Nanomaterials by Severe Plastic Deformation

Edited by
M. Zehetbauer and R. Z. Valiev

Nanomaterials by Severe Plastic Deformation

Proceedings of the Conference „Nanomaterials by Severe Plastic Deformation – NANOSPD2", December 9-13, 2002, Vienna, Austria

Edited by
Michael Zehetbauer and Ruslan Z. Valiev

Deutsche Gesellschaft
für Materialkunde e.V.

WILEY-VCH

WILEY-VCH Verlag GmbH & Co. KGaA

Editors:
Prof. Dr. Michael Zehetbauer
Institut für Materialphysik
Universität Wien
Boltzmanngasse 5
1090 Wien
Austria

Prof. Ruslan Z. Valiev
Institute of Physics of Advanced Materials
Ufa State Aviation Technical University
12 K. Marks Str.
Ufa, 450 000
Russia

This book was carefully produced. Nevertheless, editors, authors, and publisher do not warrant the information contained therein to be free of errors. Readers are advised to keep in mind that statements, data, illustrations, procedural details or otheritems may inadvertently be inaccurate.

The cover picture symbolizes the various aspects of current research in the field of severe plastic deformation (SPD). All figures have been taken from papers of the present Proceedings of "NANOSPD2".
Bottom left position
Structure of ultrafine and nanograins developed during High Pressure Torsion of Cu as investigated by Electron Back Scatter Patterning (EBSP). With a hydrostatic pressure of 8 GPa, an von Mises equivalent strain of e = 145 has been reached. The side length of image is 800 nm. (paper by T.Hebesberger, A.Vorhauer, H.P. Stüwe, R.Pippan, p. 447)
Bottom right position
HRTEM micrograph of HPT deformed Ni3Al (Cr,Zr)+B at room temperature up to a shear strain ~ 800, showing a deformation twin with atomic resolution which is typical of this deformation. The white bar represents a distance of 1nm (paper by Ch.Rentenberger, H.P.Karnthaler, R.Z.Valiev, p. 80)
Right side top
Cold rolling of Cu and Al increases their yield strength but decreases their elongation to failure (ductility). The extraordinary combination of both high strength and high ductility in nanostructured Cu and Ti processed by SPD clearly sets them apart from coarse-grained metals (paper by R. Z. Valiev, p. 109)
Right side middle
Triple junction between 3 grains, which slide significantly relative to each other because of a macroscopic deformation of 1.3 %. Molecular Dynamics Simulation of a full 3D grain boundary network yields atomic displacement vectors with the colour determined by the magnitude of the displacement (paper by H. Van Swyggenhoven, P.M. Derlet, A. Hasnaoui, p. 599)
Left side top
Repetitive Corrugation and Straightening (RCS), a new mode of Severe Plastic Deformation to achieve bulk nanomaterials (paper by N.Tsuji, Y.Saito, S.H. Lee, Y. Minamino, p. 479)
Left side middle
Strain intensity and distribution for Equal Channel Angular Pressing (ECAP) as calculated by FEM for Aluminium alloy 5083 using the MARC™ code (paper by P.A.Gonzales, C.J.Luis, p. 251)
Background
HRTEM image of nanocrystalline and amorphous phase in Ni-50.3%-Ti after HPT processing with pressure of 6 GPa (paper by T.Waitz, V.Kazykhanov, R.Z.Valiev, H.P.Karnthaler, p. 351)

Library of Congress Card No.: Applied for

British Library Cataloguing-in-Publication Data:
A catalogue record for this book is aailable from the British Library

Bibliografic information published by Die Deutsche Bibliothek
Die Deutsche Bibliothek lists this publication in the Deutsche Nationalbibliografie;
detailed bibliografic data is available in the Internet at <http://dnb.ddb.de>.

ISBN 3-527-30659-5
© 2004 WILEY-VCH Verlag GmbH & Co. KGaA, Weinheim

Printed on acid-free paper

All rights reserved (including those of translation in other languages). No part of this book may be reproduced in any form – by photoprinting, microfilm, or any other means – nor transmitted or translated into machine language without written permission from the publishers. Registered names, trademarks, etc. used in this book, even when not specifically marked as such, are not to be considered unprotected by law.

Composition: W.G.V. Verlagsdienstleistungen GmbH, Weinheim
Printing: betz-druck GmbH, Darmstadt
Bookbinding: Litges & Dopf Buchbinderei GmbH, Heppenheim
Printed in the Federal Republic of Germany

Editorial

Dear Reader,

the present Proceedings include the papers of the conference "Nanomaterials by Severe Plastic Deformation – NANOSPD 2" which was held from December 9 to 13, 2002 in Vienna, Austria, at the Institute of Materials Physics of the University of Vienna. As many as 153 participants from 22 nations testify the high interest in the field of Severe Plastic Deformation (SPD) which has been revealing as an attractive tool to achieve ultrafine grained and nanocrystalline materials in bulk shape without defective pores and impurities. The conference offered 15 sessions on themes which spanned the spectrum from general particular physical properties of nanostructured materials to the unique ones of SPD nanomaterials, the modelling of properties and of SPD production of nanocrystalline materials, the peculiarities of nanostructures evolving during SPD investigated by electron optical techniques, X-ray diffraction and other methods, the thermostability of nanocrystalline materials, and a final session on actual and futural applications of nanomaterials from SPD. Each session has been introduced by a keynote and followed by several related oral and poster contributions. In order to guarantee a high scientific level, each contribution was peer reviewed by at least one referee, revised by the authors and finally checked by us. Of course this procedure took some time but we did our best to achieve a certain level of quality without loosing too much time in publication procedure which may have been harmful to actuality of the contributions.

As a resume of the scientific outcome of the conference, at first it can be stated that since the first SPD conference held in 1999 in Moscow, nanostructured materials produced by SPD have become an extensively pursued area of research in materials science. Significantly increased research efforts can be observed in modelling and simulation of the different SPD processes, as well as in modelling of the mechanical properties of SPD materials. For the first group of tasks the simulations of strengthening under high hydrostatic pressure, the simulations of strain by several Finite Element Methods, and the texture simulations must be emphasized, while for the second type of tasks very promising simulations by Molecular Dynamics have been introduced. As a partial consequence of the improved modelling capabilities, some progress has been in bringing ECAP to a more industrial style. In this connection, other SPD techniques like Cyclic Channel Die Compression as well as Multiforging and Accumulative Roll-Bonding have been demonstrated to yield very homogeneous UFG and nanostructures. These methods may importantly complement the "classical" ones such as Equal Channel Angular Pressing (ECAP) and Torsion under Elevated Hydrostatic Pressure (HPT) especially in connection with commercial application and/or exploitation. There have been also demonstrated possibilities to combine SPD with ball milling achieving bulk samples with defect free microstructures of even nanometer scale. An increasing number of papers were concerned with the important issue being the thermostability of the SPD nanomaterials. Although this may be worse than that of conventional coarse grained materials, some specific recovery and even recrystallization treatment reveals to be highly beneficial to the ductility of material while keeping the strength on elevated level of nanomaterials. Furthermore, the research of SPD materials more and more turn to solid solutions and precipitation alloys which – in comparison with pure metals – show much smaller grain sizes as well as a markedly higher thermostability achievable by the application of SPD. There are still left some open questions like "how is fragmentation really going on ?", "what are the real reasons for enhanced ductility of SPD nanomaterials ?", "which preconditions do allow for grain boundary sliding i.e. superplasticity in SPD nanomaterials ?", which will be certainly in focus of the SPD research during the next years. Other still important tasks will be to select certain SPD methods for industrial and commercial applications, to scale them up and/or modify them in order to find the most economical solution for a continuous SPD processing.

Following the successful research activities in the field of nanostructured SPD-produced materials, various international and national projects are being proposed in this area. A number of international workshops and symposia will take place in the near future which are related to this research. During the NANO-SPD2 meeting in Vienna, an International Advisory Board ('Steering Committee') on NANO-SPD has formed which is supposed to coordinate the activities in the field of SPD nanomaterials. As a tool for this coordination, the Steering Committee has launched a special web-site under www.nanospd.org.

We do not want to close before having expressed our thanks to the Head of Institute of Materials Physics, Prof. Hans Peter Karnthaler for his general support and hospitality, and the provision of lecture halls. We are grateful to Mrs. Renate Seidl for having done most of the secretary work of the conference, and the ladies and gentlemen of the Atominstitut Wien for essential help in organization (Prof. Peter Wobrauschek – general advice, Maria Paukovits – booklets production, and Dipl.Ing. Shokufeh Zamini – homepage). Dr. Erhard Schafler (University of Vienna), Dipl.Ing. Anna Dubravina and Dr. Nariman Enikeev (both Ufa State Aviation Technical University) significantly contributed to the administration of these Proceedings. We should not forget to thank all the referees who kindly followed our requests for substantial help in improving the submitted manuscripts.

Many thanks also go to the sponsors of this conference i.e. the Austrian Federal Ministry of Education, Science and Culture, the Austrian Research Center Seibersdorf, the Vienna Business Agency, and the Department for Culture of the City of Vienna for their financial support which allowed many scientists from Eastern countries to participate in this conference. Last not least, we thank all the participants for coming and having contributed so many good and attractive papers which made possible that NANO-SPD2 became a remarkable success. We also look forward to seeing you at the NANO-SPD3 meeting in Fukuoka, Japan in 2005.

Vienna, September 2003

Michael J. Zehetbauer, Chairman   Ruslan Z. Valiev, Co-Chairman

Content

I Reasons to Use Nanostructured Materials 1

Unique Features and Properties of Nanostructured Materials 3
Hahn, H. Technische Universität Darmstadt, Institute of Materials Science, Darmstadt (D)

Properties, Benefits and Application of Nanocrystalline Structres in Magetic Materials 18
Grössinger, R., Sato, R., Holzer, D., Dahlgren, M. Technische Universität Wien, Institut für Festkörperphysik, Wien (A)

Formation of Nanostructures in Metals and Composites by Mechanical Means 30
Fecht, H.-J. University of Ulm, Center for Micro-and Nanomaterials, Ulm (D)

Scale Levels of Plastic Flow and Mechanical Properties of Nanostructured Materials 37
Panin, V. E., Panin, A. V., Derevyagina, L. S. Institute of Strength Physics and Materials Science, Tomsk (Rus); Kopylov, V. I. Physical-Technical Institute of National Academy of Sciences of Belarus, Minsk, Belarus; Valiev, R. Z. Institute of Physics of Advanced Materials associated with (USA)TU, Ufa (Rus)

Hydrogen-Induced Formation of Silver and Copper Nanoparticles in Soda-Lime Silicate Glasses 44
Suszynska, M., Krajczyk, L., Macalik, B. Institute of Low Temperature and Structure Research, Polish Academy of Sciences, Wroclaw (Pl)

II Large Strain Cold Working and Microstructure 53

Equivalent Strains in Severe Plastic Deformation 55
Stüwe, H. P. Erich Schmidt Institute of Materials Science, Austrian Academy of Sciences and Institute of Metal Physics, University of Leoben, Leoben (A)

Stage IV: Microscopic Or Mesoscopic Effect ? 65
Aldazabal, J., Alberdi, J. M., Sevillano, J. Gil CEIT (Centro de Estudios e Investigaciones Técnicas de Guipúzcoa), and TECNUN (Technological Campus, University of Navarra), San Sebastián (E)

Deformation Substructure and Mechanical Properties of BCC - Polycrystals 72
Firstov, S. A. Francevych Institute for Problems of Materials Science, Kyiv (Ua)

Micro- and Nanostructures of Large Strain-SPD deformed L12 Intermetallics 80
Rentenberger, C., Karnthaler, H. P. Institute of Materials Physics, University of Vienna, Vienna (A); Valiev, R. Z. Institute of Physics of Advanced Materials, Ufa State Aviation Technical University, Ufa (Rus).

The Nature of the Stress-Strain Relationship in Aluminum and Copper over a
Wide Range of Strain 87
*Chinh, N. Q. Eötvös University, Budapest (Hu); Horita, Z. Kyushu University,
Fukuoka (J); Langdon, T. G. University of Southern California, Los Angeles (USA).*

Strain Hardening by Formation of Nanoplatelets 95
*Han, K., Xin, Y., Ishmaku, A. National High Magnetic Field Laboratory,
Florida State University, Tallahassee, Fl (USA)*

Modelling the Draw Hardening of a Nanolamellar Composite: A Multiscale
Transition Method 101
*Krummeich-Brangier, R. Groupe de Physique des Matériaux-UMR CNRS 6634,
Université de Rouen (F); Sabar, H., Berveiller, M. Laboratoire de Physique et
Mécanique des Matériaux-UMR CNRS 7554, Université de Metz (F)*

III Unique Features of SPD – Microstructure and Properties 107

Paradoxes of Severe Plastic Deformation 109
*Valiev, R. Z. Institute of Physics of Advanced Materials, Ufa State Aviation
Technical University, Ufa (Rus)*

Investigation of Phase Transformations in Nanostructured Materials Produced by
Severe Plastic Deformation 118
*Sauvage, X., Guillet, A., Blavette, D. Groupe de Physique des matériaux,
UMR CNRS 6634, Université de Rouen, Saint Etienne du Rouvray (F)*

Structure-Phase Transformations and Properties in Nanostructured Metastable Alloys
Processed by Severe Plastic Deformation 125
*Stolyarov, V. V., Valiev, R. Z. Institute of Physics of Advanced Materials,
Ufa State Aviation Technical University, Ufa (Rus)*

On the Influence of Temperature and Strain Rate on the Flow Stress of ECAP Nickel 131
*Hollang, L., Thiele, E., Holste, C. Institut für Physikalische Metallkunde,
Technische Universität Dresden, Dresden (D); Brunner, D. Max-Planck-Institut
für Metallforschung, Stuttgart (D)*

The Effect of Second-Phase Particles on the Severe Deformation of Aluminium
Alloys during Equal Channel Angular Extrusion. 138
*Apps, P. J. UMIST, Manchester (GB); Bowen, J. R. Risø National Labs (Dk);
Prangnell, P. B. UMIST, Manchester (GB)*

Effects of ECAP Processing on Mechanical and Aging Behaviour of an AA6082 Alloy 145
*Bassani, P., Tasca, L., Vedani, M. Politecnico di Milano, Dipartimento di
Meccanica, Milano (I)*

Influence of the Thermal Anisotropy Internal Stresses on Low Temperature
Mechanical Behavior of Polycrystalline and Nanostructured Ti 151
*Bengus, V. Z, Smirnov, S. N. B. Verkin Inst. for Low Temperature Physics &
Engineering Ukraine Academy of Sciences, Kharkov (Ua)*

Nano- and Submicrocrystalline Structure Formation During High Pressure Torsion
of Al-Sc and Al-Mg-Sc Alloys 158
*Dobatkin, S. V. Baikov Institute of Metallurgy and Material Science RAS,
Moscow (Rus); Zakharov, V. V., Rostova, T. D. All-Russia Institute of Light Alloys,
Moscow (Rus); Vinogradov, A. Yu. Osaka City University, Osaka (J); Valiev, R. Z.,
Krasilnikov, N. A. Ufa State Aviation Technical University, Ufa (Rus); Bastarash, E. N.,
Trubitsyna, I. B. Moscow State Steel and Alloys Institute (Technological University),
Moscow (Rus)*

Phase Transformation in Crystalline and Amorphous Rapidly Quenched
Nd-Fe-B Alloys under SPD 165
*Gunderov, D. V., Stolyarov, V. V. Institute of Physics of Advanced Materials,
USATU, Ufa (Rus); Popov, A. G., Schegoleva, N. N. Institute of Physics of Metals,
UrD RAS, Ekaterinburg (Rus); Yavary, A. R. National Polytechnic Institute,
Grenoble (F)*

Structure and Functional Properties of Ti-Ni-Based Shape Memory Alloys
Subjected to Severe Plastic Deformation... 170
*Khmelevskaya, I. Yu., Trubitsyna, I. B., Prokoshkin, S. D. Moscow Steel and
Alloys Institute, Moscow (Rus); Dobatkin, S. V. Moscow Steel and Alloys Institute,
Moscow (Rus) and Baikov Institute of Metallurgy and Material Science, Russian
Academy of Science, Moscow (Rus); Stolyarov, V. V., Prokofjev, E. A. Ufa State
Aviation Technical University, Ufa (Rus)*

Formation of Submicrocrystalline Structure in The Hard Magnetic Alloy
Fe-15wt.%Co-25%Cr During Straining by Complex Loading... 177
*Korznikova, G. F., Korneva, A. V. Institute of Metals Superplasticity Problems,
Russian Academy of Sciences, Ufa (Rus); Korznikov, A. V. Austrian Research
Centres, Seibersdorf Research GmbH and Institute of Metals Superplasticity
Problems, Russian Academy of Sciences, Ufa (Rus)*

Experimental Investigations of the Al-Mg-Si Alloy Subjected to Equal-Channel
Angular Pressing... 183
*Krallics, G., Szeles, Z. Budapest University of Technology and Economics,
Budapest (Hu); Semenova, I. P., Dotsenko, T. V., Alexandrov, I. V. Ufa
State Aviation Technical University, Ufa (Rus)*

Mechanical Properties of AZ91 Alloy after Equal Channel Angular Pressing 190
Máthis, K. Department of Metal Physics, Charles University, Prague (Cz) and Department of General Physics, Eötvös University, Budapest (Hu); Trojanová, Z., Lukác, P. Department of Metal Physics, Charles University, Prague (Cz); Lendvai, J. Department of General Physics, Eötvös University, Budapest (Hu); Rauch, E., Mussi, A. Génie Physique et Mécanique des Matériaux, INP Grenoble - ENSPG (F)

Influence of Microstructural Heterogeneity on the Mechanical Properties of Nanocrystalline Materials Processed by Severe Plastic Deformation 194
Pakiela, Z., Sus-Ryszkowska, M. Faculty of Materials Science and Engineering, Warsaw University of Technology, Warsaw (Pl)

Creep Behaviour of Pure Aluminium Processed by Equal-Channel Angular Pressing........... 200
Sklenicka, V., Dvorak, J., Svoboda, M. Institute of Physics of Materials, Academy of Sciences of the Czech Republic, Brno (Cz)

Low Temperature Strain Rate Sensitivity of some Nanostructured Metals 207
Tabachnikova, E. D., Bengus, V. Z., Natsik, V. D., Podolskii, A. V., Smirnov, S. N. B. Verkin Inst. for Low Temperature Physics & Engineering (Ua) Academy of Sciences, Kharkov (Ua); Valiev, R. Z., Stolyarov, V. V., Alexandrov, I. V. Institute of Physics of Advanced Materials, Ufa (Rus)

IV Modelling of SPD and Mechanical Properties of SPD Materials............................... 213

Importance of Disclinatioin in Severe Plastically Deformed Materials 215
Romanov, A. E. Ioffe Physico-Technical Institute, St. Petersburg (Rus)

Disclination-Based Modelling of Grain Fragmentation during Cold Torsion and ECAP in Aluminium Polycrystals 226
Seefeldt, M., Houtte, P. Van K.U. Leuven, Heverlee (B)

Modeling of Deformation Behavior and Texture Development in Aluminium under Equal Channel Angular Pressing 233
Baik, S. C. Technical Research Labs., Pohang Iron & Steel Co. Ltd., Pohang (ROK); Estrin, Y., Hellmig, R. J. IWW, TU Clausthal, Clausthal-Zellerfeld (D); Kim, H. S. Department of Metallurgical Engineering, Chungnam National University, Daejeon (ROK) and Technical Research Labs., Pohang Iron & Steel Co. Ltd., Pohang (ROK)

Process Modeling of Equal Channel Angular Pressing 239
Kim, H. S., Hong, S. I., Lee, H. R., Chun, B. S. Chungnam National University, Daejeon (ROK)

Deformation Behaviour of ECAP Cu as described by a Dislocation-Based Model 245
Enikeev, N. A. Institute of Mechanics, Ufa Science Centre, RAS, Ufa (Rus); Alexandrov, I. V. Institute for Physics of Advanced Materials (USA)TU, Ufa (Rus); Hong, S. I., Kim, H. S. Chungnam National University, Taejon (ROK)

Severe Plastic Deformation by ECAP in an Acommercial Al-Mg-Mn-Alloy 251
González, P. A., Luis, C. Departimento de Ingeniera Mecánica, Energética y Materiales, UPNA, Campus de Arrosadía, Pamplona (E)

Evolution of Mechanical and Microstructural Properties of ECAP Deformed Copper 257
Hellmig, R. J., Estrin, Y. Technische Universität Clausthal, Clausthal-Zellerfeld (D); Bowen, J. R, Juul Jensen, D. Center for Fundamental Research: Metal Structures in Four Dimensions, Roskilde (Dk); Baik, S. C. Pohang Iron & Steel Co. Ltd., Pohang (ROK); Kim, H. S., Seo, M. H. Chungnam National University, Daejeon (ROK)

A Composite Grain Model of Strengthening for SPD Produced UFG Materials 263
Kozlov, E. V., Popova, N. A., Koneva, N. A. Tomsk State University of Architecture and Building, Tomsk (Rus); Zhdanov, A. N. Altai State Technical University, Barnaul (Rus)

Computer Simulation of Equal-Channel Angular Pressing of Tungsten by Means of the Finite Element Method 271
Krallics, G. Budapest University of Technology and Economics, Budapest (Hu); Budilov, I. N., Alexandrov, I. V., Raab, G. I., Zhernakov, V. S., Valiev, R. Z. Ufa State Aviation Technical University, Ufa (Rus)

V Texture Evolution and Simulation During SPD 279

Texture Evolution in Severe Plastic Deformation by Equal Channel Angular Extrusion 281
Tóth, L. S. Laboratoire de Physique et Méchanique des Matériaux, Université de Metz (F); Rauch, E. F., Dupuy, L. Génie Physique et Mécanique des Matériaux (GPM2), Institut National Polytechnique de Grenoble ENSPG, Saint-Martin d'Hères (F)

Textural Evolution during Equal Channel Angular Extrusion versus Planar Simple Shear 297
Rauch, E. F. and Dupuy, L. Génie Physique et Mécanique des Matériaux (GPM2), Institut National Polytechnique de Grenoble ENSPG, Saint-Martin d'Hères (F)

Grain Refinement and Texture Formation during High-Strain Torsion of NiAl 303
Skrotzki, W., Klöden, B., Tamm, R., Oertel, C.-G. Institut für Strukturphysik, Technische Universität Dresden, Dresden (D); Wcislak, L. HASYLAB at DESY, Hamburg (D); Rybacki, E. Geoforschungszentrum Potsdam, Potsdam (D)

Severely Plastically Deformed Ti from the Standpoint of Texture Changes 309
Bonarski, J. Polish Academy of Sciences, Institute of Metallurgy and Materials Science, Krakow (Pl); Alexandrov, I. V. Ufa State Aviation Technical University, Ufa (Rus)

Development of Crystallographic Texture and Microstructure in Cu and Ti, Subjected to Equal-Channel Angular Pressing 315
Bonarski, J., Tarkowski, L. Polish Academy of Sciences, Institute of Metallurgy and Materials Science, Krakow (Pl); Alexandrov, I. V. Ufa State Aviation Technical University, Ufa (Rus)

VI Details of SPD Nanostructures as Investigated by Electron Microscopy 321

Boundary Characteristics in Heavily Deformed Metals 323
Winter, G., Huang, X. Center for Fundamental Research: Metal Structures in 4D, Risø Laboratory (Dk)

Quantitative Microstructural Analysis of IF Steel Processed by Equal Channel Angular Extrusion 332
De Messemaeker, J., Verlinden, B., Van Humbeeck, J., Froyen, L. Katholieke Universiteit Leuven, Leuven (B)

HRTEM Investigations of Amorphous and Nanocrystalline NiTi Alloys Processed by HPT 339
Waitz, T., Karnthaler, H. P. Institute of Materials Physics, University of Vienna, Vienna (A); Valiev, R. Z. Institute of Physics of Advanced Materials, Ufa State Aviation Technical University, Ufa (Rus)

Effect of Grain Size on Microstructure Development during Deformation in Polycrystalline Iron 345
Kawasaki, K., Hidaka, H., Tsuchiyama, T., Takaki, S. Kyushu University, Fukuoka (J)

Microstructure and Phase Transformations of HPT NiTi 351
Waitz, T., Karnthaler, H. P. Institute of Materials Physics, University of Vienna, Wien (A); Kazykhanov, V. Institute of Physics of Advanced Materials, Ufa State Aviation Technical University, Ufa (Rus)

Types of Grains and Boundaries, Joint Disclinations and Dislocation Structures of SPD-produced UFG Materials 357
Koneva, N. A., Popova, N. A., Ignatenko, L. N., Pekarskaya, E. E., Kozlov, E. V. Tomsk State University of Architecture and Building, Tomsk (Rus); Zhdanov, A. N. Altai State Technical University, Barnaul (Rus)

Microstructure Development of Copper Single Crystal Deformed by Equal
Channel Angular Pressing.. 363
*Koyama, T., Miyamoto, H., Mimaki, T. Doshisha University, Kyotanabe (J);
Vinogradov, A., Hashimoto, S. Osaka City University, Osaka (J)*

TEM Investigations of Ti Deformed by ECAP 369
*Mingler, B., Zeipper, L., Karnthaler, H. P., Zehetbauer, M. Institute of
Materials Physics, University of Vienna, Wien (A)*

Microstructural Evolution during Severe Deformation in Austenitic Stainless
Steel with Second Phase Particles.. 375
*Miura, H., Hamaji, H., Sakai, T. Department of Mechanical Engineering
and Intelligent Systems, University of Electro-Communications, Chofu, Tokyo (J)*

Structural Models and Mechanisms for the Formation of High-Energy
Nanostructures under Severe Plastic Deformation 381
*Tyumentsev, A. N., P. Pinzhin, Yu., Litovchenko, I. Yu., Ovchinnikov, S. V.
Institute of Strength Physics and Materials Technology, RAS, Tomsk (Rus);
Korotaev, A. D., Ditenberg, I. A., Surikova, N. S., Shevchenko, N. V. Siberian
Physicotechnical Institute, Tomsk (Rus); Valiev, R. Z. Institute of Physics of
Advanced Materials Ufa State Aviation Technical University, Ufa (Rus)*

Grain Refinement and Microstructural Evolution in Nickel During
High-Pressure Torsion 387
*Zhilyaev, A. P. Universitat Autònoma de Barcelona, Bellaterra (E);
Nurislamova, G. V. Institute for Physics of Advanced Materials, Ufa State
Aviation Technical University, Ufa (Rus); Kim, B.-K., Szpunar, J. A. Department
of Metals and Materials Engineering, McGill University, Montreal (Cdn);
Baró, M. D. Universitat Autònoma de Barcelona, Bellaterra (E); Langdon, T. G. Departments
of Aerospace & Mechanical Engineering and Materials Science,
University of Southern California, Los Angeles, (USA)*

VII Analyses of SPD Materials by Selected Physical Methods... 393

The Meaning of Size Obtained from Broadened X-ray Diffraction Peaks............................ 395
Ungár, T. Department of General Physics, Eötvös University, Budapest (Hu)

Ultra Fine Grained Copper Prepared by High Pressure Torsion: Spatial
Distribution of Defects from Positron Annihilation Spectroscopy... 407
*Cizek, J., Prochazka, I., Kuzel, R., Cieslar, M. Charles University in Prague,
Faculty of Mathematics and Physics, Prague (Cz); Islamgaliev, R. K. Institute
of Physics of Advanced Materials, Ufa State Aviation Technical University,
Ufa (Rus); Anwand, W., Brauer, G. Institut für Ionenstrahlphysik und
Materialforschung, Forschungszentrum Rossendorf, Dresden (D)*

Anelastic Properties of Nanocrystalline Magnesium 413
Trojanová, Z., Lukác, P., Stanek, M. Department of Metal Physics, Charles University, Praha (Cz); Riehemann, W., Weidenfeller, B. Department of Materials Engineering and Technology, Technical University Clausthal, Clausthal-Zellerfeld (D)

X-ray Peak Profile Analysis on the Microstructure of Al-5.9%Mg-0.3%Sc-0.18%Zr Alloy Deformed by High Pressure Torsion Straining... 420
Gubicza, J. Department of General Physics, Eötvös University, Budapest (Hu) and Department of Solid State Physics, Eötvös University, Budapest (Hu); Fátay, D., Nyilas, K., Ungár, T. Department of General Physics, Eötvös University, Budapest (Hu); Bastarash, E., Dobatkin, S. Moscow State Steel and Alloys Institute (Technological University), Moscow (Rus)

Evolution of Microstructure during Thermal Treatment in SPD Titanium 426
Schafler, E. Institute of Materials Physics, University of Vienna (A) and Erich Schmid Institute of Material Science, Leoben (A); Zeipper, L. Institute of Materials Physics, University of Vienna (A) and ARC Seibersdorf Research GMBH, Seibersdorf (A); Zehetbauer, M. J. Institute of Materials Physics, University of Vienna (A)

VIII Influence of Deformation Parameters to SPD Nanostructures 433

The Role of Hydrostatic Pressure in Severe Plastic Deformation 435
Zehetbauer, M. J., Schafler, E. Institute of Materials Physics, University of Vienna, Vienna (A); Stüwe, H. P., Vorhauer, A. Erich Schmid Institute of Materials Science, Austrian Academy of Sciences, Leoben (A); Kohout, J. Department of Physics, Military Academy Brno, Brno (Cz);

Influence of the Processing parameters at High Pressure Torsion 447
Hebesberger, T. Erich Schmid Institute of Materials Science Austrian Academy of Sciences, Leoben (A), now Voest-Alpine Stahl, GmbH; Vorhauer, A. Erich Schmid Institute of Materials Science Austrian Academy of Sciences, Leoben (A); Stüwe, H. P. Erich Schmidt Institute of Materials Science, Austrian Academy of Sciences and Institute of Metal Physics, University of Leoben, Leoben (A); Pippan, R. Erich Schmid Institute of Materials Science Austrian Academy of Sciences, Leoben (A) and Christian Doppler Laboratory for Local Analysis of Deformation and Fracture, Leoben (A)

Mechanical Properties and Thermal Stability of Nano-Structured Armco
Iron Produced by High Pressure Torsion 453
*Ivanisenko, Yu., Minkow, A. Division of Materials, Ulm University, Ulm (D);
Sergueeva, A. V. Chemical Engineering and Material Science Department,
University of California, Davis (USA); Valiev, R. Z. Institute of Physics of
Advanced Materials, Ufa State Aviation Technical University, Ufa (Rus);
Fecht, H.-J. Division of Materials, Ulm University, Ulm (D) and Institute
of Nanotechnology, Research Center Karlsruhe, Karlsruhe (D)*

Properties of Aluminum Alloys Processed by Equal Channel Angular Pressing
Using a 60 Degrees Die 459
*Furukawa, M. Fukuoka University of Education, Munakata (J); Akamatsu, H.,
Horita, Z. Kyushu University, Fukuoka (J); Langdon, T. G. University of
Southern California, Los Angeles (USA)*

Deformation Behaviour of Copper Subjected to High Pressure Torsion 465
*Dubravina, A. A. Institute of Physics of Advanced Materials, Ufa State
Aviation Technical University, Ufa (Rus) and Department of Chemical
Engineering & Materials Science, University of California, Davis (USA);
Alexandrov, I. V., Valiev, R. Z. Institute of Physics of Advanced Materials,
Ufa State Aviation Technical University, Ufa (Rus); Sergueeva, A. V. Department
of Chemical Engineering and Material Science, University of California, Davis (USA)*

Features of Equal Channel Angular Pressing of Hard-to-Deform Materials 471
Raab, G. I., Soshnikova, E. P. Ufa State Aviation Technical University, Ufa (Rus)

IX New Methods of SPD 477

ARB (Accumulative Roll-Bonding) and other new Techniques to Produce
Bulk Ultrafine Grained Materials 479
*Tsuji, N., Saito, Y., Minamino, Y. Osaka University, Suita (J); Lee, S-H
Mokpo National University, Mokpo (ROK)*

Optimal SPD Processing of Plates by Constrained Groove Pressing (CGP) 491
*Alkorta, J., Sevillano, J. G. Centro de Estudios e Investigaciones Técnicas
de Guipúzcoa y Tecnun, University of Navarra, San Sebastian (E)*

Comparative Study and Texture Modeling of Accumulative Roll Bonding
(ARB) processed AA8079 and CP-Al 498
Heason, C. P., Prangnell, P. B. UMIST, Manchester (GB)

Nanocrystallization in Carbon Steels by Various Severe Plastic Deformation
Processes 505
*Todaka, Y., Umemoto, M., Tsuchiya, K. Department of Production System
Engineering, Toyohashi University of Technology, Toyohashi (J)*

Severe Plastic Deformation by Twist Extrusion 511
Beygelzimer, Y., Varyukhin, V., Orlov, D., Synkov, S., Spuskanyuk, A., Pashinska, Y. Donetsk Phys&Tech. Institute, Donetsk (Ua)

SPD Structures Associated With Shear Bands in Cold-Rolled Low SFE Metals 517
Higashida, K., Morikawa, T. Kyushu University, Fukuoka (J)

Features of Mechanical Behaviour and Structure Evolution of
Submicrocrystalline Titanium under Cold Deformation 523
Mironov, S. Yu., Salishchev, G. A. Institute of Metals Superplasticity Problems, Russian Academy of Sciences, Ufa (Rus); Myshlyaev, M. M. Baikov Institute of Metallurgy and Materials Science, Russian Academy of Sciences, Moscow (Rus)

Ultra Grain Refinement of Fe-Based Alloys by Accumulated Roll Bonding 530
Reis, A. C. C., Tolleneer, I., Barbé, L., Kestens, L., Houbaert, Y. Ghent University, Department of Metallurgy and Materials Science, Ghent (B)

X SPD with Ball Milling and Powder Consolidation 537

Mechanically Activated Powder Metallurgy : A Suitable Way To Dense
Nanostructured Materials 539
Gaffet, E. UMR 5060 CNRS / UTBM, Sévenans, Belfort (F) and GFA, GdR 2391 CNRS, Dijon (F); Paris, S., Bernard, F. LRRS UMR 5613 CNRS / Univ. Bourgogne, Dijon (F) and GFA, GdR 2391 CNRS, Dijon (F)

Characterization and Mechanical Properties of Nanostructured Copper
Obtained by Powder Metallurgy 545
Langlois, C., Hÿtch, M. J., Lartigue, S., Champion, Y. Centre d'Etudes de Chimie Métallurgique CECM-CNRS, Vitry-sur-Seine (F); Langlois, P. Laboratoire d'Ingénierie des Matériaux et des Hautes Pressions LIMHP-CNRS, Villetaneuse (F)

Densification of Magnesium Particles by ECAP with a Back-Pressure 551
Lapovok, R. Ye., Thomson, P. F. CAST CRC, School of Physics and Materials Engineering, Monash University, Melbourne (Aus)

Annealed Microstructures in Mechanically Milled Fe-0.6%O Powders 558
Belyakov, A., Sakai, Y., Hara, T., Kimura, Y., Tsuzaki, K. Steel Research Center, National Institute for Materials Science, Tsukuba (J)

Processing and Characterization of Nanocrystalline Aluminum obtained
by Hot Isostatic Pressing (HIP) 564
Billard, S., Dirras, G., Fondere, J. P., Bacroix, B. Laboratoire des Propriétés
Mécaniques et Thermodynamiques des Matériaux (LPMTM) CNRS, Institut
Galilée Université Paris 13, Villetaneuse (F)

Characteristics of Nano Grain Structure in SPD-PM Processed AISI304L
Stainless Steel Powder 571
Inomoto, H. Ritsumeikan University, Kusatsu, Shiga (J); Fujiwara, H.
Department of Environmental Systems Engineering, Kochi University of
Technology, Tosayamada, Kochi (J); Ameyama, K. Deptartment of
Mechanical Engineering, Ritsumeikan University, Kusatsu, Shiga (J)

Formation of Powder and Bulk Al-Cu-Fe Quasicrystals, and of Related
Phases During Mechanical Alloying and Sintering 579
Kaloshkin, S. D., Tcherdyntsev, V. V., Laptev, A. I., Shelekhov, E. V.
Moscow State Institute of Steel and Alloys, Moscow (Rus); Principi, G.,
Spataru, T. Settore Materiali and INFM, DIM, Padova (I)

Production and Consolidation of Nanocrystalline Fe Based Alloy Powders 585
Rombouts, M., Froyen, L. Department of Metallurgy and Materials
Engineering (MTM), Katholieke Universiteit Leuven (B); C. Reis, A. C.,
Kestens, L. Department of Metallurgy, Ghent University (B)

Strain Measurement in the ECAP Process 591
Werenskiold, J. C., Roven, H. J. Norwegian University of Science and
Technology, Trondheim (N)

XI Mechanical Properties and Thermostability of Nanocrystalline Structures 597

Atomistic Modeling of Strength of Nanocrystalline Metals 599
van Swygenhofen, H., Derlet, P. M., Hasnaoui, A. Paul Scherrer
Institute, Villigen (CH)

Multiscale Studies and Modeling of SPD Materials 609
Alexandrov, I. V. Ufa State Aviation Technical University, Ufa (Rus)

Microstructural Stability and Tensile Properties of Nanostructured
Low Carbon Steels Processed by ECAP 616
Shin, D. H. Department of Metallurgy and Materials Science, Hanyang
University, Ansan (ROK); Park, K.-T. Division of Advanced Materials
Science & Engineering, Hanbat National University, Taejon (ROK)

Microstructure and Mechanical Properties of Severely Deformed Al-3Mg
and its Evolution during subsequent Annealing Treatment 623
*Morris-Muñoz, M. A., Garcia Oca, C., Gonzalez Doncel, G., Morris, D. G.
CENIM, CSIC, Madrid (E)*

Dependence of Thermal Stability of Ultra Fine Grained Metals on Grain Size 630
*Stulíková, I., Cieslar, M., Kuzel, R., Procházka, I., Čížek, J. Faculty
of Mathematics and Physics, Charles University, Prague, (Cz);
Islamgaliev, R. K. Institut of Physics of Advanced Materials, Ufa State
Aviation Technical University, Ufa (Rus)*

Thermomechanical Properties of Electrodeposited Ultra Fine Grained
Cu-Foils for Printed Wiring Boards 636
*Betzwar-Kotas, A., Gröger, V., Weiss, B., Wottle, I., Zimprich, P. Institute
of Material Physics, University of Vienna (A); Khatibi, G. Institute of
Material Physics, University of Vienna (A) and Institute of Physical
Chemistry, Material Science, University of Vienna (A)*

Effect of Grain Boundary Phase Transitions on the Superplasticity in the Al-Zn system 642
*López, G. A., Gust, W., Mittemeijer, E. J. Max Planck Institute for Metals
Research and Institute of Physical Metallurgy, University of Stuttgart,
Stuttgart (D); Straumal, B. B. Max Planck Institute for Metals Research
and Institute of Physical Metallurgy, University of Stuttgart, Stuttgart (D)
and Institute of Solid State Physics RAS, Chernogolovka (Rus)*

Microstructure and Thermal Stability of Tungsten based Materials after
Severe Plastic Deformation.. 648
*Vorhauer, A. Erich Schmid-Institute of Materials Science of the Austrian
Academy of Sciences, Leoben (A) and Christian Doppler Laboratory of
Local Analysis of Deformation and Fracture, Leoben (A); Pippan, R.
Erich Schmid-Institute of Materials Science of the Austrian Academy of
Sciences, Leoben (A) and Christian Doppler Laboratory of Local Analysis
of Deformation and Fracture, Leoben (A); Knabl, W. Plansee AG, Reutte (A)*

Development of Microstructure and Thermal Stability of Nano-structured
Chromium Processed by Severe Plastic Deformation 654
*Wadsack, R. Erich Schmid Institute for Materials Science of the Austrian
Academy of Sciences, Leoben (A); Pippan, R. Erich Schmid Institute for
Materials Science of the Austrian Academy of Sciences, Leoben (A) and
Christian Doppler Laboratory for Local Analysis of Deformation and
Fracture, Leoben (A); Schedler, B. Plansee AG, Reutte (A)*

XII Influence of Deformation Path to Properties of SPD Materials 661

Fatigue of Severely Deformed Metals 663
Vinogradov, A., Hashimoto, S. Department of Intelligent Materials
Engineering, Osaka City University, Osaka (J)

Cyclic Deformation Behaviour and Possibilities for Enhancing the Fatigue
Properties of Ultrafine-Grained Metals 677
Höppel, H. W., Kautz, M., Barta-Schreiber, N., Mughrabi, H.
Friedrich-Alexander-Universität Erlangen-Nürnberg, Erlangen (D); Xu, C.,
Langdon, T. G. University of Southern California, Los Angeles (USA)

The Influence of Type and Path of Deformation on the Microstructural
Evolution During Severe Plastic Deformation 684
Vorhauer, A., Pippan, R. Erich Schmid Institute for Materials Science of
the Austrian Academy of Sciences, Leoben (A) and Christian Doppler
Laboratory for Local Analysis of Deformation and Fracture, Leoben (A)

Formation of a Submicrocrystalline Structure in Titanium during Successive
Uniaxial Compression in Three Orthogonal Directions 691
Salishchev, G. A., Zherebtsov, S. V., Mironov, S. Yu. Institute for Metals
Superplasticity Problems, Ufa (Rus); Myshlayev, M. M. Baikov Institute
of Metallurgy and Materials Science, Moscow (Rus); Pippan, R. Institute of
Material Science, Leoben (A)

XIII Features and Mechanisms of Superelasticity in SPD Materials 699

Achieving a Superplastic Forming Capability through Severe Plastic Deformation 701
Xu, C., Langdon, T. G. University of Southern California, Los Angeles (USA);
Horita, Z. Kyushu University, Fukuoka (J); Furukawa, M. Fukuoka University
of Education, Munakata (J)

Production of Superplastic Mg Alloys Using Severe Plastic Deformation 711
Horita, Z., Matsubara, K., Miyahara, Y. Kyushu University, Fukuoka (J);
Langdon, T. G. University of Southern California, Los Angeles (USA)

High Strain Rate Superplasticity in an Micrometer-Grained Al-Li Alloy
Produced by Equal-Channel Angular Extrusion 717
Myshlyaev, M. M., Myshlyaeva, M. M. Institute of Solid State Physics,
Russian Academy of Sciences, Chernogolovka (Rus); Kamalov, M. M.
Institute of Solid State Physics, Russian Academy of Sciences, Chernogolovka
(Rus) and Baikov Institute of Metallurgy and Material Science, Russian
Academy of Sciences, Moscow (Rus)

Diffusion-Controlled Processes and Plasticity of Submicrocrystalline Materials 722
R. Kolobov, Yu., Ivanov, K. V., Grabovetskaya, G. P., Naidenkin, E. V.
Institute of Strength Physics and Materials Science, Tomsk (Rus)

Superplastic Behavior of Deformation Processed Cu-Ag Nanocomposites 728
Hong, S. I., Kim, Y. S., Kim, H. S. Department of Metallurgical Engineering,
Chungnam National University, Taedok Science Town, Taejon (ROK)

Features of Microstructure and Phase State in an Al-Li Alloy after ECA
Pressing and High Strain Rate Superplastic Flow 734
Myshlyaev, M. M. Institute of Solid State Physics, Russian Academy
of Sciences, Chernogolovka (Rus) and Baikov Institute of Metallurgy and
Material Science, Russian Academy of Sciences, Moscow (Rus); Mazilkin, A. A.,
Kamalov, M. M. Institute of Solid State Physics, Russian Academy of Sciences,
Chernogolovka (Rus)

Microstructure Refinement and Improvement of Mechanical Properties of a
Magnesium Alloy by Severe Plastic Deformation 740
Mussi, A., Blandin, J. J., Rauch, E. F. Génie Physique et Mécanique des
Matériaux (GPM2) Saint-Martin d'Hères (F) and Institut National Polytechnique
de Grenoble (INPG) Saint-Martin d'Hères (F)

Grain Refinement and Superplastic Properties of Cu-Zn Alloys Processed
by Equal-Channel Angular Pressing 746
Neishi, K., Horita, Z. Kyushu University, Fukuoka (J); Langdon, T. G.
University of Southern California, Los Angeles (USA)

XIV Mechanisms of Diffusion Related Processes in Nanocrystalline Materials 753

Diffusion in Nanocrystalline Metals and Alloys – A Status Report 755
Würschum, R., Brossmann, U. Technische Universität Graz, Institut für
Technische Physik, Graz (A); Herth, S. Forschungszentrum Karlsruhe,
Institut für Nanotechnologie, Karlsruhe (D)

Self-Diffusion of ^{147}Nd in Nanocrystalline $Nd_2Fe_{14}B$ 767
Sprengel, W., Barbe, V., Herth, S., Wejrzanowski, T. Institut für
Theoretische und Angewandte Physik, Universität Stuttgart, Stuttgart (D)
and Faculty of Materials Science, Warsaw University of Technology, Warsaw
(Pl); Gutfleisch, O. Leibniz-Institut für Festkörper- und Werkstoffforschung
Dresden, Dresden (D); Eversheim, P. D. Helmholtz Institut für Strahlen- und
Kernphysik, Universität Bonn, Bonn (D); Würschum, R. Institut für Technische
Physik, Technische Universität Graz, Graz (A); Schaefer, H.-E. Institut für
Theoretische und Angewandte Physik, Universität Stuttgart, Stuttgart (D)

Theoretical Investigation of Nonequilibrium Grain Boundary Diffusion Properties 773
*Perevezentsev, V. N. Blagonravov Nizhni Novgorod Branch of Mechanical
Engineering Research Institute,Russian Academy of Sciences, Nizhny Novgorod (Rus)*

On Annealing Mechanisms Operating in Ultra Fine Grained Alloys 780
*Sakai, T., Miura, H. Department of Mechanical Engineering and
Intelligent Systems, University of Electro-Communications, Chofu, Tokyo (J);
Belyakov, A., Tsuzaki, K. Steel Research Center, National Institute for
Materials Science, Tsukuba (J)*

XV Application of SPD Materials 787

Commercialization of Nanostructured Metals Produced by Severe Plastic
Deformation Processing 789
Lowe, T. C., Zhu, Y. T. Los Alamos National Laboratory, Los Alamos (USA)

The Main Directions in Applied Research and Developments of SPD
Nanomaterials in Russia 798
*Fokine, V. A. Editor of the Journal "Russia and World: Science and Technology",
Director General of the F&F Consulting Co., Moscow (Rus)*

Developing of Structure and Properties in Low-Carbon Steels During Warm
and Hot Equal Channel Angular Pressing 804
*Dobatkin, S. V. Baikov Institute of Metallurgy and Material Science, Russian
Academy of Sciences, Moscow (Rus) and Moscow State Steel and Alloys
Institute (Technological University), Moscow (Rus); Odessky, P. D. Institute
of Building Constructions, Moscow (Rus); Pippan, R. Erich Schmid Institute
of Material Science, Austrian Academy of Sciences, Leoben (A); Raab, G. I.,
Krasilnikov, N. A. Ufa State Aviation Technical University, Ufa (Rus);
Arsenkin, A. M. Moscow State Steel and Alloys Institute (Technological
University), Moscow (Rus)*

Mechanical Properties of Severely Plastically Deformed Titanium 810
*Zeipper, L., Korb, G. ARC Seibersdorf Research GmbH, Seibersdorf (A);
Zehetbauer, M., Mingler, B., Schafler, E., Karnthaler, H. P. University of
Vienna, Vienna (A)*

Ways to Improve Strength of Titanium Alloys by Means of Severe
Plastic Deformation 817
Popov, A. A. Ural State Technical University, UGTU-UPI, Ekaterinburg (Rus)

Structures, Properties, and Application of Nanostructured Shape Memory
TiNi-based Alloys 822
*Pushin, V. G. Institute of Metal Physics, Ural Division of Russian Academy
of Sciences, Ekaterinburg (Rus)*

Microstructure and Properties of a Low Carbon Steel after Equal Channel
Angular Pressing 829
*Wang, J., Wang, Y., Du, Z., Zhang, Z., Wang, L., Zhao, X. School of
Metallurgical Engineering, Xi'an Univ. of Arch. & Tech., Xi'an (VRC); Xu, C.,
Langdon, T. G. Depts. of Mech. Eng. and Mater. Sci., Univ. of Southern
California, Los Angeles (USA)*

Formation of Submicrocrystalline Structure in Large-Scale Ti-6Al-4V Billets
during Warm Severe Plastic Deformation 182 835
*Zherebtsov, S. V., Salishchev, G. A., Galeyev, R. M., Valiakhmetov, O. R.
Institute for Metals Superplasticity Problems, Ufa (Rus); Semiatin, S. L. Air
Force Research Laboratory, Wright-Patterson Air Force Base (USA)*

Author Index 841

Subject Index 845

I   Reasons to use Nanostructured Materials

Unique Features and Properties of Nanostructured Materials

Horst Hahn
Technische Universität Darmstadt, Institute of Materials Science, Thin Films Division, Darmstadt, Germany

1 Abstract

In this introductory paper an attempt is made to give an overview of the area of nanostructured „materials irrespective of the synthesis process. The various microstructural features such as clusters or isolated nanoparticles, agglomerated nanopowders, consolidated nanomaterials and nanocomposite materials as well as all materials classes are considered. As an important component of modern research on nanomaterials a section describes the various characterization tools available. Based on these remarks some properties of nanostructured materials will be summarized emphasizing the property-microstructure relationships. Finally, a brief outlook on applications and initial industrial use of nanomaterials is presented.

2 Introduction

Nanostructures are plentiful in nature. In the universe nanoparticles are distributed widely and are considered to be the building blocks in planet formation processes. Biological systems have built up inorganic-organic nanocomposite structures to improve the mechanical properties or to improve the optical, magnetic and chemical sensing in living species. As an example, nacre (mother-of-pearl) from the mollusc shell is a biologically formed lamellar ceramic, which „exhibits structural robustness despite the brittle nature of its constituents. [1] Figure 1 shows an SEM imge of a fracture surface of an abalone shell exhibiting the $CaCO_3$-platelets which are se-

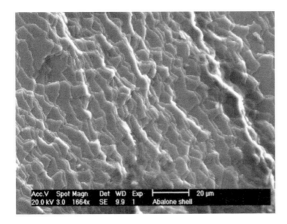

Figure 1: SEM image of a fracture surface of a Korean abalone shell showing the individual calcium-carbonate platelets separated by organic compounds.

parated by organic compounds which exhibit nanometer dimensions. These systems have evolved and been optimized by evolution over millions of years into sophisticated and complex structures. In natural systems the bottom-up approach starting from molecules and involving self organization concepts has been highly successful in building larger structural and functional components. Functional systems are characterized by complex sensing, self repair, information transmission and storage and other functions all based on molecular building blocks. Examples of these complex structures for structural purposes are teeth, such as shark teeth, which consist of a composite of biomineralized fluorapatite and organic compounds. These structures result in the unique combination of hardness, fracture toughness and sharpness, see Figure 2.

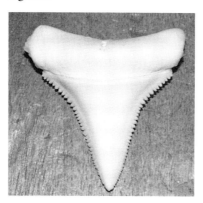

Figure 2: Example of a nanostructure found in nature: shark tooth with unique mechanical properties. The overall dimension of the tooth can reach several cm.

Another example for a biological nanostructure is opal which exhibits unique optical properties. The self cleaning effects of the surfaces of the lotus flower have been attributed to the combined micro- and nanostructure which in combination with hydrophobic groups give the surface a water and dirt repellent behavior. [2] In the past few years, numerous companies have realized products resembling the surface morphology and chemistry of the lotus flower such as paint, glass surface and ceramic tiles with dirt repellent properties. The realization that nature can provide the model for improved engineering has created a research field called „bio-„mimicking or bio-inspired materials science. It has been possible to process these ceramic-organic nanocomposite structures which provide new technological opportunities and potential for applications. [3] Other exciting results have been published such as the biomimetic growth of synthetic fluorapatite [4] in the laboratory and promising new technical applications of these nanomaterials are envisioned. [5] Other man-made nanostructures were manufactured for their attractive optical properties, such as the colloidal gold particles in glass as seen in medieval church windows.

While plentiful man made materials with nanostructures have been in use for a long time (partially without knowing it) a change of the scientific and technological approach can be identified over the past two decades. This change can be related to a few key ideas and discoveries: the idea of assembling nanostructures from atomic, molecular or nanometer sized building blocks, [6] the discovery of new forms of carbon, i.e., fullerenes [7] and carbon nanotubes, and the development of scanning probe microscopy, [8] such as scanning tunneling microscopy (STM) and atomic force microscopy (AFM). With the visionary goals many researchers worldwide have worked intensively on the development of novel or improved synthesis methods, new and better characterization techniques and the measurement and the design of the properties of nanostructured materials. In this paper some aspects of the immensely wide field will be described. However, as the field of nanostructured materials is very broad including all classes of materials as well as composites it is not possible and not attempted to consider all developments and all research groups and industries working in this area.

3   Synthesis

The microstructure and properties of nanostructured materials depend in an extreme manner on the synthesis method as well as on the processing route. Therefore, it is of utmost importance to select the most appropriate technique for preparation of nanomaterials with desired properties and property combinations. Synthesis techniques can be divided into bottom-up and top-down approaches. The top-down approach starts with materials with conventional crystalline microstructures, typically metals and alloys, and defects such as dislocations and point defects are introduced by severe plastic deformation such as in equal channel pressing. The recrystallization of the material leads to finer and finer grain sizes and under certain processing conditions to nanostructured materials. The advantage of these approaches is the fact that bulk nanostructured materials with theoretical density can be prepared. An alternative to obtain theoretical dense materials is the pulsed electrodeposition method developed by Erb and El-Sherik which yields nanocrystalline strips, however, only with thicknesses of several hundred microns. [9] The bottom-up approach includes many different techniques which are based on liquid or gas phase processes. Classically, wet chemical processes such as precipitation and sol-gel have been employed to obtain nanoparticles, however, with the disadvantage of severe agglomeration. In the gas phase metallic and ceramic nanoparticles have been synthesized by using Inert Gas Condensation, Flame Pyrolysis (Aerosol process by Degussa) and chemical vapor based processes. The major microstructural features in preparing nanoparticles for subsequent use are: nanometer sized primary particles with narrow size distributions, minimum amount of agglomeration, good crystallinity, etc.

Two techniques, chemical vapor synthesis (CVS) in the gas phase [10] and electrodeposition under oxidizing conditions (EDOC) in the liquid phase, [11] together with the resulting microstructures will be presented in more detail and the advantages and disadvantages be discussed. CVS is based on chemical vapor deposition (CVD) for the synthesis of thin films and coatings by the decomposition of metalorganic precursors. Whether thin films are deposited by heterogeneous „nucleation or nanoparticles are formed in the gas phase by homogeneous nucleation is determined by the residence time of the precursor in the hot zone of the reactor. The most important parameters determining the growth regime and the particle size are the total pressure, the precursor partial pressure and the temperature of the reaction zone. A typical reactor set-up is shown schematically in Figure 3 with one precursor source, the hot wall reactor, the thermophoretic collector, the pumping unit and the control devices for pressure and temperature. The hot wall concept operating at reduced pressures has been successfully scaled up in a cooperation project with a large German corporation involved in the synthesis of nanopowders such as carbon black, titania and silica. [12]

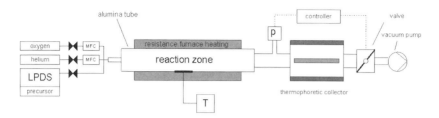

Figure 3: Schematic diagram of the major components of a CVS hot wall reactor: precursor source (liquid precursor delivery system, LPDS), hot wall reaction zone, thermophoretic particle collector, pumping system, and pressure and temperature control.

When two precursors are used, the precursor delivery can be modified in the following way:

(1) two precursors are introduced simultaneously into the reaction zone yielding doped nanoparticles (i.e. alumina doped zirconia); [13]
(2) two precursors are introduced into two concentric reaction tubes, reacted to form nanoparticles and then mixed in the gas phase to yield a nanocomposite structure (i.e., alumina mixed with zirconia) and
(3) in the first reaction zone the first precursor is decomposed to form nanoparticles by homogeneous nucleation which are subsequently coated in a second reaction zone by introducing the second precursor under conditions which favor CVD deposition (i.e., alumina surface coated zirconia). [14]

The experimental set-up of case 3) can be further modified by using a plasma reaction zone with pulse option which allows the controlled functionalization with organic molecules and polymeric shells. [15,16] Figure 4 shows a high resolution electron image of polymer coated titania nanoparticles where the crystalline titania core can be clearly distinguished from the amorphous organic shell on several grains.

Further evidence of the complete coating can be obtained by surface analysis, FTIR studies and by dispersion „experiments in different organic liquids and water. The modification of the surfaces of nanoparticles allows the improvement of dispersibility in various aqueous and organic solvents which is important for many ceramic processing steps (dip- or spin-coating, slurries for ceramic processing, etc.) and for technical applications of dispersions. Additionally, the inorganic core/polymer shell structure allows the preparation of polymer nanocomposites with excellent separation between the inorganic nanoparticles.

A further variation by exact control of all synthesis parameters allows the growth of thick nanocrystalline coatings on dense and porous substrates. Depending on the substrate temperature the porosity of the coating can be changed over a wide range up to theoretical density. This intermediate stage, called CVD/CVS, has been successfully used to deposit a nanocrystalline coating of yttrium stabilized zirconia on porous anode substrates for high temperature solid oxide fuel cell applications. Figure 5 shows a high resolution scanning electron image of a coated anode substrate. [17]

The processes leading to particle formation have been modeled and simulated by many authors. The detailed description of these efforts is beyond the scope of this paper. A comprehen-

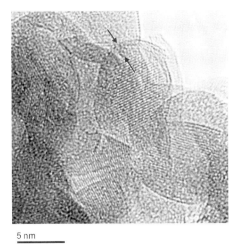

Figure 4: High resolution TEM of titania nanoparticles (crystalline core) coated with an amorphous polymeric shell.

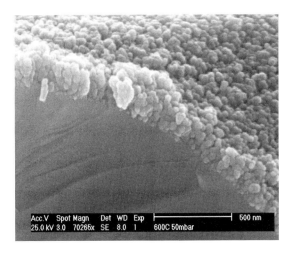

Figure 5: High resolution SEM of a nanocrystalline coating of yttrium stabilized zirconia on a porous anode substrate.

sive overview of the modeling is given by Winterer. [18] Recently, with the availability of large scale supercomputers and parallel PC clusters, the simulation of atomic processes involving millions of atoms has become available. Several authors have employed the molecular dynamics (MD) simulation technique to obtain details of the processes during particle formation, agglomeration, and sintering. As the nanoparticles contain only a limited number of atoms, it is possible to study diffusion and rearrangement processes leading to aggregation and particle growth. Atomistic simulations are extremely useful in describing the initial stages of sintering and equilibrium particle morphologies which determine the final structure and properties of nanoparticles prepared in the gas phase. Figure 6 shows the time evolution of Ge-nanoparticles, the aggregation and sintering during condensation in an Ar-gas. [19] In order to be able to simulate this process, high concentrations of atoms in the gas phase have to be employed in order to be able to adapt the time scale to the computing capabilities. A scaling law was established which allows the comparison to experimental conditions. MD-simulations have been employed for many other processes in nanocrystalline materials such as the calculation of the elastic properties [20] of nanocrystalline nickel and the „plastic and superplastic deformation of nanocrystalline metals [21–23] and the atomistic structure and energies of grain boundaries in nanocrystalline metals.

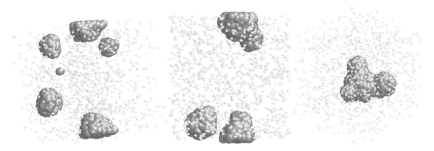

Figure 6: Result of a Molecular Dynamics (MD) simulation showing the evolution of Ge-nanoparticles in Ar-gas for 0,96 ps a), 1,29 ps b) and 3.36 ns c). The aggregation, sintering and change of morphology can be observed.

The EDOC process was developed to synthesize nanocrystalline oxides with improved properties as compared to oxides obtained by precipitation and sol-gel routes. [11,24] In particular, it was attempted to improve the control on size and size distribution, the dispersibility and the possibility to modify the surfaces of the nanocrystalline powders. In the EDOC process an anode, i.e., metallic Zn, is dissolved in an organic electrolyte with an organic conducting salt by applying a voltage. The Zn^{2+}-ions migrate to an inert cathode, are discharged and subsequently oxidized by means of bubbling air through the electrolyte, thus forming ZnO-nanoparticles. The conducting salt has the additional role to prevent continued growth of the nanoparticles and their agglomeration. After filtration, the nanoparticles can be further modified or functionalized by exchanging the organic shell by other molecules to adapt the surface chemistry to other chemical environments. Other metal oxides have been prepared by EDOC, such as SnO_2, TiO_2, and ZrO_2 as well as doped oxides. Compared to other wet chemical processes, the EDOC nanopowders exhibit superior dispersibility in solvents.

In conclusion, it can be stated that many synthesis techniques exist with individual advantages and disadvantages depending on the requirements of the material. In many processes, the microstructure, the morphology, the size and size distribution, the agglomeration and the elemental distribution on the scale of the nanoparticles can be controlled. In addition, it is possible to control the surface chemistry and thus to control the reactivity of the nanomaterials with the environment. The availability of versatile synthesis techniques is the prerequisite for materials design on the nanometer scale.

4 Characterization

The characterization of nanostructured materials in the form of thin films, nanoparticles and bulk structures demands special techniques which allow for the structural and chemical analysis with a sufficient lateral resolution. Consequently, special characterization techniques are required besides the standard techniques available to materials science. The extremely high surface and interface areas have to be considered. Experimental difficulties such as oxidation/reaction at surfaces, the disordered structure at the grain boundaries and other internal interfaces and substantial porosity can be present in these materials. The specific surface area of nanopowders with average grain sizes below 10 nm can rise to several hundred m^2/g. That is, two to three grams of nanopowders which fit into a small volume of a few cm^3 can have the same surface area as an entire football field. Therefore, „nanocrystalline powders are considered to be excellent candidates for catalysis and gas sensing devices.

The crystallinity even in materials with average particle/grain sizes below 10 nm is very good. Figure 7 shows an X-„ray diffractogram of nanocrystalline zinc oxide together with the Rietveld fit. An average particle size of 8 nm was determined independently by HRTEM and from the fit of the line broadening. The clear presence of all crystalline maxima indicates an excellent crystallinity of the nanopowders after the preparation and without any calcination treatment. The crystallinity of nanoparticles from the CVS process is even better which can be observed clearly in the high resolution TEM shown in Figure 8 for the case of zirconia.

An important aspect is the analysis of the elemental distribution on the scale of the nanoparticles. Only a few techniques provide the information with sufficient lateral resolution. HRTEM in combination with an Omega filter is a new technical solution to increase the analytical capabilities of transmission electron microscopy. Winterer describes the possibility to use EXAFS in

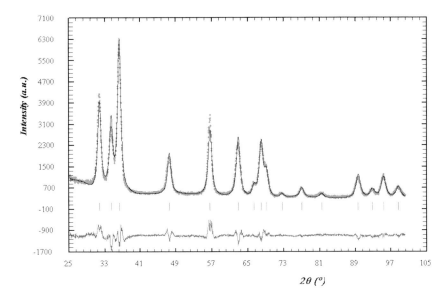

Figure 7: X-ray diffractogram of nanocrystalline ZnO powder prepared by EDOC. The experimental data, the Rietveld fit and the difference plot are shown.

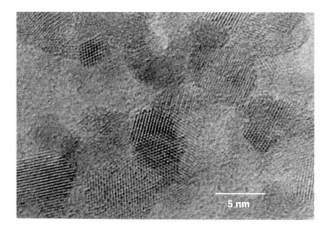

Figure 8: High Resolution TEM of as prepared CVS-zirconia nanoparticles. The lattice fringes extend to the surface proving excellent crystallinity. A narrow size distribution and crystallographic habitus planes at the surfaces are observed.

combination with Reverse Monte Carlo simulation to obtain the detailed structure and elemental distribution in nanometer sized particles. [25] By analyzing the pair distribution functions of Zr-Zr, Zr-Y, and Y-Y it was shown that depending on the synthesis parameters a surface layer rich in yttrium oxide is formed instead of the expected yttrium doped zirconia. Other standard characterization techniques were not capable to resolve the thin surface layer with a different composition. It should be pointed out that a change of the synthesis conditions resulted in a homogeneously doped oxide. The knowledge of the detailed elemental distribution is extremely important in order to establish the structure-property relationships and to understand the complex behavior of nanomaterials. Other frequently used characterization techniques include nitrogen adsorption for porosity and pore size distribution, [26] small angle neutron scattering (SANS) for particle and pore size determination, [27] surface analytical techniques such as X-ray photoelectron spectroscopy (XPS), ion scattering spectroscopy (ISS), nuclear techniques such as Mössbauer spectroscopy and many more depending on the details of interest.-

5 Properties

The interest in nanostructured materials arises from the fact that due to the small size of the building blocks and the high density of interfaces (surfaces, grain and phase boundaries) and other defects such as pores, new physical and chemical effects are expected or known properties can be improved substantially. In addition, the novel processing routes allow a bottom-up approach in materials design. In the „following sections several examples of properties which are altered dramatically in the nanometer regime are presented.-

5.1 Nanoparticles

The main reason for using nanoparticles is the large surface area which will be favorable for gas-solid interactions such as in catalysis [28] and gas sensing. [29] In addition, nanoparticles are the building blocks of bulk nanocrystalline materials, i.e., ceramics, prepared by sintering. Several authors have examined the catalytic activity of nanocrystalline materials. In catalysis, alumina with a high specific surface area is used extensively as support material for catalytically active noble metals. However, the role of the substrate material was not studied extensively. For several systems it was found that the influence of the substrate material on the catalytic performance in gas reactions, such as complete oxidation of methane, can be quite drastic. [30–32] Another application of the surface reactivity of nanocrystalline oxides is in gas sensing devices. As an example, nanocrystalline titania prepared by CVS was used for gas sensing. [29] A typical experimental arrangement consists of an alumina substrate with an interdigitized finger-like metallic structure on one side and a resistive heater on the other side. This allows the deposition of the sensing material and the measurement of the resistivity as a function of the atmosphere. As shown in Figure 9 a nanocrystalline coating was obtained by screen printing a dispersion and a moderate sintering step at 600 °C for one hour. The film is highly porous which allows the reactive gases to reach all surfaces of the sensing material in short times. All grains are in good contact to each other and sintering necks with well developed grain boundaries are formed. This

Figure 9: High resolution SEM demonstrating the nanocrystalline structure with nanoporosity of the titania layer prepared by screen printing and sintering.

is an important prerequisite for the measurement of the resistance of the film. As seen in Figure 10, the reaction to a change of atmosphere, i.e. exposure to a different concentration of oxygen, is extremely fast, with response times (between 10 and 90 % of the total signal) in the order of less than three seconds. This compares to the response times of several minutes for commercial sensors which are based on dense thin films deposited by physical vapor deposition such as sputtering. The fast response time is an attractive feature of the nanostructured gas sensors and is a direct consequence of the nanocrystalline and nanoporous structure.

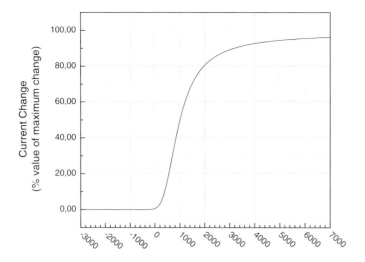

Figure 10: Change of the signal of a sensor made of nanocrystalline titania in response to a change of the oxygen concentration in the atmosphere. The sensor temperature was set at 300 °C. The reaction time of the sensor, i.e. the time between 10 % and 90 % of the final signal is less than 3 s.

Other interesting physical effects are observed in metallic nanoparticles such as FePt prepared by a wet chemical process described by Sun et al. [33] Fe50Pt50 nanoparticles in the size range from three to seven nanometers can be prepared with an organic coating of oleic acid which prevents the oxidation even when exposing the nanoparticles to air. [34] The „organic shell also serves to keep the nanoparticles at a molecular distance. Therefore, the nanoparticles are superparamagnetic and can be arranged on clean surfaces in regular arrays which opens interesting applications for magnetic storage „devices.

5.2   Nanoceramics

Nanoceramics with grain sizes in the range of 10 to 100 nm are typically prepared by consolidation, i.e., uniaxial compaction, cold isostatic pressing and sintering in air or vacuum. The green density, i.e. the density after initial compaction, reaches values of 40 to 50 % of the theoretical value for CVS nanopowders. The most striking feature in the case of CVS powders is the transparency of the compacted pellet. This is a direct consequence of the narrow size distribution of pores in the material which can only occur if no agglomerates are present in the as prepared powder or are broken up during the consolidation step. The dependence of the green density with increasing compaction pressure does not show any change in slope which would be an indication of breakage of agglomerates. The size distribution shows only pores smaller than the primary particle size. This is further evidence for the lack of agglomeration in CVS powders. This feature is a prerequisite for a good sinterability of the CVS powders which has been found to occur in all nanoceramics under investigation (TiO_2, ZrO2, Y_2O_3, Al_2O_3, various doped oxides etc.) at temperatures well below 1/2 T_M. In the intermediate sintering steps at lower temperatures, the pores disappear but no excessive pore growth is observed. However, in pure ceramics grain growth occurs when the density exceeds 90 % of the theoretical value. This undesirable effect has been reduced by doping, surface doping (core-shell nanoparticles) and composite structures. [13,14] In Figure 11 the grain size is plotted as a function of the sintering temperature for several oxides.

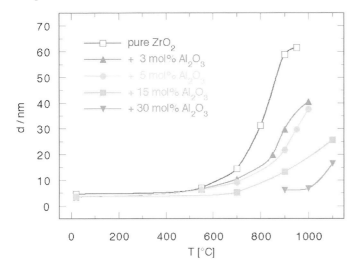

Figure 11: Grain size as a function of sintering temperature for pure zirconia and zirconia doped with four different concentrations of alumina. Only the $ZrO_2/30$ mol.-% Al_2O_3 nanopowder is amorphous in the as prepared state, all other samples are crystalline with the tetragonal structure. The crystallization of the amorphous powder occurs at a temperature of approx. 800 °C. Pure ZrO_2 shows excessive grain growth with an average grain size typically exceeding 100 nm, while the grain growth is effectively reduced by the doping.

In the pure nanocrystalline zirconia extensive grain growth is observed at temperatures and in the fully dense ceramic (100 % TD) sintered at 1000 °C the average grain size is typically above 100 nm. Fully dense ceramics are also obtained for zirconia doped with 3 to 5 mol % alumina at grain sizes in the range of 40 to 50 nm. Further increase of the doping to 15 mol % further reduces the grain growth; however at a sintering temperature of 1000 °C the theoretical density is not reached. At a higher doping level of 30 mol.-% alumina, the as prepared nanopowders are amorphous, crystallize at the sintering temperatures above 600 °C and show a phase separation. At this doping level the grain growth is suppressed most effectively as the grains grow only up to 30 nm even at 1200 °C. It should be mentioned that the crystallographic structure of as prepared pure zirconia with average grain size below 10 nm is monoclinic and tetragonal, while in the alumina doped samples the tetragonal structure is stabilized. Core-shell particles with an alumina rich surface and two phase zirconia-alumina structures prepared with the CVS method as described above also exhibit a drastic effect on the grain boundary mobility. In summary of the study of various zirconia ceramics, it can be stated that during pressureless sintering 1) pure nanocrystalline oxides exhibit extensive grain growth at high densities; 2) grain growth can be suppressed effectively by doping with different elemental distribution of the doping element; 3) depending on the concentration of the doping element fully dense and nanoporous, nanocrystalline ceramics can be obtained which are stable against pore and grain growth at temperatures exceeding 1000 °C. The latter ceramics could be of interest for high temperature gas filters and as catalyst support materials at high temperatures for gas phase reactions. The densification can be further enhanced by the application of a high uniaxial or hydrostatic pressure while grain growth is further suppressed.

The grain size can influence the properties of nanomaterials in various ways. As shown for sintering the basic processes leading to densification and grain growth are identical and consequently, the properties are enhanced according to the grain size dependence. However, it is also possible that new processes lead to a different dependence on grain size. As an example, the plastic or superplastic deformation of metals and ceramics with nanocrystalline grain sizes is described. In metallic systems the Hall-Petch relationship describes the grain size dependence as a result of the interaction of dislocations and grain boundaries. Several theoretical approaches lead to a grain size dependence of the hardness or mechanical strength. The hardness increases as the inverse of the square root of the grain size. Several authors have reported that this well established behavior changes drastically as the grain size gets below a critical value in the nanometer regime. In this case the material becomes softer as the grain size is reduced and consequently, the effect is named Inverse Hall-Petch effect. It is obvious that different processes which require a lower stress than the dislocation-grain boundary interaction are operative in the nanometer regime. Grain boundary sliding is an alternative process, however, the steric hindrance by neighboring grains has to be overcome. Many theories describe dislocation and diffusion processes which relax the high stresses at the triple junctions when sliding occurs between grains. An alternative process based on the formation of mesoscopic grain boundaries has been proposed for nanocrystalline materials. [35] By grain boundary migration the grain morphology is altered to produce a flat mesoscopic interface over the dimensions of many grain diameters. The threshold stress for the smoothening of the grain boundaries has been calculated and found to decrease with decreasing grain size, i.e., an opposite behavior as in materials with larger grain sizes. In the model the basic step of the sliding is described by local atomic rearrangements of grain boundary atoms at the stress concentration sites leading to the sliding of the neighboring grains. As dislocations in the grains do not play a role in this process mesoscopic sliding can be active in both metals and ceramics. The model of a new process responsible for the deformation of nanocrystalline metals and ceramics has been further supported by molecular dynamics simulations of nanocrystalline nickel at different grain sizes. [36] It is found that at the smallest grain sizes the dominant processes occur in the interfaces leading to grain boundary sliding, atomic shuffling in the interfaces, grain rotation etc. However, in these materials no dislocation activity is observed in agreement with the model predictions. At larger grain sizes, the effects of dislocation activity are observed as stacking faults indicate the motion of dislocations through the grains. However, MD requires very short time scales in the range of nanoseconds and therefore, high stresses which do not correspond to experimental values are applied. The consequences for the atomic processes are not clear. In summary, it can be stated that for nanocrystalline materials a transition from dislocation based mechanisms in larger grains to an interface controlled sliding mechanism seems most likely as it explains the observed change of grain size dependence. The combination of deformation experiments, theoretical models and atomistic simulations provide an insight into the complex deformation processes of nanocrystalline materials.

6   Applications

Nanotechnology is already used extensively in modern industrial products. The semiconductor industry which relies on the miniaturization of the structural components has used device structures in the nanometer range in commercial products. The time to market in this industry is extremely short as the demand by the end user is very high. An example for the fast realization of a technical product based on a new physical effect is the giant magneto resistance (GMR) effect originally discovered in 1985. A few years later the first hard disks with a GMR-based read head were in the market and a revolutionary increase in storage density leading to smaller hard disks followed.

Over decades several companies have marketed agglomerated powders in many industrial applications. Although the primary particle size is in the nanometer range, properties such as dispersibility, light scattering and sinterability are determined by the size of the agglomerates or aggregates. The continued need for better products can only be fulfilled by improved materials and consequently, many companies have intensified research in this area. In developing new products companies have to consider the costs of new materials and technologies as private and industrial customers are not willing to \(buy nano, but only improved products and processes\). For the continuing support of basic science in the field of nanomaterials and nanotechnology it is of utmost importance that the new properties and technologies also lead to commercial successes. The range of ideas for commercial products in nanomaterials is as wide as the field of materials science, covering mechanical, physical, chemical, biological, pharmaceutical, medical and cosmetic areas. Several companies have introduced into the market sun screen products based on TiO_2 and ZnO nanopowders in an attempt to reduce the amount of organic sun blockers at the same sun protection factor. An important consideration for many applications is the dispersibility of the nanopowders in various liquid (dispersions or slurries for ceramic processing) and solid media (nanocomposites) which requires surface modifications to adapt the inorganic surface to the matrix. Ferrofluids, i.e., magnetic fluids, which have been in commercial use for many years are an example of the extremely good stability of liquid dispersions despite the strong magnetic interaction forces. Dispersions with excellent stability have been synthesized and are used in many processes. The use of nanopowders in semiconductor processing for chemical mechanical polishing (CMP) is a huge market and an important component in the manufacturing of modern integrated circuits with three-dimensional (3D) architecture. In the field of catalysis and gas sensors nanocrystalline powders and porous ceramics are considered. The development of new products by transferring basic science results has led to the foundation of numerous start-up companies as well as the involvement of large international companies in many countries. Some examples in the area of nanomaterials are Nanophase Technologies, Nanopowder Enterprises in the United States, Samsung Corning in Korea, APT in Australia, Sus-Tech Darmstadt, NanoGate, and Degussa in Germany. It is anticipated that in the next few years many approaches will lead to the use of nanomaterials and only future can tell which attempt will also be a commercial success.

7 Acknowledgement

The author acknowledges the financial support of the research work by the Deutsche Forschungsgemeinschaft, the Bundesministerium für Forschung und Technologie, the Alexander-von-Humboldt Stiftung, the Deutsche Akademische Austauschdienst, the Hahn-Meitner-Institut Berlin and SusTech GmbH & Co. KG Darmstadt.

The plentiful contributions and discussions by PD Dr. habil. M. Winterer, Dr. B. Stahl, Prof. K. Albe, Dr. M. Ghafari, Prof. A. Raju, Dr. A. G. Balogh, Dipl.-Ing. J. Seydel, M. Sc. Yong-Sang Cho and other members of the Thin Films Division, and Dr. W. Miehe and Prof. H. Fuess of the Structural Research Division in the Institute of Materials Science, TU Darmstadt, and Prof. V. Srdic, Novi Sad, Serbia and Montenegro, Prof. K. A. Padmanabhan, Hyderabad, India, Prof. G. Hoflund, Gainesville (USA) are appreciated.

8 References

[1] R. Z. Wang, Z. Suo, A. G. Evans, N. Yao, I. A. Aksay, J.'Mater. Res. 2001, 16 2485.
[2] W. Barthlott, C. Neinhuis, Planta 1997, 202, 1
[3] I. A. Aksay, M. Trau, S. Manne, I. Honma, N. Yao, L.'Zhou, P. Fenter, P. M. Eisenberger, S. M. Gruner, Science 1996, 273, 892
[4] S. Busch, H. Dolhaine, A. DuChesne, S. Heinz, O. Hoch‚rein, F. Laeri, O. Podebrad, U. Vietze, Th. Weiland, R.'Kniep, Eur. J. Inorg. Chem. 1999, 1643
[5] For some examples in magnetic nanocomposites, nanodispersions, and others, see www.sustech.de
[6] H. Gleiter, Proc. of the 7th Riso Int. Symposium on Metallurgy and Materials Science (1981), Roskilde, pp. 15–21
[7] R. Smalley, R. Curl, H. Kroto, Nobel Prize in Chemistry 1996
[8] Heinrich Rohrer, Gerd Binnig, Nobel Prize in Physics 1985
[9] A. M. El-Sherik, U. Erb, J. Mater. Sci. 1995, 30, 5743
[10] W. Chang, G. Skandan, H. Hahn, S. C. Danforth, B. H. Kear, Nanostructured Mater. 1994, 4, 345
[11] A. Dierstein, H. Natter, F. Meyer, H.-O. Stephan, C.'Kropf, R. Hempelmann, Scripta Mat. 2001, 44, 2209
[12] Degussa, Hanau, http://www.degussa.com
[13] V. V. Srdic, M. Winterer, H. Hahn, J. Am. Ceram. Soc. 2000, 83, 1853
[14] V. V. Srdic, M. Winterer, A. Möller, G. Miehe, H. Hahn, J. Am. Cer. Soc. 2001, 84, 2771
[15] I. Lamparth, D. Szabo, D. Vollath, Macromol. Symp. 2002, 181, 107
[16] M. Schallehn, M. Winterer, T. Weirich, U. Keiderling, H.'Hahn, Chem. Vap. Dep. 2003, 9, 40
[17] J. Seydel, M. Winterer, H. Hahn, Mat. Res. Soc. Symp. Proc 2001, 676, Y8.14.1-5
[18] M. Winterer, Nanocrystalline Ceramics, Synthesis and Structure, Springer Series in Materials Science 53, New York 2002
[19] P. Krasnochtchekov, K. Albe, Y. Ashkenazy, R. S. Averback, J. Phys. Chem., submitted
[20] S.-J. Zhao, K. Albe, H. Hahn, Acta Mater., submitted

[21] H. Van Swygenhoven, D. Farkas, A. Caro, Phys. Rev. B 2000, 62, 831
[22] V. Yamakov, D. Wolf, M. Salazar, S. R. Phillpot, H. Gleiter, Acta. Mater. 2001, 49, 2713
[23] K. W. Jakobsen, J. Schiotz, Nature Materials 2002, 1, 115
[24] R. Hempelmann, H. Natter, EP 1 121 477, 2003
[25] M. Winterer, J. Appl. Phys. 2000, 88, 5635
[26] H. Hahn, Nanostructured Mater. 1993, 2, 251
[27] U. Keiderling, A. Wiedenmann, V. Srdic, M. Winterer, H. Hahn, J. Appl. Cryst. 2000, 33, 483
[28] H. Hahn, H. Hesemann, W. Epling, G. B. Hoflund, Mat. Res. Soc. Symp. 1998, 497, 35
[29] Yong-Sang Cho, Ph.D. Thesis, Technische Universität Darmstadt 2003
[30] G. B. Hoflund, Z. Li, W. S. Epling, T. Göbel, P. Schneider, H. Hahn, React. Kinet. Catal. Lett. 2000, 70, 97
[31] A. Tschöpe, D. Schaadt, R. Birringer, J. Y. Ying, Nanostructured Mater. 1997, 9, 423
[32] S. H. Oh, M. L. Everett, G. B. Hoflund, J. Seydel, H. Hahn, Mat. Res. Soc. Symp. Proc. 2001, Y4.6.1-6, 676
[33] S. Sun, C. B. Murray, D. Weller, L. Folks, A. Moser, Science 2000, 287, 1989
[34] B. Stahl, N. S. Gajbhiye, G. Wilde, D. Kramer, J. Ellrich, M. Ghafari, H. Hahn, H. Gleiter, J. Weißmüller, R.'Würschum, P. Schlossmacher, Adv. Mater. 2002, 14, 24
[35] H. Hahn, K. A. Padmanabhan, Phil. Mag. B 1997, 76, 559
[36] A. Hasnaoui, H. Van Swygenhoven, P. M. Derlet, Phys. Rev. B 2002, 66, 184 112

Properties, Benefits, and Application of Nanocrystalline Structures in Magnetic Materials

Roland Grössinger, Reiko Sato, David Holzer, and Mikael Dahlgren
Institut für Festkörperphysik, Technische Universität Wien, Wien (Austria)

1 Abstract

The magnetic properties of nanocrystalline hard magnetic and soft magnetic are summarized. When the grain size becomes of the order f the magnetic exchange length exchange coupling occurs. The different concepts of exchange coupling in these materials are discussed. Exchange coupling leads in "isotropic hard magnetic materials to a remanence enhancement. Soft magnetic materials exhibit due to exchange coupling a lower coercivity, lower losses and consequently also improved properties.

2 Introduction

In the last years great effort was made to improve the magnetic properties of hard and soft magnetic materials. Permanent magnets based on Nd–Fe–B with an energy product of 451 k J/m^3 were recently reported. [1] Any significant further improvement needs a new compound with a higher saturation magnetization.

Similar is the situation for soft magnetic materials. There since long time Fe–Si (about 3 % Si) was and is the most important material. New developments lead to soft magnetic amorphous materials were the best magnetic properties were found in the system (Fe,Co)80(Metalloid = Si,B)20. Even extrinsic properties such as the permeability, the coercivity and the losses were improved, never a real large scale break through was achieved because of the limited saturation magnetization of about 1.4 T (see e.g., Luborsky). [2]

This limits were overcome in the last year by the invention of nanocrystalline materials. In this case a special sample preparation procedure leads to nanocrystalls (grain size typical 10–30 nm) which exhibited new unexpected magnetic properties. The most outstanding improvement was the remanence enhancement of isotropic hard magnetic materials on one side but also the extremely improved soft magnetic properties on the other side. In the following an overview of the magnetic properties of hard and soft magnetic nanocrystalline materials will be given.

3 Production of Nanocrystalline Materials

Similar as for amorphous materials various production techniques were established to produce nanocrystalline hard magnets; melt spinning, mechanical alloying, but also direct powder production such as spark erosion. By melt spinning or rapidly quenching the nanocrystalline state can be obtained directly [3] or after a heat treatment of an over-quenched material. [4]

Also, besides melt-spinning and splat cooling, other rapid solidification techniques have been used like vapor deposition, [5–8] atomization, [9] and mechanical alloying applying a high energy ball mill [10,11] for the fabrication of nanophase magnets. Additionally, sputtering techniques have also been used to prepare nanosize Sm–Co alloys, [12,13] and CoPt and FePt alloys for high density recording media. [14,15] All kind of milling procedures forms first an amorphous powder, the nanocrystalline material is formed due to a heat treatment. [16] Recently, some nanocrystalline materials were produced by mechanical deformation. [17,18]

4 Isotropic Hard Magnetic Materials

Concerning application permanent magnets are considered to be of good quality if their energy product is as high as possible. Other favorable properties are a high coercivity, a small temperature coefficient, a high Curie temperature, T_C etc. Beside good magnetic properties magnets should be "resistant against corrosion and from commercial point they should be inexpensive.

The theoretical limit for the maximum energy product, $(BH)_{max}$ of a given magnetic material [19,20] is given by:

$$(BH)_{max} \le \frac{J_S^2}{4\mu_0} \tag{1}$$

The remanence for a polycrystalline magnet with non-interacting isotropic, uniaxial grains with an easy c-axis can be calculated to be:

$$J_r = J_S \frac{\int_0^{2\pi}\int_0^{\pi/2} \cos\vartheta \sin\vartheta \, d\vartheta \, d\varphi}{\int_0^{2\pi}\int_0^{\pi/2} \sin\vartheta \, d\vartheta \, d\varphi} = \frac{1}{2} J_S \tag{2}$$

In a nanocrystalline material due to the fact that exchange coupling between neighboring grains occur even for an isotropic material a ratio $J_r/J_S > 0.5$ can be achieved. This leads to a larger energy product which is important for applications.

5 Exchange Coupling

For nanocrystalline material the way how the small grains are coupled is of great importance for the understanding of the remanence enhancement. Therefore, in the following part the fundamental considerations are surveyed. For grains of nanosize different exchange length have to be considered:

The exchange length due to an external field:

$$\lambda_H = \sqrt{\frac{2A}{H\mu_0 M_S}} \tag{3}$$

The exchange length due to the crystal energy:

$$\lambda_K = \sqrt{\frac{A}{K}} \qquad (4)$$

The exchange length due to stray fields:

$$\lambda_S = \sqrt{\frac{2\mu_0 A}{(\mu_0 M_S)^2}} \qquad (5)$$

It depends on the considered material which type of "exchange length" is more important, however, for hard magnetic material the second type which is the result of a competition between quantum-mechanical exchange and magnetocrystalline anisotropy may be the most important one. For a soft magnetic material the stray field between the grains may polarize the neighboring grains. It turns out that the ratio between the mean grain diameter <D> grain and the range of the exchange interaction given by the exchange length κ_K mainly controls the magnetization distribution. [21] To decide if κ_K or κ_S is dominating for a certain magnet, the relation below can be used. [22]

$$\frac{K}{J_S^2/4\mu_0} \qquad (6)$$

If the Expression 6 is much smaller than 1 then the stray-field energy is dominating and κ_S is describing the magnetic state. If instead Expression 6 is much larger than 1 the magneto-crystalline anisotropy energy is dominating and thus κ_K determines the magnetic state (see Table 1).

Table 1: Different kind of exchange lengths and the criteria to decide which exchange energy is dominating. The calculated exchange length for some typical compounds are also given.

	$\lambda_{exch.}$	$\lambda_H \equiv \sqrt{\frac{2A}{HJ_S}}$	$\lambda_K \equiv \sqrt{\frac{A}{K_1}}$	$\lambda_S \equiv \sqrt{\frac{2\mu_0 A}{J_S^2}}$	$\frac{K}{J_S^2/4\mu_0}$
Nd$_2$Fe$_{14}$B [23,24]		λ_H = 3.1 nm	λ_K = 1.3 nm	λ_S = 2.7 nm	8.4
		λ_H = 4.5 nm (H = 1 MA/m)	λ_K = 2.0 nm	λ_S = 4.0 nm	8.4
Fe$_{17}$Sm$_2$N$_{2.7}$ [25]		λ_H = 2.5 nm (H = 1 MA/m)	λ_K = 0.7 nm	λ_S = 2.3 nm	7.4
Co$_5$Sm [26]		λ_H = 4.8 nm (H = 1 MA/m)	λ_K = 0.8 nm	λ_S = 5.2 nm	20
α-Fe [27]		λ_H = 4.8 nm (H = 1 MA/m)	λ_K = 23 nm	λ_S = 3.6 nm	$2.7*10^{-2}$

6 Nanocrystalline Hard Magnetic Materials

The nanocrystalline magnets do not exceed the magnetic properties of the best rare earth magnets as shown in Figure 1.

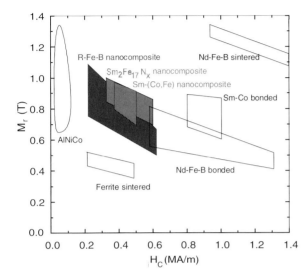

Figure 1: Remanence as a function of the coercivity for different hard magnetic materials.

The most likely way in which nanocomposite materials are to succeed in the permanent magnet market is in providing a lower cost alternative to current bonded Nd_2Fe14B_1 magnets with similar or slightly better values of maximum energy product.

Nanocrystalline hard magnetic materials with an enhanced remanence were first discovered by Coehoorn et al. [28] and by Clemente and co-workers [29] These new kind of materials are mainly characterized by an increase over the expected remanence of 50 % from the saturation magnetization that is expected for an ensemble of non-interacting isotropic magnetic uniaxial grains. [30] The remanence enhancement results in an increase of the maximum energy product $(BH)_{max}$ without a necessary magnetically alignment of the grains. In relatively low fields and in any direction these materials can be magnetized, because they are crystallographically isotropic.

The nanocrystalline magnets with an enhanced remanence can be divided into: single nanocrystalline hard magnetic phase and nanocomposites known also as spring magnets.

Generally, the phases for the nanocomposites are composed of a hard magnetic phase and a soft magnetic phase. The soft magnetic phase should have a higher saturation magnetization than the hard magnetic phase in order to "enhance the remanence even more than the "nanocrystalline single phase material. The further enhancement of the remanence is due to the fact that the hard magnetic phase is polarizing the soft magnetic phase. Higher remanence is gained in cost of a "decrease of the coercivity. The nanocomposite magnets show an increase of the (BH)max compared to the nanocrystalline single phase materials due to the higher enhancement of the remanence. However, nanocomposite magnets are limited by the low coercivity, normally less or equal to the required H_C of $0.5 \, J_r/\mu_0$.

Applying a finite element modeling allows the calculation of the optimum conditions for the formation of a nanocomposite. [31] As an example we show in Figure 2 [31] the results of such a model calculation. The right side shows the domain structure at the two different working points 1) and 2). One can also see that the total demagnetizing curve is not just the superposition of that of the Nd–Fe–B and the α-Fe phase.

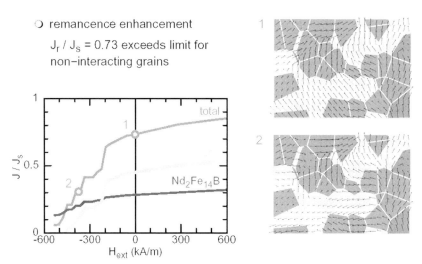

Figure 2: Finite element modeling assuming a mean grain size of 10 nm with 50 % Nd_2Fe14B and 50 % α-Fe

6.1 Magnetization and Remanence

As an example, results of magnetization and remanence obtained on nanocrystalline Nd–Fe–B and Pr–Fe–B will be here reported. [32–34] From the measurements of the hysteresis loop the coercivity and the remanence J_r were obtained. The saturation polarization J_s was determined applying the law of approach to saturation. Figure 3 shows as an example the temperature dependence of the J_s, the remanence and the ratio J_r/J_s of a sample with a Pr concentration below

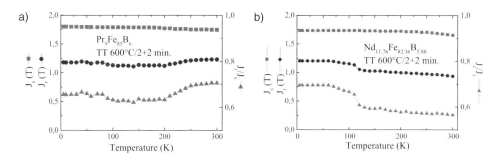

Figure 3: Temperature dependence of the saturation polarization J_s, the remanence J_r and the ratio J_r/J_s for nanocrystalline a) mechanically alloyed Pr_9Fe85B_6, and b) rapidly quenched stochiometric Nd11.76Fe82.36B5.88.

the 2-14-1 stochiometry (Pr_9Fe85B_6) as well as that of a stochiometric nanocrystalline Nd–Fe–B sample (Nd11.76Fe82.36B5.88). As a comparison in Figure 4 the same is shown for the sample with high Pr content (18 % Pr). Because of the spin reorientation which happens in Nd_2Fe14B at 135 K an easy cone is formed at low temperatures which is accompanied by a reduction and change of sign of the first order anisotropy constant. Therefore below 135 K J_r/J_S increases due to the easy cone situation (see Fig. 3b). For the Pr_9Fe85B_6 sample a decrease of J_r/J_S with decreasing temperature was found, which indicates a decoupling due to the increase of the anisotropy at low temperatures. A similar behavior was found for the stochiometric sample.

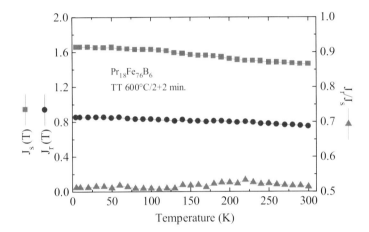

Figure 4: Temperature dependence of the saturation polarization J_s, the remanence J_r and the ratio J_r/J_s for mechanically alloyed nanocrystalline Pr18Fe76B6.

It is worth to note that for the Pr18Fe76B6 sample where the grains should be decoupled due to the existence of a nonmagnetic Pr rich phase, a J_r/J_s ratio of about 0.5 was found for the whole temperature range.

6.2 Applications

Up to now no large scale application of nanocrystalline hard magnetic materials exists. Recently Magnequench developed a new powder of nanocrystalline material, however this material is not yet on the market. The general advantage of nanocrystalline hard magnetic material can be:

a. reduced costs
b. tailoring of the magnetic properties
c. easier magnetizing process

7 Soft Magnetic Nanocrystalline Materials

The most prominent examples of Fe-based nanocrystalline alloys are devitrified Fe–Cu–Nb–Si–B alloys [35] and Fe–Zr–B–Cu alloys. [36] Nanocrystalline structures offer a new opportunity for tailoring soft magnetic materials. They are generally obtained from amorphous materials by a heat treatment at elevated temperature. So an ultrafine grain structure with a typical grain size ranging from 10 nm to 20 nm can be produced. This nanocrystalline state is characterized by a two-phase system "consisting of small crystallites embedded in an amorphous matrix. In the Fe–Cu–Nb–Si–B alloy the addition of copper is necessary to enhance the nucleation of α-FeSi grains, whereas niobium hinders the growth of the grains. [37] The

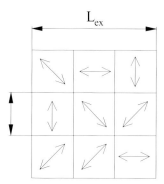

Figure 5: Schematic presentation of the random anisotropy model with grain size D and the magnetic exchange length L_{ex}. The arrows indicate the local easy axis of magnetization [38].

nanocrystalline alloys show excellent soft magnetic behavior, such as high permeability, low coercivity, and relatively high saturation magnetization.

7.1 Grain Size Dependence

An explanation of the soft magnetic behavior and also of the dependence of the coercivity on the grain size was given by Herzer, [38] by applying the Alban–Becker–Chi model, [39] the so-called random anisotropy model. Figure 5 shows the schematic representation of this model. The model shows that the magnetic exchange length has to be larger than the grain size.

The magnetic properties are influenced by the averaging effect of the ferromagnetic exchange interaction on the local magneto-crystalline anisotropy energy. For large grains the local magnetization follows the local easy magnetic directions and the magnetization process is determined by the mean value of the magneto-crystalline anisotropy.

For very small grains the magnetic exchange interaction forces the magnetic moments to align parallel. Therefore, the magnetization can not follow the easy directions of each individual grain. So the effective anisotropy is an average over several grains and therefore drastically reduced in magnitude.

$$N = (L_{ex}/D)^3 \tag{7}$$

$$\langle K \rangle = \frac{K_1}{\sqrt{N}} \tag{8}$$

The critical parameter differentiating these two cases is given by the ferromagnetic exchange length:

$$L_{ex}^0 = (A/K_1)^{1/2} \tag{9}$$

In this Equation A denotes the exchange stiffness and K_1 is the magneto-crystalline anisotropy. The averaged anisotropy can be described by the random anisotropy model. The basic idea can be deduced from Figure 5. For a three-dimensional structures the grain size dependence of the average anisotropy is given by

The coercivity as well as the initial permeability must be closely correlated to the average anisotropy <K>:

$$H_C = p_C \frac{\langle K \rangle}{J_S} \approx p_C \frac{K_1^4 D^6}{J_S A^3} \tag{10}$$

$$\mu_i = p_\mu \frac{J_S}{\mu_0 \langle K \rangle} \approx p_\mu \frac{J_S^2 A^3}{\mu_0 K_1^4 D^6} \tag{11}$$

The parameters p_c and p_μ are factors close to unity. It was found [38] that the experimental dates for coercivity and permeability are compatible with the expected -dependence for grain sizes below 40 nm.

Therefore the soft magnetic properties can be improved by reducing the grain size (only valid for D < D_{crit}). The dependence of the coercivity on the grain size is shown in Figure 6. [40] It

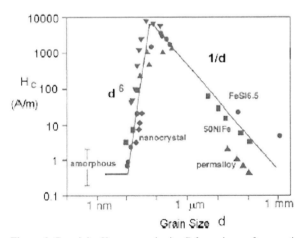

Figure 6: Coercivity H_c versus grain size D for various soft magnetic alloys [40].

must be noticed that only an average grain size D can be determined, but a real material has a grain size distribution.

Table 2: Magnetic properties of a commercial Finemet type material

Material data, magnetic properties	VITROPERM 500 Z	VITROVAC 6025 Z
Material base	nanocrystalline Fe-based	amorphous Co-based
Saturation flux density (25C), B_s	1.2 T	0.58 T
Bipolar flux density swing (25C), $\Delta B_{ss,25C}$	2.35 T	1.15 T
Bipolar flux density swing (90C), $\Delta B_{ss,90C}$	2.15 T	1.0 T
Squareness, B_r / B_s (typical value)	>94 %	>96 %
Core losses P_{Fe} (typical value, at f = 50 kHz, B = 0.8 T)	100 W/kg	60 W/kg
Static coercivity H_c	10 mA/cm	3 mA/cm
Saturation magnetostriction (25C)	< 0.5 x 10^{-6}	< 0.2 x 10^{-6}
Curie temperature, T_c	>600 °C	240 °C
Continuous upper operation temperature	120 °C	90 °C
Specific electrical resistivity 120 µm	120 µm	135 µm
Density	7.35 g/cm^3	7.70 g/cm^3

7.2 Magnetic Properties of Nanocrystalline Softmagnetic Materials

Figure 7 shows the temperature dependence of the coercivity of a typical samples of a \(Finemet\) type material. Similar curves were reported also by. [41,42] The upturn of the coercivity above 300 °C can be explained regarding the low Curie temperature of the amorphous phase

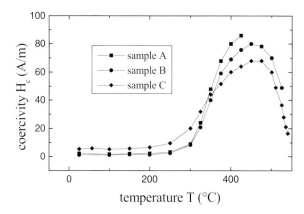

Figure 7: Temperature dependence of the H_c of different samples of nanocrystalline finemet type material.

TCa which lies close to 300 °C. Also the temperature dependence of the saturation magnetization shows a kink at this temperature. At temperatures above TCa the nanocrystalline &agr;-Fe

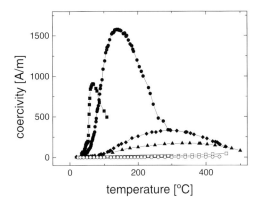

Figure 8: Temperature dependence of H_C measured on $Fe_{86}Zr_7Cu_1B_6$ after 1h annealing at different temperatures T_a $T_a = 450\ °C$ (■); $T_a = 475\ °C$ (●), $T_a = 500\ °C$ (♦), $T_a = 550\ °C$ (▲), $T_a = 600\ °C$ (□), Joule heated (○)].

grains start to decouple which causes an increase of the coercivity. Table 2 summaries the magnetic properties of a commercial Finemet type material.

Table 2: Compares the magnetic properties of a commercial nanocrystalline Fe-based material with soft magnetic Co-based amorphous samples.

Zr–Fe–Cu–B is a similar nanocrystalline system, however, there the Curie-temperature of the amorphous phase lies between 50–100 °C which means just above room temperature, where also the upturn of coercivity occurs. Figure 8 shows the temperature dependence of the coercivity on nanocrystalline Zr–Fe–Cu–B after different heat treatments. Also here one can observe the magnetic hardening above the Curie temperature of the amorphous phase. Because of this rather low temperature range but also due to production problems the Zr–Fe–Cu–B system is not so well suited for technical applications.

7.3 Application

Nanocrystalline soft magnetic materials (\(Finemet\)) is industrial used as a high permeable core for inductivities in switched power supplies for computers. There exists several industrial available product, as e.g., Vitroperm from VAC (see Table 2).

8 References

[4] W. Rodewald, B. Wall, M. Katter, K. Üstüner; S. Steinmetz, in Proc. of 17th Int. Workshop on Rare Earth Magnets (Ed: G. C. Hadjipanayis, M. J. Bonder), Rinton Press, Delaware 2002, pp. 25–36

[5] F. E. Luborsky, in Amorphous Metallic Alloys, Butterworths Monographs in Materials, London 1983

[6] G. C. Hadjipanayis, J. Magn. Magn. Mater. 1999, 200, 373
[7] A. Jha, H. A. Davies, R. A. Buckley, J. Magn. Magn. Mater.1989, 80, 109
[8] F. J. Cadieu, in Physics of Thin Films, vol. 16, Academic Press, San Diego 1992
[9] F. J. Cadieu, Int. Mater. Rev. 1995. 40, 137
[10] D. J. Sellmyer, J. Alloys Compounds 1992, 181, 397
[11] E. E. Fullerton, J. S. Jiang, M. Grimsditch, C. H. Sowers, S. D. Bader, Phys. Rev. B 1998, 58, 12 193
[12] K. Narasimhan, C. Willman, E. J. Dulis, US Patent No. 4 588 439, 1986
[13] P. G. McCormick, W. F Miao., P. A. I. Smith, J. Ding, R.'Street, J. Appl. Phys. 1998, 83, 6256
[14] L. Schultz, in Science and Technology of Nanostructured Materials (Ed: G. Prinz, G. C. Hajipanayis), NATO ASI Series 1990, 259, p. 583
[15] Lambeth D., in Magnetic Hysteresis in Novel Magnetic Materials (Ed: G. C. Hadjipanayis), NATO ASI Series 1996, 338, p. 767
[16] D. J. Sellmyer, in Magnetic Hysteresis in Novel Magnetic Materials (Ed: G. C. Hadjipanayis), NATO ASI Series 1996, 338, p. 419
[17] J. P. Liu, Y. Liu, C. P. Luo, Z. S. Shan, D. J. Sellmyer, Appl. Phys. Lett. 1998, 72, 483.
[18] S. Starroyiannis, I. Panagiotopoulos, D. Niarchos, J.'Christodoulides, Y. Zhang, G. C. Hadjipanayis, Appl. Phys. Lett. 1998, 73, 3453
[19] J. Ding, Y. Liu, P. G. McCormick, R. Street, J. Appl. Phys. 1994, 75, 1032
[20] A. Giguère, N. H. Hai, N. M. Dempsey, D. Givord, J.'Magn. Magn. Mater. 2002, 242-245, 581
[21] A. Giguère, N. M. Dempsey, M. Verdier, L. Ortega, D.'Givord, IEEE Trans. Magn. 2002, 38, 2761
[22] W. H. Meiklejohn, C. P. Bean, Phys. Rev. 1956, 102, 1413
[23] W. H. Meiklejohn, C. P. Bean, Phys. Rev. 1957, 105, 904
[24] R. Fischer, H. Kronmüller, J. of Magn. Magn. Mat. 1999, 191, 225
[25] E. F. Kneller, R. Hawig, IEEE Trans. Magn. 1991, 27, 3588
[26] S. Hock, Ph.D. Thesis, University of Stuttgart 1990.
[27] G. Ruwei, L. Hua, J. Shouting, M. Liangmo, Q. Meiying, J. Magn. Magn. Mater. 1991, 95, 205
[28] M. Katter, Ph.D. Thesis, TU Wien 1991
[29] R. Kütterer, H. R. Hilzinger, H. Kronmüller, J. Magn. Magn. Mater. 1977, 4, 1
[30] E. F. Kneller, in Ferromagnetismus, Springer-Verlag, Berlin 1962
[31] R. Coehoorn, D. B de Mooij, J. P. W. B. Duchateau, K. H. J. Buschow, J. de Physique 1988, C8, 669
[32] A. M. Kadin, R. W. McCallum, G B. Clemente, J. E. Keem, Proc. of Materials Research Society Symposia Pittsburgh, (Ed: M. Tenhover, W. L.Johnson, L. E. Tanner), Mater. Res. Soc. 1987, p. 385
[33] E. C. Stoner, W. P. Wohlfarth, Phil. Trans. Roy. Soc. 1948, A240, 5299
[34] J. Fidler, T. Schrefl, J. Physics D: Appl. Phys. 2000, 33, R135-R156
[35] M. Dahlgren, R. Grössinger, in Proc. of Rare Earth Magnets and Their Appications (Ed: L. Schultz, K. H. Müller), Dresden, 1998, pp. 253–262
[36] R. Grössinger, H. Hauser, M. Dahlgren, J. Fidler, Physica B 2000, 275, 248

[37] D. Suess, M. Dahlgren, T. Schrefl, R. Grössinger, J. Fid"ler, J. Appl. Phys. 2000, 87, 6573
[38] Y. Yoshizawa, S. Oguma, K. Yamauchi, J. Appl. Phys. 1988, 64, 6044
[39] K. Suzuki, A. Makino, A. Ihoue, T. Masumoto, J. Appl. Phys. 1991, 70, 6232
[40] Y, Yoshizawa, K. Yamauchi, Mater. Trans. JIM 1990, 31, 307
[41] G. Herzer, Mater. Sci. Eng. 1991, A 133, 1
[42] R. Alben, J. J. Becker, M. C. Chi, J. Appl. Phys. 1978, 49, 1653
[43] G. Herzer, Phys. Scr. 1993, T49, 307
[44] A. Hernando, T. Kulik; Phys. Rev. B 1994, 49(10), 7064
[45] M. Dahlgren, R. Grössinger, A. Hernando, D. Holzer, M. Knobel, P. Tiberto, J. Magn. Magn. Mater 1996, 160, 247

Formation of Nanostructures in Metals and Composites by Mechanical Means

H.-J. Fecht
University of Ulm, Center for Micro-and Nanomaterials, Ulm, Germany

1 Introduction

Nanostructured materials as a new class of engineering materials with enhanced properties and structural length scales between 1 and 100 nm can be produced by a variety of methods [1]. Besides the fabrication of clusters, thin films and coatings from the gas or liquid phase, chemical methods such as sol-gel processes and electrodeposition are common methods of processing. As a versatile alternative however, mechanical methods have been developed which allow fabricating nanostructured materials in large quantities with a broad range of chemical composition and microstructure. These methods can be applied to powder samples, thin foils and to the surface of bulk samples.

In the 1970's, the method of mechanical alloying of micrometer sized powder particles followed by high temperature sintering has been developed as an industrial process to successfully produce new alloys and phase mixtures by mechanical means. For example, this powder metallurgical process allows the preparation of alloys and composites, which can not be synthesized via conventional casting routes. This method can yield [2]

a) uniform dispersions of ceramic particles in a metallic matrix (superalloys) for use in gas turbines,
b) alloys with different compositions than alloys processed from the liquid and
c) alloys of metals with quite different melting points with the goal of improved strength and corrosion resistance.

In the 1980's, the method of high-energy ball milling has gained additional attention as a non-equilibrium solid state process resulting in materials with nanoscale microstructures. The formation of nanocrystals within initially single crystalline powder samples has been first studied systematically in pure metals and intermetallic compounds [3]. Moreover, solid state (mechanical) alloying beyond the thermodynamic equilibrium solubility limit can lead to the formation of amorphous metallic materials as observed for a broad range of alloys with a considerable atomic size mismatch and a negative enthalpy of mixing. This process is considered as a result of both mechanical alloying [4] and the incorporation of lattice defects into the crystal lattice [5]. More recent investigations demonstrate that the nanostructure formation can also occur for several unexpected cases, such as materials with positive enthalpies of mixing, brittle ceramics, ceramic / metal nanophase mixtures and polymer blends.

Whereas the deformation processes within the powder particles are important for fundamental studies of extreme mechanical deformation and the development of nanostructured states of matter with particular physical and chemical properties, similar processes control the deformation of technologically relevant surfaces. For example, the effects of work hardening and erosion during wear situations result in surface microstructures comparable to those observed during mechanical attrition [6]. Similar to mechanical attrition of powder particles this is the consequence of the formation of dislocation cell network, subgrains and consequently grain boundaries and the dissolution of second phase particles by mechanical alloying.

2   Experimental

A variety of ball mills have been developed for different purposes [7]. The basic process of mechanical attrition is illustrated in Fig. 1. Powder particles with typical particle diameters of about 20–100 μm are placed together with a number of hardened steel or WC coated balls in a sealed container, which is shaken violently. Consequently, plastic deformation at high strain rates ($\sim 10^3$–10^4 s^{-1}) occurs within the particles and the average grain size can be reduced to a few nanometers after extended milling. The temperature rise during this process is modest and is generally estimated to be 100 to 200 °C. The collision time corresponds to typically 2 ms.

For most metallic elements and intermetallic compounds a refinement of the internal grain size is observed to typically 5 to 20 nm. Figure 2 exhibits the distribution of scattering objects (Dislocations at the early stage, small and large angle grain boundaries at a later stage) after 0.5 and 30 hours of mechanical milling for Fe-particles as determined by small angle neutron scattering.

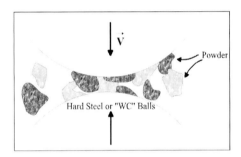
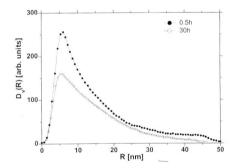

Figure 1: Schematic sketch of the process of mechanical attrition of metal powders

Figure 2: Grain/sub-grain size distribution for nanocrystalline Fe (volume weighted) calculated from small angle neutron diffraction spectra for milling times of 0.5 and 30 h

The elemental processes leading to the grain size refinement include three basic stages as found by combined X-ray, electron and neutron analysis:

1)   Initially, the deformation is localized in shear bands consisting of an array of dislocations with high density.

2) At a certain strain level, these dislocations annihilate and recombine to small angle grain boundaries separating the individual grains. The subgrains formed via this route are already in the nanometer size range.
3) The orientations of the single-crystalline grains with respect to their neighboring grains become completely random.

3   Results and Discussion

3.1   Solid Solutions and Metallic-Glass Formation

Extended solid solutions far beyond the thermodynamic equilibrium have generally been noted in the course of mechanical milling of alloys. In addition, for phase mixtures with negative enthalpies of mixing and large (> 15 %) atomic size mismatch solid-state amorphization is observed. During this process long-range solute diffusion and solute partitioning are suppressed and therefore, highly metastable amorphous and nanocrystalline states become accessible.

For example, during mechanical alloying of 75 at% Zr and 25 at% Al the formation of a supersaturated hcp (α-Zr) solid solution was observed prior to the solid-state-amorphization reaction [8]. However, in all cases of binary alloys it remained unclear weather indeed a metallic glass has been formed or just a material with "x-ray amorphous structure". More recently, a similar phase transformation sequence has been investigated in a mechanically alloyed multicomponent elemental $Zr_{60}Al_{10}Ni_9Cu_{18}Co_3$ powder mixture with an alloy composition which is known to form a bulk metallic glass when cooled from the liquid state [9]. These multi-component alloys are considerably more stable than binary alloys and can be heated above the glass transition temperature before crystallization sets in. The X-ray spectra at different stages of the milling process are characterized by the successive disappearance of the elemental Al, Co, Cu and Ni peaks and a simultaneous shift of the Zr-peaks to higher scattering angles, corresponding to a decrease in the lattice constant of the hcp-Zr as a result of the rapid dissolution of the smaller atoms, such as Cu, Ni, Co and Al in the (α-Zr) matrix [10].

3.2   Nano-Composites

Mechanical alloying is also a very versatile process to prepare new nano-composites. For example, metallic glass / ceramic composites can be obtained by mechanical alloying of multicomponent Zr-based elemental metallic powders together with SiC particles [11].

A secondary electron microscopy (SEM) image of such a $Zr_{65}Al_{7.5}Cu_{17.5}Ni_{10}$ / 10 vol% SiC metallic glass/ceramic composite after a milling time of 30 hr (Fig. 3) reveals a uniform distribution of fine SiC particles in the metallic glass powder matrix, as proven by further X-ray diffraction and EDX analysis.

The size distribution of the SiC particles ranges from 1 µm down to values below 50 nm. It is further interesting to note that the SiC particles do not act as potent heterogeneous nucleation sites when the composite is heated to the crystallization temperature above the glass transition temperature. Thus, mechanical alloying represents a convenient method to achieve dispersion-strengthened amorphous alloys with considerably improved strength and wear resistance by a powder metallurgical pathway.

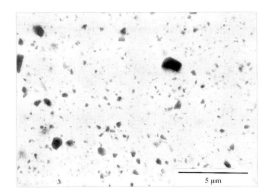

Figure 3: Scanning electron micrograph of a metal/ceramic composite of 10 vol% SiC particles in a $Zr_{60}Al_{7.5}Cu_{17.5}Ni_{10}$ metallic glass matrix.

3.3 Mechanical Properties (Nanocrystalline-Fe)

As a further consequence of the grain size reduction a drastic change in the mechanical properties has been observed. In general, nanocrystalline materials exhibit very similar mechanical behavior as amorphous materials due to shear band formation as the prevalent mechanism of deformation. Strain hardening is not observed and conventional dislocation mechanisms are not operating in nanostructured materials. Typical results for nanocrystalline (nx) Fe powder samples exhibit an increase in microhardness by a factor of 6–7 (9 GPa for nx-Fe with d about 16 nm versus 1.3 GPa for annealed px-Fe as shown in Fig. 4). In general, the microhardness MH follows a trend similar to the Hall-Petch relationship (MH $\approx d^{-1/2}$) though the dislocation based deformation mechanism in the nanocrystalline regime certainly does not apply. Furthermore, the Young's modulus can be measured by this method as well and shows a decrease by typically about 10% in comparison with the corresponding polycrystal.

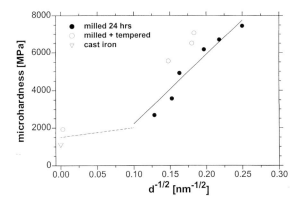

Figure 4: "Hall Petch" relationship for the hardness of nanocrystalline and polycrystalline-iron using a nanoindenter

As such, it is suggested that the mechanical properties of nanophase materials prepared by mechanical attrition after extended periods of milling are not being controlled by the plasticity of the crystal due to dislocation movement anymore but rather by the cohesion of the nanocrystalline material across its grain boundaries. From the considerable increase of hardness and the changes of the deformation mechanisms improved mechanical properties can generally be expected as attractive features for the design of advanced powder metallurgical materials.

3.4 Friction Induced Nanostructure Formation (Fe-C-Mn)

Many microscopic processes occurring during mechanical attrition and mechanical alloying of powder particles exhibit common features with processes relevant in tribology and wear. For example, the effects of work hardening and mechanical alloying result in similar microstructures of the wear surface [12,13]. In particular, during sliding wear, large plastic strains and strain-gradients are created near the surface. Typical plastic shear strain rates can correspond here to several 10^3 s^{-1}.

During sliding wear a special tribolayer develops on the surface of components being subjected to large plastic strains. This surface layer often is called the Beilby layer which for a long time was thought to be amorphous because its microstructure could not be resolved with the instruments commonly used at that time [14].

As a typical example of technical relevance, the development of high speed trains reaching velocities higher than 300 km/h is also a materials challenge concerning the mechanical integrity and safety required for the railway tracks [15,16]. In particular, the interaction and slip between wheel and rail has been optimized and is controlled by sophisticated electronics whereas the materials for the rail remained unchanged since over two decades. In particular, on the steel surface (Fe-0.8 at% C-1.3 at% Mn) the local pressure due to the wheel/rail interaction corresponds to typically 1–1.5 GPa thus exceeding the yield strength of the original (core) material of 0.8–0.9 GPa considerably.

Corresponding X-ray diffraction and TEM results indicate that the average grain size of the extremely deformed surface layer is decreased to about 20 nm whereas a gradient in grain size is observed further away from the surface where values up to 200 nm are reached. As a consequence, hardness measurements have been performed by nanoindentation and conventional methods as shown in Fig. 5. Steep hardness gradients have been found in cross section with a lateral resolution of a few micrometer which are clearly correlated with the change in microstructure. As a result it is found that the hardness is increased from typical values for the pearlitic steel of approximately 2.5–3 GPa to 12–13 GPa next to the surface which corresponds to values about double as much as for pure nanocrystalline Fe as indicated in Fig. 5.

This remarkable increase in hardness and mechanical strength (about 1/3 of the hardness values) of regions near the surface is clearly related to the fact that the average grain size is considerably decreased by the continuous deformation process. As such, the improvement of the mechanical properties clearly shows the importance of the formation of nanostructures for technologically relevant wear problems. However, at least in this situation the microstructural change also leads to embrittlement and the formation of microcracks that grow in the rail as well as the wheel and can cause dramatic failure situations.

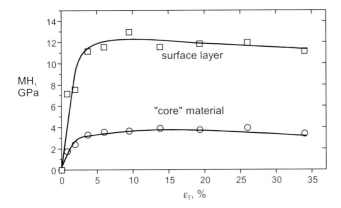

Figure 5: Microhardness (MH) of the nanostructured surface layer of a high-speed railway track and the pearlitic material (core) as a function of total deformation

4 Summary and Conclusions

Extreme plastic deformation encountered in a variety of mechanical processes may yield nanostructured and nanophase materials with a broad range of chemical composition and microstructure. This paper discusses several such examples including the deformation induced nanophase formation in powder particles, thin foil sandwich structures and at the surface of metals and alloys exposed to friction induced wear conditions. The deformation processes occurring within powder samples during mechanical attrition are important for fundamental studies on the effect of extreme mechanical deformation on the microstructure and its stability.

5 Acknowledgements

The continuous financial support by the Deutsche Forschungsgemeinschaft (G.W. Leibniz program) and BMBF (03N6020) is gratefully acknowledged.

6 References

[1] H. Gleiter, Progress in Mat. Science, 33, 223 (1989)
[2] J. S. Benjamin, Sci. American, 234, 40 (1976)
[3] H.-J. Fecht in: "Nanomaterials; Synthesis, Properties and Applications", eds. A. S. Edelstein and R.C. Cammarata, Institute of Physics Publ. (1966)
[4] C. C. Koch, O. B. Cavin, C. G. McKamey and J. O. Scarbrough, Appl. Phys. Lett. 43, 1017 (1983)
[5] H.-J Fecht, E Hellstern, Z Fu and W L Johnson, Adv. Powder Metallurgy, 1, 111 (1989)
[6] D. A. Rigney, L. H. Chen, M. G. S. Naylor and A. R. Rosenfield, Wear, 100, 195 (1984)

[7] W. E. Kuhn, I. L. Friedman, W. Summers and A. Szegvari, ASM Metals Handbook, Vol. 7, Powder Metallurgy, Metals Park (OH) p. 56 (1985)
[8] H.-J. Fecht, G. Han, Z. Fu and W. L. Johnson, J. Appl. Phys. 67, 1744 (1990)
[9] A. Sagel, R. K. Wunderlich, J. H. Perepezko and H.-J. Fecht, Appl. Phys. Lett., 70, 580 (1997)
[10] A. Sagel, N Wanderka, R. K. Wunderlich, N. Schubert-Bischoff and H.-J. Fecht, Scipta Mater., 38, 163 (1998)
[11] C. Moelle, I. R. Lu, A. Sagel, R. K. Wunderlich, J. H. Perepezko and H.-J. Fecht, Materials Science Forum, 269-272, 47 (1998)
[12] Y. V. Ivanisenko, G. Baumann, H.-J. Fecht, I. M. Safarov, A. V. Korznikov and R. Z. Valiev, The Physics of Metals and Metallography, 83, 303 (1997)
[13] H.-J. Fecht, Nanostructured Materials, 6, 33 (1995)
[14] G. Beilby, Aggregation and Flow of Solids, Macmillan (London) (1921)
[15] G. Baumann, H.-J. Fecht and S. Liebelt, Wear 191, 133 (1996)
[16] G. Baumann and H.-J. Fecht, Nanostructured Mat., 7, 237 (1996)

Scale Levels of Plastic Flow and Mechanical Properties of Nanostructured Materials

Victor E. Panin, Alexey V. Panin, Ruslan Z. Valiev[1], Ludmila S. Derevyagina, Vladimir I. Kopylov[2]

Institute of Strength Physics and Materials Science, SB, RAS, Tomsk, Russia
1 Institute of Physics of Advanced Materials associated with USATU, Ufa, Russia
2 Physical-Technical Institute of National Academy of Sciences of Belarus, Minsk, Belarus

1 Abstract

The relation of shear contributions at micro-, meso-, and macroscale levels governs the mechanical behavior and stress-strain curve of ultrafine-grained materials. Special attention must be paid to their surface layer which is the autonomous mesoscopic structural level of plastic deformation.

2 Introduction

Any solid under loading is the multilevel system where a plastic flow is developed self-consistently as shear stability loss at different scale levels: micro, meso, and macro, Fig. 1 [1]. At the micro- and meso I levels the plastic flow is realized by the motion of dislocations and dislocation cells, Fig. 1(a, b). At the meso II- and macroscale levels the main contribution to the plastic flow is related to development of shear bands and macrobands, Fig. 1(c, d).

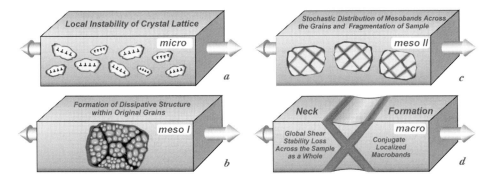

Figure 1: Schematic representation of scale levels of shear stability loss in deformed solid

The mechanical behavior features of ultrafine-grained materials are governed by the fact that two stages (micro and meso I) of plastic flow have been already realized during preliminary severe plastic deformation (spd). Therefore, the subsequent loading of SPD-materials causes mainly their shear stability loss at the meso II- and macroscale levels. Taking into account the heterogeneity of spd-materials their plastic flow develops at all scale levels of shear stability loss.

This paper is an overview of the authors' investigations of scale levels of plastic deformation of ultrafine-grained materials and their mechanical properties [1–12, etc.].

2   Correlation between Surface Strain-induced Relief and Dislocation Substructure in Ultrafine-Grained Materials

In recent works [6–12] it is found that a surface layer of a loaded solid is the autonomous mesoscopic structural level of plastic deformation. Within the surface layer the peculiar deformation mechanisms of various scale levels are developed and affect the deformation of the bulk of the material. Therefore, it is very important to consider the mechanical behavior of the bulk of ultrafine-grained materials as well as their surface layers. The ultrafine-grained structure of the bulk of the material can be produced by equal channel angular pressing, whereas such a structure in the surface layer can be obtained by ultrasonic treatment of cold-rolled specimens. By combining optical, scanning-tunneling and transmission electron microscopy (STM and TEM) it is possible to reveal very good correlations between the surface strain-induced relief and shear-bands in the bulk of the ultrafine-grained materials.

Figure 2 shows the STM- (a), optical (b), and TEM-images (c) of the tensile specimen of polycrystalline titanium which was preliminary subjected to cold-rolling and subsequent ultrasonic treatment of its surface. Two types of localized deformation bands were observed on the surface which were oriented by an angle of 45° with respect to the tension axis. The investigation with STM reveals the surface mesobands manifesting as an extruded material, Fig. 2(a).

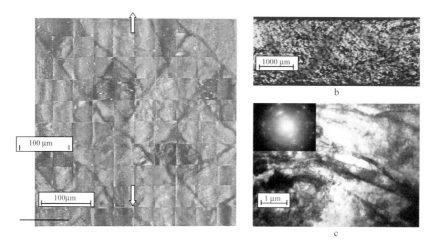

Figure 2: Strain-induced surface relief (*a, b*) and dislocation structure (*c*) in the bulk of the titanium specimen with the ultrafine-grained surface layer; tension, $\varepsilon = 16$–$18\ \%$

Mesoband width is ~ 80 mm, and height gradually rises with strain and reaches 3–4 mm at $e = 18$ %. Deformation within the mesobands is realized by multilevel consecutive shears of separate lamellas of various scales. It is much clearly expressed for the ultrafine-grained surface layer of titanium specimens subjected to preliminary hydrogenation, Fig. 3. According to Fig. 3, each lamella has finer transverse lamellar structure. The extruded material mesobands reflect the peculiarities of surface layer deformation at the mesoscale level and are accompanied by the appearance of a banded dislocation substructure in the bulk of the specimen, Fig. 2(c).

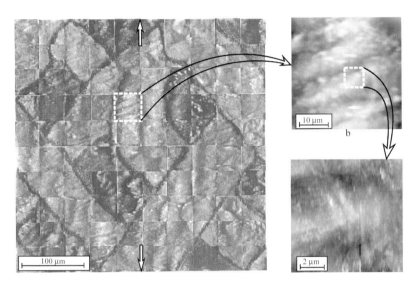

Figure 3: STM-images of Ti specimen; tension after preliminary cold-rolling, ultrasound treatment and hydrogen saturation, $\varepsilon = 16$ %

The studies by optical microscopy reveal another type of the surface bands, which propagate along the working part of the specimen according to the scheme of the wave of total internal reflection, Fig. 2(b). Such band is developed at the stage of homogeneous specimen elongation and are propagated in zig-zag fashion across the total width of the treated surface. It is deformation macrolocalization within the surface layer with ultrafine-grained structure. The width of macrobands amounts to ~300 mm. On the untreated specimen surface no localized deformation macrobands occur. Essentially, the macrobands form the extended zig-zag neck in the ultrafine-grained surface layer. Such bands emerge at $e = 2$–3 %. With increasing strain, they locally reduce the specimen thickness in the area of the extended zig-zag neck and form the first stage of deformation macrolocalization.

The development of localized deformation meso- and macrobands within the surface layer of loaded titanium specimens subjected to preliminary ultrasonic treatment greatly affects the dislocation substructure of materials' bulk. Fig. 2(c) indicates that banded dislocation substructures are formed with diffuse boundaries characterized by discrete disorientations through angles $> 15\times$. The material in between the microbands is fragmented. The fragment dimensions are 1–2 mm in length and 0.1–0.3 mm in width, their disorientation is 2–6°. The zones where the banded dislocation substructures are clearly pronounced alternate with areas of highly defected material with a dislocation density $> 10^{11}$ cm^{-2}. These areas can be homogeneously distributed as well as can form non-disoriented cellular structures with wide dislocation boundaries.

At the second stage, plastic flow macrolocalization of the titanium specimens develops in the bulk of the material and governs necking followed by specimen fracture. An analysis of the displacement vector fields makes possible to show a clear correlation of the necking mechanism in the specimens with ultrafine-grained structure within the surface layer, or in the bulk of the material and the development of localized deformation macrobands, Fig. 4. The mechanism of two conjugate macrobands interaction within the neck has been investigated in [4] for tensile spd-Cu specimens. It can be specified as a phase wave: the development of shear in one macroband is accompanied by stunted shear in the conjugate macroband, and vice versa.

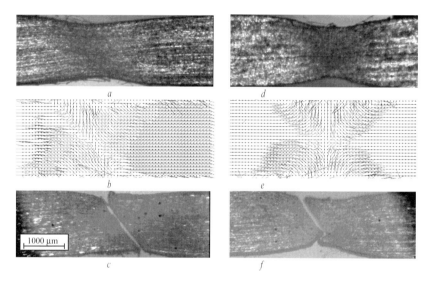

Figure 4: Neck formation in tensile Ti specimen with SPD surface layer, $\varepsilon = 18$ %, a, d — optical images; b, e — displacement-vector field; c, f — fracture in the neck region

3 Influence of Scale Levels of Plastic Flow on Stress-Strain Curves

The study of the tensile Ti, Cu, a-Fe, low-carbon steel specimens with ultrafine-grained structure within the surface layer or in the bulk allows us to separate the effect of micro-, meso-, and macrobands on the stress-strain curve. The mechanical behavior of the Ti specimen in various structural states is presented in Fig. 5 as an example.

One can say with reasonable confidence that development of micro- and mesobands causes work-hardening of tensile spd-specimens. Extended zig-zag macrobands propagated along the working part of the specimen cause moderate work-softening if the ultrafine-grained structure is formed only within the surface layer (curve 3 in Fig. 5). The development of macrobands in the bulk of the spd-materials is accompanied by an marked drop of the stress-strain curve [11]. From an engineering point of view, the plastic flow macrolocalization is always hazardous, especially if a material is in a highly non-equilibrium state.

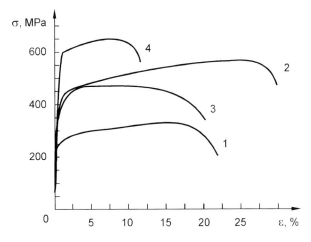

Figure 5: Stress-strain curves of the Ti specimens after different treatments: annealed (*1*); cold-rolled (*2*); subjected to ultrasonic treatment (*3*), or equal channel angular pressing (*4*) [12]

The low ductility of SPD-materials is related to the fast development of macrolocalization in tensile specimens. This process may be suppressed by low temperature annealing of SPD-materials. An example how such treatment influences the stress-strain curves of low-carbon steel is presented in Fig. 6. The high power ultrasonic treatment makes it possible to highly refine the grain structure in low-carbon steel. The surface layer itself consists of a pseudo-amorphous structure. The more getting from materials surface to the bulk, the less refined grains are found, this effect showing up as much as 180 mm below the surface.

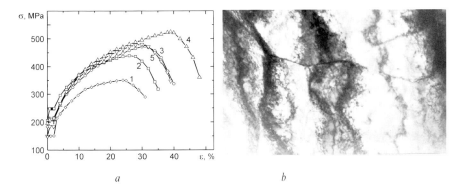

Figure 6: (a) RT stress-strain curves of flat specimens of low-carbon steel (initial state (1)), after ultrasonic treatment of the surface (2), and after subsequent annealing at different temperatures T = 1023 (3), 1103 (4), and 1173 K (5); (b) dislocation substructure in the deformed surface layer with grain size ~ 0.5 mm, TEM, × 14 000

Upon annealing at $T = 1103$ K, a submicrocrystalline structure with grain size ~ 0.5 mm develops in the surface layer, while in the bulk the grains are still quite coarse ~ 10 mm. Under tension, uniformly distributed and paired deformation mesobands propagate in the surface layer, while a common cellular dislocation structure develops in the bulk of the material. This gives rise to an increase in the strain resistance and the plasticity of the material. The same results were obtained previously for SPD-Cu [4].

In conclusion, two remarks should be done. First, rather different data for mechanical behavior of spd-materials could be found in the literature. This can be attributed to the different contributions of deformation mechanisms arising from different scale levels for ultrafine-grained structure obtained by various modes of severe plastic deformation. Second, the development of micro-, meso-, and macrobands of plastic flow localization in spd-materials must be taken into account which comes from the relaxation process which is superimposed to dislocation work-hardening. It must be importantly considered in modeling of mechanical properties of ultrafine-grained materials.

4 Summary

1. The formation of ultrafine-grained structure within thin surface layer or in the bulk of the cold-rolled materials causes an increase of stress but is accompanied by a decrease of material ductility. The latter is related to the development of plastic flow localization in SPD-materials at all levels of scale. Low-temperature annealing of SPD-materials allows to obtain high values of both strength and ductility of fine-grained materials.
2. The surface layers of SPD-materials are an autonomous mesoscopic structural level of plastic deformation. The peculiar mechanisms of deformation localization within the specimen surface layer govern the formation of a banded dislocation substructure in the bulk of the specimen.
3. The characteristics of the stress-strain curves of SPD materials is determined by individual contributions of shears at different scale levels. In order to reach high reliability of SPD-material with respect to their mechanical properties, all kinds of plastic flow macrolocalization have to be suppressed.

5 Acknowledgement

This work was carried out under financial support of the Russian Foundation for Basic Research (Grants Nos. 00-15-96174 and 02-01-01195) and the Siberian Branch of the Russian Academy of Sciences (Integration Project No. 45).

6 References

[1] V. E. Panin in NATO Advanced Research Workshop on Investigations and Applications of Severe Plastic Deformation (Ed.: T. C. Lowe, R. Z. Valiev), Kluwer Academic Publishers, Dordrecht–Boston–London, 2000
[2] R. Z. Valiev, I. V. Aleksandrov, Nanostructured Materials: Production, Structure and Properties, Nauka, Moscow, 1999, p. 244

[3] R. Z. Valiev, Mat. Sci. Eng. 1997, A234–236, 59–66
[4] V. E. Panin, L. S. Derevyagina, R. Z. Valiev, Phys. Mesomech. 1999, 2, 85–90
[5] A. N. Tyumentsev, V. E. Panin, L. S. Derevyagina, R. Z. Valiev, N. A. Dubovik, I. A. Ditenberg, Phys. Mesomech. 1999, 2, 105–112
[6] A. V. Panin, V. A. Klimenov, N. L. Abramovskaya, A. A. Son, Phys. Mesomech. 2000, 3, 83–92
[7] A. V. Panin, V. A. Klimenov, Yu. I. Pochivalov, A. A. Son, Phys. Mesomech. 2001, 4, 81–88
[8] P. V. Kuznetsov, V. E. Panin, Phys. Mesomech. 2000, 3, 85–91
[9] V. E. Panin, Phys. Mesomech. 1999, 2, 5–21
[10] V. E. Panin, Mat. Sci. Eng. 2001, A319–321, 197–200
[11] A. V. Panin, V. E. Panin, I. P. Chernov, Yu. I. Pochivalov, M. S. Kazachonok, A. A. Son, R. Z. Valiev, V. I. Kopylov, Phys. Mesomech. 2001, 4, 87–94
[12] A. V. Panin, V. E. Panin, Yu. I. Pochivalov, V. A. Klimenov, I. P. Chernov, R. Z. Valiev, M. S. Kazachonok, A. A. Son, Phys. Mesomech. 2002, 5, 73–84

Hydrogen – Induced Formation of Silver and Copper Nanoparticles in Soda-Lime Silicate Glasses

M. Suszynska, L. Krajczyk, B. Macalik

Institute of Low Temperature and Structure Research, Polish Academy of Sciences, Wrocaw, Poland

1 Abstract

It was investigated the effect of stretching upon optical and structural characteristics of the soda-lime silicate glasses doped either with silver or copper and thermally treated in the hydrogen atmosphere. It has been stated that both the hydrogenation and deformation strongly affect both the morphology of the glassy matrix and the colloidal nanoparticles of the dopant. Because the induced changes bear an anisotropic character, the deformation could be useful for new applications of such composite materials.

2 Introduction

Nanoscale metal- or semiconductor-doped oxide glasses are now attracting considerable attention for photonic applications because the non-linear optical susceptibility and the picosecond response time at high excitation intensities of such composites is comparable with those characteristic of crystalline materials [1–3]. Ion exchange has been shown to be a useful fabrication method to form these nanoclusters in the oxide glasses because of the ability to produce waveguiding configurations with a high volume fraction of the metal colloids. For instance, copper-containing waveguides have recently drawn new attention owing to their blue/green luminescent properties [4]. In addition to the optical applications, the glass composites synthesized with transition elements are important for their magnetic properties resulting – among others – in superparamagnetism and applications in the field of magnetic recording substrates for high-density information storage. Moreover, an improvement of the mechanical characteristics (the Vickers microhardness number and the crack formation resistance) together with a change of the chemical durability has been observed in these composites [5]. During the ion exchange large compressive stresses appear in the near-surface layer for dopant ions which have a larger radius than the original alkali ion, like Ag+ and K+. Pre-stressing of such composites has a practical effect of providing a protection measure against the surface damage.

From the point of view of the applied research, much work must, however, be done to develop suitable technologies for fabricating glass-composites with prescribed properties.

Preliminary investigations of commercial multicomponent soda-lime silicate (SLS) glasses exchanged either with silver or copper have shown that many the structure-sensitive properties of such composite materials could be affected by both the microstructure of the glassy matrix and the peculiar behavior of the quantum dots related with the dopant.

The aim of the work described was to investigate the effect of deformation upon the microstructure and optical absorption of hydrogenated SLS-glasses of different origin and partially substituted either by silver or copper ions.

3 Samples and Techniques

We used two types of commercial multicomponent SLS-glass the composition of which corresponds to the miscibility-gap in the SiO_2-Na_2O system [6]. The main components of both glasses were (in mol%): SiO_2/74, Na_2O/13, CaO/6.4, MgO/4.5. Different glass-making procedures resulted in different content of the Al, Fe, Mn and K-ions in samples of type I (prepared as 1 mm-thick sheets) and II (prepared as 5 mm-thick plates). Samples (I) with less potassium and no manganese have been exchanged with silver, whereas samples (II) gave the copper-exchanged material.

Ion exchange was performed either in molten $AgNO_3$ at 670 K or in Cu_2Cl_2 at 940 K; for more detail see [7, 8]. The concentration profiles of the exchanging ions were monitored by the Philips scanning microscope (SEM 515) with a roentgenographic analyzer (EDAX 9800) working at 20 kV. After exchange the specimens have been annealed in a continuous stream of dry hydrogen at temperatures ranging between 573–773 K at atmospheric pressure. The samples were quenched to room temperature (RT) without passivation.

For deformation of samples (I) a homemade apparatus was used [9]. The samples moved through a relatively narrow heating zone at 859 K under a constant tensile load (98 N) and different tension velocity (800–7200 mm/min).

Optical absorption (OA) was measured at RT in the range between 250–2500 nm by using a Varian (Carry 5E) spectrophotometer. For the deformed samples linearly polarized light in direction parallel or perpendicular to the tensile axis was used.

For electron microscopy studies a transmission electron microscope (TEM, Philips-CM 20) was used at 200 kV with 0.24 nm point-to-point resolution. Shadowed carbon and extraction replicas have been prepared from sample surfaces selectively etched in diluted water solutions of HF. The high-resolution transmission electron microscopy (HRTEM) and selected area electron diffraction (SAED) analysis have accomplished the TEM-studies.

4 Results and Discussion

In order to obtain an insight into the processes, which control the structure-sensitive properties of doped SLS glasses, TEM, SAED and HREM analyses have been performed for replicated surfaces of both types of specimens.

It has been stated that samples of type (I) are separated into two phases already during the glass-making procedure with a remarkable difference in the chemical durability. Their etching was accomplished with rather weak HF-water-solutions. From the shadow of micrograph1.A. it has been inferred that the dispersed phase is low silica, whereas the matrix phase is high silica one. The separated particles have been tentatively ascribed to Na_2O-rich particles. After hydrogenation secondary phase separation phenomena occur, and the new elements of the glass structure are etched together with the originally separated phase suggesting that both be of comparable chemical durability and probably also of similar composition.

The glass of type (II) was separated into phases with a less pronounced difference in chemical durability and stronger HF solutions were required. This difference is probably the consequence of some changes in the chemical composition of phases during the glass-making procedures. Moreover, addition of K_2O to the SLS glasses probably retards the phase separation, similar to the effect observed for coloration of these glasses [10].

4.1 Structural Characteristics from Electron Microscopy

For the sake of clarity, the TEM observations are presented as first, since they show directly the structural changes, which deeply affect the OA data of the doped SLS glasses.

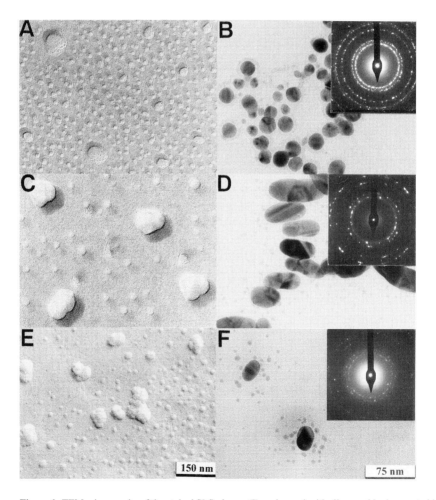

Figure 1: TEM micrographs of the etched SLS glasses (I) exchanged with silver and hydrogenated in not deformed (A, B) and deformed (C-F) state; left column presents the matrix morphology, while the right column shows the silver nanoparticles. Insets present the electron diffraction patterns.

Figure 1 (A–F) presents a typical set of TEM-micrographs of carbon replicas obtained for the specimens of type (I). There are shown the matrix morphology (left column) and the behavior of silver (right column) in these glasses after hydrogenation and deformation. Annealing in hydrogen yields a size increase of the Na_2O-rich droplets, a secondary phase separation and the reduction of silver ions to atoms. Subsequent diffusion of these atoms leads to the formation of spherical colloidal nanoparticles the size of which increases with the annealing time as consequence of Ostwald ripening. During these processes, the colorless samples become yellow-brownish.

Especially interesting are the effects of deformation of specimen (I). The largest silver particles become elongated in the direction of sample-tension, whilst the smallest remain not defor-

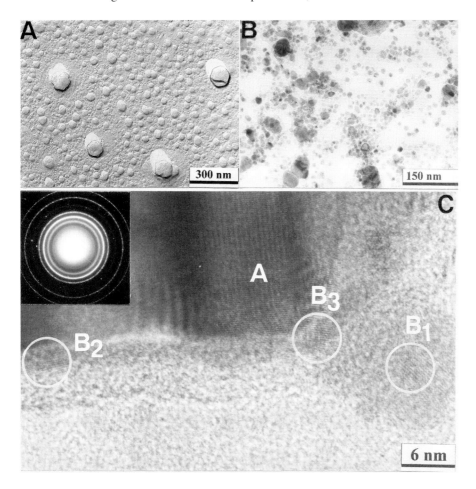

Figure 2: TEM micrographs of the etched SLS glasses (II) exchanged with copper. The matrix morphology (A), copper nanoparticles (B) and the HRTEM picture (C) with lattice elements of crystalline Cu and Cu2O nanoparticles are shown.

med in the surrounding of the deformed ones. The stretching affects only particles with a diameter larger than 5 mm, interpreted as a threshold size, at least at stresses used in the described work. This bimodal size-distribution and the change of spherical silver particles into prolate spheroids imply new applications of such glasses.

For copper-doped and hydrogenated specimen (II), the particles related to the dopant are of various shapes and size (see Fig.2 A–C). The smallest particles are nearly equisized and spherical in shape, while the largest are irregular in shape and exhibit the presence of many stacking faults. According to the electron diffraction pattern, crystalline Cu and Cu_2O particles have been evidenced in these specimens. The small particles entirely correspond to Cu_2O (see the B1/ 0.302 nm /(110) and B2/0.247 nm/(111) fringes), while the largest ones are composed of the metallic Cu-core (see the fringes A/0.209 nm related to the (111) planes) surrounded by the semiconducting Cu2O layer (see the fringes: B2/0.247 nm from (111), and B3/0.214 nm from (200)). It seemed reasonably to suppose that the colloidal Cu-particles formed during hydrogenation are highly reactive with respect to the surrounding air-atmosphere and form the nanoparticles of Cu_2O.

4.2 Optical Absorption Spectra of Microstructurally Anisotropic SLS Glass

Figure 3 gives the optical absorption spectra induced by silver particles deformed in various degrees. For comparison, the inset shows the spectrum characteristic of not-deformed specimens. In this case, the band is located at about 410 nm and corresponds to the surface-plasmon resonance of the Ag-particles [11]. For the deformed specimens the OA spectra have been measured with linearly polarized light, and the pairs of absorption curves a-d, b-e, and c-f correspond to the absorption of light polarized perpendicularly (short λ) and parallelly (large λ) to the tensile axis. The separation of the corresponding bands increases by increasing the deformation degree, i.e. the eccentricity of the prolate spheroidal silver particles. Moreover, for samples annealed in hydrogen a doubling of the long λ-band into two Lorentzian components occurs, corresponding to some periodical size and/or shape changes (modulations) of the particles; see the results of TEM. The weaker band, located at short λ's, is probably related to the small spherical silver particles, which are not affected by deformation. Since colors and polarizing effects related to the

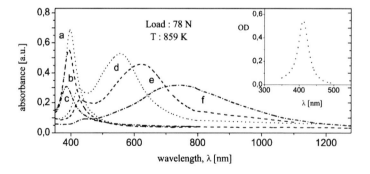

Figure 3: Absorption spectra of colloidal Ag-spheroids in the H2-annealed samples measured with polarized light: curves a,b,c are for polarization ⊥, and curves d,e,f for polarization ∥ to the tensile axis.

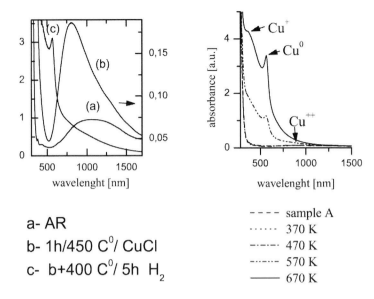

a- AR
b- 1h/450 C^0/ CuCl
c- b+400 C^0/ 5h H$_2$

---- sample A
······ 370 K
─·─·─ 470 K
─··─·· 570 K
──── 670 K

Figure 4: The effect of isochronal (5 h) hydrogenation of SLS glass (II), exchanged with monovalent copper during 1 h at 728 K, at temperatures from the range 370–670 K upon the optical absorption

particles may widely be controlled by variation of their aspect ratio, this behavior enables the fabrication of novel dichroic polarizers of high mechanical, chemical and thermal durability which have a wide range of dichroic colors accessible. Polarizers with such attributes could be useful for the production of colored liquid crystal displays [12].

To interpret the OA spectra of copper (see Fig. 4), one has to emphasize that after exchange the copper in SLS glasses is present in the form of cuprous (I) and cupric (II) ions. The cupric ions create color centers with a broad absorption band (at about 780 nm) inducing the blue/green color of these glasses. The shift of their fundamental absorption edge towards lower energies is probably due to introduction of an absorption band related to the cuprous ions. Hydrogenation of samples above 470 K results in the formation of atomic copper reduced to the metallic state in a surface layer whose thickness increases with time and temperature. The growth kinetic of the reduced layer, followed by optical spectroscopy, could be described by the tarnishing model [13] The copper nanoparticles induce a red/ruby color of samples, and this color is related with a nonsymmetrical absorption band located at about 560 nm due to the surface plasmon resonance of the colloidal nanoparticles [14]. The evolution of this band with increasing reduction temperature reflects the changes in size of these particles. The broad absorption band, located at the high-energy side of the colloidal band is probably related to the presence of some compounds of monovalent copper, e.g. to the presence of nanosized Cu_2O particles evidenced by the TEM observations.

5 Concluding Remarks

Since both the exchange and formation of the colloidal nanoparticles seem to depend upon the type of the glass involved and the history of the individual specimens, a specific effect upon the experimental data is quite difficult to interpret.

Combined X-ray diffraction and electron microscopy studies demonstrate that the color changes of the matrix are for both composites due to the formation of metallic colloids. It is suggested that the overall coloring process be controlled by the initial reduction of the dopant-ion to the atomic state and by subsequent diffusion of the atoms to growing nuclei, which increase in size with time.

The presence of a secondary phase separation instead of partial devitrification, characteristic of prolonged application of electric field [15] or irradiation [16] suggests that the permeation of hydrogen not only allows the reduction of silver but enters the glass network, probably forms hydroxyl ions i.e. a new component of the defect structure, and in this way affects the structure-sensitive properties.

For the copper-exchanged specimens it is the first evidence of the formation of colloidal Cu_2O nanoparticles in the SLS-matrix, and it could be interesting for studies of the quantum confinement effects that are expected when the semiconductor-nanoparticles are small enough.

The extraordinarily strong polarizability of the metal clusters at the surface plasmon frequency could result in marked nonlinear properties of such composites.

The deformation induces remarkable dichroism of the optical absorption, which could be exploited for new applications, e.g. to produce color-selective polarizers. Further work on these problems is in progress.

6 Acknowledgments

The work was performed in frames of a KBN project no. 7T 08D 06021.

7 References

[1] Hache, F., Ricard, D., Flytzanis, C., Kreibig, U., Appl.Phys. A 1988, 47, 347–353
[2] Myers, R.A., Mukherjee, N., Bruek, S.R.S., Optical Letters 1991, 16, 1937–1951
[3] Haglund, R.F.Jr., Yang, L., Magruder III, R.H.; Whute, C.W., Zuhr, R.A., Yang L., Dorsinville, R., Alfano, R.R., NIM B 1994, 91, 493–501
[4] Debnath, R., J. Lumin., 1989,43, 375–379
[5] Suszyska, M., Szmida, M., Grau, P., Mater.Sci. Engin. A, 2001, 319/321, 702–705
[6] Porai-Koshits, E.A., Averjanov, V.J., J.Non-Crystal.Solids 1965, 1, 29–38
[7] Suszynska, M., Krajczyk, L., Mazurkiewicz, Z., Mater.Chem.Phys. 2003, 81, 404–406
[8] Macalik, B., Krajczyk, L., Okal, J., Morawska-Kowal, T., Nierzewski, K.D., Suszynska, M., Rad. Effects and Defects in Solids 2002, 157, 887–893
[9] Drost, W.G., Dissertation Thesis, 1992, Halle, Germany,
[10] Suszynska, M., Macalik, B., Morawska-Kowal, T., Okuno, E., Yoshimura, E.M., Yukihara, E., Rad. Effects And Defects in Solids 2001, 156, 353–358

[11] Kreibig, U., Vollmer, M., Optical Properties of Metal Clusters, Springer Verlag, 25, 1995, Berlin
[12] Berg, K.J., Dehmel, A., Berg, G., Glastech.Ber., 1995, 68, 554–559
[13] Shelby, J.E., J.Appl.Phys. 1980, 51, 2589–2593
[14] Estournes, C., Cornu, N., Guille, J.L., J.Non-Cryst.Solids 1994, 170, 287–294
[15] Berg, K.-J., Capeletti, R., Krajczyk, L., Suszynska, M., Proc. of ISE 9, IEEE, eds. Zhongfu, X.; Hongyan, Z. 1996 378–383
[16] Morawska-Kowal, T., Krajczyk, L., Macalik, B., Nierzewski, K.D., Okuno, E., Suszynska, M., Szmida, M., Yoshimura, E.M., NIM B 2000, 166/167, 490–494

II Large Strain Cold Working and Microstructure

Equivalent Strains in Severe Plastic Deformation

Hein Peter Stüwe

Erich Schmid Institute of Materials Science, Austrian Academy of Sciences and Institute of Metal Physics, University Leoben, Leoben, Austria

1 Abstract

The concept of equivalent plastic strain is discussed. Definitions may be based on kinematic arguments (like change of geometry) or on equivalence of plastic work. Complications arise when strain is accumulated in increments with variable strain path.

2 Introduction

Severe plastic deformation (SPD) is defined as "intense plastic straining under high imposed pressure". [1] It has received increasing attention in the last decade because it produces extremely fine-grained structures (with grain sizes < 100 nm) and, consequently, very high strength.

Very high strains can be achieved by a number of methods which may be classified by their strain paths:

1. Continuous strain without change of strain path: Compression, extrusion, high pressure torsion (HPT).
2. Accumulated strain without change of strain path: Rolling, drawing, equal channel angular pressing (ECAP) route A.
3. Accumulated strain with reversal of strain path: Cyclic extrusion--compression (CEC), ECAP (route C), channel die compression.
4. Accumulated strain with variable strain path: swaging, ECAP (routes other than A or C), channel die compression, accumulated roll bonding.

All of these methods have in common that the material is deformed under high hydrostatic pressure. One "extrinsic" reason for this is obvious: pressure will prevent cracks and will thus permit to achieve much higher strains (e.g., in torsion). There may be also "intrinsic" effects leading to a microstructure different from that achieved without pressure. Preliminary experiments show that such an effect does, indeed, exist although it is not of overriding importance. [3]

This paper shall discuss how to compare strains achieved by different SPD methods, i.e., the question of equivalent strains.

3 Conventional Definition of Strain in Longitudinal Extension

3.1 Continuous Strain

The conventional definition of equivalent strain φ^* is based on the proposition that two specimens deformed on different strain paths should be compared when the same specific plastic work has been spent on them.

For a tensile test this can be written as

$$\frac{A}{V} = \int_{l_0}^{l} \frac{\sigma F}{V} dl = \int_{l_0}^{l} \sigma \frac{dl}{l} = \int_{0}^{\varphi^*} \sigma \, d\varphi \qquad (1)$$

where A is the total plastic work, V is the active volume of the specimen (assumed to be constant), l is the active length, F the cross section (assumed to be equal over the specimen length) and ρ is the flow stress. For an ideally plastic material (ρ = const, no work hardening) Equation 1 simplifies to

$$\frac{A}{V} = \sigma \varphi^* \qquad (2)$$

with

$$\varphi^* = \ln \frac{l}{l_0} \qquad (3)$$

This is often a sufficiently good approximation, especially at high strains where the work hardening coefficient is small.

Tensile tests may follow different strain paths with different principal strains as shown in Table 1.

Table 1: Different strain paths with different principal strains.

Principal strains	Axial symmetry	General	Plane strain
φ_1	φ^*	φ^*	φ^*
φ_2	$-\frac{\varphi^*}{2}$	$-x$	0
φ_3	$-\frac{\varphi^*}{2}$	$-\varphi^* + x$	$-\varphi^*$

The Equations 2,3 are valid only if the largest principal strain is used. On the other hand, the Equations 2,3 are also valid in compression tests where all terms of Table 1 change their sign. A more general definition of equivalent strain would then be to use the supreme absolute value of the principal strains, i.e.,

$$\varphi^* = \left| \ln \frac{l}{l_0} \right|_{sup} \qquad (4)$$

3.2 Accumulated Strain without Change of Strain Path

Equation 4 is very useful when deformation is carried out in several steps like, e.g., in successive compression tests. The accumulated strain is then

$$\varphi^* = \Sigma |\Delta\varphi| \tag{5}$$

Because they are so useful and simple, the Equations 4,5 are widely used to describe accumulated strains in technical processes such as rolling and drawing. (Length l or diameter are measured easily on the semifinished product.). This is satisfactory for many purposes but is not quite exact. Usually there is a shear component in the strains near the surface due to friction between the material and the tool. This "redundant strain" produces strain gradients in the material leading to inhomogeneities, e.g., in drawn rods (see section 6 in this paper) and in accumulated roll bonding.

4 Equivalent Strain in Simple Shear

Simple shear is a good model for torsion tests which in turn are one of the most important tools in the study of high strains. For instance, stage IV of the stress strain curve was first observed in free torsion tests [4] and the highest plastic strains up to now have been reached in HPT. [5] It is therefore, of interest to know an equivalent strain. In analogy to Equation 1 we can write

$$\frac{A}{V} = \int_0^\gamma \frac{\tau \, Fl \, d\gamma}{V} = \int_0^\gamma \tau \, d\gamma \tag{6}$$

(where χ is the tangent of the shear angle) or, for ideal plasticity

$$\frac{A}{V} = \tau \gamma \tag{7}$$

Comparing specimens that have seen equal specific plastic work, we obtain

$$\int_0^{\varphi^*} \sigma \, d\varphi = \int_0^\gamma \tau \, d\gamma \tag{8}$$

This equation is fulfilled when

$$\sigma = a\tau$$
$$d\varphi = \frac{1}{a} d\gamma \tag{9 a/b}$$

Obviously Equation 8 is fulfilled for any arbitrary value of a, so an additional assumption is needed. Many have been proposed, but only three have found wide acceptance. They are listed in Table 2.

The first two values are based on Equation 9 and a heuristic assumption on the shape of the yield locus. These assumptions have two advantages:

Table 2. Most acceptable values of a.

	a
Tresca	2
v. Mises	$\sqrt{3}$
Taylor	M

1. They are very simple mathematically.
2. Almost one century of experience has shown that predictions based on them rarely deviate very much from experiment.

They have a great disadvantage: they are designed for isotropic materials and thus cannot account for plastic anisotropy.

Taylor-type theories are based on a kinematic analysis of Equation 9b in a crystal. χ is then the accumulated shear on a selected set of co-operating slip systems in one crystal, averaged over all crystals in a polycrystal.

The advantage of such theories is that they permit to predict the plastic response of anisotropic (i.e., textured) materials. They also permit to predict the evolution of deformation textures. Their disadvantage is that they are mathematically complicated and require the use of elaborate computer programs. Only one special case is simple: in a single crystal oriented for

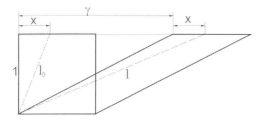

Figure 1: Maximum elongation in simple shear

single slip M is the inverse of the Schmid orientation factor and has, thus, a minimum value at $M = 2$.

Let us now see whether a combination of Equations 9b and 4 leads to a reasonable definition of a.

Fig.1 shows a square deformed in simple shear. The tangent of the shear angle is χ. The direction of maximum elongation is characterized by the distance x. The lengths l_0 and l are then given by

$$l_0^2 = 1 + x^2$$
$$l^2 = 1 + (\gamma + x)^2$$
(10a/b)

x can be found from the condition that $d(l/l_0)/dx = 0$ as

$$x = -\frac{\gamma}{2} \pm \sqrt{\frac{\gamma^2}{4} + 1} \tag{11}$$

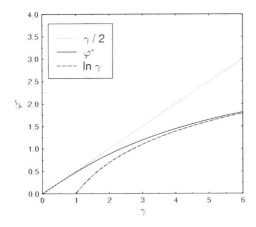

Figure 2: φ^* vs. χ according to Equation 12

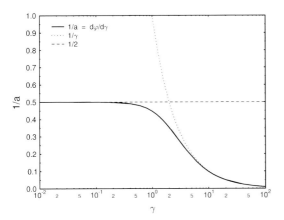

Figure 3: $1/a$ vs χ computed from Equations 12, and 9 b.

Using the positive root one obtains for Equation 4:

$$\varphi^* = \ln\left(\frac{l}{l_0}\right)_{max} = \frac{1}{2}\ln\frac{1+\left(\sqrt{\frac{\gamma^2}{4}+1}+\frac{\gamma}{2}\right)^2}{1+\left(\sqrt{\frac{\gamma^2}{4}+1}-\frac{\gamma}{2}\right)^2} \tag{12}$$

(The negative root leads to the compressive strain perpendicular to *l*). The function of Equation12 is shown in Figure2. It can be approximated by

$$\ln(l/l_0) \approx \frac{\gamma}{2} \quad \text{for } \gamma \ll 2 \quad \text{(good for } \gamma < 1\text{)} \tag{13}$$

and

$$\ln(l/l_0) \approx \ln \gamma \text{ for } \gamma \gg 2 \text{ (good for } \gamma > 4\text{)} \tag{14}$$

Equation 12 can be used to compute 1/*a* according to Equation 9b. The result is shown in Figure 3. One sees that ≈2 is a good approximation only for $\chi < 1$ and that $a \rightarrow \gamma$ for larger strains.

5   Efficiency of Simple Shear

Consider a cubic volume element of 10 μm diameter in the undeformed matrix. This element may be a grain or, better still, a one-phase volume in a dual phase alloy. Let this material be deformed by accumulated roll bonding where the specimen thickness is reduced by one half in each pass. After 15 passes the thickness of the element should (on a purely geometric basis) be reduced to 0.3 nm, i.e., to atomic dimensions. Obviously, the element will have lost its metallographic identity long before that. The corresponding true strain is 10.4.

According to Equation9b and using the Tresca criterion ($a = 2$), the same amount of work is spent for $\gamma = 20.8$. The thickness of the element is then given by Equation 14 as 0.48 μm.

This shows that simple shear is an inefficient way to change the geometry of original structural elements.

The typical SPD structure, however, is not determined by the elements of the original structure but rather by the storage of a great number of newly created dislocations. For these, the plastic work spent on the material is a better indication. The definition of Equation4 must be used with caution when shear strains are involved. (This has been hinted at already in section 2.2).

For the discussion of the microstructure it is more appropriate to compare the accumulated shear of the active slip systems, i.e., to use the Taylor factor *M*. *M*, of course, is also a function of strain but it tends to approach a constant value characteristic for each strain path. (This corresponds to the development of a specific deformation texture). For crude estimates it may be sufficient to use one of the other constants given in Table2.

6   Accumulated Strain with Reversal of Strain Path

Many methods for SPD are based on a reversal of strain path. This is because one hopes to apply very high strains without changing the dimensions of the specimen. As an example we consider CEC as shown schematically in Figure4. The specimen is pressed from a cylindrical recipient of diameter d_0 through a die of diameter d_m into another recipient of diameter d_0. It ex-

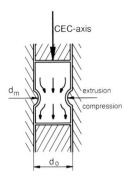

Figure 4: One cycle of CEC.

periences an extrusion with $\Delta\varphi_1 = \ln(d_0 / d_m)^2$ and a compression with $\Delta\varphi_2 = \ln(d_m / d_0)^2$. Obviously, $\Delta\varphi_1 + \Delta\varphi_2 = 0$.

For a discussion of microstructure this is not satisfactory. The material has seen two deformations in this cycle so we tend to add the absolute values of the partial strains like in Equation 5.

This has been done in Figure 5 where flow stress is plotted vs. strain accumulated in many passes of CEC. The curve shows several stages that are called A, B, C, and D. Figure 6 shows the flow curve of a similar material obtained in unidirectional strain showing the conventional stages III, IV, and V. Very extensive investigations [6,7] have shown that the development of microstructure in stages B, C, and D corresponds to the well known stages III, IV, and V. This means that the strain accumulated by reversals of strain path is not equivalent to unidirectional strain so that Equation 5 should be written as

$$\varphi^* = \eta \sum_{1}^{2n} |\Delta\varphi| \tag{15}$$

where n is the number of cycles and γ is an efficiency factor. Comparing the onset strains marked by vertical lines in Figures 5 and 6 we conclude that $0.1 < \gamma < 0.2$ in this experiment. It is no surprise that γ should be smaller than one. The increment of dislocation density φ produced by an increment in strain can be estimated as

$$\Delta\rho = \frac{2M}{\Lambda b} \Delta\varphi \tag{16}$$

where Λ is the free path of a dislocation with Burgers vector b, and M the Taylor factor. If this is followed by an equal increment in strain of the opposite direction, we can consider two extreme cases:

1. All newly created dislocations run back into their sources and disappear (like a movie that is run backwards). The efficiency for creating dislocations, γ, would then be zero if taken over the complete cycle.

2. A new set of dislocations is created according to Equation 16. Since the same set of slip systems is operating, no immediate change in M and Λ is expected. If all dislocations remain in the material, then $\gamma = 1$.

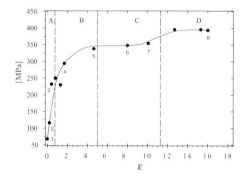

Figure 5: Flow stress vs. accumulated strain for AlMg5 deformed by CEC (from Richert et al. [7]).

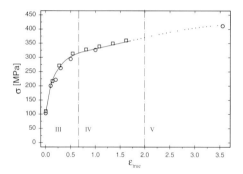

Figure 6: Flow stress vs. strain for AlMg5 deformed by rolling μ and torsion δ (from Richert et al. [7])

The second set, however, consists of dislocations of the opposite sign running in the same direction and/or of dislocations of the same sign running in the opposite direction. This strongly enhances their chance to annihilate with dislocations of opposite sign and we expect $0 < \gamma < 1$ for the complete cycle.

We can call this effect "geometric recovery" or "kinematic softening". It is a familiar phenomenon. Related classical examples are known as the "Bauschinger-effect" and as "fatigue softening" of work hardened material (see, e.g., Schrank et al.). [8]

7 Accumulated Strain with Variable Strain Path

Figure 7 shows the distribution of hardness over the diameter of copper rods. The rods have been drawn (upper curves) or swaged (lower curves) to increasing strains (defined by change of length).

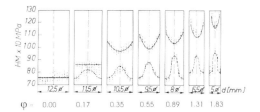

Figure 7: Distribution of hardness over the cross section of copper rods deformed by drawing (upper curves) and by swaging (lower curves). (from Grabinowski et al. [9])

The upper curves are easily understood: The core of the rod is work hardened by unidirectional strain. The outer layers are work hardened somewhat more by an additional (redundant) shear (see section 2.2).

In swaging only part of the external work is spent on lengthening the rod. During each stroke the material is also sheared in direction of the circumference. [8] If the specimens were rotated by 45° between strokes this lateral shear would be essentially reversed. In commercial machines the rotation angle is different (in the machine used by Grabianowski et al. [9] it was 57°) so the strain path for each stroke varies irregularly.

It is difficult to predict an efficiency for the production of dislocations for strains accumulated in this way, because M and Λ may change with each stroke and many kinds of dislocations are produced. The result, however, is clear from Figure 7: kinematic softening very nearly prevents work hardening in the outer layers of the rod and even the center of the rod work hardens much less than the drawn rods.

On the basis of these observations one would expect that the efficiency for producing many dislocations and, hence, SPD structures by ECA pressing should increase following routes C (reversal of strain path) < B (other changes of strain path) < A (constant strain path). This was, indeed, observed in experiments designed for this purpose. [10] Unfortunately, a literature survey presented in [11] comes to different conclusions and proposes the sequence A < B_A < C < B_C. It is possible that an "efficiency to produce fine grained and equiaxed structures with a wide spread of misorientations" is too vaguely defined to serve for quantitative arguments.

8 Conclusions

Equivalent strains are used when materials deformed by different methods are compared. Their definition may be based on the shape of the specimen or on the plastic work spent. Special care must be taken when comparing strains accumulated in steps of varying strain path. It may be helpful to introduce efficiency factors which depend on the material property under discussion.

9 Acknowledgement

The author is indebted to Dr. H. Weinhandl for computing the Figures 2 and 3, and to F.D. Fischer, R. Pippan, and W. Schwenzfeier for helpful discussions.

10 References

[1] R. Z. Valiev, Investigations and Applications of SPD (Eds. T.C. Lowe, R.Z. Valiev), Kluwer Academic Publishers, Norwell 2000, p.211
[2] Investigations and Applications of SPD (Eds. T.C. Lowe, R.Z. Valiev), Kluwer Academic Publishers, Norwell 2000
[3] T. Hebesberger, A. Vorhauer, H.P. Stüwe, R. Pippan, in Proc. of the 2nd Int. Conf. on Nanomaterials by Severe Plastic Deformation, Wien 2002, Wiley-VCH, Weinheim 2003, in press
[4] H.P. Stüwe, Z. Metallk. 1965, 56, 633
[5] T. Hebesberger, A. Vorhauer, R. Wadsack, H.P. Stüwe, R. Pippan, Berg-Hüttenmaenn. Monatsh., in press
[6] M. Richert, H.P. Stüwe, J. Richert, R. Pippan, C, Motz, Mater. Sci. Eng. 2001, A301, 237
[7] M. Richert, H.P. Stüwe, M.J. Zehetbauer, J. Richert, R.'Pippan, C. Motz, E. Schafler, Mater. Sci. Eng. A, in press
[8] J. Schrank, B. Ortner, H.P. Stüwe, A. Grabianowski, Mater. Sci. Techn. 1985, *1*, 544.
[9] A. Grabianowski, A. Danda, B. Ortner, H.P. Stüwe, Mech. Res. Comm. 1980, *7*, 125
[10] P.B. Prangnell, A. Gholinia, M.V. Martenshev, in Investigations and Applications of SPD (Eds. T.C. Lowe, R.Z. Valiev), Kluwer Academic Publishers, Norwell 2000, p.65
[11] T.C. Lowe, Y.T. Zhu, S.I. Semjatin, D.R. Berg, in Investigations and Applications of SPD (Eds. T.C. Lowe, R.Z. Valiev), Kluwer Academic Publishers, Norwell 2000, p.347

Stage IV: Microscopic Or Mesoscopic Effect ?

J. Aldazabal[1], J. M. Alberdi[2] and J. Gil Sevillano[1]

[1] CEIT (Centro de Estudios e Investigaciones Técnicas de Guipúzcoa), and TECNUN (Technological Campus, University of Navarra), San Sebastián, Spain
[2] Faculty of Chemistry, Basque Country University, San Sebastián, Spain

1 Abstract

The generality of theoretical explanations of Stage IV, the large strain deformation stage, like those available for Stages I to III, are up to date based on a balance of accumulation and annihilation dislocation processes taking place at a microscopic scale, in the sense of being reducible to the behaviour of a small sub-micrometric single-oriented volume undergoing a uniform plastic strain. This paper shows that mesoscopic plastic gradients, i.e., spatial plastic strain and crystallographic slip gradients of micrometric wavelength, can by themselves explain the emergence of Stage IV at large deformations or at least it is shown that they can make a significant contribution to it.

2 Introduction

Deformation Stage III, characterised by a decreasing work hardening rate that would seem to lead towards a saturation stress, τ_s^{III}, is followed by Stage IV, characterised by a small but persistent work hardening rate that delays the attainment of a true saturation stress much further. At moderate or low homologous temperatures (T/TM<0.5), the Stage III/Stage IV transition occurs rather abruptly for an equivalent strain of about unity. The ensuing Stage IV occupies a large strain range until its work hardening rate is exhausted or until a new deformation stage, Stage V, takes the relay leading rapidly to a structural change that annihilates the hardening capacity (e.g., continuous recrystallization) or induces ductile fracture [1,2].

The Stage III hardening rate, $\theta = \partial \tau / \partial \Gamma$, in terms of the critical resolved shear stress, τ, and the amount of crystallographic slip, Γ, at constant strain rate and temperature, is, as thoroughly demonstrated by Mecking and Kocks [2], very well described by the Voce strain hardening equation

$$\frac{\theta}{\theta_0^{III}} = 1 - \frac{(\tau - \tau_0)}{\tau_s^{III}} \qquad (1)$$

both for single crystals deforming by multiple slip or for polycrystals. The stress τ_0 is a part of the flow stress independent of the dislocation density, θ_0^{III} is an athermal asymptotic limit of the strain hardening rate for zero dislocation density (equivalent to the Stage II work hardening rate of single crystals) and τ_s^{III} is a thermally and strain rate dependent saturation stress. The limit work hardening rate θ_0^{III} has a value of the order of $5 \cdot 10^{-3} G$ (with G the appropriate shear modulus). It decreases as the stacking fault energy, SFE, decreases. For the same "homologous temperature", T/T_M or $k/Gb3$ (with k the Boltzmann constant), the saturation stress τ_s^{III} of different materials also scales with the SFE but in the opposite way, i.e., it becomes larger as SFE decreases [2].

Equation (1) makes possible to collapse Stage III work hardening data from different materials measured at different temperatures and strain rates in a single "master curve", fig. 1.

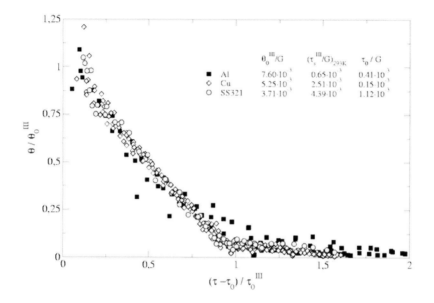

Figure 1: Torsional work hardening rate vs. flow stress data of 99.98 % Cu (77, 198, 293, 373 and 473 K), 1050 Al (77, 198, 293 and 373 K), both at $\dot{\Gamma} = 2 \cdot 10^{-2} s^{-1}$, and austenitic SS321 stainless steel (20 °C and 200 °C, $\dot{\Gamma} = 6 \cdot 10^{-3} s^{-1}$) normalised according to eq. (1). A constant orientation factor $M = 1.5$ has been used for transforming the macroscopic stress-strain data in CRSS- crystallographic slip data. Re-plotted results from Alberdi [3]. Grain size of Al and Cu is 25 µm.

In such a representation, Stage IV appears as a prominent tail that allows to harden the materials by plastic deformation up to twice their Stage III virtual saturation stress and beyond. It may also be seen in the figure that Stage IV data are, if not perfectly, remarkably well normalised by using the two Stage III fundamental parameters. This suggests that both deformation stages bear some close connection.

The CRSS for plastic flow is related to the dislocation density trough the accepted relationship

$$\frac{(\tau - \tau_0)}{G} = \alpha b \rho^{1/2} \qquad (2)$$

The non-dimensional parameter α takes a value of about 0.3 for most materials.

On account of eq. (2), Stage III behaviour is microscopically explained by the competing action of dislocation storage and annihilation processes that depend on the dislocation density as

$$\frac{\partial \rho}{\partial \Gamma} = C_1 \rho^{1/2} - C_2 \rho \qquad (3)$$

$C_1 \geq 0$ is an athermal coefficient of dislocation storage (hardening) and $C_2 \geq 0$ is a thermally and strain rate dependent coefficient of dislocation recovery (softening). Such functional dependencies are mechanistically explained in terms of percolation hardening theories and dynamic recovery models [1,2].

A consensus on the mechanistic explanation for Stage IV has not been reached. Several more or less elaborated theoretical models that lead to an extension of the hardening beyond Stage III have been produced [1]. The generality of them, as it happens with the models available for Stages I to III, are up to date based on a balance of accumulation and annihilation dislocation processes taking place at a microscopic scale, in the sense of being reducible to the behaviour of a small sub-micrometric single-oriented volume undergoing a uniform plastic strain. Such models should be particularly suited to predict the single crystal behaviour. There is however a worry: it seems that there are very few results of "clean" stress-strain curves of single crystals deforming by multiple slip and reaching very large strains. Quite often in the few cases where the work hardening of single crystals has been plotted against the flow stress [4–6], the Stage IV tail is absent or poorly developed but for high homologous temperatures, at least if compared with the many results now available for polycrystals deformed under comparable temperature and strain rate conditions. Inherent to the plastic deformation of polycrystals are the spatial heterogeneities of plastic strain and of crystallographic slip activity arising as a consequence of the orientational differences. The magnitude of such fluctuations of grain size wavelength is very big if compared with the small heterogeneities that deformation develops in single crystals because of the patterns of non-uniform dislocation density that strain induces and that are also present in polycrystals at intragranular level.

Plastic gradients imply a storage of "geometrically necessary dislocations" (a non-local or mesoscopic work hardening effect) additional to the dislocation accumulation that would occur in their absence (the local or microscopic work hardening effect of strain). The observed work hardening rate results from the superposition of both local and non-local effects. However, the extra term of dislocation accumulation induced by plastic gradients does not represent a mere quantitative modification of the work hardening law that holds in the absence of those gradients: the "geometrically necessary dislocations" are composed of sets of dislocations of the same sign at a "local" or microscopic scale and they are thus immune to short-range (low-temperature) dislocation annihilation (recovery) mechanisms. It seems thus pertinent to investigate the implication of such mesoscopic effects on the manifestation of Stage IV.

In fact, a discussion on the possible non-microscopic origin of Stage IV was quite early raised [7] and perhaps too quickly forgotten. This paper shows that without recourse to any ad-hoc mechanistic assumption for some microscopic process responsible for the existence of Stage IV, the plastic effects associated to the presence of mesoscopic plastic gradients in crystalline materials, can by themselves explain the emergence of it at large deformations or at least significantly contribute to it.

Stress-strain curves up to large strains have been numerically calculated for single crystals or polycrystals using a simple cellular automaton that takes into account spatial heterogeneities of dislocation density, orientation factors, elastic-plastic strains and internal stresses. In the model, the size establishing the micro/meso border is the cell size, below which plastic strain is assumed to take place uniformly. A microscopic Stage III work hardening law typical of multiple slip is assumed to hold inside each cell. In absence of strain gradients, it would lead to a saturation stress without any Stage IV. Strain gradients however imply a proportional storage of geometrically necessary dislocations that contribute to increase the flow stress above the Stage III saturation stress. For simulated polycrystals the calculated curves display a Stage III to Stage IV transition that occurs at a ratio of the work hardening rate to the strain-dependent part of the flow stress in reasonable agreement with experimental observations and a Stage IV extension also of the right order of magnitude. Finally, it is to be mentioned that Kok et al. [8] have quite recently published a detailed Finite Element Analysis of polycrystalline deformation using a very similar approach to that presented here. They observe the emergence of Stage IV and reach conclusions very similar those of the present paper.

3   A simple Mesoscopic Model

A very simple one-dimensional mesoscopic cellular automaton model has been developed in which a strip composed of "grains" is divided in elementary cells of size much smaller than the average grain size. Each cell is characterized by its local orientation, dislocation density (comprising two types of dislocations, statistically stored, SSD, and geometrically necessary, GND, dislocations) and a one-dimensional internal stress. The local orientation and dislocation density of a cell determine its local plastic strength. The dislocation transmittance capability of each grain boundary is also considered. The strip deforms in equilibrium under a macroscopic one-dimensional stress increasing step by step, the elastic-plastic and structural status of each cell being synchronously updated at every step. A coupled evolution of the two dislocation density types is considered (the GND density, ρ_{gn}, is determined by the local plastic strain gradient but the evolution of the SSD density, ρ_{ss}, depends on the total, SSD+GND, local dislocation density, ρ).

Starting from an initial condition, the macroscopic response and the evolution of the internal heterogeneity of the strip can be followed. Neither changes of the geometrical shape of the cells nor crystallographic rotations have been taken into account. Despite its very simple structure the model yields quite realistic results of stress-strain and Hall-Petch behaviour [9,10]. See those two references for a detailed description of the model.

The conditions and data input for the numerical simulations to be presented here are as follows. The spatial boundaries at the two sides of the strip correspond to a periodic structure. The strip is composed of N = 45 000 cells of size $l/b = 1000$. The cell length determines the resolution of the measures of the dislocation density and strain. The shear modulus G of the cells has been assumed uniform.

A low initial average dislocation density, $b^2/(\rho_i)_0 = 3 \cdot 10^{-6}$, corresponding to a mean of three dislocations per cell i has been assumed. Such initial density has been assigned to discrete SSD randomly distributed in the cells but for some simulations for which the starting dislocations have been uniformly distributed (in order to test the influence of microscopic vs. mesoscopic heterogeneity).

Calculations have been made either for a single crystalline strip, assuming a uniform orientation factor $M = 1.5$ for the whole strip, or for a "polycrystalline" strip divided in segments ("grains"), the cells of each segment having an orientation factor randomly distributed in the range $1 \leq M \leq 2$, i.e., $\langle M \rangle = 1.5$ for the polycrystal. The polycrystalline strip has been constructed locating at random 450 grain boundaries on the strip, i.e., the average grain size is $10^5 b \cong 25\ \mu m$. With the values used for the orientation factors, the calculated macroscopic stress-strain behaviour adequately represents a simple shear deformation of a cubic fcc or bcc polycrystal. An orientation-dependent factor relates the instantaneous GND density, ρ_{gn}, in a cell to the plastic strain gradient, χ, there

$$\rho_{gn} = M_R (\chi/b) \tag{4}$$

It has been set to $M_R = M$ for the present one-dimensional calculations.

Two variants of the model have been considered with respect to the storage of geometrically necessary dislocations: a lower bound, in which grain boundaries are considered fully transparent to slip (or, alternatively, perfect dislocations sinks) and an upper bound, in which grain boundaries are assumed completely impermeable to slip. In the first case, the density of geometrically necessary dislocations is computed by eq. (4). In the latter case, gliding dislocations are not free to cross grain boundaries and the cells limited by boundaries store a surplus of dislocation density of the same sign: a dislocation layer will form at both sides of the boundary.

The dislocation density independent part of the CRSS has been assumed $s_0 = 0$, thus the simulations are representative of pure FCC metals. The proportionality constant between the (normalised) CRSS and the square root of the dislocation density has been taken to be $\alpha = 0.3$, in agreement with most experimental or theoretical estimations for cubic metals [1]. The macroscopic stress is increased in $\Delta\sigma/G = 10^{-5}$ steps and the simulations correspond to stress control.

The statistically originated term, ρ_{ss}, is the dislocation storage not directly linked to the presence of mesoscopic plastic gradients. However, its evolution most probably depends on the instantaneous value of the dislocation density in the cell whatever its origin, thus it will indirectly depend on the plastic gradient after any finite plastic deformation. Consequently, it has been calculated assuming a Voce hardening law adapted for the presence of geometrically necessary dislocations. The expression of it to be used in incremental form is, for each cell i

$$\frac{\Delta(\rho_{ss})_i}{(\Delta\Gamma)_i} = C_1 \rho_i^{1/2} - C_2 (\rho_{ss})_i \tag{5}$$

Any dislocations, statistically or necessarily stored, are assumed to induce dislocation accumulation (first term in eq. 5) but geometrically necessary dislocations are assumed to be immune to annihilation by recovery at moderate temperatures (second term in eq. 5).

The two coefficients of the previous equation are related to the parameters of the Voce hardening equation as,

$$\frac{\theta_0^{III}}{G} = \frac{\alpha(C_1 b)}{2}$$

$$\frac{\tau_s^{III}}{G} = \frac{\alpha(C_1 b)}{C_2} \tag{6}$$

4 Results and Discussion

Calculations of stress-strain curves using a range of parameters C1 and C2 corresponding to Stage III of FCC metals of high to low SFE and a wide range of temperatures (of the order of those corresponding to the Stage III region of the experimental results in fig. 1) have been made for the 105b grain size polycrystals. Some of them, after proper normalising, are shown in fig. 2, where the experimental points of fig. 1 are also included. Given the simplicity of the model and its one-dimensional character, the agreement with the experimental behaviour is surprisingly good. Similar calculations for single crystals (notice however that no account has been made for strain-induced tessellation of the single crystals in misoriented domains) do not show but Stage III until saturation close to the corresponding τ_s^{III}.

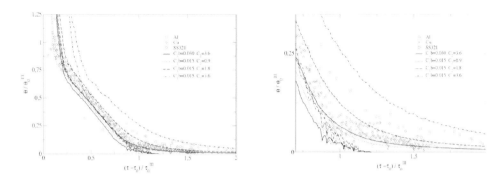

Figure 2: a) Results of the model (C_1 and C_2 parameters that have been used in the computations are shown in the figures) superposed to the experimental points previously shown in fig. 1 b) Zoom of the Stage IV region.

Without any parameter tuning, using the Stage III microscopic hardening law of a material but accounting for the extra accumulation of dislocations necessary to absorb the mesoscopic plastic gradients that built-up during the deformation process in a heterogeneous material that mimics a polycrystal, Stage IV appears after Stage III is nearly completed (roughly just after the work hardening rate has decayed to 10% of the Stage III limit hardening rate θ_0). Of course, total agreement cannot be pretended and other contributions to Stage IV hardening coming from microscopic origins (debris, Stage I type hardening contributions, etc.) could also form part of the observed behaviour.

5 Conclusions

- In polycrystals, mesoscopic plastic gradients, i.e., spatial plastic strain and crystallographic slip gradients of grain size wavelength, can by themselves explain the emergence of Stage IV at large strains or at least it is shown that they can make a significant contribution to it.
- In single crystals, Stage IV could arise because of the formation of increasingly misoriented regions as deformation goes on.

- The model used for simulating the non-uniform deformation of a polycrystalline aggregate is extremely simple but it seems to capture the essence of the Stage IV process. However, other contributions to it from microscopic processes can not at present be excluded

6 References

[1] J. Gil Sevillano, in H. Mughrabi (ed.), "Materials Science and Engineering. A Comprehensive Treatment", vol. 6, p. 19. VCH, Weinheim, Germany, 1993.
[2] U. F. Kocks and H. Mecking, "Physics and Phenomenology of Strain Hardening, the FCC Case". Progr. Mater. Sci., in press.
[3] J. M. Alberdi Garitaonaindía, Doctoral thesis, Faculty of Sciences, University of Navarra, San Sebastián, Spain, 1984.
[4] H. Mecking, B. Nicklas, N. Zarubova and U. F. Kocks, Acta mater., 34, 1986, 527–535.
[5] P. N. B. Anongba, J. Bonneville and J. L. Martin, Acta Metall. Mater., 41, 1993, 2897 and 2907.
[6] A. Mecif, B. Bacroix and P. Franciosi, Acta Mater., 45, 1997, 371.
[7] H. Mecking and A. Grimberg, Proc. ICSMA 5, P. Haasen, V. Gerold and G. Kostorz, eds., vol.1, p. 289, 1979.
[8] S. Kok, A. J. Beaudoin and D. A. Tortorelli, Acta Materialia, 50, 2002, 1653.
[9] J. Aldazábal and J. Gil Sevillano, Z. Metallkde., 93, 2002, 681.
[10] J. Aldazábal and J. Gil Sevillano, Mater. Sci. Eng., in press.

Deformation Substructure and Mechanical Properties of BCC-Polycrystals

S.A. Firstov
Francevych Institute for Problems of Materials Science, Kyiv, Ukraine

1 Introduction

BCC transition metals are very good objects to study regularities of strain hardening at high plastic deformations. The data on evolution of dislocation substructures enable an essential change of the strain hardening mechanism at high plastic deformations to be predicted as well as a non-monotonous influence of plastic deformation on temperature of brittle-to-ductile transition to be explained [1–3, etc.]. Recently, unquestionable evidences for deformation substructure effects on changes of fracture micro-mechanisms have been obtained [4–6] which also explain the non-monotonous dependence of fracture toughness on strain degree.

2 Dislocation Substructure and Strain Hardening

The most considerable differences of bcc metals from other metals are: i) an existence of abrupt temperature dependence of yield stress; ii) a high value of stacking fault energy. The last factor facilitates the cross-slip and climb processes promoting formation of cell structures. At the same time, such yield stress behavior complicates a dislocation rearrangement in energetically more favorable configurations such as cell structures or sub-grains as temperature decreases. This can

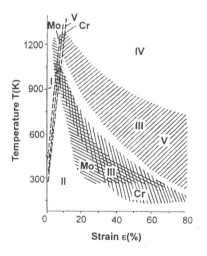

Figure 1: Structural stage diagram. I – tangles; II – homogeneous distribution of dislocations; III – intermediate structure; IV – disoriented cellular structure.

be seen on the diagram of structural states (Fig.1), comprising the data of electron microscopic investigations. It follows from the diagram that a decrease of temperature results in expanding the area of existence of relatively homogeneous dislocation distributions. One can see also that the area disappears as deformation temperature increases while cell structure appears practically at once, due to facilitation of both climb and cross-slip processes.

The term „cell-structure" often denotes distributions of dislocations with areas of a high dislocation density appearing in slip bands, surrounding areas which are almost free of dislocations. Boundaries of these cells usually have a considerable thickness. However, this structure is typical only of the first stages of deformation. As strain increases cells boundaries become much narrower and similar to grain boundaries; within them, dislocations can not be longer observed separately. A comparison with grain boundaries becomes even more pertinent since a continuous increase of neighbor cells disorientation is the important feature of cellular substructure appearing at high deformations. For this reason, an introduction of the term „disoriented (off-oriented) cell-structure" is required [1–3] to describe these structures.

Generalization of their own and literature data let authors [6] obtain the following dependence of average disorientation angle w on true strain :

$$\omega(e) = \alpha \cdot \varepsilon^{3/2}. \tag{1}$$

Thus, the important feature of structural changes at increase of the strain degree is a transition from structures which do not exhibit disorientation, to such which do. The transition means an essential change of the original strain hardening mechanism following the well-known dependence

$$\Delta\sigma(\varepsilon) = \alpha G b \sqrt{\rho(\varepsilon)} \tag{2}$$

In eq. (2), is the deformation, G is the shear modulus, b is the Burgers vector, ρ is the dislocation density. The new hardening is associated with formation of a disoriented cellular and almost superfine-grained structures and amounts to

$$\Delta\sigma(\varepsilon) = k \cdot d^{-m} \tag{3}$$

where $m = 0.5$ or 1, depending on deformation temperature, size of structural elements d, and disorientation.

Thus, the data of structural investigations obviously reveal a considerable change of the deformation mechanism and the hardening law at transition from low to high plastic deformations. However, experimental curves of strain hardening have usually not any features. In case of a homogeneous deformation the hardening is usually monotonous and consists of two parts, i.e. a parabolic and a linear one. The parabolic stage of hardening is described by one of the following two dependencies, i.e. either

$$\sigma = K_1 \varepsilon^n \tag{4}$$

or

$$\sigma = \sigma_s + K_2 \varepsilon^n, \qquad (5)$$

where σ_s is the yield stress, K_1, K_2 are the coefficients and n is the strain hardening exponent.

The value of n is in the range of 0.3–0.6. After the parabolic stage an extended linear hardening is usually observed, being

$$\sigma = \sigma_3 + \theta \varepsilon, \qquad (6)$$

where σ_3 is the hardening corresponding to the end of parabolic stage, θ is the modulus of plasticity.

Note that at least three stages of strain hardening have to be observed according to the structural state diagram. These were revealed by analysis of experimental strain hardening curves in coordinates of σ vs $\varepsilon^{0.5}$ [7]. This way, a „simple" parabolic curve disintegrates in three rectilinear parts displaying a stage character of structural changes in accordance with the structural state diagram. A comparison of dislocation substructures and the strain hardening parts have allowed the conclusion to be made that only the first stage of hardening can be described by dependence (2); when dealing with the following stages one have to consider a hardening contribution by the boundaries of cell structures.

The next extended stage of linear hardening is associated with continuous decrease of structural element size. It is shown in [3,8 etc.] that the dependence of sample cross-section area on deformation suggests a relation of cell size d as

$$d(\varepsilon) = d_0 \cdot \exp(-\varepsilon/2) \qquad (7)$$

while in fact the cell size diminishes with deformation more slowly, obeying at high deformations the following law

$$d(\varepsilon) = \frac{d_0}{1 + \alpha \varepsilon} \ . \qquad (8)$$

Schematically this distinction is illustrated in Fig. 2. It follows from this observation that as strain increases, not only a simple decrease of cell size occurs. A significant part of cell boundaries disappears, reducing the rate of the average cell size decrease. This can happen due to a change of the cells' disorientation up to zero either by cell slipping or by neighbor cell coalescence.

Authors [9,10] arrived at the conclusion on the significant role of so-called rotational modes at high plastic deformations, which can be described by the movement of specific defects like partial disclinations, by forming such boundaries and changing the disorientation of neighbor cells).

Assuming that even in ideal case when a hardening material does not fail up to reaching theoretical strength, failure strain can be evaluated from formula (6), using parameters derived from an experimental hardening curve. Ultimate size of structural elements can be obtained from empirical $d(\varepsilon)$ dependencies. Such an evaluation was carried out for molybdenum and iron [6]. Achievement of theoretical strength was found to require the realization of extremely high

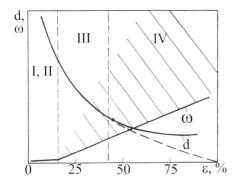

Figure 2: Schematic dependencies of d and ω vs. ε. As concerns d, the dashed line corresponds to eq. (7), the solid one represents eq. (8).

true strain being 40 for iron and 25 for molybdenum. The ultimate size of cells was shown to be 0.02 mm.

Of course, these estimations are correct under the condition that mechanisms of deformation and hardening remain unchanged. However, in spite of increasing number of structural investigations, the nature and operation of deformation mechanisms in these structures are still matters of dispute.

We have carried out an analysis of temperature dependence of yield stress and critical deformations corresponding to different stages of hardening and strengthening coefficients [11]. It was shown for polycrystalline molybdenum that at least before the transition to the final (linear) stage, the activation energy of the process-controlling temperature dependence of the above parameters remains unchanged.

Note that during the decrease of sizes of structural elements (SSE) we found some critical sizes at which a change of hardening mechanism occurs. Namely, if SSE becomes $\approx 1\mu m$ or slightly less, then the parameter m in expression (3) changes from $-1/2$ to -1 and size dependence of flow stress becomes much stronger. Further decrease of SSE up to nano-sizes brings about a situation at which transgranular slip can occur only at a very high stress level (up to a theoretical strength). Due to this reason another change of deformation and fracture mechanism occurs. The nature of these mechanisms is not clear yet. In one-component materials many investigators observed that strength (or hardness) decreases as SSE is reduced to the nano-scale level („negative Hall-Petch dependence phenomenon"). In case of two- or multi-component materials one can obtain a type of „healing" of SSE-boundaries due to segregation of „useful" doping agents and, hence, an increase of strength (further discussion see below).

3 Brittle-to-Ductile Transition and Fracture Toughness

It was shown [1–3] that the formation of a mainly homogeneous dislocation distribution leads to an increase of the temperature of brittle-to-ductile transition T_x, while a decrease of SSE at high plastic deformations decreases T_x. For deformed chromium and molybdenum, T_x can be lowered to temperatures below 200 K where at the same time significant hardening occurs.

It was also found that the dependence of fracture toughness on deformation is a non-monotonous one [4,5]. The typical dependencies of $K_{1C}(\varepsilon)$ are shown in Fig. 3.

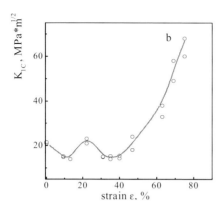

Figure 3: Influence of deformation power on K_{1c} for Mo [5]

Here, one can see that, after some decrease of fracture toughness in the early stages of deformation, a distinct increase of fracture toughness occurs for deformation powers of 20–30%. At these deformations a transition starts from dislocation distribution without disorientation (see stages 1 and 11, Figs. 1, 2) to disoriented structures. It was established by fractographical investigations that fracture mechanisms are of quasi-cleavage type. Data of $K_{1C}(\varepsilon)$ from Chromium for various test temperatures are presented in Fig.4. It is evident that the dependence of $K_{1C}(\varepsilon)$ becomes more weak with decreasing test temperature.

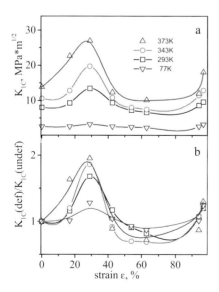

Figure 4: Influence of deformation power on K_{1c} for Cr tested at different temperatures [5]

At high values of plastic deformation, > 60 %, a superfine granular structure is formed, and the fracture toughness increases again. It is seen that the increase of fracture toughness in molybdenum essentially exceeds the increase of that in chromium. This effect, in our opinion, is associated with different tendencies of chromium and molybdenum to form intercrystalline (intercellular) fracture and is governed by the influence of interstitial impurities, i.e. oxygen, on processes of fracture along internal interfaces.

Some examples of different effect of oxygen on mechanical behaviour of Cr, Mo and W are discussed in [12]. If molybdenum and tungsten contain a small amount of oxygen, then they exhibit the intergranular fracture in the recrystallized condition and delaminate after a severe plastic deformation (SPD). On the contrary, Cr cannot be fractured via intergranular mechanism and is not susceptible to delamination after SPD. For deformed Mo-based alloys the delamination is usually observed on grain and cell boundaries. Fig. 5 shows the fractograms of chromium and molybdenum wires produced by drawing up to a strain of more than 98%. On the fracture surface of molybdenum, deep delaminating cracks have been observed, while chromium fails by cleavage and does not demonstrate any attributes of delamination. One of the reasons why delamination does not take place in Cr-based alloys is the presence of impurities (oxygen) in grain and cell boundaries [12]. The melting temperatures of some refractory metals and their stable oxides are presented in Table 1.

Table 1: Melting temperature of metals and their oxides

Element	Cr	Mo	W	Cr_2O_3	MoO_2	WO_2
Tmelt [K]	2048	2883	3653	2607	2200	1843

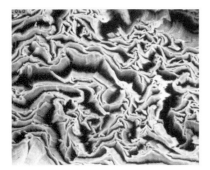 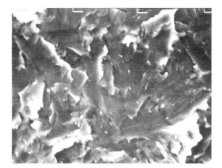

Figure 5: Typical fractograms of Mo (left) and Cr (right) wires. × 10 000

It is evident that in sequence the bonding energies of Cr, Mo, W increase. Considering their oxides, however, the bonding energy of them in the same sequence decreases because of the changed melting temperatures. Accordingly, the segregation of oxygen on internal boundaries of Mo and W brings about a decrease of bonding energy whereas its segregation on internal boundaries of Cr causes the bonding energy to increase. This fits to the fact that molybdenum oxide films can be easily separated from metal surface whereas chromium oxide films are strongly bonded to pure chromium.

Earlier it was noted by Glickman as well as in our researches [1,13] that the elements enriching interfaces can be subdivided in "harmful" and "useful" ones. Obviously, oxygen can be signified as an useful impurity in chromium but as a harmful one in molybdenum and tungsten.

In summarizing, it should be noted that a doping of grain boundaries with useful impurities is one of the ways to additionally increase strength of bcc nanocrystalline materials. Fig. 6 shows results of our cooperation with T. Rogul [15] on influence of grain size decrease on hardness and yield stress (calculated using Marsh formula) in films of Chromium. After annealing under temperatures up to 1873 K, the grain size increased to 450–500 mm and the microhardness decreased to 1750 MPa. It was established that $m = 1/2$ is valid for grain sizes $d > 1$ mm, corresponding to the Hall-Petch law. In a range of 0.1 mm $< d < 1$ mm, however, $m = 1$ is valid. Further decrease of grain size does not result in stabilisation of hardness (yield stress) or a „negative" Hall-Petch-ratio but in amore intense increase of hardness. It should be noted that high values of hardness (up to 26000 MPa) were observed in chromium coatings [14]. After recrystallization, a small amount of Cr_2O_3 (not more than 3 vol.%) was detected. The observed high hardness values can be explained neither by the presence of small oxides nor by mechanisms of dispersion hardening.

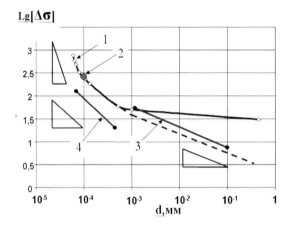

Figure 6: Influence of grain size on hardening of chromium. Curves 2, 3 and 4 correspond to Fe alloys according to [15]

Probably, the hardening is a result of oxygen atoms embedded in nanocrystal boundaries, i.e. „healing" of weak sites on grain boundaries.

4 References

[1] Trefilov, V.; Milman, Yu.; Firstov, S., Physical base of refractory metals strength, Naukova Dumka, Kyiv, 1975, 315
[2] Trefilov, V.; Firstov, S., Luft, A.; Schlaubitz, K.; in: Problems of Solid State Physics and Materials Science, Nauka, Moscow, 1976, pp.97–112
[3] Firstov, S.; Sarzhan, G.; Izv. vuz. Fiz. , 1991, 34, 23–34

[4] Danylenko, M. et al., Soviet Powder Metallurgy and Metal Ceramics, 1991, 30, 780–784
[5] Danylenko, M.; Podrezov, Yu.; Firstov, S.; Theoretical and applied fracture mechanics, 1999, 32, 9–14
[6] Graivoronsky, N.; Sarzhan, G.; Firstov, S.; Phys. Metals Mod. Techn. 1998, 17, 105–118
[7] Trefilov, V.; Moiseev V.; Pechkovsky E., Deformation hardening and fracture of polycrystalline metals, Naukova Dumka, Kiev, 1989, p.256
[8] Firstov, S., in Cooperative deformation processes and localization of deformation, Naukova Dumka, Kiev, 1989, pp.196–219
[9] Rybin, V.; Lichachev, V.; Vergazov, A., Phys. metall. metallov. 1974, 37 N3, 620–624
[10] Rybin,V., High plastic deform. & fracture of metals, 1980, Moscow, Metallurgy, p.224
[11] Firstov, S; Pechkovsky E., in Voprosy materialovedenija, 2002, 29, N1, 70–84
[12] Vasiliev.A.; Perepiolkin, A.; Firstov, S.; Ukrainian Phys. J. 1985, 30 N4, 603–606
[13] Glickman, E.; Bruver, R., Metallophysica, Naukova Dumka, Kiev, 1972, 43, 42–57
[14] Dudko, D.; Barg, A.; Milman, Yu., Poroshkovaja metallurgija, 1993, N8, pp.70–75
[15] Firstov S.A.,Rogul T.G., Marushko V.T., Sagaydak V.A., in Voprosy materialovedenija, 2003, v.33, N1 , pp.201–205

Micro- and Nanostructures of Large Strain-SPD deformed L1$_2$ Intermetallics

C. Rentenberger[1], H. P. Karnthaler[1] and R. Z. Valiev[2]
[1] Institute of Materials Physics, University of Vienna, Wien, Austria
[2] Institute of Physics of Advanced Materials, Ufa State Aviation Technical University, Ufa, Russia

1 Introduction

Intense plastic straining using various techniques (e. g. equal-channel angular pressing ECAP, high pressure torsion HPT) enable the formation of nanostructured materials. Even intermetallic alloys which are rather brittle can be deformed to a large strain and processed to grain-refined materials [1] showing unique properties. The severe plastic deformation of binary Ni3Al of the ordered L12 structure leads to the loss of order. It is the purpose of this paper to study the refinement of the structure and the transformation into the nanostructure of L12 ordered materials.

Therefore, transmission electron microscopy (TEM) methods are used to investigate the micro- and nanostructure of intense strained Ni$_3$Al and Cu$_3$Au samples at different strain levels (distance from the center of HPT samples).

2 Experimental Procedure

A Ni$_3$Al based alloy (Ni-18at.%Al-8at.%Cr-1at%Zr-0.15at%B) with an initial grain size of about 6 µm was severely deformed by the HPT technique at room temperature using a quasi hydrostatic pressure QHP of about 6 GPa. The HPT specimen of 12 mm in diameter was torsion strained (5 revolutions) to a shear strain of about 800 at a distance of 5 mm from the centre. A L12 ordered Cu$_3$Au sample (∅ ~ 8 mm, grain size of about 500 µm) was strained by HPT at RT using a QHP of about 4 GPa. The shear strain at a distance of 3 mm reached a value of about 80 after 2.5 turns.

For the microscopic investigations 2.3 mm discs were punched out mechanically or by spark erosion from different areas of the torsional plane of the HPT processed material (Fig. 1). Discs from the central area, from the near-central area and from the periphery of the HPT sample were designated C, NC and P, respectively. Some of the punched samples were marked along the radial direction of the HPT specimen that is perpendicular to the shear direction. The marks were eroded with an accuracy of about ±10°. Thin foils were prepared using the twin-jet polisher Tenupol 3. For the thinning process of the Ni$_3$Al(Cr,Zr)+B sample an electrolyte consisting of 10 % perchloric acid and 15 % acetic acid in methanol was used at 10 °C and 30 V. The Cu$_3$Au discs were dimpled mechanically with the Model2000 (E. A. Fischione Instruments) to a thickness of about 50 µm and then electropolished to thin foils using a solution of chromic acid and acetic acid at 20 °C and 25 V [2].

The TEM transmission electron microscopic studies were carried out using a Philips CM200 electron microscope operating at 200 kV and a Philips CM30 at 300 kV. Diffraction contrast under bright-field and dark-field conditions and high resolution transmission electron microscopy (HRTEM) were used to analyse the microstructure. For the selected area electron diffraction (SAED) patterns the illuminated area was about 1 µm^2. The study of the surface of the thinned discs were carried out using an optical microscope Zeiss Axioplan.

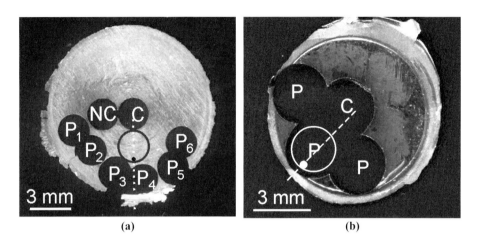

Figure 1: Ni$_3$Al(Cr,Zr)+B sample (a) and Cu$_3$Au sample (b) after deformation by high pressure torsion. From different areas of the HPT samples TEM discs marked along the radial direction were punched out (periphery P, near centre NC, centre C)

3   Experimental Results

3.1   HPT Deformation with a Shear Strain of about 80

Figure 2 shows the electropolished surface of TEM discs punched from different areas of the Ni$_3$Al(Cr,Zr)+B sample deformed by the HPT process. The disc belonging to the peripheral area (Fig. 2(a)) shows slightly curved striations on the surface. The superposition of images of both sides of a TEM disc shows that the striations on both surfaces are running parallel having the same curvature. The curvature of a circle with a diameter of 10 mm is indicated in Figure 2(a). The direction of the striations is perpendicular to the marked radial direction of the HPT sample (cf. Fig. 1). Figure 2(b) shows the surface of a TEM disc from the central area of the HPT sample. Two different structures are observed: Near the edge of the TEM disc striations with a radius of about 0.6 mm are observed. The striations are running round an area with a more isotropic surface structure located near the top of the indicated sector.

Figure 3 shows a bright field TEM image of the central area of the Ni$_3$Al(Cr,Zr)+B sample (TEM disc C of Fig. 1(a) and 2(b)). The observed area corresponds to a shear strain of about 80. Since two completely different structures are coexisting the observed characteristic structure is called duplex structure consisting of a nanocrystalline structure and a band-type structure. The transition between the two structures is rather sharp and is in this case nearly along the direction of shear (indicated in Fig.3). The two insets show the diffraction patterns corresponding to the

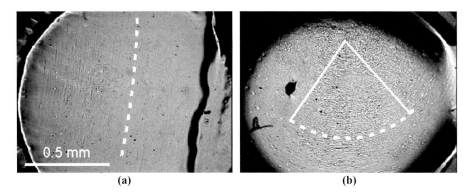

Figure 2: Optical micrographs of the surfaces of electropolished TEM discs from to the peripheral and central area of the $Ni_3Al(Cr,Zr)+B$ HPT sample are shown in (a) and (b). The curvature of striations visible on the surface increases with decreasing distance from the centre.

two structures. The illuminated area was in both cases about 1.2 µm in diameter. The diffraction pattern (top left) from the nanocrystalline structure shows continuous rings indicating small grains with many different orientations. Since the diffraction rings from the superlattice reflections are missing it is concluded that the $L1_2$ order is destroyed. The other diffraction pattern (top right) corresponds to the band-type structure and shows a spotpattern containing the superlattice reflections as in the undeformed sample. The spreading of the spots of about 10° indicates an orientation variation by small-angle subboundaries within the band-type structure.

Figure 4 shows a TEM bright field image of the band-type structure when two sets of {111} planes are end on. The traces of the two sets of {111} planes of the large grain are indicated as dashed lines. The extended boundaries parallel to these traces are sharp when the corresponding {111} planes are end on but diffuse when they are inclined. From these tilting experiments it is concluded that they are boundaries parallel to {111} planes (termed crystallographic boundaries as in [3]). The smallest observed distance between these boundaries was about 100 nm. The neighbouring grain on the left hand side (grain boundary along the dotted curve) shows under the present orientation condition only one set of crystallographic boundaries. The occurrence of superlattice reflections in the diffraction pattern indicates the presence of the $L1_2$ structure.

Figure 5 shows a TEM image of the peripheral area of a HPT processed Cu_3Au sample under dark-field condition. The observed area corresponds to a shear strain of 80. The submicrocrystalline structure is confined to a band of about 1 µm in width and the transition to the coarse grained structure is sharp. The diffraction pattern of the indicated area shows continuous rings within the band; rings corresponding to superlattice reflections are missing.

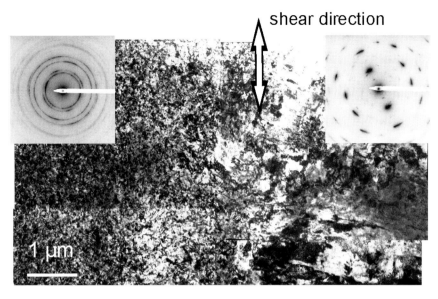

Figure 3: Ni$_3$Al(Cr,Zr)+B, shear strain ~ 80: TEM bright field image of a duplex structure consisting of a nanocrystalline structure (left) and a band-type structure (right). The transition between the two structures is parallel to the shear direction SD. The two insets show the corresponding diffraction patterns.

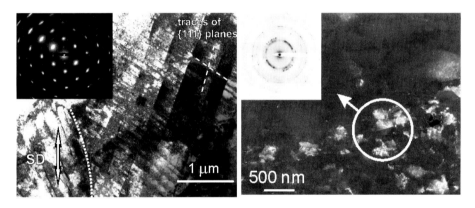

Figure 4: Ni$_3$Al(Cr,Zr)+B, shear strain ~ 80 (disc C), shear direction SD. Crystallographic boundaries in two neighbouring grains are imaged when {111} planes are end on (the SAED pattern of the larger grain is shown).

Figure 5: Cu$_3$Au, shear strain ~ 80 (disc P). The submicrocrystalline structure is confined to a band. The corresponding SAED pattern shows diffraction rings.

3.2 HPT Deformation with a Shear Strain of about 800

Figure 6 shows a TEM dark-field image of the nanocrystalline structure that is characteristic for the peripheral area of the Ni$_3$Al(Cr,Zr)+B sample (Pi in Fig. 1(a)). A broad distribution of the

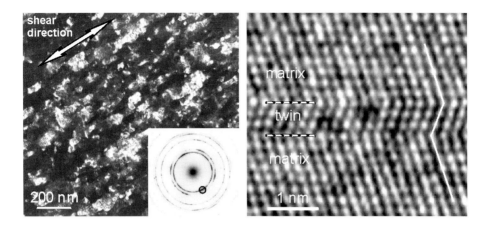

Figure 6: Ni$_3$Al(Cr,Zr)+B, shear strain ~ 800. A high volume fraction of nanograins whose {111} planes are nearly parallel to SD are elongated along SD.

Figure 7: Ni$_3$Al(Cr,Zr)+B, shear strain ~ 800. High resolution TEM image showing a deformation twin with atomic resolution

grain size (from several nm up to 200 nm) is observed. When the aperture is positioned at the highest intensity of the {111} ring (as indicated in the diffraction pattern) grains elongated along the shear direction SD are observed having a large volume fraction and the traces of {111} planes parallel to SD. Diffraction rings corresponding to superlattice reflections are missing in the diffraction pattern.

In Figure 7 a deformation twin of nanometer size is shown under high-resolution imaging conditions. These twins are mainly observed in grains elongated along SD. The twin being formed during the HPT process has a {111} twinning plane and a width of 3 atomic layers. The twinning planes of most of the twins lie nearly parallel to the shear direction.

4 Discussion

The electropolished surface of TEM samples punched out from different areas of a HPT deformed Ni$_3$Al(Cr,Zr)+B sample shows striations parallel to the direction of shear. Since the thickness of the discs is reduced considerably when TEM foils are prepared the striations correspond to structures in the middle of the HPT sample. The striations can be used to identify the direction of shear and to estimate the local shear strain from the curvature since the curvature decreases with increasing distance from the centre of the HPT sample.

At shear strains below 100 a duplex structure consisting of a band-type structure and a nanocrystalline structure is observed. In the coarse grained area the grains are fragmented by boundaries leading to the band-type structure being similar to that observed after rolling. The occurrence of superlattice reflections indicate that the ordered L1$_2$ structure is still present. Boundaries parallel to {111} planes that are characteristic at this strain level are also observed in cold-rolled Ni$_3$Al+B when the samples are rolled to a thickness reduction of about 40% [4].

They were designated as microbands and their width was similar to the distance between the crystallographic boundaries observed in this study (100 nm as seen in Fig. 4). It is interesting to note that in binary Ni3Al deformed by HPT the smallest boundary spacings are about 20 nm [5]. The fragmentation by crystallographic boundaries seems to occur homogeneously across the whole grain. According to Winther et al. [3] crystallographic boundaries are formed when two glide systems in the same slip plane account for a large fraction of the total slip. In addition to boundaries parallel {111} planes non-crystallographic boundaries aligned parallel to the shear direction are observed in some grains that are fragmented by crystallographic boundaries. The transition into the nanocrystalline structure seems to occur locally by a shear instability at intersecting boundaries leading to bands consisting of nanograins in the case of $Ni_3Al(Cr,Zr)+B$ and submicrograins in the case of Cu_3Au. These 'nanostructured' bands are comparable to brass-type shear bands observed in cold-rolled Ni_3Al+B [4] that carry large strains and consists of small equiaxed crystallites.

The volume fraction of the nanocrystalline structure increases with increasing strain at the expense of the band-type structure until the whole volume is transformed. Within the nanostructured volume the ordered structure is destroyed. The deformation and fragmentation of the grains by the motion of dislocations seems to be replaced by other deformation modes; the nanograins rotate at further shear and a further fragmentation of the nanograins is enabled by deformation twinning. Contrary to other studies of binary Ni_3Al [5] in the present case the grains are elongated along the direction of shear (Fig. 6) leading to a different texture.

5   Summary

- The surface of electropolished HPT deformed $Ni_3Al(Cr,Zr)+B$ samples shows striations parallel to the direction of the shear when the nanocrystalline structure is formed.
- At shear strains < 100 a duplex structure consisting of a band-type structure and a nanocrystalline structure was observed by TEM methods.
- The band-type structure shows bands that are comparable to cold-worked structures.
- With increasing strain the volume fraction consisting of nanocrystallites increases at the expense of the band-type structure.
- The precursor of the transformation into the nanocrystalline structure is a homogeneous fragmentation of grains by crystallographic boundaries parallel {111} planes.
- The transformation itself is heterogeneous and starts locally at boundaries forming nanocrystalline bands.
- The nanocrystalline structure shows a high fraction of elongated grains. The atomic structure of deformation twins of nanometer size was observed by HRTEM methods.

6 Acknowledgments

The authors thank Dipl. Ing. A. Vorhauer from the Erich-Schmid Institute of Materials Science (Leoben, Austria) for the HPT deformation of the Cu_3Au sample.

7 References

[1] Dimitrov, O., Korznikov, A. V., Korznikova, G. F., Tram G., J. Phys. IV, 2000, 10, 33
[2] Fisher, R. M, Marcinkowski, M. J., Phil. Mag. 1961, 6, 1385
[3] Winther, G., Jensen, D. J., Hansen, N., Acta mater. 1997, 12, 5059–5068
[4] Ball J., Gottstein, G., Intermetallics 1993, 1, 171–185
[5] Korznikov, A. V., Tram, G., Dimitrov, O., Korznikova, G. F., Idirisova, S. R., Pakiela, Z., Acta mater. 2001, 49, 663–671

The Nature of the Stress-Strain Relationship in Aluminum and Copper over a Wide Range of Strain

Nguyen Q. Chinh[1], Zenji Horita[2] and Terence G. Langdon[3]
[1] Eotvos University, Budapest, Hungary (e-mail: chinh@metal.elte.hu)
[2] Kyushu University, Fukuoka, Japan
[3] University of Southern California, Los Angeles, U.S.A.

1 Abstract

The relationships between true stress and true strain for pure Al and Cu were investigated both experimentally and theoretically. Mathematical analysis of the data obtained on the samples after annealing or ECAP showed that the relationship between true stress and true strain may be accurately described by an exponential-power law constitutive relationship over a wide range of strain. The significance of this new relationship has been examined theoretically using a physical model to develop equations which describe the evolution of both the mobile and the forest dislocation densities during plastic deformation. The results of numerical calculations show that, under certain conditions appropriate to the main micro-mechanisms of plastic deformation, the solution of these equations leads to an exponential-power law stress-strain relationship. The new constitutive equation has been considered also for interpretation of steady state creep for the situation where there is a strong recovery during plastic deformation.

2 Introduction

The work hardening of polycrystalline metals has been the subject of extensive investigations over a period of many years. However, the understanding of this behavior, particularly at large strains, is far from complete. It is well known that in the case of single crystals the work hardening behavior, as represented by the true stress-true strain (σ-ε) curves, can be divided into three stages, I, II, and III, in which the basic mechanisms of the deformation processes are reasonably well established [1–4]. In the case of polycrystals, the presence of grain boundaries complicates the deformation processes even in pure metals so that the work hardening characteristics of these materials are generally not well understood. This may be the reason that no physically-based functional, rather than empirical, description of the stress-strain curves of polycrystalline metals can be found in the litterature. The theoretical afforts are mainly restricted to the limiting cases of low strains at low temperarutes (below $T_m/2$) and of high strains at high temperatures (above $T_m/2$, where T_m is the absolute melting temperature).

In the treatment of work hardening behavior of metals the most commonly used empirical relationship, the Hollomon power-law type [5]

$$\sigma = \sigma_0 + K \cdot \varepsilon^m \tag{1}$$

can be regarded as the first useful equation in which the constants σ_0, K and m are mainly material and temperature dependent. It can be seen that power-law relationships of the form given in equation (1) can physically describe the stress-strain behavior only within a limited range because they imply there is no upper limit in stress at large strains so that, in principle at least, $\sigma \to \infty$ as $\varepsilon \to \infty$.

It appears the problem may be solved by using an essentially different relationship introduced by Voce [6]:

$$\frac{\sigma_m - \sigma}{\sigma_m - \sigma_0} = \exp\left(-\frac{\varepsilon}{\varepsilon_c}\right) \tag{2}$$

where σ_m, σ_0 and ε_c are also material and temperature dependent empirical parameters. This equation implies that the stress, σ, approaches exponentially the asymptotic value σ_m at high strains. In spite of this advantage, the Voce equation does not provide an accurate description at small strains.

In this paper, the relationships between true stress and true strain (σ-ε) for pure Al and Cu were investigated both experimentally and theoretically. A new constitutive equation is proposed for the description of σ-ε curves. This new formula has been considered also for interpretation of steady state creep for the situation where there is a strong recovery during plastic deformation.

3 Experimental Materials and Procedures

The experiments were conducted using high purity (4N) Al and Cu which were given an annealing treatment for 30 minutes at 400 °C and 600 °C, respectively. Samples were processed for different numbers of passes using processing route B_c in ECAP. Further details on ECAP processing are given elsewhere [7–9].

Samples were machined for compression tests in the form of cylinders having diameters and heights of 4 and 7 mm, respectively. The compression tests were conducted using an MTS testing machine operating at an initial strain rate of 10^{-3} s^{-1}. Earlier results obtained in tension [9] were also used in the present analysis, where the tensile tests were performed at room temperature using an initial strain rate of $3.3 \cdot 10^{-4}$ s^{-1}.

4 Experimental Results

4.1 Experimental Evidence on Plastic Deformation

4.1.1 Stress-Strain Curves

Figs. 1 and 2 show the stress-strain relationships of Al and Cu, respectively, obtained at room temperature (RT) in the region of small (left hand side) and high (right hand side) strains. Mathematical analysis shows that in both metals the rate of work hardening ($d\sigma/d\varepsilon$) decreases mo-

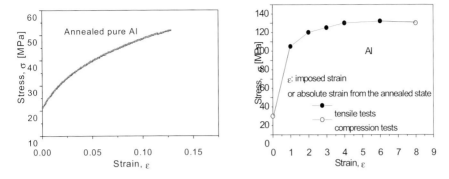

Figure 1: Stress-strain relationships of Al obtained at RT for small (left) and high (right) strains

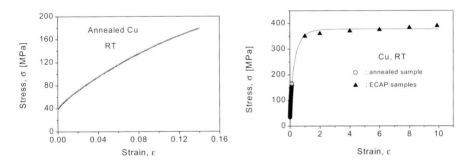

Figure 2: Stress-strain relationships of Cu obtained at RT for small (left) and high (right) strains

notonously practically from the beginning of plastic deformation, and it tends to 0 at high strains.

4.1.2 Evolution of Microstructures During Deformation

Fig. 3 shows the microstructures of the Al samples deformed to different amounts by compression and by ECAP at RT. It can be seen that a cellular structure is formed in the very early stages ($\varepsilon < 0.05$) of deformation and this leads to a stable microstructure with an average grain-size

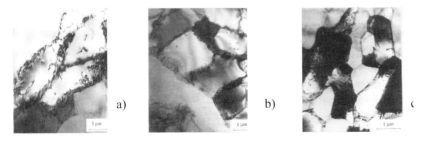

Figure 3: Microstructure of pure Al after deformation at RT by a) $\varepsilon = 0.04$, b) $\varepsilon = 0.2$ (by compression) and c) $\varepsilon = 8$ (by ECAP)

of ~1 μm during ECAP. Earlier results showed that in both metals a stable grain structure is established after ~4 passes when using route B_c in ECAP. More details of the microstructures achieved in ECAP are given in other reports [7–9].

4.2 A new Constitutive Description of the Stress-Strain Relationship

Mathematical analysis shows that an exponential-power law eqution as

$$\sigma = \sigma_0 + \sigma_1 \left[1 - \exp\left(-\frac{\varepsilon^n}{\varepsilon_c} \right) \right] \tag{3}$$

describes best the relationship between true stress and true strain over a wide range of strain. Here σ_0, σ_1, ε_c and the exponent n are constant fitting parameters. It can be seen that this constitutive relationship reflects the main features of the more conventional Hollomon-type power law and the Voce-type exponential equations in the regions of small and high strains, respectively.

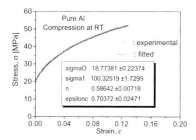
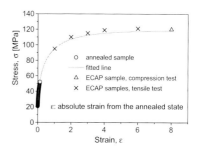

Figure 4: Experimental σ- ε data and a fitted line based on equation (3) for a) small and b) a wide range of strain in the case of Al

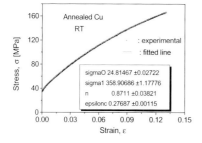
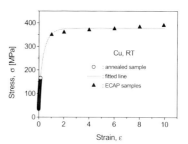

Figure 5: Experimental σ- ε data and a fitted line based on equation (3) for a) small and b) a wide range of strain

Figs. 4 and 5 show experimental σ-ε data obtained at room temperature and a fitted line based on equation (3) for the two situations of small strains and a wide range of strain, respectively. In both cases the fitted lines match very well to the experimental datum points, thereby confirming the validity of the new constitutive equation (3). More details on the analysis of the stress-strain curves can be seen in a previous paper [10].

5 Discussion

5.1 Theoretical Background of the new Equation

It is well known that according to the Taylor formula:

$$\sigma_p = \alpha \cdot G \cdot b \sqrt{\rho} \tag{4}$$

where σ_p is the part of flow stress occurring during plastic deformation ($\sigma = \sigma_0 + \sigma_p$) and ρ is the average dislocation density, the σ_p-ε can be obtained theoretically if we know the ρ-ε function during plastic deformation.

In order to predict the ρ-ε connection, some experimental evidence has to be taken into account as follows:

1. The rate of work hardening ($d\sigma/d\varepsilon$) decreases monotonously practically from the beginning of plastic deformation, and it tends to 0 at high strains (see Figs.1 and 2),
2. Together with the feature 1) the cell- and subgrain structures (see Fig.3) are formed very quickly from the early stage of strain.

These facts lead to the conclusion that even at RT the recovery takes place strongly from the early stage of strain (at least in the case of Al). Furthermore, because of the early formed cell-structure, both types of dislocation densities: ρ_m for mobile dislocations, and ρ_f for forest dislocations, have to be taken into account for describing the dislocation development during plastic deformation.

According to these conclusions the Kubin-Estrin model [11] is used to describe the development of dislocation densities during plastic deformation. The following equation system has to be solved numerically:

$$\frac{d\rho_m}{d\varepsilon} = C_1 - C_2 \cdot \rho_m - C_s \cdot \rho_f^{1/2}$$
$$\frac{d\rho_f}{d\varepsilon} = -C_2 \cdot \rho_m + C_s \cdot \rho_f^{1/2} - C_4 \cdot \rho_f \tag{5}$$

where the parameters C_i are related to the multiplication of mobile dislocations (C_1), their mutual annihilation and trapping (C_2), their immobilization through interaction with forest dislocation (C_3), and to dynamic recovery (C_4).

More details on these numerical calculation are given elsewhere [12]. Primary resshow that using the solution of eq. system (5), taking the average dislocation density as $\rho = \rho_p + \rho_f$, in the case of pure Al, for instance, according to the maximum average dislocation density, ρ measured by X-ray of about $1.2 \cdot 10^{14} \text{m}^{-2}$ (after 8 passes in route B_c), with the parameters:

$\rho_m(0) = 10^{10} \text{m}^{-2}$ and $\rho_m(0) = 10^{10} \text{m}^{-2}$, as well as

$C_1 = 2 \cdot 10^{14} \text{ m}^{-2}$, $C_2 = 2$, $C_3 = 3 \cdot 10^6 \text{ m}^{-1}$ and $C_4 = 2$, $\alpha = 0.38$ \hfill (6)

the theoretical $\sigma_p - \varepsilon$ approximately gives the experimental curve (see Figs. 6–8). Note that the experimental curve means the flow stress subtracting the σ_0. Furthermore, the theoretical $\sigma_p - \varepsilon$ relationship can be fitted very well – see Fig.9 – with the exponencial-power law function:

$$\sigma_p = \sigma_1 \left[1 - \exp\left(-\frac{\varepsilon^n}{\varepsilon_c} \right) \right] \qquad (7)$$

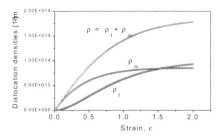

Figure 6: A solution of Eq. sys (5) with parameters given in (6)

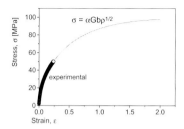

Figure 7: A theoretical $\sigma_p - \varepsilon$ curve and experimental data for Al with parameters given in (6)

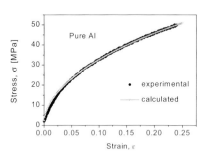

Figure 8: The curves shown in Fig.7 at smaller strains

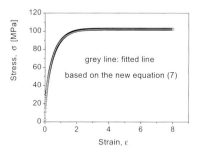

Figure 9: A theoretical $\sigma_p - \varepsilon$ connection and fitted line based on formula (7)

5.2 A further Application of the new Formula

During high temperature deformation due to strong recovery the saturated stress, $\sigma_{sat} = \sigma_0 + \sigma_1$ in formula (3), may be reached at relatively small strain, and for the subsequent deformation σ_{sat} will give the value of the steady state (stacioner) stress, σ_s.

Using the usual formulas for steady state flow:

$$\sigma_s = K \cdot \dot{\varepsilon}^m \quad \text{and} \quad \dot{\varepsilon}^m = B \cdot \exp(-Q/kT), \tag{7}$$

we get

$$\sigma_{sat} = K^* \cdot \exp(-mQ/kT) \tag{8}$$

where K, B, K^* are material-dependent constant, $\dot{\varepsilon}$ is strain rate, m is the value of strain rate sensitivity and Q is the activation energy of the deformation process. According to the eq. (7) from the $\ln(\sigma_{sat})$-$1/T$ relationship the value of Q can be determined. Experimental results show that on the basis of this relationship, from the point of view of the deformation processes, two (low and high) temperature regions can clearly be distinguished with the boundary at 200 °C, i.e. at 0.51 T_m (see Fig. 10). Calculations based on eqs. (7) and (8) lead to the values characterizing the deformation process of Al at high temperarutes (see Figs. 11 and 12). More details on the determination of these parameters will be given elsewhere [12].

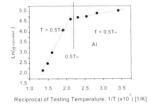
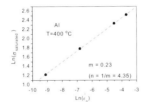
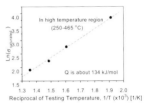

Figure 10: Low and high temperatureregions of plastic deformation

Figure 11: Determination of strain rate sensit. parameter (*m*)

Figure 12: Determination of the activation energy (*Q*)

5 References

[1] F. R. N. Nabarro, Z. S. Basinski and D.B. Holt: Adv. Phys., 1964, 13, 193–202
[2] R. Berner and H. Kronmüller: "Moderne Probleme der Metallphysik", Ed.: A Seeger, Springer, Berlin, 1965
[3] H. Mecking: Proc. Symp. on "Work Hardening in Tension and Fatigue", 1975, p.67–85
[4] I. Kovács: Def. Beh. Mat., 1977, 2, 211–219
[5] J. H. Hollomon: Trans. AIME, 1945, 162, 268–276
[6] E. Voce: J. Inst. Met., 1948, 74, 537–552
[7] M. Furukawa, Y. Iwahashi, Z. Horita, M. Nemoto and T.G. Langdon: Mater. Sci. Eng., 1998, A257, 328–332
[8] Y. Iwahashi, Z. Horita, M. Nemoto, T.G. Langdon: Acta Mater., 1998, 46, 3317–3331

[9] Y. Iwahashi, Z. Horita, M. Nemoto, T. G. Langdon: Metal. Mater. Trans., 1998, 29A, 2503–2510
[10] N. Q. Chinh, G. Vörös, Z. Horita, T. G. Langdon: Ultrafine Grained Materials II, The Minerals Metals and Materials Society, Warrendale, PA, 2002, p. 567–572
[11] L. P. Kubin and Y. Estrin, Acta Metall., 1990, 38, 697–709
[12] N. Q. Chinh, Z. Horita and T. G. Langdon: to be published

Strain Hardening by Formation of Nanoplatelets

K. Han, Y. Xin, and A. Ishmaku
National High Magnetic Field Laboratory, Florida State University, 1800 E Paul Dirac Dr., Tallahassee, Fl, USA

1 Abstract

The strengthening mechanisms of MP35N (35wt%Co-35wt%Ni-20wt%Cr-10wt%Mo) were studied in order to explore the maximum strength achievable in such alloys. Room temperature rolling introduces both high density of dislocations and nanoplatelets in the fcc matrix. The formation of the nanoplatelets is considered diffusionless. The main mechanism is the very high rate of strain hardening that develops dynamically from the high-dislocation- density substructure and high density of nanoplatelets. The deformation also introduces $\{110\}<1\bar{1}2>$ texture and the rotation of the annealing twin boundaries. The habit planes of the nanoplatelets are on $\{111\}$ and the thickness is about a few atomic layers. The formation of the platelets and dislocations strengths the fcc matrix significantly. After materials were deformed and aged, the MP35N can reach hardness of 5647±78 MPa and yield strength of 2125 MPa at room temperature. The further strengthening by aging can be related to the increase of the total length of the nanoplatelets without significantly increasing the thickness of the platelets. Deformation at 77 K increases the work hardening rate, indicating further formation of nanoplatelets and dislocations.

2 Introduction

MP35N (35wt%Co-35wt%Ni-20wt%Cr-10wt%Mo) is one of the high strength, high modulus and high corrosion resistance materials with various applications. In addition, MP35N is considered as one of the biomaterials. For instance, MP35N can be used as fasteners for repair of fractures of bones [1–3]. Because of its commercial values, researches have been undertaken to improve the properties and to understand the strengthening mechanisms and phase transformation in MP35N.

MP35N can be used in an annealed condition, which provides a solid solution with a face-centered-cubic (fcc) structure. MP35N is more usually used in as-rolled or as-drawn conditions. In such conditions, MP35N is described as a multiphase cobalt-nickel alloy because some cobalt-nickel alloys are work-hardened by formation of the stress-induced hexagonal-close-packed (hcp) platelets in fcc matrix [1–4]. However, the lack of evidence for hcp phase in cold deformed MP35N suggests that the strengthening mechanisms and phase transformation of MP35N alloys are different from the alloys with only nickel and cobalt [1]. Even for cobalt-nickel alloys, when the alloys with Co:Ni ratios (wt%) less than 45:25, it was not possible to detect the stress-induced hcp phase by the x-ray diffraction techniques[1, 6]. Instead, the materials were considered to be strengthened by the production of deformation twins in the fcc matrix [1, 2]. In MP159, some researchers reported the absence of hcp and found the dispersed γ' phase particles of several nanometers in diameter[1]. Recently, it has been recognized that deformation introduces mainly the planer defects or stacking faults in the fcc matrix of the MP35N [1, 2].

The cold deformed MP35N multiphase alloys can be further hardened by aging [1, 2]. After the materials reach the maximum strength at a defined temperature range, additional aging time contributed no further increase of the strength. Most applications use the materials aged at between 813 to 873 K for 4 hours [1]. However, it appears that it is necessary to age the deformed MP35N in order to get aging strengthening effects. Thus, in order to achieve higher strength in this material, it is necessary to understand the connection between the deformation, aging and phase transformation. It is also found that this material has good cryogenic properties although no detailed research was undertaken in this area. This paper will report some finding of the nanoplatelets formed in MP35N and to relate the strengthening to the nanoplatelets. In addition, the cryogenic properties of MP35N will be considered. All the tests of MP35N have been done at both room temperatures and at 77 K. In addition, the microstructure of the materials is examined and related to the properties of the materials.

3 Experimental Techniques

The alloys were fabricated by annealing, cold rolling and aging. The detailed fabrication procedures were reported before. The materials were received from H.C.Starck, USA in cold rolled condition. The typical chemical content is shown in Table 1. Further annealing and aging were performed in the NHMFL.

Table 1: Typical Chemical Composition (wt%) of MP35N

Elements	B	C	Co	Cr	Fe	Mn
Content	0.005	0.003	32.93	20.07	0.27	0.03
Elements	Mo	Ni	P	S	Si	Ti
Content	9.81	36.03	0.003	0.002	0.04	0.81

The MP35N samples for environmental scanning electron microscopy examinations (ESEM) were electro-polished in an electrolyte of 35vol% H_2SO_4 in methanol. The electro-polish was undertaken using a current of about 0.5A for 60 to 260 sec at room temperature. The samples were then cleaned with ethanol and etched by modified Aqua Regia (67 vol% HCl, 33 vol% HNO3) for 3–4.5 minutes at room temperature. The samples were examined in an Electroscan Model E-3 ESEM operated at 20 kV.

Discs of 3-mm in diameter were sectioned from the MP35N sheets for Transmission Electron Microscopy (TEM) examinations. The discs were polished using a Fischione Instruments Model 110 Twin-Jet Electropolisher. The electrolyte contained 19 vol% H2SO4, 76 vol% methanol, and 5 vol% H_3PO_4. The electropolisher was operated at a current of 28–32 mA and temperatures of 3 °C–5 °C. The foils were examined in a JEOL JEM-2010 Electron Microscope operated at 200V.

A Tukon 200 Micro-hardness tester was used to perform the hardness test. The hardness measurements were taken from surfaces polished for optical microscopy without being etched. The load was applied mainly in the transverse direction with respect to the rolling direction. Twenty indentations were obtained from each specimen using a load of 300 grams. The minimum impression spacing (center to edge of adjacent impression) was about 3 times of the diagonal of the impression and at least 0.02mm from the edge of the specimen.

Tensile tests were performed on the 100 kN MTS machine in displacement control at a rate of 0.5 mm/min. Both tensile strength (TS) and yield strength (YS) were recorded. The elastic modulus for sheet was determined by the stress-strain curves in an unload/reload cycle after the onset of plastic deformation (at 0.01 strain).

4   Results

Both X-Ray diffraction and transmission electron microscopy were used to study the crystallographic structures of the MP35N rolled to different reduction-in-areas. In the majority of the areas of the materials rolled to 65 %, the textures are close to $\{110\} <1\bar{1}2>$, where $\{110\}$ is parallel to the sheet surface whereas $<112>$ is approximately parallel to the RD, as shown in Figure 1. In addition, the $\{110\}<112>$ component can also be found.

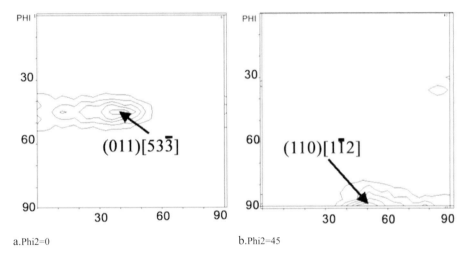

Figure 1: Orientation distribution function for MP35N cold-rolled to 65 % showing the texture of the materials close to $\{110\} <1\bar{1}2>$ in section of Phi2 = 0 and Phi2 = 45.

The microstructure of the annealed MP35N are composed of equiaxial grains and annealing twins. The cold rolling introduces rolling lines that are parallel to the rolling direction (RD). Some of those rolling lines were still visible after the samples were polished by electrolyte as shown in Figure 2a. Nevertheless, electro-polishing removed the majority of the rolling lines, and revealed continuous streaks, as shown in Figure 2a. The streaks intersect with the rolling lines. The majorities of the streaks represented the corrugation of traces of the $\{111\}_{fcc}$ planes in MP35N, as shown by the TEM results. The aging didn't change the microstructure significantly. The majority of the $\{111\}_{fcc}$ planes acted as the habit planes for the formation of the nanoplatelets induced by deformations, as demonstrated by TEM results which are discussed in the following text.

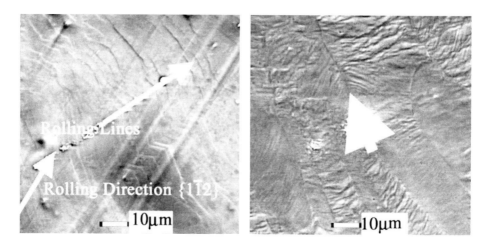

Figure 2: Environmental scanning electron microscopy images of MP35N, which was cold-rolled to 65 % showing the microstructure variations, revealed by electropolishing (a) and by etching (b). The electro-polish was undertaken with 35 vol% H_2SO_4 in methanol at 3–5 °C for 4 minutes. The etching was undertaken with modified Aqua Regia (67 vol% HCl, 33 vol% HNO_3) for 3–4.5 minutes at room temperature after polishing. The arrow in (b) indicates the rolling direction and a midrib.

Close examination of the SEM images of various magnifications showed detailed microstructure resulted from a combination of rolling and polishing. In lower magnifications, some streaks are bent slightly. However, high magnification images indicate that these bent streaks result from the overlapping of different steps, which are generally in {111} orientations. The intersection of the {111} steps and steps which are approximately parallel to the RD contributed to the streaks in various orientations in low magnifications. In some areas, the lengths of the streaks or steps are so small that they appear in irrational orientations.

Etching revealed different microstructure of the deformed MP35N compared to the electropolished samples, as shown in Figure 2b. Midribs can be seen in some areas. Those midribs are thought to be attribute to the deformed twin boundaries. In addition, one can also see other types of boundaries, which are not exactly parallel to RD.

Typical TEM image of MP35N rolled to 65 % revealed grain boundaries, high density of dislocations and parallel nanoplatelets. The majority of the images are imaged in the zone axis of <110>, which is parallel to the foil normal, as shown in the texture results. The grain boundaries are shown as the midribs in the SEM images (cf. figure 2b and figure 3a). Significant amount of such boundaries along the rolling direction are identified as incoherent twin boundaries. In addition, the twin planes appear to deviate from {111} due to the heavy deformation.

The nanoplatelets have the habit planes of {111}, as shown in Figure 3b. Occasionally, some of the nanoplatelets are intercepted. The distance between the parallel platelets is from a few nanometers to 50 nanometers. The thickness of the platelets is below 1 nanometer. Because two types of {111} planes within one grain are visible under the imaging condition, the platelets are formed at least on one of the habit planes. These two types of {111} habit planes are perpendicular to the rolling planes {110}. These fine platelets form after the deformation and strengthen the material by providing barriers for dislocation motion. As the deformation was undertaken at room temperatures, and the majority of the alloying elements diffuse by substitution, the formation of the platelets is considered as diffusionless. Therefore, the platelets should be the defects similar to stacking faults on {111}.

Both hardness and tensile tests show that the strength or hardness of the materials increase as deformation strain increase, primarily because of the formation of the nanoplatelets and dislocations. Aging further increases the strength of the materials. After the materials were deformed and aged at optimum conditions, the hardness and the yield strength of the materials reach 5647±78 MPa and yield strength of 2125 MPa, respectively. Tensile tests at 77 K increase the strain-hardening rate of the materials. Therefore, the material reaches even higher strength at 77 K. This is considered due to the further formation of the nanoplatelets at cryogenic temperatures. Currently, the maximum strength achievable at 77 K is about 2700 MPa. This is probably the limit of the strength of this material.

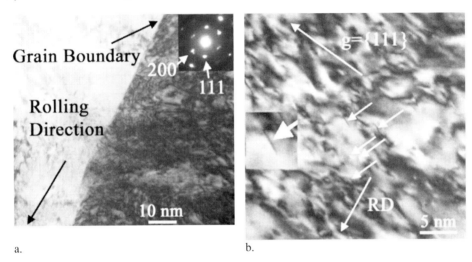

Figure 3: Transmission electron microscopy image showing the microstructure and nanoplatelets formed by deformation of the MP35N to 65 % of reduction-in-area. The longer arrows in both figures indicates the rolling direction (RD). A midrib observed in ESEM is shown to be a grain boundary along <112>. Nanoplatelets are indicated by small arrows (which are in {111}) in figure (b) nucleated on {111} habit planes in MP35N. An inset shows an enlarged nanoplatelet.

5 Conclusions

The strengthening of MP35N alloys is achieved by formation of the nanoplatelets induced by deformation. The nanoplatelets have the habit planes of {111}, which is perpendicular to the rolling plane {110}. The majorities of the habit planes of the nanoplatelets have an angle of 60-70 with respect to the rolling direction of $<1\bar{1}2>$. The primary factor to determine the formation of the nanoplatelets is the strain of deformation and the reaction is diffusionless.

6   Acknowledgments

We thank Mr. Goddard, Mr. Walsh and Mr. Toplosky for assistance in doing some experiments. This work was carried out at the NHMFL under cooperative agreement DMR-0084173. The Microscopy facilities are supported by NSF Grant No. DMR-9625692. NHMFL.

7   References

[1] J. L. Gilbert, C.A. Buckley, and E.P. Lautenschlager (Northwestern Univ.), Compatibility of Biomedical Implants , San Francisco, CA, (May 23-25, 1994), PV94-15, The Electrochemical Society, Inc., Pennington, NJ, (1994), 319–330
[2] M. Donachie, Advanced Materials & Processes , vol. 154, no. 1, pp. 63–65, July 1998
[3] S.F. Cogan, G.S. Jones, D.V. Hills, J.S. Walter, and L.W. Riedy, J. Biomed. Mater. Res., (February 1994), 28, (2), 232–240
[4] Multiphase MP35N alloy technical data, Latrobe Steel Company Subsidiary of the Timken Company Latrobe, Pennsylvania, (1994)
[5] R.P.Singh and R.D. Doherty, Metall. Trans., 23A: .321–334 (1992)
[6] A.H. Graham and J.L. Youngblood, Metall. Trans., 1: 423–430 (1970)
[7] J.M. Drapier, P. Viatour, D. Coutsouradis and L. Habraken, Cobalt, No.49:171–186, (1970)
[8] M. Raghavan, B. J. Berkovitz, and R. D. Kane, Metall. Trans., 11 A: 203–207 (1980)
[9] Lu S.Q., Huang B.Y., He Y.H., Tang J.C., Transactions of the Nonferrous Metals Society of China (China), 12 (2), 256–259 April 2002
[10] S. Asgari, E.El-Danaf, S.R. Kalidindi, and R.D. Doherty, Metall. and Mater. Trans., 28A:1781–1795 (1997)
[11] E. El-Danaf, S. R. Kalidini, and R. D. Doherty, Metall. and Mater. Trans., 30A:1 223–1233 (1999)
[12] Lu S.Q., Huang B.Y., He Y.H., Tang J.C., Transactions of the Nonferrous Metals Society of China (China), 12 (2), 256–259, April 2002
[13] A. Ishmaku and K. Han, Materials Characterization, 47, (2), 139–148, August (2001)
[14] K. Han, A. Ishmaku, Y. Xin, H. Garmestani, V. J. Toplosky, R. Walsh, C. Swenson, B. Lesch, H. Ledbetter, S. Kim, M. Hundley and J. R. Sims, Jr., IEEE Transactions of Applied Superconductivity, 12: (1) 1244–1250, MAR 2002
[15] R.P.Singh and R.D. Doherty, Metallurgical Transaction, 23A: 307–319 (1992)
[16] L.A.Pugliese and J.P.Stroup, Cobalt, 43:80–86 (1969)
[17] B. Fultz, A. DuBois, H.J.Kim, and J.W.Morris, Jr, Cryogenics, 24(12): 687–690 (1984)

Modelling the Draw Hardening of a Nanolamellar Composite: A Multiscale Transition Method

R. Krummeich-Brangier [1], H. Sabar [2], M. Berveiller [2]

[1] Groupe de Physique des Matériaux-UMR CNRS 6634-Université de Rouen-France
[2] Laboratoire de Physique et Mécanique des Matériaux-UMR CNRS 7554-Université de Metz-France

1 Introduction

Pearlitic steel cords may reach a tensile strength up to 4 GPa [1]. This property is related to the process of cold drawing which reduces the size of the lamellar substructure, in proportion to the wire diameter (Embury's similitude principle [2]) and induces the formation of nanocrystalline and morphological textures.

Using a multiscale transition method [3] based on a non pile-up micromechanical model for dislocations the aim of this work is to model and simulate the draw hardening of pearlitic lamellar nanocomposite.

The multi-scale transition method is described in the next paragraph and applied to wire drawing of a pearlitic industrial steel in a third paragraph.

2 Multiscale Transition Scheme

When looking at the microstructure of a pearlitic nano-lamellar material (see Figure 1), many scales are to be considered.

The finest scale is related to the elementary mechanism of dislocation storage at the interface between two adjacent lamellae. In the framework of micromechanical modelling, we may introduce the morphological anisotropy of single crystals with the critical resolved shear stress τ_g^* on a glide plane g, leading to plastic flow, as a function of the interlamellar spacing (i. s. l). The relationship established in [3] is:

$$\tau_g^* = \tau_o + \frac{k^g}{\lambda} \qquad (1)$$

where τ_0 is related to the lattice friction and k^g is related to the local plastic flow on a glide plane g in relation with the orientation of the interface between the two single crystals.

The second scale is relative to single crystal elasto-plasticity. The classical finite strain analysis (see for example [4–5]) is used here to take into account the spin of single crystals.

The meso-scale is attached to a set of alternate lamellae, defined as the representative elementary volume (REV) and named colony (see figure 1b). The colony is submitted to a mesoscopic velocity gradient G which verifies:

$$\mathbf{G} = <\mathbf{g}>_{REV} = f<\mathbf{g}>_{V^\beta} + (1-f)<\mathbf{g}>_{V^\alpha} \qquad (2)$$

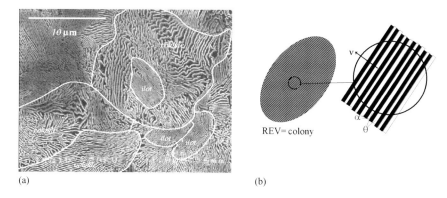

Figure 1: (a) Microstructure of a patented industrial pearlitic steel obtained with a scanning electron microscope (SEM). The dark phase is ferrite; the light grey is cementite. The morphology of grains takes the form of alternate parallel lamellae with a width of some tens nanometers. A set of parallel alternate lamellae is a colony. The typical value of the interlamellar spacing is around 100 nm. (b) Schematic representation of the representative elementary volume (REV). α respectively θ defines a ferrite respectively a cementite lamella. n is the unit normal at the interlamellar interface.

where $<\mathbf{x}>_V$ is the average value of the tensor $\mathbf{x}$ over the volume V and f is the volume fraction of cementite.

The homogenisation step uses interfacial operators [6] to relate stress and strain rate fields on each side of the material interface [7], knowing orientation relationships from experimental data of the literature (see for example [8]).

The basic assumptions at this scale are as follows:

- the velocity gradient $\mathbf{g}^\varphi$, $\varphi = \alpha, \theta$ is supposed to be homogeneous inside each single crystal (when compared to the velocity gradient jump at the interlamellar interface). Thus, equation (2) becomes $\mathbf{G} = f\mathbf{g}^\theta + (1-f)\mathbf{g}^\alpha$;
- the interface is supposed to be coherent and remains compatible;
- the interface verifies equilibrium conditions.

The following basic equations define the set to be solved:

$$\begin{cases} [\mathbf{g}] = \dot{\xi} \cdot \mathbf{v} \\ [\dot{\mathbf{n}}] \cdot \mathbf{v} = \mathbf{0} \\ \dot{\mathbf{n}}^\varphi = \mathbf{l}^\varphi : \mathbf{g}^\varphi , \varphi = \alpha, \theta \end{cases} \quad (3)$$

where $[\mathbf{x}]$ is the jump of the tensor $\mathbf{x}$ across the interface, v is the unit normal to the interface, $\mathbf{g}^\varphi$ and $\dot{\mathbf{n}}^\varphi$ respectively the local velocity gradient and the local nominal stress rate and $\dot{\xi}$ a vector characterising the amplitude of the velocity gradient jump at the interface. $\dot{\xi}$ has to be determined. The first equation in the set (3) is the compatibility of the interface, the second one is its equilibrium and the last equation is the tangent formulation of the elastoplastic behaviour of each single crystal.

Consider a reference isotropic elastic medium with the compliance tensor $\mathbf{C}^o$. Combining equations (3i) and (3ii) leads to:

$$[\mathbf{g}] = -\mathbf{v} \cdot \mathbf{C}^{o^{-1}} \cdot \mathbf{v} : [\delta \mathbf{l} : \mathbf{g}] \tag{4}$$

where $\delta \mathbf{l}^\varphi = \mathbf{l}^\varphi - \mathbf{C}^o$ and $\mathbf{C}^o = \mathbf{v} \cdot \mathbf{C}^o \cdot \mathbf{v}$ is the Christoffel matrix.

The interfacial operator $\mathbf{P}^o$ is defined in the following expression:

$$\mathbf{P}^o = \mathbf{v} \cdot \mathbf{C}^{o^{-1}} \cdot \mathbf{v} \tag{5}$$

Thus, rewriting equation (4) with $\mathbf{A}^\varphi = \mathbf{I} + \mathbf{P}^o : \delta \mathbf{I}^\varphi$ leads to:

$$\mathbf{A}^\alpha : \mathbf{g}^\alpha = \mathbf{A}^\theta : \mathbf{g}^\theta \tag{6}$$

Combining this last equation with the average assumption (2) on the velocity gradient leads to the following expression of the localisation tensor $\mathbf{a}^\varphi$, $\varphi = \alpha, \theta$:

$$\mathbf{g}^\varphi = \mathbf{a}^\varphi : \mathbf{G} \quad \text{with} \quad \mathbf{a}^\varphi = \mathbf{A}^{\varphi^{-1}} : \left\langle \mathbf{A}^{\varphi^{-1}} \right\rangle^{-1} \tag{7}$$

At this step, it is possible to relate local velocity gradient (in each constitutive phase of pearlite) to the mesoscopic (imposed) one. Assuming the average equation on stress rates - with $\dot{\mathbf{N}}$ the mesoscopic nominal stress rate of the REV- allows estimating the effective tangent modulus L of the colony with the Hashin-Strikmann lower bound LHS:

$$\dot{\mathbf{N}} = \mathbf{L} : \mathbf{G} \quad \text{with} \quad \mathbf{L} \approx \mathbf{L}^{HS} = \left\langle \mathbf{l}^\varphi : \mathbf{a}^\varphi \right\rangle \tag{8}$$

The material interlamellar interface is supposed to evolve in the homogenised state of the REV through its unit normal ν using the classical equation:

$$\dot{\nu} = -\mathbf{G} : \nu \tag{9}$$

where $\dot{\nu}$ is the increment of the unit normal n in the actual configuration of the solid.

In this context, a last average step is needed to validate the simulation with experimental data. It is possible to identify the local behaviour of cementite in the context of a nanolamellar microstructure (see figure 2 or [9]). An implicit assumption is the non-correlation of colonies, considering the pearlite as a poly-colonies composite.

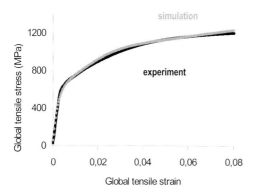

Figure 2: Experimental and simulated tensile curve of an industrial pearlite. The simulation is made for a set of a hundred colonies under Taylor assumption. By inverse method, fitting the experimental data leads to an overestimated value of the lattice friction shear stress τ_0 in the bulk of cementite of 1 GPa.

3 Draw Hardening of a Pearlitic Industrial Steel: Results and Discussion

Embury's principle of similitued states [2]:

$$\frac{\lambda}{\lambda_0} = \frac{r}{r_0} \qquad (10)$$

where λ_0 is the initial average i. s. of the patented steel wire, r_0 its initial radius. When drawing, the wire section reduces from $S_0 = \pi r_0^2$ to $S = \pi r^2$ with the drawing true strain $\varepsilon_t = \ln(S_0/S)$. Considering morphological texture as a determining microstructural evolution during the process of cold drawing, we may relate i.s. with ε_t and simulate a tensile test in order to obtain the draw hardening curve. Results are shown in figure 3. Comparison between simulation and experimental data allows concluding that a draw-hardening slope of ¼ does not mean a pile-up mechanism at the nanoscale: in this model, the elementary mechanism is related to the discrete behaviour of dislocations.

More generally, it is in the average process that one may accurately define determining mechanism or proper scale to describe the nanomaterial local behaviour. Moreover, in such a long process of modelling (at least three scales are under consideration) the whole local and global behaviour of the nanomaterial may be followed for any type of load. Such a method for modelling nanomaterials should be useful in assessing physical hypothesis on main flow mechanism at the finest scale.

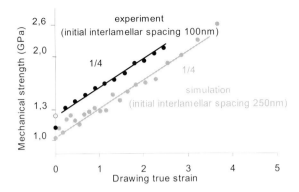

Figure 3: Comparison between simulation and experiment for the draw-hardening curve of a pearlitic wire. The drawing strain is related to the interlamellar spacing (i.s.) using Embury 's similitude principle and tensile test simulation are made for a set of a hundred colonies. The good agreement between experimental draw-hardening and simulated one tends to show that the main mechanism responsible for such a behaviour is related to the reduction of the dislocation mean free path (related to i.s.) for the dislocation behaviour.

4   Perspectives

The original assumption of the model is the ductility of the hard phase (cementite) leading to a good correlation between simulation and experiment. The morphological mesoscale texture has consequences on the second order stresses [7].

The estimation of internal stresses in the lamellar nanocomposite under shear and tensile test lead to some saturation of the internal stress of cementite after reaching the elastoplastic domain [9]. Moreover, internal stresses reached in local single crystals during the process of drawing might contribute for a large part to the driving forces leading to the process of cementite dissolution. Further development is needed to assess these points.

5   Acknowledgements

The author would like to thank IRSID/Usinor for financial support during the PhD thesis. Also, the author would like to thank the Groupe de Physique des Matériaux at Rouen University and especially P. Pareige for financing of the conference fees. The author gratefully acknowledges the scientific committee and Professor Yuri Estrin who supported successfully the author's participation at the conference.

6   References

[1] J. Gil Sevillano, J. Phys. III, 1991, 1, 967–988
[2] J. D. Embury, Scripta Metallurgica et Materialia, 27, 1992, 981–986
[3] R. Krummeich, H. Sabar, M. Berveiller, Journal of Eng. Mat. and Tech., Trans. ASME, 2001, 123, 216–220
[4] R. J. Asaro, Journal of Applied Mechanics, 1983, 50, 921
[5] R. Hill, J.R. Rice, Journal of Mechanics and Physics of Solids, 1972, 20, 401
[6] R. Hill, Journal of Mechanics and Physics of Solids, 1983, 31, 4, 347
[7] R. Krummeich-Brangier, H. Sabar, M. Berveiller, International Journal of Plasticity, to be submitted
[8] M. X. Zhang, P.M. Kelly, Scripta materialia, 1997, 37, 12, 2009
[9] R. Krummeich-Brangier, PhD thesis, Université de Metz, April 2001

III Unique Features of SPD - Microstructure and Properties

Paradoxes of Severe Plastic Deformation

Ruslan Z. Valiev
Institute of Physics of Advanced Materials_Ufa State Aviation Technical University, 450 000 Ufa, Russia

1 Abstract

Severe plastic deformation (SPD) can lead to emergence of microstructural features and properties in materials which are fundamentally different from the ones well known for conventional cold deformation. In particular, the instances of unusual phase transformations resulting in development of highly metastable states associated with formation of supersaturated solid solutions, disordering or amorphization and their further decomposition during heating, high thermal stability of the SPD-produced nanostructures, and the paradox of strength and ductility in some SPD-processed metals and alloys are discussed.

2 Introduction

In recent years SPD processing is getting to be established as the innovative technique for producing bulk nanostructured metals and alloys. [1–3] SPD materials are viewed as "advanced structural and functional materials of the next generation of metals and alloys. In this connection a deeper understanding of the physical nature of SPD and its differences from ordinary plastic straining is very topical. In certain cases these differences are not only essential, but also paradoxical. Three of such new paradoxical phenomena, which we found recently during our research, are considered in this paper. They are unusual phase transformations and formation of highly metastable states, enhanced thermostability of some SPD-produced nanostructures, and high strength and ductility revealed in several SPD-processed metals and alloys.

3 Unusual Phase Transformations and Metastable SPD Alloys

Although SPD by high pressure torsion (HPT) or equal-channel angular pressing (ECAP) is usually implemented at relatively low temperatures, as a rule at less than 0.3 T_m, [1] many of its characteristic features are typical of warm and even hot deformation. It can be seen, for instance, in development during SPD in alloys of phase transformations typically occurring at elevated temperatures. These transformations can lead to development of highly metastable states in alloys associated with formation of supersaturated solid solutions, disordering or amorphization. This issue is considered in more detail in the paper [4] of the present edition. Here we would like only to mention recent investigations of formation at SPD of supersaturated solid solutions in immiscible Al–Fe alloys, [5,6] complete dissolution of cementite during high pressure torsion in high-carbon steel, [7] disordering and even amorphization in SPD intermetallics. [8,9]

However, in order to attain the objective set for the present paper, it is especially important to address to nanocrystallization during severe straining in the amorphous alloy Ti50Ni25Cu25 taking place already at room temperature. [10] This shape-memory alloy can be produced in the amorphous state using melt spinning. [11] Following that, this amorphous alloy was subjected to HPT at RT under imposed pressure of 6 GPa.

X-ray investigations have shown that in both states – in the'initial state after melt-spinning and in the state after HPT–X-ray diffraction patterns look amorphous-like (Fig. 1). According to the data from DSC, crystallization of the initial amorphous alloy occurs at heating up to 450 °C

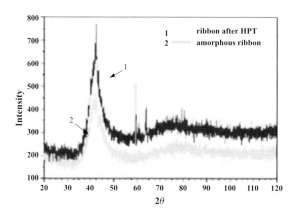

Figure 1: X-ray patterns from two states of the Ti50Ni25Cu25 alloy. 1: ribbon after HPT, 2: amorphous ribbon.

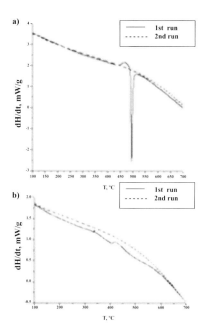

Figure 2: DSC curves for the two states of the Ti50Ni25Cu25 alloy: a) amorphous ribbon; b) the ribbon after HPT (heating rate 40° / min).

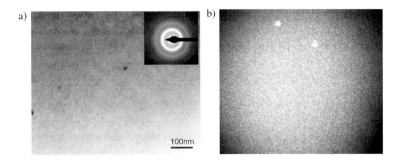

Figure 3: TEM micrograph of the amorphous Ti50Ni25Cu25 alloy after HPT: bright field image with SAE b) dark field image.

(Fig. 2a). However, there was observed no peak of crystallization after heating of the HPT-processed alloy (Fig. 2b), which suggests the possibility of its nanocrystallization during severe straining already at room temperature.

This assumption was evidenced by direct TEM/HREM "investigations. At TEM bright field observations a typical amorphous state was observed (Fig. 3a), but the dark field image revealed formation of very small nanocrystals of 2–3 nm in size (Fig. 3b). This was also verified by HREM analysis of the alloy (Fig. 4) which allowed to observe clearly the lattice fringes of small crystals.

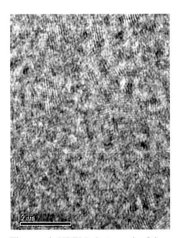

Figure 4: HREM photograph of the amorphous Ti–Ni–Cu alloy after HPT

Thus it has been demonstrated that nanocrystallization of this amorphous alloy during severe straining occurs already at room temperature which is more that 400 °C lower that the temperature of nanocrystallization during ordinary heating. Moreover, the nanocrystals created in the alloy are very small and are close in size to the nuclei of crystallization.

From the point of view of the crystallization theory, the revealed sharp decrease in the nanocrystallization temperature resulting from severe plastic deformation can be accounted for by introduction into the amorphous material of high "excess free energy [12] leading to excess free volume and enhanced dynamic activity of atoms which are typical mainly for elevated temperatures.

4   Enhanced Themostability of SPD-Processed Nanostructures

It is well established that during conventional cold deformation with an increase in the strain amount the temperature of recrystallization decreases. However, in contrast to cold deformation, during severe plastic deformation the nanostructures developing at late stages can be thermally more stable than the microstructures from earlier stages of SPD or cold straining.

This unusual phenomenon was revealed in the course of experiments we carried out using Armco iron [13,14] and commercially pure (CP) titanium [15] subjected to HPT under high pressure of 5–6 GPa with a various amount of revolutions ranging from 1 to 5.

In both materials we observed a strong microstructure "refinement, and after five turns formation of ultrafine grains with a mean size of about 150 nm was found. It was shown by HREM and orientation imaging microscopy that these "materials possess an UFG structure with predominantly high-angle grain boundaries (totalling more that 80 %) and these grain boundaries contain a high density of dislocations [14,15] indicating their non-equilibrium structure. During heating there is some grain coarsening starting both for Armco Fe and CP Ti after annealing at temperatures higher than 400 °C (Figs. 5, 6). However, even after annealing at 520 °C and 500 °C the grain size was uniform and still quite small with a mean size of 400 nm and 800 nm for Armco Fe and CP Ti, respectively.

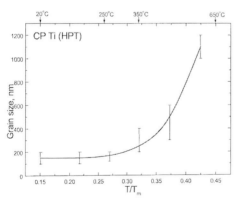

Figure 5: Variation of grain size with temperature for CP Ti after HPT (annealed for 5 h at certain temperature)

Such enhanced thermostability of UFG Armco Fe and CP Ti is very interesting, because even for low-carbon steel after cold rolling the recrystallization temperature is essentially lower and after annealing at 520 °C in the microstructure there are many large grains with a mean size of about 10 µm. [16] The same behavior is well known for CP Ti as well.

These paradoxical features of enhanced thermal stability of UFG Fe and Ti can be understood using the modeling of grain coarsening in SPD-produced ultrafine-grained metals. [17] In accordance with these modelling results, the kinetics of grain growth in UFG materials is close-

Figure 6: Typical TEM microstructure of Armco Fe after HPT (5 revolution) and annealing at 520 °C

ly connected with the non-equilibrium grain boundaries which predetermines both the driving force of grain growth and the mobility of grain boundaries, following the relation

$$d \sim M \cdot F^p, \quad p \geq 1 \tag{1}$$

where d is the grain size, $M \sim D_b/kT$ represents the mobility of atoms in the grain boundary with diffusion coefficient D_b at temperature T (k being Boltzmann's constant), $F \sim (\Omega/\delta)(2\Gamma/d)$ is the driving force with Ω as the atomic volume, δ as grain boundary thickness, and Γ as the energy of grain boundaries (interface tension).

In their turn, enhanced diffusivity and excess energy of non-equilibrium boundaries containing a high density of grain boundary defects provide for a high rate of grain growth. However, during heating the recovery of the non-equilibrium boundaries structure may proceed quite rapidly due to high grain boundary diffusion in UFG metals, [1] which leads to a drastic decrease in the driving force and especially in mobility of grain boundries and consequently results in an increase of the UFG structures thermostability.

5 Strength and Ductility in SPD-Produced Nanomaterials

It is well known that plastic deformation induced by conventional forming methods such as rolling, drawing or extrusion can significantly increase the strength of metals. However, this increase is usually accompanied by a loss of ductility. For example, Figure 7 shows that with increasing plastic deformation, the yield strength of Cu and Al monotonically increases while their elongation to failure (ductility) decreases. The same trend is also true for other metals and "alloys. Here, we report and discuss an extraordinary combination of high strength and high ductility produced in metals subject to severe plastic deformation. [18–20]

In this investigation, pure Cu (99.996 %) was processed using ECAP with 90° clockwise rotations along the billet axis between consecutive passes, while pure Ti (99.98 %) was processed using HPT. [18] All processes were performed at room temperature.

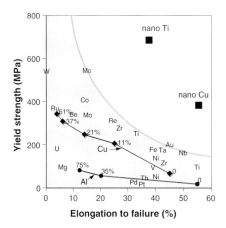

Figure 7: Cold rolling (the reduction in thickness is marked by each data point) of Cu and Al increases their yield strength but decreases their elongation to failure (ductility). The extraordinary combination of both high strength and high ductility in nanostructured Cu and Ti processed by SPD clearly sets them apart from coarse-grained metals.

Strength and ductility were measured by uniaxial tensile tests performed using samples with gauge dimensions of 5 × 2 × 1 mm. Resulting engineering stress-strain curves are shown in Figure 8. Results for Cu tested at room temperature in its initial and three processed states are shown in Fig"ure 8a. The initial coarse-grained Cu, with a grain size of about 30 µm, has a low yield stress but exhibits significant strain hardening and a large elongation to failure. This behavior is typical of coarse-grained metals. The elongation to failure is a quantitative measure of ductility, and is taken as the engineering strain at which the sample broke. Cold rolling of the copper to a thickness reduction of 60 % significantly "increased the strength (curve 2 in Fig. 8a) but dramatically decreased the elongation to failure. This is consistent with the classical mechanical behavior of metals that are deformed plastically. [21,22]

This tendency is also true for Cu subjected to two passes of ECA pressing (curve 3 in Fig. 8a). However, further deforming the Cu to 16 ECA passes simultaneously increased both the strength and ductility (curve 4 in Fig. 8a). Furthermore, the increase in ductility is much more significant than the "increase in strength. Such results have never been observed before and challenge our current understanding of mechanical properties of metals processed by plastic deformation.

Similar results were also observed in Ti samples subjected to HPT which were tested in tension at 250 ∘C. The coarse-grained Ti with a grain size of 20 µm exhibits a low strength and a large elongation to failure (curve 1 in Fig. 8b). After being processed by HPT for 1 revolution, the Ti material had a very high strength but significantly decreased ductility. Further HPT processing to 5 revolutions dramatically increased the ductility and slightly increased the strength (curve 3 in Fig. 8b).

Figure 8 shows that small SPD strains (2 ECA passes or 1'HPT revolution) significantly increase the strength at the expense of ductility, while very large SPD strains (16 ECA passes or 5 HPT revolutions) dramatically increase the ductility and at the same time further increase the strength. This is contrary to the classical mechanical behavior of metals that are deformed plas-

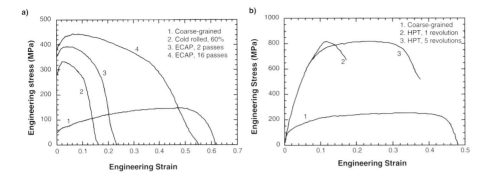

Figure 8: Tensile engineering stress-strain curves of a) Cu tested at 22 °C and b) Ti tested at 250 °C. Both were tested at a strain rate of 10⁻³ s⁻¹. The processing conditions for each curve are listed in the Figure.

tically. Greater plastic deformation by conventional techniques such as rolling, drawing or extrusion introduces greater strain hardening, which in turn increases the strength, but decreases the ductility of the metal (see Fig. 7).

The strain-rate sensitivity of stress, defined as

$$m = (\delta \ln \sigma / \delta \ln \dot{\varepsilon})_\varepsilon \qquad (2)$$

where ρ is flow rate, $\dot{\varepsilon}$ is strain rate, was measured using the standard jump-test method. [21] The samples with high ductility were found to have higher strain rate sensitivity. For "instance, the value m was equal to 0.14 for ECAP Cu (16 passes) in contrast to $m = 0.06$ for ECAP Cu (2 passes). Higher strain rate sensitivity renders the materials more resistant to necking. [23,24]

The extraordinary mechanical behavior in metals processed by SPD suggests a fundamental change in deformation mechanisms after the metals have been processed by SPD to very large strains.

For coarse-grained metals, dislocation movement and twinning are the primary deformation mechanisms. Ultrafine, equiaxed grains with high-angle grain boundaries impede the motion of dislocations and consequently enhance strength. At the same time, these grains may also facilitate other deformation mechanisms such as grain boundary sliding and enhanced grain rotation, [25] which improves ductility. We experimentally observed significant grain boundary sliding in ultrafine-grained copper deformed at room temperature. [25] The enhanced strain rate sensitivity observed in this work also indicates an active role of grain-boundary sliding. [26,27]

This work demonstrates the possibility of tailoring microstructures by SPD techniques to produce ultrafine nanostructured metals and alloys that have a combination of high strength and high ductility. Figure 7 shows that the trend of higher strength accompanied by lower ductility is not only followed individually by the work-hardened Cu and Al, but also followed collectively by 21 other coarse-grained metals. The nanostructured Cu and Ti are clearly separated from coarse-grained metals by their coexisting high strength and high ductility. Furthermore, the ECAP technique has the potential to provide nanostructured materials in sufficiently large product forms to enable their use in advanced structural and functional applications.

6 Summary

In this paper. we have demonstrated that severe plastic "deformation can lead to tailoring in metals and alloys of "microstructural features and properties which are fundamentally different from the ones well known for conventional cold deformation. Using several examples from our recent "investigations we have considered and discussed unusual phase transformations during SPD, in particular, nanocrystallization in the amorphous alloy which takes place already at room temperature, enhanced thermal stability of the SPD-produced UFG structures, and exceptionally high strength and ductility revealed in several SPD-processed metals. The "unusual behaviour of these materials is originated from the possibility to produce by SPD ultrafine nanostructured metals and alloys with extremely high density of high-angle grain boundaries, dislocations and obviously point defects.

7 Acknowledgement

The author is deeply grateful to the friends and colleagues who collaborated in the joint works presented in the references. This paper was prepared during the author\9s stay in Germany under the Alexander v. Humboldt Research Award and I thank H.'Gleiter, H.-E. Schäfer, H. Rössner, G. Wilde, and Yu. Ivanisenko for useful discussions and cooperation.

8 References

[1] R. Z. Valiev, R. K. Islamgaliev, I. V. Alexandrov, Prog. Mater. Sci. 2000, 45, 103
[2] R. Z. Valiev, in Ultrafine-Grained Materials II, Proc. of a Symposium Held during the 2002 TMS Annual Meeting (Eds: Y. T. Zhu, T. G. Langdon et al.), TMS Publ., Warrendale, PA 2002, pp. 313–322
[3] T. G. Langdon, M. Furukawa, M. Nemoto, Z. Horita, JOM 2000, 52(4), 30
[4] V. V. Stolyarov, R. Z. Valiev, this issue
[5] O. N. Senkov, F. H. Froes, V. V. Stolyarov, R. Z. Valiev, J. Liu, Scripta Mater. 1998, 38, 1511
[6] V. V. Stolyarov, I. G. Brodova, D. V. Gunderov, Phys. Met. Metall., in press
[7] Yu. V. Ivanisenko, W. Lojkowski, R. Z. Valiev, H. Fecht, Acta Mater., submitted
[8] A. V. Korznikov, G. Tram, O. Dimitrov, G. F. Korznikova, S. R. Idrisova, Z. Pakiela., Acta Mater. 2001, 49, 663
[9] A. V. Sergueeva, R. Z Valiev, A. K Mukherjee, Mater. Sci. Eng. A, in press
[10] R. Z. Valiev, A. P. Zhilyaev, H. Rösner, V. G. Pushin, to be published
[11] H. Rösner, A. V. Shelyakov, A. M. Glezer, P. Schloss"macher, Mater. Sci. Eng. A 2001, 307,188
[12] Physical Metallurgy (Ed: R. Cahn), Russian translation, Mir, Moscow, 1983, p. 330
[13] R. Z. Valiev, Yu. V. Ivanisenko, E. F. Rauch, B. Baudelet, Acta Mater. 1996, 44, 4705
[14] Yu. V. Ivanisenko, A. V. Sergueeva, R. Z. Valiev, H.'Fecht, Proc. of the NanoSPD$_2$ Conf., in press
[15] A. V. Sergueeva, R. Z. Valiev, A. K. Mukherjee, Acta Mater., to be published

[16] Steel. A Handbook for Materials Research and Engineering (Ed: Verein Deutscher Eisenhüttenleute), vol. 1, Springer-Verlag, Berlin, Germany 1992, p. 737
[17] J. Lian, R. Z. Valiev, B. Baudelet, Acta Metall. Mater. 1995, 43, 4165
[18] R. Z. Valiev, I. V. Alexandrov, Y. T. Zhu, T. C. Lowe, J.'Mater. Res. 2002, 17(1), 5
[19] R. Z. Valiev, Nature 2002, 419, 887
[20] Y. M. Wang, M. W. Chen, F. H. Zhou, E. Ma, Nature 2002, 419, 912
[21] E. A. Brandes, G. B. Brock, Smithells Metals Reference Book, 7th ed., Butterworth-Heinemann, Oxford 1992, Ch. 22
[22] E. R. Parker, Materials Data Book for Engineers and Scientists, McGraw-Hill, New York 1967
[23] D. Jia, Y. M. Wang, K. T. Ramesh, E. Ma, Y. T. Zhu, R. Z. Valiev, Appl. Phys. Lett. 2001, 79, 611
[24] E. W. Hart, Acta Metall. 1967, 15, 351
[25] R. Z. Valiev, E. V. Kozlov, Yu. F. Ivanov, J. Lian, A. A. Nazarov, B. Baudelet, Acta Metall. Mater. 1994, 42, 2467
[26] T. G. Nie, J. Wadsworth, O. D. Sherby, Superplasticity in Metals and Ceramics, Cambridge University Press, Cambridge, UK 1997
[27] O. A. Kaibyshev, Superplasticity in Commercial Alloys and Ceramics, Springer-Verlag, Berlin, 1993

Investigation of Phase Transformations in Nanostructured Materials Produced by Severe Plastic Deformation

X. Sauvage, A. Guillet and D. Blavette
Groupe de Physique des matériaux, UMR CNRS 6634, Université de Rouen, site du Madrillet, Avenue de l'Université, Saint Etienne du Rouvray – Cedex, France

1   Introduction

Severe plastic deformation is now a widely used technique to produce nanostructured metallic materials [1]. Many different techniques have been developed during the past two decades : Equal Channel Angular pressing (ECAP) [1,2], Severe plastic torsion straining (SPTS)[3,4], cold rolling [5] and drawing [6,7]. The resulting nanoscaled structures strongly depend on the technique used. However, the severe plastic deformation always push the material far away from its thermodymical equilibrium because a large amount of defects are created. The dislocation density strongly increases and has been estimated up to 10^{13} cm^{-2} [7]. The vacancy density may increase as well [8]. The level of internal stresses in the as–deformed material might be extremely high, up to a few GPa [9, 10]. Grains are elongated along the flow direction [5,6,7] and/ or high angle grain boundaries are created [1] which leads to a dramatic increase of the interface area. All these defects may promote phase transformations in nanostructured materials produced by severe plastic deformation. Many examples of such phenomenon have already been reported : solid state amorphization [11, 12], cluster dissolution [2,13, 14] or disordering [3].

Such phase transformations might have a significant influence on the properties of these nanostructured materials. By the way, severe plastic deformation might be an appropriate technique to prepare new materials containing metastable phases with unique properties.

The aim of this study is to get further understanding on the mechanisms of such phase transformations. Nanoscaled structures of materials prepared by drawing were investigated because they exhibit quite a simple grain morphology : most of interfaces are aligned along the wire axis [6]. The microstructures were investigated thanks to transmission electron microscopy (TEM), high resolution TEM (HRTEM), Field Ion Microscopy (FIM) and 3D Atom Probe [15].

2   Experimental Procedures

Two kinds of metal matrix composites (MMC) were investigated : cold drawn pearlitic steels provided by IRSID USINOR (ARCELOR) and Cu/Nb nanocomposite wires provided by the "Laboratoire des Champ magnetique pulsés", Toulouse (France).

The investigated steel wire was produced from a fully pearlitic steel, 3.69C-0.78Mn-0.52Si-0.28Cr-0.66Cu at%. This steel was hot rolled to 11 mm in diameter, austenitised and continuously cooled down at about 5 °C / s to get an interlamellar spacing of about 100 nm. Then, this wire was cold drawn to 3.3 mm in diameter. The true strain is therefore about $\varepsilon = 2.4$ ($\varepsilon = 2 \ln (d_i/d_f)$ where d_i and d_f are the initial and final diameter of the wire).

The Cu/Nb metal matrix composite wire was produced by drawing a cast ingot of Cu-Nb$_{18}$-Hf$_{0.2}$ (%$_{vol.}$) up to a true strain of 11. More details about the wire production process are given elsewhere [16]. The average width of the niobium fibers in the drawn wire was in a range of 10 to 60 nm.

3D-AP analyses were carried out on the new generation of optical tomographic atom-probe (OTAP) developed in the GPM (University of Rouen). TEM observation were performed with a Philips CM200 microscope operating at 200 kV and with a JEOL JEM-4000EX microscope operating at 400 kV for high resolution observations. More details, especially on specimen preparation, are given in previously published papers [17, 18].

3 Results

3.1 Cold Drawn Pearlitic Steel [17]

The FIM micrograph of the cold drawn pearlitic steel (figure 1 (a)) shows two cementite lamellae. The interlamellar spacing is about 20 nm. As expected, the ferrite exhibits a strong <110> texture parallel to the wire axis. The lamella located in the center of the image was analyzed with the 3D-AP along the <110>$_{\alpha\text{-Fe}}$ direction (white square, figure 1 (b)). The tracks of (110)$_{\alpha\text{-Fe}}$ planes appear slightly curved because of a disorientation of about 9° between the ferrite grains located on each side of the lamella. Almost the same angle was measured between the two (021) poles on the FIM micrograph (figure 1(a)). The composition profile computed across the cementite lamella shows that the maximum amount of carbon is about 15 at% (figure 1 (c)). Diffuse zones that appear in the ferrite close to interfaces indicate that cementite has partly dissolved during drawing.

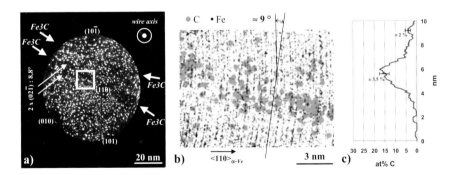

Figure 1: a) Neon field ion micrograph of the cold drawn pearlitic steel. Two cementite lamella are arrowed. b) Carbon and iron map of the volume analysed in the area indicated in the FIM image. The tracks of (110)$_{\alpha\text{-Fe}}$ planes and a cementite lamella are exhibited. c) Composition profile computed through the analysed lamella showing the dissolution of cementite [17].

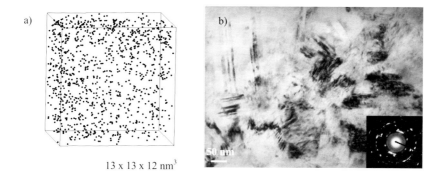

Figure 2: (a) 3D-AP carbon map showing the presence of a homogenous high-carbon content region. (b) Cross-sectional TEM bright field image of a drawn wire exhibiting a martensite-like microstructure [17].

Surprisingly, homogeneous regions were also found with a carbon level close to the nominal composition (figure 2(a), carbon concentration 3.99 ± 0.14 at.%). This indicates that cementite may completely dissolve during drawing. TEM observations were performed to determine the crystallographic structure of such zones. TEM specimens exhibited frequently the typical nanostructure of cold drawn pearlitic steels with cementite lamellae aligned along the wire axis. However, some regions with a martensitic-like microstructure were as well exhibited (figure 2(b)). This phase may result from the complete dissolution of cementite such as illustrated in figure 2(a).

3.2 Cu-Nb Nanocomposite Wire [18]

The HRTEM micrograph of figure 3 exhibits the cross-sectional nanostructure of the Cu/Nb MMC. This pictures shows two very thin Nb filaments and Cu/Nb interfaces layered by a 2 nm thick amorphous layer. This solid state amorphization occurs only in the region where the microstructure is the thinnest and might be the result of the severe plastic deformation.

In order to get more information about this new phase, 3D AP experiments were performed. Figure 4(a) shows the 3D reconstruction of an analysed volume. In this picture, only copper atoms are plotted. Between two niobium rich zones, a thin copper lamella is aligned along the wire axis. The concentration depth profile (figure 4(b)) was obtained in the perpendicular direction to this lamella. The copper concentration is plotted as a function of the sampling box position used for the computation. The concentration of copper detected in the niobium rich zones is about 12 ± 4 at.% and the 3D-AP data indicates also that the copper lath contains only 30 ± 2.5 at.% copper. So, this analysis reveals that in the regions where the microstructure is the finest, the interdiffusion of Cu and Nb has occurred. This mechanical alloying might be related to the solid state amorphization pointed out by HRTEM observations (figure 3).

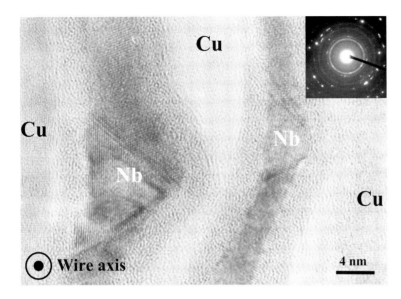

Figure 3: HRTEM image of a cross-sectional specimen showing an amorphous phase located at Cu/Nb interfaces [18]

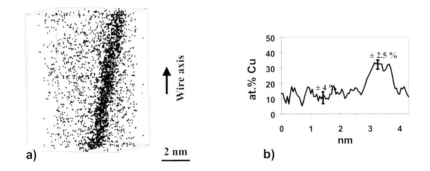

Figure 4: a) 3D reconstruction of a small analysed volume ($7.2 \times 7.2 \times 10$ nm^3). Only copper atoms are plotted. b) Composition profile across the copper rich channel [18].

4   Discussion

Our experimental investigations clearly show that because of severe plastic deformation during the drawing process of the pearlitic steel wire and of the Cu-Nb nanocomposite wire, phase transformations occur. Indeed, the cementite dissolution was clearly pointed out in the first case while a solid state amorphization along with the interdiffusion of Cu and Nb was reaveled in the Cu-Nb MMC. Several interpretations already exist for this kind of phenomena, especially for the cementite dissolution in steel cords [19,20,21]. Dislocations play an important role and the great proportion of interface area as well. In this paper, we will focus only on the effect on internal stresses. Such stresses result from the plastic deformation and are more pronounced in multiphase materials [22]. In cold drawn pearlitic steels they have been measured up to 2 GPa in the cementite phase [10], while in the Nb fibers of Cu-Nb MMC they can reach 5 GPa [9].

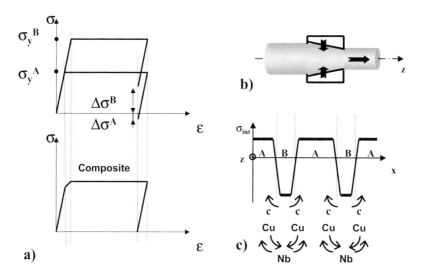

Figure 5: a) Origins of internal stresses in a multiphase material submitted to tensile plastic deformation. Material with a perfect elasto-plastic behavior and rule of mixture, from [22]. b) Stress configuration during drawing. c) Internal stress gradients along a direction perpendicular to the interfaces in MMC : A = α-Fe and B = Fe$_3$C (pearlitic steel) or A = Cu and B = Nb (Cu-Nb MMC). The stress field promote the dissolution of Fe$_3$C, and the interdiffusion of Cu and Nb.

Figure 5(a) shows the schematic representation of the tensile curves of a composite material containing two phases with different yield stresses. Because of the higher yield stress of the B phase, this phase remains in tension along the tensile axis after plastic deformation while the A phase remains in compression. During drawing, grains are elongated along the wire axis, so that the stress field during the process is very close to a compression perpendicular to these interfaces (figure 5(b)). The internal stress field after drawing might be therefore very similar to the sketch given figure 5(c). Along a direction perpendicular to the wire axis, Fe$_3$C or Nb phases are in compression, while α-Fe and Cu are in tension. Since carbon atoms diffuse in interstitial sites in the ferrite, this stress field will enhance both the diffusion and the solubility and may promote the destabilization of cementite. For the Cu-Nb case, one should note that Nb atoms are much bigger than Cu atoms. So, big atoms are in compression and small atoms in tension. That is why the solid state amorphization along with the interdiffusion of Cu and Nb is favorable: some internal stresses can be relaxed.

5 Conclusions

Thanks to 3D atom probe, TEM and HRTEM, phase transformations in two kinds of nanocomposite wires were investigated. Severe plastic deformations induce first the dissolution of cementite in pearlitic steels and as well the solid state amorphization of Cu-Nb nanocomposite wires. Internal stresses seem to play an important role in these transformations : the elastic energy promote the destabilization of the original phases while the atomic mobility is significantly increased along stress fields.

6 Acknowledgements

The authors would like to gratefully acknowledge F. Lecouturier and L. Thilly from the LNCMP and N. Guelton from IRSID for providing materials and fruitful discussions.

7 References

[1] Valiev, R.Z.; Islamgaliev, R.K.; Alexandrov, I.V., Progress in Material Science, 45 (2000), 103–189
[2] Morris, D.G.; Munoz-Morris, M.A., Acta Materialia, 50 (2002), 4047–4060
[3] Korznikov, A.V.; Dimitrov, O.; Korznikova, G.F.; Dallas, J.P.; Quivy, A.; Valiev, R.Z.; Mukherjee, A., Nanostructured Materials, vol. 11, No. 1, (1999), 17–23
[4] Korznikov A.V., Ivanisenko Y.V., Laptionok D.V., Safarov I.M., Pilyugin V.P. and Valiev R.Z., Nanostruct. Mater., 4 (2), (1994), 159–167
[5] Gholinia, A.; Humphreys, F.J.; Prangnell P.B., Acta Materialia, 50, (2002), 4461–4476
[6] Russel, A.M.; Chumbley, L.S.; Tian, Y., Adv. Eng. Mat., 2, (2000), 11–22
[7] Bevk, J., Ann. Rev. Mater. Sci., 13, (1983), 319–338
[8] Ruoff A.L. and Balluffi R.W., Jour. Appl. Phys., 34 (1963), 2862
[9] Hong S.I., Hill M.A., Sakai Y., Wood J.T. and Embury J.D., Acta Metall., 43, (1995), 3313–3323
[10] Van Acker K., Root J., Van Houtte P. and Aernoudt E., Acta Mater. 44, (1996), 4039–4044.
[11] Atzmon M., Unruh K.M. and Johnson W.L., J. Appl. Phys., 58 (1985), 3865
[12] Koike J. and Parkin D.M., J. Mater. Res., 5 (1990), 1414
[13] Sagaradze V.V. and Shabashov V.A., Nanostruc. Mater., 9, (1997), 681
[14] Murayama M., Hono K. and Horita Z., Mater. Trans. – JIM., 40, (1999), 938
[15] Blavette, D.; Bostel, A.; Sarrau, J.M.; Deconihout, B.; Menand, A., Nature, 363, (1993), 432
[16] Snoeck, E., Lecouturier, F., Thilly, L., Casanove, M.J., Rakoto, H., Coffe, G., Askénazy, S., Peyrade, J.P., Roucau, C., Pantsyrny, V., Shikov, A. and Nikulin, A., Scripta Met., 1998, 38, 1643
[17] Sauvage,X.; Guelton,N.; Blavette, D., Scripta mater., 46, (2002), 459–464
[18] Sauvage,X., Ping, D.H.; Blavette, D.; Hono, K., Acta. mater., 49, (2001), 389–394

[19] Gridnev, V.N. ; Gavrilyuk, V.G., Phys. Metals, 4 (1982), 531
[20] Languillaume, J. ; Kapelski, G. ; Baudelet, B., Acta Mater., 45 (1997), 1201
[21] Sauvage, X.; Copreaux, J.; Danoix, F.; Blavette, D., Phil. Mag. A, 80 (2000), 781
[22] Biselli, C. ; Morris, D.G., Acta Metall., 44 (1996), 493–504

Structure-Phase Transformations and Properties in Nanostructured Metastable Alloys Processed by Severe Plastic Deformation

V. V. Stolyarov, R. Z. Valiev
Institute of Physics of Advanced Materials, Ufa State Aviation Technical University, Ufa, Russia

1 Introduction

During last years materials researchers exhibit a great interest to ultrafine-grained (UFG) metals and alloys processed by severe plastic deformation (SPD). These materials are viewed as advanced materials with unique properties that can be applied in various areas [1,2]. Unusual and often extraordinary properties of SPD materials are originated with high microstructure refinement and high density of various crystal lattice defects, elastic long-distance stresses and other factors under investigations. However, as compared to pure metals or solid solutions based on them, multiphase alloys represent more complex objects, because during SPD processing different structure-phase transformations, resulting in new phases formations, creation of supersaturated solid solutions, disordering, amorphization etc., can occur in them. As it was shown in [3, 4] due to this such unique properties as superplasticity, enhanced fatigue strength, record coercivity can be revealed, in particular, in such metastable nanostructured alloys, processed by SPD methods under extreme conditions. At the same time the SPD influence on structure-phase state and properties of such alloys is associated with accumulated strain and depends on the chosen processing methods and regimes [5].

In the present paper we consider using our recent experimental results the potentials of SPD methods for achieving metastable states as well as the comparison of some structure feature and mechanical properties of several Ti and Al based alloys, subjected to two SPD methods which are presently the most widely used, namely, high pressure torsion (HPT) and equal channel angular pressing (ECAP).

2 Materials and Procedure

The immiscible Al-5 wt.%Fe alloy and such titanium materials as CP Ti, Ti-6Al-4V and Ti-50.5 at.%Ni were used as materials to be investigated. All alloys were supplied by the Russian manufacturers. At that the aluminum rod $\varnothing 30$ mm was produced as chill mold with rapid cooling rate (10^2 K/s), and titanium alloys were subjected to additional hot thermo-mechanical treatment (hot rolling or rotary forging) in order to obtain the final rod size $\varnothing 25$–30 mm.

SPD was performed by HPT and ECAP methods. Their principle scheme and strain regimes for the materials under investigation are presented in Fig.1 and Table 1, as well as described in more detail in [6, 7, 8]. The calculation of accumulated (true) strain was made in accordance with the ratios given in our previous publications.

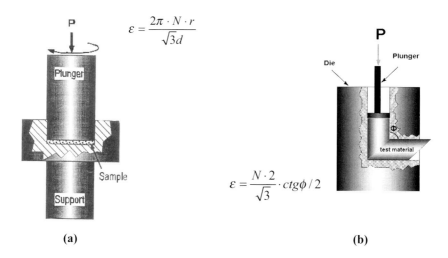

Figure 1: Principles of HPT (a) and ECAP (b) methods. N- number of passes, r- sample radius, d- sample thickness, ϕ -crossing channel angle.

Table 1: Processing parameters for HPT and ECAP methods, used for the investigated alloys.

Method	Alloy	Forward (back) pressure, GPa	Strain rate, mm/s	T_{spd}, °C (T_{spd}/T_{melt}^{**})	Sample sizes, mm	Turns (passes) N	Equivalent strain, e
HPT	CP Ti, Ti-64, TiNi, Al-Fe	P = 6–7 GPa	0.26 0.26 0.26 0.26	20 (0.15) 20 (0.16) 20 (0.17) 20 (0.30)	⌀10 × 0.2 ⌀10 × 0.2 ⌀10 × 0.2 ⌀10 × 0.2	5 5 5 5	225 225 225 225
ECAP Route Bc	CP Ti Ti-64 TiNi Al-Fe	P_f = 1–2 P_f = 1–2 P_f = 1–2 P_f = 0.5; P_b = 0.25	6 6 6 2	450 (0.37) 700 (0.5)* 400 (0.39) 20 (0.30)	⌀20 × 100 ⌀20 × 100 ⌀20 × 100 20 × 0 × 80	8 13 8 16	9.2 (ϕ = 90°) 10 (ϕ = 120°) 6.4 (ϕ = 110°) 18.4 (ϕ = 90°)

Note: * - nonisothermal conditions; ** - $T_{melt} \equiv$ melting temp. in [K]

The simple shear straining realized at relatively low temperatures under high imposed pressure is common for these two SPD methods of microstructure refinement. However, a number of used processing parameters differs considerably and has an influence on UFG structure formation and properties. To such parameters we refer: ratio T_{spd}/T_{melt}; (non) isothermal conditions; strain speed; forward (P_f) and backward pressure (P_b); accumulated equivalent strain e; strain localization. The type of forming crystallographic texture also differs from these methods. As it can be seen in Table 1, judging by the parameters e and T_{spd}/T_{melt} for HPT, the used strain conditions are more severe than for ECAP.

3 Experimental Results

3.1 Microstructures

TEM images of nanostructures of investigated alloys obtained after HPT and ECAP:

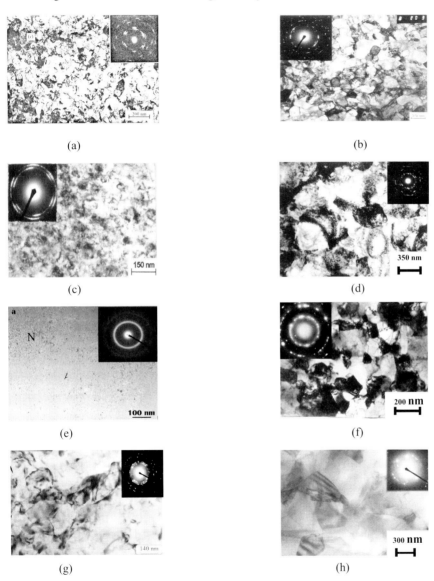

Figure 2: TEM microstructures of CPTi (a, b), Ti-64 (c, d), TiNi (e, f) and Al-5%Fe (g, h) as processed by HPT (a, c, e, g) and ECAP (b, d, f, h)

The very high refinement of initial microstructures is common for them. However, the mean size of the formed grains differs considerably for HPT and ECAP samples (Table 2). The processed UFG structures are characterized by high level of internal stresses and elastic distortions of crystal lattice that is revealed in spreading of spots in TEM pictures, curved grain boundaries and absence of their well-defined contrast.

The analysis of phase composition of the processed alloys shows that SPD results not only in nanostructure formation but also in creation of metastable phase states connected with formation of supersaturated solid solutions. The higher the accumulated strain and applied pressure are, the higher metastability we obtain (Table 2). For example, in the Al-Fe alloy, processed by HPT, the maximum dissolution of Fe in Al is higher (0.8 %) than in ECAP alloy (0.6 %). Two-phase Ti-64 alloy after HPT is transformed into single-phase alloy due to dissolution of β phase that is not revealed during ECAP. One more SPD effect, observed during HPT of intermetallic TiNi alloy, is its partial or complete amorphization [7,12,17].

Table 2: Structure parameters and mechanical properties of the HPT and ECAP processed alloys.

Alloy	State	T_{def}	Grain size, nm	Phase composition	UTS, MPa	YS, MPa	EL, %	Hv, MPa	Ref.
CP Ti	Hot-rolled		>1000	α-Ti	460	380	27	1800	[6]
	HPT	RT	120	α-Ti	950	790	14	2800	[10]
	ECAP	450	300	α-Ti	720	640	14	2800	[6]
Ti-64	Hot-rolled		>1000	α + β	1000	800	32	3100	[10]
	HPT	RT	80	α	1750	1750	1	5100	[10]
	ECAP	700	400	α + β	1160	1110	12	4000	[11]
$Ti_{49.5}Ni_{50.5}$	Hot-rolled		>1000	TiNi	840	500	19	2450	[13]
	HPT	RT	20	A + TiNi	1570		1	5500	[12]
	ECAP	400	300	TiNi+Ti_2Ni	1140	1050	9	2960	[13]
Al-5%Fe	Cast		>1000	Al+Al_3Fe_{14}	58	52	6.8	310	[14]
	HPT	RT	100	s.s (0.8 %)	590	473	1.9	1830	[14]
	ECAP	RT-	300	s.s (0.6 %)	244	216	5.8	660	[8]

Note: s.s. – solid solution (solubility of Fe in Al)

3.2 Mechanical and other Properties

In all the alloys under investigation SPD results in considerable strengthening (Table 2). During tensile tests strength properties increase 1.2–2 times, and for aluminum alloy –considerably as compared to the initial coarse-grained (CG) state. The highest increase of strength is observed while using HPT method. It is connected both with higher structure refinement as compared to ECAP method and with the role of second phases (amorphous or nanocrystalline, their volume fraction and distribution), as the contribution of second phases can be comparable or even higher than from decreasing of grain sizes. The fact, that at quite similar mean grain size of matrix phase the ultimate tensile strength and yield stress of ECAP and HPT processed Al-Fe alloy differ by more than 2 times, is obviously associated with this.

It was shown [10,14,18] that the strength of SPD metastable alloys can be considerably improved by additional thermal and mechanical treatment (CP Ti) or ageing (Ti-Ni, Al-Fe). For example, using cold rolling and intermediate annealings one can enhance ultimate tensile strength and fatigue limit of ECAP Ti up to 1100 and 500 MPa, correspondingly, without decreasing of elongation [10]. Such a high strength exceeds even the strength of HPT samples, while the grain size is similar. Low temperature annealings after HPT enabled to enhance record values of strength to 2650 MPa in TiNi [17] and 750 MPa in Al-Fe [14]. We have shown, that these greatest changes of strength are caused by the transformation of phase composition as a result of ageing. Not only mechanical, but also some unusual physical properties of the investigated alloys occurred to be sensitive both to the SPD methods and strains applied. For instance, the temperatures of forward martensite transformation (Ms and Mf) in ultrafine-grained $Ti_{49.6}Ni_{50.4}$ alloy as compared to coarse- grained one (0 °C and –17 °C) decreased and amounted to –38 °C and –77 °C for HPT and –17 °C and –47 °C for ECAP [7]. The magnetic susceptibility χ in paramagnetic CP Ti, processed by ECAP and HPT methods, was higher by 5 % and 18 % as compared to coarse-grained Ti [18], respectively.

4 Concluding Remarks

We have compared the microstructural changes and properties of several Ti and Al alloys, subjected to two SPD methods: i.e. HPT and ECAP. It was established that both methods can result in high refinement of microstructure up to nanoscale as well as in formation of high metastable states, associated with fabrication of oversaturated solid solutions or even amorphization. However, the level of metastability in HPT samples is, as a rule, considerably higher than in ECAP ones. It is obviously connected with more severe conditions of deformation during HPT, i.e. lower temperatures and higher applied pressures. Higher metastability in the investigated alloys enables to increase considerably their properties.

At present the SPD methods, in particular, modification and improvement of ECAP processing are being intensively developed. To improve the nanostructured parameters and to enhance further properties in metals and alloys especially the performed analysis demonstrates that future ECAP developments should be aimed at further decreasing of deformation temperature and increasing of the applied forward and backpressures.

5 Acknowledgement

The work in part was supported by the INTAS grant # 01NANO-0320, ISTC projects ## 2070 and 2398.

6 References

[1] Lowe T.C and Valiev R.Z., in Proceedings of the NATO ARW on Investigation and applications of severe plastic deformation, Moscow, 1999, Kluwer Academic Publishers (2000) 367–372

[2] Valiev R.Z., Islamgaliev R.K., Alexandrov I.V., Progr. Mater. Sci. 45, (2000), 103.

[3] Stolyarov V.V., Valiev R.Z. Journal of Metastable and Nanocrystalline Material, ed A.R. Yavari, Trans Tech. Publications, Materials Science Forum, v. 307, v.1 (1999) 185–190

[4] Stolyarov V.V., Valiev R.Z., in book "Ultrafine-grained Materials" edited by Mishra R.S., Semiatin S. L., Suryanarayana C., Thadhani N.N., T.C. Lowe, TMS society (2000) 351–360

[5] Utyashev F.Z., Enikeev F.U., Latysh V.V., in Ann. Chim. France., 21, 1996, 379–389.

[6] Stolyarov V.V., Zhu Y.T., Alexandrov I.V., Lowe T.C., Valiev R.Z., Mater. Sci&Eng., 299, 1–2 (2001) 59–67

[7] Pushin V.G., Stolyarov V.V., Valiev R.Z., Kourov N.I., Kuranova N.N., Prokofiev E.A., Yurchenko L.I., in Annales de Chimie – Science des Materiaux, 27, No.3, (2002) 77–88

[8] V.V. Stolyarov, R. Lapovok, I.G. Brodova, P. F. Thomson, Mater.Sci&Eng (2003), in press

[9] Sergueeva A.V., Stolyarov V.V., Valiev R.Z., Mukherjee A.K., Scripta Materialia 45 (2001) 747–752

[10] Stolyarov V.V., Shestakova L.O., Zharikov A.I., Latysh V.V., Valiev R.Z., Zhu Y.T., Lowe T. C., Proceedings of 9th Int. Conf. Titanium-99 eds Gorynin I.V. and Ushkov S.S., Nauka, v.1 (2000) 466–472

[11] The Quarter Report on ISTC project #2125

[12] Stolyarov V., Prokofiev E., Sergueeva ., Mukherjee A.K, Valier, R. Z. Proceedings of 37^{th}International Symposium"Actual problems of strength. Shape memory effect alloys and other perspective materials" September 24–27 2001 St. Petersburg, (2001), 108–113 (in Russian)

[13] current result

[14] Stolyarov V.V., Soshnikova E.P.,Brodova I.G., Bashlykov D.V., Kilmametov A.R., The Physics of Metals and Metallography, 93, 6,(2002), 74–81

[15] Valiev R.Z., Alexandrov I.V., Zhu Y.T., Lowe T.C., J.Mater.Res.,17,1 Jan (2002) 5–8

[16] Alexandrov I.V., Dubravina A.A., Kim H.S., Defect and Diffusion Forum, 208–209 (2002) 229–232.

[17] A.V. Sergueeva, C. Song, R.Z. Valiev, A.K. Mukherjee, Materials Science and Engineering: A, 339, 1–2 (2003) 159–165

[18] S. Z. Nazarova, A. A. Rempel, A. I. Gusev, V. V. Stolyarov, Proceedings of 6th UDS Conf., August, (2002) Tomsk, in press

On the Influence of Temperature and Strain Rate on the Flow Stress of ECAP Nickel

L. Hollang[1], E. Thiele[1], C. Holste[1], and D. Brunner[2]

[1] Institut für Physikalische Metallkunde, Technische Universität Dresden, Germany
[2] Max-Planck-Institut für Metallforschung, Stuttgart, Germany

1   Motivation and Theoretical Background

Severe plastic deformation (SPD) by equal-channel angular pressing (ECAP) is one of the most efficient methods to produce sub-microcrystalline materials with grain-sizes in the range 0.1 to 1µm. ECAP materials are dense bulk materials whose structural parameters can be adjusted widely by varying the processing parameter as processing temperature, die geometry, and processing route. Accordingly, the structure and the properties of ECAP materials have been the subject of numerous investigations. However, there are little attempts to use the thermal activation analysis to determine the elementary processes governing the plastic behavior of ECAP materials. The aim of the present paper is to show that thermal activation analysis is an useful tool to characterize the SPD state.

At low enough temperatures the plastic deformation of metals is governed by the conservative motion of dislocations. The dislocation motion is impeded by different interaction mechanisms which are effective at different length scales. Therefore, the flow stress measured in a deformation experiment may be decomposed into two contributions [1]

$$\sigma = \sigma_\mu + \sigma^* \quad . \tag{1}$$

The athermal stress contribution σ_μ is result of long-range internal stresses impeding the glide dislocations. σ_μ is related to the average dislocation density N_z and, in the case of polycrystals, additionally to a stress contribution σ_δ describing the influence of the grain size δ, by [2]

$$\sigma_\mu = \alpha \mu b N_z^{1/2} + \sigma_\delta \quad , \tag{2}$$

where μ is the elastic modulus, b the Burgers vector, and α is a constant. The temperature dependence of σ_μ is entirely due to that of μ. The *thermal stress contribution* σ^*, which accounts for the temperature and strain-rate dependence of σ, arises from glide dislocations *overcoming short-range obstacles* with the help of thermal activation.

Let us assume these obstacles to be randomly distributed. The average distance between the obstacles in the glide plane is given by l. If ΔG is the free activation enthalpy for cutting through the obstacles and v_0 the attempt frequency, the plastic strain rate at high stresses is

$$\dot{\varepsilon}_{pl} = \dot{\varepsilon}_0 \exp\left\{-\frac{\Delta G(\sigma^*, \hat{\sigma}^*)}{k_B T}\right\} \quad , \tag{3}$$

with $\dot{\varepsilon}_0 = b\,v_0\,l^2\,N$, where N is the number of glide dislocations per unit volume pressed against the obstacles. The Boltzmann term (k_B is Boltzmann's constant, T is the absolute temperature) describes the stress and temperature dependent probability of an activation event. The obstacle strength $\hat{\sigma}^*$ is the stress needed to surmount the obstacles without thermal activation and is assumed to follow $\hat{\sigma}^* \sim l^{-1}$. Provided N_z, δ and λ are *constant* during a strain-rate-change experiment (and thus σ_μ, $\hat{\sigma}^*$, and $\dot{\varepsilon}_0$), differentiating of ΔG in (3) with respect to σ^* gives [3]

$$\Delta a(\sigma^*) := -\frac{1}{b}\left(\frac{\partial \Delta G}{\partial \sigma^*}\right)_T = \frac{k_B T}{b}\left(\frac{\partial \sigma}{\partial \ln \dot{\varepsilon}_{pl}}\right)^{-1}_{T,l} = l\Delta x(\sigma^*). \tag{4}$$

The activation area Δa is important for applying the activation theory as it has a special geometrical meaning: it is the area swept out by $\dot{\varepsilon}_0$ a dislocation segment *during* cutting through the obstacles over an activation distance Δx. Δa can be determined experimentally since the strain-rate sensitivity $\lambda := \left(\partial \sigma / \partial \ln \dot{\varepsilon}_{pl}\right)_{T,l}$ appearing in (4) is a measurable quantity.

During a deformation experiment N_z, l, and δ can change considerably. Of importance is the process of dislocation multiplication and storage leading to work-hardening. In a first approximation such changes may be accounted for over a wide stress range by combining N_z and l by

$$l = N_z^{-1/2}. \tag{5}$$

From (5) follows that both, the athermal stress difference $\sigma_\mu - \sigma_\delta$ and the obstacle strength $\hat{\sigma}^*$ result from the same type of obstacles, namely dislocations with density N_z. Involving changes of N_z in the course of a deformation experiment (4) can be rewritten as

$$\Delta a(\sigma) = \frac{k_B T}{b}\left(\frac{\partial \sigma}{\partial \ln \dot{\varepsilon}_{pl}}\right)^{-1}_{T,l} = \frac{k_B T}{b}\frac{m(T)}{(\sigma - \sigma_\delta)}. \tag{6}$$

Here m depends only on temperature. Equation (6) is equivalent to the Cottrell-Stokes (CS) law which predicts a straight line in a λ-σ plot with a temperature dependent slope m^{-1}.

As the preceding discussion shows, Δa is sensitively dependent on l. We suppose significant changes of Δa if δ is decreased to values where it directly affects l. Moreover, the appearance of a competing thermally activated glide mechanism, involving, e.g., grain-boundary migration, should lead to a drastic change of the Δa-T dependence. We will apply the theory sketched above in order to characterize ECAP nickel and, for comparison, finecrystalline nickel.

2 Experimental Procedure and Evaluation of Data

Regarding the deformation state prior to tensile testing four different classes of Nickel specimens have been investigated in this paper (Tab.1). *ECAP nickel* was produced at *room temperature (ECAP-RT)* [4] and at an *elevated temperature (ECAP-ET)* of 250 °C [5]. *Fine-crystalline (FC) Nickel* billets were produced from rolled plates. The *annealed state (FC-AS)* was obtained by annealing the billets in vacuum. The *pre-deformed state (FC-PS)* was result of tensile deformation of FC-AS billets at 296 K to a stress level of about 390 MPa.

From the billets rods of 35 mm length and 3 mm in diameter were machined by spark erosion. The rod axis was chosen parallel to the pressing axis or tensile axis, respectively. From the rods tensile specimens with gauge length 10 mm and about 1 mm in diameter were produced by electrochemical polishing. The specimens were isothermally deformed in tension with an applied strain rate $\dot{\varepsilon}_a = 4.3 \cdot 10^{-4} \mathrm{s}^{-1}$ at different temperatures $4\,\mathrm{K} \leq T \leq 320\,\mathrm{K}$ in a helium cryostat permitting to keep the temperature constant within less than 0.05 K [6].

In standard experiments the specimen were deformed at constant temperature in isothermal tensile straining (ITS) tests. One specimen was subjected two times to an incremental-strain–temperature-lowering (ISTL) test, i.e. the specimen was repeatedly strained in a small strain interval beyond the yield stress σ_y at successively decreasing temperatures. During the ITS and ISTL tests stress-relaxation (SR) tests were performed in order to determine the experimental strain-rate sensitivity λ as the slope of a curve σ vs. $\ln(-\dot{\sigma})$ [7].

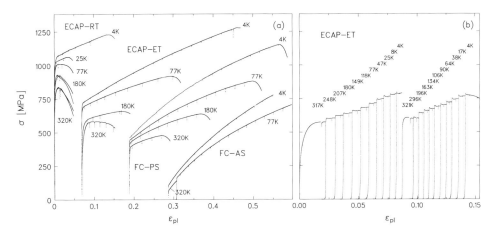

Figure 1: Selected stress–strain curves of the four different classes of nickel specimens investigated. (a) ITS tests at different temperatures. The curves are shifted along the ε_{pl}-axis in order to avoid overlaps. The stress drops refer to SR tests. (b) ISTL tests with *one* ECAP-ET specimen. The deformation temperature was lowered stepwise (as indicated) from 317 K to 4 K and, after heating up, again from 321 K to 4 K.

Table 1: Classes of nickel specimen investigated

	ECAP-RT	ECAP-ET	FC-AS	FC-PS
ingot purity [wt.%]	99.99	99.99	99.9	99.9
treatment	ECAP at 296 K 10 passes	ECAP at 523 K 8 passes	annealing	annealing and pre-deformation
grain size [μm]	0.5	0.8	35	35
References	[9], [10]	[9], [10]	[11]	[11]

Throughout this paper the flow stress σ means the true stress. The yield stress σ_y was determined by back-extrapolation of the stress–strain curves to zero strain in the case of work-hardening, or by taking the flow-stress maximum in the case of work-softening, respectively. In the ISTL tests the effect of cumulated hardening or softening on σ_y was corrected [8].

3 Experimental Results

3.1 Yield Stress and Work-hardening Behavior

Figure 1a shows stress–strain curves of all four classes of nickel specimens investigated. The stress–strain curves of *ECAP-RT nickel* can be subdivided into two types: at temperatures below 77 K the curves display work-hardening until necking whereas above 77 K short work-hardening at the beginning is followed by pronounced work-softening. Generally, the tendency to work-hardening is the lower the higher the temperature is. ECAP-RT nickel shows a rather limited ductility, with the exception of the curve at 4 K. σ_y decreases linearly with increasing temperature (Fig. 2a). The *ECAP-ET nickel* exhibits work-hardening at all temperatures. Compared to ECAP-RT nickel the level of σ_y is markedly decreased (Fig. 2a), but the σ_y-T relation is still linear and equals that obtained from the ISTL tests shown in Fig.1b. Finally, all *FC nickel* specimens show pronounced, temperature dependent work-hardening until necking in the entire temperature range investigated with a slightly temperature dependent σ_y as it is shown in Fig.2a, too. Figure 2b displays the work-hardening coefficient $\theta := \partial\sigma / \partial\varepsilon_{pl}$ derived from the stress-strain curves at 4 K and at 77 K (Fig.1a) as a function of σ. All four classes of Nickel specimen exhibit an almost linear θ-σ-relationship at high stresses.

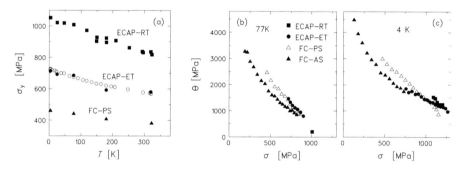

Figure 2: (a) Yield stress vs. temperature from the ITS tests (full symbols) and from ISTL tests (open symbols). (b), (c) Work-hardening coefficient vs. flow stress from the stress-strain curves at 77 K and at 4 K in Fig.1a.

3.2 Activation Area and Activation Enthalpy

The strain-rate sensitivity λ determined along the stress–strain curves of Fig.1a is shown in Fig.3a as a function of σ. At constant temperature λ depends linearly on σ with deviations at lower stresses which are the more pronounced the more the temperature is decreased. The slopes of the straight lines represent m^{-1} in (6). These findings hold for both types of deformation behavior observed: work-hardening *and* softening. However, whereas at constant temperature all λ data of FC nickel (AS and PS, resp.) coincide very well, the λ values of ECAP nickel (RT and ET, resp.) differ from each other and from that of FC nickel in a characteristic manner: At temperatures below 180K the λ-σ relationship of ECAP nickel is shifted parallel to that of FC nickel by a stress which we ascribe to σ_δ (eq.(6)). σ_δ is always 180 MPa for ECAP-RT nickel (shown in Fig.3a) and is between 120 MPa and 60 MPa for ECAP-ET nickel, respectively, is 120 MPa at the beginning of plastic deformation, and is decreased to 60 MPa (short lines in Fig.3a) during further deformation if the specimens exhibit work-hardening (Fig.1a). At temperatures above 180 K the slope m^{-1} of ECAP nickel is increased compared to that of FC nickel at the same temperature. Fig.3b shows the activation area Δa vs. σ-σ_δ. In Fig.4a Δa vs. σ_y-σ_δ obtained from the ISTL tests (Fig.1b) is shown together with the corresponding data of the ITS tests at the onset of plastic deformation. In Fig.4a the good agreement is attained between the data, resulting from both types of tests, when choosing (in both cases) the σ_δ values typical for the onset of plastic deformation in the initial state of ECAP nickel, namely 180 MPa (ECAP-RT) and 120 MPa (ECAP-ET). The strong influence of temperature on Δa at constant stress σ-σ_δ in Figs. 3b and 4a is described by the quantity $k_B Tm$, given by (6); its temperature dependence is shown in Fig.4b. For FC nickel $k_B Tm$ increases linearly in the entire temperature

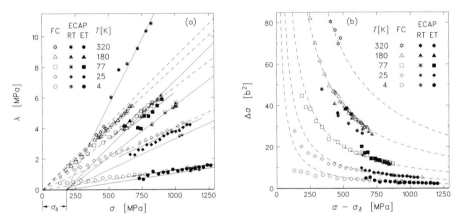

Figure 3: (a) Strain-rate sensitivity λ vs. σ from ITS tests (cf. Fig.1a). Straight lines corresponding to the Cottrell-Stokes law are drawn for ECAP-RT (full lines through σ=180 MPa), ECAP-ET (short full lines), and FC nickel (broken lines through origin). (b) Activation area Δa vs. σ-σ_δ; for σ_δ see text; broken lines as in (a).

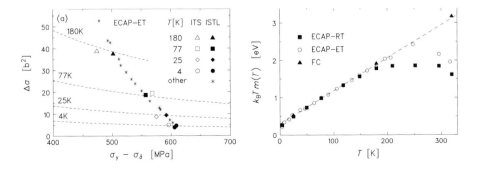

Figure 4: (a) Activation area Δa vs. σ-σ_δ of ECAP-ET nickel at low temperatures, determined from λ of the ISTL tests, σ_δ =120 MPa. The corresponding data of the ITS tests in Fig. 3b (open symbols, σ_δ =180 MPa) are shown for comparison. Dotted lines from Fig.3b. (b) $k_B Tm(T)$ vs. T for ECAP and FC nickel.

range investigated. ECAP nickel displays the same behavior at low temperatures, but deviates above 150 K (ECAP-RT) or 250 K (ECAP-ET).

4 Discussion and Conclusions

A striking feature of ECAP nickel is the tendency to pronounced work-softening above a certain temperature. It must be questioned why the ECAP nickel softens and also why the transition from work-hardening to work-softening can not be recognized in the σ_y-T curves. A qualitative argument which accounts for both observations offers the fact that the strain path is changed from ECAP (multiaxial stress) to tensile testing (uniaxial stress). Changing the strain path gives rise to local instabilities in connection with strain localization. However, at the onset of plastic deformation we find the smooth σ_y-T relationship caused by the *initial* ECAP structure. The initial ECAP structure is characterized by the dislocation density N_z, resulting from the ECAP process. If N_z is too high to permit work hardening by tensile deformation at the given temperature, the dislocation structure will be destroyed, at least locally, upon further straining. The resulting decrease of N_z leads to the observed macroscopic softening effects. Since $N_z = 10.0 \cdot 10^{14}$ m^{-2} of ECAP-RT nickel is about two times that of ECAP-ET nickel [10], ECAP-RT nickel softens already at 77 K instead of 320 K.

If deformation instabilities occur the question arises, whether SR tests can further be used to determine λ. If a SR test starts in the presence of instabilities, the immediate decay of σ will stop the activity of the dislocation sources responsible for strain localization. Provided, the volume fraction of the zones of lowered N_z is not too high, we will measure the relaxation behavior of the surrounding matrix and, therefore, a nearly unaltered λ–σ relationship. This accounts for the observation that $k_B Tm$ vs. T of ECAP-RT nickel follows the behavior of FC nickel up to 150 K, despite the fact that ECAP-RT nickel softens already at 77 K.

If ECAP nickel exhibits work-hardening during tensile deformation, it hardens in work-hardening Stage III of fcc metals. This is indicated by the almost linear, strongly temperature dependent decrease of θ with increasing σ. Stage III is the result of dynamic recovery processes reducing the dislocation-storage rate due to competing enhanced annihilation of dislocations as a result of cross slip and climb. This accounts for the strong influence of temperature and strain rate on the flow stress of ECAP nickel. Moreover, the dynamic processes in work-hardening Stage III occur in such a way that for both, FC *and* ECAP nickel, the CS law for strain-rate-changes is found to be valid.

At low temperatures the differences between FC and ECAP nickel can be completely ascribed to differences of σ_δ, which differ already at the beginning and may change during tensile deformation. Assuming $\sigma_\delta \sim \delta^{-1}$ we can ascribe the different σ_δ to differences of grain size δ (Tab.1). Taking into account the grain-size effect, the activation volume Δa at low temperatures is found to be independent on the deformation history of the specimens as demonstrated in Figs. 3b and 4a. Therefore we conclude that the same thermally activated low-temperature process governs the plastic behavior of FC *and* ECAP nickel. The almost linear increase of $k_B Tm$ vs. T in Fig.4b suggests for this process that $m = m_0(1+T_0/T)$ with $m_0 \approx 100$ and $T_0 \approx 35$ K. At high temperatures the deviations from the linear $k_B Tm$-T relationship indicate the appearance of a competing thermally activated high-temperature process in ECAP nickel. The nature of both processes will be discussed in a forthcoming paper.

From our results and from the discussion we conclude:

(1) The influence of grain size on the plastic behavior of submicrocrystalline nickel, produced by ECAP, can be analyzed successfully by using the thermal activation analysis.
(2) During uniaxial deformation ECAP nickel exhibits Stage III hardening at low temperatures. Above a critical temperature the deformation becomes unstable due to (local) lowering of the average dislocation density after changing the strain path from ECAP to uniaxial deformation.
(3) The strain-rate sensitivity of ECAP nickel follows the Cottrell-Stokes law.
(4) For ECAP nickel an additional athermal stress contribution, σ_δ, has to be considered which is caused by a grain-size effect.
(5) At low temperatures the temperature and strain-rate sensitivity of the flow stress of ECAP nickel is determined by the thermally activated interaction of dislocations, which occurs in the same way as in finecrystalline nickel. At higher temperatures an additional thermally activated process appears, which is not present in finecrystalline nickel.

5 References

[1] Seeger, A., Phil. Mag. 1954, 45, 711
[2] Nes, E. and Marthinsen, K., Mat. Sci. Eng. 2002, A 322, 176
[3] Kocks, U.F.; Argon, A.S.; Ashby, M.F., Progr. Mat. Sci. 1975, 19, 1
[4] produced by V.I. Kopylov, Belorussian Academy of Sciences, Minsk, Belorussia
[5] produced by R.Z. Valiev, Ufa State Aviation Technical University, Russia
[6] Brunner, D. and Diehl, J., Z. Metallkde. 1992, 83, 828
[7] Guiu, F. and Pratt, P.L., phys. stat. sol. 15, 1966, 539
[8] Brunner, D. and Diehl, J., phys. stat. sol. (a) 1987, 104, 145
[9] Klemm, R.; Thiele, E.; Baum, H.; Holste, C., Proceedings of the 8th International Fatigue Congress, Stockholm, 2002, EMAS, Vol. 3, p. 1609
[10] Klemm, R., Dr. rer. nat. thesis 2003, submitted to Technical University Dresden
[11] Buque, C.; Bretschneider, J.; Schwab, A.; Holste, C., Mat. Sci. Eng. A 300, 254

The Effect of Second-Phase Particles on the Severe Deformation of Aluminium Alloys during Equal Channel Angular Extrusion

P. J. Apps[1], J. R. Bowen[2], P. B. Prangnell[1]
[1]UMIST, Manchester, UK
[2]Risø National Labs, DK

1 Abstract

To date there has been little systematic work published on the effect of a material's initial microstructural variables on the evolution of the deformed state during severe deformation processing. In this work, the rate of grain refinement in an AA8079 alloy (Al-1.3%Fe-0.09%Si) containing coarse second-phase particles has been compared to that in a single-phase Al-0.13%Mg alloy, during ECAE processing to a strain of ten, using high resolution EBSD analysis. Coarse particles have been shown to greatly enhance the rate of grain refinement, and the mechanisms by which this occurs are discussed.

2 Introduction

A large body of work has now been published on the effect of severe plastic deformation (SPD) on the microstructure and properties of a variety of model and commercial alloys [e.g. 1,2]. However, to date only a few researchers have studied the evolution of the deformed state during SPD, with a view to determining the mechanisms that control grain subdivision and refinement to the sub-micron scale [3,4,5]. This work has shown that grain refinement occurs by the extension and compression of grain boundaries with strain, combined with grains subdividing by new high angle grain boundaries (HAGBs) being formed discontinuously, on finer and finer lengthscales, until ultimately a limit is reached where the HAGB spacing converges with the subgrain size. Research has also shown grain refinement is encouraged by heterogeneity in plastic deformation [6]. In conventional deformation it has been found that the presence of hard, micron-scale, second-phase particles can increase the rate of dislocation generation and promote HAGB formation at relatively low strains [7,8]. However, few systematic studies of the effects of a material's initial microstructural variables, and in particular the presence of second-phase particles, on the rate of grain refinement during severe deformation processing, have been carried out.

The objective of the current work was, therefore, to compare the rate and mechanisms of HAGB generation during the SPD of an Al-alloy containing a significant volume fraction of coarse second-phase particles, to previous work on a single-phase alloy deformed under identical conditions [3,4]. Detailed analysis of the deformed samples was carried out by high-resolution electron backscattered diffraction (EBSD) in order to quantify the differences between the two materials.

3 Experimental

An AA8079 Al-alloy (Al-1.3%Fe-0.09%Si) was chosen for the current investigation, as it can be processed to give a reasonably uniform distribution of coarse second phase particles, with no fine dispersoids present and a minimal solute level in the matrix. An 18 mm thick, DC cast and hot rolled, slab of AA8079, containing approximately 2.5 vol% of θ-phase ($Al_{13}Fe_4$) particles, was supplied by Alcan International. The mean particle diameter was ~2 μm and the inter-particle spacing ~10 μm. The plate was recrystallised for 1 hour at 500 °C, giving an initial grain size of 0.5–1.0 mm. 15 mm diameter cylindrical rods were machined from the plate, in the rolling direction, and processed by Equal Channel Angular Extrusion (ECAE), at room temperature, by up to 15 ECAE passes with no billet rotation between extrusion cycles (Route A). A die angle of 120° was used, giving a von Mises' effective strain per pass of ~ ε_{vm} = 0.7. Full details of the ECAE press and deformation behavior in the die used can be found elsewhere [8]. Microstructural characterization was carried out using high-resolution EBSD in the center of the ND-ED plane of the billet (the symmetry plane of the die). The EBSD maps typically contained several thousand grains and allowed statistically significant quantitative data to be obtained. The results were compared to previous work on a single-phase Al-0.13%Mg alloy deformed under identical conditions [3–5].

4 Results and Discussion

4.1 Evolution of Microstructural Parameters

Fig. 1 shows the transverse HAGB spacing (mean linear intercept) and the percentage of HAGB area (HAGBs; defined as the fraction of boundary area with misorientations greater than 15°), as a function of ECAE process strain. In the low strain regime, the EBSD data is somewhat unreliable due to wide variations in boundary spacing and a large difference between the subgrain

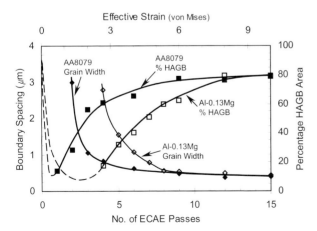

Figure 1: EBSD data showing average transverse HAGB spacing, and HAGB fraction, in Al-0.13%Mg and AA8079, as a function of strain during ECAE processing

and grain sizes. For this reason, data is only presented above two ECAE passes for alloy AA8079 and above four passes for the Al-0.13%Mg alloy.

From Fig. 1, it is clear that at low strains the rate of grain refinement in the particle-containing alloy is much greater than in the single-phase material. A transverse HAGB spacing of ~1 micron was achieved in the AA8079 material, at an effective strain of only 2 (3 ECAE passes), whereas the same transverse HAGB spacing was not reached in the single-phase Al-0.13%Mg alloy until an effective strain of $\varepsilon_{vm} = 4$ (6 ECAE passes), and even then the grain aspect ratio was still significantly greater in the Al-0.13%Mg sample. This much increased rate of grain refinement in the AA8079 alloy is also reflected in the more rapid generation of a large fraction of HAGB area at relatively low strains. In comparison, in the single-phase alloy, a significant increase in the percentage of HAGB area was not observed until an effective strain in excess of 3. Furthermore, in the particle-containing alloy the HAGB area fraction reached a stable maximum level of ~ 80% at a strain of $\varepsilon_{vm} = 6$, while this did not occur until a strain of $\varepsilon_{vm} \sim 10$ in the single-phase alloy, although the same maximum level was ultimately reached in both materials. Deformation to strains higher than ten showed that the fraction of HAGB area did not further increase in either material, and this appears to be the maximum level that can be obtained by severe deformation with this deformation mode. It should be noted that this level is still considerably less than that expected for a random grain assembly (97%).

In Fig. 2, the boundary misorientations measured form the EBSD maps at different strain levels have been subdivided into four 15° misorientation classes; Low Angle Boundaries (LABs), less than 15°, Low High Angle Boundaries (LHABs), 15° to 30°, Medium High Angle Boundaries (MHABs), 30° to 45°, and Very High Angle Boundaries (VHABs), defined as having misorientations in excess of 45°. Although these divisions are somewhat arbitrary, they give useful information on the differences in microstructural evolution between the two materials. In the particle-containing alloy, AA8079, Fig. 2b shows the formation of significant quantities of highly misorientated MHABs and VHABs from the outset of deformation. In contrast, boundary misorientations in the single-phase alloy develop much more slowly. At low strains only new HAGBs with misorientations just larger than 15° are formed (LHABs Fig. 2a) and significant densities of MHABs and VHABs are not seen until strains greater than $\varepsilon_{vm} = 3$. This data

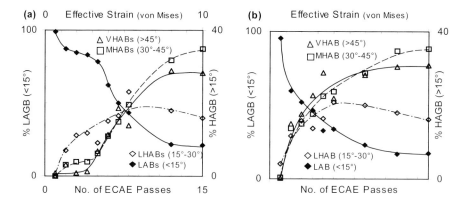

Figure 2: Comparison of the rates of evolution of boundaries of differing misorientation class, in (a) the Al-0.13%Mg and (b) the AA8079 alloy, as a function of strain. The boundaries have been divided into 15° misorientation bins giving four boundary classes, as described in the text.

clearly demonstrates that a different mechanism of grain subdivision and refinement must be occurring in the particle-containing alloy to that seen in single-phase alloys [3,4], particularly at low strains.

4.2 Local Deformation around Particles

Fig. 3a shows an example of a high resolution EBSD map (step size 0.05µm) around a single second-phase particle, in the AA8079 alloy, after only one ECAE pass (strain ~ 0.7). The black region in the center of the map corresponds to the un-indexed particle and the misorientations (in degrees) of selected boundaries in the surrounding matrix are annotated. At this strain, the matrix consists mainly of low-angle boundaries forming a subgrain cell structure throughout the alloy. However, close to the second-phase particles local zones of a size similar to one particle diameter were observed of much higher misorientation; e.g. the region labeled A in Fig 3a. A corresponding <111> pole figure in Fig. 3b indicates that region A has rotated about the transverse direction (normal to the symmetry plane of the die) by about 20° relative to the matrix

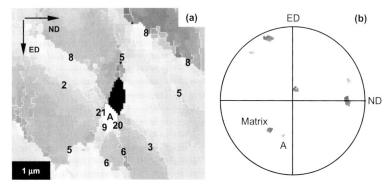

Figure 3: (a) High-resolution EBSD map (with HAGBs as black lines) around a particle in the AA8079 alloy after deformation to a strain of ~0.7 (1 ECAE pass) and (b) a corresponding <111> pole Fig

and, despite the low strain, is already delineated by new HAGBs of ~ 20° in misorientation. This rotated region has formed due to the influence of the particle, where strain gradients and local variations in the active slip systems have resulted in the development of high local matrix orientation gradients that readily collapse into new high-angle boundaries. Similar observations have been made during conventional deformation [7].

4.3 Microstructural Evolution with Strain

Fig. 4 shows EBSD maps from the Al-0.13%Mg and AA8079 alloys after deformation to a strain of ε_{vm}~1.3 (a & d) at a low magnification and at higher magnifications at strains of ε_{vm} ~4 and ε_{vm} ~10. In the low magnification images 4a and d, for clarity, only the HAGBs are shown. At a strain of ε_{vm} = 1.3, both primary deformation bands (PDBs), on the granular scale, and secondary deformation bands (SDBs), within the primary bands, are clearly visible in the single-phase alloy and these features are labeled in Fig. 4a. In contrast, in the AA8079 alloy, at the same strain (Fig. 4d) coarse deformation bands cannot be seen and grain subdivision is occurring much more homogeneously on the scale of the second-phase particle spacing. At higher strains, further grain subdivision and rotation of the deformation band boundaries leads to the evolution of a lamellar HAGB structure in the single phase Al-0.13%Mg alloy, of large elongated grains containing relatively equiaxed subgrains (Fig. 4b). With increasing strain, the transverse HAGB spacing has been observed to converge with the subgrain dimensions and produce a 'bamboo' structure of narrow elongated grains of high aspect ratio [3–5]. This structure has been found to be very stable and further deformation to ultra-high strains only results in a gradual reduction of grain aspect ratio (Fig. 4c). In contrast, at the same strain (ε_{vm} = 4) a deformation structure of submicron HAGB dimensions was seen to be already starting to form in the AA8079 alloy (Fig. 4e). Although some grains of high aspect ratio were still present in the matrix, regions of fine equiaxed grains already existed around the second-phase particles, which expanded with strain, and the inter-particle spacing was low enough so that these zones had already begun to impinge at this strain level. The formation of local deformation zones around coarse particles, which produces fine equiaxed grains at low strains, thus results in rapid grain refinement and helps prevent the formation of the lamella stable 'bamboo' structure observed in single-phase alloys by randomizing the matrix flow behavior.

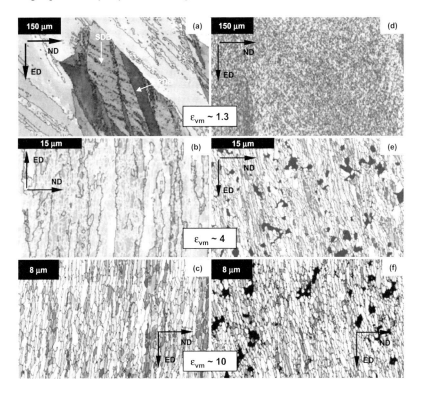

Figure 4: EBSD maps showing the evolution of the deformation structures in the single phase Al-0.13%Mg alloy, for strains of ε_{vm}~1.3, 4 and 10, (a) (c) and (d), and in the AA8079 particle containing alloy at the same strain levels (c), (e) and (f). Black and white lines represent HAGB and LAGBs, respectively.

At ultra-high strains ($\varepsilon_{vm} \sim 10$), the deformation structures of the two alloys were remarkably similar (Fig. 4c and f). Table 1 gives statistical EBSD data from both alloys after deformation to a strain of ~ 10 (15 ECAE passes) and shows virtually identical transverse HAGB spacings and percentages of HAGB area. Hence, the presence of coarse second-phase particles has no effect on the limiting cell size formed in alloys with similar matrix compositions. The grain aspect ratio in AA8079 is slightly lower than that in the single-phase alloy, and this reflects the greater disruption of the lamella boundary structures due to the matrix flow around the particles. However, it must be remembered that an ultra-fine grain structure was formed at a much lower strain in the particle containing alloy, which exhibited little further refinement for strains greater than $\varepsilon_{vm} = 6$. As the final average HAGB spacings are less than 1 µm in both materials when grain refinement saturates, they can both be considered to be 'submicron grained', although it must be remembered that they are in fact deformation structures and still contain substantial densities of low angle boundaries.

Table 1: Comparative statistical EBSD data of the Al-0.13%Mg and AA8079 alloys deformed to an effective strain of ~10 by ECAE.

	Al-0.13%Mg	AA 8079
Percentage HAGB	79%	82%
Transverse HAGB spacing	0.44 µm	0.40 µm
Longitudinal HAGB spacing	0.93 µm	0.63 µm
Grain Aspect Ratio	2.1	1.6
Mean LAGB misorientation	6.0°	5,9°
Mean HAGB misorientation	40°	41°

5 Conclusions

The presence of coarse second-phase particles dramatically increases the rate of grain refinement during SPD. This has been attributed to the local rapid generation of new HAGBs in deformation zones surrounding the particles, which develop into micron-scale grains at relatively low strains. It has been demonstrated that the formation of a submicron-grained structure, with in excess of 70 % HAGB, can be achieved at an effective strain as low as 4. In comparison, the production of an equivalent microstructure in a single-phase material requires severe deformation to a strain approaching ten.

6 Acknowledgments

The authors would like to acknowledge the support of the EPSRC (grant GR/L96779) and Alcan Int. (Banbury Labs. UK).

7 References

[1] M.V. Markushev, M.Yu. Murashkin, Phys of Met and Metall 2000; 90(5): p. 506–515
[2] S. Ferrasse, V.M. Segal, K.T. Hartwig, R.E. Goforth, Metall and Mater Trans A 199728A: p. 1047–1057
[3] J. R. Bowen, P.B. Prangnell, F.J. Humphreys, Mat Sci and Tech 2000; 16: p. 1246–1250
[4] J. R. Bowen, The formation of UFG model Al and steel alloys. PhD thesis, UMIST (2000)
[5] P. B. Prangnell, J. R. Bowen, A. Gholinia, in 22nd Risø Int. Symp. on Mat Sci – Science of Metastable and Nanocrystalline Alloys, Structure Properties and Modelling (Eds.: A. R. Dinesen, M. Eldrup, D. Juul Jensen, S. Linderoth, T. B. Pedersen, N. H. Pyrds, A. Schrøder Pedersen, J. A. Wert.) Risø, DK, 2001, p. 105.
[6] B. Hutchinson, Phil Trans of the Royal Society A 1999; A357: p. 1471–1485
[7] F. J. Humphreys, M.G. Ardakani, Acta Metall et Mater 1994; 42(3): p. 749–761
[8] F. J. Humphreys, Acta Metall 1979; 27: p. 1801–1814
[9] J. R. Bowen, A. Gholinia, S. M. Roberts, P. B. Prangnell, Mat. Sci. and Eng 2000; A287: p. 87–99

Effects of ECAP Processing on Mechanical and Aging Behaviour of an AA6082 Alloy

P. Bassani, L. Tasca, M. Vedani
Politecnico di Milano, Dipartimento di Meccanica, Milano, Italy

1 Introduction

The physical and microstructural properties of alloys processed by severe plastic deformation (SPD) techniques have been widely studied during recent years. In particular, many model and commercial Aluminium alloys were investigated after Equal Channel Angular Pressing (ECAP) and Accumulative Roll Bonding (ARB) to cast light on fundamental aspects as well as on industrial feasibility of this processing routes [1–11]. Among commercial Al alloys, the 6XXX series have received a great attention, being it largely used in aerospace and automotive industries and owing to its hardenability by aging treatments.

Recent published works showed that the AA6061 alloy, is well suited to ECAP and ARB processing [7,12–16]. Data are available on pre-ECAP solution treated and aged samples as a function of aging time (i.e. peak aging and overaging) and on samples that underwent the peak aging treatment only after ECAP processing. The combination of pre-ECAP solid solution treatment with post-ECAP low-temperature (e.g. 100 °C) aging resulted to be more effective in increasing the strength of the AA 6061 alloy with respect to pre-peak aged samples subjected to the same number of ECAP passes. An extensive work published by Kim and co-workers [12,14] also showed that the above mentioned increase in strength with aging was achievable in the ECAP processed AA6061 alloy only at low aging temperatures (i.e. at 100 and only partially at 140 °C). At higher aging temperatures (i.e. at 175 °C) softening and/or grain growth effects became more pronounced and dominated over the precipitation hardening effect, especially when large strains were accumulated in the materials by a relatively large number of ECAP passes.

From TEM investigations it was also reported that modification occurred in hardening precipitate morphology. The usually fine spherical or needle-like G.P. zones found in 6061 aged alloys are replaced by rather coarse spherical particles mainly located around the dislocation substructure in severely deformed alloys [12]. The increased diffusion rate and the strong stress field induced in the heavily deformed microstructure by ECAP might therefore significantly alter the precipitation behaviour of the alloy.

To better elucidate some of the aspects related to aging behaviour of ECAP processed 6XXX series alloys, a study was undertaken on aging behaviour of an AA6082 alloy processed by ECAP at different level of accumulated strains. The results of DSC analyses are presented here as a basis for further investigations aimed at defining optimised heat treatment conditions and the corresponding mechanical properties achievable.

2 Materials and Experimental Procedures

Samples cut from extruded bars of an AA6082 Al alloy having a diameter of 10 mm were solution treated at 530 °C for 2 hours and water quenched before being severely deformed by ECAP. The ECAP facility was based on a system of two symmetrical half dies machined from two blocks of tool steel. Cylindrical channels having a diameter of 10 mm were machined with a configuration characterised by angle values of $\Phi = 90°$ and $\Psi = 20°$. With the aim of reducing friction and increasing wear properties, the two half die were ion nitrided.

ECAP testing was carried out at room temperature by repeatedly pressing several specimens in order to accumulate the desired plastic deformation. The samples were processed by route C (rotation by 180° of the sample between each pressing) for up to 6 passes. According to the well accepted Iwahashi equation [17], it can be estimated that for the present ECAP die geometry, the strain given at each pass is 1,055. Further detail on experimental setup and testing procedures are given elsewhere [18].

After processing, the samples were longitudinally sectioned for optical microscopy and microhardness analyses. Transmission electron microscope (TEM) samples were also prepared starting from 1 mm thick disks transversally cut from the samples.

Mechanical testing was carried out on the processed samples at room temperature at a engineering strain rate of $1,5 \cdot 10^{-3}$ s^{-1}. The tensile specimens had a gauge length of 30 mm and a gauge diameter of 6 mm.

Aging behaviour of the pre-solution treated and ECAP processed materials was studied by differential scanning calorimetry (DSC). Heating cycles from room to 550 °C were set at a heating rate of 30 K/min in a Ar atmosphere.

3 Results and Discussion

Results on microstructure evolution of the AA 6082 alloy as a function of passes during ECAP were presented elsewhere [18]. From TEM analyses it was shown that a fine equiaxed grain structure with high-angle grain boundaries formed after 4 passes. The severe plastic deformation induced at room temperature allowed to reduce the starting grains size from 6 mm down to 250 nm.

In this paper, the results based on mechanical testing of the as ECAP processed materials and their aging behaviour are described. Figure 1 shows the tensile curves of the pre-solution treated and ECAP processed samples while in Figure 2 the tensile data are summarised as a function of the number of passes. It is apparent from the above figures that a significant increase in strength occurred after the first pressing. Subsequent pressings further increased the strength but to a lower rate. Eventually, a saturation was attained after four passes.

The high level of yield and ultimate tensile strength values achieved by the solution treated and ECAP processed samples demonstrated the remarkable hardening induced by grain structure modification, even before the aging treatment.

From Figure 2 it is also noticed that fracture elongation underwent a significant reduction after the firs pressing. Further accumulation of strain did not lead to substantial losses in ductility, the fracture elongation ranging from 6 % to 10 % for all the ECAP processed samples. The loss in ductility over the as solution treated AA 6061 alloy is supposed to be due to the relatively

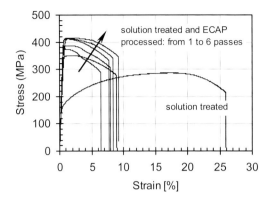

Figure 1: Stress vs. strain curves of the ECAP processed AA 6082 alloy at different passes

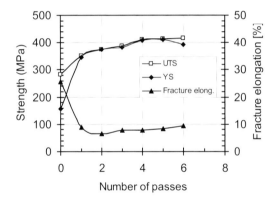

Figure 2: Evolution of tensile data as a function of number of passes

small strain hardening after yielding of the ECAP processed materials, which leads to high sensitivity to strain localization (necking).

DSC scans of the materials investigated, starting from the solution treated condition, are gathered in Figure 3. For the interpretation of the heat flow evolution, the known precipitation sequence during aging must be recalled [19-21]: SSS → independent clusters of Si and Mg, co-clusters of Si and Mg → GP zones → Si rich phase → β" coherent metastable phase → β' semicoherent metastable phase → Si precipitates → β stable Mg_2Si phase. Depending on specific alloy composition, on heating rate and on calorimetric system sensitivity, some of the transformation peaks might be not visible or superimposed to other broader peaks.

From the DSC profile of the solution treated AA 6082 alloy given in Figure 3, the broad peak at about 295 °C is supposed to represent the β" precipitation. In literature it is reported that a second peak, usually located at about 330 °C and corresponding to β' formation, might be superimposed to a shoulder of the former peak or even completely hidden by it [19]. Finally, the

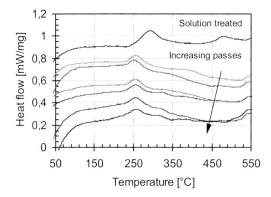

Figure 3: DSC curves of the solution treated AA 6082 alloy and of the pre-solution treated and ECAP processed samples

peak at 485 °C visible in the solution treated DSC profile is to be associated to formation of stable β phase.

When comparing the aging behaviour of the solution treated sample to that of the solution treated and ECAP processed samples, it is apparent that significant modification occurred due to the severe plastic deformation experienced by the alloy. There is clear evidence that the broad peak at 295 °C is shifted to lower temperatures in the ECAP processed samples. The position of this β'' peak is shown in Figure 4 as a function of ECAP passes. It can be inferred that after the first ECAP pressing the acceleration of the aging kinetics saturates, the peak being located at about 250–260 °C from the first up to the sixth pass.

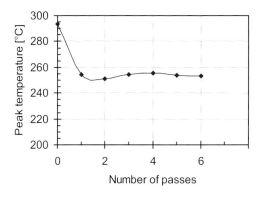

Figure 4: Temperature of the β'' peak as a function of plastic strain accumulated during ECAP (each pass theoretically corresponds to 1,055 strain)

It is also interesting to note that the peak at 485 °C, visible for the solution treated alloy, does not appear any more in the ECAP processed samples. Presumably, the corresponding β phase might form at lower temperatures, as suggested by the peak visible at 390 °C in the pass 1

ECAP processed sample curve. Secondly, a careful observation of the curves corresponding to passes 4 to 6 of the ECAP processed samples suggests that heavy plastic deformation has separated the two originally superimposed peaks of β" and β' phases. The shift to lower temperatures of the broader β" peak apparently made more visible the latter peak, whose presence in the solution treated samples was only presumed on the basis of literature data.

4   Conclusions

An investigation was carried out on mechanical properties and aging behaviour of a pre-solution treated and ECAP processed AA6082 alloy. Room temperature tests showed that the considered alloy was suitable to processing by severe plastic deformation, at least up to six ECAP passes under route C.

The tensile properties of the processed specimens suggested that a significant increase in strength occurred after the first pressing. The subsequent passes induced a further hardening even though to a lower degree. Eventually a saturation after four passes occurred. Fracture elongation underwent a significant reduction after the firs pressing. Further accumulation of strain did not lead to substantial losses in ductility.

The results of a thermal analysis study were also presented as a basis for further investigations aimed at defining optimised heat treatment conditions of the ECAP processes materials. When comparing the aging behaviour of the solution treated sample to that of the solution treated and ECAP processed materials, it was apparent that significant modifications occurred. There was clear evidence that the peak corresponding to β" phase formation was shifted to lower temperatures in the ECAP processed samples. Several observations contributed to state that precipitation sequence and aging kinetics of the ECAP alloy differed to that of the solution treated standard alloy.

5   References

[1]   R.Z. Valiev. Mater. Sci. Engng., A234-236 (1997) 59–66
[2]   M. Furukawa, Z. Horita, M. Nemoto, T.G. Langdon. J. Mater. Sci., 36 (2001) 2835–2843
[3]   Y. Iwahashi, Z. Horita, M. Nemoto, T.G. Langdon. Acta Mater., 45 (1997) 4733–4741
[4]   R.Z. Valiev, R.K. Islamgaliev, I.V. Alexandrov. Prog. Mater. Sci., 45 (2000) 103–189
[5]   F. H. Froes, O.N. Senkov, E.G. Baburaj. Mater. Sci. Engng., A301 (2001) 44–53
[6]   Y.T. Zhu, T.C. Lowe. Mater. Sci. Engng., A291 (2000) 46–53
[7]   Z. Horita, T. Fujinami, M. Nemoto, T.G. Langdon. J. Mater. Proc. Techn., 117 (2001) 288–292
[8]   J.C. Lee, H.K. Seok, J.Y. Suh. Acta Mater., 50 (2002) 4005–4019
[9]   J.C. Lee, H.K. Seok, J.Y. Suh, J.H. Han, Y.H. Chung. Metall. Mater. Trans., 33A (2002) 665–673
[10] N. Tsuji, Y. Ito, Y. Saito, Y. Minamino. Scripta Mater., 47 (2002) 893–899
[11] X. Huang, N. Tsuji, N. Hansen, Y, Minamino. Mater. Sci. Engng., A340 (2003) 265–271

[12] J. K. Kim, H. G. Jeong, S. I. Hong, Y. S. Kim, W. J. Kim. Scripta Mater., 45 (2001) 901–907
[13] K.-T. Park, H.-J. Kwon, W.-J. Kim, Y.-S. Kim. Mater. Sci. Engng., A316 (2001) 145–152
[14] W. J. Kim, J. K. Kim, T. Y. Park, S. I. Hong, D. I. Kim, Y. S. Kim, J. D. Lee. Metall. Mater. Trans., 33A (2002) 3155–3164
[15] C. S. Chung, J. K. Kim, H. K. Kim, W. J. Kim. Mater. Sci. Engng., A337 (2002) 39–44
[16] S. Ferrasse, V. M. Segal, K. T. Hartwig, R. E. Goforth. J. Mater. Res., 12 (1997) 1253–1261
[17] Y. Iwahashi, J. Wang, M. Horita, M. Nemoto, T.G. Langdon. Scripta Mater., 35 (1996) 143–146
[18] M. Vedani, P. Bassani, M. Cabibbo, V. Latini, E. Evangelista. Scripta Mater., submitted for pubblication (2002)
[19] G. Biroli, G. Caglioti, L. Martini, G. Riontino. Scripta Mater., 39 (1998) 197–203
[20] M. Takeda, F. Ohkubo, T. Shirai, K. Fukui. Mater. Sci. Forum., 217–222 (1996) 815–820
[21] S. B. Kang, L. Zhen, H. W. Kim, S. T. Lee. Mater. Sci. Forum., 217–222 (1996) 827–832

Influence of the Thermal Anisotropy Internal Stresses on Low Temperature Mechanical Behavior of Polycrystalline and Nanostructured Ti

V. Z Bengus, S. N. Smirnov

B. Verkin Inst. for Low Temperature Physics & Engineering, Ukraine Academy of Sciences, Kharkov, Ukraine

1 Introduction

In polycrystalline non-cubic materials temperature variations can cause internal thermoelastic microstresses due to the anisotropy of the thermal expansion coefficient in individual grains. These stresses were called [1] microstructural thermal anisotropy stresses (MTAS). If the MTAS exceed the critical stress at which slip (or twinning) starts in any of the crystallographic systems of plastic shear in an individual grain, this grain experiences plastic deformation. This really occurs on heating, cooling and thermocycling of soft polycrystalline metals. The phenomenon was first detected by Boas and Honeycombe in the 1944-47ies [2], who tried to estimate roughly the order of MTAS magnitude.

V.A. Likhatchev obtained formulae for the order of magnitude calculation of MTAS in polycrystals of all non-cubic symmetries. He used the model of an elastically and thermally anisotropic grain immersed into isotropic medium [1]. We employ Likhatchev's model to calculate the MTAS in the grains of coarse- and ultrafine grain (nanostructured) titanium on cooling below room temperature. The MTAS effect on the mechanical behavior of titanium is estimated.

2 Thermal Anisotropy Stresses in Titanium

In Likhatchev's model the MTAS tensor $\hat{\sigma}^{th}$ has a diagonal form in the Cartesian coordinates $x^{(c)}$, $y^{(c)}$, $z^{(c)}$ connected with the crystallographic axes of each grain (the $x^{(c)}$ and $y^{(c)}$ axes are along $[2\bar{1}\bar{1}0]$ and $[01\bar{1}0]$, respectively; the $z^{(c)}$ axis is oriented along $[0001]$, i.e. the six-fold axis of a hcp crystal). At the temperature T the $\hat{\sigma}^{th(c)}$ components are determined as

$$\sigma_{11}^{th(c)} = \sigma_{22}^{th(c)} = -\int_{T}^{T_0} \frac{(c_{11}+c_{12}-2c_{13})(\alpha_3 - \alpha_1)}{3[1+(c_{11}+c_{12}+c_{13})(c_{11}^* - c_{12}^*)^{-1}/2]} dT + \sigma_{22}^{th(c)}(T_0), \quad (1)$$

$$\sigma_{33}^{th(c)} = \frac{4}{3}\int_{T}^{T_0} \frac{(c_{33}-c_{13})(\alpha_3 - \alpha_1)}{[1+(c_{11}+c_{12}+c_{13})(c_{11}^* - c_{12}^*)^{-1}/2]} dT + \sigma_{33}^{th(c)}(T_0) \quad (2)$$

where T_0 is temperature of the 'initial" state of the sample from which counting is started; $\sigma_{11}^{th(c)}(T_0) = \sigma_{22}^{th(c)}(T_0)$, $\sigma_{33}^{th(c)}(T_0)$ are the MTAS tensor components at T_0; c_{11}, c_{12}, c_{13}, c_{33} are the elastic constants of the hcp single crystal; c_{11}^*, c_{12}^* are the averaged elastic constants of the polycrystalline environment of the grains; α_3, α_1 are the thermal expansion coefficients parallel and perpendicular to the six-fold axis, respectively.

It is shown below that the MTAS effect on the shear processes in the grains is dependent on the difference $\sigma_{th} \equiv \sigma_{33}^{th(c)} - \sigma_{22}^{th(c)}$. Using Eq. (1), we can find

$$\sigma_{th} \equiv \sigma_{33}^{th(c)} - \sigma_{22}^{th(c)} = \int_{T}^{T_0} \frac{(4c_{33} + c_{11} + c_{12} - 6c_{13})(\alpha_3 - \alpha_1)}{3[1 + (c_{11} + c_{12} + c_{13})(c_{11}^* - c_{12}^*)^{-1}/2]} dT + \sigma_{33}^{th(c)}(T_0) - \sigma_{22}^{th(c)}(T_0) \quad . \tag{3}$$

The dependences $\sigma_{22}^{th(c)}(T)$, $\sigma_{33}^{th(c)}(T)$ and $\sigma_{th}(T)$ are estimated numerically from c_{ik} [3], c_{ik}^* [4] and the dependences $\alpha_i(T)$ for titanium iodide [5] are shown in Fig.1. It is seen that the highest changes in these values occur in the interval 300–40 K. The further cooling produces only slight effect. Since $\sigma_{33}^{th(c)} < 0$, while $\sigma_{11}^{th(c)}$ and $\sigma_{22}^{th(c)} > 0$, below 300 K each grain in iodide titanium experiences compressive stress along the six-fold axis and tensile strength in the perpendicular direction. It is known [6] that such stresses do not stimulate twinning in hcp single crystals with $c/a < 1.633$. Titanium belongs to this type of single crystals ($c/a \approx 1.585$), and no twinning occurs when titanium iodide is cooled.

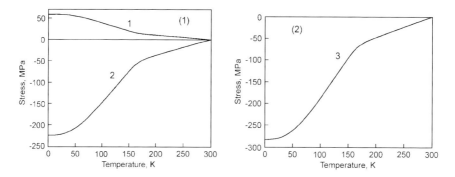

Figure 1: Temperature dependences of MTAS of the iodide titanium by $T_0 = 300$ K. 1: $\sigma_{22}^{th(c)}(T) - \sigma_{22}^{th(c)}(T_0)$, 2: $\sigma_{33}^{th(c)}(T) - \sigma_{33}^{th(c)}(T_0)$, 3: $\sigma_{th}(T) - \sigma_{th}(T_0) \equiv \sigma_{33}^{th(c)}(T) - \sigma_{22}^{th(c)}(T) - \sigma_{33}^{th(c)}(T_0) + \sigma_{22}^{th(c)}(T_0)$.

3 Effect of Thermal Anisotropy Stresses on Twinning and Slip in Titanium

Let us assume that our sample is loaded uniaxially and experiences uniform elastic deformation. In this case the tensor of elastic stresses $\hat{\sigma}^e$ induced by the applied force is uniform too. In the Cartesian coordinates $x^{(0)}$, $y^{(0)}$, $z^{(0)}$ related to the sample (the $z^{(0)}$ axis is along the sample axis) the tensor $\hat{\sigma}^{e(0)}$ has only one non-zero component $\sigma_{33}^{e(0)} = \sigma$.

The total stress initiating plastic deformation in the sample is described by the tensor $\hat{\sigma}$, which is the sum of the tensors $\hat{\sigma}^e$ and $\hat{\sigma}^{th}$:

$$\hat{\sigma} = \hat{\sigma}^e + \hat{\sigma}^{th} . \tag{4}$$

The tensor $\hat{\sigma}$ is uniform within each grain. Besides, in the $x^{(c)}\ y^{(c)}\ z^{(c)}$ coordinate system the tensor $\hat{\sigma}^{th(c)}$ is the same for all grains while the tensor $\hat{\sigma}^{e(c)}$ varies because it is dependent on the grain (single crystal) orientation with respect to the sample axes.

The components of the tensor $\hat{\sigma}^{e(c)}$ in the coordinates $x^{(c)}$, $y^{(c)}$, $z^{(c)}$ is given by the matrix:

$$\sigma_{ik}^{e(c)} = \sigma \begin{vmatrix} \sin^2\varphi \sin^2\theta & \sin\varphi \cos\varphi \sin^2\theta & \sin\varphi\sin\theta\cos\theta \\ \sin\varphi \cos\varphi \sin^2\theta & \cos^2\varphi \sin^2\theta & \cos\varphi\sin\theta\cos\theta \\ \sin\varphi\sin\theta\cos\theta & \cos\varphi\sin\theta\cos\theta & \cos^2\theta \end{vmatrix} \qquad (5)$$

where θ and φ are the Eulerian angles between the $z^{(0)}$ - $z^{(c)}$ axes and $x^{(0)}$ - $x^{(c)}$ axes, respectively. We assume that φ, θ are random quantities having the distribution density $f(\varphi, \theta)$. The φ and θ angles take values from region Ω_1: $0 \leq \varphi \leq \pi/3$, $0 \leq \theta \leq \pi/2$ (since the c-axis of the hexagonal lattice is the six-fold axis). The function $f(\varphi, \theta)$ is normalized by 1 in the region Ω_1.

We denote the slip and twinning systems as "deformation systems". The Cartesian coordinates $x^{(s)}$, $y^{(s)}$, $z^{(s)}$ can be introduced into each of these systems by directing the $y^{(s)}$ axis along plastic shear and the $z^{(s)}$ axis perpendicular to the deformation plane. Eqs. (1), (2), (4), (5) determine the tensor $\hat{\sigma}$ in the $x^{(c)}$, $y^{(c)}$, $z^{(c)}$ coordinates. The shear stresses of any deformation system can be calculated using the standard transformation rules for the stress tensor components and the known crystallogeometric coupling for the coordinate systems $x^{(c)}\ y^{(c)}\ z^{(c)}$ and $x^{(s)}\ y^{(s)}\ z^{(s)}$.

3.1 $<1\bar{1}01>\{1\bar{1}02\}$ Type Twinning

The explicit expression $\sigma_{23}^{(tw)}$ for the component inducing twinning in the systems of the $<1\bar{1}01>\{1\bar{1}02\}$ type can be obtained using the tensor components $\hat{\sigma}^{(c)}$. In this case the rotation of the $x^{(tw)}$, $y^{(tw)}$, $z^{(tw)}$ coordinates with respect to the $x^{(c)}$, $y^{(c)}$, $z^{(c)}$ coordinates is determined only by one Eulerian angle $\theta_{tw} \approx 42,4616°$ ($\varphi_{tw} = 0$). As a result,

$$\sigma_{23}^{(tw)} = \frac{1}{2}\left\{ \sin 2\theta_{tw}\ [(\cos^2\theta - \cos^2\varphi \sin^2\theta)\ \sigma + \sigma_{th}] + \cos 2\theta_{tw}\ \cos\varphi\ \sin 2\theta \cdot \sigma \right\}. \qquad (6)$$

We assume that twinning in a particular twinning system only starts when $\sigma_{23}^{(tw)}$ exceeds the stress of the onset of twinning $\sigma_{tw} \geq 0$. Let σ_{tw} be the same for all $<1\bar{1}01>\{1\bar{1}02\}$ systems in the grain and for all grains, i.e. σ_{tw} characterized the twinning system in the whole sample. The criterion of initiating twinning in a particular grain in a certain twinning system can be written as $\sigma_{23}^{(tw)} \geq \sigma_{tw}$. Using Eq.(6), this criterion is

$$\left[\cos^2\theta - \cos^2\varphi \sin^2\theta + \text{ctg} 2\theta_{tw} \cdot \cos\varphi \sin 2\theta\right]\sigma + \sigma_{th} \geq 2\sigma_{tw}/\sin 2\theta_{tw}. \qquad (7)$$

With $\sigma \neq 0$, the criterion of Eq. (7) can be written conveniently as

$$\begin{aligned} F_{tw}(\varphi,\theta) &\geq p \quad \text{for } \sigma > 0, \\ F_{tw}(\varphi,\theta) &\leq -p \quad \text{for } \sigma < 0 \end{aligned} \qquad (8)$$

where $F_{tw}(\varphi, \theta) \equiv \cos^2\theta - \cos^2\varphi \sin^2\theta + \text{ctg} 2\theta_{tw} \cos\varphi \sin 2\theta$,

$$p = \frac{1}{|\sigma|}\left(\frac{2\sigma_{tw}}{\sin 2\theta_{tw}} - \sigma_{th}\right). \quad (9)$$

It is seen in Eqs. (8)–(9), that only the parameter p allows for three factors σ, σ_{th} and σ_{tw} responsible for initiation of twinning. Twinning can be stimulated only by MTAS (at $\sigma = 0$) if the following condition is obeyed $\sigma_{th} \geq 2\sigma_{tw}/\sin 2\theta_{tw}$. In this case we should have $\sigma_{th} > 0$. With the accuracy sufficient for the practical applications, we can assume that $\sigma_{th} \geq 2\sigma_{tw}$.

Each Ti grain has six identical twinning systems of the $<1\overline{1}01>\{1\overline{1}02\}$ type. The system $<1\overline{1}01>\{1\overline{1}02\}$ on whose plane contains the grain axis $x^{(c)}$ $[2\overline{1}\overline{1}0]$ can be denote as "system 1" (φ is the angle between this axis and the $x^{(0)}$-axis of the sample). Then, to calculate the stresses in any of the six twinning systems of a particular grain, φ in Eq. (6) should be replaced with the sequence $\varphi_k = \varphi + (k-1)\pi/3$, $k = 1, 2, ..., 6$. Here, k is the number of the twinning system in a given grain. The systems are thus numbered in the sequence of increasing φ_k angles. The whole set of φ_k is within the interval $[0, 2\pi]$. To analyse the twinning criterion in all the twinning systems of all grains, we should therefore consider the function $F_{tw}(\varphi, \theta)$ in the angle region Ω_2: $0 \leq \varphi \leq 2\pi$ and $0 \leq \theta \leq \pi/2$. Since in Ω_2 the function $F_{tw}(\varphi, \theta)$ is symmetric with respect to the straight line $\varphi = \pi$, the sets of stresses $\sigma_{23}^{(tw)}$ in the grains characterized by the angles (φ, θ) and ($\pi/3 - \varphi, \theta$) are identical. Besides, the stresses in the twinning systems of these grains form equal pairs: in system 1 of one grain and in system 6 of the other grain, as well as in systems 2 and 5, 3 and 4, 4 and 3, 5 and 2, 6 and 1, respectively.

The equal level lines of the function F_{tw} in the region Ω_2 determined by the requirement $F_{tw}(\varphi, \theta) = C$ are shown in Fig. 2. These lines can be considered as a family of functions $\theta = \theta(\varphi, C)$, where $|C| \leq \sqrt{1 + \text{ctg}^2 2\theta_{tw}}$. The maximum of the function $F_{tw}(\varphi, \theta)$ is achieved at $\varphi = 0$ and $\theta = \theta_{max} \approx 2{,}5384°$. The absolute minimum is at $\varphi = \pi$ and $\theta = \theta_{min} \approx 87{,}4616°$. Twinning is thus possible on both tension and compression if

$$|p| \leq p_{tw} \equiv \sqrt{1 + \text{ctg}^2 2\theta_{tw}} \approx 1{,}003839.$$

The grains in which twinning occurred at least in one twinning system are here referred to as twin grains. To find out whether the grain is twinned at a preassigned parameter p, the fulfilment of the criterion of Eqs. (8) should be verified for each of the six twinning systems available in this grain. The grain whose orientation is determined by the angles φ and θ is twinned if there is at least one point of the set (φ_k, θ) in the region Ω_2 which does not overlie the constant level line $\theta(\varphi, p)$. For each p this condition dictates a certain region of φ and θ angles Ω_p, in Ω_1, where the grains undergo twinning.

The relative number of twinned grains is given by the integral

$$n(p) = \int_{\Omega_p} f(\varphi, \theta) d\theta \, d\varphi. \quad (10)$$

As the analysis shows, the Ω_p region is essentially dependent on the p-value. The random distribution of the grain orientations is considered below with the density $f(\varphi, \theta) = 3\pi^{-1}\sin\theta$.

The twinning evolution in the sample subjected to uniaxial compression ($p > 0$) is considered here only from the standpoint of fulfilment of the twinning criterion for each of the six systems in each grain. For definiteness, we assume that the external load σ onto the sample is increasing

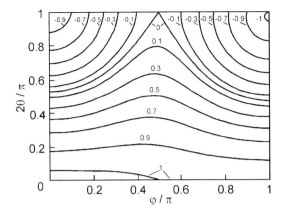

Figure 2: Equal level lines $F_{tw}(\varphi, \theta) = C$. C values are indicated near the corresponding curves.

gradually. As a result, the formal model parameter p describing the criterion of the twinning onset decreases. These regularities also hold in our model if variations of p are caused by all three factors σ, σ_{tw} and σ_{th}.

(1) $p > p_{tw}$. Twinning in impossible in any twinning system, in any grain.
(2) $p_1 \approx 1{,}003655 \leq p \leq p_{tw}$. The criterion is fulfilled in the favourably oriented grains in the only one twinning system ($k = 1$, $k = 6$).
(3) $p_2 \approx 1{,}002626 \leq p \leq p_1$. The number of grains undergoing twinning in the systems 1 or 6 increases. Some grains twinned by one of these systems undergo twinning again by the other system. The total relative number of twinned grains $n(p)$ is given by the formulae:

$$n(p) = \frac{6}{\pi} \int_0^{\pi/6} [\cos\theta_-(\varphi,p) - \cos\theta_+(\varphi,p)] \, d\varphi \tag{11}$$

where

$$\theta_\pm(\varphi,p) = \arcsin\left[\frac{(1-p)(1+\cos^2\varphi) + 2\operatorname{ctg}^2 2\theta_{tw}\cos^2\varphi \pm \sqrt{D}}{(1+\cos^2\varphi)^2 + 4\operatorname{ctg}^2 2\theta_{tw}\cdot\cos^2\varphi}\right]^{1/2} \tag{12}$$

$$D = 4\operatorname{ctg}^2 2\theta_{tw}\cos^2\varphi\,[\,p(1-p) + (1-p+\operatorname{ctg}^2 2\theta_{tw})\cos^2\varphi\,]. \tag{13}$$

(4) $1 \leq p \leq p_2$. Some grains are twinned by system 1 or 6, in some grains twinning occurs in systems 1 and 6. In some grains twinned by systems 1 and 6, one more system 2 or 5 comes into effect. The total relative number of twin grains is determined by the same Eqs. (11)-(13).
For example, at $p = 1$ we have $n(1) = 0{,}0039$.
(5) $p_3 \approx 0{,}997 \leq p < 1$. In this interval of p new grains become twinned by one of the systems (1 or 6). In addition to the grains twinned by two (1 and 6), three (1, 6, 2 or 1, 6, 5) systems, there appear grains in which twinning involves more systems: four (1, 2, 5, 6), five (1, 2, 3, 5, 6 or 1, 2, 4, 5, 6) and six (1–6) systems. The number of twinned grains $n(p)$ is given by

$$n(p) = 1 - \frac{6}{\pi} \int_0^{\pi/6} \cos\theta_+(\varphi, p) d\varphi$$

where $\theta_+(\varphi, p)$ is determined by Eqs. (12)–(13).

(6) $p_4 \approx 0{,}9596 \leq p < p_3$. The new grains are twinned by one of systems 1, 2, 5, 6.

(7) $0 \leq p < p_4$. In this p-interval grains are twinned by one system (2 or 5). In this region

$$n(p) = 1 - \frac{6}{\pi} \int_{\pi/3}^{\pi/2} \cos\theta_+(\varphi, p) d\varphi$$

where $\theta_+(\varphi, p)$ is determined by Eqs. (12)–(13).

The highest possible relative number of twinned grains under compression corresponds to the $p = 0$ case with $n(0) = 0.7772$.

4   Conclusions

The MTAS effect on initiation of twinning is dependent on their value and the sign. Positive σ_{th} decrease the parameter p at assignment σ and σ_{tw}; hence, they increase the number of twinned grains in the sample. Negative σ_{th} increase the parameter p and hence they decrease the number of twinned grains. As follows from Eqs. (1)–(2), the signs of $\sigma_{33}^{th(c)}$, $\sigma_{22}^{th(c)}$, σ_{th} are directly dependent on the sign of the difference between the thermal expansion coefficients ($\alpha_3 - \alpha_1$). As was shown above, in iodide titanium $\alpha_3 < \alpha_1$ (a similar correlation of α_3 and α_1 was found in [7]) and $\sigma_{th} < 0$. At the same time, in Ti presumably containing oxygen and nitrogen impurities we have $\alpha_3 > \alpha_1$ [4, 7]. We can expect that in such titanium $\sigma_{th} > 0$ and MTAS should be favourable for (and even stimulating) twinning on cooling below 300 K.

In the frame of [1] the MTAS values are independent on grain size d if elastic constants of a material and thermal expansion coefficients are not dependent on d. Therefore in a nanostructured and in a polycrystalline material these stresses would be the same.

But the defect structure of grain boundaries of nanostructured material processed by severe deformation is substantially different from an equilibrium grain boundary structure [8]. Repeated crossing of grain boundaries by plastic shears in the course of severe deformation lead to arising of a large density of glide and sessile grain boundary dislocations. Free shifting of glide grain boundary dislocations along grain boundaries, which bordering each grain, can ensure partial relaxation of thermal anisotropy stresses arising under a temperature change. Therefore in a nanostructured material (specifically in Ti) processed by severe deformation one, generally speaking, can expect some decreasing of values of thermal anisotropy stresses (at equal other conditions) comparative to polycrystals.

But final judgement about values of thermal anisotropy stresses in Ti processed by severe deformation needs experimental data on its elastic constants and thermal expansion coefficients because in nanostructured metals these characteristics can be substantially different from polycrystalline ones [8].

5 Acknowledgement

Authors are thankful to Dr. E. D. Tabachnikova for promoting this work.
This work has been partially supported by the INTAS 01 - 0320 Project.

6 References

[1] V. A. Likhatchev, Fizika tverdogo tela 1961, 3, 1827–1834
[2] W. Boas, R. W. K. Honeycombe, Proc. Royal Society 1947, A188, 427–439; J. Inst. Metals 1946–1947, 73, 433–444
[3] E. S. Fisher, C. J. Renken, Phys. Rev. 1964, 135, A482–A494
[4] U. Zwicker, Titan und Titanlegierungen, Springer-Verlag Berlin Heidelberg New York, 1974, p. 511
[5] V. I. Nizhankovsky, M. I. Katznelson, G. V. Peschanskikh, A. V. Trefilov, Pisma v ZhETF 1994, 59, 693–696
[6] E. Schmid, W. Boas, Plasticity of crystals, Hughes F. A. and Co., London, 1950
[7] Ram Rao Pawar, V. T. Deshpande, Acta Cryst. 1968, A24, 316–317
[8] R. Z. Valiev, R. K. Islamgaliev, I. V. Alexandrov, Progress in Materials Science 2000, 45, 103–189

Nano- and Submicrocrystalline Structure Formation During High Pressure Torsion of Al-Sc and Al-Mg-Sc Alloys

Dobatkin S.V.[1,2], Zakharov V.V.[3], Valiev R.Z.[4], Vinogradov A.Yu.[5], Rostova T.D.[3], Krasilnikov N.A.[4], Bastarash E.N.[2], Trubitsyna I.B.[2]

[1]Baikov Institute of Metallurgy and Material Science RAS, Moscow, Russia
[2]Moscow State Steel and Alloys Institute (Technological University), Moscow, Russia
[3]All-Russia Institute of Light Alloys, Moscow, Russia
[4]Ufa State Aviation Technical University, Ufa, Russia
[5]Osaka City University, Osaka, Japan

1　　Abstract

The structure and microhardness after severe plastic deformation (SPD) by torsion under high pressure of binary Al-Sc alloys with contents of Sc-0.14 wt%, 0.4 % and 0.55 %, as well as of industrial Al-Mg-Sc alloys 01515, 01535, 01545 and 01570 are investigated. By SPD of binary Al-Sc alloys, only submicrocrystalline structure is formed. Addition of more than 4 % Mg in Al-Sc alloy results in formation of nanostructure during room temperature SPD. The minimum size of a grain – 40 nm – is revealed in alloy Al-5.9 % Mg-0.3%Sc. The maximum hardening of both Al-Sc and Al-Sc-Mg alloys after SPD by torsion is reached after annealing at 300 °C.

2　　Introduction

It is known that severe plastic deformation (SPD) of aluminum alloys allows to obtain rather massive samples with submicrocrystalline structure with the grain size of more than 100 nm [1–3]. Thus, there is a task of further refinement of structure in aluminum alloys down to nano-scale and increase of their thermal stability. In aluminum alloys, this is especially important for realization of low-temperature and high-speed superplasticity. Probably, the task can be solved by means of phase transformations during SPD [4, 5] and additional alloying, in particular, with Sc [6–11] Alloying of aluminum alloys with Sc has a strong influence on their structure and properties.

The purpose of the present work is to study the structure and properties of aluminum alloys alloyed with Sc after SPD by high pressure torsion.

3　　Experimental Procedure

The object of study were binary Al-Sc and industrial Al-Mg-Sc alloys. The binary alloys were produced by continuous casting from aluminum 99.99 % with 0.14; 0.4 and 0.55 wt% Sc (Table 1). The impurity content was as follows: 0.02 % Fe; 0.01 % Si; 0.06 % Ti. The following industrial Al-Mg-Sc alloys were also investigated [8]: 01570, 01545, 01535 and 01515 (Table 2). The alloys basically differ by the content of Mg, though the content of Sc and Zr also differ a little. The initial state of Al-Mg-Sc alloys: cold-rolled sheets after subsequent annealing.

Deformation was performed by torsion under 6 GPa pressure in a Bridgman anvil. The samples of 10 mm in diameter and 0.6 mm in thickness were deformed by torsion to 5 revolutions which corresponds to strain of $\varepsilon = 5.8$ at different temperatures (Fig.1, left hand side) and subsequent heating (Fig.1, right hand side). Structure analysis was carried out using the transmission electron microscope JEM-100CX. Microhardness was defined using the PMT-3 device with 20 g and 50 g load.

Table 1: Chemical composition of Al-Sc alloys (wt%)

Alloy	Sc	Fe	Si	Ti
Al-0,14%Sc	0,14	0,02	0,01	0,06
Al-0,4%Sc	0,4	0,02	0,01	0,06
Al-0,55%Sc	0,55	0,02	0,01	0,06

Table 2: Chemical composition of Al-Mg-Sc-Zr alloys (wt%)

Alloy	Mg	Sc	Zr
01515	3,1	0,31	0,08
01535	4,1	0,16	0,13
01545	5,1	0,22	0,18
01570	5,9	0,3	0,08

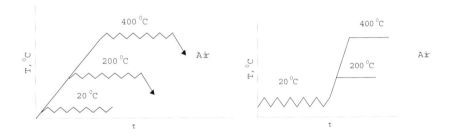

Figure 1: Processing routes for the alloys investigated

4 Results and Discussions

In pure aluminum after severe plastic deformation (SPD) by torsion at room temperature the grain structure is formed with the average grain size of 660 nm (Fig.2). This structure is usually defined as submicron one. High angle misorienation of grains was observed using the micrographs showing double contrast on grain boundaries.

In binary Al-Sc alloys, with increase of the Sc contents the grain size after SPD monotonously decreases: 570 nm for Al-0.14%Sc, 370 nm for Al-0.4%Sc and 280 nm for Al-0.55%Sc (Fig.2, 3).

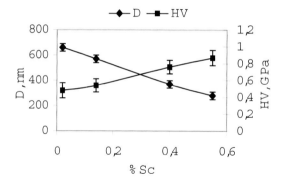

Figure 2: Grain size and microhardness of Al-Sc alloys after SPD at 20 °C

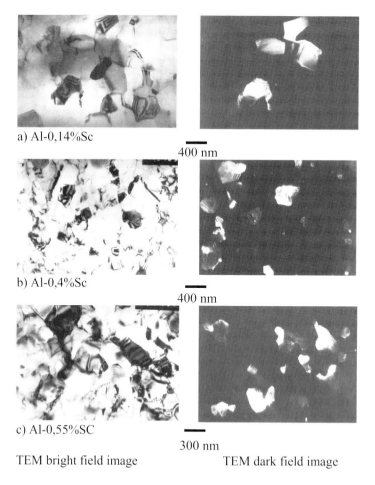

a) Al-0,14%Sc

400 nm

b) Al-0,4%Sc

400 nm

c) Al-0,55%SC

300 nm

TEM bright field image   TEM dark field image

Figure 3: Structure of Al-Sc alloys after SPD at 20 °C

But, as can be seen, the grain size is in submicron range. As primary particles Al$_3$Sc after crystallization in a cast state were absent, and precipitation of particles Al$_3$Sc during deformation could not be determined by electron microscopy, it can be assumed the reduction of the grain size is connected with increasing Sc concentration in a solid solution of Al. However, the possibility of partial precipitation of highly disperse particles Al$_3$Sc during deformation can not be ruled out.

The microhardness of binary Al-Sc alloys after SPD is much higher than in the as cast state and it increases with increasing Sc content (Fig.2).

There is evidence of an aging process of binary Al-Sc alloys when they are heated, as suggested by an increase of microhardness due to precipitation dispersion particles of Al$_3$Sc (Fig.4). In the Al-0.14%Sc alloy the effect of aging is small. The aging is more effective in alloys with a higher content of Sc. It reaches a maximum for heating at 300 °C.

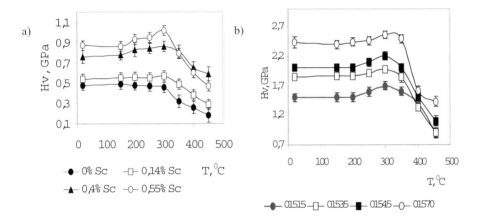

Figure 4: Dependence of microhardness on the annealing temperature after SPD: a) Al-Sc alloys, b) Al-Mg-Sc alloys

Despite a fine grain after SPD and the aging effect, binary Al-Sc alloys have low hardness. For example, in strongest Al-0.55%Sc alloy the microhardness is hardly more than 1 GPa (Fig.4). Therefore Al-Mg-Sc alloys with various Mg content processed by SPD were investigated.

The microhardness of all Al-Mg-Sc alloys after SPD is higher than in Al-Sc alloys. It increases with increase of the Mg content (Fig.4). The values of microhardness after SPD are higher than in the initial state by a factor of 1.5. Aging during annealing occurs in the same interval as in binary Al-Sc alloys: it begins at 250 °C and the maximum effect is observed at 300 °C. Heating of Al-Mg-Sc alloys in the initial state also revealed aging in the same temperature interval.

The observed effect of aging during annealing both for the as received material, and after SPD in torsion at room temperature testifies, first, that in heating of as-cast structure, heating under hot-rolling and heating after cold rolling the decomposition of a solid solution of Sc is not complete, part of Sc remaining in solid solution, and, secondly, that during SPD at a room temperature there is no or small precipitation of Al$_3$Sc particles, and basically Sc remains in solid solution in Al.

Electron microscopy investigation has revealed nanocrystalline structure in the alloy 01570 after SPD by torsion at $T_{def} = 20\,°C$ and $\varepsilon = 5.8$ (5 revolutions) with the average grain size of 40 nm and mainly high angle boundaries. On reduction of the Mg content in alloys the grain size increased to 60 nm in 01545 alloy, 100 nm in 01535 alloy and 150 nm in 01515 alloy (Fig.5). Earlier, a reduction of average grain size within the range of 0.7 to 0.2 µm was observed for Al-0.2% Sc alloys containing from 0 to 3 % Mg severely deformed by equal channel angular pressing [9].

The heating of Al-Mg-Sc alloys up to 200 °C after SPD does not result in any appreciable change of the grain size and the microhardness. For example, for the alloy 01570 grains grow up from 50 to 90 nm, remaining in the nanoscale range.

For heating at 300 °C grains in all alloys grow up to 400–600 nm. This temperature corresponds to the maximum hardening. This means that hardening by means of precipitation of dispersion particles Al$_3$Sc compensates for the softening by grain growth. Annealing at 400 °C causes a further grain growth up to 800–1000 nm. Hence, it is proved that nanocrystalline structure (the grain size < 100 nm) is formed by SPD at $T_{def} = 20\,°C$ in Al-Mg-Sc alloys with the content of Mg not less than 4 %.

For obtaining maximum hardening it is thus necessary to deform at room temperature and to heat up on 300 °C for the occurrence of aging. Increase of the content of Sc and Mg in the investigated limits promotes hardening.

To obtain an ultra fine grained (UFG) structure, stable to the subsequent heating under superplasticity, we have tried warm and hot large strain deformation of alloy 01570. After SPD at 200 °C the grain size in alloy 01570 is increased a little, but remains in the nanoscale range. The microhardness practically does not vary. After deformation at 400 °C the grain size of about 150 nm was obtained, while annealing at this temperature of the material deformed at room temperature (D = 40 nm), resulted in an increase of the grain size to 800 nm. Probably the small grain size for deformation with $T = 400\,°C$ can only be explained by decomposition of the solid solution during deformation.

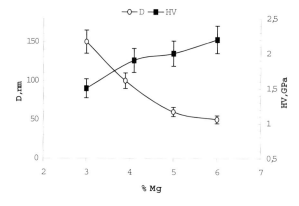

Figure 5: Grain size and microhardness of Al-Mg-Sc alloys after SPD at 20 °C

Preliminary hot large strain deformation tests at the temperature of superplastic forming allow to obtain a grain structure that should be stable at short-term reheats at the same temperatures. The grain size in this case passes from nano-scale to submicron-scale, but remains smaller than after heating of nanostructure obtained by cold SPD to the same temperature. Decomposition of supersaturated solid solution could give additional grain refinement. The goal is to introduce an increased quantity of Sc for making anomalous supersatureted solid solution. Decomposition of this solid solution can be initiated both during the deformation and during the subsequent heating. One of processing routes for obtaining ultrafine grained structure at high temperature includes solution treatment at a temperature near the melting point in order to maximize the amount of scandium contained in solid solution, cold SPD and subsequent heating [10,11]. We could suggest another schedule: rapid crystallization for retaining Sc in solid solution, deformation of as-cast structure and hot large strain deformation with precipitation of Al_3Sc particles.

5 Conclusions

(1) In binary Al-Sc alloys subjected to SPD by torsion at a room temperature a submicrocrystalline structure is formed. The grain size decreases with increase of the Sc content and reaches a minimum value of 280 nm at 0.55 % Sc.
(2) Addition of Mg to binary Al-Sc alloy results in nanocrystalline structure formation by SPD in torsion with a minimum grain size of 40 nm in Al-5.9 % Mg-0.3 % Sc alloy.
(3) The maximum hardening of both Al-Sc and Al-Sc-Mg alloys due to SPD by torsion is reached after aging by annealing at 300 °C.
(4) During hot SPD by torsion of Al-Mg-Sc alloy 01570 at = 400 °C an UFG structure with grain size of 150 nm is formed, apparently by virtue of decomposition of a solid solution.

6 References

[1] R.Z.Valiev, N.A.Krasilnikov, N.K.Tsenev, Mater.Sci.Eng., 1991,A137, 35–40
[2] J.Wang, Z.Horita, M.Furukawa and T.G.Langdon., J.Mater.Res., 1993,Vol.8, 11, 2810–2821
[3] Y.Iwahashi, Z.Horita, M.Nemoto, and T.G.Langdon, Acta Mater, 1998,.Vol.46, 3317–3331
[4] A.V.Korznikov, Yu.V.Ivanisenko, D.V.Laptionok et al., Nanostructured Materials, 1994, Vol. 4, 159–168
[5] V.V.Sagaradze, V.A.Shabashov, A.G.Mukoseev, N.L.Petcherkina, in "The Structure and Properties of Nanocrystalline Materials",(Eds.N.Noskova and G.Taluts), RAS, Ekaterinburg, Russia, 1999,46–63 (in Russian)
[6] T.D. Rostova, V.G. Davydov, V.I. Yelagin and V.V. Zakharov, Materials Science Forum, 2000,Vol.331-337, Part 2, 793–798
[7] V.G. Davydov, T.D. Rostova, V.V. Zakharov et al., Mat. Sci. Eng., 2000, A 280, 30-36.
[8] Yu.A.Filatov, J.of Advanced Mat., 1995, 5, 386–390

[9] M.Furukawa, A.Utsunomiya, K.Matsubara, Z.Horita and T.G.Langdon, Acta Mater. 2001, vol.49, pp. 3829–3838

[10] Z.Horita, M.Furukawa, M.Nemoto, A.J.Barnes and T.G.Langdon, Acta Mater. 2000, 48, 3633–3640

[11] S.Komura, M.Furukawa, Z.Horita, M.Nemoto, T.G.Langdon, Mat. Sci. Eng., 2001,A 297, 111–118

Phase Transformation in Crystalline and Amorphous Rapidly Quenched Nd-Fe-B Alloys under SPD

D.V. Gunderov [1], A.G. Popov [2], N.N. Schegoleva[2], V.V. Stolyarov[1], A.R. Yavary [3]

[1]Institute of physics of advanced materials, USATU, Ufa, Russia
[2]Institute of physics of metals, UrD RAS, Ekaterinburg, Russia
[3]National polytechnic institute, Grenoble, France

1 Introduction

In recent years we studied the influence of severe plastic deformation (SPD) by means of high-pressure torsion (HPT) on the structure and magnetic hysteresis properties of R-Fe-B (R-Nd, Pr) alloys with the main intermetallic phase $R_2Fe_{14}B$ [1–3]. These alloys have a great scientific and practical importance as the materials for permanent magnets with record properties [4,5]. The investigations showed that HPT leads to the decomposition of $R_2Fe_{14}B$ phase into amorphous Nd-rich phase and N -Fe phase [1–3]. On the other hand in the result of rapid quenching (RQ) one can succeed in making the homogeneous amorphous structure in these alloys [4]. In this connection we consider that studying structures of RQ amorphous alloy Nd-Fe-B subjected to SPD is of great interest. This will allow to define the dependence of the structure forming during SPD on the initial structure of the deformed material.

2 Materials and Methods

The coarse-grain $Nd_{11.8}Fe_{82.3}B_{5.9}$ alloy of stoichiometric $Nd_2Fe_{14}B$ composition was cast in the smelting furnace and then homogenized at $T = 1050$ °C during 10 hours. The amorphous alloy of the similar composition was obtained by rapid quenching on a rotating disk at the cooling rate of 10^6 K/s.

The powder prepared from RQ material and the plate cut from the coarse-grained alloy were subjected to HPT at room temperature and pressure $p = 5$ GPa. As a result of such processing 100%-density samples with 6–10 mm in diameter and 0,2 mm in thickness were obtained both from the cast alloy and RQ powder. The strain degree was changed by the angle of anvils rotation φ from 0 to 16π. Annealing of samples was performed in vacuum furnace at 10^{-3} Pa and at temperatures from 300 to 900 °C. Transmission electron microscopy (TEM) was carried out using the electronic microscope JEM-200CX. Demagnetization curves were measured by means of vibration sample magnetometer in the field up to 1,6 MA/m. Before the measurements the samples were magnetized in a pulse field of 5 MA/m. Temperature dependence of magnetization was measured in the field of 0,72 MA/m and in the temperature range from 20 to 800 °C.

3 Results and Discussion

The coarse-grained alloy in the initial state has the microstructure of $Nd_2Fe_{14}B$ grains with the mean size of 100 µm. After HPT the regions of refined $Nd_2Fe_{14}B$ grains co-exist with the regions where α-Fe grains of 20 nm, surrounded by amorphous phase are observed (Fig. 1). In these regions $Nd_2Fe_{14}B$ phase decomposed.

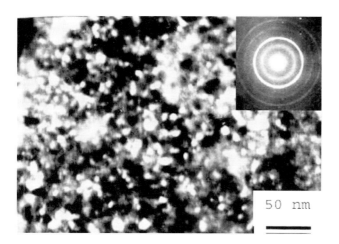

Figure 1: Dark field image in α-Fe reflex of the coarse-grained Nd-Fe-B alloy, HPT $\varphi = 16\pi$

Homogeneous amorphous structure (Fig. 2a) is typical of the rapidly quenched alloy. Crystallites are not observed in the dark field image. The microstructure of RQ alloy after HPT at $\varphi = 16\pi$ changes considerably (Figure 2b). Nanocrystalline -Fe grains with the size of about 10 nm appear in amorphous phase.

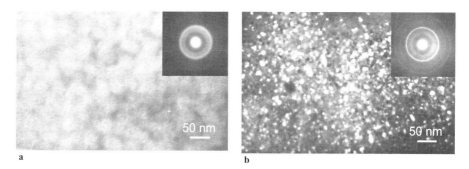

Figure 2: Dark field image in α-Fe reflex of the RQ $Nd_{11.8}Fe_{82.3}B_{5.9}$ alloy, a) initial RQ state b) HPT $\varphi = 16\pi$

Fig. 3 shows the temperature dependence of specific magnetization σ for coarse-grained alloy before (1) and after HPT (2,3). The increase of strain results in the growth of contribution

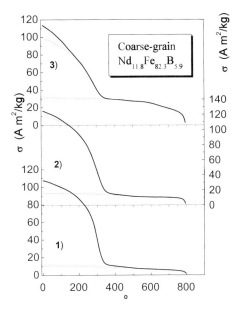
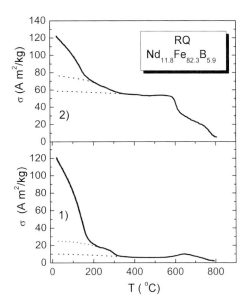

Figure 3: Temperature dependence of magnetization σ for the coarse-grained Nd-Fe-B alloy 1) initial state; 2) HPT, $\varphi = 5\pi$; 3) HPT, $\varphi = 16\pi$

Figure 4: Temperature dependence of magnetization σ for RQ Nd-Fe-B alloy 1) as-RQ state; 2) HPT, $\varphi = 12\pi$

into total magnetization from -Fe (Curie temperature $T_C = 773$ °C) and in the reducing contribution from $Nd_2Fe_{14}B$ phase ($T_C = 310$ °C). The additional kink appears on the curves σ(T) at a temperature about 150 °C corresponding to the Curie temperature of amorphous Nd-Fe phase [4,5]. Thus, TMA testifies that in result of HPT $Nd_2Fe_{14}B$ phase decomposes into α-Fe and amorphous phase.

The phase composition of the samples was calculated according to the additive contribution of the phases into total magnetization. The next simple equations are true:

$$\sigma = \sigma_T V_T + \sigma_{Fe} V_{Fe} + \sigma_A V_A ; V + V_{Fe} + V = 1$$

where σ, σ_{Fe}, σ_A are the values of magnetization at 20 ° and V_T, V_{Fe}, V_A are the volume fractions of $Nd_2Fe_{14}B$, α-Fe and amorphous phases, correspondingly. The σ_{Fe} and σ_T values are equal to 220 and 100 Am²/kg for the measurement field = 720 kA/m [5].

Figure 4 shows dependence of magnetization on temperature for the alloy after RQ and HPT at $\varphi = 12\pi$. The calculations the phase compositions of the samples are presented in Table 1. One can see HPT increases the volume fraction of of -Fe and decreases the volume fraction of the amorphous phase (Table 1).

Table 1: Phase composition of the coarse-grained and of RQ $Nd_{11.8}Fe_{82.3}B_{5.9}$ alloys subjected to HPT.

Alloy	Rotation angle (rad/π)	Phase composition, wt %		
		$Nd_2Fe_{14}B$	α-Fe	Amorphous phase
Cast	0	97	3	-
	5	90	7	3
	16	45	22	33
Rapidly quenched	0	14	3	83
	12	10	25	65

Thus, as a result of SPD of both RQ amorphous and coarse-grained Nd-Fe-B alloys the similar microstructures of the composite of the amorphous phase and NC α-Fe phase mixture are formed.

The experiments [6] showed that $R_2Fe_{14}B$ phase during the deformation by intensive milling decomposes into NC α-Fe and amorphous phase. During the intensive milling of the mixture of powders of pure Nd, Fe and B the structure of amorphous phase and NC α-Fe are also formed [7]. Thus, the influence of different methods of SPD on the system of Nd-Fe-B elements in different initial structural states leads to the formation of similar structures – composite out of the amorphous phase and NC α-Fe. We can ascertain that in case of NdFeB alloy the microstructure formed during SPD doesn't depend on the initial structural state of the material.

These phase transformations can be explained in the following way: the increase of defect density and refinement of $Nd_2Fe_{14}B$ grains up to NC-dimensions under SPD raises free energy (F). As a result F becomes larger than the free energy of the amorphous phase (F_a) of the same alloy composition. However, F_a of homogeneous-amorphous phase is larger than the free energy of the composite out of the amorphous phase and NC α-Fe phase. SPD of the coarse-grained alloy results in the decomposition of $Nd_2Fe_{14}B$ phase in accordance with the following scheme:

$Nd_2Fe_{14}B$ phase $\xrightarrow{SPD}$ amorphous phase + NC α-Fe.

At the same time SPD causes crystallization of α-Fe phase in the amorphous rapidly quenched alloy in accordance with the following scheme:

amorphous phase $\xrightarrow{SPD}$ amorphous phase + NC α-F

Probably, SPD leads to introduction energy that provokes nanocrystallization of α-Fe out of amorphous phase both in decomposed $Nd_2Fe_{14}B$ phase and amorphous phase in RQ alloy.

As it was shown earlier [1,2], annealing at $T = 600$ °C HPT samples of coarse-grained alloys results in H_c increasing due to the recovery of $Nd_2Fe_{14}B$ phase in NC-state from the decomposition products.

It is known that annealing of amorphous rapidly quenched Nd-Fe-B alloy leads also to the formation of NC $Nd_2Fe_{14}B$ phase and increasing H_c and formation of hard-magnetic state [4].

Our result shows that the values of H_c and σ_r are 10–30 % higher at annealing of RQ alloys after SPD as compared to non-deformed material (Table 2). Probably SPD changes the dynamics of crystallization of RQ alloys at annealing. It leads to the formation of more homogeneous and fine NC structure of $Nd_2Fe_{14}B$ phase that results in the growth of hysteresis properties. This phenomenon can be important from the practical point of view.

Table 2: Hysteresis properties of coarse-grained and RQ $Nd_{11.8}Fe_{82.3}B_{5.9}$ alloy after HPT and annealing.

State		HPT j, p	HPT			HPT + annealing at 600 °C 10 min		
			Hc (kOe)	sm (Am²/kg)	sr (Am²/kg)	Hc (kOe)	sm (Am²/kg)	sr (Am²/kg)
Coarse-grained	Initial		0.1	110		0.1	110	
	16	0.8		129	59	4.6	121	87
RQ	Initial			122	54	4.2	109	72
	12	0.33		130	29	6.1	119	91

4 Conclusions

In the coarse-grained $Nd_{11.8}Fe_{82.3}B_{5.9}$ alloy SPD results in $Nd_2Fe_{14}B$ phase decomposition into NC -Fe and amorphous phase. The effect of SPD on the initial amorphous rapidly quenched Nd-Fe-B alloy results in crystallization of NC α-Fe and decrease of volume fraction of amorphous phase. Thus, under SPD the similar structures are formed both in coarse-grained Nd-Fe-B alloy and in rapidly quenched amorphous one. SPD of the amorphous RQ alloy enables to increase its magnetic properties after annealing.

5 Acknowledgements

The performed work has been supported by INTAS grant No 99-01741

6 References

[1] V. V. Stolyarov, D. V. Gunderov, R. Z. Valiev, A. G. Popov, V. S. Gaviko, A. S. Ermolenko, JMMM, 1999. Vol. 196–197, 166–168
[2] A. G.Popov, A. S.Ermolenko, V. S.Gaviko, N. N.Schegoleva, V. V.Stolyarov, D. V. Gunderov, Proc. on 16th Int. Workshop on Rare-Earth Magnets and Their Application, Sendai, Japan, 2000, 621–630
[3] V. S. Gaviko, A. G. Popov, A. S. Ermolenko, N. N. Shegoleva, V. V. Stolyarov, D. V. Gunderov, Physics of Metals and Metallography, Vol. 92, 2001, No. 2, 58–66
[4] K. H. J. Buschow, Materials Science Rep. 1986, v. 1
[5] J. Fidler and T. Schrefl, J. Appl. Phys., 1996, v. 79, 5029–5034
[6] V. V. Maikov, A. E. Ermakov at al, JMMM, 1995, v. 151, 167–172
[7] P. Crespo, V. Neu, L. J. Schultz, Appl. Phys., 1997, v. 30, pp. 2298–2303

Structure and Functional Properties of Ti-Ni-Based Shape Memory Alloys Subjected to Severe Plastic Deformation

I.Yu. Khmelevskaya[1], I.B. Trubitsyna[1], S.D. Prokoshkin[1], S.V. Dobatkin[1,2], V.V. Stolyarov[3], E.A. Prokofjev[3]

[1]Moscow Steel and Alloys Institute, Moscow, Russia
[2]Baikov Institute of Metallurgy and Material Science, Russian Academy of Science, Moscow, Russia
[3]Ufa State Aviation Technical University, Ufa, Russia

1 Introduction

The possibilities to regulate the functional properties of shape memory alloys by means of thermomechanical treatment can be enlarged by using a nanocrystalline structure formed under conditions of severe plastic deformation. For practical purposes, it is important to obtain nano- or submicrocrystalline structures [1–3] in rather large samples. In various alloys this is achieved with the method of equal-channel angular pressing (ECAP). Hence, it was interesting to apply this method to titanium-nickel-based shape memory alloys. In addition, structure formation and its dependence on strain and pressure under conditions of high-pressure torsion (HPT) was studied.

2 Experimental Procedure

Ti-50.0 at.% Ni (alloy 1), Ti-50.7 % Ni (alloy 2) and Ti-47 % Ni-3 %Fe (alloy 3) alloys were subjected to HPT at RT and to ECAP in the temperature range 500–400 °C (the last one for alloys 1 and 2 only) (Table 1). To determine the conditions of nano- or submicrocrystalline structure formation, the samples Ø 10 × 0.6 mm in size (Ø: diameter) were deformed by the HPT (N = 5 revolutions) as well as Ø 3 × 0.2 mm samples (N = 1, 3, 5, 10, 15). ECAP true strains of Ø 16 × 70 mm samples were 2.5 after 3 passes, 6.5 after 8 passes and 9.7 after 12 passes. After HPT, the samples of Ø 10 mm were annealed in the temperature range from 150–500 °C. Structure formation was studied using X-ray diffraction and transmission electron microscopy (TEM). Functional properties (temperature range of shape recovery, recovery strain and recovery strain rate) were determined by heating of the samples after bending at 0 °C around mandrels.

Table 1: Alloys studied

Alloy	Chemical Composition, at.%			Transformation temperatures*, °C					Main phase composition before deformation
	Ti	Ni	Fe	M_s	M_f	A_s	A_f	T_R	
1	50.0	50.0	–	68	55	86	98	–	B19'
2	49.3	50.7	–	–20	–35	5	15	–2	B2 (metastable)
3	50.0	47.0	3.0	–160	–196	–130	–60	–60	B2 (stable)

*M_s, M_f: martensitic transformation starting and finishing temperatures; A_s, A_f: reverse martensitic transformation starting and finishing temperatures; T_R -R-phase formation starting temperature.

3 Results and Discussion

Before HPT, stable martensite and stable austenite structures were observed in alloys 1 and 3, respectively, and the austenite structure in a premartensitic state in alloy 2 (between M_s and M_d, where M_d is the highest martensitic transformation starting temperature under deformation).

Electron microscopy study was conducted first after HPT and after HPT with following annealing at 450 °C of Ø 10 mm samples ($N = 5$) of alloys 1 and 3 (Figs. 1, 2). The following regularities were observed.

In the case of HPT of alloy 1 with initial martensite structure, amorphization takes place. Characteristic contrast (Fig. 1 a) and amorphous haloes in microdiffraction pattern (Fig. 1 b) are observed.

With annealing lower than 450 °C, the amorphized alloy 1 crystallizes, and the nanocrystalline (10–20 nm) structure forms (Figs. 1 c-e).

Figure 1: TEM micrographs and electron diffraction patterns of Ti-50.0 % Ni alloy after HPT. a: directly after HPT; b: electron diffraction from "a"; c: bright field image after HPT and annealing at 450 °C; d: electron diffraction from "c"; e: dark field image of "c" in {110} reflection of austenite

In the case of HPT of alloy 3 with the initial 2-austenite structure, a nanocrystalline (10–20 nm) structure forms as a result of deformation (Figs. 2 a-c). After annealing at 450 °C, the nanocrystalline structure of alloy 3 transforms to the submicrocrystalline (100–150 nm) structure (Figs. 2 d, e).

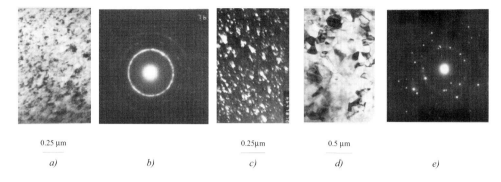

Figure 2: TEM micrographs and diffraction patterns of Ti -47 % Ni –3 % Fe alloy after HPT. *a–c:* directly after HPT; d, e: after HPT and annealing at 450°C; b, e: electron diffraction; a, d: bright field images; c: dark field image of "*a*" in {110} reflection of austenite

ECAP of the alloy 2 at 450 °C was limited by 3 passes corresponding to a true strain of 2.5. As a consequence, only a highly dislocated dynamically recovered substructure formed. In this case, the ECAP strain was limited by strain aging defining brittleness of the material. However, already in this condition, interesting functional properties were obtained [4].

On the contrary, the non-aging alloy 1 sustained ECAP in 12 passes at 500 °C and in 8 passes at 400 °C. Electron microscopy study shows that submicrocrystalline structures (0.3 to 0.5 μm and 0.1 to 0.3 μm, respectively) form containing high dislocation densities (Fig. 3 *a–d*).

Figure 3: TEM micrographs and electron diffraction patterns of Ti-50.0 % Ni alloy after ECAP. a, b: after ECA at 500 °C; c,d: after ECAP at 400 °C

The diffraction pattern after ECAP at 400 °C is similar to the one typical for Ti-Ni alloys in the case of annealing after cold deformation in the transitional stage from static polygonization to static recrystallization (compare Fig. 3 *d* with Fig. 4 *b*). Thus, it seems that the mechanism of submicrocrystalline structure formation is recrystallization, i.e. gradual accumulation of sub-grain misorientations up to high-angle ones.

A fully recoverable strain of 7 % obtained after ECAP at 400 °C (Fig. 5) corresponds to the best results obtained after cold deformation followed by polygonizing heating at 400–450 °C for the same Ti–50.0%Ni alloy. The temperature range of shape recovery (Fig. 5) is lower and narrower than that after the latter reference treatment.

Figure 4: TEM micrographs (*a*) and electron diffraction pattern (*b*) of Ti–50.7 % Ni alloy subjected to cold rolling with 25 % reduction and followed by post-deformation annealing at 550 °C (partially polygonized and recrystallized structure) [4]

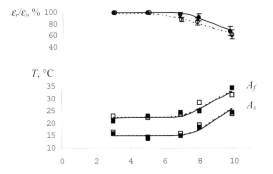

Figure 5: ε_r to ε_i ratio (above) and temperatures A_s and A_f (below) vs induced strain ε_i (ε_r is the recovery strain). ■,● - ECAP at 400 °C; □, ○ - ECAP at 500 °C

For estimating the effects of strain and pressure on nanocrystalline and amorphized structures formation, X-ray and TEM studies of the alloys 1, 2, and 3 have been performed using HPT of Ø 3 × 0.2 mm samples at N = 1, 3, 5, 10 and 15 revolutions and 4 to 8 GPa pressures.

X-ray line width B in $\{110\}_{B2}$ neighbourhood vs N is shown in Fig. 6. B increases as strain increases, and reaches a certain limit $B = 9 - 10°$ which is equal for all three alloys (see Fig. 6). It means that finally, after HPT in 10 to 15 revolutions, the alloys obtain the same structure. However, the kinetics of B changing in the different alloys are different: stabilization of B is reached at N = 3, or 5–10, or 10–15 in alloys 1, 2 and 3, respectively.

It is logical to suppose that the regularities of the kinetics of B changing with strain changing (see Fig. 6) are determined by the structure changes in the following sequence: strain-hardened dislocation substructure-nanocrystalline structure-amorphized structure. The structure changes were studied by TEM in more details after HPT in 5 and 10 revolutions. TEM study shows that although the kinetics of the same structure formation processes are different in different alloys, finally, at rather large strain, they lead to the formation of the amorphized structure which correlated with the results of the X-ray study (see Fig. 6).

After HPT, N = 5, under 4 GPa, the amorphized structure occupies large continuous fields in alloy 1 (Figs. 7 *a–c*). The gaps between these fields contain mixed amorphized and nanocrystal-

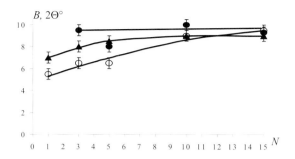

Figure 6: X-ray line width vs N. Full circles and triangles, and empty circles represent the results received in alloys 1, 2, and 3, respectively

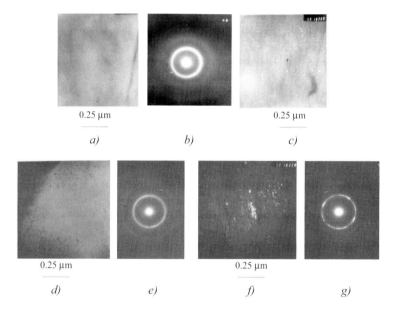

Figure 7: TEM micrographs and electron diffraction patterns of alloys 1 (a–c), 2 (d–f) and 3 (g) after HPT, $N = 5$, $P = 4$ GPa. a, d: bright field images, c, f: dark field images

line structures. The amorphized fields and regions contain also fine dispersed nanocrystalline inclusions (Fig. 7 c).

In alloy 2, under the same conditions of $N = 5$, under 4 GPa, mixed nanocrystalline and amorphized structures are observed in approximately equal fractions (Figs.7 d–f). In alloy 3, after HPT, $N = 5$, only nanocrystalline structure is observed together with traces of strain-hardened substructure (Fig. 7g, compare with Figs.2 a–c).

The structure of all three alloys after HPT, $N = 10$, under 4 GPa, is the amorphized one containing nanocrystalline inclusions (Fig. 8). Thus, a tendency to form an amorphized structure in-

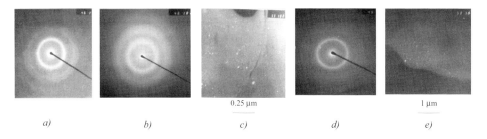

Figure 8: TEM micrographs and electron diffraction patterns of alloys 1 (a), 2 (b, c) and 3 (d, e) after HPT, $N = 10$, $P = 4$ GPa. c, e: dark field images

creases from alloy 3 to alloy 2 and alloy 1. It seems that this depends on relative positions of M_S and deformation temperatures, as follows: first, initially martensitic alloy 1 amorphizes, then metastable austenitic alloy 2, and, finally, the stable austenitic alloy 3.

Change of the pressure from 4 to 8 GPa at HPT, $N = 5$ leads to suppression of the amorphized structure formation in alloy 1: the amorphized areas become narrower and the nanocrystalline structure fraction increases considerably (Figs. 9 a, b). From place to place, the strain-hardened substructure regions are observed. In alloy 2, after HPT, $N = 5$, increasing of the pressure from 4 to 8 GPa reveals the same tendency as for the alloy 1 and after HPT under 8 GPa, only a nanocrystalline structure plus strain-hardened dislocation substructure are observed

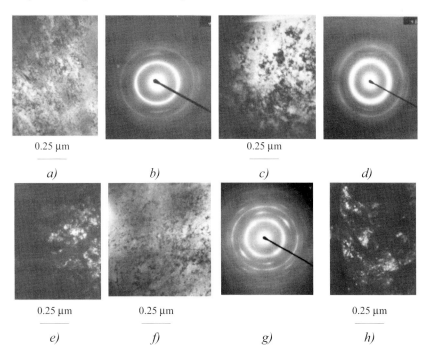

Figure 9: TEM micrographs and electron diffraction patterns of alloys 1 (a, b), 2 (c, d, e) and 3 (f, g, h) after HPT, $N = 5$, $P = 8$ GPa. a, c, f: bright field images, e, h: dark field images

(Figs. 9 c–e). In alloy 3, the tendency manifests itself in a transition from the completely nanocrystalline structure after HPT under 4 GPa to the mixed nanocrystalline structure plus the strain-hardened dislocation substructure after HPT under 8 GPa (Figs. 9 f–h).

4 Conclusions

1. In Ti-Ni-based shape memory alloys, nanocrystalline or amorphized structures can be obtained by severe plastic deformation at room temperature. It seems that the tendency to form amorphized and nanocrystalline structures depends on the relative positions of M_s and deformation temperatures: it is stronger when the deformation temperature is lower than M_s, intermediate when deformation temperature lies between M_s and M_d, and weaker when deformation temperature is higher than M_d.
2. During post-deformation annealing at 450 °C, an amorphized structure forming as a result of HPT crystallizes into a nanocrystalline structure. The nanocrystalline structure forming as a result of HPT transforms to a submicrocrystalline one during post-deformation annealing at 450 °C.
3. Increasing of pressure at HPT suppresses the tendency of amorphized structure formation from the nanocrystalline structure, and of the nanocrystalline structure formation from the strain-hardened dislocation substructure.
4. Equal-channel angular pressing of Ti-50.0 % Ni alloy with true strains up to 9.7 in the range 400 to 500 ºC leads to a submicrocrystalline structure formation. In this case, shape recovery properties are comparable to the ones obtained by polygonizing annealing after cold deformation.

5 Acknowledgements

The present work was performed under support of the INTAS grant 01-0320 and grant 00-15-99083 of Russian Foundation for Basic Research.

6 References

[1] Fedorov V.B., Kurdyumov V.G.., Khakimova D.K. et. al., Doklady AN SSSR 269 (1983) 885–888
[2] Tatyanin E.V., Kurdyumov V.G. and Fedorov V.B., Fiz. Metalloved. 62 (1986) 133–137
[3] Structure and properties of nanocrystalline materials (UrO RAN, Ekaterinburg, 1999) pp. 77–82
[4] Prokoshkin S.D., Khmelevskaya I.Yu., Brailovski V. et. al. Proc. Int. Conf. ICOMAT'02, in press

Formation of Submicrocrystalline Structure in the Hard Magnetic Alloy Fe-15wt.%Co-25%Cr during Straining by Complex Loading

G. F. Korznikova[1], A. V. Korneva[1], A. V. Korznikov[2,1]
[1] Institute of Metals Superplasticity Problems, Russian Academy of Sciences, Ufa, Russia.
[2] Austrian research centre, Seibersdorf research GmbH

1 Introduction

Ultrafine-grained materials attract great attention of investigators because of their specific physical and mechanical properties [1, 2]. Severe plastic deformation (SPD) is one of the efficient techniques for processing bulk submicrocrystalline (SMC) materials. In particular, torsion under quasi-hydrostatic pressure [2, 3] provides deformation without failure of both pure metals and brittle intermetallics [4]. In ductile and less hardenable metals and alloys a structure with a grain size of 100–200 nm is formed [4], whereas the size of grains that are formed during plastic deformation in intermetallics, alloys with insoluble elements, and metal-metalloid alloys is 10–20 nm [2, 4].

Severe plastic deformation also strongly influences the phase composition of alloys [2, 4]. It was shown [4–6] that the second phases often dissolve during severe plastic deformation to form solute concentrations in the solid solutions significantly exceeding the solubility limits of these elements at low temperatures.

The Fe-15wt.%Co-25%Cr alloy belongs to hard magnetic materials of the precipitation-hardening class [7]. Magnets are produced from this alloy by both casting and plastic working methods. The formation of a high-coercivity state during spinodal decomposition leads to a sharp decrease in the strength and plasticity characteristics due to a modulated structure consisting of coherent ordered precipitates of the α_1 phase in the α_2 matrix. It is known that the plasticity characteristics of industrial alloys can be significantly increased by changing the size and morphology of ordered phases [4, 8]. Previously it was reported that room temperature severe plastic deformation by high-pressure torsion on Bridgman anvils of Fe-15wt.%Co-25%Cr alloy in its high-coercivity state results in the dissolution of the α_1 and α_2 phases and the formation of a supersaturated solid solution with a grain size of about 50 nm [9]. It was also revealed that the dependence of the mechanical properties on the degree of deformation in the high-coercivity state is nonmonotonic. The maximum plasticity is characteristic of the alloy with a mixed SMC and cellular structure. The formation of a nanocrystalline structure results in a decrease in the plasticity [9].

The drawback of the samples processed by means of Bridgman anvil techniques is their small thickness (less than 1 mm) whereas complex loading at elevated temperatures enables to produce massive samples. The purpose of this work was to study the evolution of the structure and mechanical properties of the Fe-15wt.%Co-25%Cr alloy during complex loading at elevated temperatures by upsetting and subsequent torsion.

2   Experimental

The Fe-15wt.%Co-25%Cr alloy consists of 25 wt% Cr, 15 % Co, 1 % Ti, 1 % V, 0.4 % Si, 1 % Al, 1 % Nb, and Fe balance. The ingot of the cast alloy was water-quenched from 1200°C to obtain a single-phase α solid solution. Four cylindrical samples 12 mm in diameter and 10 mm in height were cut from the ingot and then subjected to severe plastic deformation by complex two-step loading at constant temperature. Different temperatures of deformation (700, 750, 800 and 850 °C) were chosen from the range of two phase (α+γ) domain. At the first step of loading (figure 1a) the samples were subjected to upsetting at the strain rate $4 \cdot 10^{-4}$ s^{-1} and at the second step (figure 1b) the torsion of a lower anvil was applied to the samples (strain rate $4 \cdot 10^{-2}$ s^{-1} at a distance $R/2$ from the center). The amount of strain in upsetting and torsion was estimated using the equation (1) proposed by Degtyarev [10].

$$e = e_1 + e_2 = \ln(h_o / h_{iR}) + \ln(1 + (\varphi^* R / h_{iR})^2)^{1/2} \quad (1)$$

where h_o is the initial height, h_{iR} is the height of the sample after processing at a distance R from the center, φ is an angle of rotation of a mobile anvil.

Pure compression of the sample results in a strain $e_1 = 1.4$. The contribution of torsion to the total amount of strain at a distance $R/2$ from the center is $e_2 = 4.5$.

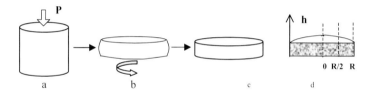

Figure 1: Schematic illustration of the processing of the samples

Microhardness was measured according to the Vickers method with a load of 0.2 kg on the cross section of the samples along the dashed lines indicated in figure 1d. The microstructure of the samples was examined in JXA 6400 scanning electron microscope (SEM). Thin foils were examined by JSM 2000 EX transmission electron microscope (TEM), operating at 150 kV.

3   Results and Discussion

3.1. The Structure of the Fe-15wt.%Co-25%Cr Alloy

The alloy Fe-15wt.%Co-25%Cr after annealing in the temperature range 750–950 °C represents two-phase structure with γ lamellae in coarse α grains. An example of non-deformed structure is given in figure 2.

The application of plastic deformation by upsetting and consequent torsion at all the temperatures under study causes transformation of the lamella structure into a granular one. It is necessary to note that the structure of the processed samples is not uniform (figure 3a). The degree of

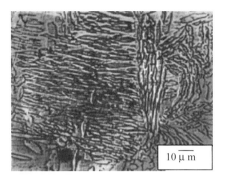

Figure 2: The microstructure of non-deformed Fe-15wt.%Co-25%Cr alloy after annealing at 850 °C

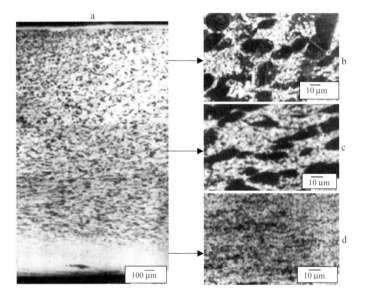

Figure 3: Panoramic image of the cross-section of the sample of Fe-15wt.%Co-25%Cr alloy after complex loading at 700 °C (a) and detailed images of microstructure of the top (b), middle (c) and bottom (d) parts of the sample

deformation is lowest obviously in the top part of the sample which was close to immobile anvil (figure 3b). The grains of α-phase are slightly elongated, some of γ-phase lamellae are broken. The structure in the middle part of the sample consists of markedly elongated coarse grains of α-phase surrounded by small grains and subgrains of γ-phase.

The most significant refinement of the structure occurs in the layer located in the vicinity of the mobile as illustrated in figure 3 (d) for a sample deformed at 700 °C (the bottom array). TEM observation of the foil, prepared from the bottom part of the sample revealed microduplex type SMC structure with equaxial globular grains of α and γ-phases with a mean size less than 500 nm (figure 4).

Figure 4: TEM image of the layer with SMC structure of the sample shown on figure 2 d

Essentially the structure of the samples processed at higher temperatures is similar to that described above. The maximum size of α grains (about 100 μm) was observed on the top part of the sample processed at 850 °C.

The thickness of the layer with SMC structure in the bottom part of the sample was measured as varies from 570 μm at 700 °C to 700 μm at 850 °C (table 1). As is seen from the table the volume fraction of SMC structure is increasing with the rise of the temperature of loading.

Table 1: The thickness of SMC layer in the samples of Fe-15wt.%Co-25%Cr alloy after loading at different temperatures

Temperature of deformation	The thickness of SMC layer (mm)	The thickness of the sample (mm)	The fraction of SMC structure
700°C	570	3800	0.15
750°C	700	3800	0.19
800°C	700	3000	0.23
850°C	700	2700	0.26

3.2 Microhardness of the Fe-15wt.%Co-25%Cr Alloy

On the whole the value of microhardness is not uniform in all the samples (figure 5). The maximum value of Hv was observed in the array close to the bottom of all the samples (H from 0 to 1 mm) especially that deformed at 700 and 750 °C. According structure analysis these areas are characterized by SMC microduplex structure (figure 5) suggesting that amount of strain was maximum near the mobile anvil. In the middle part of the samples (H from 1 to 2 mm), where the microstructure comprises a mixture of SMC γ and significantly elongated α grains, the Hv value decreases. In the top layer of the samples located near immobile anvil the Hv value is lowest and caused obviously by a coarse structure with a slight effect of deformation. In the samples deformed at 800 and 850 °C the distribution of Hv seems to be more homogeneous that is probably caused by more uniform deformation of the samples due to some rise of ductility on temperature increasing [4].

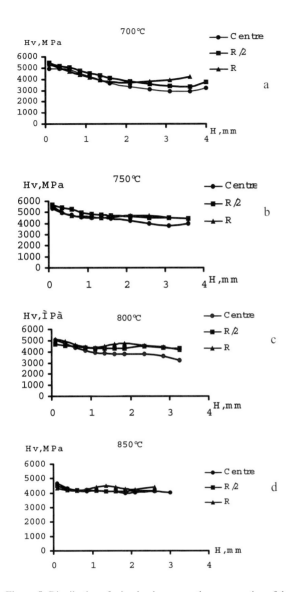

Figure 5: Distribution of microhardness over the cross section of the samples of Fe-15wt.%Co-25%Cr alloy, processed at different temperatures: 700 °C (a), 750 °C (b), 800 °C (c), 850 °C (d). H is the distance from the bottom of the sample.

It is necessary to note that microhardness is dependent also on the distance from the mobile anvil and only slightly on the distance from the axis of the sample, i. e. the curves $Hv(H)$ corresponding the center of the sample, $R/2$ and R differ only slightly. By contrast, as it was previously reported [9] the microhardness value of the Fe-15wt.%Co-25%Cr samples processed by Bridgman anvil type unit at room temperature depends significantly on the amount of deformation and on the distance from the center of the sample. Such a difference is probably a result of lower value of yield stresses at elevated temperatures. This fact gives no way of reducing the grain size down to nanometer scale as in the case of processing at room temperature but makes it possible to produce massive samples with uniform SMC layer.

4 Conclusions

Complex two-step loading of alloy Fe-15wt% Co-25% Cr including upsetting and torsion provides transformation of a coarse lamella structure into granular submicrocrystalline one.

The maximum structure refinement occurs in the layer in the vicinity of mobile anvil. The thickness of SMC layer slightly depends on temperature.

The process may be used for surface treatment (hardening) of bulk specimens.

5 Acknowledgements

The financial support of Austrian Research Centre Seibersdorf is gratefully acknowledged by A. Korznikov.

6 References

[1] Gleiter, H., Nanostructured Materials: Basic Concept and Microstructure, Acta Mater, 2000, 48, p.1–29
[2] Valiev R.Z., Islamgaliev R.K., Alexandrov I.V., Prog.Mater.Sci. 2000, 46, p.103
[3] Bridgman, P.W., Studies in Large Plastic Flow and Fracture, New York: McGraw-Hill, 1952; Effect of High Shearing Stress Combined with High Hydrostatic Pressure, Phys. Rev., 1935, 48, no. 15, p. 825–847
[4] Korznikov, A.V., Structure and Mechanical Properties of Metals and Alloys Subjected to Severe Plastic Deformation, Doctoral (Tech.) Dissertation, Ufa: Inst. of Problems of Metals Superplasticity, 2000
[5] Murayama M, Horita Z. and Hono K. Acta Materialia, 2001, 49, p.21–29
[6] Jang J.S.C., Koch C.C., Journal Mater. Res. 1990, 5 N3, p. 498–510
[7] Sergeev, V.V. and Bulygina, T.I., Magnitotverdye materialy (Magnetically Hard Materials), Moscow: Energiya, 1980
[8] Korznikova G., Dimitrov O., Korznikov A. Annale de Chemie, Sci. Mat., 2002,.27(3) p..35–44
[9] Korznikova G. F., Korneva A. V., Pakiela Z., Korznikov A. V. Physics of Met. and Metallogr. 2002, 94 Suppl 2, S69–S74
[10] Degtyarev M.V., Chashuhina T.I., Voronova L.M., Davudova L.C., Pilyugin V.P. Physics of Met. and Metallogr. 2000, 90 , p.83–90

Experimental Investigations of the Al-Mg-Si Alloy Subjected to Equal-Channel Angular Pressing

G. Krallics[1], Z. Szeles[1], I.P. Semenova[2], T.V. Dotsenko[2], I.V. Alexandrov[2]
[1]Budapest University of Technology and Economics, Budapest, Hungary
[2]Ufa State Aviation Technical University, Ufa, Russia

1 Introduction

The severe plastic deformation (SPD) method has demonstrated its advantages when obtaining bulk ultrafine-grained (UFG) nanostructured states in different metals and alloys. The question concerning the optimal conditions under which SPD method creates nanostructured states is not answered yet. The answer is of importance especially in such a widely spread SPD technique as the equal-channel angular pressing (ECAP).

The main idea of ECA pressing is that the metal is deformed through a process of simple shear taking place in the cross sectional area without any geometrical change of the ingot [1]. Several studies dealt with different aspects of this process from structure change to process modeling [2-5]. The main attention during these studies was connected with the UFG structure formation and stability in pure Al [6-8], Al-Mg alloys [2, 3, 4, 9, 10], Al-Mg-Li-Zr alloys [2, 4, 5]. Using such materials for research can be explained by their low strength and relatively large ductility. As a result, the ECAP process can be performed at room temperature using different routes and a large number of passes. At the same time, there is no need to use high strength tools. Although a large number of studies were done in this field, systematic researches concerning the process of structure formation, changes of mechanical properties during ECAP and subsequent thermo-mechanical treatment are still topical.

In the present report the results of systematic investigations of the influence of SPD on the microstructure and mechanical properties of Al-Mg-Si alloy billets, processed by the ECAP technique are examined.

2 Material and Procedures

The as-received alloy used in this study was a commercial Al-Mg-Si alloy (Table 1). Before ECAP the as-received alloy was heat treated at 643 K for 30 minutes. Cylindrical billets of 15 mm in diameter and 50 mm in length were pressed through the die with 90° intersecting channels using a 500 kN testing machine. Eight passes by the route C (the 180° rotation of the billet around its longitudinal axis after each pass) at 293 K at the constant displacement rate of 1 mm/s were completed. The set-up of the tool is shown in Figure 1 and the geometry of the die is given in Figure 2.

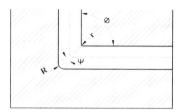

Figure 1: Die geometry (D = 15 mm, R = 3 mm, r = 0.5 mm, $\Phi = 90°$, $\Psi = 12.5°$)

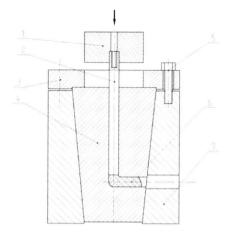

Figure 2: Die set-up (1 – punch holder; 2 – punch; 3 – blankholder; 4 – die; 5 – blank holder bolts (6 pieces); 6 – specimen; 7 – die holder)

Table 1: Chemical composition of the Al-Mg-Si alloy (mass%)

Mg	Si	Cu	Mn	Fe	Zn	Ti	Cr	Al
0.7	0.5	0.1	0.3	0.5	0.2	0.2	0.1	bal.

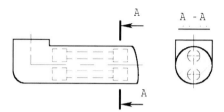

Figure 3: Tensile samples

Tensile specimens were machined parallel to the direction of ECAP (Figure 3). The gauge length of a specimen was 15 mm and the diameter was 3 mm. A ZWICK 2020 testing machine was used, at the 5 mm/min constant displacement rate at room temperature.

Vickers hardness tests were conducted with a KLEIN – HARTEPRUFER DURIMET installation. The specimens were cut perpendicular to the longitudinal axis of the ECAP billet. The tests were done at load of 3 N during 15 s. Nine series with 24 measurements at each series were done. The analysed cross section is located at the middle length of the specimen.

The TEM foils were prepared from slices cut perpendicular and parallel to the longitudinal axes from the as-received and ECAP ingots. The thickness of slices was 0.8-1.0 mm. All slices were mechanically ground to a thickness of 0.15 mm and subjected to twin-jet electropolishing in a mixture of 200 ml CH_3COOH, 150 ml H_3PO_4, 100 ml HNO_3, 50 ml H_2O at room temperature.

Transmission electron microscopy (TEM) was conducted at an accelerating voltage of 200 kV using a JEM-100B. Grain and subgrain sizes were measured on TEM images and averaged (minimum at 10 different places on a foil). The maximum values of width and length of subgrains that have maximal visible boundaries were averaged.

3 Experimental

3.1 Mechanical Testing

The results of the tensile tests are indicated in Figure 4. The yield stress (YS) and the ultimate tensile stress (UTS) increased after the first ECAP pass while the reduction of area (Z) decreased. The subsequent passes caused no significant changes of the mechanical parameters. The increase in the yield stress after the first ECAP pass is very pronounced. This increase is nearly three times of the initial value.

The mean values of the hardness increase sharply after the first pass and become almost constant with the higher number of passes (Figure 5). This tendency is similar to that of the strength properties (Figure 4). As the scatters show the hardness distribution in the cross section is not uniform at each pressing.

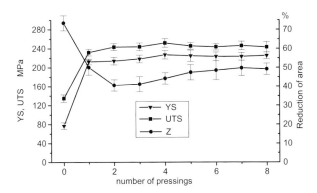

Figure 4: Mechanical properties at ambient temperature of the Al-Mg-Si alloy after various ECAP passes

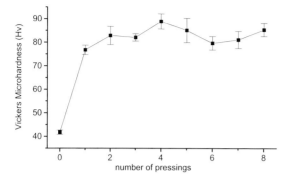

Figure 5: Mean value and the scatter of the cross-section Vickers hardness vs. the number of ECAP passes

3.2 Transmission Electron Microscopy

The as-received and annealed microstructure of the Al-Mg-Si alloy is given in Figure 6 in the longitudinal and transverse directions, respectively. The average size of the grains in the as-received and annealed condition is about 3.5 m. The typical microstructures after 1, 4 and 7 passes of ECAP are shown in Figures 7, 8 and 9, respectively.

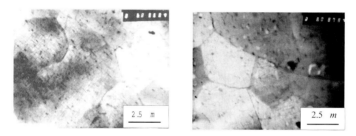

Figure 6: Microstructure at initial state in transverse (left) and longitudinal (right) directions

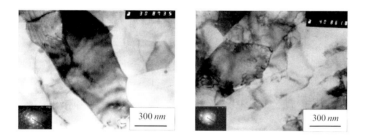

Figure 7: Microstructure after the first ECAP pass in transverse (left) and longitudinal (right) directions

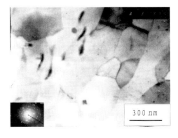 

Figure 8: Microstructure after the fourth ECAP pass in transverse (left) and longitudinal (right) directions

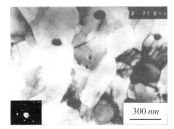 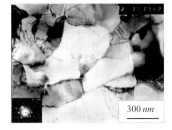

Figure 9: Microstructure after the seventh ECAP pass in transverse (left) and longitudinal (right) directions

After the first and fourth passes, the microstructure consists of parallel groups of elongated subgrains both in the longitudinal and transverse directions. The diffraction pattern suggests that subgrains have small angles of misorientation. The average length of subgrains after the first pass is ~800 nm and the average width is ~420 nm. After the fourth pass the corresponding length decreased to ~500 nm with an unchanged width. After the seventh pass of ECAP the shape of grains and subgrains is close to equiaxial (Figure 9). The dimensions of grains after the seventh pass are almost the same ~500 nm in length and ~410 nm in width. The diffraction patterns indicate an increase in the misorientation angles of the subgrains (Figures 7–9).

4   Discussion

In the experiments, the tendencies in the evolution of the microstructure of the Al-Mg-Si alloy were found to be the same as for pure Al [6-8] and its alloys containing Mg [2, 3, 9]. Namely, after the first pass the microstructure both in the longitudinal and transverse cross sections is inhomogeneous and consists of parallel groups of elongated subgrains divided by low angle boundaries. As reported for pure Al [8], the subgrains in the Al-Mg-Si alloy under investigation are elongated in the direction of pure shear (45° with respect to the longitudinal axis of the ECAP ingot). After the forth pass the elongation of subgrains with small angle boundaries is still retained. Only the seventh pass resulted in the development of equiaxed grains with high angle grain boundaries.

The observed case of the Al-Mg-Si alloy's tendency in the evolution of the microstructure is, in principle, in agreement with the phenomena characteristic of pure Al and its alloys with Mg, as observed in the literature. In pure Al, however, homogeneous equiaxed microstructure has developed already after four ECAP passes [6-8]. Besides, the addition of Mg to the Al matrix when the Mg concentration is between 1 % to 3 % resulted in a decrease of the evolution rate of the low angle cell boundaries into high angle grain boundaries [3]. Namely, 1 % addition of Mg decelerated such a transformation up to the seventh ECAP pass. Similar effect took place in the Al-Mg-Si alloy under study when the concentration of Mg was 0.7 %. Iwahashi et al. [3] related these observations to the strengthening of the Al matrix by Mg atoms. The latter hampers the dislocation mobility. As a result, the transformation rate of the low angle cell boundaries into the high angle grain boundaries in Al-Mg-Si alloy is decelerated during ECAP.

As well as in the case of pure Al and its alloys investigated earlier, the microstructure refinement effect in the Al-Mg-Si alloy in the present study was observed already after the first ECAP pass. At this, the length and width of the subgrains were estimated as ~800 nm and ~420 nm, respectively. The increase in the number of ECAP passes up to 4 and 7 resulted in the average length of subgrains and then grains of ~500 nm and almost unchanged width of ~410 nm. The average over the length and width of the grain size in the homogeneous microstructure after 7 ECAP passes is equal to ~460 nm. This size is comparable with the grain size of ~450 nm observed in an Al based alloy with addition of 1 % Mg after 6 ECAP passes [3]. At the same time in pure Al after 10 ECAP passes the minimum grain size reached ~100 nm [8]. The observed more intense effect of the microstructure refinement in alloys rather than in pure metals is related (as well as in the case of the boundaries transformation) to the dislocations pinned in the crystal lattice by the alloying elements. It decelerates the dislocations mobility and probably influences to a great extent the process of the new subgrain boundary development.

In principle, the revealed tendency in the hardness evolution at ECAP of the Al-Mg-Si alloy is similar to results observed in the case of pure Al and its alloys with Mg [6–10]. After two ECAP passes the hardness is doubled. The following passes do not change the hardness values and they are stabilized (although, a second small peak was observed after 4 passes for hardness. Similar tendency was observed in the case of mechanical properties such as tensile yield stress and ultimate tensile stress. They tend to be stable after the first ECAP pass.

5   Conclusion

A commercial Al-Mg-Si alloy was cold deformed by ECAP in eight passes (route C). The effect of cumulative strain hardening was monitored, and the hardness, the yield and the tensile strengths increased significantly at the first pass and almost remained unchanged when the process continued. The ductility also decreased at the first pass and practically remained constant during the following passes.

The refinement of the microstructure up to the subgrain size and low angle subgrain microstructure transformation to the high angle grain boundaries at ECAP of the commercial Al-Mg-Si alloy possess the same features as observed in the literature for pure Al and its alloys with Mg. It was revealed that at chosen experimental conditions the replication of seven ECAP passes results in the homogeneous microstructure development with the average grain size of ~ 460 nm.

6 References

[1] Valiev, R.Z.; Islamgaliev, R.K.; Alexandrov, I.V., Progr. Mater. Sci. 2000, 45, p. 103–189
[2] Berbon, P.B.; Tsenev, N.K.; Valiev, R.Z.; Furukawa, M.; Horita, Z.; Nemoto, M.; Langdon, T.G., Metall. Mater. Trans. 1998, 29A, p. 2237–2243
[3] Iwahashi, Y.; Horita, Z.; Nemoto, M.; Langdon, T.G., Metall. Mater. Trans. 1998, 29A, p. 2053–2510
[4] Horita, Z.; Smith, D.; Nemoto, M.; Valiev, R.Z.; Langdon, T.G., Mater. Research Soc. 1998, 13, p. 446–450
[5] Berbon, P.B.; Furukawa, M.; Horita, Z.; Nemoto, M.; Tsenev, N.K.; Valiev, R.Z.; Langdon, T.G., Mater. Sci. 1996, 217–222, p. 1013–1018
[6] Iwahashi, Y.; Horita, Z.; Nemoto, M.; Langdon, T.G., Acta Mater. 1997, 45, p. 4733–4741
[7] Nakashima, K.; Horita, Z.; Nemoto, M.; Langdon, T.G., Acta Metal. 1998, Vol.No5, p. 1589–1599
[8] McNelley, T.R.; Swisher, D.L.; Horita, Z.; Langdon, T.G., Ultrafine Grained Materials II, TMS, Warrendale, PA, 2002, p. 15–23
[9] Patlan, V.; Vinogradov, A.; Higashi, K.; Kitagawa, K., Mater. Sci. Engng. 2001, A300, p. 171–182
[10] Furukawa, M.; Horita, Z.; Nemoto, M.; Valiev, R.Z.; Langdon, T.G., Acta mater. 1996, 44, p. 4619–4629
[11] Berbon, P.B.; Furukawa, M.; Horita, Z.; Nemoto, M.; Langdon, T.G., Scr. mater. 1996, 35, p. 143

Mechanical Properties of AZ91 Alloy after Equal Channel Angular Pressing

Kristian Máthis[1,2], Alexandre Mussi[3], Zuzanka Trojanová[1], Pavel Lukáč[1], Edgar Rauch[3], János Lendvai[2]

[1]Department of Metal Physics, Charles University, Prague, Czech Republic
[2]Department of General Physics, Eötvös University, Budapest, Hungary
[3]Génie Physique et Mécanique des Matériaux, INP Grenoble – ENSPG, France

1 Abstract

The deformation behavior of a fine-grained AZ91 alloy has been investigated. The specimens were prepared using equal channel angular pressing (ECAP). Mechanical properties of the AZ91 alloy after multi-angular pressing were estimated at testing temperatures from 20 °C to 300 °C. The values of the yield stress and the maximum stress of ECAP specimens are higher than those of initial specimens. Elongation to fracture (ductility) increases with increasing test temperature and with number of extrusion. The values of the strain rate sensitivity of the flow stress suggest a superplastic behavior of the ECAP processed material above 200 °C.

2 Introduction

Over the last decades, magnesium alloys have been widely used as structural components in the automotive and aircraft industry due to their excellent strength-weight ratio. However, the disadvantages of these materials are the poor workability because of their h.c.p. structure and a low corrosion resistance. A significant improvement in the workability is expected by refining the grain size of materials and by achieving materials for superplastic forming.

It is well known that the materials with a small grain size have excellent mechanical properties. A reduction of the mean grain size is expected to increase the yield stress at room temperature and the toughness of the materials. This may promote superplastic properties at higher strain rates and/or lower temperatures than those conventionally used in superplastic forming. A very fine structure may be produced by equal channel angular pressing (ECAP) technique (in the literature the process is also termed equal channel angular extrusion) Horita, Z., Furukawa, M. Nemoto, M., Langdon T.G., Mater. Sci. Techn., 2000, 16, 1239 . In the ECAP technique a die containing a channel bent within the die into an L shaped configuration is used. The pressed sample does not change in the cross-sectional dimensions.

One of the most important parameters of ECAP process is the number of pressings. The most important refinement of the microstructure takes place during the first pressing (pass). The development of an ultrafine grain size leads to an increase in the yield stress. With increasing number of passes the elongation to the fracture is increasing but, on the other hand, the yield stress or the maximum stress shows only a small change.

3 Experimental Procedure

The cast AZ91 alloy (Mg-9wt.%Al-1wt.%Zn-0.2wt.%Mn) was solution-treated at 413 °C for 18 hours. The heat-treated samples with dimensions of $10 \times 10 \times 60$ mm^3 were ECAP processed in a right-angled channel via route C [4] (i.e. the samples were rotated by 180° after each pressing). The temperature during the first pass was 270 °C and it was decreased by 20 °C at each pass. The number of passes was 1, 4 and 8. The total strain was 1.15, 4.6 and 9.2; these values were calculated by the relation for the total strain given in [5]:

$$\varepsilon_n = 1.15 \, N \, \mathrm{ctg} \frac{\phi}{2} \qquad (1)$$

where N is the number of passes, ϕ is the angle between the channels. Tensile specimens were machined from the rods produced by ECAP and they had a rectangular cross section of 1×2 mm^2 and a 10 mm gage length. Tensile tests were carried out in a MTS machine at a constant strain rate of $5 \cdot 10^{-4}$ s^{-1} and in the temperature range from 20 °C to 300 °C.

4 Results

Figure 1 shows the true stress-true strain curves of both as-cast AZ91 and ECAP-processed AZ91 samples at various temperatures. It can be seen that ECAP processing improves the ductility in the whole temperature range.

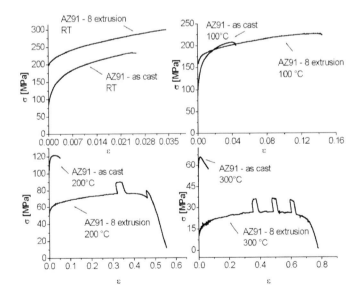

Figure 1: True stress-true strain curves of as-cast AZ91 and ECAP-processed AZ91 samples at various temperatures

In addition, strain rate step tests were conducted. An increase in the flow stress due to an increase in the strain rate twice at 200 °C and 300 °C is shown in Fig.1. The strain rate sensitivity of the flow stress (m value) was obtained using the equation:

$$m = \frac{\partial \ln \sigma}{\partial \ln \dot{\varepsilon}} \qquad (2)$$

where σ is the flow stress and $\dot{\varepsilon}$ is the strain rate. The mean value of m is 0.24 at 200 °C and 0.45 at 300°C suggesting superplastic behavior above 200 °C ($\sim 0.51 T_m$, T_m being melting point of pure Mg in K). The yield stress σ_{02} and the maximum stress σ_{max} are strongly depending on the test temperature. It can be seen (Figs. 2 and 3) that up to 100 °C both the yield stress and the maximum stress are higher for the ECAP processed samples. Above this temperature a strong degradation of mechanical properties takes place and the values of σ_{02} and σ_{max} are higher for the as cast material. This is most probably due to grain boundary sliding in ECAP -processed samples at higher temperatures.

The dependence of yield and maximum stress on the number of passes at 300 °C was also investigated. It can be seen (Figs 4 and 5) that the elongation to the fracture increases, on the other

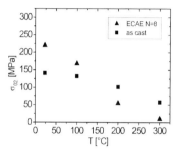

Figure 2: Temperature dependence of the flow stress for the as-cast and ECAP-processed sample

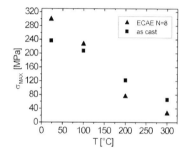

Figure 3: Temperature dependence of the maximum stress for the as-cast and ECAP-processed sample

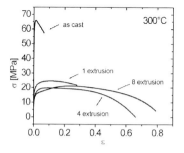

Figure 4: Dependence of true stress-true strain curves on the number of extrusion at 300 °C

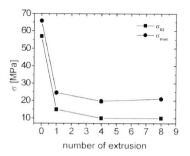

Figure 5: Dependence of the yield stress and the maximum stress on the number of extrusion at 300 °C

side both the yield stress and the maximum stress decrease with the first ECAP pass and remain practically constant with increasing number of passes.

5 Discussion

The grain size of the samples after pressing is about 1 µm. An increase of both the yield stress and maximum flow stress of the samples ECAP deformed at room temperature is due to the refinement in grain size. The lower values of the yield stress (and the maximum flow stress) of the ECAP alloys measured at 200 and 300 °C in comparison with those of the unpressed material indicate some recovery process(es). Rapid diffusion is expected at these temperatures. The ECAP procedure has a significant influence on the elongation to fracture of specimens if pressed at higher temperatures. The elongations to failure of the pressed AZ91 alloy are much higher than those of the unpressed alloys. The elongation to failure of the specimens pressed at 300°C increases with the number of the passes.

The values of the strain rate sensitivity parameters and the elongation to failure for the ECAP materials deformed at 200 and 300 °C may indicate the necessary conditions required for superplasticity in the AZ91 magnesium alloy.

6 Conclusion

Magnesium alloy AZ91 subjected to ECAP exhibits microstructure with ultrafine grain size. The grain refinement leads to an increase in the strength of the material at room temperature. At temperatures above 200 °C the yield stress and the maximum stress of samples after ECAP are lower than those for as-cast AZ91. The values of the strain rate sensitivity of the flow stress (parameter m) are 0.24 at 200 °C and 0.45 at 300 °C. At these high testing temperatures the elongation to failure increases markedly. These experimental data demonstrate that it is possible to achieve superplastic behavior.

7 Acknowledgements

This work was supported in part by the Program Barrande No 2000-013-3.

8 References

[1] Horita, Z., Furukawa, M. Nemoto, M., Langdon T.G., Mater. Sci. Techn., 2000, 16, 1239
[2] Furukawa, M., Horita, Z., Langdon, T.G., Adv. Eng. Mater. 2001, 3, 121.
[3] Mabuchi, M., Iwasaki, H., Yanase, K. and Highasi, K., Sci. Mat., 1997, 36, 681–686
[4] Segal, V.M., Mat. Sci. Eng. A, 1995, 27A, 157
[5] Iwahasi, Y., Wang, S., Horita, Z. Nemoto, M., Langdon, T. G., Scripta Mater., 1996, 35, 143-146.
[6] Aida, T., Matsuki, K., Horita, Z., Langdon, T.G., Scripta Mater. 2001, 44, 575

Influence of Microstructural Heterogeneity on the Mechanical Properties of Nanocrystalline Materials Processed by Severe Plastic Deformation

Z. Pakiela, M. Sus-Ryszkowska
Faculty of Materials Science and Engineering, Warsaw University of Technology, Warsaw, Poland.

1 Introduction

The great interest in the mechanical behavior of nanocrystalline materials has been inspired by the observation that they have unique mechanical properties. Materials with very small (including sub-micron) grains can be produced by several methods, such as crystallization from the amorphous state, powder metallurgy or vapor condensation. Most of these methods however, allow for producing only relatively small quantities of the material. A new group of methods developed recently is based on severe plastic deformation (SPD) of bulk metals. The nanostructured materials processed by SPD have very advantageous properties, such as a high strength combined with acceptable plasticity [1]. As a result these methods are extensively studied at the present.

An important part of these studies is focused on plastic deformation of SPD nano-materials. Some investigators suggest the possibility of grain boundary sliding and a grain rotations [2]. Mechanical twinning has been also pointed out as relevant process [3]. Strain localization within the shear bands has been also suggested.

Theoretical studies confirm that these mechanisms can play a significant role in deformation of nanocrystalline materials. Besides however these mechanisms, there are also other phenomena that can affect significantly the deformation of nanocrystalline materials, processed by SPD. The aim of this paper is to show that these phenomena occur in the scale of tens and hundreds of micrometers and not in the nanometer scale.

In the present study, the severe plastic deformation and the refinement accompanying it was effected by the two methods: high-pressure torsion (HPT) and equal channel angular extrusion (ECAP). The processed materials were examined in terms of microstructure heterogeneity on their mechanical properties. The aim was to find how various macro- and microscale parameters affect the mechanical properties of the material.

2 The Specimens

Specimens were obtained via ECAP and HPT. The volume of the material processed by the ECAP method is sufficiently large for machining standard tensile tests specimens. This is not possible with the material processed by HPT. In order to compare the properties of the materials processed by these two methods, micro-specimens were prepared with a length of 8–12mm, gauge length of 2–3 mm, and thickness of 0.1–0.2 mm. Examination of these small specimens brings about various difficulties.

An important difficulty appears when one measures the deformation of such a small specimen. Small specimens require sensitive methods for measuring strain. A 0.2 % elongation of a specimen with a gauge length of 2 mm results in 4μm displacement. To measure this elongation with an accuracy of ±5 %, we should have a measuring device with a resolution of ±0.2 μm. Such high resolution can be achieved with a tensometric extensometre. A typical small standard extensometre has a gauge length of 5 mm, which permits achieving a resolution of 0.1 μm. If the gauge length of the specimen is below 5 mm, the extensometre is too large to be installed. If this is the case, the extensometre may be fixed to the specimen holders. Knowing the gauge length of the specimen, we can determine its deformation, but the measurements then bear a considerable error. This is due to the fact that the strain is localized in stress concentration regions, which in the specimens prepared for tensile tests lie between the holding portion of the specimen and the gauge length as shown in Figs. 1 and 2. Fig. 1 shows the strain field in microspecimen calculated by the finite element method (FEM). We can see that the stress induced within the transition regions between the gauge length and the holder portion of the specimen is several times as great as that in the gauge region. The consequences of this stress concentration can also be seen in Fig.2, which is a photograph of an ECAP processed iron specimen after a tensile test. As a result of the stress concentration in the transition zone, the material here becomes plastic earlier than that in the gauge region. If the specimen deformation is measured indirectly by measuring the displacement of the holders, the flow stress of the material is underestimated and the specimen elongation overestimated. For this reason, when dealing with

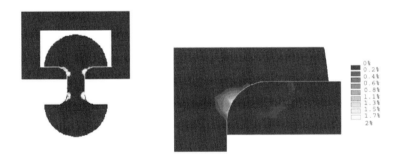

Figure 1: Strain field in the specimen, as calculated by the finite element method FEM

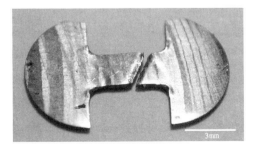

Figure 2: ECAP processed iron specimen after a tensile test

small specimens on which an extensometer cannot be installed, the measurements should be performed by optical or other direct methods.

3 Surface and Microstructure

It has been known that the state of the specimen surface can affect the mechanical properties of the material. In the nanocrystalline materials obtained by the SPD methods, the mean free dislocation path is by definition very short. The technique used for cutting the test specimens may here strongly affect the properties of the material. Mechanical cutting techniques generate stresses that can reduce the plasticity of the surface zone to zero. The electric spark cutting may in turn relax stresses in the surface zone and thereby increase the plasticity of the specimen. This effect is the stronger the thinner is the specimen. Although the surface layer is routinely removed by mechanical polishing of the specimen, the disturbance of its bulk properties may remain, since the stress relaxation could go down deeper than the observable change of the specimen microstructure.

Another reason of the difference in the behavior between the individual specimens may be the heterogeneity of the material from which they were cut out. The HPT process yields discs in which the material deformation degree and, thus, the material microstructure depends on the distance from the disc center. In the central portion of the disc, the deformation is the smallest. Therefore, the tensile test specimens will differ from one another in their properties according to the disc portion of which they were cut out. Examples of these differences are shown in Fig.3. Fig. 3a shows the microstructure of the Ni_3Al intermetallic phase in the central region of the disc. One can see the shear bands characteristic of a plastically deformed material. The diffraction image reveals the presence of long range ordering (LRO), characteristic of this material.

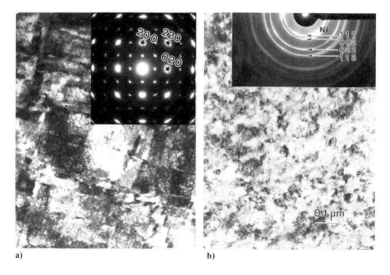

Figure 3: Microstructure of the intermetallic compound Ni3Al processed by HPT. a) central region of the sample. Notice presence of the LRO diffraction spots. b) microstructure of the same sample observed at a distance of 3mm from its center. Notice the lack in LRO diffraction rings.

Fig.3b shows the microstructure of the same disc, but observed at a distance of 3mm from its center. The material here has the structure of equiaxial crystallites with an average diameter of about 20 nm. The diffraction image shows no reflexes due to long range ordering, which means that the plastic deformation has led to the disordering. This results in considerable differences in the material properties between the central portion of the disc and the region positioned at a certain distance from the disc center. Hence, the samples cut out of various portions of the disc will have different mechanical strengths and show differing plasticity. The measurements of these parameters were described in detail in our earlier papers [4,5].

4   Micro and Macro Strain Localization

In the SPD materials strongly hardened by plastic deformation, the range within which the deformation is homogeneous is very small. At relatively small deformation, the specimen develops a neck and the strain is localized within this region. In specimens of the same material with the same thickness and gauge width, the neck has similar widths. Thus, the proportion of the strain localized in the neck to the strain in the entire specimen depends on the gauge length. In the samples with a very short gauge length, this proportion is great, which apparently suggests that the specimen elongation is high. As a result, when determining the plasticity of small size specimens, it is reasonable to measure uniform deformation rather than the total deformation.

Proper understanding properties of SPD materials requires identification of the leading mechanism of deformation. Many investigators suggest that the essential mechanism is the grain boundary sliding. In fact, the results of certain examinations, e.g. observations by atomic force microscopy (AFM), indicate that this deformation mechanism really occur [6]. If the material is homogeneous and the grain size in the whole volume of the specimen is very small, the mechanism of grain boundary sliding may predominate in the development of plastic deformation. When, however, the material is not homogeneous, it is worth considering also the possibility of the occurrence of other dislocation mechanisms.

In the SPD processed materials, the microstructure is often heterogeneous. Fig.4 shows the microstructure of iron subjected to ECAP. Multiple pressing through intersecting at 90° round channels (20 mm in diameter) was performed twelve times at room temperature to give an imposed strain of $\varepsilon \approx 13$, via the so-called 'route C', when the sample was rotated through 180°

Figure 4: Microstructure of iron subjected to ECAP

clockwise about working axis between subsequent passes. Polished section were prepared following standard metallographic procedures followed by etching in 2 % nital.

One can see shear bands of different size and orientations whose width amounts to several tens of µm. It is characteristic of these materials that their microhardness and plastic properties are heterogeneous. Fig.5 shows the results of the measurements of the deformation of the iron specimen mentioned above, made in a randomly selected measurement area. The measurement was performed using grid interferometry. The details can be found in [7]. As can be seen in Fig.5, the observed area sized at 535×770 µm contains regions that differ manifold from one another in the deformation degree. The size of the regions that have undergone similar deformation ranges from several to several tens of µm, and it greatly exceeds the average grain size, which, in this material, was about 500 nm. This may suggest that certain regions in the specimen have been hardened in a weaker degree, and that the deformation is localized just in them. If these regions are distributed uniformly throughout the specimen, it will behave as a composite material that is composed of a plastic phase and a hard phase.

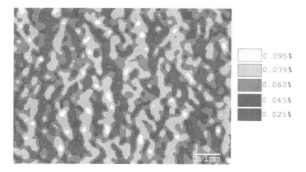

Figure 5: Deformation field of the iron specimen measured by diffraction grid method

5   Summary

The mechanisms of deformation of nanocrystalline materials are still subject of many theoretical and experimental studies. The behavior of these materials during deformation depends on various phenomena that occur on various microstructure levels. In addition to the phenomena occurring on the nanometric scale, such as grain boundary sliding, and also other phenomena, that take place in mezoscopic scale, can affect the mechanical properties of SPD materials.

The present paper provides examples showing that the plastic deformation may be affected by phenomena that occur in the scale of tens and hundreds of µm. The effect of surface zone of the material, deformation in the stress concentration regions, and strain localization in regions that have been less hardened than the rest of the material have been discussed.

6 Acknowledgements

The authors thank Prof. K.J. Kurzydlowski (Warsaw University of Technology, Poland) for helpful discussion and Prof. R.A. Valiev (Ufa State Aviation Technical University, Russia) for material supply and encouragement.

This paper is a part of a research program currently executed at the Faculty of Materials Science and Engineering of Warsaw University of Technology and was supported by the Polish Committee of Scientific Research under grants no. 4T08A00322 and grant no. PBZ-KBN-013/T08/05.

7 References

[1] R.Z. Valiev, E.V. Kozlov, Y.F. Ivanov, J. Lian, A.A. Nazarov, B.Baudelet, Acta Metall. 42(1994)2467
[2] R.W. Siegel, Materials Science Forum, 235–238(1997)851
[3] J.Y. Huang, Y.K. Wu, H.Q. Ye, Acta Mater. 44(1996)1211
[4] A.V. Korznikov, Z. Pakiela, K.J. Kurzydlowski, Scripta Mater. 45(2001)309
[5] Z. Pakiela, W. Zielinski, A.V. Korznikov, K.J. Kurzydlowski, Inzynieria Materialowa, 5(2001)698
[6] A. Vinogradov, S. Hashimoto, V. Patlan, K. Kitagawa, Mater. Sci. Eng. A319–321(2001)862
[7] M. Sus-Ryszkowska, Z. Pakiela, R.A. Valiev, K.J. Kurzydlowski, to be published

Creep Behaviour of Pure Aluminium Processed by Equal-Channel Angular Pressing

Vaclav Sklenicka, Jiri Dvorak, Milan Svoboda
Institute of Physics of Materials, Academy of Sciences of the Czech Republic, Brno, Czech Republic

1 Introduction

It is well known that physical and mechanical properties of metallic materials are very sensitive to their grain sizes. Equal-channel angular pressing (ECAP) is a processing procedure in which a material is subjected to a very severe plastic strain without any concomitant change in the cross-sectional dimensions of the work-piece [1–4]. Numerous reports over the last decade have firmly established ECAP processing as a technique for achieving very significant grain refinement in bulk polycrystalline materials with the grain sizes typically reduced to the submicrometer and/or nanometer level and thus to levels that are generally not attainable in conventional thermomechanical processing. There are many reports demonstrating that these remarkable grain refinements lead to unusual properties in the as-processed materials [5, 6]. It is surprising to note that there have been no published results to date describing the creep properties of materials processed by ECAP. Accordingly, the present investigation was undertaken to address this deficiency. Specifically, the creep behaviour and microstructural characteristics of pure aluminium after ECAP were examined.

2 Experimental Material and Procedures

High purity (99.99%) aluminium with an initial grain size of ~ 5 mm in the as-received state was subjected to ECAP. The ECAP pressing was conducted at room temperature with a die that had a 90° angle between the die channels and one or repetitive pressing followed either route A, B or C [7] - Fig. 1. In routes B ≡ B_c and C, the rotation was always 90° and 180°, respectively, in the same sense. Constant stress tensile creep tests were conducted at 473 K and at applied stress of 15 MPa on the ECAP billets. For comparison purposes, some creep tests were also conducted on unpressed coarse-grained aluminium. Following ECAP passes and creep testing, samples were prepared for examination by means of light microscopy, and both scanning (Philips SEM 505) and transmission (Philips CM 12) electron microscopy.

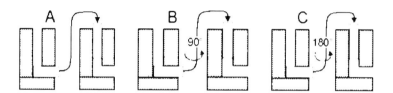

Figure 1: The three processing routes for ECAP [7]

3 Experimental Results

3.1 The ECAP Procedures and Microstructural Observations

All ECAP pressings were conducted in air at room temperature with the samples coated with a thin film of a lubricant based on molybdenum disulfide. Several samples were pressed consecutively to avoid any difficulties in removing samples from the die. Samples were pressed, using routes A, B or C, from a minimum of 1 to a maximum of 12 passes through the die. Figure 2 gives an example of the microstructure and associated SAED pattern in the cross-section normal to the pressing direction after a single and repetitive pressings using different routes. Examination shows that a single pressing leads to a substantial reduction in the (sub)grain size (~ 1.4 µm) and the microstructure now consists of parallel bands of subgrains in the shearing direction, the grain boundaries have low angles of misorientation. These subgrains subsequently evolve with further pressings into a reasonably equiaxed and homogeneous microstructure with an average grain size of the order of ~ 1 µm independently of the processing routes. The SAED patterns indicate that the subgrain misorientation increases during repetitive ECAP.

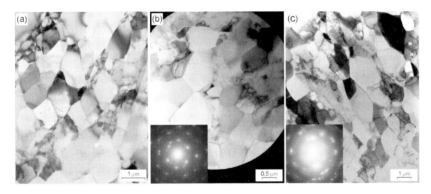

Figure 2: Typical microstructures and associated SAED patterns after passage through the die for (a) single pressing, (b) 4 pressings, route B, and (c) 8 pressings, route C.

To explore the thermal stability of the ECAP material the post-ECAP annealing at temperature of 473 K for different periods of time were performed. Microscopic examination revealed that the post-ECAP annealing makes the ECAP structure quite unstable and a noticeable grain growth occurs at the very beginning of annealing (Tab. 1.). Simultaneously, annealing at 473 K gives measurable change in the Vickers microhardness.

3.2 Creep Results

Selected creep curves for coarse-grained aluminium and the ECAP material after different routes of pressing are shown in Figs. 3a, 4a, 5a and 6a in the form of strain ε, versus time, t, for an absolute testing temperature, T, of 473 K. These standard ε vs. t creep curves can be easily replotted in the form of the instantaneous strain rate $\dot{\varepsilon}$, versus time, t, as shown in Figs. 3b, 4b, 5b and 6b. As demonstrated by the figures, significant differences were found in the creep behaviour of the ECAP material when compared to its coarse-grained counterpart. First, the ECAP material exhibits markedly longer creep life than coarse grained aluminium. Second, the minimum creep rate for the ECAP material is about one to two orders of magnitude less than that of coarse-grained material. Third, the shapes of creep curves for the ECAP material for different process routes (A, B and C) used after and different number of pressings differ considerably. However, this difference in the shapes of the ECAP creep curves is more clearly illustrated for the tests conducted at small number of the ECAP pressings (Figs. 3 and 4).

Table 1: Thermal stability and Vickers microhardness of the ECAP material

Annealing conditions	ECAP 4 passes route A		ECAP 4 passes route B	
	grain size [mm]	microhardness HV5	grain size [mm]	microhardness HV5
no treatment	0.9	37	0.9	38
473 K/ 0.5 h	6.6	27	4.5	32
473 K/ 1 h	7.9	23	4.8	32
473 K/ 2 h	7.3	23	4.8	27
473 K/ 5 h	7.3	21	5.3	27
473 K/ 24 h	12.2	19	5.0	23
473 K/ 168 h	13.4	18	10.4	21

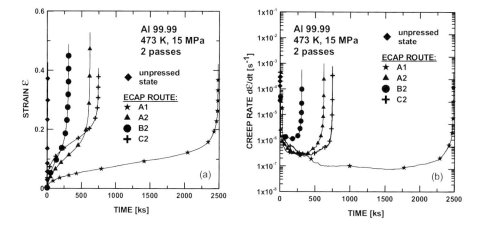

Figure 3: Creep curves and creep rate versus time for unpressed state and different ECAP routes after two passes (for comparison purposes also one pass (A1) was plotted)

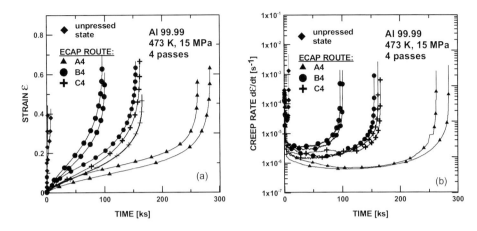

Figure 4: Creep curves and creep rate versus time for unpressed state and different ECAP routes after four passes

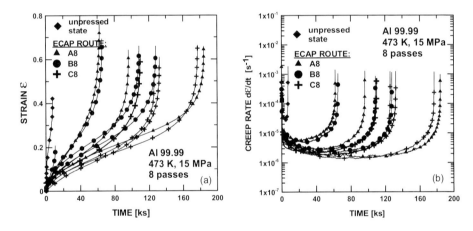

Figure 5: Creep curves and creep rate versus time for unpressed state and different ECAP routes after eight passes

The effect of the number of ECAP passes by route B is shown in Fig. 7 for two, four, eight and twelve passes. Inspection shows that the creep life of the ECAP material decreases with increasing number of ECAP passes. Further, a comparison between creep fracture strain of unpressed coarse-grained material and the ECAP material shows that the ECAP pressing leads to an increase in the creep plasticity. This difference consistently increases with increasing number of ECAP passes.

A metallographic investigation currently in progress has not revealed substantial grain growth due to creep. The observed grain sizes were within the range of 9 μm to 14 μm. These values agree with the results of the grain size measurements of the ECAP material after post-

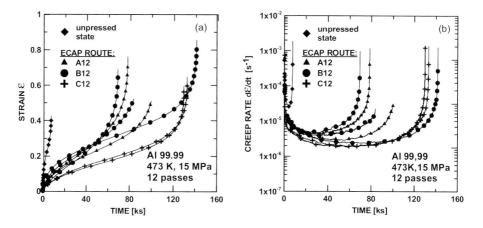

Figure 6: Creep curves and creep rate versus time for unpressed state and different ECAP routes after twelve passes

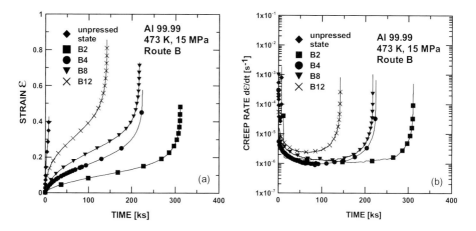

Figure 7: Creep curves and creep rate versus time for unpressed state and various number of ECAP passes via route B

ECAP annealing (cf. Tab. 1.). Thus, it may be concluded that grain growth of the ECAP material during creep test is caused by thermal exposure rather than by creep deformation.

A question naturally arises about the reason(s) for a gradual decrease of creep strength of the ECAP materials with increasing number of ECAP passes (Fig. 7). As successive pressing is performed, the microstructure evolves from one consisting of subgrains and low-angle boundaries to a state, with grains surrounded by high-angle boundaries. It should be stressed that high-angle boundaries are needed to achieve grain-boundary sliding in creep deformation. Further, the inhomogeneous nature of a banded structure following small number ECAP passes can reflect a coarse initial grain size and inhomogeneous deformation due to differences in initial orientations of the grains prior to the ECAP pressing operations. Finally, the texture of the ECAP material produced by the different process routes due to different rotation of the billets may account for the differences in creep behaviour of the ECAP material after the same number of passes. Thus, further investigation is required to determine the dependence of the texture development on ECAP passes.

4 Discussion

The present results provide an initial report on the creep characteristics after ECAP of pure aluminium, a material wherein it appears to be relatively easy to attain a uniform, essentially equiaxed grain structure and a high population of high-angle boundaries.

In the study of Oh-ishi et al. [4], pure aluminium (99.99 %) was deformed at room temperature by different ECAP routes with a die that had a 90° angle between the die channels, and the microstructure was observed by using TEM and SAED patterns. The effectiveness of producing equiaxed ultrafine grains was found in the route order B > C > A. Now it is generally accepted that an internal angle of 90° is needed in order to most readily establish an array of equiaxed and ultrafine grains separated by high-angle grain boundaries [9]. Systematic research of the development and evolution of the microstructure of pure aluminium by transmission electron microscopy (TEM) and orientation imaging microscopy (OIM) methods after one, four and twelve ECAP passes and different pressing routes was carried out in very recent works [8, 10]. The microstructure for all cases examined indicated little difference in the grain size produced via the various routes. There was little apparent dependence of the disorientation distribution on the process route for an ECAP die having an internal angle equal to 90°. The disorientation data revealed that repetitive pressing results in a progressive increase in the fraction of high-angle boundaries. By contrast, the texture of the material by the various process routes was greatly influenced by the processing route. Thus, the various textures may be attributed to the different rotations of the billets between subsequent ECAP passes.

5 Conclusions

The creep resistance of an ultrafine-grained pure aluminium after the ECAP pressing is shown to be considerably improved compared to an unpressed coarse-grained material. However, the creep behaviour of an ultrafine-grained material indicates a dependence on the number of ECAP passes resulting in homogenisation of the microstructure and microtexture and the influence of the ECAP process route which may be attributed to the different rotations of the billets via the ECAP pressing.

6 Acknowledgements

Financial support for this work was provided by the Grant Agency of the Academy of Sciences of the Czech Republic under Grant A 2041301.

7 References

[1] V.M. Segal, Mat. Sci. Engng. 1995, A197, 157–164
[2] R.Z. Valiev, R.K. Islamgaliev, I.V. Alexandrov, Prog. Mater. Sci. 2000, 45, 103–189
[3] Y. Iwahashi, Z. Horita, M. Nemoto, T.G. Langdon, Acta Mater. 1997, 45, 4733–4741
[4] K. Oh-ishi, Z. Horita, M. Furukawa, M. Nemoto, T.G. Langdon, Metall. Mater. Trans. A 1998, 29A, 2011–2013
[5] R.Z. Valiev, T.C. Lowe, K. Mukherjee, JOM 2000, 52, 37–40
[6] Y. Lu, P.K. Liaw, JOM 2001, 53, 31–35
[7] T.G. Langdon, M. Furukawa, M. Nemoto, Z. Horita, JOM 2000, 52, 30–33
[8] T.R. McNelley, D.L. Swisher, Z. Horita, T.G. Langdon, in Ultrafine Grained Materials II (Eds.: Y.T. Zhu et al.) TMS, Warrendale, USA, 2002, pp. 15–24
[9] N. Nakashima, Z. Horita, M. Nemoto, T.G. Langdon, Acta Mater. 1998, 46, 1589–1599
[10] S.D. Terhune, D.L. Swisher, K. Oh-ishi, Z. Horita, T.G. Langdon, Metall. Mater. Trans. A 2002, 33A, 2173–2184

Low Temperature Strain Rate Sensitivity of some Nanostructured Metals

E.D.Tabachnikova[1], V. Z Bengus[1], V.D.Natsik[1], A.V.Podolskii[1], S. N. Smirnov[1], R.Z.Valiev[2], V.V.Stolyarov[2], I.V.Alexandrov[2]

[1]B. Verkin Inst. for Low Temperature Physics & Engineering, Ukraine Academy of Sciences, Kharkov, Ukraine
[2]Institute of Physics of Advanced Materials, Ufa, RUSSIA

1 Introduction

Ultrafine-grained (nanostructured) metals (NSM) with 100–300 nm mean grains manufactured under intensive deformation by the equal-channel angular pressing (ECAP) are known for their high strength and plasticity at low temperatures (300–4.2 K) and are therefore promising structural materials for cryogenic applications [1]. It was found [2,3] that under low temperature deformation the grain sizes decreased in nanostructured (NS) Ti and NS Cu from ~15 μm to 100 nm and the yield stresses increased 5 to 6 times. The work hardening was less pronounced (as compared to polycrystalline samples) in the stress-strain curves. These are qualitative indications of change in the NSM mechanism of deformation: along with intragrain dislocation slip, grain boundary slip develops even at relatively low temperatures. The intragrain slip in compressed NS Ti shows up as ductile shear rupture at the end of unstable intragrain shear for large distances at nitrogen and helium temperature [2,3]. No such fracture is observed in Ti polycrystals with traditional grain size ~ 15 μm under the same conditions. However, the micromechanisms of low temperature plastic deformation in NSM are not clear yet. One of the parameters characterizing the slip micromechanism is the activation volume V of dislocation slip during work hardening. The commonly accepted indirect method of finding the v-value is to measure the rate sensitivity of the deforming stress [4]. The presently available data on activation volumes of dislocation slip under low temperature deformation of NSM are not sufficient. This study was therefore intended to measure the deformation dependence of the activation volume V for nanostructured Ti and Cu at room and liquid nitrogen temperatures.

2 Materials and Methods

The compression specimens (2 × 2 × 7 mm) of commercial Cu and Ti metals were cut by spark erosion along the billet axis (14 × 14 × 160 mm rods), prepared using ECAP. The Cu samples used had three structural states: the initial state - 99,9 % pure polycrystalline Cu annealed (600 °C, 10 hours) samples; route A – obtained by ECAP with number of deformation passes n = 8, route B – obtained by ECAP with n = 8 (after each pass the billet was rotated by 90^0 around the ECAP axis with reversing the shear direction). The Ti samples had the following states: 1- initial (polycrystalline VT-1-0 Ti, the average grain size d~ 15 μm); 2 - ECAP, n = 8, d ~ 0,3 μm, 3 - ECAP, cold rolling followed by 1 hour annealing at 300 °C. The samples were uniaxially compressed in the deformation machine (1,3 · 10^7 N/m rigidity) at 300 and 77 K. The measurement accuracy was ± 3 MPa (stress) and ± 0.1% (strain). The Ti specimens were cut parallel

(∥) and perpendicular (⊥) to the axis of ECAP. The following mechanical characteristics were measured: stress-strain curves (σ-ε), macroscopic yield stress σ_0 (this value was found by the method of intersection of tangents to the deformation curve at its initial and steady parts), deformation ε_0, corresponding to σ_0, estimated through averaging the results for five and more samples tested The strain-rate sensitivity of the yield strength was determined by increasing the strain-rate from the basic strain rate $9{,}1 \cdot 10^{-2}$ mm/min to $3{,}8 \cdot 10^{-1}$ mm/min. To find the activation volume V, we used the known Arrhenius equation interpreting the deformation rate as a rate of a thermally activated process

$$\dot\varepsilon = v\exp\left\{\frac{-H-V\sigma}{KT}\right\} \quad (1)$$

where

$$v = \varepsilon_0 v_0 N \exp\left\{\frac{S}{K}\right\},$$

v_0 is the frequency factor, ε_0 is the microscopic deformation corresponding to the process when a dislocation overcomes the obstacle resistant to the dislocation motion, N is the number of activated elements, S is the activation entropy, H is the activation energy, V is the activation volume. Assuming that H and V are the functions of stress and temperature in the experiment, the activation volume V can be found as

$$V = KT\left(\frac{\ln\left(\dot\varepsilon_1/\dot\varepsilon_2\right)}{\Delta\sigma}\right)_T, \quad (2)$$

where $\dot\varepsilon_1$ and $\dot\varepsilon_2$ are deformation strain rates during the rate cycling, $\Delta\sigma$ is the difference between the flow stresses in the stress-stain curve which corresponds to each value of deformation strain rate.

3 Results and discussion

3.1 Strain rate sensitivity of nanostructured Cu

The principal regularities of low temperature stress-strain curves for NS Cu were discussed elsewhere [3]. It was found that fragmenting the grains from $r \sim 30$ mcm (starting polycrystalline Cu) to 200 nm by ECA pressing leads to a 5 fold increase in the yield strength both at room and lower temperatures (77 K and 4.2 K). Besides that the compressed NS Cu retains its high plasticity even 30 % in the whole temperature interval. The curves taken at 300 K and 77 K have the region of appreciable microplastic strain (~5 %), then the distinct yield strength occurs, which is followed by practically one stage of deformation. The strain rate of loading the samples was changed at the stage of steady-state flow. It is seen that the change in the strain rate from

$9.1 \cdot 10^{-2}$ mm/min to $3.8 \cdot 10^{-1}$ mm/min does not affect the work hardening coefficient θ. θ is close to zero at 300 K and is 1,6 kg/mm² at 77 K for routes A and B. The strain dependence of activation volumes corresponding to the dislocation motion in all studied cases (pure polycrystalline Cu (1), route A (2), route B (3)) are shown in Fig. 1 (a, b).

It appears that:

1. V decreases with growing ε in all the routes;
2. for all ε and T, V is smaller in states 2 and 3 than in state 1.

The decrease in $V(\varepsilon)$ may be attributed to the intragrain dislocation density which increases in the process of deformation. Indeed, the electron-microscopic examination of this material [5] shows that the intragrain dislocation density increases with deformation from $5 \cdot 10^{14}$ to 10^{15} m^{-2}.

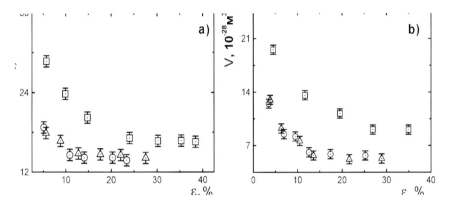

Figure 1: (a, b) Strain dependence of activation volume for Cu for 300 K (a) and for 77 K (b) in different structural states $\square$ –1, O – 2, Δ – 3

The obtained V-values can be used to estimate the mean dislocation segment length for dislocations involved in the thermally activated motion as [4]:

$$L = \frac{V}{bd}, \quad (3)$$

assuming $b \sim d$, where b is Burgers vector and d is the potential barrier width.

The L-values for $\varepsilon \sim 5$ % obtained by Eq. (3) are listed in Table 1. They can be used to test the model [6] explaining the mechanical behaviour of NS Cu in terms of the dislocation bending mechanism during intragrain slip. It is assumed that on the onset of intragrain plastic deformation, the dislocation loops bend outward between the fixed points taking the form of a semicircle. The critical stress for this to occur is given by the following equation [7]:

$$\sigma = \frac{MGb}{2\pi L(1-\mu)}\left[\left(1-\frac{3\mu}{2}\right)\ln\left(\frac{L}{b}\right) - 1 + \frac{\mu}{2}\right], \quad (4)$$

where M is the Taylor factor, μ is the Poisson ratio and L is the mean dislocation length. The values obtained from Eq. (4) with the L data of Table 1 are about 15–20 % lower than those measured at 300 K and 77 K (see Table 1). This discrepancy can be attributed to the internal stresses disregarded in Eq. (4) and to the intragrain slip contribution. According to [5], in NS Cu this contributions is 15–20 %.

The above estimates permit us to conclude that in the complex processes occurring under low temperature deformation of NS Cu [5] (simultaneous development of intragrain and grain-boundary slip, grain boundary migration) the measured activation volume of dislocation slip corresponds to the intragrain termoactivated slip at 300 K and 77K.

Table 1: Value of L found by Eq. (3); comparison of experimental σ and σ_T; σ_T is calculated for $\varepsilon = 5$ % by Eq. (4).

	T, K	$L \cdot 10^{-8}$, m	σ, GPa	σ_t, GPa
Cu, Route A	300	3,59	0,45 ± 0,03	0, 38
	77	2,98	0,52 ± 0,03	0,43
Cu, Route B	300	3,36	0,43 ± 0,03	0,40
	77	2,45	0,56 ± 0,04	0,49

3.2 Strain-rate sensitivity of nanostructured Ti

The principal regularities of low temperature stress-strain curves for NS Ti were discussed elsewhere [2]. The studies of low temperature plasticity and fracture show that the yield strength of NS Ti (BT-1-0) is 1.5–2 times higher than in coarse-grained polycrystals. The same increase in the yield strength is observed on cooling from 300 K to 4.2 K. The temperature dependences of the yield strength of coarse-grained and NS Ti [2] are typical of thermally activated plasticity. It was therefore interesting to measure the rate sensitivity of the deforming stress for three states of Ti (sample cut off parallel (||) and perpendicular ($\perp$) to the ECA pressing axis). Strain dependence of the activated volumes of the dislocation motion are shown in Figs.2 (a, b) for samples cut off parallel to the ECA pressing axis at 300 K (Fig.2a) and 77 K (Fig.2b) for three states of Ti.

It is seen that in state 1 the V-value is independent of ε and is sensitive to the structure at 77 K at the early stages of deformation. These results coincide with the previously obtained data for polycrystalline Ti BT-1-0 [8]. At 300 K they can be attributed to the thermoactivated motion of dislocations through the obstacles formed by the impurity atoms, which control plastic deformation under these conditions. At 77 K the dependence $V(\varepsilon)$ is attributed to new obstacles formed by dislocations of other slip and twinning systems which are activated in Ti at lower temperatures [9].

It is found for NS Ti at 300 K that:

1. under fixed deformation, the V-value are lower in states 2 and 3 than in state 1 and the lowest in state 3;
2. the dependence $V(\varepsilon)$ is observed in all the cases;
3. the V-values practically coincide for (||) – and ($\perp$)- samples.

It is found at 77 K that:

1. the absolute values of V are lower than those of the corresponding states at 300 K;
2. the V-value for states 2 and 3 and their deformation strain dependences practically coincide;
3. no difference was detected between these dependence for (ǁ) – and (⊥)- samples.

Using the obtained values of V for 300K and Eq. (3), we can estimate the dislocation density ρ for states 2 and 3. For $\varepsilon \sim 2\%$ we obtained $\rho = 6{,}5 \cdot 10^{-10}$ sm^{-2} (State 2), $\rho = 7{,}25 \cdot 10^{-10}$ sm^{-2} (State 3). For $\varepsilon \sim 20\%$ $\rho = 8{,}1 \cdot 10^{-10}$ sm^{-2} (State 2) and $\rho = 9 \cdot 10^{-10}$ sm^{-2} (State 3, $\varepsilon \sim 12\%$).

These estimates agree with the electron-microscopic result on the dislocation density ρ [11] for submicrocrystalline BT1-00 titanium. In [11] for 300 K ρ was observed to increase nearly by an order of magnitude during plastic deformation. The discrepancy in V-value found in this study for states 2 and 3 correlated well with electron-microscopic measurements of dislocation density in NS Ti after ECA pressing [5]. It was found [5] that in Ti in state 3 the dislocation density in the grains is an order of magnitude higher than in state 2.

The changes in V observed during deformation suggest that in NS Ti the thermally activated motion at 300 K and 77 K is controlled by the obstacles formed by dislocations of other slip and twinning systems rather than by impurity atoms. We would like to emphasize that the authors of [11] have arrived at a similar conclusion. Their findings (growth of ρ with ε, essential changes in the intragrain dislocation substructure, formation of metallographic texture, loss of deformation-caused striation contrast at the intragrain boundaries at the earliest stages of plastic flow numerous lattice dislocations trapped by grain boundaries) lead to the conclusion that the intragrain dislocation slip is the main mechanism of deformation at 300 K.

Thus, despite of the small grain size, the thermally activated motion of intragrain dislocations in NS Cu and NS Ti can be considered as a controlling factor.

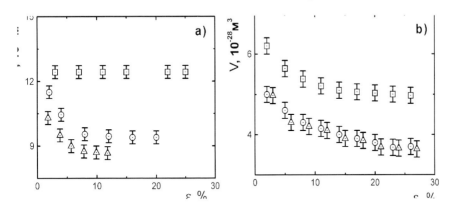

Figure 2: (a, b) Strain dependence of activation volume for Ti for 300 K (a) and for 77 K (b) in different structural states ☐–1, O – 2, Δ– 3

Finally, we note that in this study we were able to measure the low temperature strain rate sensitivity of the deforming stress of NS titanium and cupper only for two temperatures. This gave us no way of carrying out a detailed thermo activation analysis of plasticity and specifying unambiguously the microscopic mechanisms responsible for plastic flow in the material, such as the type of dislocations, the mechanism of their nucleation, the nature and parameters of barriers impeding slip and so on. Such a study is planned for the future. It will include detailed measurements of the temperature dependences of yield stress and strain-rate sensitivity of the flow stress, as well as the thermo activation analysis of these dependences within the procedure used earlier for CG titanium [12].

4 Acknowledgements

The authors are indebted to V. S. Kovaleva and V. A. Moskalenko for helpful discussions and to S. N. Khomenko for assistance in experiments and preparation of the paper. This study has been partially supported by INTAS-2001 within project 0320.

5 References

[1] T. C. Lowe, R. Z. Valiev, Investigations and Applications of Severe Plastic Deformation, NATO Science Series 3. High Technology, Kluwer Academic Publishers, Dordrecht, 2000, 80, 394
[2] V. Z. Bengus, E. D. Tabachnikova, V. D. Natsik, Phys. Phisika nizkih temperatur (in russian). 2002, 28, 11
[3] V. Z. Bengus, E. D. Tabachnikova, K. Csach, J. Miskuf, V. V. Stolyarov, R. Z. Valiev, V. Ocelik, J. Th. M. De Hosson, Proceedings of the 22nd Riso International Symposium on Materials Science, 2001, 217–222
[4] A. G. Evans, R. D. Rawlings, Phys. Status Solid, 1969, 34, 1, 9–32
[5] R. Z. Valiev, I. V. Alexandrov, The nanostructurals materials manufactured under intensive plastic deformation, Moscow, Logos, 2000, p. 271
[6] J. Lian, B. Baudelet, A. A. Nazarov, Mater. Sci. Engin., 1993, A172, 23–30
[7] J. P. Hirth, J. Lothe, Theory of Dislocations, New York: Willey, 1982, p. 331
[8] G. Baur and P. J. Lehr, Less-Cpmmon Met., 1980, 69, 203
[9] V. N. Kovaleva, V. A. Moskalenko and V. D. Natsik, Philosophical Magazine A, 1994, 70, No.3, 432–438
[10] V. A. Moskalenko, A. R. Smirnov, Materials Scince and Eng., 1998, A246, 282-288
[11] S. Y. Mironov, G. A. Salichev, M. M. Mishlaev, Phisika metalov i metalovedenie (in Russian), 2002, 93, No. 4, 75–87
[12] V. N.Kovaleva, V. A. Moskalenko, V. D Natsik, Philosophical Magazine, 1994, A 70, 423–438

IV Modelling of SPD and of Mechanical Properties of SPD Materials

Importance of Disclinations in Severe Plastically Deformed Materials

A. E. Romanov
Ioffe Physico-Technical Institute, St. Petersburg, Russia

1 Abstract

A short overview of recent achievements in disclination approach application to severe plastically deformed materials is given. Necessary definitions and designations: Volterra rotational and translational dislocations, disclination Frank (rotation) vector, wedge and twist disclinations, are introduced and explained. The properties of screened disclination configurations (loops, dipoles, defects in small particles etc.) with relatively small energies are considered. Disclination models for the processes in the structure of plastically deformed materials are reviewed. The bands with misorientated crystal lattice in metals and other materials are described as a result of partial wedge disclination dipole motion. The disclination approach is applied to the description of workhardening at large strains and to the analysis of grain boundaries and their junctions in conventional polycrystals and nanocrystals.

2 Introduction

It was shown over the past two decades that nanostructured materials demonstrate a unique set of physical and mechanical properties. [1–3] Their unique properties originate mainly from two factors:

1. size effects [2] and
2. presence of high density of interfaces in the material interior. [1,3]

Additional factors are the elastic distortions, which may also be present in materials prepared by severe plastic deformation. [3] The interfaces (grain or phase boundaries) separate neighboring regions (grains) with otherwise perfect crystal lattice. In mono-phase nanostructured materials (nanocrystals) the grain boundaries (boundaries of crystal lattice misorientation) form a three-dimensional network with linear junctions and point nodes. Usually these are triple junctions and four-fold nodes. The density of junctions and nodes drastically increases with decreasing grain size. [4] In nanocrystals prepared by severe plastic deformation technique the refinement of grains takes place during material processing. [3] Therefore, both the properties of nanocrystals and the nanocrystalline material processing itself cannot be understood without experimental and theoretical analysis of grain boundary network.

It was suggested that in bulk nanocrystals produced by extreme plastic deformation grain boundaries are in metastable, non-equilibrium state. [3] Non-equilibrium grain boundaries generate long-range elastic stress fileds, which are responsible for modified properties of nanocrystals. [3,5] The structure of non-equilibrium grain boundaries has been modeled in terms of disordered arrays of grain boundary dislocations and/or disclinations. [6] The other important sources of internal stresses in nanocrystals are junction disclinations. These junction disclinations have been proved to be necessary elements of defect structure at large strains [7,8] They appear as a result of the incompatibility of plastic strain in neighboring misoriented grains. [9] Disclinations contribute to the latent energy of nanocrystals and volume expansion of materials produced by severe deformation [10] They also influence work-hardening in fine grain materials. [2,11] Nucleation and motion of disclinations were shown to be a relevant mechanism for grain subdivision in the course of plastic deformation. [12]

The present article gives a brief introduction to the disclination concept together with the list of relevant references. The properties of disclinations –their geometry, elastic fields, and energies are then discussed and the examples of disclination observation in the structure of deformed materials are given. The disclination models of rotational structures in strongly deformed materials are introduced and explained The effect of screening of disclination elastic field in nanostructured materials and nanoparticles is considered in connection with relaxation processes in such materials.

3 Definitions of Disclinations. Perfect and Partial Disclinations

The concept of disclinations was originally introduced in the mechanics of deformed solids (see references in Nabarro and Romanov et al. [13,14]) Vito Volterra considered disclinations as the rotational dislocations along with the translational dislocations [15] The latter ones are now simply referred to as dislocations. [16] Similar to dislocations, disclinations are linear defects with a long-range distortion field. [14] It is known that the strength of a dislocation is determined by its Burgers vector b. In the same manner, the strength of a disclination is related to an axial vector ω (Frank vector). The Frank vector gives the relative displacement $[u]$ of the cut faces bound by the defect line in the following form

$$[\mathbf{u}] = \omega \times (\rho_\omega - \mathbf{r}) \tag{1}$$

where r is the radius-vector and r_ω is the position of the axis for ω. The displacement defined by Equation 1 corresponds to rigid body rotation. This explains why sometimes ω is cited as the vector of rotation.

With regard to the angle between the vector ω and the line vector l, analogously to the screw and edge dislocations two fundamental types of disclinations can be defined: wedge disclinations with $\omega \parallel l$ and twist disclinations with $\omega \perp l$. Wedge disclination can be most easily imagined as an additional wedge of the material inserted in or taken out from the elastic body. Figure 1 gives an example of so-called negative wedge disclination where the wedge of angle ω is inserted into the region between the cut faces. As a result, near the negative disclination line (i.e., near the vertex of the wedge) the material is under compression. For positive wedge disclinations the wedge has to be taken out and the material near the disclination line is under tension. The physical distinction between positive and negative wedge disclinations is related to the fact that ω is a pseudo-vector. [14]

In crystals the vectors b or ω of perfect dislocations or disclinations, respectively, must be compatible with the lattice symmetry. That means, that the module of b has to be a translation period of the crystal lattice, while that of ω has to correspond to the multiplicity n of a symmetry

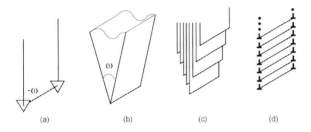

Figure 1: Relation between wedge disclinations and terminated edge dislocation walls. a) negative wedge disclination with strength ξ; b) wedge of angle ξ, which has to be inserted to create the disclination; c) a set of half-planes modeling the wedge; d) terminated wall of equidistant edge dislocations.

axis of the lattice: $\xi = 2\pi/n$ ($n = 2,3,4,6$). Such Frank vectors cannot be realized without distortion of the whole crystal structure. The elastic stresses of straight disclination diverge with the radial distance r from the defect line as $\ln r$. [14] Therefore, it was assumed for a long time that disclinations cannot play any role in bulk crystals. [13]

However, as illustrated in Figures 1c,d for the case of a wedge disclination, an equivalent dislocation configuration can be introduced. In this case, straight wedge disclination is modeled as terminated dislocation wall. Such a model gives rise to the idea of partial disclinations, i.e., linear rotational defects with smaller vectors ω but which are also associated with boundaries of misorientation. It is well documented that boundaries of misorientation observed in crystals cover a wide range of angles regardless of the crystal lattice type. [8] So it is not crystallographic considerations that determine the nature of the rotations. In the same manner, we may conclude that disclinations in crystals are partial, i.e., imperfect; they inevitably entail surface defects such as subgrain, grain or twin boundaries. Therefore, the partial disclination strength ξ can vary over a wide range. The disclination line characteristic scale also may have a spectrum of length values. We may expect, however, that typical partial disclinations will have their segment sizes comparable to the elements of material structure, i.e., grain, subgrains, or fragments. Disclination dipoles or quadrupoles can be formed by partial disclinations of opposite sign. [8,14] Defects of this kind (disclination dipoles) are compatible with the crystal structure and can be considered as superdislocations having an effective Burgers vector of magnitude $B_{eff} = 2a\omega$ ($2a$, dipole arm), and as in the case of straight dislocations their strain and stress far fields decrease as $1/r$.

4 Rotational Structures in Deformed Metals: Relation to Partial Disclinations

The examples of defect configurations, which can be interpreted in terms of disclinations, can be easily found in the structure of strongly deformed crystals. Figure 2 presents the typical microband pattern observed in rolled copper single crystal face-centered cubic (fcc) lattice. [17,18] Microbands are regions of the crystal with misoriented crystal lattice; they may give rise to a dark contrast in the transmission electron microscopy (TEM) images (as can be seen in Fig. 2a) obtained under primary beam conditions. The misorientation in microbands is up to a few degrees. In the material region shown, microbands form junctions of cell walls. Special analysis based on the experimental technique developed in [17,19] proves the presence of disclination defects in two neighboring junctions. It follows from the schematics of the Figure 2b that disclinations have a strong dipole component (the vectors ω_1 and ω_2 are practically antiparallel). The absolute value of disclination strength in this case is approx. one degree.

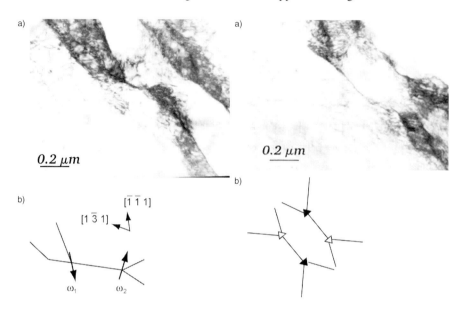

Figure 2: Microbands in a copper single crystal rolled down to 70 % thickness reduction at room temperature [18]. a) TEM image of typical defect structure with cell wall junctions; b) schematics of junction disclinations with Frank vectors ξ_1 and ξ_2; characteristic crystallographic directions are shown in the region above the junctions.

Figure 3: Disclination quadrupole configuration in strongly deformed copper [18]. TEM image a) and schematics b) for disclination configuration. The deformation conditions are the same as for Figure 2. Filled and empty triangles designate positive and negative disclinations respectively.

A further example of partial disclinations in rolled Cu, which is given in Figure 3, demonstrates a quadrupole configuration with similar magnitude of disclination strength, i. e., of the order of one degree. When disclinations form a quadrupole their elastic fields are mainly localized in the "region between disclinations and the energy depends only on geometrical parameters of the quadrupole.

The presence of disclinations and related terminated boundaries of misorientation does not depend on the crystal lattice type. [8,20] Examples supporting the above statement are given in Figures 4,5 for tungsten, body-centered cubic (bcc), and titanium, hexagonal closed-packed (hcp), correspondingly. The examples of disclination related structures in "deformed metals shown in Figures 2 to 5 reveal one common feature: at the place of cell walls termination and in nearby material's regions a non-uniform TEM contrast can be detected. We assume that such a diffraction contrast can be associated with the elastic distortions of crystal lattice caused by disclinations. Modeling the diffraction contrast can provide a tool for disclination identification in deformed materials. [21]

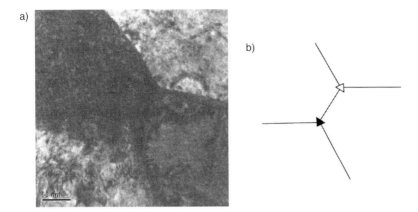

Figure 4: A pair of non-compensated triple junctions (disclination dipole) in tungsten rolled down to 70 % thickness reduction at 600 °C (after V. Klemm and P. Klimanek). TEM image a) and schematics b) for dipole configuration.

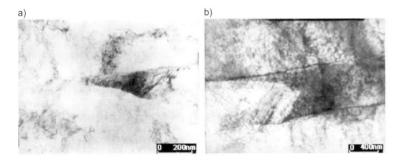

Figure 5: Terminated boundaries of misorientation in α-Ti deformed at room temperature [20]. a) Non-compensated triple junction; b) dipole of terminated dislocation walls.

5 Disclinations Properties: Geometry, Elasticity, Screening

To apply disclination approach to the modeling of deformation behavior of materials (in particular, SPD materials) the properties of individual disclination defects and their ensembles have to be well understood.

Geometrical laws are the foundation of disclination theory. The geometrical approach is based on the analysis of distortions introduced by a disclination when it is created in initially defectless material. Obviously, the material response depends both on the character of a disclination and the material properties. For example, geometrical laws for disclinations in liquid crystals are completely different from those for structureless solid body. [22] Geometrical consideration includes the study of the relation between dislocations and disclinations and the analysis of disclination motion. An important branch of geometrical approach relates continuous distributions of defects with the properties of curved space. These and other details of the disclination geometry can be found by Romanov and Vladimirov [14] and Anthony. [23]

In the elastic continuum disclinations generate strong elastic fields. We already mentioned that elastic strains and stresses have a logarithmic divergence at large distances from the line of a straight disclination. As a result, the elastic energy associated with a single disclination scales as L^2, where L is a characteristic length of the elasticity problem. For example in linear isotropic elastic solid the energy of disclinated cylinder (per unit length) is [14]

$$E = \frac{G\omega^2}{16\pi(1-v)} R^2 \qquad (2)$$

where R is cylinder radius, ω is wedge disclination strength, G is the shear modulus, and μ is Poisson's ratio. For a wedge disclination placed at the distance d parallel to the free surface of a half-space the energy is [14]

$$E = \frac{G\omega^2}{4\pi(1-v)} d^2 \qquad 3)$$

At present a set of expressions similar to Equations 2,3 have been obtained for various geometry of elastic bodies, various configurations of disclination lines and various materials properties, i.e., elastically anisotropic solids, elastically nonlinear solids etc. Relevant references may be found in. [14,24]

The above examples for disclination energies lead to the important concept of disclination elastic field screening. [25] When disclinations appear in small volumes or close to free surfaces their energy can be rather low and it can be comparable with the energies of other defects characteristic of deformed materials. In such cases one can speak of screened disclination configurations. Screening can be achieved not only with the help of free surfaces, but also with the aid of other defects. Self-screening is realized for dipole, quadrupole and disclination loop configurations. As it was already shown, in the process of plastic deformation the screened configurations of partial disclinations come into play.

6 Disclination Modeling of SPD Materials

6.1 Modeling of Rotational Structures in Strongly Deformed Materials

By definition partial disclinations are associated with terminated boundaries of misorientations in otherwise perfect crystals. [14] Wedge disclination, as we have shown above, can be considered as a mesoscopic model for terminated tilt boundary. Boundaries of misorienation related to disclinations have a different nature. They can be low angle dislocation walls (Fig. 6a) with the misorientation angle ξ_{DW}, high angle grain boundaries (Fig. 6b) with misorientations ξ_{GB}, or twin boundaries with misorientation ξ_{TB}. The geometry of terminated boundaries of misorientation can also be different. For single terminated tilt boundaries the disclination strength is exactly equal to the angle of misorientation. In Figure 6a the negative wedge disclination ξ with the strength

$$w = jDW \qquad (4)$$

is shown. At the junction of several boundaries (Fig. 6b) the resulting disclination with the strength ω_j can appear under the condition

$$\omega_j = \sum_i \omega_i = \sum_i \varphi_{GB_i} \neq 0 \quad (i = 1, 2, 3) \tag{5}$$

i. e., in the case of so-called non-compensated junctions. Disclination dipole configurations can be associated with terminated lamellae with misoriented crystal lattice: misorientation and kink bands or twins (as shown in Fig. 6c). [14] In the latter case two terminated boundaries of misorientation belong to the disclination dipole.

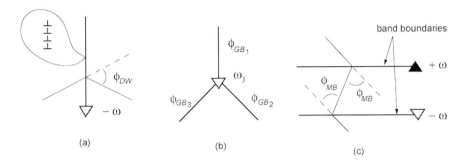

Figure 6: Disclination models for terminated tilt boundaries in structure of deformed materials. a) single terminated dislocation wall with misorientation φ_{DW}; b) triple junction of high angle grain boundaries; φ_{GBi} (i = 1,2,3) are misorientations at individual grain boundaries; (c) misorientation (kink) band with misorientation angle φ.

The concept of screened disclinations enables one to describe properties of misorientation boundaries themselves. The most promising here is the application of the disclination model to high-angle grain boundaries where the dislocation description does not work. [26,27] Within the framework of this model and using a limited number of experimentally measured parameters the dependence $E_g(\varphi)$ of grain boundary energy on misorientation φ was evaluated. [27] It was shown that calculated dependences $E_g(\varphi)$ for symmetrical equilibrium tilt boundaries in Al are in good agreement with experiment. More recently disclination models were applied for the multiscale analysis of grain boundary properties in polycrystalline diamond [28] including different orientation of grain boundary plane. [29] Disclination models enable one to investigate the so-called non-equilibrium state of grain boundaries. [6] For this purpose a random distribution of disclination dipoles should be considered. Due to the non-uniformity in the dipole arrangement the grain boundary contains an additional energy which may be comparable with the grain boundary energy in the lowest state. Additionally, grain boundaries with so-called quasiperiodic structure have been analyzed. [30]

6.2 Disclination Mechanism for Plastic Deformation and Work-Hardening

When moving, disclinations contribute to the plastic deformation. To calculate the amount of strain ε the geometry of disclination motion must be known in detail. In "order to obtain an elementary estimate of strain rate one can apply the following relation: [25]

$$\frac{d\varepsilon}{dt} = \theta \omega l_r V \qquad (6)$$

where θ is the density of screened disclination systems (dipoles) with the strength ω, V their velocity, and l_r the characteristic size (the dipole arm $2a$). The motion of the dipoles of wedge partial disclination (as it is shown in Fig. 7) is the main mechanism of rotational plastic deformation. [14,25] The terminated band of misorientation is modeled as a disclination dipole. The dipole movement occurs when edge dislocations travel from the bulk of the grain to the planes where partial disclinations are located [31] (see Fig. 7b). The disclination approach permits a "description of the details of the development of misorientation bands in crystals and other solids. It provides a connection of the parameters of a band (the thickness $2a$ and misorientation $\varphi = \omega$) with the density of mobile dislocations ρ and applied stress σ_e. [31]

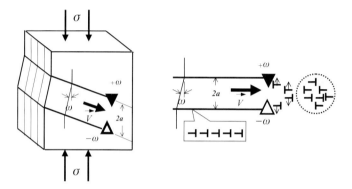

Figure 7: Disclination contribution to the plastic strain. a) macroscopic change in the sample shape to due to the disclination dipole motion; b) microscopic mechanism of disclination dipole motion related to redistribution of dislocations in front of the dipole.

The nucleation of disclination dipoles and corresponding misorientation bands usually takes place near stress concentrators. [8,14] Grain boundary junctions and other imperfections of grain boundary structure serve as such concentrators. The models considering the wedge disclination dipole generation near triple junctions predict the value for critical external shear stress σ_g, above which an event of disclination nucleation takes place. [32] The numerical estimates for σ_g give the values of $G/1000$. These estimates correspond to the level of deforming stress at the stage III of ρ-ε diagram for initially monocrystalline materials. The critical stress σ_g is shown to be strongly dependent on the geometry and strength of initial disclinations at grain boundary faults. The further development of disclination structure at an isolated grain boundary fault demonstrates two main regimes of misorientation band development: stable and unstable propagation. The transition between the above two regimes of misorientation band development is controlled by the other critical value for external stress σ_p. [33]

Work-hardening at the stage of rotational structure formation is related to the resistance to dislocation motion in the elastic fields of existing disclinations. [14, 34] At the stage of active rotational deformation the hardening appears to be due to interaction between disclinations. Typical for disclination hardening is the following dependence of the deforming stress: [2,14]

$$s = \beta G w \qquad (7)$$

where β is a geometry factor determined by the actual type of interacting disclinations. Consideration of the above relation helps to explain deviations from the well-known dependence of the deforming stress on the dislocation density ρ:

$$\sigma = \alpha G b \sqrt{\rho} \qquad (8)$$

For the disclination mechanisms of hardening one can obtain $\rho \sim \rho^n$ with $1/2 < n \leq 1$. Such "anomalies are often observed from the onset of stage III of the $\rho(\varepsilon)$ curve on.

6.3 Screened Disclinations in SPD Materials and Nanoparticles

It was reported that disclination mechanisms of hardening play an important role for nanocrystals. [11] Other characteristic values were also estimated for nanocrystalline materials in the framework of disclination approach. [10] They included root mean square (rms) strain, dilatation and stored energy. It was found, for example, that disclination contribution to dilatation δ for ultra fine grain Al (grain size 100 nm) can be of the order of 10^{-4}. One may expect that disclinations will become more important with grain size reduction.

Another important area for the application of the disclination approach can be found in the description of the properties of so-called pentagonal nanoparticles, [35] which are frequently observed for fcc materials. Such particles have characteristic shape of decahedron or icosahedron and possess axes with five-fold symmetry. It was proven that each five-fold axis appears at the junction of five twin boundaries and contains a positive wedge disclination of the strength

$$\omega = 2\pi - 10\sin^{-1}\left(\frac{1}{\sqrt{3}}\right) \approx 7° 20' \qquad (9)$$

Owing to the presence of disclinations, non-uniform elastic distortions are generated in the particle interior. Elastic stresses related to these distortions may then relax in various ways. It was proposed, [36] that the principal relaxation channels in pentagonal particles are the creation of additional lattice dislocations, displacement of the pentagonal axis (which is equivalent to the motion of the disclination towards the surface of the nanoparticle, splitting of the pentagonal axis (i.e., formation of new partial disclinations) and the emergence of pores in the disclination core. These findings are in good agreement with the available experimental results. [35]

7 Conclusions

Disclination defects are important elements of the structure of materials obtained by severe plastic deformation. Disclinations contribute to high level of internal stresses and to the occurrence of anomalies in deformation behavior of strongly deformed materials. Disclinations defects are found in junctions of several grain or subgrain boundaries and they can be identified by applying diffraction methods of the structure analysis of distorted crystals.

In connection with materials structure research it is important to stress that the disclination concept is able to

1. simultaneously explain the origin of significant lattice rotations and long-range stresses in strongly deformed metals;
2. simplify the description of cooperative dislocation slip and patterning in terms of diclination generation and movement;
3. couple the micro- and the mesoscales of plastic deformation.

8 Acknowledgement

This work was supported by Volkswagen Foundation (Research Project I/74 645) and by the Physics of Solid State Nanostructures Program of the Ministry of Industry and Sciences of "Russia. The author is grateful to Prof. P. Klimanek and Dr. V. Klemm for fruitful cooperation and for the possibility to present the TEM images of deformed metals.

9 References

[4] H. Gleiter, Prog. Mater. Sci. 1989, 33, 223
[5] V. G. Gryaznov, L. I. Trusov, Prog. Mater. Sci. 1993, 37, 289
[6] R. Z. Valiev, R. K. Islamgaliev, I. V. Alexandrov, Prog. Mater. Sci. 2000, 45, 103
[7] W. Ning, G. Palumbo, W. Zhirui, U. Erb, K. T. Aust, Scr. Metall. Mater. 1993, 28, 253
[8] A. A. Nazarov, A. E. Romanov, R. Z. Valiev, Nanostruct. Mater. 1995, 6, 775
[9] A. A. Nazarov, A. E. Romanov, R. Z. Valiev, Acta Metall. Mater. 1993, 41 1033
[10] V. A. Likhachev, V. V. Rybin, Izvestia AN SSSR, Ser. Fiz 1973, 37, 2433
[11] V. V. Rybin, Large Plastic Deformation and Fracture of Metals, Metallurgia, Moscow 1986
[12] V. V. Rybin, A. A. Zisman, N. Yu. Zolotorevsky, Acta Metall. Mater. 1993, 41, 2211
[13] A. A. Nazarov, A. E. Romanov, R. Z. Valiev, Scr. Mater. 1996, 34, 729
[14] V. G. Gryaznov, M. Yu. Gutkin, A. E. Romanov, L. I. Trusov, J. Mater. Sci. 1993, 28, 4359
[15] M. Seefeldt, L. Delannay, B. Peeters, S. R. Kalidini, P.'van Houtte, Mater. Sci. Eng., A 2001, 319–321, 192
[16] F. R. N. Nabarro, Theory of Crystal Dislocations, Clarendon Press, Oxford 1967
[17] A. E. Romanov, V. I. Vladimirov, Dislocat. Solids 1992, 9, 191
[18] V. Volterra, Ann. Ecole Normale Sup., Paris 1907, 24, 401
[19] J. P. Hirth, J. Lothe, Theory of Dislocations, McGraw-Hill, New York 1982
[20] V. Klemm, P. Klimanek, M. Seefeldt, Phys. Status Solidi A 1999, 175, 569
[21] V. Klemm, P. Klimanek, M. Motylenko, Mater. Sci. Eng. A 2001, 324, 174
[22] V. Klemm, P. Klimanek, M. Motylenko, Solid State Phenom. 2002, 87, 57
[23] P. Klimanek, V. Klemm, A. E. Romanov, M. Seefeldt, Adv. Eng. Mater. 2001, 3, 877
[24] A. L. Kolesnikova, V. Klemm, P. Klimanek, A. E. Romanov, Physica Status Solidi A 2002, 191, 467
[25] M. Kleman, Points, Lines and Walls, Wiley, New York 1983

[26] K.-H. Anthony, Solid State Phenom. 2002, 87, 15
[27] F. Kroupa, Lejcek, Solid State Phenom. 2002, 87, 1
[28] A. E. Romanov, Mater. Sci. Eng. 1993, A 164, 58
[29] J. C. M. Li, Surf. Sci. 1972, 31, 12
[30] V. Yu. Gertsman, A. A. Nazarov, A. E. Romanov, R. Z. Valiev, V. I. Vladimirov, Philosophical Magazine 1989, A 59, 1113
[31] O. A. Shenderova, D. W. Brenner, A. A. Nazarov, A. E. Romanov, L. Yang, Phys. Rev. B 1998, 57, 3181
[32] A. A. Nazarov, O. A. Shenderova, D. W. Brenner, Phys. Rev. B 2000, 61, 928
[33] K. N. Mikaelyan, I. A. Ovid\9ko, A. E. Romanov, Mater. Sci. Eng. A 2000, 288, 61
[34] V. I. Vladimirov, A. E. Romanov, Sov. Phys. Solid State 1978, 20, 1795
[35] M. Yu. Gutkin, A. E. Romanov, P. Klimanek, Solid State Phenom. 2002, 87, 113
[36] M. Yu. Gutkin, K. N. Mikelyan, A. E. Romanov, P. Klimanek, Physica Status Solidi A 2002, 193, 35
[37] M. Seefeldt, Rev. Adv. Mater. Sci. 2001, 2, 44
[38] V. G. Gryaznov, J. Heydenreich, A. M. Kaprelov, S. A Nepijko, A. E. Romanov, J. Urban, Cryst. Res. Technol. 1999, 34, 1091
[39] V. G. Gryaznov, A. M. Kaprelov, A. E. Romanov, I. A. Polonsky, Physica Status Solidi B 1991, 167, 441

Disclination-Based Modelling of Grain Fragmentation during Cold Torsion and ECAP in Aluminium Polycrystals

M. Seefeldt, P. Van Houtte,
K.U. Leuven, Heverlee; Belgium

1 Introduction

The present work considers the ultra-fine microstructures resulting from severe plastic deformation at low temperatures (i. e. with negligibly slow diffusion) as deformation micro-structures, their formation as a process of continuous fragmentation or grain subdivision, and this fragmentation as a process of nucleation and growth of fragment boundary segments. In terms of defect theory, this approach is best reflected by the disclination concept. Fragmentation by nucleation and growth of fragment boundary segments translates into generation, propagation and immobilization of (partial) disclinations.

Disclinations are rotational line defects, lines of discontinuity of lattice rotation – like dislocations are translational line defects, lines of discontinuity of shear [1,2]. For example, the two ends of a both-side terminating wall of edge dislocations with the same sign are lines of discontinuity of misorientation of opposite signs and thus form a partial disclination dipole. The end stresses around the tips of the terminating wall are equal to the the long-range stress field of a partial disclination dipole. These stresses can attract additional mobile dislocations and attach them to the terminating wall. That means that the end lines of the wall move or, in disclination language, that the partial disclinations propagate. This propagation is a non-conservative motion which has to be maintained by dislocations – like non-conservative climb of edge dislocations has to be maintained by vacancies. The propagation of a partial disclination thus leaves a dislocation wall, a low-angle grain boundary, behind – like a moving partial dislocation leaves a stacking fault behind. The theory of disclinations in crystals has been developed since the 1960s (see e.g. [1,2]), but has only recently found indirect as well as direct experimental confirmation.

2 Coupled Substructure and Texture Evolution Model

The structure of the coupled substructure-texture model proposed in this paper is as follows: The substructure is represented in terms of defect densities. The defects undergo interactions with each other, so that the balance equations for the defect densities include generation, immobilization and annihilation rates and are driven by the mobile dislocation currents, i.e. the slip rates. These slip rates are obtained from the texture evolution model. On the other hand, the substructure evolution results in increasing critical resolved shear stresses and in increasing misorientations which are passed to the texture code. Starting from a certain critical intra-granular misorientation, the „old" grain is replaced by the „new" fragments in the texture.

2.1 Representation of the Substructure

A preexisting cell structure is used as a starting-point for the substructure model. Six cell wall families are defined as follows: two intersecting dislocations from different slip systems can undergo two different types of interaction: attractive interaction resulting in the formation of Lomer-Cottrell locks, Hirth locks and other barriers, or repulsive interaction resulting in cutting processes, possibly leaving dislocation loops or debris behind. Since there are four slip systems in two intersecting slip planes which undergo attractive interaction and form sessile locks and since these barriers are supposed to be the nuclei of cell walls, the six possible combinations of two intersecting slip planes (corresponding to the edges of Thompson's tetrahedron) correspond to six possible families of cell walls.

On the one hand, it is assumed that dislocations can get stored in the cell walls through both types of interaction (with different storage rates though). On the other hand, it is proposed that only attractive, junction-forming interaction can contribute to the formation of misorientations because the dislocation loops and debris left behind from cutting processes would always include both signs.

The fragment or cell block structure is superimposed on the cell structure on a larger scale [3,4]. It is assumed to evolve through excess dislocation layers arising at and expanding along existing cell walls. The tips of these expanding layers, the partial disclinations, thus use the cell walls to propagate – like dislocations use the slip planes to glide.

2.2 Elementary Processes and Balance Equations

The cell structure is assumed to evolve in the interplay of dislocation storage and recovery in the cell walls. The underlying elementary processes are individual trapping of mobile dislocations into cell walls and annihilation of mobile with wall dislocations of opposite sign. The fragment structure is assumed to develop on the background of the existing cell structure through creation and storage of disclinations. The underlying elementary processes are collective trapping of groups of excess dislocations of the same sign at cell walls (i. e. generation of partial disclination dipoles), capturing of mobile dislocations by the end stresses around the ends of the excess dislocation groups (i. e. propagation of partial disclination dipoles), and immobilization of propagating partial disclinations at fragment boundaries.

2.2.1 Cell Structure - Dislocations

The elementary processes which control the evolution of the cell structure can be quantified as follows (one slip system):

1. The rate of immobilization of individual dislocations at cell walls scales with the probability P of an individual dislocation getting trapped, with the inverse cell diameter $1/d_c$ and with the mobile dislocation current $v\rho_m = \dot{y}/b$. Immobilization due to either reaction-type, lock-forming or cutting-type, loop- and debris-forming interaction is represented by two different trapping probabilities P_l, and P_d, respectively.
2. The rate of pairwise annihilation of mobile with cell wall dislocations scales with the annihilation length y_a, with the cell wall dislocation density ρ_c and with the mobile dislocation

current. If it is either assumed that edge and screw dislocations exhibit similar annihilation lengths (e. g. Cu at RT) or that one character dominates the population of cell wall dislocations (e. g. Al and Ni at RT), the recovery behaviour of both characters can be summarized in one annihilation rate. In the present case, y_e is the annihilation length for disintegration of edges and, c_e is the fraction of edges among the mobile dislocations.

2.2.2 Fragment Structure - Disclinations

The elementary processes which control the evolution of the fragment structure can be quantified as follows (one slip system and one fragment boundary family):

1. The rate of generation of partial disclination dipoles (i.e. of groups of excess dislocations, idealised as both-side terminating low-angle grain boundary segments) scales with the probability of a dislocation getting immobilized, together with neighbouring dislocations of the same sign, in a collective trapping event. In order to end up with a both-side terminating fragment boundary segment, the counterparts of the collectively trapped dislocations must not get immobilized at the other side of the cell wall. In the resulting probability $(P_1(1-P_1))^N$ that a dislocation participates in a PDD-forming event, the number N of involved dislocations in the exponent reflects the required collective behaviour, while the product reflects the non-balanced character of the event.
2. The rate of capture of mobile dislocations by partial disclinations (i.e. by the end stresses around the tips of the terminating fragment boundary segments) scales with a capturing length y_c up to which the partial disclination stresses (or end stresses) are strong enough to capture a mobile dislocation from a random distribution.
3. The rate of immobilization of propagating partial disclinations is assumed to scale only with the inverse fragment diameter $1/d_f$ and with the propagating disclination current $V\theta_m$ where the disclination mean velocity V is controlled by the capturing process and depends on the PDD width $2a_p$ and strength ω_p. The probability of a propagating partial disclination getting stopped at a fragment boundary is thus assumed to be 1.

2.2.3 Balance or Evolution Equations

Putting all the generation, immobilization and annihilation rates together gives the following set of balance or evolution equations for the defect densities. The index (i) is for the twelve f.c.c. slip systems, the index (w) for the six cell wall and fragment boundary families; $(i_l(w))$ runs over the slip systems (i) which contribute through lock-forming interaction to the evolution of cell wall and fragment boundary family (w), $(w_l(i))$ and $(w_d(i))$ run over the cell wall families which act as obstacles for lock-forming or debris-forming interaction, respectively, for mobile dislocations in the slip system (i). For details, see [5].

$$\frac{d\rho_c^{(i)}}{dt} = \left(\left(\sum_{w_l(i)} \frac{P_l f_{geom}}{d_c^{(w_l(i))}} + \sum_{w_c(i)} \frac{P_c f_{geom}}{d_c^{(w_d(i))}} \right) - 2 y_e c_e \rho_c^{(i)} \right) \frac{\dot{\gamma}^{(i)}}{b} \tag{1}$$

$$\frac{d\rho_{exc}^{(w)}}{dt} = \left(\frac{(P_l(1-P_l))^N f_{geom}}{d_c^{(w)}} + \frac{(P_f(1-P_f))^N}{d_f^{(w)}} + y_c c_e \theta_m^{(w)} \right) \sum_{i_l(w)} \frac{\dot{\gamma}^{(i_l(w))}}{b} \quad (2)$$

$$\frac{d\theta_m^{(w)}}{dt} = \left(\frac{(P_l(1-P_l))^N f_{geom}}{Nbd_c^{(w)}} + \frac{(P_f(1-P_f))^N}{Nbd_f^{(w)}} - \frac{10 a_p c_e}{4\omega_p^{(w)}} \sqrt{\theta_i^{(w)}} \theta_m^{(w)} \right) \sum_{i_l(w)} \dot{\gamma}^{(i_l(w))} \quad (3)$$

$$\frac{d\theta_i^{(w)}}{dt} = \left(\frac{10 a_p c_e}{4\omega_p^{(w)}} \sqrt{\theta_i^{(w)}} \theta_m^{(w)} \right) \sum_{i_l(w)} \dot{\gamma}^{(i_l(w))} \quad (4)$$

Note that during numerical integration of these evolution equations, the dislocation content of each cell wall and fragment boundary family is registered according to their Burgers vector.

2.2.4 Cell Sizes, Fragment Sizes, Fragment Boundary Orientations and Misorientations

The spacings $d_c^{(w)}$ between the cell walls are derived from the wall dislocation densities by using a Holt-type scaling law and a principle of similitude separately for each family *(w)*. The spacings between the fragment boundaries $d_f^{(w)}$ are estimated from the density of fragment boundary triple junctions $\theta_i^{(w)}$, assuming a parallelepipedal structure of the fragment boundary mosaic with the partial disclinations in the mosaic nodes,

$$d_c^{(w)} = \frac{K}{\sqrt{\sum_{i_l(w)} f_l^{(i \to w)} \rho_c^{(i_l(w))} + \sum_{i_d(w)} f_d^{(i \to w)} \rho_c^{(i_d(w))}}} \quad \text{and} \quad d_f^{(w)} = \frac{1}{\sqrt{2\theta_i^{(w)}}} \quad (5)$$

where K is the Holt constant and $f_l^{(i \to w)}$ denotes the fraction of cell wall dislocations in slip system *(i)* which got stuck in cell walls of family *(w)*. The average misorientation $\varphi_f^{(w)}$ across fragment boundaries of family *(w)* is calculated by distributing the excess dislocation density on the total available boundary surface. For simplicity, a homogeneous distribution on all boundaries of one family as well as within all boundaries is assumed,

$$\varphi_f^{(w)} = \frac{b \rho_{exc}^{(w)}}{2 d_f^{(w)}} \quad . \quad (6)$$

Since the Burgers vectors of all dislocations in the fragment boundary families have been registered during the numerical integration of eqs. (1)–(4), the normal and misorientation axis vectors of the boundaries can be calculated, in case if it is assumed that the boundaries are tilt boundaries, i.e. composed of edge dislocations. One has to sum up the Burgers and line vectors of all dislocations in a wall segment,

$$\hat{n}^{(w)} = \frac{\sum_{i_l(w)} N^{(i_l(w))} b^{(i_l(w))} + \sum_{i_d(w)} N^{(i_d(w))} b^{(i_d(w))}}{\left| \sum_{i_l(w)} N^{(i_l(w))} b^{(i_l(w))} + \sum_{i_d(w)} N^{(i_d(w))} b^{(i_d(w))} \right|} \quad , \quad (7)$$

$$\underline{\hat{s}}_{sub}^{(w)} = \frac{\sum_{i_j(w)} N^{(i_j(w))} \underline{s}^{(i_j(w))}}{\left|\sum_{i_j(w)} N^{(i_j(w))} \underline{s}^{(i_j(w))}\right|} = \frac{\sum_{i_j(w)} N^{(i_j(w))} (\underline{\hat{n}}^{(i_j(w))} \times \underline{b}^{(i_j(w))})}{\left|\sum_{i_j(w)} N^{(i_j(w))} (\underline{\hat{n}}^{(i_j(w))} \times \underline{b}^{(i_j(w))})\right|} . \tag{8}$$

The Euler angles of the new fragment orientations are then calculated by rotating the old grain orientations by plus and minus the half misorientation vector around the misorientation axis. Please note that at the moment of splitting the „old" grain orientation into the „new" fragment orientations, the excess dislocation density of the corresponding family *(w)* has to be reset.

2.2 Texture Development

For the texture part of the model, a standard FC Taylor code has been used (see e.g. [6]).

3 Application to Torsion and ECAP

In a case study, the model outlined above is applied to a simple example of severe plastic deformation. Equal-channel angular pressing with perpendicular channels and low friction can be approximated as a sequence of simple shear deformations, see figure below [7,8].

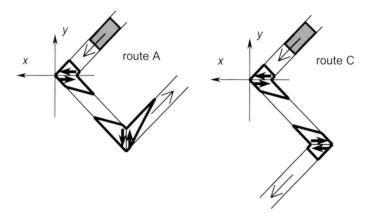

Figure 1: Schematic representation of ECAP routes *A* and *C*, approximated as sequences of simple shear deformations.

If the reference system is chosen as indicated in Figure 2, the first pass can, as a rough approximation, be represented by the same velocity gradient tensor as for simple shear and for torsion. The second pass can again be approximated as simple shear, but now with rotated shear planes and shear directions, depending on the route,

$$\underline{\underline{L}}_1 = \begin{pmatrix} 0 & \dot{\gamma}_\beta & 0 \\ 0 & 0 & 0 \\ 0 & 0 & 0 \end{pmatrix} \quad \underline{\underline{L}}_{2A} = \begin{pmatrix} 0 & 0 & 0 \\ -\dot{\gamma}_\beta & 0 & 0 \\ 0 & 0 & 0 \end{pmatrix} \quad \underline{\underline{L}}_{2C} = \begin{pmatrix} 0 & -\dot{\gamma}_\beta & 0 \\ 0 & 0 & 0 \\ 0 & 0 & 0 \end{pmatrix}. \tag{9}$$

The velocity gradient tensors are then used as input for the Taylor FC texture code which, for his part, gives as output the slip rates which drive the substructure balance equations (1)–(4).

Grain subdivision has been studied for eight orientations along the two partial fibres in a copper-type shear texture [9]. All these orientations concentrate slip on few slip systems and fragmentation on few fragment boundary families. In the first pass, all orientations except for $\varphi_1 = 40°$, $\Phi = 54.7°$, $\varphi_2 = 45°$ develop significant misorientations only in two families.

In the following, the orientation $\varphi_1 = 0°$, $\Phi = 45°$, $\varphi_2 = 45°$ is considered as an example. If one „old" orientation is replaced by two „new" fragment orientations when one family reaches a critical intra-granular misorientation of 2.5° (the choice of the critical misorientation only slightly affects the orientation path after splittting), then the first, second and third splitting events in the two active families occur at equivalent strains of 0.37 and 0.39, between 0.63 and 0.65, and between 0.89 and 0.94, respectively. After the first pass, the two fragments with the most diverging orientations have the Euler angles (–19°/45°/50°) and (3°/46°/43°), respectively. After only one pass or an equivalent strain of 1.15, one can thus obtain misorientations of about 20° and fragment boundary spacings of about 0.1 µm.

In the second pass (see Table 1), routes B+ (rotation of the specimen between the two passes with +90°) and B– (rotation with –90°) lead to the next splitting event in one fragment boundary family a little bit earlier than routes A and C. However, first of all they activate more fragment boundary families, i.e. let more fragment boundary families develop significant misorientations and reach the critical misorientation during the second pass.

Table 1: Fragmentation of the orientation (–19°/45°/50°) during the second ECAP pass for the routes A, B+, B– and C

	A	B+	B–	C
equivalent strain for the next subdivision step	1.35	1.33	1.32	1.35
number of active fragment boundary families (i.e. number of families reaching the critical misorientation during the second pass)	3	4	5	3

These results hold qualitatively for all eight studied orientations. Although they refer to states during rather than after the second pass, they indicate that the routes B+ and B- promote the activation of more fragment boundary families during the second pass and thus a more equi-axed microstructure, as reported in the literature [10]. A detailed study up to the ends of the second and further passes is of course desirable, but faces numerical difficulties because many active fragment boundary families lead to many splitting events in all fragments and let the number of texture components „explode".

4　Acknowledgements

M.S. gratefully acknowledges a postdoctoral fellowship granted by the Fonds voor Wetenschappelijk Onderzoek – Vlaanderen, Brussels.

5　References

[1]　Romanov, A.E., Vladimirov, V.I., in: Dislocations in Solids, vol. 9 (Ed.: F.R.N. Nabarro), North Holland, Amsterdam, The Netherlands, 1992, p. 191–422
[2]　Nabarro, F.R.N., Theory of Crystal Dislocations, Clarendon, Oxford, U.K., 1967
[3]　Rybin, V.V., Bolshie plasticheskie deformatsii i razrushenie metallov, Metallurgiya, Moscow, R.F., 1986
[4]　Bay, B., Hansen, N., Hughes, D.A., Kuhlmann-Wilsdorf, D., Acta metall. mater., 1992, 40, 205–219
[5]　Seefeldt, M., Delannay, L., Peeters, B., Aernoudt, E., Van Houtte, P., Acta mater., 2001, 49, 2129–2143
[6]　Van Houtte, P., Textures Microstruct., 1988, 8/9, 313
[7]　Segal, V.M., Mater. Sci. Engng. A, 1995, 197, 157–164
[8]　Segal, V.M., Mater. Sci. Engng. A, 1999, 271, 322–333
[9]　Van Houtte, P., Acta metall., 1978, 26, 591–604
[10]　Langdon, T.G., Furukawa, M., Horita, Z., Nemoto, M., in: Investigations and Applications of Severe Plastic Deformation (Eds.: T.C. Lowe , R.Z. Valiev), Proceedings of the NATO Advanced Research Workshop, 2-7 August, 1999, NATO Science Series 3. High Technology, vol. 80, Kluwer, Dordrecht, The Netherlands, 2000, p. 149–154

Modeling of Deformation Behavior and Texture Development in Aluminium under Equal Channel Angular Pressing

Seung Chul Baik[1], Yuri Estrin[2], Ralph Jörg Hellmig[2], and Hyoung Seop Kim[3]
[1] Technical Research Labs., Pohang Iron & Steel Co. Ltd., Pohang, Korea
[2] IWW, TU Clausthal, Clausthal-Zellerfeld, Germany
[3] Department of Metallurgical Engineering, Chungnam National University, Daejeon, Korea

1 Introduction

The development of nanocrystalline materials for structural applications hinges on the availability of techniques for manufacturing fully dense bulk material with the average grain size in the nanometer range. Equal channel angular pressing (ECAP) [1–3] has emerged as such a technique, see e.g. [4]. Even though the smallest grain size achievable by ECAP lies typically in the range of several hundred nanometers, a relative ease of operation and the possibility of repeated pressings with almost no cross-sectional change between the passes makes ECAP a very promising method of producing materials with a sub-micron grain size [5].

The mechanisms of this extreme grain refinement is far from being understood. There are reasons to believe, however, that it is related to a decrease of the average dislocation cell size, possibly accompanied with transformation of cell walls into large angle grain boundaries due to accumulation of misorientations between cells with strain [6]. A model that accounts for dislocation density and cell size evolution and is also capable of desribing the deformation behavior at large strains was developed in Refs. [7,8]. Here we report the results of modelling ECAP of Aluminium with a three-dimensional version of the model [8]. The report is based on some results presented earlier [9] as well as on new simulations relating to texture development under ECAP.

2 Dislocation Density Based Strain Hardening Model

The model used in the present simulations was discussed in previous publications [7–9] in great detail. Here it suffices to review it briefly. The model is based on the notion that the material can be partitioned in two 'phases': the dislocation cell walls with dislocation density ρ_w and the cell interiors with dislocation density ρ_c, which is much lower than ρ_w. The mechanisms governing the evolution of the average cell size are not considered. Instead, a certain scaling relation is assumed, see below. The dislocation densities in the two 'phases' are introduced as *scalar* internal variables of the model. Their evolution with strain determines the overall strain hardening behavior. This approach is combined with crystal plasticity considerations, thus making it possible to simultaneously trace strain hardening and texture evolution. The total dislocation density, ρ_t, is given by a rule of mixtures:

$$\rho_t = f\rho_w + (1-f)\rho_c , \qquad (1)$$

where f denotes the volume fraction of the cell walls. An important element of the model is the consideration of the evolution of the volume fraction of the cell walls, which is based on the experimental observations. According to reports in literature [10,11], f decreases with strain gradually tending towards saturation. The evolution of f was approximated in [7–9] by the following empirical function:

$$f = f_\infty + (f_0 - f_\infty)\exp(-\gamma^r/\tilde{\gamma}^r), \qquad (2)$$

where f_0 is the initial value of f, f_∞ is its saturation value at large strains and the quantity $\tilde{\gamma}^r$ describes the rate of variation of f with resolved shear strain γ^r. A scaling relation between the average cell size d and the inverse of the square root of the total dislocation density is assumed to hold:

$$d = K/\sqrt{\rho_t}, \qquad (3)$$

where K is a proportionality constant. Obviously, d decreases with the accumulation of the total dislocation density in the course of ECAP. Given that the total cell wall area increases, the reduction of the wall volume fraction implies that the walls get 'sharper' as deformation proceeds. A similar trend was observed in Ref. [12]: the cell wall thickness decreased with the number of fatigue cycles, presumably due to dislocation annihilation.

Strain hardening is accounted for within the modeling framework of Refs. [14-16]. The equivalent (scalar) resolved shear stresses τ_c^r and τ_w^r are related to the equivalent resolved plastic shear rates $\dot{\gamma}_c^r$ and $\dot{\gamma}_w^r$ and the dislocation densities in the cell interiors and the cell walls, respectively:

$$\tau_c^r = \alpha G b\sqrt{\rho_c}\left(\frac{\dot{\gamma}_c^r}{\dot{\gamma}_0}\right)^{1/m}, \qquad (4)$$

$$\tau_w^r = \alpha G b\sqrt{\rho_w}\left(\frac{\dot{\gamma}_w^r}{\dot{\gamma}_0}\right)^{1/m}. \qquad (5)$$

Here G is the shear modulus, b is the magnitude of the Burgers vector, $\dot{\gamma}_0^r$ is a reference strain rate, $1/m$ is the strain rate sensitivity parameter and α is a constant, about 0.25, say. The overall behavior of the 'composite' structure is defined by a scalar quantity, τ^r, that is obtained using the *rule of mixtures* applied to the two 'phases':

$$\tau^r = f\tau_w^r + (1-f)\tau_c^r. \qquad (6)$$

The kinetic equations, Eqs. (4) and (5), are combined with evolution equations for the two dislocation densities that include dislocation generation and annihilation terms and incorporate various dislocation reactions [9]. Strain compatibility between cell interiors and cell walls is imposed through the following relation:

$$\dot{\gamma}_c^r = \dot{\gamma}_w^r = \dot{\gamma}^r. \qquad (7)$$

The above scalar model is further combined with a *crystal plasticity model* thus providing a tool for modelling texture evolution in a polycrystal alongside the strain hardening behavior [8,9].

The misorientations between the dislocation cells within a grain are disregarded, and the grain is characterized by a unique equivalent resolved shear strain rate that can be expressed in terms of the individual resolved shear strain rates $\dot{\gamma}_s^r$ on slip systems s as [8]

$$\dot{\gamma}^r = \left[\sum_{s=1}^{N} \dot{\gamma}_s^{r\frac{m+1}{m}} \right]^{\frac{m}{1+m}} \qquad (8)$$

where N is the number of active slip plains in the grain. The full constraint Taylor model was used. The random choice method [17] was applied to calculate the sum in Eq. (8). That is to say, 5 slip systems were selected at random from among the 12 potentially active slip systems in each grain. In finite element calculations with ABAQUS [18], each element was assumed to consist of 300 initially randomly oriented grains. The evolving pole figures were extracted by aggregating all grain orientations in a mesh. Details of the FE simulation procedure as well as the parameter values used are available from Ref. [9].

3 Experimental

ECAP experiments were conducted at room temperature using coarse grained samples of technical purity (99.9 %) aluminium (grain size: 2 mm) polished prior to pressing. A split-design die with a square-shaped 12 mm Þ 12 mm channel was used. A description of the experimental set-up is given in another paper in these Proceedings [19]. The pressing speed for the aluminium specimens was 4 mm/min. A modified Route C, referred to as Route C_r, was used: between two consecutive pressings, the sample was rotated by 180° about its long axis and additionally turned upside down.

The average dislocation cell size and the corresponding yield strength of the material were measured by transmission electron microscopy (Philips CM200) and tensile testing of ECAP processed material. Texture measurements were done on a Siemens D5000 X-ray diffractometer. Complete pole figures for {111} reflections were measured. The samples with the length of 20 mm for measuring the pole figures were cut from the uniformly deformed middle portion of the workpieces.

4 Results

The effect of the number of ECAP passes on the tensile strength is shown in Fig. 1. The agreement between the experimental data and the simulation results is remarkably good. It is seen that, as common to ECAP results, the strongest effect is achieved after the first pressing. Subsequent deformation does lead to a gradual increase in strength, but the curve tends to level off very quickly. This strong effect of the first ECAP pass, followed by a gradual variation towards saturation with further pressings, is paralleled by the cell size evolution, Fig. 2. Again, the simulation results provide a very good description of the observed behaviour. While no significant gain in grain refinement is achieved with an increase in the number of ECAP passes, the overall uniformity of structure does improve, cf. Fig. 3.

It is interesting to note that the general trends in strain hardening behavior are similar for Cu and Al, cf. Ref. [19]. However, there are significant quantitative differences. Thus, the modulus compensated stress, σ/G, of Al is lower than that of Cu [20]. Similarly, the cell refinement effect in Cu is also more pronounced, the average cell size dropping down to about 200 nm already after a single ECAP passes [19,20]. This difference was attributed to a higher dynamic recovery rate for Al owing to its larger stacking fault energy [20].

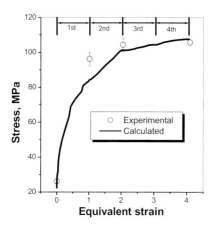

Figure 1: Variation of the tensile strength of Al with the number of ECAP passes

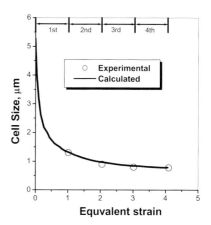

Figure 2: Variation of the cell size in Al with the number of ECAP passes

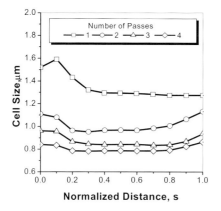

Figure 3: Variation of the cell size with the distance from the bottom of the workpiece (normalized with respect to the workpiece thickness) for different numbers of ECAP passes

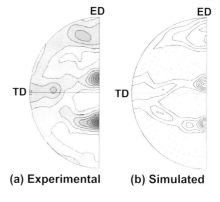

Figure 4: Measured [21] and calculated (111) pole figures after 4 ECAP passes (Route A)

The orientations of grains in every mesh were calculated using the above model at every time step. To verify the reliability of the calculations, they were compared with the experimental pole figures [21]. Figure 4 shows a calculated (111) pole figure vis-à-vis a measured one. This experimental pole figure, available from literature, cf. Ref. [21], corresponds to the texture after the fourth ECAP pass (Route A). Accordingly, the simulation was made for Route A. A good agreement between the two pole figures is evident.

5   Conclusion

The deformation behavior of Al under equal channel angular pressing was studied experimentally and by finite element simulations based on the model [7,8]. A comparison of the simulated evolution of the dislocation cell size, yield strength in tension and texture with experimental data showed a good capability of the model to account for major features of the deformation behavior of Al under equal channel angular pressing.

6   Acknowledgements

This work was funded by the Deutsche Forschungsgemeinschaft within grant ES74-9/1. Support from the Ministry of Science & Culture of Lower Saxony is appreciated. One of the authors (HSK) acknowledges support from the Center for Advanced Materials Processing (CAMP) of the 21th Century Frontier R&D Program funded by the Ministry of Science & Technology in Korea.

7   References

[1]   V. M. Segal: USSR Patent, No575892 (1977)
[2]   V.M. Segal, V. Reznikov, A. Drobyshevkiy and V. Kopylov: Russ. Metall., 1 (1981) 99
[3]   V.M. Segal: Mater. Sci. and Eng., A197 (1995) 157
[4]   R.Goforth, K. Hartwig and L. Cornwell, in "Investigations and applications of severe plastic deformation", T.Lowe and R. Valiev, eds., Kluwer Academic Publishers, Dordrecht, (2000) p. 3
[5]   R.Z. Valiev, R.K. Islamgaliev and I.V. Alexandrov: Progress Mater. Sci., 45 (2000) 103
[6]   A. Belyakov, T. Sakai, H. Miura and K. Tsuzaki, Phil. Mag. 81 (2001) 2629
[7]   Y.Estrin, L.S.Tóth, A. Molinari and Y. Bréchet, Acta Mater., 46 (1998) 5509
[8]   L.S.Tóth , A. Molinari and Y.Estrin: J. Eng. Mater. Techn. 124 (2002) 71
[9]   S.C. Baik, Y. Estrin, H.S. Kim and R.J. Hellmig, Mater. Sci. Eng. A 351 (2003) 86
[10] M. Müller, M. Zehetbauer, A. Borbély and T. Ungár: Z. Metallkunde, 86 (1995) 827
[11] M. Müller, M. Zehetbauer, A. Borbély and T. Ungár: Scripta Mater., 35 (1996) 1461
[12] H. Yaguchi, H. Mitani, K. Nagano, T. Fujii and M. Kato: Mat. Sci. Eng., A315 (2001) 189
[13] U.F. Kocks: J. Eng. Mater. Tech., 98 (1976) 76
[14] H. Mecking and U.F. Kocks: Acta Metall., 28 (1981) 1865

[15] Y. Estrin and H. Mecking: Acta Metall., 32 (1984) 57
[16] Y. Estrin: ìUnified Constitutive Laws of Plastic Deformationî, ed. A.S. Krausz and K. Krausz, Academic Press (1995) 69
[17] P. Van Houtte: Proceeding of ICOTOM 6, Tokyo (1981) p. 128
[18] ABAQUS/Standard, Version 5.8, Hibbitt, Karlsson & Sorensen, Inc. (1998)
[19] R.J. Hellmig, S.C. Baik, J.R. Bowen, Y. Estrin, D. Juul Jensen, H.S. Kim and M.H. Seo, this volume
[20] S.C. Baik, R.J. Hellmig, Y. Estrin and H.S. Kim, Z. Metallkunde 94 (2003) 754
[21] A. Gholinia, P. Bate and P.B. Prangnell: Acta Mater., 50, 2121 (2002)

Process Modeling of Equal Channel Angular Pressing

H. S. Kim, S. I. Hong, H. R. Lee and B. S. Chun
Chungnam National University, Daejeon, Korea

1 Introduction

Recently several methods of severe plastic deformation (SPD), such as equal channel angular pressing (ECAP), high pressure torsion straining (HPT), accumulated roll bonding (ARB), multiple forging, etc., have been developed to process bulk materials with ultrafine-grained microstructures [1,2]. Among them, the ECAP process is a convenient forming procedure to extrude material by use of specially designed channel dies without a substantial change in geometry and to make an ultrafine grained material. Figure 1(a) shows the principle of ECAP, where two channels of equal cross-section intersect at an oblique angle (or channel angle) Φ. With reference to the workpiece, three perpendicular directions denoted as pressing direction, width direction and thickness direction, respectively, are introduced. In a rectangular workpiece with the width of W and length of L, the thickness direction is perpendicular to the width and length directions, so that the strain along the thickness direction is zero, i.e. plane strain condition prevails. Therefore, the deformation during the ECAP process of rectangular specimens becomes two-dimensional. However, unlike the idealized case of Fig. 1(a), the real deformation characteristics is inhomogeneous due to the effects of friction, a curved die corner or incomplete filling of the die, see Fig. 1(b). The die corner angle Ψ is defined as the angle subtended by the arc curvature and lies between $\Psi = 0$ and $\Psi = \pi - \Phi$.

The properties of the materials are strongly dependent on the plastic deformation behaviour during pressing, which is governed mainly by die geometry (channel angle Φ and corner angle Ψ) [3–7], material properties (strength and hardening behaviour) [5,8,9], and process variables (lubrication, deformation speed and temperature) [10–12]. Because the evolution of microstructures and the mechanical properties of the deformed material are directly related to the amount and history of plastic deformation, the understanding of the phenomenon associated with strain development is very important in the ECAP process.

It is clear that there is a need for modeling techniques which may permit a wider study of the effects observed for better process control and the understanding of process related phenomena. In this study, we describe a range of our continuum modeling results of the ECAP process in order to illustrate the modeling applicability. Firstly, the general finite element results of ECAP modeling are described. Secondly, the inhomogeneous deformation due to the hardening property of the material is explained.

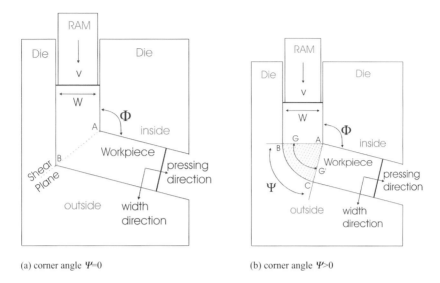

(a) corner angle $\Psi=0$ (b) corner angle $\Psi>0$

Figure 1: Schematic illustration of ECAP showing the channel angle Φ and the corner angle Ψ. (a) corner angle $\Psi = 0$ and (b) corner angle $\Psi > 0$

2   Deformation of the Workpiece during ECAP

A finite element analysis using the rigid-plastic DEFORM2D code [13] was carried out in order to investigate the plastic deformation behaviour of the workpiece during the ECAP process with a die channel angle of $\Phi = 90°$ and a die corner angle of $\Psi = 90°$. The imaginary model material that exhibits elasto-perfect-plastic behavior was used in the calculations, in order to exclude the effect of strain hardening and strain rate on the deformation behavior and to make investigation into a pure geometric effect. . The detailed calculation procedures can be found in reference [5]. The coefficient of friction between the die channel inside and the specimen was assumed to be zero – frictionless condition. Although the effect of friction was not considered in the present study in order to focus on the intrinsic geometry effect of the ECAP process, friction is an important process variable which retards the flow of the surface material and results in high shear strain at the bottom region and lower shear strain at the top region of the workpiece. A detailed effect of friction on the deformation behavior during ECAP will be published elsewhere. The mean stress distribution of the workpiece at the steady state of the load during the ECAP process is shown in Fig. 2. It can be seen from the mean stress distribution that the stress in the inner side of the exit channel is different from that in the outer side, not just in its value, but even in the sign. The inner part of the entry side receives compressive stress due to the compression of the ram, and the maximum compressive stress (contour line A) appears in the inside corner point. The main deformation zone (area ABC in Fig. 1(b)) is in the compressive stress state. On the other hand, the outer part of the entry side of the workpiece is in the tensile stress state (contour line F in Fig. 2) since the outer part is elongated in the pressing direction ahead of and in the front part of the main deformation zone. However, since the elongated elements are compressed again from around the centre of the main deformation zone, they acquire a compressive stress state. After exiting the main deformation zone, the compressed elements of the inner part are in tensile stress state, while the tensioned elements of the outer part turn to the compressive stress state because the elements between the inner part and the outer part have to maintain continuity. Of course, the stress states and the deformation behaviour of the front part and the rear part are different due to the difference in the deformation history.

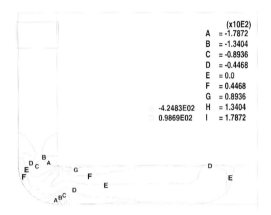

Figure 2: Distribution of mean stress during the ECAP process

3   Formation of Corner and Exit Channel Gaps

Figure 3 shows the calculated deformed meshes for (a) the strain hardening material (1100Al) and (b) the quasi-perfect plastic material (6061Al-T6). It can obviously be found that the die corner filling behaviour of the strain hardening and quasi-perfect plastic materials are quite different. The corner gap between the die and the workpiece is large in 1100Al which has high strain hardening behaviour and relatively small in 6061Al-T6 with low strain hardening behaviour. The corner gap is expected to disappear completely for real perfect plastic materials with zero hardening rate. Although the die corner angle Ψ was 0°, the arc curvatures generated on the surface of the workpiece are $\Psi = 51°$ for 1100Al and $\Psi = 22°$ for 6061Al-T6, due to the corner gap formation between the die and the workpiece. Therefore, it can be concluded that a sharp die corner angle is not essential for the high strain generation in the workpiece because the workpiece does not exactly follow the corner shape of the die. An interesting observation in Fig. 3 is that the shape of the corner gap is not symmetrical. That is, the corner gap along the entry side is longer that that along the exit side. This can be attributed to the asymmetry of the deformation strain and the loading conditions with respect to the central shear plane.

In order to investigate the effects of strain hardening and strain rate sensitivity of materials, non-hardening (perfect plastic) materials of $n = 0$, strain hardening materials of $n = 0.5$, strain rate insensitive materials of $m = 0$, and rate sensitive materials of $m = 0.5$ were considered. The workpiece material used in the calculations was imaginary non-dimensional model materials with various strain hardening exponents and strain rate sensitivities. The stress-strain curves for the model materials can be expressed by the following equation; $\sigma = \varepsilon^n \dot{\varepsilon}^m$, where σ is the flow stress, ε is the strain, n is the work hardening exponent and m is the strain rate sensitivity (see also ref. [5]).

Examples of (nearly) non-hardening materials are peak-aged and overaged Al alloys with high dynamic recovery rate due to easy cross slip in the peak-aged and overaged matrix and fully pre-strained (e.g. pre-ECAP experienced) materials which show the saturation of the strength. The example for the strain rate sensitive materials may correspond to ECAP processing of materials at high temperatures. The result on the effect of the strain rate sensitivity, however, can be used to predict the relative homogeneity of plastic flow during ECAP of materials with slightly different rate sensitivities.

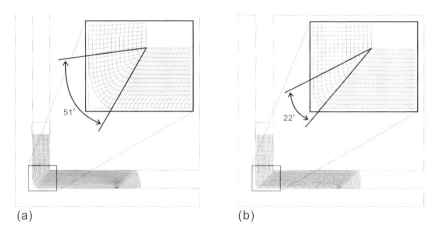

Figure 3: FEM-predicted grid distortions for (a) a strain hardening material (1100Al) and (b) a quasi-perfect plastic material (6061Al-T6)

Figure 4 shows the deformed geometries (flow net) for various materials of different material parameters at two different pressing speeds. In case of (a) the non-hardening and rate insensitive materials ($n = 0$ and $m = 0$), the deformed geometry represents almost ideal and homogeneous behavior within the workpiece. The workpiece flows along the die filling all the inside channel. On the other hand, in case of (b) the strain hardening and rate insensitive materials ($n = 0.5$, $m = 0$), two important points should be noted. First, the corner gap between the die and the workpiece develops. The size of the corner gap was found to increase with increase of the hardening exponent n. Second, the center part and the inner part of the workpiece are heavily sheared whereas the outer part appears to be much less sheared. It is clear, from this study, that the less sheared zone is attributed to the flow path of the workpiece controlled by the die geometry.

The strain rate sensitive ($m = 0.5$) materials in Fig. 4(c) shows quite different deformation distribution in ECAP compared to the perfect plastic (Fig. 4(a)) or strain hardening (Fig. 4(b)) materials. The outer (bottom after pressed) part of the workpiece experiences more severe deformation and the inner (top after pressed) part of the workpiece shows less shear deformation. Contrary to the strain hardening materials, the formation of the corner gap is insignificant. However, the upper channel gap instantly develops upon shearing of the billet from the vertical entrance channel to the horizontal exit channel. The formation of the upper channel gap be-

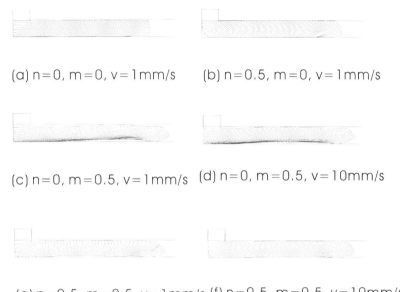

Figure 4: Deformed geometry changes during ECAP. (a) $n = 0$, $m = 0$; (b) $n = 0.5$, $m = 0$; (c) $n = 0$, $m = 0.5$, $v = 1$ mm/s; (d) $n = 0$, $m = 0.5$, $v = 10$ mm/s; (e) $n = 0.5$, $m = 0.5$, $v = 1$ mm/s; (f) $n = 0.5$, $m = 0.5$, $v = 10$ mm/s

tween the billet top and die exit channel results from the same effect as increasing the die channel angle. The reaction force from the bottom wall would promote the bending of the billet, resulting in the formation of the lower channel gap between the billet bottom and the die. Since the shearing occurs approximately 45 degrees from the billet axis as shown in Fig. 4, the strain rate in the lower part of the billet is lower in the horizontal channel and the strength is lower in the bottom part of the billet. Therefore, the local strain in the bottom part would increase in the horizontal channel. From the engineering point of view, the bent workpiece needs to be straightened for the usage and the bending should be avoided. It should be noted that less sheared zone is formed in the top part of the workpiece in the strain rate sensitive materials with $n = 0$ and $m = 0.5$, while it is formed at the bottom part in the strain hardening materials with $n = 0.5$ and $m = 0$.

In Fig. 4(d), the deformation distribution for the strain rate sensitive materials with $n = 0$ and $m = 0.5$ at a higher pressing speed is shown. The strain distribution and the shape are similar to those at a lower pressing speed (Fig. 4(c)). However, the rate dependent deformation characteristics (i.e. local inhomogeneity, less shear zone and exit channel gaps) are reduced at high deformation speeds. In case of strain hardening and rate sensitive materials with $n = 0.5$ and $m = 0.5$ (Fig. 4(d) and 4(f)), the corner gap develops due to the strain hardening effect and the upper channel and exit channel gaps develop due to the strain rate sensitivity effect. It should be noted, however, that the strain rate sensitivity acts to reduce the corner gap promoted by the hardening exponent (compare Fig. 4(b) and 4(e)) and the hardening exponent tends to reduce the upper channel and the bottom exit channel gaps (compare Fig. 4(d) and 4(f)). It should also be stressed that the inhomogeneity of the shear deformation strongly depends on the relative effects of strain hardening exponent and strain rate sensitivity.

4 Acknowledgements

This research was supported by a grant from the Center for Advanced Materials Processing (CAMP) of the 21st Century Frontier R&D Program funded by the Ministry of Science and Technology, Republic of Korea. The authors acknowledge the help of Mr. M. H. Seo for performing the DEFORM calculation.

5 References

[1] V. M. Segal, V. I. Reznikov, A. E. Drobyshevkii, V. I. Kopylov, Metally 1 (1981) 115
[2] R. Z. Valiev, R. K. Islamgaliev, I. V. Alexandrov, Prog. Mater. Sci. 45 (2000) 103
[3] V. M. Segal, Mater. Sci. Eng. A271 (1999) 322
[4] Y. Iwahashi, J. Wang, Z. Horita, M. Nemoto, T. G. Langdon, Scripta Mater. 35, (1996) 143
[5] H. S. Kim, M. H. Seo, S. I Hong, Mater. Sci. Eng. 291A (2000) 86
[6] H. S.Kim, Mater. Sci. Eng. A315 (2001) 122-128
[7] H.S. Kim, Mater. Sci. Eng. A328 (1–2) (2002) 317–323
[8] B. S. Moon, H. S. Kim, S. I. Hong, Scripta Mater. 46 (2) (2002) 131–136
[9] H. S. Kim, S. I. Hong, M. H. Seo, J. Mater. Res. 16 (2001) 856-864
[10] Y. Wu, I. Baker, Scripta Mater. 37 (1997) 437
[11] H. S. Kim, Mater. Trans. JIM 42 (2001) 536-538
[12] H. S. Kim, J. Mater. Res. 17 (2002) 172-179
[13] DEFORM2D, vers. 5.1, Scientific Forming Technology Corporation, Columbus, OH 1997

Deformation Behaviour of ECAP Cu as described by a Dislocation-Based Model

N.A. Enikeev[1], H.S. Kim[2], I.V.Alexandrov[3], S.I. Hong[2]
[1]-Institute of Mechanics, Ufa Science Centre, RAS, Ufa, Russia
[2]-Chungnam National University, Taejon, South Korea
[3]-Institute for Physics of Advanced Materials, USATU, Ufa, Russia}

1 Introduction

Material science community experiences a growing interest to severe plastic deformation (SPD) due to its capability to produce bulk nanostructured materials which exhibit outstanding mechanical and changed physical properties [1]. The properties of SPD nanomaterials are strongly determined by their microstructure parameters, such as small grain size, big fraction of high—angle grain boundaries and high defects density. Meanwhile, the deformation mechanisms leading to structure refinement and high average misorientation angle are still not evident enough. Recent years there have been fulfilled many attempts to set up various models which are dealing with large imposed strains (see, for example [2–6] and others). However, the case of SPD still gives a large amount of unanswered questions. For example, it was recently shown that dislocation density evolution during SPD has a certain peculiarity: with strain increasing it reaches its maximum, then drops down and stabilizes [12–14]. Meanwhile, the dislocation structure is a very important parameter of SPD nanomaterials which strongly affects their properties. So the aim of current research is to modify current approach to be applicable for SPD describing. Namely, let us concentrate on the above-mentioned effect of dislocation density evolution, typical precisely for SPD.

Having conducted an analysis of various deformation models we chose the Estrin-Toth dislocation model [28] as guidance. Let us briefly summarize the essence of the model. It is a 2D two-phase dislocation model for the late deformation stages. It deals with the cell structure consideration. The macroscopic stress can be calculated according to the rule of mixtures for the stresses in the cell walls τ_w^r and in cell interiors τ_c^r.

$$\tau = f\, M\, \tau_w^r + (1-f) M \tau_c^r \tag{1}$$

where f is a cell wall volume fracture, M is the Taylor factor.

For the each of two phases constitutive equations are written to relate the resolved shear stress to the resolved shear strain. The main part of the model are the evolution equations for dislocation densities in cell walls and cell interior which determine the deformation process. The equations are written under assumption that dislocation generation takes place in the cell walls while their annihilation is possible both in cell walls and cell interior. The decrease in cell size d and cell wall fraction f is set phenomenologically according to experimental data available. A comprehensive description of the model one can find in [3]. Later on the authors had reformulated the model for the 3D case [4].

This model allowed to describe the late (III, IV and V) hardening stages at large plastic deformation within one general approach with very good agreement to experiment. It also provided the data on microstructure parameters evolution and mechanical response at large plastic strains. Thus, Estrin-Toth model proved to be a very useful basis for the description of large deformations. In the present research it was modified to meet the conception of SPD features.

2 Experimental

In order to validate the obtained theoretical results the experiments on ECAP of Cu and further microstructure investigations have been carried out. Pure Cu (99.9 %) samples annealed at 600 °C for 10 hours and having an initial bar shape with the 10 mm square cross-section and a 100 mm length have been treated by ECAP technique at room temperature. The angle between channels was 90°. For the current demonstration purposes the results on samples treated by B_A ECAP route for 1, 2, 4 and 12 passes have been selected.

The X-ray analysis of the obtained samples was done with the Rigaku X-ray diffractometer under the following settings: 2θ technique, CuKα radiation, the accelerating voltage 40 kV, the current 30 mA. The X-ray diffraction patterns were obtained in the 2θ angular range from 40 to 142. The 2θ step was equal to 0.02° and exposition time was 20 s while scanning X-ray peaks. Samples for the X-ray analysis have been mechanically polished and electropolished afterwards. For the further treatment the (111), (222) and (200), (400) peaks were chosen. For the each pair of peaks belonging to the same crystallographic direction the grain size and microstrain values were calculated using `MISIS' X-ray software package.

Using the obtained crystallite size and elastic microdistortions the dislocation density was evaluated as for the case of random dislocation distribution:

$$\rho_{hkl} = 2\sqrt{3} <\varepsilon_{hkl}^2>^{1/2} \tag{2}$$

The samples with nanostructures obtained after 12 passes ECAP have been subjected to low-temperature annealing (150 °C for 60 minutes) in order that recovery of non-equilibrium grain boundary structure take place [7]. The X-ray diffraction patterns of the annealed samples were used for determination of extrinsic grain boundary dislocation density using a recently developed method [8,9].

3 Model Description

As it was already mentioned above the Estrin-Toth dislocation model has been chosen as a basis to develop a modified one adjusted for the case of SPD process due to its attractive capabilities for the large straining description. However, the application of this model to the case of SPD seems to be somewhat restricted due to the several points. For example, the dislocation evolution equations here assume that nucleation of dislocations by Frank-Read sources takes place in cell walls, however in case of SPD materials it is assumed to occur in the grain interior [10,11] and all. This could lead, in particular, to the fact that the resulting predicted curve by Estrin-Toth model for dislocation density evolution in cell interior (growing and saturation) does not fit experimental data for the case of SPD materials (growing, noticeable dropping down and then stabilizing) observed by X-ray in ECAP Cu [12], HPT Cu [14], by TEM for multiple forging [13].

To adjust the Estrin-Toth model for the SPD, some modifications are proposed to be introduced. Our approach is based on the ideas worked out in papers by Valiev and colleagues [10,11]. The general idea is that during SPD there happens evolution of cell structure to grain boundary one. The dislocations generated by Frank-Read sources are absorbed by cell walls transforming them to grain boundaries (Fig.1a). The captured lattice dislocations are decomposed into a number of grain boundary sessile and glide dislocations. After annihilation of grain boundary dislocations with opposite signs, the defect structure of grain boundaries is represented by ensembles of glide dislocations which create long-range stress fields (Fig.1c) and sessile dislocations which contribute to grain boundary misorientation increasing (Fig.1b).

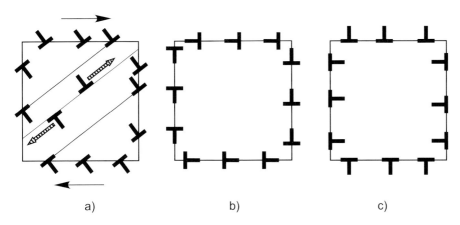

Figure 1: Nucleation of extrinsic grain boundary dislocations ensembles

So, the main idea of this approach for the dislocation reactions is quite different from the Estrin-Toth model and is described by the following relationships:

$$\dot{\rho}_g = \dot{\rho}_{FR} - \dot{\rho}_a - \dot{\rho}_{wall} \tag{3}$$

$$\dot{\rho}_{gb} = \dot{\rho}_{gr} - \dot{\rho}_m - \dot{\rho}_a \tag{4}$$

where:

$\dot{\rho}_g$ is the rate of density variation in a grain

$\dot{\rho}_{FR}$ are dislocations, generated by Frank-Read sources in grains

$\dot{\rho}_a$ is the annihilation of dislocations by cross slip

$\dot{\rho}_{wall}$ are dislocations, moving form cell body to cell walls

$\dot{\rho}_{gb}$ the rate of density variation of dislocations in grain boundaries

$\dot{\rho}_{gr}$ dislocations, transformed to grain boundaries from grain bodies

$\dot{\rho}_m$ the quantity of sessile dislocations contributing to the increase in grain boundary misorientations

The total number of Frank-Read sources in the grain interior was counted as a number of possible dislocations segments between interceptions in the dislocation net at given dislocation density. A coefficient was used to determine which part of them is actually active. The rest parts of dislocation reaction equations are taken to be the same as in the Estrin-Toth model. The final modified dislocation equations then have a form:

$$\dot{\rho}_{gb} = \frac{3\beta^* d\rho_g}{2b}\dot{\gamma}_g - \frac{4\beta^*\dot{\gamma}_g}{bd(1-f)^{1/2}} - k_0\left(\frac{\dot{\gamma}_g}{\dot{\gamma}_0}\right)^{-1/n}\dot{\gamma}_g \rho_{gb} \qquad (5)$$

$$\dot{\rho}_{gb} = \frac{4\beta^*\dot{\gamma}_{gb}(1-f)^{1/2}}{bdf} + \frac{2\alpha^*(1-f)\dot{\gamma}_{gb}\sqrt{\rho_{gb}}}{\sqrt{3}fb} - k_0\left(\frac{\dot{\gamma}_w}{\dot{\gamma}_0}\right)^{-1/n}\dot{\gamma}_{gb}\rho_{gb} \qquad 6)$$

A misorientation accumulation was fulfilled following [15], where estimation of contribution of excess dislocation density captured by grain boundaries to misorientation growth was conducted:

$$\Theta = \int_0^t bPd(\dot{\Delta\rho})dt \qquad (7)$$

where P is the possibility of dislocation immobilization, $P = 2d/L = 1/3$

4 Results and Discussion

The above mentioned equations have been formed a set of constitutive equations for description of SPD process. Model constants were used to be the same as obtained in the Estrin-Toth model [3] for the case of pure Cu. With agreement to the aim of current publication, the main attention is focused on the average dislocation density evolution during SPD. The results of calculations are shown at Fig.2 (solid line). For the comparison the dislocation density values derived by X-ray analysis for the Cu subjected to ECAP (route Ba, passes 1,2,4 and 12) are plotted at the same figure as empty squares. The similar data earlier obtained in [14] for the HPT Cu are shown here as full circles. For the purpose of better presentation dislocation density is given having been normalized.

At Fig.3 one can see the results of calculations for average misorientation growth as a function of imposed strain. The experimental data, obtained in [13] for the multiple forging have been used as a reference.

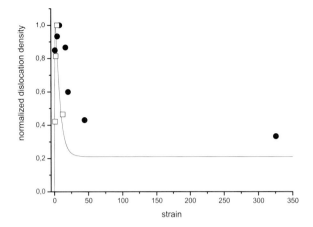

Figure 2: Evolution of average dislocation density at SPD: calculated from the modified dislocation model (solid line); measurements in ECAP Cu (□), and in HPT Cu (●) [14]

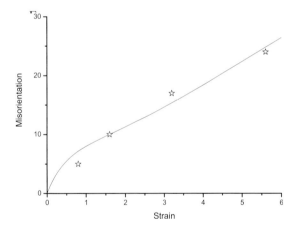

Figure 3: Misorientation evolution at large plastic strains compared to experimental data (★) by Belyakov el al [13] for Cu, subjected to SPD by multiple forging

rom both figures it is clearly seen that the suggested approach allowed reproducing the character of experimentally observed dependence of dislocation density on strain at SPD. Thus, the modified Estrin-Toth model is capable to describe the SPD process and its further development promise to be very helpful in prediction of microstructure parameters evolution during SPD. For instance, it is still not possible to solve the most challenging problem facing SPD investigation – the description of grain refinement mechanism. The recent publications [6] allow to hope that the promising way here is using of dislocation-disclination consideration of grain subdivision under certain energetic criterion.

5 Summary

A deformation-based model was put forward in order to describe deformation behaviour of materials at severe plastic deformation (SPD). The 2D Estrin-Toth approach [3] has been chosen as a guidance, into which some modification was introduced in accordance with speculations on dislocation evolution typical for SPD. As a preliminary result of conducted calculations for the case of ECAP Cu, experimentally observed facts of dislocation density and misorientation evolution have been matched. The presented approach is capable to help in describing of the deformation processes and predicting of microstructure parameters evolution during SPD.

6 References

[1] R.Z. Valiev, R.K. Islamgaliev, I.V. Alexandrov, Progr. Mater. Sci. 2000, V. 45, 103 P
[2] M. Zehetbauer, Acta Mater. 1993, V. 41, No. 2, 557—588
[3] Y. Estrin, L.S. Toth, A. Molinari, Y. Brechet, Acta Mater. 1998, V. 46, No. 15, 5509–5522
[4] L.S. Toth, A. Molinari, Y. Estrin, Special Issue of J. Eng. Mat. Techn. 2002, V. 124, 71–77
[5] G.A. Malygin, Phys. -- Uspekhi 1999, V. 42, No. 9, 887–916
[6] M. Seefeldt, L. Delannay, B. Peeters, E. Aernouldt, P. Van Houtte, Acta Mater. 2001, V. 49, 2129–2141
[7] Yu. Gertsman, R. Birringer, R.Z. Valiev, H. Gleiter, Scripta Met. Mater. 1994, V.30, 229–234
[8] I.V. Alexandrov, R.Z. Valiev, Phil.Mag. 1996, B73, 861
[9] N.A. Enikeev, I.V. Alexandrov, R.Z. Valiev, Phys. Met. Metlgf. 2002, V. 93, No. 6, 515–524
[10] A.A. Nazarov, A.E. Romanov, R.Z. Valiev, Acta Metal. Mater. 1993, V. 41, No. 4, 1033–1040
[11] R.Z. Valiev, Yu.V. Ivanisenko, E.F. Rauch, B. Baudelet, Acta Metal. Mater. 1996, V. 44, No. 12, 4705–4712
[12] T. Ungar, I.V. Alexandrov, P. Hanak, Investigation & Applications of Severe Plastic Deformation (Eds.: T. Lowe and R.Z. Valiev), NATO Sci Series, 3. High Technology: Kluwer Academic Publishers, 2000, V. 80, 293–299
[13] A. Belyakov, T. Sakai, H. Miura, K. Tsuzaki, Phil. Mag. A 2001, V. 81, No. 11, 2629–2643
[14] [14] I.V. Alexandrov, A.A. Dubravina, H.S. Kim, Defect and Diffusion Forum 2002, V. 208–209, 229–232
[15] W. Pantleon, Acta Mater. 1998, V. 46, 451–462

Severe Plastic Deformation by ECAP in a Commercial Al-Mg-Mn Alloy

P.A. González, C.Luis

Departamento de Ingeniería Mecánica, Energética y de Materiales. UPNA. Campus de Arrosadía, Pamplona, Spain

1   Introduction

In recent years there has been an increasing interest in the production of ultrafine-grained material both for commercial as well as for investigative purposes. These materials exhibit mechanical and other physical properties, which are out of the ordinary and of great interest, in particular their remarkable strength and toughness and their potential for superplasticity at low temperatures [1]

Recent investigations show that determined techniques of severe or intense plastic deformation (SPD) can be applied for the production of submicron grain materials with great advantages over other conventional methods. The severe plastic deformation is a powerful medium to process materials and for controlling its properties which constitutes a very important area of investigation in physical metallurgy and has been the object of important studies such as Ref [2]. The equal channel angular pressing or extrusion (ECAP/E, from now on) is an ingenious method of severe plastic deformation which can be applied to a great variety of materials with enormous advantages over other methods of deformation [3].

2   Finite Element Modelling

The ECAP/E process has been numerically modelled by several authors, using the Finite Element Method (FEM), to simulate the plastic flow and the strain conditions of the pressed materials [4,5].

FEM simulations of our alloy AA5083 have been carried out using MARC™ code. Isothermal 2-D plane strain conditions were supposed to simplify the analysis. It is assumed also that the die and the ram deform only elastically. In order to investigate the effect of hardening behaviour on the mechanical flow of the material during ECAP, the stress-strain room temperature response of the alloy was used. With regard to friction conditions it was considered low friction (shear friction coefficient $m = 0,05$) .

2.1   FEM Analysis

Figure 1 shows the strain levels for partial passage ($n = 1$) for the 90° and 120° die, respectively and a detail of the corner filling as well. The shear deformation zone is the central one and therefore the ends of the billet must be discarded. "Secondary deformations" at the end of the billet can be seen because of the flow and rotation of material, causing "facets" to emerge from the end edge.

Five numbered nodes were randomly marked into the mesh. In agreement with Figure 2, the effective strain is sufficient homogeneous in both cases, but the maximun strain ($\varepsilon = 1,2$) is reached for the 90° die according to Segal´s analysis [3].

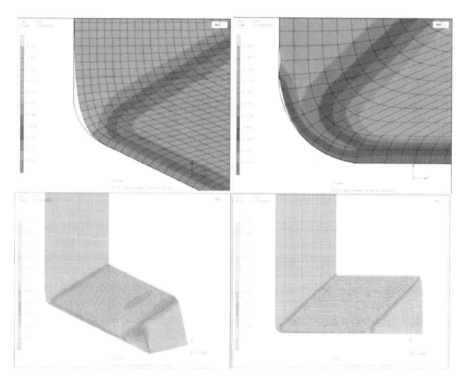

Figure 1: Partial passage $n = 1$ for 120° angle and 90° angle die and corner filling

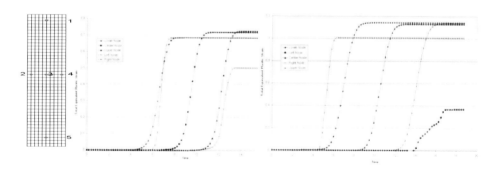

Figure 2: Nodes for the analysis and effective plastic strain in the nodes versus time for 120° and 90°

3 Experimental Conditions

The AA5083 (93,69 wt%Al; 4,67 % Mg; 0,70 % Mn) was received in cast bars with a homogenization treatment and was machined to 15 mm diameter × 100 mm in length billets with a grain diameter of about 50 µm.

The extrusions were carried out in an 400 kN ECAE press with a speed of 50 mm/min at 473K . The billets were extruded through two equal channel dies of 90° and 120° (true strains $e = 1,15$ and e = 0,68 per pass respectively) up to 8 passes in constant strain path with low friction conditions , using PTFE and MoS_2 sprays as lubricants (see half-extruded billets in Fig. 3).

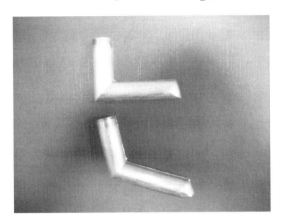

Figure 3: Cylindrical billets of 5083 aluminum alloy extruded halfway through the 90° and 120° die

Specimens were cut in the extrusion axis direction. Vickers microhardness was measured with a load of 4,9N (500gf). The tensile tests were performed in a Hounsfield universal testing machine using machined tensile specimens by "wire electroerosion machining".

SEM samples were produced by metalographic polishing followed by light electropolished with perchloric acid.

The microstructures were characterised by electron back scattered diffraction (EBSD) in the transverse direction in a Philips XL30 SEM (FEG).

4 Experimental Results

The ECAP deformation increases the hardness, but saturates (at a value of HV 113 at 90° and at 107 at 120°) after four passes (figure 4). These results are consistent with a higher accumulated strain for the 90° die, although the saturation level reached does not have the same value for the two dies. This, in spite of the fact that theoretically similar accumulated strains could be reached in both dies after sufficient extrusion passes . In any case the values of hardness reached are similar to those results in [6] although perhaps slightly lower due to certain recovery from the higher processing temperature (473K versus 423K).

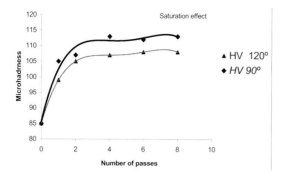

Figure 4: Variation of Vickers microhardness

Tension tests give different results. The yield stress of AA5083-110 MPa as received- increases up to 8 passes (400 MPa and 260 MPa respectively) and does not reach a limit. About 25 MPa hardening is obtained between every two passes. (figure 5).

The stability of those mechanical properties was examined carrying out annealing processes between 400–723 K during 1h. The microhardness variation of ECAP 5083 alloy is shown (figure 6). It is a drop in hardness of about 550 K , a result comparable to ref [7].

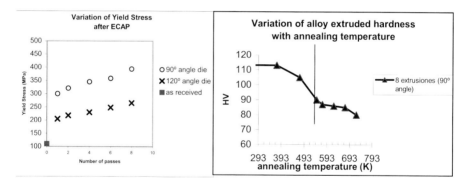

Figure 5: Variation of yield stress with increasing number of passes

Figure 6: Microhardness variation of the extruded alloy (8 passes) with annealing

As the drop in hardness at over 550 K seems to show, the alloy undergoes recovery and recrystalization processes. The Al_6Mn dispersoids are very coarse and too heterogeneously distributed to restrict the grain boundary mobility (see Figure 7).

Despite of the poor grain stability, the ECAE process produces grain refinement from 50 μm in the alloy before extrusion to elongated grains of 3-4 μm size after 8 passes, as can be seen in EBSD mapping area of Figure 8. The dark lines are high angle boundaries and the light ones are low angles boundaries.

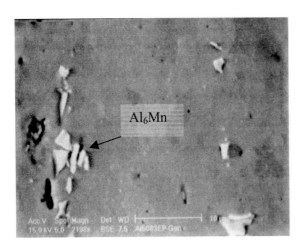

Figure 7: SEM image of AA5083

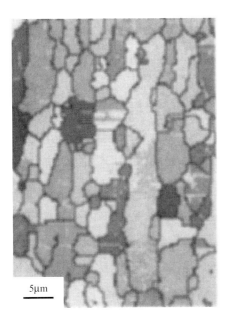

Figure 8: EBSD mapping for $e = 9.2$ ($n = 8$)

In any case it has not reached a sub-micron grain structure probably because the temperature at which the ECAE process was carried out. That temperature prevents a higher grain reduction due to the feasible recovery processes at around 473 K.

5 Conclusions

1. The ECAP process enhances the room temperature mechanical properties of the extruded alloy, better with the 90° die than the 120° one in agreement with Segal's analysis and the FEM simulation.
2. The severe plastic deformation increases the hardness, but saturates (at a value of HV 113 at 90° and at 107 at 120°) after a few passes. Tension tests appear more sensitive. The yield stress increases progressively up to 8 passes. About 25 MPa hardening is obtained between each two passes.
3. A grain refinement is obtained from 50 μm in the alloy before extrusion to 3-4 μm elongated grains after 8 passes. It is likely that the temperature of the process prevents more grain reduction due to recovery processes around 473 K.
4. The alloy exhibits a poor grain stability as the drop in hardness at around 550 K seems to show. The Al_6Mn dispersoids are very coarse and too heterogeneously distributed to restrict the grain boundary mobility.

6 Acknowledgements

The authors are grateful for the support of the research project :MAT2002-04343-C03-02 of MCYT and also to the company "Alcoa Extrusion Navarra".

7 References

[1] R.Z.Valiev, N.A.Krasilnikov, N.K.Tsenev, Mat. Sci. Eng A137, 1991, 35–40
[2] J. Gil-Sevillano, P.V.Houtte, E. Aernoudt, Prog. Mat. Sci. 25 ,1980, 69–410
[3] V. M. Segal, Mat. Sci. Eng. A197, 1995, 157–164
[4] J. Bowen, A. Gholinia , S. M. Roberts, P. B. Prangnell, Mat. Sci. Eng. A287 ,2000, 87–89
[5] V.Zhernakov, I.Budilov, G.Raab, I.Alexandrov, R.Valiev, Scr. Mater, 44,2001, 1765–1769
[6] L.Dupuy, E.Rauch, J.Blandin Proc.Advanced Research Workshop in applications of SPD Kluwer Academic Publisher 2000, Vol 80, 189
[7] K. T. Park, D. Y. Hwang, Y. K. Lee, Y. K. Kim, Mat. Sci. Eng A341,2003, 273–281

Evolution of Mechanical and Microstructural Properties of ECAP Deformed Copper

R.J. Hellmig[1], S.C. Baik[2], J.R. Bowen[3], Y. Estrin[1], D. Juul Jensen[3], H.S. Kim[4] and M.H. Seo[4]
[1] Technische Universität Clausthal, Clausthal-Zellerfeld, Germany
[2] Pohang Iron & Steel Co. Ltd., Pohang, Korea
[3] Center for Fundamental Research: Metal Structures in Four Dimensions, Roskilde, Denmark
[4] Chungnam National University, Daejeon, Korea

1 Introduction

Over the last few years [1], equal channel angular pressing (ECAP) has become a popular method for producing bulk ultra-fine structured materials. It allows repetitive pressing of a billet through a die having two intersecting channels without changing the billet's cross-sectional dimensions. Therefore, very high shear strains can be introduced into the workpiece [2,3].

Grain refinement by ECAP is related to the formation and evolution of dislocation cell boundaries under this process. The evolution of the cell size distribution with increasing equivalent strain has been determined using transmission electron microscopy (TEM). For a single pass specimen the dislocation cell misorientation distribution was determined. Recently, it was shown in case of an Al-0.13%Mg alloy that even after a deformation to an effective strain of 10, a correlation between neighboring orientations is still retained [4], induced mainly by low angle grain boundaries. Another recent report demonstrated that in the case of copper a lamellar structure of the cell boundaries as well as a large proportion of low angle grain boundaries were still observed even after an effective strain of approximately 8 [5].

It is possible to describe the microstructural changes and the macroscopic deformation behavior by modelling the evolution of the dislocation density and the cell size. A suitable constitutive model [6] predicting the strain hardening behavior of materials forming dislocation cells was recently extended from a two-dimensional to a three-dimensional cell structure [7]. The model was applied to the experimental data, including those on the mechanical properties of the ECAP processed material as obtained by tensile testing. In this paper, the results of an experimental study of the effects of ECAP on the microstructure, thermal stability and mechanical properties of Cu will be presented, along with the outcomes of the constitutive modeling.

2 Experimental

Specimens produced from 99.95 % purity copper having an initial grain size of 30 µm (after annealing of the as received material for 2 hours at 450 °C) were severely deformed by ECAP according to route C_r (which means that the specimens were rotated by 180° about their axis and turned upside down between two consecutive pressings). The initial specimen size was 12 mm × 12 mm × 60 mm. The ECAP rig used (built from a common use tool steel X38 CrMoV 5 1) had a split design where one part contained the full channel and the other part was used to close the die. The angle Θ between the two intersecting channels is 90° and the die corner angle

Ψ was 20°. The diameter of the exit channel was slightly decreased in comparision to the entrance channel to facilitate multiple pass operation [8]. The die was placed into an INSTRON 8502 machine allowing a maximum applicable load of 200 kN (see Figure 1). A molybdenum disulphide grease was used as lubricant, the pressing speed was chosen between 4 mm and 8 mm per minute.

Figure 1: The ECAP setup used

From the ECAP processed workpieces, tensile specimens having a gauge length of 1 cm were prepared by plasma wire cutting and mechanical polishing. The tensile tests were performed with the strain rate of 10^{-3} s^{-1} using a MTS 810 machine. The strain was determined from the crosshead displacement.

TEM studies were performed with a JEOL 2000 FX microscope. For the one pass specimens, the pictures taken were combined with the measurements of grain misorientation angles using the convergent beam technique to analyse the Kikuchi patterns produced by single dislocation cells [9]. TEM mappings of specimens with a higher strain were also produced. The TEM foils were prepared from the homogeneous middle parts of the deformed workpieces. In addition, a heat treatment for 1 h at 150 °C was applied to investigate thermal stability of the cell structure of the ECAP deformed specimens.

3   Results of TEM Investigations

A typical TEM micrograph of a one pass copper specimen cut perpendicular to the extrusion direction can be seen in Figure 2. It shows a typical deformation structure. Two kinds of boundaries can be recognized in this picture. The thick dark lines can be identified as dense dislocation walls (DDWs) [10] while the thinner, lighter ones are the individual cell boundaries. The cell block structure is of a lamellar type, the distance between the DDWs being basically the width of a single cell.

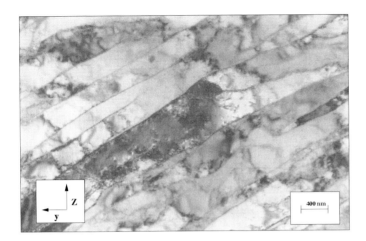

Figure 2: TEM micrograph of a single pass copper specimen, the directions y and z denote the transverse and the normal direction, respectively, x (not shown) being the extrusion direction

By processing a number of such pictures, cell size distributions were obtained. The distributions of misorientation angles between adjacent cells were measured using single crystallite orientations in TEM. As this is a very time consuming procedure, the results are concentrated on certain areas of the specimen only.

Figure 3 (left) shows the misorientation distribution determined in an area between DDWs (measurement parallel to DDWs). It can be seen that this area consists of low angle grain boundaries only, as expected. If an area also containing DDWs is considered, a bimodal misorientation distribution is observed, as can be seen on the right.

The bimodality of this distribution can be expected as the low angle grain boundaries originate from the area between the DDWs, while the observed high angle grain boundaries are the DDWs. In this structure a very large number of dislocation cells within the DDWs was observed, which explains the shape of the distribution. Some boundaries with an angle of up to 25 degrees were observed as well, which is remarkably high for a one pass ECAP deformation. It cannot be ruled out that this high degree of misorientation is attributable to an original grain

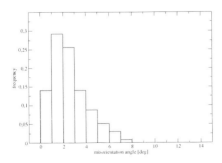 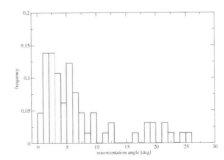

Figure 3: Misorientation distribution between adjacent cells of a single pass copper specimen in an area containing only low angle grain boundaries (left, 137 misorientations measured), containing low and high angle grain boundaries (right, 65 misorientations measured)

boundary of the non-deformed material. It should be noted that the measurement of approximately 100 misorientation in this study is not sufficient to yield fully adequate statistics. Significantly better statistics, albeit with slightly lower spatial and angular resolution, can be obtained using high resolution electron backscattered diffraction (HR-EBSD) [11,12].

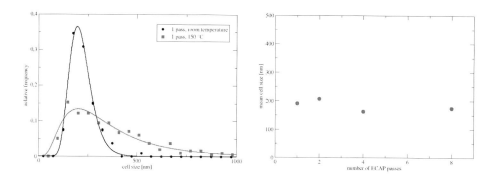

Figure 4: Cell size distribution for a single pass specimen before and after annealing at 150 °C (left), mean cell size of ECAP deformed specimens without subsequent annealing (right)

The cell size distributions were determined by the analysis of a set of TEM pictures. Using the image analysis software package ImageC, the areas of the apparent dislocation cells measured were converted to the equivalent circle diameter. Between 300 and 1500 cells were taken into account for the determination of the cell size distribution for each specimen. To calculate the most frequent cell size, a fit was performed to the original data using the lognormal distribution. For up to 8 ECAP passes it was found that the most frequent cell size was around 200 nm. After a rapid decrease to approximately 200 nm the average dislocation cell size does hardly change with increasing strain (compare Figure 4 right). The concomitant microstructure chang-

Figure 5: Montage of TEM micrographs for an annealed single pass specimen

es due to the evolution of misorientation angle distribution of adjacent cells lead to an increased number of high angle grain boundaries.

Annealing for 1 hour at 150 °C led to a change in the cell size distribution function. While the most frequent cell size did not change, the distribution broadened showing a pronounced structure coarsening (see Figure 4 left).

Figure 5 shows parts of the TEM montage for an annealed single pass specimen that was used to determine the cell size graphs presented in Figure 4 (left). As can be seen, the lamellar cell block structure observed in Figure 2 (no annealing) is still partly present here. Regions of increased growth in the annealed specimen can be observed for example in the right top corner of Figure 5. It is also seen that the lamellar structure has become less regular. Similar results were obtained for the investigated two and four pass specimens not shown here.

4   Mechanical Properties and Modelling

The tensile tests performed after ECAP reveal a rapid increase of the equivalent stress after the first pressing and only little hardening after the subsequent passes (see Figure 6). The same trend is observed for the decrease in cell size as a fuction of strain (compare Figure 4 right).

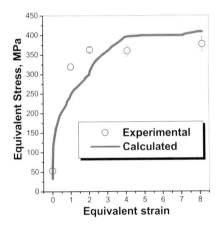

Figure 6: Simulated equivalent stress versus equivalent strain curve for ECAP copper [14]

A dislocation density based strain hardening model was used to describe the deformation behavior of the ultrafine grain material in terms of the evolution of the dislocation density [6]. In the model, a mixture of two 'phases' is considered. One is related to the cell interior with its relatively low dislocation density, the other is made up of cell walls with a high dislocation density. The total dislocation density is calculated using a rule of mixtures, with a weight factor associated with the volume fraction of the cell walls: a quantity that was assumed to decrease monotonically with increasing strain, as observed experimentally [13]. The average cell size is considered to be inversely proportional to the square root of the total dislocation density. This model was implemented in the ABAQUS software for FEM calculations. Details of the simulation procedure can be found elsewhere [14,15]. The model parameters were mainly adopted from previous work [6,7].

The simulated stress versus equivalent strain shows a good agreement with experimental data, cf. Figure 6.

The model used does not yet include the evolution of the grain structure (such as growing average misorientation angle and an increase of the proportion of high angle boundaries). The conversion of dislocation cells into grains due to an increasing misorientation between neighboring cells as well as formation and growth of new grain would be useful additions to the existing modelling framework.

5   Summary

In this work copper specimens were deformed using ECAP. TEM investigations were performed to analyze the microstructure and to determine the dislocation cell size distribution. For a single pass specimen, the distribution of misorientation angles between the adjacent cells was determined. It was found that misorientation angles of up to 25 degrees existed which may be attributed to an original grain boundary of undeformed copper. Annealing of ECAP processed specimens for 1 hour at 150 °C leads to the onset of grain coarsening, yet the most frequent cell size did not change under these conditions. A good agreement between model predictions and the experimental results-was obtained.

6   References

[1] Valiev, R.Z.; Islamgaliev, R.K. ; Alexandrov, I.V., Progress Mater. Sci. 2000, 45, 103
[2] Iwahashi, Y.; Wang, J.; Horita, Z.; Nemoto, M.; Langdon, T.G.: Scripta Mater., 1996, 35, 143
[3] Segal, V.M.: Mater. Sci. Eng. 1995, A197, 157
[4] Bowen, J.R.; Mishin, O.V.; Pragnell, P.B.; Juul Jensen, D.: Scripta Mater. 2002, 47, 289
[5] Mishin, O.V.; Juul Jensen, D.; Hansen, N: Mat. Sci. Eng. 2003, A342, 320
[6] Estrin, Y.; Tóth, L.S. ; Molinari, A.; Bréchet, Y.: Acta. Mater. 1998, 46, 5509
[7] Tóth, L.S.; Molinari, A.; Estrin, Y.; J. Eng. Mater. Techn. 2002, 124, 71
[8] Iwahashi, Y.; Horita, Z.; Remoto, M; Langdon, T.G.: Acta Mater. 1998, 46, 3317
[9] Liu,Q.:Ultramicroscopy 1995, 60, 81
[10] Hughes, D.A.; Hansen, N: Acta Mater. 1997, 45, 3871
[11] Bowen, J.R.; Pragnell, P.B.; Humphreys, F.J.: Mat. Sci. Tech. 2000, 16, 1246
[12] Bowen, J.R.; Pragnell, P.B.; Humphreys, F.J.: Mat. Sci. Forum 2000, 331–3, 545
[13] Müller, M; Zehetbauer, M; Borbély, A.; Ungár, T.: Z. Metallkunde 1995, 86, 827
[14] Baik, S.C.; Estrin, Y.; Kim, H.S. ; Hellmig, R.J. : Mater. Sci. Eng. A 2003, in press
[15] Baik, S.C.; Hellmig, R.J.; Estrin, Y.; Kim, H.S.: Z. Metallkunde 2003, 94, 754

A Composite Grain Model of Strengthening for SPD Produced UFG Materials

E.V. Kozlov, A.N. Zhdanov*, N.A. Popova, N.A. Koneva
Tomsk State University of Architecture and Building, Tomsk, Russia
*Altai State Technical University, Barnaul, Russia

1 Introduction

By now the problem of strengthening of ultrafine grained (UFG) materials obtained by the severe plastic deformation (SPD) is highly actual. Basically the solving of this problem concentrates on the analysis of a applicability of the well-known Hall-Petch (H-P) relationship

$$\sigma = \sigma_0 + k\, d^{-1/2} \qquad (1)$$

(σ_0 and k are relationship parameters, d is a grain size), the behavior of k parameter in the large grain size interval, consideration of the experimental results and deformation mechanisms. A number of models of the k parameter behavior were suggested. Some of these models are considered to be highly exotic. The present work is devoted to the study of these questions in UFG copper and is the direct continuation and development of our investigations [1].

2 The Hall-Petch Relationship Parameters: The Value of k Coefficient

Using the effect of grain refining for the increase of strengthening of metallic materials is closely connected with the σ_0 and k parameters in the H-P relation. The problem of constancy or change of σ_0 and k values in different grain size intervals is, still a matter of debate. One of the widespread variants of the k parameter behavior with the grain size decrease is the following: (1) its jump decreases [2, 3] or (2) the change of sign of k value at attaining a certain grain size [4]. In present work only the experimental data for copper should be considered. Tables 1 and 2 give k and σ_0 values for copper with the ordinary grain size (Table 1) and the grain size that varies over a range from microns to nanometers (Table 2). The data are given for tests at a room temperature (T_{room}). The behavior of the H-P parameters for the copper grain values $d > 25$nm is basically analyzed. It should be noted that the parameter k becomes negative at $d < 25$nm [3,14]. As it was expected the parameter σ_0, characterizing a strength to deformation in the grain body, varies over a wide range. For coarse grains it varies from 10 to 90 MPa, for the grain size range including the nanometric grains it varies from 10 to 150 MPa. The change of σ_0 value for pure copper is caused by impurities in the crystal lattice and by defects in its structure, basically by dislocations. It should be noted that the intervals of σ_0 variations for different grain sizes are in substantial agreement. In the main this may confirm the universal nature of the phenomenon.

The systemized data for the coefficient k are still more interesting. For copper samples the necessary grain size in Table 1 and 2 was attained by SPD followed by annealing treatment. The

k parameter varies in the narrow range of values 0.10...0.15 MPa · m$^{1/2}$. At any rate when the grain sizes are decreased from 250 μm to 0.02 μm, i.e. are different in 10^4 times, the coefficient k for copper is close to constant value and may characterize the material.

Table 1: The H-P parameters of coarse grained copper

Interval of grain sizes μm	k, MPa × m$^{1/2}$	σ_0, MPa	Reference
3...29	0.10	60	[5]
2,5...20	0.10	10	The author's data
10...60	0.10	25	[6]
12...178	0.10	90	[7]
20...250	0.11	23	[8]
–	0.12	4.0	[9]
12....250	0.14	95	[10]
3.....150	0.14	10	[11]
10...100	0.21	11,4	[13]
17...1000	0.24	13	[12]

It should be noted that the value of k depends on purity of the metal, crystallographic and structural texture and its dispersion, the grain boundary type, possibility of sliding along the boundaries and emitting of dislocations, the triple line structure and some other factors.

Table 2: The H-P parameters of copper for interval sizes from coarse grains to UFG ones

Interval of grain sizes, nm...μm	k, MPa × m$^{1/2}$	σ_0, MPa	Reference
200...50	0.11...0.15	10	The author's data
11...100	0.22	11	[15]
300..100	0.15	30	[13]
25 ...2,5	0.10	100	[3]
200...90	0.11	150	[3]
16...10	0.01	75	[16]
16...10	0.02	130	[17]
25...20	0.01	80	[18]
26...20	0.06	70	[19]
35....40	0.11	25	[14]

Strong dependence of the copper yield stress on the boundary state was revealed in [20]. The grain boundary state was changed by annealing treatment at different temperatures. The grain size after annealing treatment was not changed and therefore the influence of the boundary state on the parameter k value was not find out at present. It is known the parameter k is also depends on the test temperature and, consequently, the thermally activated processes take part in the formation of its value. Nevertheless in the very wide range of grain sizes the k value is practically constant.

3   The Composite Grain Model

The "composite" model stands out against different models used for derivation of the relation (1). In this model the grain area adjoining the boundaries is distinguished by the increased resistance to dislocation movement. For the first time the model was suggested in the known J.P. Hirth's review [21]. Later this model was developed by H. Mughrabi [22, 23] for the cell structure. It was shown that the grain sections adjoining the subboundaries are distinguished by the increased internal stress. The authors of the present paper verified these models by direct measurements of the internal stresses in different sections of grains and subgrains [24]. It was established that the main cause of composite configuration of the dislocation slip resistance consists in the increased level of internal stresses in the grains or cells sections near boundaries and their joints. The experimental data for the suitability of the composite model for nanomaterials as well as the model by itself were considered in detail by Valiev and Alexandrov [25, 26]. Using X-ray investigations it was shown that near the nonequilibrium boundaries and joint disclinations the crystal lattice parameters are decreased, the compressing stresses and the change in the character of heat atom oscillations take place. The width and free volume of the grain boundaries of nanocrystalline materials is larger than in ordinary materials. Direct measurements of the internal stresses and the bending-torsion of the crystal lattice were carried out in our works [27, 28]. The investigations showed the existence of the near boundary zone of finite width with the increased level of the tangent stresses. The scheme of the stress field distribution in a typical grain (or subgrain) is given in Fig.1.

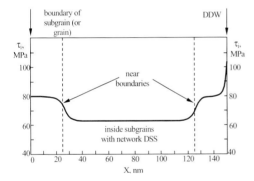

Figure 1: The distributions of the internal stresses τ_i in subgrains and grains of UFG copper. Designations DDW and DSS mean a dense dislocation wall and dislocation substructure correspondingly.

One can observe three levels of microstresses in the grain: (1) in a distance from the grain boundaries, i.e. in the grain body; (2) in areas near the boundaries; (3) immediately in the boundaries. Since the grains and subgrains may have the boundaries of different type giving rise to different level of internal stresses, this should be shown on the scheme. The scheme for copper microcrystal given in Fig.1 is much similar to the scheme for the grains of UFG nickel presented by the authors earlier [27]. However, in the case of UFG nickel the level of stresses appears to be 2–3 times higher than in the case of UFG copper. The amplitude of the internal stress field in all sections of grains was measured by the dislocation line curvature [27]. Therefore on the scheme of Fig.1 the tangent stress values averaged over the data are presented. One can see well that the results obtained not only confirm but also fit to the grain composite model in a quantitative way.

It should be borne in mind that the internal stresses are not always homogeneous. Fig.2 shows the distributions of the internal stress field amplitude for UFG copper obtained by equal channel angular (ECA) pressing.

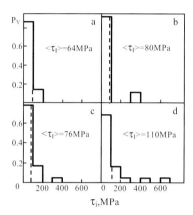

Figure 2: Distribution of the internal stress fields: (a) in the grain body, in the distance from the boundaries; (b) near the boundaries; (c) immediately in low-and high angle boundaries; (d) inside the cell boundaries and the DDWs.

The level, corresponding to 50 MPa, covers the larger part of the material volume (up to 80 %). However in local parts, stresses up to 600 MPa are found. From the data in Fig.2 one can also see that different boundaries give rise to different levels of stresses both immediately in the boundaries and in their neighborhood. The largest stresses arise near the cell boundaries and the dense dislocation walls (DDW). The rest of low and high angle boundaries give rise to smaller levels of stress. The chaotic or network dislocation substructures achieve still lower levels of stress in the grain or subgrain bodies. The simple scheme given in Fig.1 is related to the grains with network or chaotic dislocation substructures. If the substructure is a cellular one, additional sources of the internal stresses arise within the grains. In this case the scheme of internal stress distribution over the grain appears to be still more complex.

4   Deformation Resistance of UFG Copper

As shown in our work [1] the plastic deformation in UFG copper with the average grain size about 200 nm begins at T_{room} by intense sliding at the grain boundaries. The sliding rapidly increases (from 2.0 to 10.0 nm) with the increase of deformation. Simultaneously the quasi-heterogeneous slip is developed by small dislocation groups (3...7) within the grains. In separate favourably oriented grains one can observe shears up to 6 nm. Consequently the main deformation mechanism of UFG copper at T_{room} are the gliding of lattice dislocations and the sliding of grain boundaries.

In such conditions the breaking of an active dislocation source, the intersection of dislocations, cross slip, the overcoming of internal stress elastic fields and dislocation walls (boundaries of cells and, probably, fragments) as well as different grain boundary processes appear to be the mechanisms defining the deformation resistance. These processes are mostly thermally activated, the internal stress field is an exception. The existence of grain parts with increased internal stress (Fig.1) requires the application of the composite model to every type of micrograin, for the evaluation of the deformation resistance.

There are three type of grains in copper produced by ECA pressing: (1) dislocation free grains, (2) grains with chaotic or network structures and (3) grains with cellular or fragment dislocation structures. The grain size distribution with allowance for grain type is presented in Fig.3.

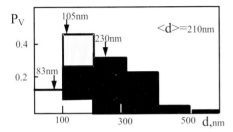

Figure 3: The distribution of grains on the sizes of UFG copper: (□) free dislocation grains, (▨) grains with dislocations, (■) grains with cells and fragments. The arrows show the average sizes of these grains

The presence of three different grain types in the investigated UFG copper allows using the complex model, whose principles were described in [1]. In [1] the formulas (3)–(5) for the complex strengthening model are given in approximation of the homogeneous stress field in each grain type. In the present work the effect of the inhomogeneous stress field is accounted for by the composite model approach described in part 3.

For intra-grain sliding for the UFG copper, the main problem of the contribution to the flow stress is the emission of dislocations from GB's. In this case the stress for dislocation emission from GB source is determined by the following relation [29]:

$$\sigma_s = \frac{m\mu b}{d} \qquad (2)$$

where m is inverse value of Schmid's factor ($m = 2,2$), μ is the shear modulus, b is Burger's vector. The internal stresses for the grain composite model are estimated according to the following relation [30]:

$$\sigma_e = m\sum_{i=1}^{3} \tau_i \frac{\delta l_i}{d}, \qquad (3)$$

where δl_i is the length of segment intra the grain body and near boundaries of grains, cells and fragments (see Fig.1). The reverse stresses acting at the source from one or n-dislocations can be estimated by the following relation [31]:

$$\sigma_{rev} = \frac{m\,\mu\,n\,b}{\pi\,(1-\nu)\,d}, \qquad (4)$$

where ν is the Poisson's number. The intra-grain dislocation contribution (σ_d) consists of contributions of the dislocations spread over a grain:

$$\sigma_d = m\,\alpha\,\mu\,b\,\rho^{1/2}, \qquad (5)$$

where α is a constant. Here the contributions from cells and fragments (H. Holt's formula [32]) were estimated by their dislocation density in accord with (5). Choosing the value α in (5) we considered that the internal stresses had been taken into account by formula (3). For different grain types one should consider different contributions to the yield stress $\sigma_{y.s}$ [1]. The value $\sigma_{y.s}$ for the dislocation glide within grains of each type consists of the following contributions:

$$\sigma_{y.s} = \sigma_s + \sigma_e + \sigma_d + \sigma_{rev}. \qquad (6)$$

Let us consider a behavior of the average size grains of every type when the yield stress is reached. The experimental value of the yield stress of UFG copper is $\sigma_{y.s}$ = 358 MPa [1]. The average grain size of dislocation free grains is d = 83nm (Fig.3). The action of dislocation source from the grain boundary requires the stress σ_s = 330 MPa. Because of the small grain size (83nm), up to 2/3 of its space are the strengthened near-boundary parts. The contribution of the internal elastic field stresses reaches the value of σ_e = 160 MPa. Since these contributions are almost not thermally activated, the dislocation free grains whose sizes are less than the average grain size are not deformed at all. Dislocation free grains with grain sizes more than the average are deformed by emitting separate dislocation. The next dislocation is emitted only when the previous dislocation has entered the opposite grain boundary and dissolved there.

The average grain size with chaotic and network substructures is 105nm, $\rho = 6 \cdot 10^{14}\,\text{m}^{-2}$ is the average dislocation density within them. In these grains the strengthening near boundary part occupies about one half of the grain. For these grains σ_s = 170 MPa, σ_e = 175 MPa. The dislocation contribution σ_d at $\rho = 6 \cdot 10^{14}\,\text{m}^{-2}$ is equal to 330 MPa. The latter contribution is thermally activated. Undoubtedly, such grains on the yield stress at T_{room} are deformed with the rate that allows realizing the thermal activation.

The average grain size with cells and fragments is equal to 230 nm, the dislocation density taking into account the cell walls is equal to $\rho = 13 \cdot 10^{14}\,\text{m}^{-2}$. For these grains σ_s =120 MPa, σ_e = 145 MPa and σ_d = 500 MPa. The latter contribution is rather thermally activated. It is in these grains i.e. their boundaries where the slip begins at deformation. Then it is actively proceeded in the grain body with the average shear P = 6.0 nm [1]. A fraction of these grains on the yield stress reaches 90 %. The development of processes on the yield stress is completely due to the contribution of these grains. Probably the grains with the size exceeding the average value mostly participate in deformation. At least the grain boundary slip is repeated over the interval of 380 nm [1]. Then the slip in the grains with network substructure is developed.

The consideration given allows to return to the origin of the H-P dependence at the grain sizes of 50…500 nm. The contributions of $\sigma_s \sim d^{-1}$, $\sigma_e \sim d^{-n}$ (n = 0.5…1 [27]), $\sigma_d \sim d^{-0.5}$. It should be stated that though the k parameter value in this interval of grain sizes coincides with that for the coarse material, the mechanisms forming the H-P dependence for UFG materials are quite different.

The contributions of elastic field stresses from strengthening areas near boundaries and the stress necessary for dislocation emission by grain boundaries are the most significant contributions for the yield stress formation of SPD produced UFG copper. The contribution of elastic field stresses achieves 160 MPa and as result $\sigma_s + \sigma_e = 490$ MPa. These contributions are not almost thermally activated. So on the yield stress dislocation free grains, which are smaller average sizes, are not deformed at all and grains which are larger average sizes are deformed by emission single dislocation on the yield stress.

5 References

[1] E.V. Kozlov, A.N. Zhdanov, L.N. Ignatenko et al., Ultrafine Grained Materials 2 (Eds.: Y.T. Zhu, T.G. Langdon, R.S. Mishra, S.L. Semiatin, M.J. Saran, T.C. Lowe), TMS, Warrendale, USA, 2002, 419–428

[2] C.J. Youngdahl, P. Sanders, J.A. Eastman, J.R. Weertman, Scr. Mater., 1997, 37, 809–813

[3] R.A. Masumura, P.M. Hazzledine, C.S. Pande, Acta. Mater., 1998, 46, 4527–4534

[4] E. Arzt, Acta. Mater., 1998, 46, 5611–5626

[5] Von E. Dick, Z. Metallkde, 1970, 61, 451–454

[6] N. Ono, S. Karashima, Scr. Metall., 1982, 16, 318–384

[7] V.Y. Gertsman, M. Hoffmann, H. Gleiter, R. Birringer, Acta. Metall. Mater., 1994, 42, 3539–3544

[8] P. Feltham, J.D. Meakin, Phil. Mag., 1957, 2, 105

[9] R.W. Armstrong, R.M. Douthwaite, Mat. Res. Soc. Symp. Proceed., 1995, 362, 41–47

[10] N. Hansen, B. Ralph, Acta. Metall. Mater., 1982, 30, 411–417

[11] A.W. Thompson, W.A. Backofen, 1971, 2, 2004–2005

[12] T.L. Johnston, C.E. Feltner, Metall. Trans., 1970, 1, 1161–1167

[13] G.T. Gray III, T.C. Lowe, C.M. Cady et al., NanoStructured Materials, 1997, 9, 477–480

[14] J.R. Weertman, D. Farkas, K. Hemker et al., MRS Bulletin, 1999, 24, 44–50

[15] E.D. Tabachnikova, V.Z. Bengus, R.Z. Valiev et al., The Structure and Properties of Nanocrystalline Materials (Eds.: G.G. Taluz, N.I. Noskova), Ekaterinburg, UD RAS, RF, 1999, 103–107

[16] C. Suryanarayanan, F.N. Froes, Met. Trans., 1992, 23A, 1071–1081

[17] P.G. Sanders, J.A. Eastman, J.R. Weertman, Acta. Mater., 1997, 45, 4019–4025

[18] G.W. Nieman, J.R. Weertman, R.W. Siegel, NanoStructured Materials, 1992, 1, 185–190

[19] R. Suryanarayanan, C.A. Frey, S.M.L. Sastry, J. Mater. Res. Soc., 1996, 11, 439–442

[20] V.Y. Gertsman, R. Birringer, R.Z. Valiev, Phys. Stat. Sol. (a), 1995, 149, 243–252

[21] J.P. Hirth, Met. Trans., 1972, 3, 3047–3067

[22] H. Mughrabi, T. Ungar, W. Keinle, M. Wilkens, Phil. Mag. A., 1986, 53, 793–813

[23] H. Mughrabi, Mater. Sci. Eng., 1987, 85, 15–31

[24] N.A. Koneva, E.V. Kozlov, L.I. Trishkina, Mater. Sci. Eng., 2001, A 319–321, 156–159

[25] R.Z. Valiev, R.K. Islamgaliev, I.V. Alexandrov, Progr. Mater. Sci., 2000, 45, 103–189

[26] R.Z. Valiev, I.V. Alexandrov, NanoStructured Materials produced by Severe Plastic Deformations, 2000, Moskow, Logos, RF, P. 272

[27] E.V. Kozlov, N.A. Popova, Yu.F. Ivanov et al., Ann. de Chimie, 1996, 21, 427–442

[28] N.A. Koneva, N.A. Popova, L.N. Ignatenko et al., Investigations and Applications of Severe Plastic Deformations (Eds.: T.C. Lowe, R.Z. Valiev), 3. High Technology, Kluwer Academic Publishers, The Netherlands, 2000, 121–126
[29] J.P. Hirth, J. Lothe, Theory of Dislocations, 2d ed., N-Y, J. Wiley, USA, 1981
[30] H. Mughrabi, Acta metal. mater., 1983, 31, 1367– 1379
[31] N.A. Koneva, E.V. Kozlov, Izvestia Vuzov. Fizika, 1982, 8, 3–14
[32] M.R. Staker, H. Holt., Acta Met., 1972, 20, 569–579

Computer Simulation of Equal-Channel Angular Pressing of Tungsten by Means of the Finite Element Method

G. Krallics[1], I.N. Budilov[2], I.V. Alexandrov[2], G.I. Raab[2], V.S. Zhernakov[2], R.Z. Valiev[2]

[1] Budapest University of Technology and Economics, Budapest, Hungary
[2] Ufa State Aviation Technical University, Ufa, Russia

1 Introduction

Recent investigations successfully demonstrated great potential of the severe plastic deformation (SPD) technique, conducted by means of equal-channel angular pressing (ECAP), to form bulk nanostructured states in different metals and alloys. However, practical application of ECAP seems to be a difficult task due to the fact, that its performance is a many-factor process [1].

The main task of ECAP is processing of homogeneous ultra-fine grained nanostructured states in the deformed billets. Meanwhile, an important factor is the homogeneity of the accumulated plastic deformation fields. The latter one depends on the conditions' invariance in the contact area, temperature changing and the character of the plastic flow in the bulk of the billet.

Numerical methods take an important place at developing the most optimal geometry of the die-set as well as regimes of ECAP. In particular, with the use the given approach there was studied the influence of such ECAP parameters as: friction coefficient, an angle of channels intersection, pressing route and a number of passes, channel's diameter, radius of channels' adjunction, strain rate and a pattern of the billet's material deformation, backpressure [2-5]. The whole sum of these factors leads to ambiguous conclusions concerning the regularities of the plastic flow, uniformity of the channel's filling up and the values of the accumulated summary plastic deformation.

It should be noted, that the majority of numerical investigations was performed applying 2D models. It diminishes their reliability and accuracy of calculations when characterizing a real plastic deformation behavior at ECAP. One more question, requiring further discussions, is the problem of considering the heterogeneity in the contact area with the influence of non-stationarity of the temperature field along the bulk of the billet and in the contact area.

It should be noticed, that up to now, simulation of ECAP has been conducted in respect to ductile metallic materials. At the same time, applying of ECAP to form ultra-fine grained (UFG) structures, hard-to-deform metals and alloys included, is still topical.

First experimental attempts to solve this problem were successful. However, they were quite expensive regarding labor efforts and material costs [6]. In this connection, development of ECAP with respect to hard-to-deform metals and alloys inevitably leads to a conduction of simulation applying numerical methods.

In the given report one may observe the results of investigation of ECAP process of CP tungsten with the use of the numerical simulation methods.

2 Simulation Method and the Material Model

The calculation of the material plastic flow at ECA pressing was performed on the basis of the flow theory [7]. The complete equation system of visco-plastic non-isothermic metal flow in Euler coordinate system includes: motion equation, kinematic proportions, incompressibility equation, thermal conductivity equations, mechanical and thermal boundary and initial conditions, defining relations, connecting deviators of tensors of strain and stress's rates [7]

$$\sigma'_{ij} = \frac{2}{3}\frac{\bar{\sigma}}{\dot{\bar{\varepsilon}}}\dot{\varepsilon}_{ij} ; \qquad (1)$$

thermal-conductivity equation

$$C\rho T = k_1 T_{ii} + \beta \sigma \varepsilon ; \qquad (2)$$

rheological equation

$$\sigma = \sigma(\dot{\varepsilon}, \varepsilon, T) , \qquad (3)$$

where

$\sigma'_{ij}, \dot{\varepsilon}'_{ij}$ deviators of stress and strain's rates

$\dot{\varepsilon}_{ij}$ elastic deformation rate

$\dot{\varepsilon}$ intensity of deformation rates

σ deformation resistance (stress intensity)

To consider friction in program [7] we use the dependence, which combines two friction laws – of Coulomb and Zibel.
To consider friction the following dependence was drawn [7]:

$$F = m\frac{\sigma}{\sqrt{3}}\left(1 - e^{-1.25(\sigma_n/\bar{\sigma})}\right) , \qquad (4)$$

where m – friction factor; $\bar{\sigma}$ – deformation resistance; σ_n – normal contact pressure.

Equation (4) describes very well as areas of high contact pressures as well as low ones.

Finite-element simulation of ECAP process was performed applying well-known code QForm [7]. These codes are widely spread. They have a scope of facilities and are intensively applied at simulation.

In the given computer simulation the condition of the real experimental ECAP conditions were taken into considerations. The angle of channels intersection was equal to 135° (Figure 1). Different deformation routes (A and C) were considered. But in the case of simulation two deformation passes were considered at the same time.

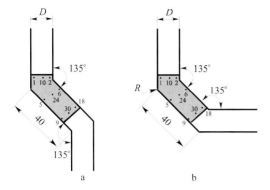

Figure 1: Channel geometry, considering the deformation route C (a) and A (b)

CP tungsten was chosen as a material for investigation. Its mechanical characteristics were: Young modulus $E = 2.1 \cdot 10^5$ MPa; Poisson coefficient $\mu = 0.3$; yield stress $\sigma_{0.2} = 220$ MPa; ultimate strength $\sigma_u = 600$ MPa. The die-set was manufactured out of instrumental high-strength steel R6MT (Russian nomenclature).

When solving the thermal task, not only the properties and the temperature of heating of the billet ($T_b = 900°$) and of the die ($T_d = 500°$) were taken into account, but we also consider properties of the contact lubricant layer. The calculation of the contact heat exchange of the billet and the tool was conducted by the following formula [7]:

$$k_1 \frac{\partial T}{\partial n} = -h_{\text{lub}} (T - T_d), \quad (5)$$

where h_{lub} – is a heat-transfer coefficient; T_d – is an average temperature of the tool; T – is a contact temperature; n - is the coordinate of the axis normal to contact layer; k_1 – thermal conductivity of the tool.

The experimental deformation curve of tungsten, which was used in calculations, is given in Figure 2. It considers the temperature of the technological process as well as the strain rate. During the process of the deformation curve has not changed. Deformation curves for W in the whole the temperature-rate interval of the calculated process of were taken from the materials' database, which was installed to the QForm [7].

As a material model we used the dependence, characterizing isotropic strengthening of the billet's material of the following type:

$$\sigma_y \left(\varepsilon_{\text{eff}}^p, \dot{\varepsilon}_{\text{eff}}^p \right) = \sigma_y^s \left(\varepsilon_{\text{eff}}^p \right) \left[1 + \left(\frac{\dot{\varepsilon}_{\text{eff}}^p}{C} \right)^{1/p} \right], \quad (6)$$

where

σ_y^s the yield stress of the material
$\varepsilon_{\text{eff}}^p, \dot{\varepsilon}_{\text{eff}}^p$ the effective strain and the strain rate correspondingly
C, p the material constants

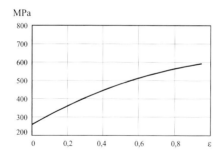

Figure 2: Curve of the deformation strengthening of CP tungsten at tension. $T = 1000°$ deformation rate = 6 mm s^{-1}

3 Analysis and Discussion of the Simulation Results

3.1 Fields of the Accumulated Shear Strains

In Figure 3 one may see a distribution of the accumulated plastic deformation in the process of ECAP in the form of isolines, with the influence of the route of the technological process being considered. The beginning of the plastic flow, when the billet just starts to pass the first angle of the channel, is shown in Figure 3. It was found, that one kernel of maximal plastic deformations appears at the initial stage. While the billet is passing along the channel, the deformation kernel shifts to the right along the direction of passing with a simultaneous growing of the accumulated plastic deformation. In Figures 3b and 3c one can observe isolines in the end of the billet's deformation after the second pass (the routes A and C correspondingly). It was established, that the channel's route influences significantly the value of the accumulated plastic deformation: on the route A – $\varepsilon_{pl}^{max} = 0.85$, and on the route C – $\varepsilon_{pl}^{max} = 1.1$ for two passes. Character of the isolines distribution ε_{pl} along the bulk of the billet in the middle and in the end of the route is changing simultaneously with a change of the deformation route.

In case of the route one can observe a presence of two specific kernels with $\varepsilon_{pl}^{max} = 0.85$. The area of uniform plastic deformations decreases significantly (twice), as compared to the area of

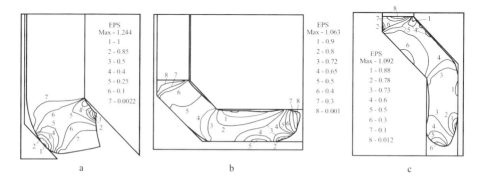

Figure 3: Isolines of the accumulated plastic deformation in the beginning of ECAP (a) and in the end of second pass (route A - b, route C - c). QForm software.

the accumulated plastic deformation, applying the route , where we can observe one local maximum with ε_{pl}^{max} = 1.1. The value of the accumulated plastic deformations applying the route C is by 23 % higher, than with the use of the route A.

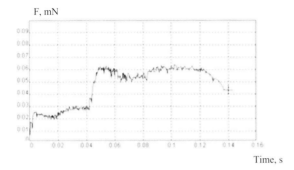

Figure 4: Punch efforts against the time at ECAP (the 1st pass)

Figure 4 shows, how changes the pressing effort (on the flank of the billet) at ECAP. The form of the curve changes significantly for different points of the billet. A sudden growth of loading in the diagram means that the billet has passed the first channel and that the deformation has started in the second channel. Maximal press efforts reach 60 kN.

Diagrams in Figure 5 show, how the accumulated plastic deformations change in the 3D model of tungsten billet at ECAP. Application of 3D FE-models allowed analyzing in more detail the distribution of the stress-strain state's components as on the surface, as in the billet's section. It also makes possible to obtain more reliable quantitative characteristics' data of the stress and strain components. Analysis of Figures 3 and 5 makes it possible to conclude, that a change of the stress state from the flat type to the bulk one leads to a considerably nonuniform

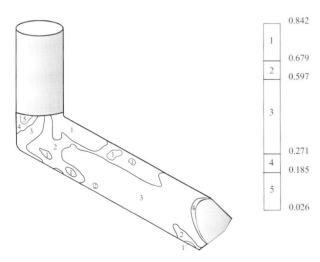

Figure 5: Accumulated equivalent plastic strain in 3D model. Superform software

distribution of summary plastic deformation along the billet's surface. Meanwhile, deformation character of the middle section (symmetry planes) has not changed significantly. But the absolute value of the accumulated plastic deformation in specific areas of the sample changes by 20–30 %.

3.2 Analysis of the Stress-strain State of the Tool

In the work the stress-strain state in the tool at ECAP for one pass of the billet was analyzed. Such an analysis is very important when choosing the material and the geometry of the die-set and the tool.

Characteristics of liquid glass EVT-24 being used as a lubricant were taken as parameters at simulation. It has been revealed, that the sizes and form of distribution areas of maximal stress intensity values in the tool matrix undergo changes during the technological process (Fig. 6). Analysis of the component distribution of the stress state shows, that there is a change of the sign of stresses on some surface areas of the channel. It means that not only the values of stresses are being changed, but the directions of them also undergo changes. It was shown, that in case of the channel under consideration the area of the largest pressures and possible exhaustion sides with the point of the outer radius of channels intersection.

The influence of the friction factor on the stress-strain state of matrix was analyzed as well. There were revealed the following: a) at first iterations changing of the friction factor from $m = 0.005$ to $m = 0.1$ and $m = 0.2$ – leads to a growth of stresses in the most loaded area by 40 %; b) at further iterations one can observe monotonous decreasing of max values σ_i with increasing of m.

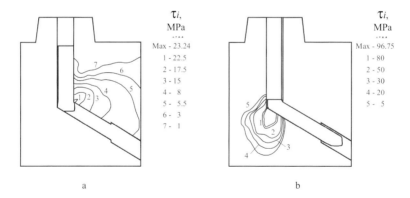

Figure 6: Change of stress intensity τ_i in the tool in the beginning (a) and in the end (b) of ECA pressing, taking into account friction factor 0.05

4 Conclusion

Finite-element analysis of the plastic flow of the material of the CP tungsten billet during ECAP at the billet temperature 1100 °C and at the die-set temperature 500 °C was performed. The influence of the friction factor and the route on the evolution kinetics of the summary plastic strain fields in various areas of the cylindrical tungsten billet was investigated.

It was shown that the plastic flow in CP tungsten in the conditions of ECAP at 1100 °C is mostly heterogeneous and depends both on the number of passes and on the type of lubricant used. It was found that an angle recess was formed. The magnitude of the recess depends on the friction factor and the type of the material deformation pattern. It was established that alongside with growing of the friction factor, the magnitude of the recess, siding with the outer angle of the channels intersection, was decreasing. It allowed increasing the accumulated plastic strain. It also had a positive influence on the uniformity of plastic deformation distributions along the bulk of the billet.

5 References

[1] Segal, V.M., Mater. Sci. Engng. 1995, A197, p. 157–164
[2] Bowen, J.R.; Gholinia, A.S.; Roberts, M.; Prangnell, P.B. Mater. Sci. Engng. 2000, A287, p. 87–99
[3] Prangnell, P.B.; Harris, C.; Roberts, S.M. Scr. mater. 1997, 37, p. 983-989
[4] Kim, H.S. Mater. Sci. Engng. 2002, A328, p. 317–323
[5] Zhernakov, V.S.; Budilov, I.N.; Raab, G.I.; Alexandrov, I.V.; Valiev, R.Z., Scr. mater. 2001, 44, p.1765–1769
[6] Aleksandrov, I.V.; Raab, G.I.; Shestakova, L.O.; Kil'mametov, A.R.; Valiev, R.Z., Phys. Met. Metallogr. 2002, 93, p. 493–500
[7] Q-Form. User's Guide. Quantor Ltd. Version 2.1, 2000

V   Texture Evolution and Simulation During SPD

Texture Evolution in Severe Plastic Deformation by Equal Channel Angular Extrusion

L. S. Tóth
Laboratoire de Physique et Mécanique des Matériaux, Université de Metz, Metz, France

The Author dedicates this paper to Prof. H. P. Stüwe for his friendship to people and his lifetime scientific achievements.

1 Abstract

The majority of the techniques of severe plastic deformation calls for simple shear deformation mode. This is why a special interest is given in this paper to textures that develop in simple shear. The evolution of texture in Equal Channel Angular Extrusion (ECAE) is also discussed in detail. The classical simple shear model of ECAE is examined as well as a new, more precise flow field which uses an analytical flow function. The proposed function is inspired from finite element calculations. The velocity gradient that follows from the analysis is incorporated into the self consistent viscoplastic polycrystal code. The evolution of deformation texture is predicted up to two passes in the A-route ECAE deformation of copper polycrystal when the extrusion angle is 90°.

2 Introduction

Large plastic strains are unavoidable in forming operations of materials and have several consequences on the properties of the product. They are: strain hardening, evolution of plastic anisotropy, and refinement of the microstructure. The equal channel angular extrusion (ECAE) process is one of the new techniques that receives particular attention these years. [1,2] In this process, the material is pushed through two channels of equal diameter connected to each other at an angle of Y, usually 90° or 120° (Fig. 1). All the above three phenomena take place in the ECAE deformation mode:

- Strain hardening strengthens the material near to its theoretical limit. [3–5]
- Ultra fine grain structure develops. [3–5]
- A particular crystallographic texture is formed. [6]

While the first two effects are examined quite extensively in recent years, the characteristics of the crystallographic textures are not yet studied sufficiently. Only one important aspect of the texture was recognized, namely, that it is similar to that of simple shear. Actually, the proposed strain mode for this deformation path is simple shear, which takes place in the in intersecting plane of the two channels if their connection is not rounded (see Fig. 1). On the basis of this model, the following formula is used for the von Mises equivalent strain. [7]

$$\bar{\varepsilon} = \frac{2}{\sqrt{3}} \cot\left(\frac{\Phi}{2}\right).\tag{1}$$

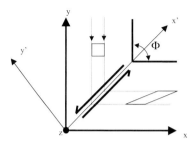

Figure 1: Scheme of the ECAE test

For a 90° die, this formula gives an equivalent strain of $\bar{\varepsilon} = 2/\sqrt{3} = 1.1547$ When the intersection is rounded, the strain is less. (For a perfectly rounded 90° die, it is $\bar{\varepsilon} = (\pi/2)/\sqrt{3} = 0.9069$). In any case, the outgoing material shows a sheared form where the shear seems to be applied on the y plane in the x-direction (see Fig. 1). In this simplified model, when the die is not rounded, the material is instantaneously sheared at the discontinuity plane; this model was called in [8] as the "discontinuous shear" model. It can also be called as the "simple shear model".

When the discontinuous shear model is used for the simulation of crystallographic texture evolution, it can be carried out in two ways. One way is to apply a negative simple shear on the discontinuity plane and then rotate the obtained texture around axis z by –45° into the x–y laboratory system of Figure 1. This rotation is necessary because the texture is usually measured in the latter reference system. The other way is first to transform the velocity gradient of negative "simple shear to the x–y laboratory system and then carry out the deformation texture simulation with this rotated velocity gradient in the laboratory system. In this way, the obtained texture can be directly compared to the experimental one. The first technique was employed by Agnew et al. [9] to simulate the texture evolution in ECAE deformed Cu and Fe. The second method is proposed independently by Tóth et al. [8] and Beyerlein et al. [10]

A more refined approach to the deformation mode in the ECAE die is to consider the flow lines – which are shown to be rounded in experiments even in the case of a non-rounded intersection of the canals. [6] A simplified model based on rounded flow lines was presented by Tóth et al. [11] and "Kopacz et al. [12] with good success in comparison to the experimental textures. That approach, however, still assumed simple shear tangent to the streamlines and neglected other strain components. This model is strictly valid only when the flow lines are perfectly rounded implying that the materiel element does not change its cross section in the direction of the flow. Gholinia et al. [6] have measured the shape of the flow lines in Aluminium and used them directly in a Taylor type polycrystal model to predict the texture evolution. Their predicted textures have reproduced some important features of the experimental textures, namely, the rotation of the ideal components with respect to the symmetry positions.

After about two passes, the process of grain subdivision is so advanced that it alters the deformation mode of individual grains and affects the evolution of the texture significantly. A model to simulate such a process has been formulated by "Beyerlein et al. [10]

The purpose of the present work is first to analyze the similarities and the differences between the crystallographic textures of large strain simple shear and ECAE textures. Then the new flow line model presented by Tóth et al. [8] will be used to predict the evolution of the crystallographic texture. It uses an analytical flow function which has only one free "parameter. The adjustment can be done using the observation of experimental streamlines or using the results of finite element calculations. In order to take into account the evolution of grain shape, the self sonsistent viscoplastic polycrystal model is employed in this work. Route A is examined in Cu. The die angle is 90° and the die-intersection is not rounded.

3 Main Characteristics of Simple Shear Textures

Simple shear textures can be readily produced by large strain torsion of a bar if length changes of the bar are small with respect to the shear strain. Tóth and co-workers have "extensively studied the texture evolution during simple shear; [13–17] here their main findings are summarized.

The ideal components of simple shear textures display small "tilts" with respect to their ideal positions and their intensity also varies significantly with strain. As an illustration, Figure 2 shows the texture evolution in torsion for the case of OFHC copper, up to very large strains. Only the first ODF sections are shown, where the A1*, C, and A2* components are present. (For more information, see Tóth et al.). [14] It can be seen that at a shear of $\chi = 2$, all three components are well developed, the C and A2* components, however, appear in rotated positions, opposite to the direction of rigid body rotation, indicated by an arrow in Figure 2. At $\chi = 5.5$, the A1* and A2* components are week, while C is further strengthened. At this strain level, all components reach their symmetry positions in orientation space. Finally, at the very large strain of $\chi = 11$, the A1* component increases and the C component is slightly rotated now in the direction of the rigid body rotation.

The above observations can be interpreted with the help of the rotation field, [16] which is very characteristic in simple shear. The rotation field is a representation of the lattice spin ($\dot{\omega}$), which is obtained as the difference between the rigid body spin ($\dot{\Omega}$) and the plastic spin ($\dot{\beta}$): [17]

$$\dot{\omega} = \dot{\Omega} - \dot{\beta} \qquad (2)$$

The rigid body rotation is inherent in shear deformation and it is the same amount for all grains of the polycrystal (in a Taylor model). It is acting as a common rotation component of the crystals. The plastic spin has to balance the rigid body rotation to obtain zero lattice spin for an ideal orientation. Large plastic spin can only be obtained if the number of operating slip systems is low. This is the reason why there are only maximum two operating slip systems in the ideal orientations of simple shear textures. [16] When strain rate sensitivity of slip is relevant, the number of operating slip systems "increases as the strain rate sensitivity index (m) increases. In such a case, exactly zero lattice spin cannot be obtained in simple shear and the ideal texture

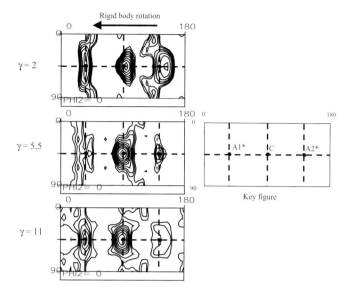

Figure 2: Evolution of the simple shear texture of copper as a function of shear deformation

components must slowly rotate in the direction of the rigid body rotation. [17] This is why we can say that the ideal components of simple shear textures are not perfectly "stable".

The general picture of the rotation field in orientation space is illustrated in Figure 3 on the example of the C ideal component. As can be seen, the arrows representing the lattice spin in this figure are convergent from the right and "divergent on the left from the C position. This asymmetric "nature of the rotation field is responsible for the accumulation of grain orientations first on the right side, and appears as a ‚tilt' of the maximum of the ODF with respect to the exact C position. Thus, at lower strains the tilts must be opposite to the rigid body rotation. At larger strains, because of the non exactly zero lattice spin at the C component, orientations con-

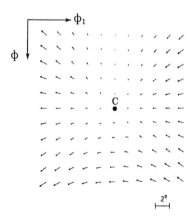

Figure 3: Rotation field around the C ideal orientation

tinue moving, arriving to the symmetry position at around a shear of 5.5, then increasing strain further rotates them in the direction of the rigid body rotation. When a component is rotated in the direction of rigid body rotation, it arrives into a divergent rotation field area and can be dissolved rapidly.

The above interpretation of the rotation field explains readily the tilted positions as well as the variations of the "intensities of the components shown in Figure 2. For the comparison below with the textures that develop in ECAE testing, it should be emphasized that the 'tilts' of the components are in the sense opposite to rigid body rotation, unless the strain is very large

4 Comparison Between Simple Shear and ECAE Textures

A sample extruded in an ECAE device leaves the die as if it was sheared on the y plane in the × direction (see Fig. 1). It is, therefore, perfectly legitimate to think that ECAE textures would be similar to that of simple shear textures. For this purpose, a comparison is made in Figure 4 between the textures of the two different tests for the case of OFHC copper. (In "order to save space the comparison is only done in the first section of the ODF, nevertheless, other sections show similar tendencies.)

As can be seen in Figure 4, the similarities are very much convincing concerning the relative intensities of the components. After the first pass, all three ideal components appearing in this section are well developed. The A1* and A2* components show some fiber nature in both deformation modes. After the second pass, the major component is the C in both textures, its intensity, however, is less in ECAE compared to simple shear. This difference can be attributed to the difference in the amount of shear, which is about $\chi = 4$ in ECAE and $\chi = 5.5$ in the simple shear test. Close inspection of the figures, however, shows important differences in the exact position of the components. The latter observation is better seen in the $\upsilon_2 = 45°$ section of the ODF, displayed in Figure 5. This section contains all ideal components of simple shear textures

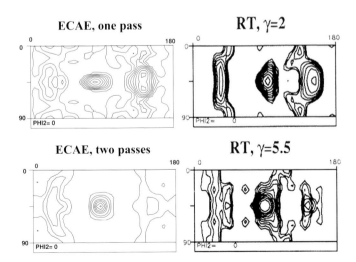

Figure 4: Comparison of ECAE (Route A) and simple shear textures for OFHC copper in the $\upsilon_2 = 0°$ section of

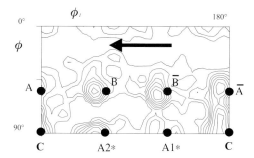

Figure 5: The $\upsilon_2 = 45°$ section of the ODF of copper after one pass in the ECAE die. Bold points are the ideal positions of simple shear textures. The arrow indicates the rigid body rotation in case of positive simple shear.

(for the definition of the ideal orientations see Tóth et al.). [17] It can be seen that – in the first pass – all ideal components appear in rotated positions by about 10° with respect to the ideal orientations. This rotation is less for further passes (see later in Fig. 7); it decreases to a constant value of about 5°. Moreover, which is a major difference when it is compared to simple shear; these rotations are in the sense of the rigid body rotation. The "tilts" are also much higher than in simple shear. These characteristics must be validated when one tries to simulate texture evolution, see the analysis of texture simulations below.

5 The Simple Shear Model of ECAE

Although the sample-form would suggest that shear "happened to it on the *y* plane in "direction *x*, this is just appearance. The flow happens in a kinked channel and Segal correctly proposed an idealized approximation of the deformation path which is negative simple shear, taking place on the intersection plane of the two channels (see Fig. 1). When a temporary *x'–y'* reference system is fixed in this plane, the velocity gradient is the following:

$$L' = \begin{pmatrix} 0 & -\dot{\gamma} & 0 \\ 0 & 0 & 0 \\ 0 & 0 & 0 \end{pmatrix} \quad (3)$$

with $\dot{\gamma}$ positive. One could use this velocity gradient to simulate the texture evolution in the rotated x'–y' reference system and just rotate the resulting texture to express it in the sample x–y reference system for comparison with experimental textures. This process, however, hides a serious problem concerning the evolution of the grain shape, very much relevant in self-consistent modelling. Namely, it would produce a wrong grain shape after simple shear and rotation (because not only the grain orientations but also their shapes have to be rotated). Tóth et al. [8] and Beyerlein et al. [10] have overcome this problem by expressing the velocity gradient L' in the sample reference system:

$$L = \frac{\dot{\gamma}}{2}\begin{pmatrix} 1 & -1 & 0 \\ 1 & -1 & 0 \\ 0 & 0 & 0 \end{pmatrix} \qquad (4)$$

This expression of the velocity gradient allows to obtain a correct form of the grain after extrusion. Formula (4), however, is very much different from simple shear! Thus, one cannot expect any more that simple shear textures in the x–y reference system could be similar to ECAE textures. Actually, expression (4) shows tension in direction x, compression in direction y and a large rigid body rotation. The latter rotation is positive around axis z, while simple shear on the y plane in direction × would represent a negative rigid body rotation around axis z.

Authors have used already the simple shear model to predict the texture development in ECAE deformation. [9–10] An example is shown also in this paper with the aim to compare the predicted textures precisely with the experimental ones. The self-consistent viscoplastic model of polycrystal plasticity was used in its isotropic version of the finite-element tuned interaction equation. [18] The material was OFHC copper; the tests were done at room temperature in a 20 mm diameter 90° die. [11] The initial texture was not perfectly random, thus it was represented by a suitable set of 500-grain orientations. Hardening was also modelled using self and latent hardening coefficients, in the same way as it is described in. [13] To simulate Route A of ECAE testing, the grain orientations as well as the grain shape matrices were rotated between passes around axis z by –90°. The results obtained after the first and second passes are presented in Figure 6.

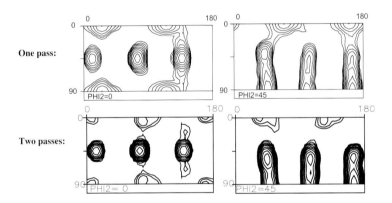

Figure 6: ECAE textures (Route A) predicted by the simple shear model with the self-consistent viscoplastic polycrystal approach

The predicted textures of Figure 6 can be readily compared to the experimental ones which are displayed in Figure 7. [11] They are quite similar, but a careful analysis shows that the ideal components appear in excessive rotated positions with respect to the experimental ones; by about 15°. When the simulations are followed for more passes, the texture remains nearly the same, mostly strengthens (not shown here). We recall that this rotation angle is 10° in the first pass and only 5° in all following passes. Thus, it can be concluded, that the "simple shear model" – although it is simple – lacks to reproduce the ideal positions of ECAE textures correctly.

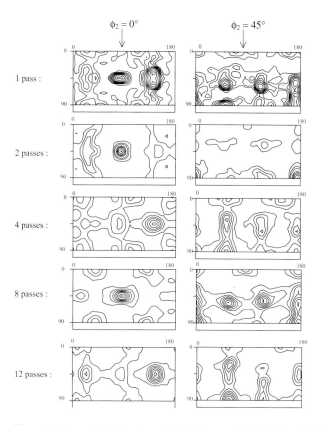

Figure 7: Measured ECAE textures in OFHC copper [11]. (Isolines: 1, 2, 3, 4, 5, 6, 7, 8, 9 ,10.)

Another difficult point in the simple shear model is pointed out by Tóth et al. [8] These authors called the simple shear model as the "discontinuous approach". The reason for this naming is that in this model the flow lines turn from vertical to horizontal instantaneously, i.e., in a discontinuous way. It means that shear is taking place only on one single plane, on the plane of intersection between the channels. As the grain dimension is finite, individual grains cross the discontinuity plane progressively. Therefore, plastic deformation is not simultaneous within a grain of a polycrystal. This situation is quite unusual in plasticity and is difficult to handle. Actually, in such an idealised case the hypotheses made for averaging in polycrystal plasticity are not valid. Special difficulty arises when one uses the self-consistent model. In that approach, the main element is the interaction of a grain with its homogenised surrounding, which is obviously not in a complete plastic state. This difficulty of the modelling can be overcome when the discontinuity of the flow lines is smoothed. Such a model was presented by Tóth et al., [8] and is called the "flow-line model" which will be described below.

6 The Flow Line Model of ECAE

The main argument of the flow line model is that in practice the flow lines are rounded even if the two channels are connected without any rounding. It is seen in "finite element modelling of the ECAE process that there is a "dead metal zone" in the outer corner region of the die. This zone can be also seen when the test is stopped with the sample just halfway through. Gholinia et al. [6] have actually measured the flow lines in such a test and showed that they are quite rounded. They were able to derive a strain field from the measured flow lines and used it in a Taylor polycrystal model. Their predicted textures are in quite good agreement with the experiments.

The flow line model of Tóth et al. [8] uses an analytical flow function; it is defined by:

$$\phi = (d-x)^n + (d-y)^n = (d-x_0)^n .\qquad(5)$$

This function contains only geometrical parameters: d is the diameter of the die and x_0 defines the incoming (and outgoing) position of the flow line. The shape of the flow line is controlled by the n parameter (see Fig. 8). Function υ satisfies the boundary conditions in the working part of the die, i.e. between the lines defined by $y = d$ and $x = d$. At these positions, the flow lines are perfectly parallel to the compression direction and the outgoing flow direction, respectively. For $n = 2$ the line is circular, for higher n values it approximates the flow better within a non-rounded die. For n $\to \infty$ the flow line function returns the above discontinuous shear model.

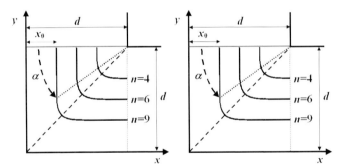

Figure 8: The shapes of the flow lines for different values of n

From the flow function, an admissible velocity field now can be defined by:

$$v_x = \lambda \frac{\partial \phi}{\partial y}, \quad v_y = -\lambda \frac{\partial \phi}{\partial x} .\qquad(6)$$

The κ parameter is determined by the incoming velocity of the material, so the velocity field is:

$$v_x = v_0 \left(\frac{d-y}{d-x_0}\right)^{n-1}, \quad v_y = -v_0 \left(\frac{d-x}{d-x_0}\right)^{n-1} \qquad(7)$$

For the purpose of texture simulations, the velocity gradient field is needed, which can be obtained by simple partial derivation of (7), after fully expressing v_x and v_y as a function of the coordinates $\times$ and y (by using Eq. (5) in (7)):

$$L_{xx} = \frac{\partial v_x}{\partial x} = -v_0(1-n)(d-x)^{n-1}(d-y)^{n-1}(d-x_0)^{1-2n},$$

$$L_{yy} = -L_{xx},$$

$$L_{xy} = \frac{\partial v_x}{\partial y} = v_0(1-n)(d-x)^n(d-y)^{n-2}(d-x_0)^{1-2n}, \quad (8)$$

$$L_{yx} = \frac{\partial v_y}{\partial x} = -v_0(1-n)(d-y)^n(d-x)^{n-2}(d-x_0)^{1-2n},$$

As can be seen from (8), the velocity gradient field describes compression along axis y, tension in direction $\times$ and shear on both the y and $\times$ planes. It gives also a large rigid body rotation, maximum at the symmetry plane of the flow (at 45°). When (8) is compared to the simple shear model, one can see that flow line model gives a similar velocity gradient as in Eq. (4). This model, however, is not discontinuous and it may also be nearer to experimental conditions. The strain mode varies continuously along the flow line.

The strain rate tensor is the symmetrical part of the velocity gradient:

$$\begin{aligned}\dot{\varepsilon}_{xx} &= -v_0(1-n)(d-x)^{n-1}(d-y)^{n-1}(d-x_0)^{1-2n}, \\ \dot{\varepsilon}_{yy} &= v_0(1-n)(d-x)^{n-1}(d-y)^{n-1}(d-x_0)^{1-2n}, \\ \dot{\varepsilon}_{xy} &= \frac{1}{2}v_0(1-n)(d-x_0)^{1-2n}\left[(d-x)^n(d-y)^{n-2}-(d-y)^n(d-x)^{n-2}\right]\end{aligned} \quad (9)$$

Now a further important difference can be seen between the present flow line model and the simple shear model: there is a non-zero x-y shear strain in the proposed flow field, which is absent in the classical model.

Important measure in ECAE testing is the total plastic strain experienced by the work piece in one pass. For this purpose, the von Mises equivalent strain rate is first expressed

$$d\bar{\varepsilon} = \sqrt{\frac{2}{3}\dot{\varepsilon}_{ij}\dot{\varepsilon}_{ij}} = \frac{1}{\sqrt{3}}v_0(n-1)(d-x_0)^{1-2n}(d-x)^{n-2}$$
$$(d-y)^{n-2}\left[(d-x)^2+(d-y)^2\right] \quad (10)$$

Then Eq. (10) is integrated along a flow line to obtain the total von Mises equivalent strain in one pass:

$$\bar{\varepsilon} = \frac{2}{\sqrt{3}}\frac{\pi(n-1)}{n^2\sin(\pi/n)} \quad (11)$$

As can be seen from this formula, the total accumulated equivalent strain is independent of the d and x_0 parameters, meaning that the total strain is the same in the whole cross section of the die, as long as the n parameter is kept constant.

The smallest physically possible value of n is 2. The flow line has a circular shape for $n = 2$. For that case, the equivalent strain is $\bar{\varepsilon}(n=2) = \pi/(2\sqrt{3}) = 0.9068$ When n approaches infinity, we obtain the Segal idealized flow field (simple shear model) with $\bar{\varepsilon} = 2/\sqrt{3} = 1.1547$. As can be seen, the accumulated strain decreases monotonically when n is decreased. Although the dependence of $\bar{\varepsilon}$ on n is quite small for reasonable n values ($3 \leq n \leq 10$), the strain field variation is much more significant. Fig. 9 displays the strain field for three different n values as a function of the angular position on the flow line. This figure also shows the strain rate variations obtained from finite element simulations. [8] Actually, the n parameter was identified by matching the results of the two simulations. Although the n parameter is not constant according to the results of finite element calculations, its variation in the cross section seems to be physically correct. Namely, it is obvious that the left corner region of the die has an immediate effect on the form of the flow lines in its vicinity. That is, as the corner shape corresponds to a flow line shape of $n \to \infty$, the n value is expected to increase as the lower left corner is approached. This is exactly the case in Figures 8,9. On the contrary, in the upper corner zone – far from the left corner – the flow lines just have to turn around the corner; in the limiting case they are just circular (a situation corresponding to $n = 2$).

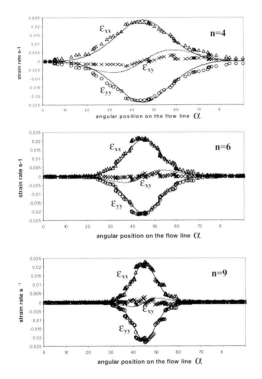

Figure 9: The variations of the three strain rate components along the flow lines for different values of *n* (continuous lines). Symbols indicate the results of finite element simulations [8]

According to the above analysis, the accumulated equivalent strain is not constant in the whole cross section of the specimen; it varies from the left to the upper corner between 1.1547 and 0.9068, respectively. This gradient, if it is confirmed by experiments, must have an effect on the evolution of the deformation texture. This question will be studied in the next Section with the help of simulations.

7 Textures Otained from the Fow line Model of ECAE

When polycrystal modeling is carried out along a flow line, the basic hypothesis is that a very small volume element contains a whole set of crystal orientations (500 in the present case) and is subjected to the strain field given by Equation 8. It travels along the flow line with the speed defined by Equations 7, which is done in the simulations incrementally. In each increment it is assured that the material element remains on the very same flow line. This condition can be readily satisfied by a radial return to the flow line defined by Equation 5 after each increment. In order to avoid overshooting, the "increments have to be quite small (around 0.001) in terms of equivalent strain. The following simulations were done in this way using the self-consistent polycrystal model as already defined in Section 4 above.

The first question to be examined is the dependence of the textures on the n parameter of the flow field. For that purpose, a series of simulations has been carried out for one pass, by only varying the n value. The results are presented in forms of {111} pole figures in Figure 10. In order to read more easily Figure 10, a key figure for the ideal orientations of "simple shear textures and the reference system are given in Figure 11.

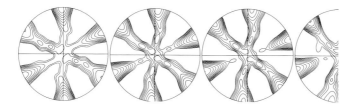

Figure 10: {111} pole figures showing the textures that develop after one pass as a function of the n parameter. Isodensities: 0.8, 1.0, 1.3, 1.6, 2.0, 2.5, 3.2, 4.0, 5.0, 6.4

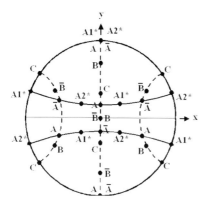

Figure 11: Key {111} pole figure on the z-projection plane

As can be seen in Figure 10, the obtained ECAE texture for $n = 2$ is a standard simple shear texture with slight positive rotation around axis z. As the n value is increased to more realistic values, however, the texture is rotated in the negative direction around axis z; an effect that is characteristic for ECAE textures and has already been pointed out in Section 3 above. The rotation angle increases monotonically with the n parameter. An interesting observation in Figure 10 also is that not all the ideal components rotate with the same amount. It is the C component, which shows the most systematic behavior. Thus, this component will be used to identify the rotation angle of the texture. When the n value becomes very high, the rotation angle approaches 15°. (Note that the $n = \infty$ case in Figure 10 was obtained by using the simple shear model which is the limiting case of the flow line model.) The rotation angles of the C component around the z-axis in measurements (using the ODF sections in Figues 5 and 7) were measured to be 10° in the first pass and 5° after all subsequent passes. According to the pole figures (Fig. 10), such rotations correspond to n values of 6 and 4, respectively. The n parameter; even after several passes it remains the same, an effect can be used to tune the values of the n parameter more precisely.

The experimental fact, that the rotation of the texture is twice higher in the first pass than in further passes, can be "related to the shape of the flow lines. It is actually observed in finite element calculations that when hardening is taken into account, the shape of the flow line becomes more rounded [19] after the first pass. It means that higher n value is to be used in the first pass as compared to later passes. In the following, simulation results will be presented that are obtained in this way: $n = 6$ in the first pass and $n = 4$ in all subsequent passes. The results of the simulations carried out up to four passes are shown in Figures 12–13. Pole figures are shown to ease comparisons with previous published pole figures.

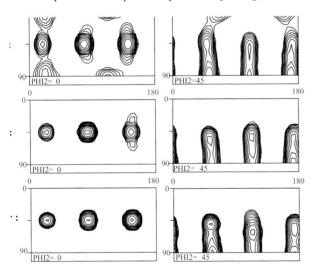

Figure 12: Textures predicted by the flow line model up to four passes in Route A deformation of copper. Isodensities: 0.7 (for pass one only), 1, 1.4, 2, 2.8, 4, 5.6, 8, 11 16, 22

The obtained textures in Figure 12 are to be compared with the experimental ones shown in the same sections of the ODF ($\upsilon_2 = 0°$ and 45°) and for the same passes in Figure 7. One can see that the flow line model reproduces all features of the experimental ODF for the first pass. The rotations from the ideal positions, the relative intensities of the components as well as the fiber characteristics around the A1* and A2* components are the same in the simulation as in the ex-

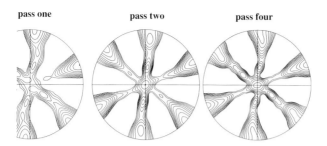

Figure 13: The simulation results of Figure 12 in forms of {111} pole figures. Isodensities: 0.8, 1.0, 1.3, 1.6, 2.0, 2.5, 3.2, 4.0, 5.0, 6.4

periment. The agreement in the second and forth passes are less satisfactory. In pass two the simulation returns correctly the rotated positions and the experimental fact that the main component is the C, however, it over-predicts significantly the intensities of the B/B° and A2* components. At pass number four, the simulation fails to reproduce the correct intensities of the A1* and C components. Only the tendencies in the intensity changes are well simulated together with the right positions of the components. It is also interesting that the "rotated cube component is only predicted in the first pass and disappears in subsequent passes. These results do not change significantly if the hardening law is replaced with another one (a dislocation cell hardening model [20] was also tested), or if the shape evolution is suppressed. Interestingly, some more rotated cube component can be obtained (unfortunately, not in the right positions) when the Taylor approach is used in the flow line model.

8 Concluding Remarks

In the present work, first a comparison is made between the textures that develop in ECAE testing of copper and the textures known in large strain simple shear. It is demonstrated that there is a good similarity between them as for the main ideal components and their intensity variations. A particularity of ECAE textures has been pointed out: its ideal orientations appear in significantly rotated positions; 10° after the first pass, which value decreases to 5° in the subsequent passes. An important difference between simple shear and ECAE textures is that the tilts of the components are in opposite sense.

The second part of the paper discussed the so-called simple shear model of ECAE deformation. Simulations have been carried out to obtain the textures that develop in Route A ECAE testing using the self-consistent polycrystal model. The main feature of the obtained ODFs is that they are rotated by about 15° around the z/TD direction in the positive sense, irrespectively of the number of passes. This rotation is too high with respect to the experimental values. Another difficulty in this classical model of ECAE deformation is that it represents a discontinuous deformation field for which averaging techniques of polycrystal plasticity cannot be applied. There is difficulty especially in the self-consistent model, which is based on an interaction law of an inclusion (the grain).

In the third part of this work, the new flow line description of the ECAE deformation field proposed by Tóth et al. [8] has been presented. This model uses an analytical flow field, which contains rounded flow lines; its roundness is controlled by one single parameter (n). It is shown that the rotation of the ideal texture components can be faithfully reproduced by choosing properly the value of the n parameter. The predicted textures are also in relative good agreement with experiments.

The weaknesses of the simulations were also emphasized in this paper. Although the flow line model is physically sounder than the simple shear model, its main contribution is only the reproduction of the right positions of the ideal components. The intensities are nearly perfectly reproduced in the first pass but not in the subsequent passes. There can be two reasons for this:
1. A more refined self-consistent model should be used which takes into account the anisotropy in the interaction equation (the isotropic interaction model was employed in the present work).
2. After the first pass there is significant grain refinement with such microstructural effects that are not accounted for by simply polycrystal modeling. The elaboration of this second point needs more knowledge about the plastic behavior of small grained material and will be studied more in the near future development of the plastic modeling of the textures in ECAE.

9 Acknowledgement

The Author acknowledges with gratitude the contribution of his co-workers who helped in the development of this research work in various ways: I. Kopacz, M. Zehetbauer, A. Eberhardt, A. Molinari, R. Massion, S. Suwas, S. C. Baik, and L. Germain.

10 References

[1] V. M. Segal, Mat. Sci. Eng. 1995, A197, 157
[2] R. Z. Valiev, I. V. Islamgaliev, I. V. Alexandrov, Prog. Mat. Sci. 2000, 45, 103
[3] S. V. Ferrasse, M. Segal, K. T. Hartwig, R. E. Goforth, Metall. Mater. Trans. 1997, A28, 1047
[4] K. Nakashima, Z. Horita, M. Nemoto, T. G. Langdon, Acta Mater. 1998, 46, 1589
[5] R. Z. Valiev, I. V. Alexandrov, Y. T. Zhu, T. C. Lowe, J.'Mater. Res. 2002, 17, 5
[6] A. Gholinia, P. B. Prangnell, V. M. Markushev, Acta Mater. 2000, 48, 1115
[7] Y. Iwahashi, J. Wang, Z. Horita, M. Nemoto, T. G. Langdon, Scripta Mater. 1996, 35, 143
[8] L. S. Tóth, R. Massion, L. Germain, S. C. Baik, Acta Mater., submitted
[9] S. R. Agnew, U. F. Kocks, K. T. Hartwig, J. R. Weertman, in Proc. 19th Riso. Int. Symp., Mat. Sci., Denmark 1998, p. 201
[10] I. J. Beyerlein, R. A. Lebensohn, C. N. Tomé, Mater. Sci. Eng., in press
[11] L. S. Tóth, I. Kopacz, M. Zehetbauer, I. V. Alexandrov, in Proc. THERMEC-2000 (Eds: Chandra, T. et al.), Las Vegas, Dec. 2000, on CD
[12] I. Kopacz, M. Zehetbauer, L. S. Tóth, I. V. Alexandrov, in Proc. 22nd Riso Int. Symp. on Mech. Sci. (Eds: A. R. Dinesen, et al.), Roskilde, Denmark, Sept., 2001, p. 295
[13] L. S. Tóth, A. Molinari, Acta Met. Mat. 1994, 42, 2459
[14] L. S. Tóth, J. J. Jonas, D. Daniel, J. A. Bailey, Textures Microstruct. 1992, 19, 245

[15] K. W. Neale, L. S. Tóth, J. J. Jonas, Int. J. Plasticity 1990, *6*, 45
[16] L. S. Tóth, K. W. Neale, J. J. Jonas, Acta Metall. 1989, 37, 2197
[17] L. S. Tóth, P. Gilormini, J. J. Jonas, Acta Metall. 1988, 36, 3077
[18] A. Molinari, L. S. Tóth, Acta Met. Mat. 1994, 42, 2453
[19] S. C. Baik, private communication
[20] L. S. Tóth, A. Molinari, Y. Estrin, J. Eng. Mat. Techn. 2002, 124, 71

Textural Evolution during Equal Channel Angular Extrusion versus Planar Simple Shear

E.F. Rauch and L. Dupuy

Génie Physique et Mécanique des Matériaux (GPM2), Institut National Polytechnique de Grenoble ENSPG, Saint-Martin d'Hères, France

1 Abstract

The texture developed in Equal Channel Angular Extrusion is compared to the one obtained in planar simple shear for an AA5083. The same features are observed with the same shifts with respect the ideal torsion components. These deviations are attributed to rate sensitive effects. The sample orientations were chosen in order to make the simple shear direction parallel to the die's intersecting plane and the results prove that this deformation path is effective. The alternative view, in which shear is considered to occur along the streamlines, is shown to be equivalent to the die's shear plane approach when additional dilatation terms are taken into account.

2 Introduction

Equal Channel Angular Extrusion has become the subject of numerous studies. This is because it has been recognized that the very large strains induced by successive extrusions modify the properties of the material. Especially the mechanical properties are known to be improved at room temperature as well as at high temperature [1]. Despite these advantages, it is remarkable to note that only few attempts are performed to apply the process for direct industrial applications and this is indicative of the difficulties that exist to scale up the extrusion device. By contrast the small ECAE units available in different laboratories offer a rather easy way to obtain very large level of strains. Therefore, these devices may be considered as very useful tools to investigate the physical mechanisms involved in alternative processes that promote also severe plastic deformation (extensive rolling, conventional extrusion …).

In order to transpose relevant results extracted from the analysis of extruded samples to other deformation modes, it is necessary to characterise unambiguously the strain path experienced by the material in the ECAE device. While all the authors agree with the fact that this type of extrusion promote mainly simple shear, there is still some doubts about the real nature of the strain tensor. Especially the shear plane and shear direction are subjects of controversial interpretations (Fig.1.a). In several works, it is mentioned that the shear occurs in a plane parallel to the channels intersection plane [2–4] (hereafter referred as die's shear plane view). Alternatively, it was proposed that shear may take place along the streamlines that progressively rotates from one channel to the other [5, 6] (streamline view). In both cases, the textural evolutions were used to check the validity of the suggested deformation paths and in all cases rather good agreements were obtained.

It is the purpose of the present work to analyse these agreements. In a first part the pole figures associated to simple shear are commented. In the second part the experimental textures of extruded samples are compared to the one obtained with the help of a planar simple shear device. Then the strain path involved in ECAE is discussed.

3 Simple Shear Texture

One of the difficulties related to texture analysis in ECAE is that the observed components are frequently deviated with respect to the expected orientations. There are two main reasons that lead to these deviations:

- the orientation of the pole figures does not always correspond to a symmetry axis of the process,
- experimental simple shear textures do not coincide with the so called ideal torsion (simple shear) orientations.

These deviations, described in more details below, are responsible for some inaccuracies in the relationship between the observed texture components and the suggested deformation mode, whatever the authors' hypothesis.

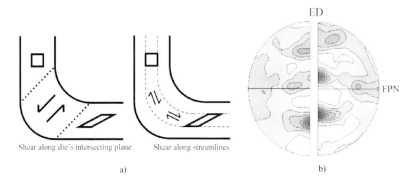

Figure 1: a) the two types of shear that are supposed to occur during extrusion, b) The 111 pole figures for aluminium after 8 extrusions in a 120° die (left) or after 4 extrusions in a 90° die (right). ED = Extrusion Direction, FPN = Flow Plane Normal (From Gholinia et al [6]). Note the difference in shift of the main textural components for the two cases.

3.1 Pole Figures Orientation

For billets extruded with the rolling direction parallel to the inlet channel (and therefore its transformed parallel to the outlet channel), the pole figures are commonly plotted in rolling convention i.e.: with the (transformed) normal direction plotted in the centre of the pole figure and the billet axis (the rolling direction) as one of the reference axes (Fig. 1.b). Besides, the process has one plane of symmetry that contains the two channel axes (the flow plane). This plane will be apparent on the pole figures as a mirror plane. This is the only symmetry element that will be observed and there is no reason why the pole figures should appear orthotropic even if shear is considered to occur along the streamlines. When compared, as sometimes done, to rolling textures, the components that emerge from extrusions seem always deviated with respect to some (symmetric) rolling components [7].

Fig.1.b extracted from [6] is a typical example of an ambiguity that arises from such projection. The two half 111 pole figures are related to ECAE devices with channels at respectively 120° and 90°. From this figure, it may either be conclude that both pole figures exhibit the same major features (a common and important deviation with respect to some symmetric orientations) or by contrast that the components are dissimilar because there is a small rotation around the flow plane normal from one pole figure to the other. In the first case it is presumed that the die angle is of no importance for the texture development (streamline view), while the reverse will be deduce from the second observation (die's shear plane view). Huang et al. [3] have shown how that these rotations are compatible with shear parallel to the channel intersecting plane.

3.2   Ideal simple shear orientations

The texture development in simple shear was theoretically analysed by Canova et al [8] with particular reference to torsional deformation. Some ideal simple shear orientations were derived for f.c.c crystals and referred as 'A' or 'B' partial fibres (respectively {111}<uvw> and {hkl}<110>) and 'C' orientation ({001}<110>). In this work the torsion projection convention is considered (i.e.: plane of shear/ direction of shear) that gives rise to centro-symmetric pole figures.

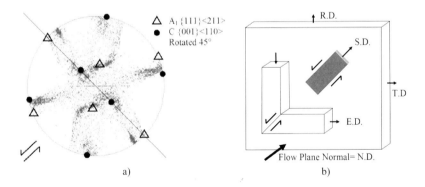

Figure 2: a) the 111 pole figures for AA5083 after one extrusion plotted in torsion convention and calculated with a Lebensohn-Tomé self consistent visco-plastic software [11]. Some textural components are close to but not exactly located on ideal orientations A_1 and C, b) orientation of extruded and sheared samples used in the present work.

The experimental pole figures obtained in torsion are known to exhibit components that are displaced with respect to these ideal orientations (see fig 9 in [8] or fig. 5.7 in [9]). This was interpreted as the effect of the strain rate sensitivity of the material [10]. Indeed, a clear shift is obtained when the texture is predicted with a self consistent model that includes visco-plasticity [11]. The 111 pole figure depicted on figure 2.a is calculated for a shear strain $\gamma=2$ with the initial texture being the rolling one for the AA5083 used in the present work. The flow plane coincides with the sheet plane and the shear direction is rotated 45° counter clockwise as shown on Fig. 2.b. The predicted pole figure appears tilted with respect to ideal components.

The same deviation is therefore expected for ECAE even if it is accepted that the process implies simple shear solely. Such deviation makes the analysis of the texture components in extrusion less straightforward.

4 Experimental Simple Shear and ECAE Textures

Even when rate sensitivity is taken into account, the texture predictions are hardly in perfect agreement with the experimental results. The details of the simulated pole figures differ with the polycrystal plasticity model that is used and the calculation parameters. It is safer to compare the texture after extrusion directly with the one developed during simple shear. This would be possible with the torsion test except that the initial orientation and symmetry would be different for twisted and extruded samples. Therefore the planar simple shear test described in [12] is used to perform shear on the same material and in the same orientation than during Equal Channel Angular Extrusion. The samples are oriented as shown in figure 2.b, i.e.: with the die's shear plane view.

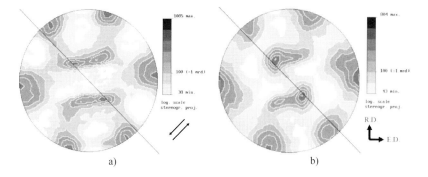

Figure 3: The 111 pole figures for AA5083 after a) simple shear up to $\gamma = 0.7$ and b) one extrusion that leads to $\gamma = 2$. Note the similitude in the orientations for these two deformation modes

Figures 3 show the 111 pole figures for an AA5083 deformed at room temperature through planar simple shear (Fig. 3.a) and Equal Channel Angular Extrusion (Fig.3.b). It can be seen that the same textural components are observed with the same shifts with respect to the shear direction. The main differences lay in the components respective intensities. This is due to the fact that it was not possible to shear the sample above $\gamma = 0.7$ with the planar simple shear apparatus because of ruptures that initiate at the free surfaces. This direct comparison strongly suggests that during extrusion shear is promoted along the channels intersection plane. Moreover, it can be noted that there is a good agreement between the experimental poles figures (Fig. 3.a and b) and the predicted one that also involve shear along the die's channels intersection (Fig.2.a).

5 Discussion

The above experimental evidences are in favour of the die's shear plane view. In this approach, the strain tensor contains uniquely a simple shear component promoted in a constant direction. In case of the streamline view it was shown that additional compression and tension components must be considered. This was deduced from direct measurements of the strain history from scribed lines on the sample [6]. The progressive distorsions of a reference frame parallel to the streamlines are shown on figure 4.a and reveal the additional terms. It is of interest to plot the distorsions encountered by the same reference frame but in case of simple shear at 45° from the extrusion direction (Fig.4.b). The latter is obtained by i) writing the related deformation gradient in the streamlines coordinates, ii) rotating the tensor in order to re-align the reference frame with the streamline coordinates (this vanishes the FnS component) and iii) integrating the values over the whole deformation path. The latter step requires an estimate of the shear strain distribution over the thickness of the deformed volume. The exact position of the inflexion points will depend on this distribution but with a reasonable guess, the curves will always have the overall shapes shown on the figure. It comes out that the grid distorsions deduced from the die's shear plane view are similar to the measured ones. It is concluded that the strain path invoked in the completed streamline approach (with the additional terms taken into account) is equivalent to a simple shear occurring along the channels intersection plane. From a textural point of view a good agreement with experimental results is expected in both cases.

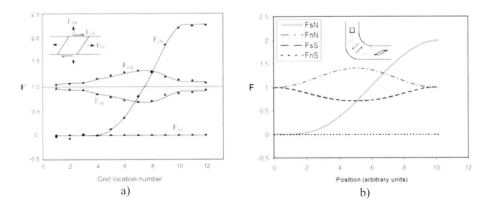

Figure 4: Deformation history through a 90° die a) measured from scribe lines on the extruded sample (from [6]), and b) obtained by rotating the deformation gradient tensor related to simple shear parallel to the channels intersection plane

6 Conclusions

It is shown that the shifts frequently observed on pole figures after Equal Channel Angular Extrusions are expected when simple shear occurs along the die's intersecting plane. The texture of an AA5083 billet extruded in a 90° die is compared to the ones obtained after a planar simple shear test: despite the lower strain that may be reached in planar simple shear, the pole figures exhibit the same features. These two facts demonstrated that the deformation mode encountered by the material in ECAE is of simple shear type and that the shear direction is parallel to the channels intersecting planes. It is argued that the streamline view is equivalent to the die's shear plane approach when the strain tensor is completed with adequate dilatation terms.

7   References

[1]   L. Dupuy, J.-J. Blandin and E.F. Rauch, Mater. Sci. Tech. 2000, 16, 1256–1258
[2]   Agnew, S.R., Kocks, U.F., Hartwig, K.T. and Weertman, J.R., in Proceedings of the 19th Risø Symposium on Materials Science: Modelling of Structure and Mechanics of Materials from Microscale to Product, eds.: J.V. Cartensen et al, Risø National Laboratory, Roskilde, Denmark 1998, p. 201–206
[3]   Huang, W.H., Chang, L., Kao, P.W. and Chang, C.P., Mater. Sci. Eng. 2001, A307, 113–118
[4]   Zhu, Y.T. and Lowe, T.C. Mater. Sci. Eng. 2000, A291, 46–53
[5]   Kopacz, I., Zehetbauer, M., Toth, L.S., Alexandrov, I.V. and Ortner,B., in Proceedings of the 22th Risø Symposium on Materials Science: Science of Metastable and Nanocrystalline Alloys, Structure, Properties and Modelling, eds.: A.R. Dinesen et al, Risø National Laboratory, Roskilde, Denmark 2001, p. 295–300
[6]   Gholinia, A., Bate, P. and Prangnell, P.B., Acta Mater., 2002, 50, 2121–2136
[7]   Pithan, C., Hashimoto, T. Kawazoe, M. Nagahora, J. and Higashi, K., Mater. Sci. Eng. 2000, A280, 62–68
[8]   Canova, G.R., Kocks, U.F. anf Jonas, J.J., Acta Metall., 1984, 32, 211–226
[9]   Kocks, U.F., Tomé C. and Wenk H.-R. (eds) in Texture and Anisotropy. Preferred Orientations in Polycrystals and Their Effect on Material Properties, 1998, Cambridge University Press, p. 189
[10] Toth, L.S., Gilormini, P. and Jonas, J.J., Acta Metall., 1988, 36, 3077–3091
[11] Lebensohn, R.A. and Tomé, C., Acta Metall. And Mater. 1993, 9, 2611–2624
[12] Rauch, E.F. in Non-Linear Phenomena in Materials Science II, G. Martin et L. Kubin eds., 1992, Solid State Phenomena., 23-24 pp. 317–334

Grain Refinement and Texture Formation during High-Strain Torsion of NiAl

W. Skrotzki[1], B. Klöden[1], R. Tamm[1], C.-G. Oertel[1], L. Wcislak[2], E. Rybacki[3]
[1] Institut für Strukturphysik, Technische Universität Dresden, Dresden, Germany
[2] HASYLAB at DESY, Hamburg, Germany
[3] Geoforschungszentrum Potsdam, Potsdam, Germany

1   Introduction

The intermetallic compound NiAl has special properties, like high melting point, low density, good corrosion resistance and moderate creep strength, making this material interesting for high-temperature applications. However, a widespread acceptance as structural material was hindered so far by the inadequate low-temperature toughness and ductility. Attempts to increase the ductility of NiAl by alloying have not been successful [1]. Another way to approach this problem is by changing the grain structure and texture on purpose. As the yield stress increases much slower with decreasing grain size than the fracture stress, grain refinement should help to reach the ductile field (Fig. 1) [2]. This field can be extended by changing the texture in such a way that the Taylor factor is reduced. Thus, in order to optimize the desired mechanical properties by thermomechanical treatment it is of particular interest to understand the mechanism of grain structure and texture formation during deformation and annealing. Therefore, it was the aim of the present work to strongly deform NiAl polycrystals in torsion, a deformation mode allowing the study of microstructure and texture with strain. To do this, methods with high spatial resolution had to be used which are given by electron back-scatter diffraction (EBSD) in the scanning electron microscope (SEM) and by texture measurements with high-energy synchrotron radiation.

2   Experimental

High-temperature extruded polycrystalline NiAl samples with two different initial textures were deformed in torsion at 1000 K in a Paterson-type rock deformation machine [3] under 400 MPa argon confining pressure (Fig. 2). The diameter and height of the cylindrical samples were 10 mm. The specimens have been deformed between alumina pistons, the whole assembly being enclosed in a thin steel jacket. The applied torsion leads to simple shear in the tangential direction in a plane normal to the torsion axis (Fig. 3). The shear strain γ and the shear strain rate $\dot{\gamma}$ in the samples increase linearly from zero at the torsion axis to a maximum $\gamma_{max} = 3.14$, ($\dot{\gamma}_{max} \cong 2 \cdot 10^{-4} s^{-1}$) at the sample edge, corresponding to one revolution. To investigate the local textures between the torsion axis and the edge, small pins with a diameter of 1 mm were prepared in the radial direction for each of the deformed samples.

To investigate the microstructure, orientation and misorientation, maps have been constructed from EBSD scans made with a SEM (Zeiss DSM 962) on longitudinal axial sections. The orientation analysis was done with the software Channel 5 of HKL, Hobro, Denmark.

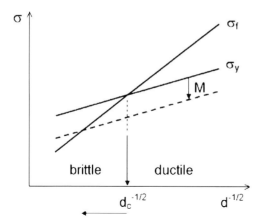

Figure 1: Grain size d dependence of the fracture stress σ_f and yield stress σ_y defining the brittle-to-ductile-transition grain size d_c. Dashed line shows the effect of texture on σ_y in the case of a decreasing Taylor factor M. In NiAl, σ_y is strongly determined by the harder secondary slip sytems.

Quantitative texture measurements were performed with high-energy (100 keV) synchrotron radiation at the beamline BW5 at DESY-HASYLAB (Hamburg, Germany) [4, 5]. The synchrotron radiation opens a possibility to investigate small sample volumes (local studies). The incident monochromatic beam was defined by a slit system to 1 mm x 2 mm. The small pins were mounted in the Eulerian cradle parallel to the rotation axis ω. An image plate detector was positioned perpendicularly to the incoming beam at a distance from the sample of about 1.3 m. Thus, the Debye-Scherrer rings with the indices (100), (110) and (111) could be registered simultaneously. The pins were rotated around the ω-axis from 0° to 180° in steps of 3°, resulting in 61 diffraction images. The integral intensities along the diffraction rings were determined using peak profile analysis [6] and then transformed into complete pole figures. The initial global texture of the two samples has been measured by neutron diffraction (Figs. 2 and 6) [7].

3 Results and Discussion

The torsion behaviour of the NiAl polycrystals is shown by the shear stress - shear strain curves in Fig. 2. These curves have been calculated from the measured torque and twist data according to the procedure given in [3]. The curves are characterized by a peak stress followed by moderate strain softening. As will be shown below, this softening is correlated with textural changes during straining. One sample has a <111> fibre parallel to the torsion axis, in the following denoted A, the other one has a texture with orthorhombic symmetry with the <100> preferred orientation parallel to the torsion axis, denoted B. Due to differences in the initial texture, sample A behaves softer than sample B.

Sample A deformed in torsion is shown in Fig. 3. The deformation geometry is illustrated by rolling traces on the steel jacket serving as passive markers, which were parallel to the cylinder axis before the experiment. Anywhere within the cylinder the shear plane lies normal to the cylinder axis. The shear direction curves tangentially around and points perpendicular to the cylin-

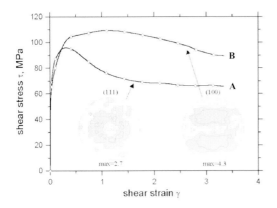

Figure 2: Shear stress - shear strain curve of torsion deformed samples A and B. The initial textures are shown as (111) and (100) pole figures, respectively. The torsion axis coincides with the center of the pole figures, texture maxima are given in multiples of a random distribution.

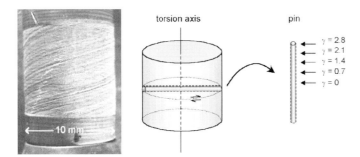

Figure 3: Sample A deformed in torsion. The rolling traces on the steel jacket were initially parallel to the cylinder axis and serve as passive strain markers to illustrate the simple shear deformation in any tangential plane. A sketch of the position of the pin taken for texture analysis is shown with different shear strains indicated.

der axis as well as to the radius. The marker lines indicate that the shear deformation was homogeneous along the torsion axis.

The change of microstructure with strain is shown in Figs. 4 and 5. The initial misorientation (θ) distribution in the recrystallized state of the extruded sample mainly consists of high angle grain boundaries (HAGB, $\theta \geq 15°$). With increasing shear strain, low angle grain boundaries (LAGB, $\theta < 15°$) are formed. The fraction of LAGB goes over a maximum, i.e. there is a change in the misorientation distribution towards HAGB (Fig. 4). The maximum fraction is about 30 %. The grain size distribution narrows with inceasing shear strain (Fig. 5a). The average grain size (arithmetical mean) rapidly decreases with shear strain and reaches a steady state value (Fig. 5b). The grains are elongated (maximum aspect ratio $\approx$ 6) with the long axis of the grains defining a steady state foliation which is inclined to the shear plane and shear direction at an angle of about 27°.

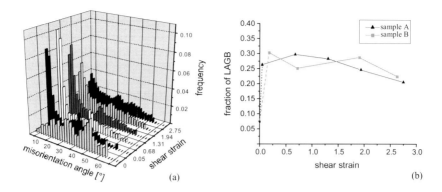

Figure 4: Shear strain dependence of the misorientation distribution in sample A (a) and fraction of low angle grain boundaries in samples A and B (b)

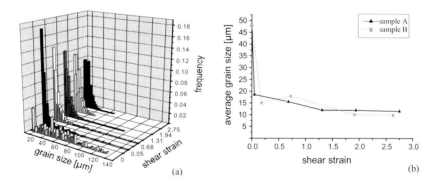

Figure 5: Shear strain dependence of the grain size distribution in sample A (a) and average grain size in samples A and B (b)

The texture after torsion was measured as a function of the shear strain γ at five different positions between $\gamma = 0$ and $\gamma_{max} = 3.14$ along the pin axis shown in Fig. 3. The corresponding (100) pole figures are shown for $\gamma = 1.4 \pm 0.3$ and 2.8 ± 0.3 and two different initial textures (Fig. 6). With increasing shear strain, both samples develop a {110}<100> shear texture (notation: {shear plane}<shear direction>). This corresponds to an alignment of the primary slip system with the slip plane {110} parallel to the shear plane and the Burgers vector <100> parallel to the shear direction (steady state texture). As the shear strain increases, for sample A the <100> maximum in shear direction becomes sharper. The texture of sample B is characterized by a decrease of the intensity of the central <100> maximum with increasing shear strain, while the other <100> maxima perpendicular to that are moving on a great circle inward corresponding to an alignment of {110} parallel to the shear plane. Apparently, the steady state texture of sample B will be reached for higher shear strains than observed for sample A.

The microstructure development with shear strain is characterized by subgrain formation. With increasing shear strain more and more dislocations are incorporated into the grain bounda-

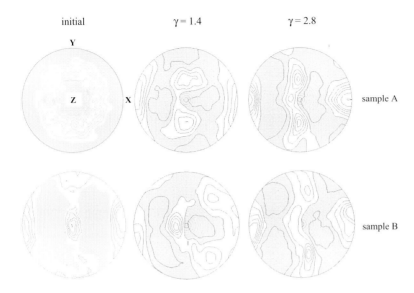

Figure 6: Shear textures (recalculated (100) pole figures) of samples A and B for different shear strains γ (X = shear direction, Z = shear plane normal)

ries increasing the misorientation. This process of continuous dynamic recrystallization leads to a change from a LAGB to HAGB structure. As grain boundary migration is negligible the grain size remains about the subgrain size. It is a general observation that the steady state subgrain size d is inversely related to the stress σ, i. e. $d = K/\sigma$ [8]. The constant K is independent of temperature and has a value of about 1000 MPa µm for NiAl [9]. Taking the maximum flow stress at room temperature of about 1500 MPa measured in compression [10] results in an extrapolated grain size of about 700 nm. Thus, this grain size seems to be the lower bound of grain refinement by deformation unless the mechanism of microstructure development changes with decreasing temperature. Investigations of the substructure in room temperature compressed NiAl are under way.

The strain softening observed in torsion is the result of texture development leading to the preferred alignment of the primary slip system. Grain refinement, unless produced by discontinuous dynamic recrystallization leading to dislocation-free grains, should not affect dislocation creep.

4 Conclusions

Continuous dynamic recrystallization leads to considerable grain refinement during high-temperature torsion of NiAl. Further refinement up to two orders of magnitude may be possible by high-pressure torsion at room temperature. Whether this helps to ductilize NiAl has to be proven.

During torsion the primary slip system preferentially aligns with {110} parallel to the shear plane and <100> parallel to the shear direction.

Torsion texture formation leads to strain softening depending on the initial texture.

5   References

[1] R.D. Noebe, R.R. Bowman, M.V. Nathal, in Physical Metallurgy and Processing of Intermetallic Compounds (Eds.: N.S. Stoloff, V.K. Sikka), Chapman & Hall, New York, USA, 1996, Chapter 7
[2] A. H. Cottrell, Trans. AIME, 1958, 212, 192–203
[3] M.S. Paterson, D.L. Olgaard, J. Struct. Geol., 2000, 22, 1341–1358
[4] L. Wcislak, H. Klein, H.J. Bunge, U. Garbe, T. Tschentscher, J.R. Schneider, J. Appl. Cryst., 2001, 35, 82–95
[5] R. Bouchard, D. Hupfeld, T. Lippmann, J. Neuefeind, H.B. Neumann, H.F. Poulsen, U. Rütt, T. Schmidt, J.R. Schneider, J. Süssenbach, M. von Zimmermann, J. Synchrotron Rad., 1998, 5, 90–101
[6] L. Wcislak, H.J. Bunge, Texture Analysis with a Position Sensitive Detector, Cuvillier Verlag, Göttingen, 1996
[7] R. Tamm. M. Lemke, C.-G. Oertel, W. Skrotzki, Mater. Sci. Forum, 1998, 273 - 275, 411–416
[8] W. Skrotzki, The Geological Significance of Microstructural analyses by transmission electron microscopy, Habilitation Thesis, Göttingen University, 1989
[9] J. Fischer-Bühner, Mechanismen der Mikrostruktur- und Texturentwicklung von polykristallinem NiAl, Shaker Verlag, Aachen , 1998
[10] W. Skrotzki, R. Tamm, C.-G. Oertel, B. Beckers, H.-G. Brokmeier, E. Rybacki, Mater. Sci. Eng., 2002, A329 - 331, 235–240

Severely Plastically Deformed Ti from the Standpoint of Texture Changes

J. Bonarski* & I.V. Alexandrov**
*Polish Academy of Sciences, Institute of Metallurgy and Materials Science, Krakow, Poland
**Ufa State Aviation Technical University, Ufa, Russia

1 Abstract

Crystallographic texture developed in Ti subjected to severe plastic deformation (SPD) by the equal-channel angular pressing (ECAP) has been experimentally investigated, and analyzed in terms of the orientation distribution function (ODF) in the Euler angles space. The analysis focused on the study of the texture inhomogeneity and its distribution along chosen radial directions of the cross-section of an ECAP Ti ingot. From the detailed texture analysis, based on the X-ray diffraction technique, some important structural parameters, like the mean deviation of crystallographic axis of crystals from the sample axis, the range of inhomogeneities, and the topography of dominating texture components have been revealed.

2 Introduction

From viewpoint of structure changes occurred in highly defected materials, like the severely plastically deformed (SPD) metals, microstructure and crystallographic texture belong to the base parameters. Favored by the equal channel angular pressing (ECAP) deformation mode, the adequate microstructure changes are accompanied with the lattice rotations and crystallographic slips. Such microstructure elements like grain boundaries, shearing bands as well as the lattice defects, and stress fields, created in these processes, cumulate most energy stored in so deformed material. It manifests by increasing the hardness, the strength, and the material density [1].

Observations of the metallic materials subjected to the SPD with an extremely high accumulated strain $\varepsilon > 10\text{-}100$ evidence that the developed ultra-fine-grained (UFG) nanostructures with the average grain size of dozens nanometers are a rather typical feature of SPD materials [2]. Such a small grains possess as a rule high angle grain boundaries. The last one contain a high density of extrinsic grain boundary dislocations (EGBDs) creating long-range stress fields and, as a result, the atoms are considerably shifted from their positions in the perfect crystal lattice. During that drastically hard deformation conditions, also a specific texture is developed [1-3]. The high deformation degree as well as the texture inhomogeneity make, however, that description of the structure changes is significantly complicated. This is also the challenge for the researchers and technologists, who tend to know and then to control the processes of formation of material by means of the SPD technique. It seems that crystallographic texture is one of the structure parameters, which gives such a chance. Especially, the advanced, non destructive methods of texture analysis become more and more popular in laboratory practice [4-6]. The present work is focused on describing the texture changes in the Ti ingot processed in the ECAP, by means of texture analysis.

3 Experimental Details

Severely plastically deformed Ti with *hcp* lattice symmetry has been investigated. ECAP was conducted by the route B_C (90° clock-wise rotations of the ingots between the passes). The die angle was 90°. The Ti ingot was subjected to 8 passes at temperature of 450–400 °C. Accumulated strain was about 9.

A sample for X-ray testing was prepared in form of slice by cutting off from the Ti ingot by means of the diamond saw from the ending part of the ingot, ca. 1 cm before the head of the ingot, measuring in its longitudinal axis. The top damaged layers on the cross-section surface were removed by a water proof paper abrasion. The sample surface was prepared by a non-preferential electro-polishing. Microstructure of the sample observed by means of the scanning electro-microscope if presented in Fig.1.

The texture analysis with reference to deformed materials is routinely based on data collected by means of X-ray diffraction technique. The measurements were carried out by means of X-ray diffractometer system of X'Pert of Philips equipped with a textural goniometer ATC-3. Filtered radiation $CoK\alpha$ ($\lambda = 1.79026$ Å) was used. A new Silicon Strip Position-Sensitive Detector, for registering the diffraction reflexes was applied [7]. Relation between the sample and the pole figure coordinates is shown in Fig. 2.

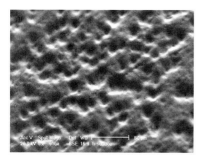

Figure 1: Microstructure of Ti sample observed by means of the SEM

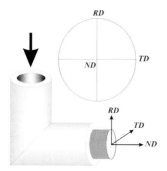

Figure 2: Relation between the sample (the slice cut off the Ti ingot) and the pole figure coordinates: ND – normal direction, TD – transverse direction, RD – radial direction

4 Results and Discussion

For recognizing the texture topography on the Ti-ingot cross-section, X-ray measurements in 14 chosen points on the sample surface have been performed. Then, the experimental pole figures (100), (102) and (110), registered by the back-reflection Schulz method were processed for the analysis of texture inhomogeneity related to the sample (cross-section of the ingot) areas [4]. Chosen complete (100) pole figures, calculated on basis of the experimental ones, correspond to the measurement points are presented in Fig. 3.

Because the applied collimated incident beam (diameter of 0.6mm), each measurement point on the Ti ingot cross-section (diameter of 20mm), marked as 1, 2,3,...14 (see Fig. 3) have not overlapped one to the other. The measurement points were arranged along two radiuses on the sample surface creating the angle of 135° to be much representative regarding the deformation route of the applied ECAP. In this way we expected obtain the texture data coming from a different concentric zones of the sample, as it was marked in Fig. 3. For example, measuring points 7 and 9 belong into the same zone. Similarly points 6 and 10, 5 and 11, 4 and 12, and so on, represent various zones, on which the sample area was divided.

The complete pole figures presented in Fig. 3, undersigned as 1i, 2i, 3i,...14i, are related to the individual zones, and show an essential texture differences between the distinguished zones. Adequate number of the pole figures registered for each of the measurement points enabled to calculate the orientation distribution functions ODF's, and to check the texture changes on the cross-section of Ti ingot. Quantitative analysis of the calculate texture functions confirms relatively sharp texture of the tested sample.

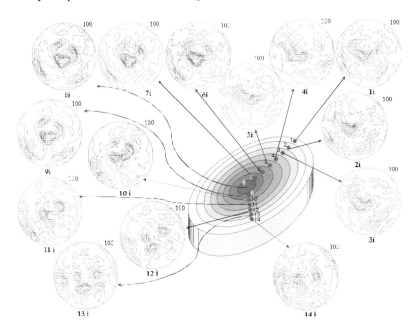

Figure 3: Complete (100) pole figures for the Ti ingot after ECAP, registered in the marked points on its cross-section

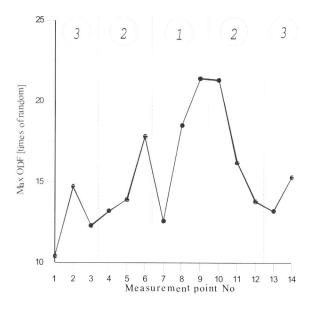

Figure 4: The line of maximum values of the texture function for the Ti ingot after ECAP deformation, calculated in measurement points on its cross-section. Three zones of texture changing have been distinguished: 1 – central zone, 2 – medium zone, 3 – border zone

Based on the calculated ODF's a line of maximum values of the texture function, illustrating the texture changes, has been constructed and given in Fig. 4. As it can be noticed, some zones of the sample characterized by diverse texture intensity can be distinguished:

- **border zone** - covered the border sample area (points of 1,2,3 and 13,14 ones) characterized by relatively the smallest ODF intensities, and by large irregularities of the texture,
- **medium zone** – covered an internal ring of the cross-section, limited by the 4–6 and 10–12 measurement points, in which sharply changed texture appears as local maxima of the ODF,
- **central zone** – relatively the narrowest sample area, presenting rather weak texture and the ODF values, localized in the centre of sample (points 7 and 8 in Fig. 3).

The above distinguished zones have been formed during deformation process. The medium zone covers some sample area in which the tendency to texturisation is the strongest. It manifests by the highest values of the ODF. Crystallographic orientations dominated in the zone can be recognized as the shear texture components [8]. Because of the compression-tension nature of the ECAP, the microstructure of materials subjected to such deformation mode is always accompanied shear bands in which defined texture components are generated. However, identification of the effects is possible only by texture analysis in a local scale, like described above.

Globally measured texture of the sample becomes less sharp and less informative from inhomogeneity viewpoint, comparing to the locally identified texture. This is due to averaging the texture effects characteristic of different sample areas as it is shown in Fig. 5.

The above described texture topography is an indispensable procedure in analysis of microstructure of materials subjected to the SPD. Localization of the components as well as it type are valuable parameter, which can be used for characterization of the ECAP process.

5 Conclusions

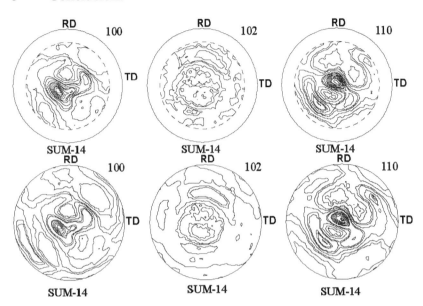

Figure 5: Experimental (top) and recalculated (bottom) pole figures for the Ti ingot after the ECAP deformation, summarized over all of 14 measurement points. Max. = 2.3 times of random. Levels: 0.5, 1.0, 1.2, 1.5, 1.8, 2.0, 2.2.

This paper documents one of the first treatises to carefully measure the homogeneity of texture across an ultrafine grained material achieved by Severe Plastic Deformation. By the example of ECAP deformed Ti, it has been found that

- there occur strong irregularities of crystallographic texture related to three spatial zones at the **border, medium and central** areas of the cross-section;
- the intensity of the texture, as well as of the misorientation angle between the main texture components and the ingot axis, hint at the localization of highly deformed (sheared) areas in processed material.

6 References

[1] R.Z. Valiev, R.K. Islamgaliev and I.V. Alexandrov, Progress in Materials Science. 2000, 45, 2

[2] S. Ch. Baik, Y. Estrin, H. S. Kim, H.-T. Jeong and R.J. Hellmig. Proc. ICOTOM-13, Ed. Dong Nyung Lee, Material Science Forum. 2002, 408-412, 697-702, Trans. Tech. Publ., Switzerland
[3] J. Kusnierz, M.H. Mathon, T. Baudin, Z. Jasienski and R. Penelle. ibid. p. 703-708
[4] K. Pawlik, Phys. Stat. Sol. 1986, (b), 134, 477
[5] J.T.Bonarski, H.J.Bunge, L.Wcislak, K.Pawlik, Textures and Microstructures, 1998, Vol. 31, No 1-2, pp. 21–41
[6] J. Pospiech, M. Wróbel, J.Bonarski, M. Blicharski, T. Moskalewicz, Archives of Metallurgy, 2002, 47/2, 185–195
[7] Detector constructed in 2001 by research groups from University of Mining and Metallurgy, and Polish Academy of Sciences in Krakow
[8] B. Major, Arch. Metall. 1986, **31**, 117–128

Development of Crystallographic Texture and Microstructure in Cu and Ti, Subjected to Equal-Channel Angular Pressing

J. Bonarski*, I.V. Alexandrov**, L.Tarkowski*
*Polish Academy of Sciences, Institute of Metallurgy and Materials Science, Krakow, Poland
**Ufa State Aviation Technical University, Ufa, Russia

1   Abstract

Severe plastic deformation (SPD) by means of equal-channel angular pressing (ECAP) allows to form ultrafine grained and nanostructured states being responsible for the enhanced mechanical properties in different metals and alloys. Meanwhile, processing of bulk billets with homogeneous structure and properties is an important issue. However, till these days investigations of the microstructure and crystallographic texture heterogeneity in ECAP metallic materials have not been practically conducted.

In the present report the results of experimental researches of crystallographic texture by means of X-ray diffraction analysis are presented. Copper samples with *cubic*-lattice and titanium ones with *hexagonal*-lattice have been studied.

2   Introduction

It is well known, that large plastic deformations result in a considerable microstructure and crystallographic texture evolution in different metals and alloys [1]. The observations of the metallic materials subjected to severe plastic deformation (SPD) with an extremely high accumulated strain $\varepsilon >10$-100 evidence that the developed ultrafine-grained (UFG) nanostructures with the average grain size of dozens nanometers are a rather typical feature of SPD materials [2]. Such a small grains possess as a rule high angle grain boundaries. The last one contain a high density of extrinsic grain boundary dislocations (EGBDs) creating long-range stress fields and, as a result, the atoms are considerably shifted from their positions in the perfect crystal lattice. Such a specific defect microstructure results in novel mechanical and physical properties of SPD metals and alloys such as, for instance, the increased hardness and strength [1].

The SPD method unlike other methods resulted in obtaining bulk nanostructured ingots without any porosity and impurities. However, the problem of the microstructure and crystallographic texture homogeneity in SPD materials is still topical. This problem is of a special interest in the case of bulk nanostructured materials processed by the equal channel angular pressing (ECAP) technique. Multiple pressing of the bulk ingot through the two intersecting channels fulfils the ECAP. In that way, material deforms by a simple shear but in a reality, ECAP is a rather complicated multifactor experiment. During ECAP, a material flow in different parts of the ingot is inhomogeneous. Therefore, it is very important to control the microstructure and texture heterogeneity in the ECAP processed ingots. At the same time in such a drastically hard deformation conditions as in the case at ECAP, a specific texture is developed [1–3]. Moreover, the texture heterogeneity makes the description of the microstructure changes significantly complicated. So, the first step in the study of the microstructure and crystallographic texture heterogeneity is the correct description of the crystallographic texture developed at ECAP in typical metals with the different crystal lattice.

The present work is an attempt to describe the microstructure of ECA-pressed FCC Cu and HCP Ti samples by means of the detailed texture analysis.

3 Experimental Details

Two samples of SPD materials, namely FCC Cu and HCP Ti were tested. ECAP was conducted by the route B_C (90° clock-wise rotations of the ingots between the passes). The die angle was 90° (see Fig.1). The Cu and Ti ingots were subjected to 8 passes at ambient temperature and at 450–400 °C, respectively. According to estimates, the accumulated true strain was about 9.

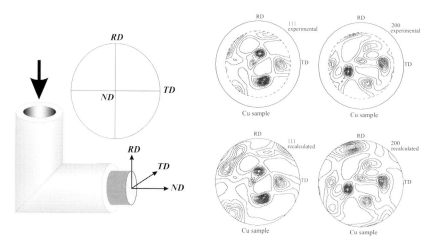

Figure 1: Schematic sketch of the ECAP process with relation between the sample and the pole figure coordinates: ND – normal direction, TD – transverse direction, RD – radial direction

Figure 2: Experimental (top) and recalculated (bottom) pole figures for the Cu sample deformed by the ECAP. Max.= 7.3 times of random. Levels: 0.5, 1.0, 2.0, 3.0, 4.0, 5.0, 6.0, 7.0.

The samples for X-ray investigations were prepared in form of slices by cutting off from the Cu and Ti ingots by means of the diamond saw from the ending part of the ingots, ca. 1 cm before the head of the ingot, measuring in its longitudinal axis. The top damaged layers on the cross-section surface have been removed by a waterproof paper abrasion. Such sample surfaces were carefully prepared by a non-preferential electro-polishing for the X-ray and electron-microscope investigations. The microstructure observations of the samples have been performed by means of SEM technique.

The texture analysis with reference to deformed materials is routinely based on data collected by means of X-ray diffraction (XRD) technique or electron back-scattering diffraction (EBSD) one.

The measurements were carried out by means of X-ray diffractometer system of X'Pert of Philips, equipped with a 4-axis textural goniometer ATC-3. Filtered radiation $CoK\alpha$ ($\lambda = 1.79026$ Å) was used. A new Silicon Strip Position-Sensitive Detector, for registering the diffraction reflexes was applied [4]. The incomplete pole figures were registered by the back-reflection Shulz method for a circular area (ca. 6 mm in diameter) located centrally on the sample surface.

The experimental data obtained from the X-ray measurement were processed for quantitative texture analysis, and for the study of the relations between the texture and the ECAP deformation modes. Texture analysis was performed by use of the three-dimensional orientation distribution functions (ODFs) in the Euler angles space, calculated by an iterative, arbitrarily defined cells (ADC) method [5].

4   Results and Discussion

The excellent agreement between the experimental (Fig.2, top; Fig.4, top) and the recalculated (Fig.2, bottom; Fig.4, bottom) pole figures confirms the correctness of performed calculating procedures, as well as the reliability of applied experimental technique in the presented texture analysis.

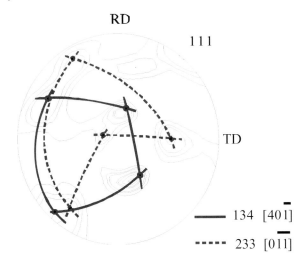

Figure 3: Example - drawing. The complete (111) pole figure for the ECA-pressed Cu sample.

4.1   Texture Development in Cu

A strongly shaped texture (max. of the ODF above 38) of the ECAP Cu ingot was observed. The given pole figures (Fig.2) and the ODF (Fig.5) illustrate its irregular character. Two main components: (134)[4 0 –1] and (233)[0 1 –1] were identified in the texture (see Tab.1). Its mutual position (see Fig.3) reveals a tendency to existing some rotating axis, close to the ⟨111⟩ di-

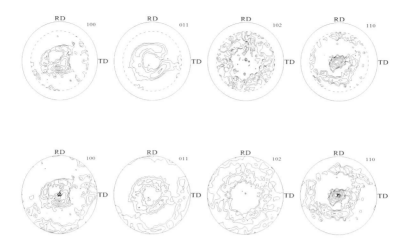

Figure 4: Experimental (top) and recalculated (bottom) pole figures for the ECA-pressed Ti sample, registered for a central part (ca. 4 x 4 mm) its cross-section (diameter ca. 18 mm). Max. = 3.9 times of random. Levels: 1.0, 1.2, 1.5, 2.0, 2.5, 3.0, 3.5.

rection, which is common for the two areas, on which divides the central part of the sample bulk during the ECAP.

The values collected in Tab.1 indicate a high disproportion in the volume fractions of those areas (ca. 22 vol.% and above 5.6 vol.%, respectively). Its localisation in frame of the pole figure is determined by the applied deformation mode, which leads to heterogeneous microstructure of the final material. This manifests in strong texture asymmetry on a cross-section of deformed Cu ingot, and have the determined consequences in anisotropic distribution of properties of the processed material.

4.2 Texture Development in Ti

Registered pole figures for investigated cross-section of the Ti ingot revealed the axis type texture, what was confirmed also by the corresponding ODF (see Figs 4 and 6). The texture shows irregular character, close to the ⟨130⟩ axis, in which dominates one, relatively strong (130)[3 -1 1] orientation (max. of the ODF above 12). Above 39 vol.% of the sample bulk have taken the preferred orientation close to the mentioned ⟨130⟩ axis (see Tab.1). A distinctly observed deviation (ca. 8°) of the ⟨130⟩ axis from the position of the ingot's longitudinal axis suggests - similarly to the Cu sample case - the presence of essential texture asymmetry on the cross-section of deformed Ti.

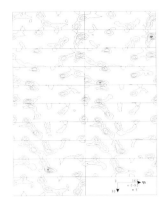
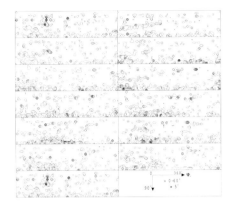

Figure 5: Orientation distribution function (j2 sections) for the ECA-pressed sample of Cu, calculated on the basis of experimental pole figures from Fig.2. Max. = 38.3 times of random. Levels: 0.5, 1.0, 2.0, 5.0, 10.0, 15.0, 20.0, 25.0, 30.0.

Figure 6.: Orientation distribution function (j2 sections) for the ECA-pressed sample of Ti, calculated on the basis of experimental pole figures from Fig.4. Max. = 12.8 times of random. Levels: 1, 2, 5, 10, 12

Table 1: Results of the quantitative texture analysis by the discrete ADC method [5]

Sample	Texture components	Component spread [°]			Volume fractions [% vol.]		
		$\Delta\varphi_1$	$\Delta\Phi$	$\Delta\varphi_2$	Components	Background	Rest
Cu	(134)[4 0–1]	17.50	12.50	15.00	21.96	0.01	72.39
	(233)[0 1–1]	10.00	10.00	10.00	5.64		
Ti	(130)[3–1 1]	45.00	45.00	20.00	39.28	0.00	60.72

3.3 Microstructure Development

It was found that, depending on the crystallographic lattice, the microstructures developed in deformed Cu and Ti samples, were typical of the applied mode of the ECAP process. Significant structure heterogeneities were presented in the both samples. Typical very fine grains structure, high density of shear bands, and numerous dislocation accumulations were observed. The linkage of texture analysis of crystalline sub-areas and microstructural elements goes far beyond the traditional concept analysis of materials processed by severe plastic deformation.

4 Conclusions

The results reported here have clearly demonstrated that strong textures in both samples of Cu and Ti, develop due to severe plastic deformation. One of the texture components highly dominates, and turns out to be characteristic of the ECAP deformation:

- In case of the Cu, the identified significant component {134}⟨401⟩ can be derived from rotation of the shear plane during the ingot deformation by B_C route [6]. Taking into account the twelve {111}⟨110⟩ slip systems in the *fcc* lattice, and a possible rotation of the shear plane in the ingot's central part (a middle angle from range of 0÷45°), the expected orientation of its cross section is close to the {134} one. As for the second, essentially weaker texture component of {233}⟨011⟩, additional material/geometrical conditions have to be considered. Moreover, the both components can be ascribed to the two areas, on which the central part of the sample bulk divides during the ECAP. The areas are mutually rotated around the common axis, close to the ⟨111⟩ crystallographic direction.
- Conclusions about the orientation distribution functions being specific of the lattice type chosen, and about their strong asymmetry over the cross-section of the deformed ingots have been drawn. The asymmetry plays an important role from viewpoint of usable properties of processed materials.
- It has been shown that the process of severe plastic deformation can be controlled by parameters like the localisation of the common rotation axis and the volume fraction of dominating texture components (for Cu sample), and the misorientation angle between the axis of main texture component and that of the ingot (for Ti sample). The selection of controlling parameters depends on material, deformation mode and type of developed texture.

5   References

[1] R.Z. Valiev, R.K. Islamgaliev and I.V. Alexandrov, Progress in Materials Science. 2000, 45, 2
[2] S. Ch. Baik, Y. Estrin, H. S. Kim, H.-T. Jeong and R.J. Hellmig. Proc. ICOTOM-13, Ed. Dong Nyung Lee, Material Science Forum. 2002, 408–412, 697–702, Trans. Tech. Publ., Switzerland
[3] J. Kusnierz, M.H. Mathon, T. Baudin, Z. Jasienski and R. Penelle. ibid. p. 703–708.
[4] Detector constructed in 2001 by research groups from Academy of Mining and Metallurgy and Polish Academy of Sciences in Kraków
[5] K. Pawlik, Phys. Stat. Sol. 1986, (b), 134, 477
[6] I. Kopacz, M. Zehetbauer, L.S. Toth, I.V. Alexandrov, B. Ortner, Proc. of the 22nd Riso
[7] Int. Symposium, Ed. A.R. Dinesen et al. Roskilde, Denmark 2001, p.295–300

VI Details of SPD Nanostructures as Investigated by Electron Microscopy

Boundary Characteristics in Heavily Deformed Metals

Grethe Winther and Xiaoxu Huang
Center for Fundamental Research: Metal Structures in 4D, Risø National Laboratory, Risø, Denmark

1 Abstract

The potential of creating nanostructured metals by plastic deformation to very high strains is currently the subject of intensive research. An important part of this research concerns evolution of the characteristics of deformation induced boundaries, in particular boundary spacing and boundary misorientation. The aim of this paper is to give an overview of the present understanding of the relations between these characteristics, the microscopic deformation mechanisms and the macroscopic deformation mode.

2 Introduction

It is well known that plastic deformation of metals induces subdivision of the initial grain structure by formation of boundaries, which evolve further during continued deformation. The potential of creating nanostructured metals by plastic deformation to very high strains is therefore currently the subject of intensive research. An important part of this "research concerns evolution of the characteristics of deformation induced boundaries, in particular boundary spacing and boundary misorientation. The aim of this paper is to give an overview of the present understanding of the relations between these characteristics, the microscopic deformation mechanisms and the macroscopic deformation mode. This understanding has been developed for fcc metals of medium to high stacking fault energy processed by different deformation modes in the low strain regime, i.e., up to von Mises strains of about one.

Experimental data on the evolution of the boundary characteristics up to a strain of 12 are presented, mainly using nickel deformed in high pressure torsion as an example. Based on the experimental data, it is discussed to what extent the correlations established for low strains may be extrapolated to high strains.

3 Deformation Induced Boundaries at Low Strain

The typical morphology of a deformation induced dislocation structure at low strains ($\varepsilon vM < 1$) is shown in Figure 1. The structure consists of extended planar dislocation boundaries seen as almost parallel lines. In between these, randomly oriented cell boundaries are found.

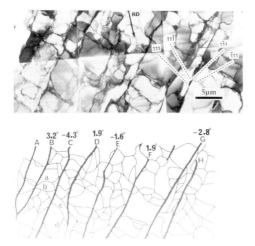

Figure 1: TEM image and corresponding tracing of dislocation boundaries for aluminum rolled to a von Mises strain of 0.12 (from Liu and Hansen [4]). The structure is seen in the plane containing the rolling direction and the rolling plane normal. The rolling "direction is indicated at the image along with the trace of crystallographic {111} slip planes. b) Misorientation angles across extended planar boundaries (labeled A-H) are given in the tracing. Examples of individual cell boundaries are also marked (a–d).

3.1 Boundary Planes

It is well established [1–3] that at low strains extended planar boundaries tend to be roughly aligned with the macroscopically most stressed planes in the sample. In rolling, these planes are inclined ± 45° to the rolling direction in the plane spanned by the rolling direction and the rolling plane normal, and they are perpendicular to the rolling direction in the rolling plane. In tension the macroscopically most stressed planes are inclined 45° to the tensile axis. In torsion, all planes parallel to the torsion axis or the shearing plane are macroscopically most stressed.

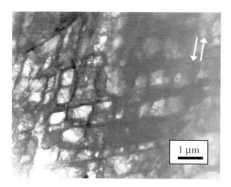

Figure 2: TEM image of nickel deformed by high pressure torsion to a von Mises strain of 0.5 at a pressure of 4 GPa. The torsion axis lies in the plane of the micrograph perpendicular to the double arrows which mark the trace of the shearing plane (from Winther et'al. [5]).

The approximate alignment of extended planar boundaries with these planes is illustrated for rolling and torsion in Figures 1 and 2, respectively.

With respect to the coordinate system given by the crystallographic lattice of the grain, boundaries have been separated into two categories, namely those that roughly align with slip planes and those that lie far from any slip plane. The occurrence of the two boundary types correlates with the crystallographic orientation of the grain as illustrated for tension in Figure 3. Classification of boundaries according to their crystallographic plane reflects fundamental differences also in other boundary characteristics as demonstrated next for the crystallographic misorientation across the boundaries and the spacing between them.

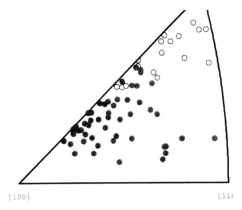

Figure 3: Stereographic triangle showing the orientation of the tensile axis for grains in aluminum deformed to von Mises strains ranging from 0.5 to 0.34 (from Winther et al. [6]). Open and filled symbols represent grains with boundaries aligned and not aligned with slip planes, respectively.

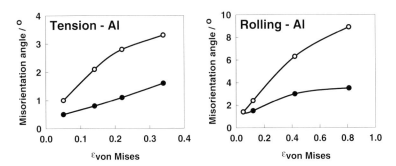

Figure 4: Average misorientation angles across extended planar boundaries for aluminum deformed in tension and rolling. Boundaries have been separated into two groups "according to whether the boundary plane is roughly aligned with a slip plane (filled symbols) or not (open symbols) (data from Liu et al. and Hansen et al. [7]).

3.2 Misorientation Angles

The average misorientation angles across the extended planar boundaries which are roughly aligned with slip planes are smaller than for boundaries not aligned with slip planes. This is demonstrated for aluminum deformed in tension and rolling in Figure 4. The misorientation angles for the two types of boundaries differ by a factor of two over the investigated strain range. A similar difference is found for the misorientation across the randomly oriented cell boundaries in between the two types of extended planar boundaries. [7,8]

3.3 Boundary Spacing

As an example, the spacing between the extended planar boundaries for tensile deformed aluminum is shown in Fig"ure 5 as a function of strain. It is seen that the curves for boundaries aligned with slip planes and boundaries not aligned with slip planes have the same shape with an initial rapid decrease, which then slows down dramatically. Over the whole strain range, the spacing between boundaries roughly aligned with slip planes is about 2 µm higher than for boundaries not aligned with slip planes. The spacing "between the ordinary randomly oriented cell boundaries in grains which have extended planar boundaries roughly aligned with slip planes is also higher than in grains where these boundaries lie far from slip planes. [7,8]

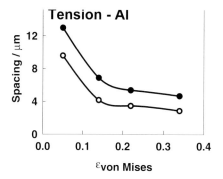

Figure 5: Average boundary spacing for boundaries which approximately align with slip planes (filled symbols) and boundaries lying far from slip planes (open symbols) for aluminum deformed in tension (data from Hansen et al. [8]).

4 Deformation Mode, Slip, and Boundary Characteristics at Low Strains

The clear development of different boundary characteristics in grains with different crystallographic orientation can be traced to the microscopic deformation mechanisms in the grain: i) Slip generates the dislocations from which the boundaries are constructed and ii) the identity of the active slip systems in each grain depends on the crystallographic orientation of the grain with respect to the macroscopic deformation mode. Therefore, the boundary characteristics are "determined by the ensemble of active slip systems, often "referred to as the slip pattern.

Recently, general correlations between the slip pattern and the closeness of the extended planar boundaries to the slip plane have been established. These correlations apply to "several macroscopic deformation modes. The link between macroscopic deformation mode, microscopic deformation mechanisms and characteristics of the induced dislocation boundaries have thus been quantitatively established. [6]

The most important single parameter which must be considered to predict the boundary characteristics from a slip pattern is the degree of slip concentration. Grains which have more than about 45 % of the slip concentrated on a single slip plane form boundaries that lie within about 10° of the slip plane. Boundaries where the slip is distributed among many slip planes lie far from any of these. This relation is illustrated in Figure 6 for selected grain orientations deformed by rolling or tension. Figure 7 demonstrates the predictive capability of this relation for rolled aluminum. It is seen to be quite good. Only in the vicinity of the borderline between the

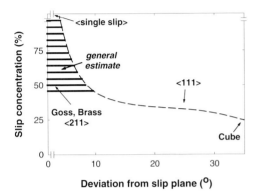

Figure 6: Experimentally observed angles between dislocation boundary planes and the nearest slip plane as a function of the fraction of slip predicted to occur on a single slip plane for grains deformed in rolling and tension to von Mises strains below 1 (from Winther et al. [6]).

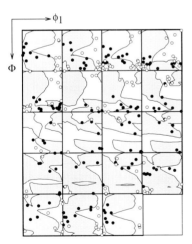

Figure 7: Orientation distribution function showing experimental grain orientations observed to have extended planar dislocation boundaries roughly aligned with slip planes or lying far from slip planes as filled and open symbols, respectively. These data are for aluminum rolled to von Mises strains of up to 0.8. Grey and white areas represent grain orientations predicted to form boundaries aligned or not aligned with slip planes, respectively (from Winther et al. [9]).

gray and white areas, i.e., grain orientations predicted to have slip plane aligned boundaries and not slip plane aligned boundaries, a small overlap of experimental data points is observed.

The tendency of extended planar boundaries to lie close to most stressed planes determined by the macroscopic deformation mode as mentioned in section 2.1 is also explained by this correlation. When slip is highly concentrated, the most active slip plane with which the boundaries align lies close to a macroscopically most stressed plane. When slip is highly distributed and boundaries lie far from any slip plane, no slip plane lies close to a macroscopically most stressed plane. Instead the slip direction is close to a macroscopically most stressed plane.

With respect to production of nanostructured metals by plastic deformation such correlations between macroscopic deformation mode, slip, crystallographic grain orientation and boundary characteristics provide a tool to evaluate the efficiency of a given deformation mode. The ideal deformation mode for generation of nanostructures should induce a slip pattern which promotes subdivision and structural refinement (high misorientation and low boundary spacing). In evaluating a "deformation mode, one should keep in mind that structural refinement is invariably accompanied by evolution of a deformation texture. As the crystallographic orientation of each grain rotates during deformation, the slip pattern also changes.

Extension of the correlations established for low strains to the high strain regime are needed before they can be fully exploited. Key questions in this process are whether boundary characteristics evolve continuously with strain and whether slip is still the dominant microscopic deformation mechanism when the strain increases. In order to answer these questions, evolution of the boundary characteristics must be followed closely over a very large strain range.

5 Deformation Induced Boundaries at High Strain

High pressure torsion is a process well suited to investigate evolution of boundary characteristics over a wide strain range because it is a monotonic, single pass process and because it is expected to resemble conventional torsion. In this section, results from pure nickel deformed by high pressure torsion at 4 GPa to von Mises strains up to 12 are presented. The samples are disks with a diameter of 7–8 mm and a thickness of 0.7 mm. The foils investigated by transmission electron microscopy are taken from the same position on disks deformed to different maximum strains. These foils are from a section parallel to the torsion axis as in Figure 2, not from the shearing plane as in many previous studies. This sample section allows observation of boundaries roughly parallel to the shearing plane. These boundaries are not resolved when looking at foils parallel to the shearing plane. For further discussion of the importance of the sample section studied see Huang et al. [10]

5.1 Morphology and Boundary Planes

As already shown in Figure 2, the morphology of the dislocation structure at a von Mises strain of 0.5 closely resembles that of rolled metals (compare with Figure 1) and the "extended planar boundaries are roughly aligned with the macroscopically most stressed planes.

At a strain of 5, only one set of boundaries roughly parallel to the shearing plane are seen (Fig. 8). Short randomly oriented cell boundaries are still found in between these. At this strain, however, very thin twins are occasionally observed. An analogous transition from two sets of

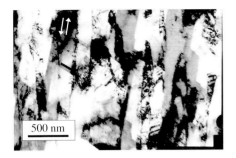

Figure 8: TEM image of nickel deformed by high pressure torsion to a von Mises strain of 1 at a pressure of 4 GPa. The torsion axis lies in the plane of the micrograph perpendicular to the double arrows, which mark the trace of the shearing plane.

intersecting boundaries to one set is also seen in rolling at a comparable strain, where boundaries become approximately aligned with the rolling plane. [11]

At a von Mises strain of 12, the overall structure is also "lamellar. As seen in Figure 9 extended planar boundaries aligned with the shearing plane still dominate. Twins are "observed more frequently than at a strain of 5. In the upper micrograph of Figure 9 twins are marked as lines. Occasionally, isolated equiaxed subgrains surrounded by high angle boundaries are seen as shown in the lower part of Figure 9.

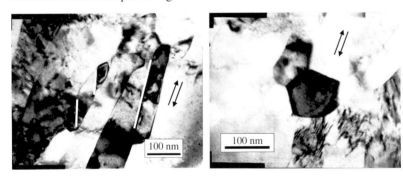

Figure 9: TEM micrographs of nickel deformed by high pressure torsion to a von Mises strain of 12 at a pressure of 4 GPa. The torsion axis lies in the plane of the micrograph perpendicular to the double arrows, which mark the trace of the shearing plane. The upper micrograph shows the typical lamellar structure. Twins are marked as white lines;. The lower micrograph shows an example of the occasionally occurring isolated equiaxed subgrains (from Huang et al.[10]).

5.2 Spacing and Misorientation

The average spacing between extended planar dislocation boundaries for the nickel samples deformed by high pressure torsion are shown as a function of strain in Figure 10. It is seen that the rate of refinement constantly decreases but "refinement definitely continues over the entire strain range with no sign of saturation.

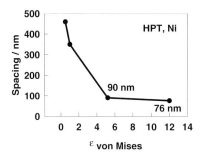

Figure 10: Average spacing between extended planar dislocation boundaries in nickel deformed by high pressure torsion at 4 GPa.

At a von Mises strain of 5, about a third of the extended planar dislocation boundaries are high angle boundaries, i.e., the misorientation angle across them are more than 15°. Similar low fractions of high angle boundaries are observed in nickel rolled to a von Mises strain of 4.5. [12] At higher strains, a detailed misorientation distribution has not yet been measured. Selected area diffraction patterns at a von Mises strain of 12 show the presence of many high angle boundaries but the relative frequency of high and low angle boundaries has not yet been determined.

6 Microscopic Deformation Mechanisms at High Strains

The continuous evolution of both morphology and boundary spacing also at high strains indicates that slip is still the dominant microscopic deformation mechanism. Nevertheless, it is clear that deformation twinning also occurs with "increasing frequency as the strain increases. In this context, it should be mentioned that the flow stress of the samples deformed by high pressure torsion as judged from hardness measurements at a strain of 5 greatly exceeds the critical stress for twinning as discussed by Huang et al. [10]

The fraction of all boundaries that are twin boundaries and the volume fraction of twinned material have not been quantified yet. This quantification must be completed before any conclusions can be drawn regarding the importance of twinning as a mechanism which must be considered to adequately model the deformation process and structural evolution. The most important contribution from twinning mechanisms to the boundary characteristics is probably the high misorientation angle of 60° across twin boundaries. Also the fact that twin boundaries contain coincident lattice sites may be important.

7 Conclusion and Outlook

An overview of the evolution of deformation induced boundary characteristics in face-centered cubic (fcc) metals of medium to high stacking fault energy over a very wide strain range has been presented. The characteristics considered are boundary planes, boundary misorientation and boundary spacing. The following conclusions are drawn:

- For low strains ($\varepsilon vM < 1$), correlations linking macroscopic deformation mode, slip and boundary characteristics are well established.
- At a von Mises strain of about one, a transition in the deformation induced structure occurs where the plane of the "induced boundaries changes. Apart from this, boundary characteristics are observed to evolve continuously up to von Mises strains of at least 12. Especially, the morphology is maintained, and the boundary spacing continuously decreases with no sign of saturation.
- Slip appears to continue as the dominant microscopic "deformation mode over the entire strain range even when the boundary spacing drops below 100 nm. However, twinning occurs more frequently with increasing strain.
- These findings are promising for the perspective of establishing correlations between boundary characteristics, "microscopic deformation mechanisms and macroscopic deformation mode valid for very high strains based on the relations found for low strains. Such correlations will provide theoretical understanding of nanostructured metals as well as guidelines for design of efficient processes aiming at these.

8 Acknowledgement

The authors gratefully acknowledge the Danish National Research Foundation for supporting the Center for Fundamental Research: Metal Structures in Four Dimensions, within which this work was performed. The authors also thank Dr. N. Hansen for scientific discussions and Prof. M. Zehetbauer, University of Vienna, and A. Vorhauer, T. Hebesberger and R. Pippan, Erich Schmid Institute, Austrian Academy of Sciences, for their collaboration and for supplying the nickel samples "deformed by high pressure torsion.

9 References

[1] B. Bay, N. Hansen, D. Kuhlmann-Wilsdorf, Mater. Sci. Eng. 1989, A113, 385.
[2] D. V. Wilson, P. S. Bate, Acta Mater. 1996, 44, 3371.
[3] G. H. Akbari, C. M. Sellars, J. A. Whiteman, Acta Mater. 1997, 45, 5047.
[4] Q. Liu, N. Hansen, Scr. Metall. Mater. 1995, 32, 1289.
[5] G. Winther, N. Hansen, T. Hebesberger, X. Huang, R.'Pippan, M. Zehetbauer, in Proc. of 22nd Risø Int. Symposium on Materials Science: Science of metastable and nanocrystalline alloys: Structure, properties and modelling, Risø National Laboratory, Roskilde, Denmark 2001.
[6] G. Winther, Acta Mater. 2003, 51, 417.
[7] Q. Liu, D. Juul-Jensen, N. Hansen, Acta Mater. 1998, 46, 5819.
[8] N. Hansen, X. Huang, D. Hughes, Mater. Sci. Eng., A 2001, 317, 3.
[9] G. Winther, D. Juul-Jensen, N. Hansen, Acta Mater. 1997, 45, 5059.
[10] X. Huang, G. Winther, N. Hansen, T. Hebesberger, A.'Vorhauer, R. Pippan, M. Zehetbauer, Mater. Sci. "Forum, in press.
[11] D. A. Hughes, N. Hansen, Metall. Trans. A 1993, 24, 2021.
[12] D. Hughes, N. Hansen, Acta Mater. 2000, 48, 2985.

Quantitative Microstructural Analysis of IF Steel Processed by Equal Channel Angular Extrusion

J. De Messemaeker, B. Verlinden, J. Van Humbeeck, L. Froyen
Katholieke Universiteit Leuven, Leuven

1 Introduction

The objective of this study was twofold. On one hand the evolution of structure and hardness of IF steel with strain was followed up to an effective strain of 9.2 for route B_A, in order to study the grain fragmentation at large strains. On the other hand the 4 classical ECAE routes A, B_A, B_C and C were compared at an effective strain of 9.2, in order to study the influence of the strain path followed during severe plastic deformation. Automated EBSD measurements were used because they allow to obtain a large collection of orientation data in a relatively short time with a fair resolution.

2 Experimental

The hot rolled IF (Interstitial Free) steel used for the ECAE samples had an equiaxed grain size of 100 μm and a hardness of (103 ± 2) H_V. The principle of ECAE has been described elsewhere [1]. The die used in the present study had an angle of 90° and a circular cross section with a diameter of 12 mm. The extrusion temperature was (200 ± 10) °C. EBSD and hardness measurements were made on the section of plane strain (perpendicular to TD) which is indicated in Fig. 1 together with the reference directions ED, TD and ND. The Vickers hardness was measured with a load of 100 gf.

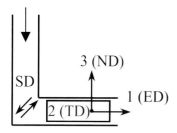

Figure 1: Sketch of the ECAE die showing the plane of section (shaded), the reference directions ED, TD and ND and the shear direction SD

The automated EBSD measurements were made with a Philips XL30 SEM-FEG at 20 kV. In order to obtain sufficient EBSD pattern quality the samples were first annealed for 3 hours at 500 °C. The raw data contained points that were indexed badly, especially in the vicinity of the

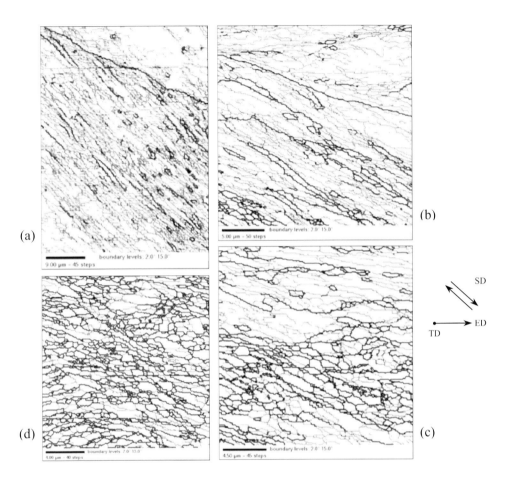

Figure 2: Boundary maps with low angle boundaries (2°–15°, thin black line) and high angle boundaries (15° or higher, bold black line) for IF steel after (a) 1 pass, (b) 2 passes via route B, (c) 4 passes via route B_A and (d) 8 passes via route B_A. Note the scale differences between the maps

subgrain boundaries. The orientation of these points was replaced using the cleanup procedures of the commercial package (OIM, TSL-EDAX) used to collect the data.

3 Results and Discussion

Because the angular resolution of EBSD measurements is about ± 1.5°, only misorientations exceeding 2° have been considered in the analysis. Subgrains are defined by boundaries with a misorientation of 2° or higher; grains by boundaries of 15° or higher. (Sub)grains were identified by a grain reconstruction algorithm, and average (sub)grain sizes were calculated as the diameter of a circle with an area equal to the average area of the (sub)grains in a given scan. The orientation data is represented by means of the boundary maps in Fig. 2 and Fig. 4. Low angle boundaries are defined as having a misorientation between 2° and 15°, and are indicated with a

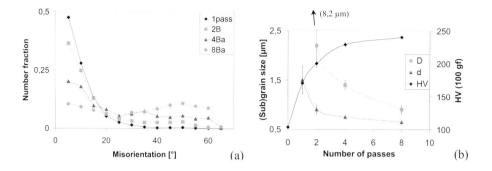

Figure 3: (a) Misorientation distributions of IF steel after 1 pass, 2, 4 and 8 passes via route B_A ("1pass", "2B", "4Ba" and "8Ba" respecively). (b) Evolution of grain size D, subgrain size d, and Vickers hardness HV with the number of ECAE passes via route B_A for IF steel

thin black line. High angle boundaries carrying a misorientation of 15° or higher, are indicated with a bold black line. For each map shown in Fig. 2 and Fig. 4, at least one additional scan was made, confirming the observations described below.

3.1 Evolution of Structure and Hardness with Strain

After 1 pass through the ECAE die ($\varepsilon_{eq} \sim 1.15$), the grains have subdivided into bands which are parallel to the shear direction and which are about 0.8–2 µm wide. The boundaries visible in Fig. 2a are mostly band boundaries (parallel to SD), and hardly any transverse boundaries (approximately perpendicular to the band boundaries) cross the bands. The misorientation distribution (Fig. 3a) reflects a typical deformation structure, with a high fraction (88 %) of low angle boundaries. The average subgrain size is 1.9 µm.

The fragmentation is not uniform and differs considerably from one grain to another. In the lower grain in Fig. 2a for example, some band boundaries carry a misorientation larger than 15°, whereas hardly any low angle boundaries have been formed in the upper grain. TEM observations revealed however that all grains contained a banded structure, though this is not always clear from the boundary maps due to their limited angular resolution. The differences in fragmentation behavior of the grains are reflected both by the magnitude of the misorientations present in the substructure, and by the fragment size. This is in agreement with the results of Seefeldt and Van Houtte [2] who modeled the fragmentation in aluminum during ECAE and found that for some orientations the structural fragmentation into new texture elements was counteracted by the textural rotation during the deformation, whereas for other orientations they assist each other, resulting in an orientation dependence of fragmentation.

After 2 passes ($\varepsilon_{eq} \sim 2.3$) via route B the bands are inclined at an angle of about 30° with the extrusion direction. The fraction of high angle boundaries has increased from 12 % after 1 pass to 36 % after 2 passes, and the subgrain size has decreased to 1.0 µm, indicating that the grain fragmentation has globally progressed. However, there is still a remarkable difference in fragmentation behavior of different grains (compare for example the grain to the upper right with the grain to the lower left in Fig. 2b, which is reflected by a bimodal subgrain size distribution).

Almost all high angle boundaries are band boundaries, and hardly any high angle transverse boundaries have developed.

With increasing strain the fragmentation continues to develop; after 4 passes via route B_A ($\varepsilon_{eq} \sim 4.6$) the subgrain size has decreased to 0.8 μm and the high angle boundary fraction has increased to 52 %. The difference in fragmentation behavior still exists, but the weakly fragmenting grains are catching up with the strongly fragmenting ones, at least as far as subgrain size is concerned. The morphological orientation of the bands has not significantly changed; in the strongly fragmenting grains the band boundaries are now all high angle boundaries.

After 8 passes ($\varepsilon_{eq} \sim 9.2$) via route B_A the fragmentation differences are leveling out as far as subgrain size is concerned; the subgrain size distribution has become unimodal about an average of 0.6 μm. However, there are still differences as far as misorientation is concerned. Fig. 2d for example shows some areas with clusters of low angle boundaries. The fraction of high angle boundaries has increased to 72 %.

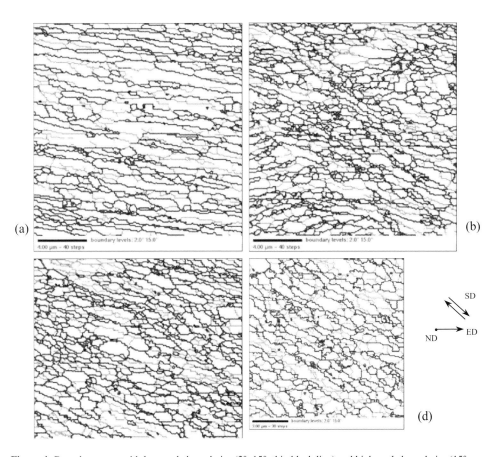

Figure 4: Boundary maps with low angle boundaries (2°–15°, thin black line) and high angle boundaries (15° or higher, bold black line) for IF steel after 8 passes via (a) route A, (b) route B_A, (c) route B_C and (d) route C. The maps are similarly scaled.

The evolution of structure and hardness is summarized in Fig. 3b. The hardness increases with increasing strain, but tends to saturate at large strains, indicating that the IF steel shows a parabolic hardening under ECAE deformation conditions. As strain increases, the grain size approaches the subgrain size, which tends to saturate at high strains.

3.2 Influence of Strain Path

The microstructures of IF steel after 8 passes ($\varepsilon_{eq} \sim 9.2$) via the 4 classical routes have significantly different morphologies and slightly different sizes (Fig. 4), even though they have the same level of strain. The structure of the route A, B_A and B_C samples consists of clearly marked bands, whereas the structure of the route C sample shows diffuse bands and has an equiaxed aspect. In the route A sample the bands tend to lie parallel to the extrusion direction; for both route B samples the bands are inclined at an angle of about 30° to the extrusion direction. In the route B_A sample a wavy structure is superimposed on the bands: a macroscopic S-shape which returns at regular intervals in the microstructure and which may be a shear band. This phenomena locally alters the morphology, as bands become aligned with the S-shape and the subgrains are locally refined with respect to the rest of the microstructure (Fig. 5d).

The route A sample has the largest grain and subgrain size, and the highest subgrain aspect ratio, because the fragmentation across the bands is less developed. As transverse boundaries on average carry a lower misorientation than the boundaries in between bands, this agrees well with the fact that the route A sample has the highest fraction of high angle boundaries (Table 1). The route B_A, B_C and C samples have the same subgrain size, but the fraction of high angle boundaries decreases in this order, while the grain size correspondingly increases.

Table 1: Grain size D, subgrain size d, percentage of high angle grain boundaries %HAGB and hardness ($H_V \pm \Delta H_V$) of IF steel after 8 passes via routes A, C, B_A and B_C (D, d and %HAGB from the EBSD data in Fig. 4)

Route	A	C	B_A	B_C
D [µm]	1.0	0.9	0.7	0.8
d [µm]	0.8	0.6	0.6	0.6
%HAGB	73 %	63 %	72 %	70 %
H_V	229	215	240	235
ΔH_V	2	3	3	2

The hardnesses of the 4 samples are significantly different, and can be classified as follows: $B_A > B_C > A > C$ (Table 1). Dupuy [3] found the same order for the flow stress at room temperature of aluminum after 4 and 8 passes through a 90° die. He calculated the α parameter introduced by Schmitt [4] to quantitatively express the change of strain path from one pass to another:

$$\alpha = \frac{\dot{\varepsilon}_p : \dot{\varepsilon}}{\|\dot{\varepsilon}_p\| \cdot \|\dot{\varepsilon}\|} \tag{1}$$

where $\dot{\varepsilon}$ and $\dot{\varepsilon}_p$ are the strain rate tensors of the consecutive passes. According to Dupuy, for a die angle of 90°, routes A and C correspond to reverse loading ($\alpha = -1$), and both routes B correspond to cross loading ($\alpha = -0.5$). However, according to the deformation mechanism that was recently proposed by Gholinia et al. [5], route A corresponds to an approximately continuous strain path ($\alpha = 1$) regardless of the die angle. For a die angle of 90°, the interpretations for the other routes remain the same ($\alpha = 0$ for both B routes). These values for α correlate with the hardness and structure of the 4 routes. Reverse loading during route C leads to an inferior hardness with respect to the continuous loading during route A, because in each pass the same slip systems are activated as in the previous pass, but in the reverse sense. The resistance to glide is therefore smaller, and fewer dislocations need to be generated to accomplish the applied strain. Cross loading during routes B_A and B_C necessitates the activation of new slip systems, and the dislocations have to cut through the existing dislocation pattern. This may explain the higher fraction of transverse boundaries in the route B samples, when compared to the route A sample. The difference between the two route B samples may be explained by the fact that the initial grain boundaries are elongated during route B_A, leading to an extra increase in high angle boundary fraction. After 4n (where n is an integer) passes via route B_C on the other hand, the original volume elements are restored to their initial shape, and no extra grain boundary area is introduced by this geometrical mechanism.

4 Conclusions

The evolution of structure and hardness of IF steel with strain, imposed by ECAE via route B_A has been investigated. It was found that the degree of fragmentation differs from grain to grain even up to a strain of 9.2. This can be explained by a model of Seefeldt and Van Houtte which shows that the structural fragmentation and the textural rotation assist each other for highly fragmenting grains and counteract for weakly fragmenting grains. The orientation dependence of fragmentation will be further investigated experimentally.

From the comparison of IF steel samples processed to a strain of 9.2 by ECAE via the 4 classical routes, a significant influence of the strain path was established. Routes B_A and B_C, corresponding to cross loading, produce a higher hardness and finer structure than routes A and C which correspond to continuous and reverse loading respectively.

5 Acknowledgements

The authors would like to thank the FWO-Vlaanderen for financial support. In addition J.D.M. would like to thank the FWO-Vlaanderen for its Ph.D. Grant.

6 References

[1] Segal, V.M., Materials Science and Engineering A 1995, 197, 157–164
[2] Seefeldt, M.; Van Houtte, P., this conference
[3] Dupuy, L., Ph.D. Thesis, Génie Physique et Mécanique des Matériaux - Institut National Polytechnique de Grenoble, 2000, p. 44 & 85
[4] Schmitt, J.H.; Aernoudt, E.; Baudelet, B., Materials Science and Engineering 1985, 75, 13–20
[5] Gholinia, A.; Bate, P.; Prangnell, P.B., Acta Materialia 2002, 50, 2121–2136

HRTEM Investigations of Amorphous and Nanocrystalline NiTi Alloys Processed by HPT

T. Waitz [1], H. P. Karnthaler [1] and R. Z. Valiev [2]
[1] Institute of Materials Physics, University of Vienna, Vienna, Austria.
[2] Institute of Physics of Advanced Materials, Ufa State Aviation Technical University, Ufa, Russia.

1 Introduction

In the intermetallic compound NiTi, functional thermomechanical properties, such as the shape memory effect and superelasticity are related to a martensitic phase transformation which is of considerable interest from both a scientific and a technological point of view. In addition, in NiTi amorphization can be induced applying various solid state processes such as particle irradiation [1], strong mechanical deformation by cold rolling [2] or mechanical alloying [3]. Severe plastic deformation by high pressure torsion (HPT) methods can be used to obtain almost complete amorphization [4].

It is the purpose of the present paper to apply different levels of strain (by HPT) to investigate the microstructural changes of bulk NiTi during the amorphization with transmission electron microscopy (TEM) methods including those of atomic resolution using high resolution transmission electron microscopy (HRTEM).

2 Experimental Procedure

In the present study, a binary NiTi alloy with a nominal composition of Ni-50 at.% Ti was used. The alloy was quenched from 800°C in water and used to prepare HPT discs at a pressure of 6 GPa applying 6 turns. The as-processed HPT discs had a diameter of 12 mm and a thickness of about 0.2 mm. From the HPT discs specimens with a diameter of 2.3 mm were punched by spark erosion using very low power to avoid any heating. Specimens were taken at the center of the HPT disc and at distances of about 2.5 and 4.3 mm from the center. Since they correspond to the central area, the middle area and the periphery of the HPT sample they were designated C, M and P, respectively. The true strain ε at center of specimens C is estimated to be less than 10; in the case of specimens M and P ε is about 500 and 800, respectively.

The specimens were used to prepare TEM foils (foil normal **FN** parallel to the normal of the HPT disc) by twin-jet polishing in a Tenupol 3. The thinning was done by electropolishing in a solution of 75 % CH_3OH and 25 % HNO_3 (using 15 V at –22 °C) or 10 % $HClO_4$ and 90 % CH_3COOH (using 20 V at room temperature (RT)). In a first step, the microstructure of samples prior to the HPT deformation (prepared at RT) were analysed. The phase structures of specimens C, M and P were analysed by selected area diffraction (SAD) applying different beam directions (**BD**) in a transmission electron microscope (Philips CM 200, operating at 200 kV). A HRTEM analysis was carried out using a Philips CM30 ST (operating at 300 kV) equipped with a Gatan slow scan CCD camera.

3 Experimental Results

Prior to HPT, the TEM analysis shows a phase structure containing mainly cubic B2 austenite at RT. Frequently, B2 has transformed to an incommensurate transition phase since diffuse streaks running parallel to $<\bar{1}10>$ were observed in the diffraction patterns (**BD** ≈ <111>); in addition, minor volume fractions (<10%) contain the trigonal R-phase and the monoclinic B19´ martensite.

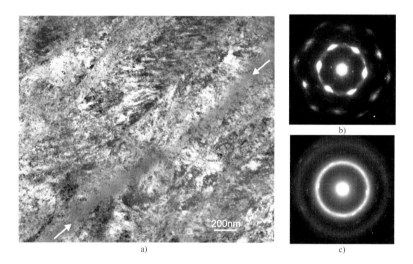

Figure 1: Ni-50 at.% Ti; HPT specimen C. a) TEM bright-field image. A thin amorphous band (marked by arrows) is contained in a matrix of strain induced martensite and residual austenite. The amorphous band has a thickness of 50 to 100 nm and a length exceeding 10 µm. The matrix contains a high density of dislocations. b) SAD pattern of the crystalline matrix; diffuse spots are caused by the lattice strains. c) SAD pattern corresponding to an amorphous band showing the diffuse rings of the amorphous phase. The rings are superimposed by weak diffraction spots of nanocrystals embedded in the amorphous band.

3.1 Specimens C

After HPT, samples C contain mainly martensite and some residual austenite (cf. figures 1a and b). An orientation relationship $[1\,1\,0]_{B19´} = [1\,1\,1]_{B2}$ between martensite and austenite was observed. The elongated grains were analysed using SAD and TEM dark-field methods. The grains have a length and a width in the range of 200 to 1000 and 50 to 200 nm, respectively. Within the grains, strong strain contrast occurs both in TEM bright-field and dark-field images. In addition, severe tilting of the lattice planes is observed in HRTEM images and locally, a high density of terminating lattice planes corresponding to edge dislocations occurs. Based on the HRTEM analysis, a local dislocation density up about $10^{16}\,\mathrm{m}^{-2}$ is estimated. Even highly strained regions seemd to be chemically ordered since the weak $[001]_{B2}$ superlattice reflections were observed. Finally, twins of the B19´ phase were encountered locally.

It is interesting that several bands of diffuse contrast were observed (marked by arrows in figure 1a). The bands have a width of 50 to 100 nm and most of them are longer than 10 µm.

SAD patterns taken at these bands contain diffuse rings corresponding to an amorphous phase (see figure 1c). In HRTEM images, nanocrystals embedded in the amorphous bands were frequently observed (cf. figure 2).

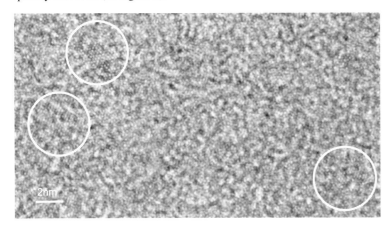

Figure 2: Ni-50 at.% Ti; HPT specimen C. HRTEM image. Amorphous band. The amorphous phase shows a granular structure. The circles mark small nanocrystals (size of about 5 nm); their atomic columns and lattice planes are resolved and correspond to the cubic B2 austenite.

3.2   Specimens M

With increasing strain, the volume fraction of the amorphous phase increases (cf. figure 3). In the specimens M, a heterogeneous nanostructured phase occurs since a high density of nanocrystals having a size between 5 and 50 nm is embedded in the amorphous phase. Most of the larger nanocrystals contain the B19´ phase whereas the smaller ones are B2. As analysed by HRTEM, large lattice strains are observed in the nanocrystals. Within the nanocrystals, thin fringes of diffuse contrast similar to an amorphous phase are frequently observed. The crystalline/amorphous interfaces are strongly curved (cf. figure 4). In addition, there is a gradual transition from the crystalline to the amorphous phase since even at a distance up to about 30 nm from the interface, lattice fringes of weak contrast are observed in the amorphous phase. It should be noted that these weak lattice fringes are running almost parallel to lattice planes of the adjacent nanocrystal.

3.3   Specimens P

In the specimens P, only a few very small nanocrystals (size less than 15 nm) are embedded heterogenously in the amorphous matrix (cf. figure 5). Using HRTEM methods, nanocrystals with a minimum size of 3 nm were encountered.

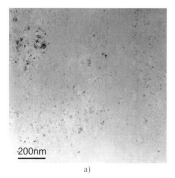

Figure 3: Ni-50 at.% Ti; HPT specimen M. a) Bright field image. Spots of dark contrast are nanocrystals that have a size of 5 to 50 nm and are embedded in an amorphous matrix (uniform contrast). b) In the corresponding dark field image most of the nanocrystals show bright contrast since the reflections $[\bar{1}10]_{B19'}$ and $[111]_{B2}$ were allowed to pass the objective aperture.

4 Discussion

The HPT deformation of Ni-50 at.% Ti causes almost complete amorphization when the strain is about 800. With increasing strain, three consecutive steps of amorphization are observed in the samples C, M and P. Firstly, in the samples C, the plastic straining leads to the transformation B2 to B19´ (cf. figure 1). The orientation relationship of the strain-induced martensite ($[\bar{1}10]_{B19'} \approx [111]_{B2}$) equals that of the thermally induced martensite. Subsequently, B19´ martensite deforms by twinning and dislocation glide. During the course of the deformation, a high density of dislocations is accumulated causing strong crystal refinement and leading to grain sizes of less than about 1 µm. In addition, a small volume fraction is subjected to amorphization already in the first stage of the HPT deformation when $\varepsilon \approx 10$. The present observations of thin amorphous bands lead to the conclusion that the amorphization occurs locally within shear bands, converting them into amorphous bands. Similar amorphous shear bands were observed in the case of cold rolled NiTi [5]. It was suggested that in NiTi, the driving force for strain induced amorphization is dislocation accumulation; a critical dislocation density ρ_c of 10^{18} m^{-2} was proposed where dislocation cores overlapp [2]. In addition, it has been proposed that a plastic shear instability can arise by dislocation accumulation [6]. In the present case, as shown by the HRTEM analysis, severely strained areas contain a dislocation density as much as 10^{16} m^{-2}. Although this is about 2 orders in magnitude lower than ρ_c, local accumulation of dislocations could give rise to a shear instability leading to the nucleation of an amorphous band. Since the amorphous bands contain nanocrystals it is concluded that crystalline debris is left during the amorphization (cf. figure 2).

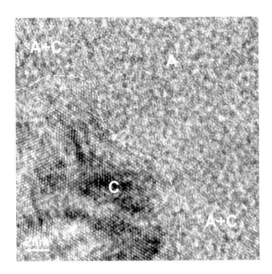

Figure 4: Ni-50 at.% Ti; HPT specimen M. a) HRTEM image. Strong strain contrast and bending of lattice planes occurs in the crystalline phase C. The transition between the crystalline area C (marked by a dotted line) and the amorphous phase A occurs gradually since lattice fringes are still observed up to a distance of about 30nm from the nanocrystal. The areas marked A+C contain both amorphous phase and fragments of crystalline phase since weak lattice fringes are visible embedded in the granular contrast of the amorphous phase.

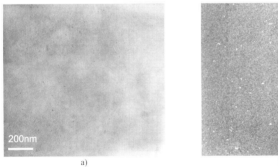 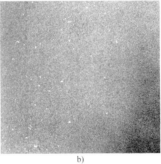

Figure 5: Ni- 50.0at.%Ti; HPT specimen P. a) Bright field image. Spots of dark contrast are nanocrystals having a size of less than 15nm embedded in an amorphous matrix. b) Dark field image of the nanocrystals (bright contrast).

In a second step, with increasing HPT strain, the amorphous volume fraction gradually increases by the accumulation of amorphous bands in the crystalline phase (cf. figure 3). This leads to the fragmentation of the crystalline phase until, at an intermediate strain ($\varepsilon \approx 500$), mainly isolated nanocrystals are left embedded in an amorphous matrix. Since the crystalline volume fraction is small (less than 30 %) and the slip distances of the dislocations are very limited by the diameter of the nanocrystals (sizes less than 50 nm) the plastic deformation must occur predominantly in the amorphous phase for geometrical reasons. In amorphous alloys strain localization is observed and failure occurs by fracture along a single shear band with little global plasticity [7]. However, in the present case failure of the amorphous NiTi is avoided by the superimposed hydrostatic pressure of the HPT deformation.

In a third step, small fragments are detached from severely strained nanocrystals (cf. figure 4). Therefore, the isolated nanocrystals are fragmented and gradually dissolve in the amorphous matrix. Finally, at a high strain ($\varepsilon \approx 800$) only very small fragments are left, embedded heterogeneously in the amorphous matrix (cf. figure 5).

5   Summary

- Ni-50 at.% Ti containing the B2 austenite phase at RT was subjected to HPT deformation. Samples corresponding to three different strains ($\varepsilon \approx 10$, 500 and 800) were investigated in detail using TEM and HRTEM methods.
- The results show that the HPT deformation leads to a nanostructured amorphous phase. With increasing HPT strain, the amorphization proceeds in three consecutive steps.
- In a first step, the HPT deformation leads to a strain-induced transformation B2 to B19´. Dislocation accumulation and deformation twinning induce crystallite refinement. In addition, at a shear strain $\varepsilon \approx 10$, a small volume fraction of amorphous bands is formed by a plastic shear instability caused by the accumulation of dislocations. Locally, dislocation densities up to about $10^{16} m^{-2}$ were measured by HRTEM methods.
- At intermediate strains $\varepsilon \approx 500$, the crystalline phase is fragmented by the accumulation of shear bands. This leads to isolated nanocrystals of a size of ≤ 50nm, which are embedded in an amorphous matrix.
- With increasing HPT strain, the nanocrystals gradually dissolve in the amorphous phase. Finally, at a strain $\varepsilon \approx 800$, only a small volume fraction of nanocrystalline debris survives the HTP deformation.

6   References

[1]   Thomas, G.; Mori, H., Fujita, H., Sinclair, R., Scripta Metall. 1982, 16, 589
[2]   Koike, J.; Parkin D. M., Nastasi, M., J. Mater. Res. 1990, 5, 1414
[3]   Schwarz, R. B.; Koch, C. C., Appl. Phys. Lett. 1986, 49, 146
[4]   Sergueeva, A. V.; Song, C., Valiev, R. Z., Mukherjee, A. K., Mater. Sci. Eng. A 2003, A339, 159
[5]   Nakayama, H.; Tsuchiya, K., Liu, Z. G., Umemoto, M., Morii, K., Shimizu, T., Mater. Trans. 2001, 42, 1987
[6]   Koike, J.; Parkin D. M., Nastasi, M., Phil. Mag. Lett. 1990, 62, 257
[7]   Leamy, H. J., Chen H. S., Wang, T. T., Metall. Trans. 1972, 3, 699

Effect of Grain Size on Microstructure Development during Deformation in Polycrystalline Iron

K.Kawasaki, H. Hidaka, T. Tsuchiyama and S.Takaki
Kyushu University, Fukuoka, Japan

1 Introduction

Grain refinement is an effective way for strengthening structural materials because ductility and toughness of materials are not reduced so much in comparison with the other strengthening ways such as dislocation hardening and precipitation hardening. In particular, it is known that ductile-brittle transition temperature (DBTT) of steel is depressed with decreasing grain size and the DBT behavior is disappeared in the ultra fine-grain size region below 1μm. Thus, fabrication technique, thermodynamics and mechanical properties of ultra fine-grained steels have been focused in recent years. The authors have succeeded to fabricate the bulk iron with ultra fine-grained structure by mechanical milling (MM) treatment using commercial pure iron powders [1]. MM treatment can introduce huge strain energy into iron powders and make the grain size within the powder particles decrease to nano-size [2,3]. The ultra fine-grained iron powder can be easily consolidated to a bulk material through warm-rolling without marked grain growth. The study on the mechanical properties in the ultra fine-grained bulk iron revealed that yield stress at room temperature increases to 1.6 GPa in the specimen with 0.2 μm yielding to the Hall-Petch relationship [1].

However, it was found that the elongation decreases by such a ultra grain refining, especially the uniform elongation almost disappears in the materials with the grain size below 0.5 μm [4]. Although the work hardening behavior should be closely related to microstructure development during deformation, the microstructure of deformed ultra fine-grained materials has not been cleared yet.

In this study, the effect of grain size on the microstructure development during cold-rolling was investigated using a commercial pure iron with conventional grain size and an ultra fine-grained iron fabricated by consolidation of mechanically milled iron powder. The strengthening mechanism was also discussed in the deformed ultra fine-grained iron.

2 Experimental Procedure

Ultra fine-grained iron used in this study is fabricated through mechanical milling (MM) and consolidation process. Commercial reduced iron powder (KIP255M, –100mesh, produced by Kawasaki Steel Co. LTD) was used for the MM treatment. The powder was put into a stainless pot (SUS304, 3 litter volume) with steel balls (SUJ2, 15 mm in diameter) and then sealed under Ar gas atmosphere. The ball/powder weight ratio was 12. Mechanical milling treatment was performed with a high-energy vibration ball mill for 720 ks. The MM iron powder was degassed at 873 K for 1.8 ks in hydrogen atmosphere and then put into a stainless steel tube of 16 mm in outer diameter and 14 mm in inner diameter. The tube was sealed in a vacuum and then hot-rolled to a 1.2 mm thick plate at 923 K or 973 K to consolidate the MM iron powder within the steel tube. The consolidated bulk iron was annealed at 973 K for 3.6 ks to remove dislocation induced during the consolidation. Commercial pure bulk iron produced by ingot metallurgy was also prepared as a referential coarse-grained material. The specimens were cold-rolled at various reductions up to 40 % and then subjected to transmission electron microscopy (TEM), X-ray diffraction (XRD) and tensile testing (initial strain rate of 0.0025 s^{-1}). TEM observation was performed at the acceleration voltage of 200 kV for films thinned by jet-polishing in the electrolyte solution of 90 % acetic and 10 % perchloric acid under the electrolytic current of about 90 mA. XRD was performed with a diffractometer using CoKα radiation ($\lambda = 0.179$ nm). Dislocation density was estimated on basis of the Hall-Williamson plot of the full width at the half maximum height of Kα_1 line.

3 Results and Discussions

3.1 Microstructure Change with Cold Rolling

Figure 1 represents TEM images of the bulk iron consolidated from the MM iron powder. The ferrite grains have an equiaxed shape, and the grain size is about 0.25 μm. In the dark field image (b), very small particles (10~20 nm in diameter) are found in the matrix. These particles were identified as Fe_3O_4 oxide[5] and its volume fraction was theoretically estimated at about 2 vol.% from the oxygen content. Evaluation of grain size based on Zener's grain boundary pinning theory indicated that the size and volume fraction of oxide particles are enough to keep the ferrite grain size fine to be 0.25 μm.

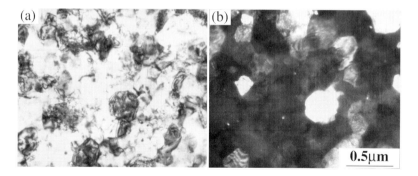

Figure 1: TEM images of ultra fine-grained iron annealed at 973 K for 3.6 ks; bright field(a) and dark field(b). The iron was fabricated from mechanically milled iron powder consolidated at 923 K.

Figure 2 represents TEM images which show substructure development in the coarse-grained iron with mean grain size of 20μm ((a), (b) and (c)) and ultra fine-grained iron with the mean grain size of 0.25 μm ((d), (e) and (f)). The specimens were cold-rolled by 3 % ((a) and (d)), 20 % ((b) and (e)) and 40 % ((c) and (f)). The coarse-grained iron (20 μm) exhibited normal microstructure change by deformation: A small amount of dislocations are scattered within ferrite grains after slight deformation (Figure 2(a)), and then dislocation density is significantly increased with deformation and a typical dislocation cell structure is formed after 20 % deformation (Figure 2(b)). The cell structure is characterized by inhomogeneouslydispersed dislocations, and consists of densely dislocation-concentrated cell walls and sparsely dislocation-scattered cell interior. With increasing reduction, the cell size is gradually decreased and it becomes around 0.5 μm in the 40 % cold-rolled iron (Figure 2(c)). It should be noted here that the cell size is larger than the grain size of the ultra fine-grained iron (0.25 μm). On the other hand, in the ultra fine-grained iron, a large number of dislocations is introduced even in the early stage of deformation and is dispersed homogeneously in the 3 % cold-rolled iron. The typical dislocation cell structure cannot be formed because the grain size is too small to form dislocation cell structure (Figure 2(d)). In the specimen cold-rolled by 40 %, no significant change in the dislocation density and morphology of dislocation substructure was found in comparison with the 20 % deformed specimen (Figure 2(e)).

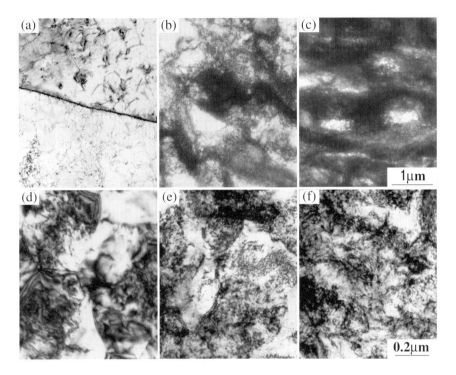

Figure 2: TEM images of bulk iron cold-rolled by about 3 % ((a) and (d)), 20 % ((b) and (e)) and 40 % ((c) and (f)); commercial pure iron(mean grain size; 20 μm((a), (b) and (c))) and ultra fine-grained iron (mean grain size; 0.25 μm((d), (e) and (f))).

3.2 Grain Size Dependence of Increase in Dislocation Density During Deformation

Figure 3 shows change in dislocation densities with cold-working in ultra fine-grained iron (0.25 μm) and coarse-grained iron (20 μm). In the coarse grained-iron, the dislocation density is gradually increased with cold-working as usual manner. On the other hand, dislocation density of ultra fine-grained iron is remarkably increased at the early stage of deformation, and then sa-

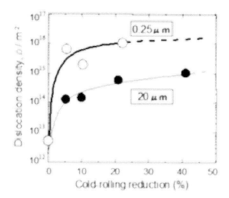

Figure 3: Change in dislocation density with cold-rolling in ultra fine-grained iron (0.25 μm) and coarse-grained iron (20 μm).

turated at about 10^{16} m^{-2}. This results in a significant difference of dislocation substructure development during deformation between the materials with different grain sizes.

The grain size dependence of dislocation multiplication could be explained by the effect of slip distance of dislocations on deformation. In general, when crystalline materials are deformed, the strain ε is expressed by the following equation as a function of dislocation density ρ [6].

$$\varepsilon = \rho b X \qquad (1)$$

where b is Burgers vector, and X is the average distance of dislocation slip. According to the equation 1, dislocation density induced by deformation is markedly enlarged with refining grains. Figure 4 shows changes in dislocation density in the ultra fine-grained and coarse-grained irons as a function of strain. The average distance of dislocation slip; X was assumed to be a half of grain size. It is clearly expressed that the dislocation density is increased with strain much earlier in the ultra fine-grained iron than in the coarse-grained iron. The dislocation densi-

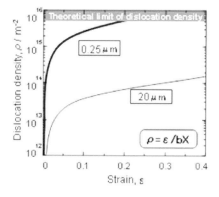

Figure 4: Change in dislocation density as a function of strain in ultra fine-grained iron (0.25 μm) and coarse-grained iron (20 μm). The dislocation densities were calculated by equation 1

ty reaches the theoretical limit ($\sim 10^{16}\,\mathrm{m}^{-2}$) [7] by 20 % deformation in the ultra fine-grained iron (0.25 μm), while is two orders of magnitude smaller ($10^{13}\sim 10^{14}\,\mathrm{m}^{-2}$) in the coarse-grained iron (20μm). These tendencies fairly correspond to results by means of XRD shown as Figure 3.

3.3 Behavior of Work Hardening with Cold-Rolling in Ultra Fine-Grained and Coarse-Grained Irons

Figure 5(a) represents the relation between 0.2 % proof stress and reduction by cold-rolling in irons with grain size of 0.25, 0.35 and 20 μm. The calculated strength in deformed specimen σ_d from Bailey-Hirsch equation with their dislocation densities are also shown in Figure 5(b). The Bailey-Hirsch equation was described as follow

$$\sigma_d = \sigma_i + \alpha G b \sqrt{\rho} \qquad (2)$$

where σ_i is friction stress, α is a constant on the order of 0.6, G is shear module (80 GPa) and b is Burgers vector (0.25 nm). The grain refinement increases strength of iron, and the 0.2 % proof stress of the ultra fine-grained iron (0.25 μm) is five times higher than that of the coarse-grained iron (20μm). However, the work hardening rate tends to decreases with ultra grain refining. The coarse grained iron exhibits normal work hardening behavior, that is, the proof stress dose not increase in the early stage where dislocations cannot be stored, and after straining by several percents, it gradually increases with increment of reduction of cold-rolling. On the other hand, the iron with grain size of 0.35 μm exhibits smaller work hardening in comparison with the coarse-grained iron. Further grain refinement to 0.25 μm results in disappearance of work hardening. The proof stress of the ultra fine-grained iron keeps a constant value even after 40 % deformation, although the deformed ultra fine-grained iron has a quite high dislocation density and fine dislocation substructure. It is well known that shear localization can lead to the lack of strain hardening in nanocrystalline and ultra fine-grained Fe alloys[8]. In terms of present study, dislocation substructure is well-developed with deformation, and its dislocation density remarkably increases, and reaches $10^{16}\,\mathrm{m}^{-2}$ from the early stage of deformation. Thus, shear bands are not too effective for stress relaxation in present study, even if shear bands devel-

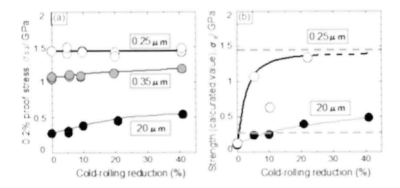

Figure 5: Relationship between stress and cold-rolling reduction in deformed iron; 0.2 % proof stress obtained by tensile testing (a) and strength calculated by the equation 2 (b)

op in the specimen. The diminishing of strain hardening seems to be reasonably explained from the view point of strengthening mechanism in each iron.

In the coarse-grained iron, the strength calculated with dislocation density (Figure 5(b)) exceeds its lower yield stress, and increases in the same manner as the work-hardening behavior from tensile testing (Figure 5(a)). This similarity reveals that the strength in coarse grained iron is subjected to dislocation strengthening. On the other hand, in the iron with grain size of 0.25 µm, the strength estimated from Bailey-Hirsh equation does not reach the same level as the proof stress of ultra fine-grained iron, even though it indicates above 1 GPa from the early stage of deformation. This result demonstrates that the proof stress cannot be estimated by simple addition of dislocation strengthening and grain refining strengthening, and the proof stress in deformed specimen was subjected to mainly grain refining strengthening in the case of ultra fine-grained iron. It is thought that the dislocation density should be increased much higher in order to work-harden the ultra fine-grained iron, but the dislocation density is saturated at the limited value shown in Figures 3 and 4. This means that it is substantially impossible to improve the uniform elongation of ultra fine-grained iron.

4 Conclusions

1. Dislocation density of the ultra fine-grained iron is markedly increased even by a slight deformation owing to the grain size dependence of increase in dislocation density. The introduced dislocations in the ultra fine-grained iron are dispersed homogeneously within grains without forming typical dislocation cell structure.
2. Work hardening rate is decreased with ultra grain refining. The iron with grain size of 0.25 µm exhibits no work hardening and keeps proof stress constant even after 40 % cold rolling.

5 References

[1] Y.Kimura, S.Takaki in Proc. of 1998 Powder Metallurgy World Congress & Exhibition, Granada, Spain, 1998, p.573
[2] J.S.C.Jang, C.C.Koch, Scr.Metar., 1990, 24, p.1599
[3] L.Daroczi, D.L.Beke, G.F.Zhou, H.Bakker, Nanostructured Materials, 1993, 2, p.515
[4] S.Takaki, K.Kawasaki, Y.Kimura, in Ultrafine Grained Materials (Ed.: R.S.Mishra et al.), TMS, New York, 2000, p.247
[5] Y.Kimura, S.Nakamyo, H.Hidaka, H.Goto, S.Takaki, in Ultrafine Grained Materials, (Ed.: R.S.Mishra et al.), TMS, New York, 2000, p.277
[6] J.C.M.Li, Y.T.Chou, Metal. Trans., 1970, 1, p.1145
[7] S.Takaki, NMS-ISIJ, 2002,177-178, p.131
[8] see, e.g.; Q. Wei, D.Jia, K.T.Ramesh and E.Ma, Appl. Phys. Lett., 2002, 81, p.1240

Microstructure and Phase Transformations of HPT NiTi

T. Waitz [1], V. Kazykhanov [2] and H. P. Karnthaler [1]
[1] Institute of Materials Physics, University of Vienna, Wien, Austria.
[2] Institute of Physics of Advanced Materials, Ufa State Aviation Technical University, Ufa, Russia.

1 Introduction

The intermetallic shape memory alloy NiTi is of considerable interest from both a scientific and technological point of view. The shape memory properties are caused by a martensitic phase transformation. The transformation occurs from the cubic B2 austenite to the monoclinic B19′ martensite; depending on the composition of the alloy, the trigonal R-phase can occur as a transition structure. Nanocrystalline NiTi alloys can be processed by mechanical amorphization followed by a suitable heat treatment [1]. In the present investigation, Ni-50.3 at.% Ti, completely transformed to martensite in a single step, was subjected to severe plastic deformation by high pressure torsion (HPT) followed by annealing, yielding a nanocrystalline structure. It is the aim of the present paper to investigate the impact of the grain size on the martensitic phase transformation.

2 Experimental Procedure

A binary NiTi alloy with a nominal composition of Ni-50.3 at.% Ti was used. As received, the coarse grained alloy shows a single-step transformation from B2 austenite to monoclinic B19′ martensite; at room temperature (RT) the alloy contains martensite (the martensite finish temperature M_f is about 45 °C). The alloy was annealed at 800 °C, quenched in water, and disc shaped specimens (diameter ~10 mm, 0.45 mm initial thickness) were used for HPT at RT applying 10 turns at a pressure of about 6 GPa. Small discs (diameter ~ 2.3 mm) were punched out at a distance of 2.5 and 4.3mm from the centre of the HPT specimens corresponding to a true logarithmic strain of about 6.7 and 7.3 (labeled M and P, respectively). Afterwards the discs were annealed in vacuum at 300, 340 and 450 °C for 1 to 5 h followed by cooling to RT. Finally, to obtain the martensitic transformation of the HPT specimens, they were cooled to –25 °C or quenched in liquid nitrogen.

The discs were used to prepare transmission electron microscopy (TEM) foils (foil normal **FN** parallel to the normal of the HPT disc) by twin-jet polishing in a Tenupol 3. The thinning was done with a solution of 75 % CH_3OH and 25 % HNO_3 (–22 °C and 15 V) or 10 % $HClO_4$ and 90 % CH_3COOH (20 V) which was used at RT. The phase structures were analyzed by selected area (SA) diffraction applying different beam directions (**BD**) in a Philips CM 200 operating at 200 kV. Grain sizes were measured by TEM dark-field methods. The impact of the grain size on the occurrence of the martensitic phase transformation was analysed in detail using SA diffraction and high-resolution TEM (HRTEM) methods using a Philips CM30 ST (operating at 250 kV and 300 kV) equipped with a Gatan slow scan CCD camera.

3 Experimental Results

Prior to the HPT deformation, the alloy is coarse-grained and contains B19′ martensite at RT. The transformation shape strains are accommodated by groups of twinned martensite habit plane variants. Mainly <110> type II and $(11\bar{1})$ type I twinning is observed. For the <110> type II twins, the width was measured and was found to be in the range of about 30 to 80 nm. After HPT, a heterogeneous nanostructured amorphous phase is formed. Both the size and the density of the nanocrystalline debris retained in an amorphous matrix are smaller in specimens P than in specimens M.

Figure 1 shows the nanocrystalline structures obtained after isothermal annealing followed by cooling to RT and quenching to –25 °C. Figures 1a and b are TEM bright-field images of specimens M and P, respectively, annealed at 300 °C for 5 h. Dark-field images were used to measure the distribution of the grain sizes (shown in figure 2). In the case of specimens P, a high density of tiny nanocrystallites is observed, and most of them have a size of less than 20 nm (cf. figures 1a and 2a). However, a rather high volume fraction of amorphous phase is still retained; in the diffraction pattern the diffuse rings corresponding to the amorphous phase are superimposed by rather sharp rings due to the nanocrystals (cf. figure 3a). Contrary to this, in the case of specimen M, crystallization is complete; all grains have a size of less than 40 nm

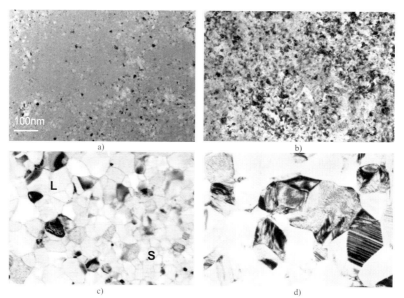

Figure 1: Ni-50.3 at.% Ti; HPT. TEM bright-field and dark-field images of specimens M and P after different heat treatments followed by cooling to RT and quenching to –25 °C. a) Specimen P after annealing at 300 °C for 5 h; TEM bright-field image. Tiny nanocrystals containing B2 austenite are embedded in an amorphous phase. b) Specimen M after annealing at 300 °C for 5 h; TEM bright-field image. The crystallization is complete. The grains contain both B2 and R-phase. c) Specimen P after annealing at 340 °C for 5 h; TEM dark-field image. A heterogeneous distribution of grain sizes is observed. Near S B2 and R-phase are observed, whereas near L mainly R-phase occurs. In addition some grains larger than about 60 nm contain B19′ martensite. d) Specimen P after annealing at 450 °C for 1 h; TEM dark field image. Most of the grains are larger than about 100 nm and contain martensite. In addition, R-phase is observed in grains up to a diameter of about 150 nm.

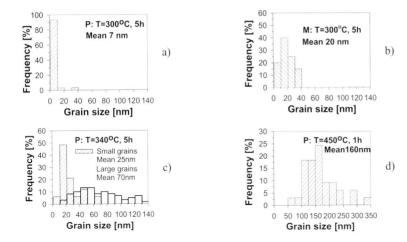

Figure 2: Ni-50.3 at.% Ti. Histograms of the sizes of nanograins arising after different heat treatments a) Specimen P after annealing at 300 °C for 5 h; most of the nanocrystals are smaller than 20 nm. b) Specimen M after annealing at 300 °C for 5 h; the nanograins have a size of less than 40 nm. c) Specimen P after heating at 340 °C for 5 h; the grain size distribution of areas that contain mainly smaller and larger grains are shown by dashed and white bars, respectively. All grains are smaller than 40nm. d) Specimen P after annealing at 450 °C for 1 h; most of the grains are larger than about 100 nm; the maximum grain size is ~350 nm.

(cf. figures 1b and 2b). In the TEM dark-field image of figure 1c, the grain structure of a specimen P after isothermal annealing at a temperature of 340 °C for 5 h is shown. The grain size is in the range of 10 to 140 nm; a heterogeneous distribution of the grains is observed since frequently areas containing mainly small grains (mean size of 25 nm, near S) are adjacent to areas containing mainly larger grains (mean size of 70 nm, near L). For these two areas, the histograms of the grain size are shown in figure 2c. Sharp and rather flat grain boundaries were observed both by TEM bright-field and HRTEM images. Within the grains, almost no dislocations were observed, and most of the grains show only weak strain contrast. Finally, in specimens annealed at 450 °C for 1 h the grains are mostly larger than about 100 nm and no grains smaller than about 50 nm were encountered; the maximum grain size is about 350 nm (cf. figures 1d and 2d).

The phase structure in grains of different diameters was analyzed in detail using both SA diffraction and HRTEM methods. In the case of P specimens annealed at 300 °C for 5 h the nanocrystals contained in an amorphous matrix give rise to reflections of B2 austenite only; it should be noted that even in the case of the very small crystallites (size less than 20 nm) weak <100> superlattice reflections were observed (cf. figure 3a). In the case of M specimens annealed at the same temperature, reflections of both B2 and the R-phase were encountered. In this case grains smaller than 15 nm contain B2 phase only (cf. figure 3b). In the case of P specimens annealed at 340 °C for 5 h, reflections of B2, R-phase and B19´ were observed (cf. figure 3c). Grains smaller than 60 nm contain B2 and R-phase but do not contain martensite even after the specimens had been quenched in liquid nitrogen. When the grain size is in the range between 60 and 140 nm both R-phase and martensite are observed by SA diffraction and HRTEM methods. In this case B2 is hardly found. Finally, in the specimens P annealed at 450 °C for 1 h, the grains

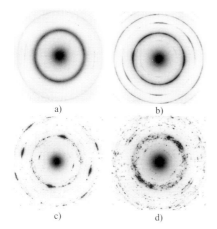

Figure 3: Ni-50.3 at.% Ti; HPT. SA diffraction patterns of specimens M and P after different heat treatments followed by cooling to RT and quenching to −25 °C. a) Specimen P after annealing at 300 °C for 5 h. Sharp diffraction rings of the nanocrystals a corresponding to the B2 phase are superimposed by rather diffuse rings of the amorphous phase. b) Specimen M after annealing at 300 °C for 5 h leading to complete crystallization. The nanograins contain both B2 and R-phase. c) Specimen P after annealing at 340 °C for 5 h. Reflections of all three phases B2, R and B19′ martensite are encountered. d) Specimen P after annealing at 450 °C for 1 h. Reflections of B19′ are dominating; in addition, reflections arising from the R-phase are observed.

contain mainly martensite (cf. figure 3d). Frequently, in grains smaller than about 150 nm R-phase is observed.

In the small grains, very fine (001) compound twins B19′ were observed by HRTEM methods (mean width of about 2 nm). In addition, as shown in figure 4, twins of the R-phase were frequently encountered that have a width of about 20 nm.

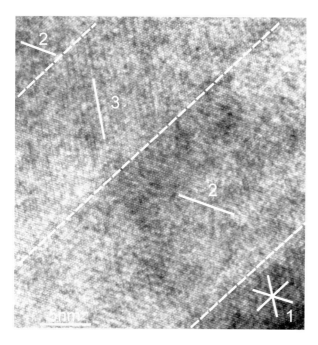

Figure 4: Ni-50.3 at.% Ti; HPT. HRTEM image of twins of the R-phase contained in a grain of about 120nm diameter. Three out of four possible compound twins are marked by 1, 2 and 3. The planar twin boundaries are indicated by dashed lines. The width of the twins is about 20 nm only.

4 Discussion

In the coarse-grained Ni-50.3 at.% Ti the occurrence of <110> type II and $(11\bar{1})$ type I twinning is in agreement with the formation of invariant habit planes during the B2 to B19´ transformation; groups of different habit-plane variants lead to an almost complete accommodation of the transformation shape strains [2]. After the HPT deformation, a nanostructured amorphous phase arises. Nanocrystalline debris embedded heterogenously in an amorphous phase survives the HPT deformation, and both density and size of the nanocrystals decrease with increasing HPT strain. These crystallites trigger the nanocrystallization of the B2 phase during isothermal annealing. Therefore, based on the experimental results, it is concluded that the grain size depends both on the local density of the nuclei and on the annealing temperature (cf. figure 1). When the crystallization is complete, a nanocrystalline structure containing weak lattice strains and almost no dislocations is found.

With decreasing grain size, the martensitic phase transformation is gradually suppressed (cf. figures 2 and 3). Since the nanograins contain only small lattice strains and almost no dislocations, it is concluded that the grain boundaries act as obstacles and suppress the transformation. Caused by the constraints of the grain boundaries, the autocatalytic formation of self-accommodating plate groups is hindered. Therefore, in the nanograins a different path of the transformation occurs. Since the transformation strains are reduced by decreasing the width of the twins, very fine (001) compound twins are formed [4, 5]; however, a decrease of the twin width leads to an increase of the twin boundary area. From the experimental results, it is concluded that with decreasing grain size both strain energy and twin boundary energy contribute to an increase of the energy barrier to be overcome upon the transformation. Therefore, at a grain size below 60 nm no martensitic transformation can occur at all.

It is concluded that the grain size should have much less impact on the B2 to R-phase transformation since the transformation strains of the R-phase are considerably smaller (about 1 %) than for the B2 to B19´ transformation (about 10 %). Therefore, in grains smaller than about 150 nm the energy barrier suppresses the martensitic start temperature (M_s) below the start temperature of the R-phase (T_R; $T_R < M_s$), in agreement with the present investigations (cf. figures 3 and 4). However, as the very small grains (<15 nm) have a size comparable to the critical diameter of nuclei of the R-phase [3], the transformation is completely suppressed. In larger grains, the shape strains can be accommodated by small twins to decrease the transformation strain energy (cf. figure 4). To reduce the strains in the nanograins, the width of the twins (20–50 nm) is significantly smaller than that of about 300nm measured in coarse grained R phase [6].

5 Summary

- Ni-50.3 at.% Ti containing B19´ martensite at RT was subjected to HPT deformation followed by isothermal annealing between 300 and 450 °C and cooling to RT. Afterwards, the samples were quenched to –25 °C or in liquid nitrogen to induce the martensitic phase transformation in the nanograined HPT specimens.
- The HPT deformation leads to a nanostructured amorphous phase. Nanocrystalline debris embedded in an amorphous phase survives the HPT deformation. The size and the density of the nanocrystals decreases with increasing HPT strain.

- During isothermal annealing, nanocrystallization occurs. The grain size depends both on the HPT strain and on the annealing temperature.
- Caused by constraints of the grain boundaries, self accommodation of the B2 to B19´ transformation is hindered; both M_s and the transformed volume fraction decrease with decreasing grain size. Within the nanograins, a different path of the transformation is favoured, leading to the formation of very thin (001) compound twins. Caused by an increasing energy barrier, the transformation to B19´ is completely suppressed in grains smaller than 60 nm.
- The grain size has less impact on the B2 to R-phase transformation since the transformation strains are considerably smaller. Therefore, in grains smaller than about 150 nm M_s is suppressed below T_R, and a two-step transformation from B2 via the R-phase to B19´ arises.
- When the grain size is of the same order as the critical radius of the nuclei of the R-phase, no transformation occurs at all.

6 References

[1] Sergueeva, A. V.; Song, C., Valiev, R. Z., Mukherjee, A. K., Mater. Sci. Eng. A 2003, A339, 159
[2] Madangopal, K.; Acta Mater. 1997, 45, 1997
[3] Murakami, Y.; Shindo, D.; Phil. Mag. Lett. 2001, 81, 631
[4] Neckel, W.; Waitz, T., Karnthaler, H. P., Proc. 6th Multinational Congress on Micr., Pula, Croatia, 193 (2003)
[5] Waitz, T.; Kazykhanov, V., Karnthaler, H. P., Acta Mater., 2004, 52, 137
[6] Miyazaki, S.; Wayman, C. M., Acta Metall. 1988, 36, 181

Types of Grains and Boundaries, Joint Disclinations and Dislocation Structures of SPD-produced UFG Materials

N.A. Koneva, A.N. Zhdanov*, N.A. Popova, L.N. Ignatenko, E.E. Pekarskaya, E.V. Kozlov
Tomsk State University of Architecture and Building, Tomsk, Russia
*Altai State Technical University, Barnaul, Russia

1   Introduction

Undoubtedly, the use of UFG materials are recognized to be promising. Their structures have been intensively investigated. Publications testify to this fact (see, for example, the proceedings of the conferences [1,2], the review [3], and the works [4,5]).

For some years the present authors have paid special attention to detailed structure of UFG materials, produced by severe plastic deformation (SPD). The grains are inhomogeneous as to structure and sizes. Three types of grains are observed: (1) almost dislocation free ones, (2) grains with chaotic or network dislocation structure, (3) grains with cell and fragment substructure [6,7]. The first two types of grains have been reported in [8]. The boundary structure of grains in UFG materials appeared to be sufficiently complex, too. In this connection it should be noted that in the composite model of strengthening of UFG materials introduced in [9] the structure and boundary resistance of each subgrain type affects significantly the yield stress.

In the following description of dislocation structure and characteristics of triple joints in UFG materials much attention should be paid to a quantitative analysis of the internal stress fields. Already in our previous works [9–11] we showed that the yield stress of bulk UFG materials depends on the amplitude of internal stress fields, their homogeneity and the size distributions of different grain types [9].

In the present paper the influence of the internal structure and the size of grains on the intra grain dislocation density is investigated. A study of the inhomogeneity of UFG material is also carried out. Attention is paid to the distribution functions of the sizes of dislocation free subgrains and their limiting sizes, since interactions of dislocations with boundaries forces the dislocations to move out from fine grains to boundaries. The grain boundary density of different types and their angles of disorientation were investigated, too.

2   Materials and Investigation Procedure

The results given in this paper were obtained for SPD produced UFG nickel, copper and the Cu-Al-O alloy. The UFG materials were made by different modes of SPD: (1) equal channel angular pressing (ECAP), (2) high temperature extension, (3) high pressure torsion (HPT). Sample annealing treatment was carried out after SPD in the case of the Ni (125 °C, 3 h). The electron microscopy (TEM) and X-ray analysis were the main methods of investigation. The transmission electron microscopes of TESLA BS-540 and EM-125K type supplied with goniometer at voltage of 120 and 125 kV correspondingly as well as automatic X-ray diffractometer DRON-3M were used. At TEM investigations the magnification was 50 000. To identify grains and their boundaries several images of one and the same part of a foil at different orientations as related to incident electron beam were investigated. Using these data and measurement results of the microdiffraction patterns the average disorientation parameters and their distributions were determined. The foils were prepared from massive samples by cutting them using the spark method followed by thinning in cold electrolytes combining chemical and electrolytical methods. Quality of preparing the foils allowed studying their large areas. Quantitative measurements of the structure parameters were carried out by the secant method followed by statistical processing of the results. The internal stresses (τ_i) and bending-torsion ($æ$) of the crystal lattice were measured by three methods: (1) the average values of τ_i were obtained by X-ray method; (2) the local values of internal stresses were determined from TEM data by the curvature of dislocation lines; (3) bending-torsion of the crystal lattice was measured using the parameters of curved extinction contours observed on the electron microscopy images. The methods have been described in more detail by the authors in [11].

3 Results and Discussion

3.1 Grains and their Boundaries

Fig.1 shows typical TEM images of grains and their boundaries. One and the same group of grains is given in Fig.1 a, b. The images were obtained at different diffraction conditions (the rotation angle of the goniometer axis in Fig.1a is different from that in Fig.1b by 6°). In these figures, 10 grains were signified for better examination. One can identify different types of grains and their boundaries. Fig.1c shows a sample area with significant density of the stripped dislo-

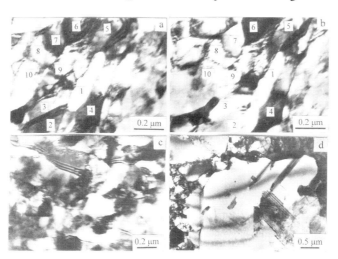

Figure 1: TEM micrographs of grains of different types observed in the investigated UFG Ni (a,b), UFG Cu (c) and UFG Cu-Al-O alloy: (a,b,c) structure of ordinary subgrains, (d) recrystallized grain. Fig. 1a,b shows one and same part of a foil at different inclinations of goniometer. One can see different types of grains numbered of figures: (1,2,4) grains with cells and fragment, (3,5,7,10) grains with chaotic or network dislocation structure, (8,9) dislocation free grains.

cation boundaries. Because of the local curvature of boundaries one cannot find lattice dislocations but the grain boundary dislocations are undoubtedly observed here. The investigation showed that the classification of grains in three of typically observed grain types is not sufficient. In the general case one can observe a fourth grain type- the nucleuses of recrystallization or recrystallized grains. They are found locally and in small numbers. They are characterized by significant average sizes and low dislocation density (a little higher than in almost dislocation free grains). The presence of annealing twins is another significant distinction of them. Fig.1d shows a typical image of such grain.

3.2 The Dependence of Dislocation Density on the Grain Size, and the Problem of Fine Dislocation Free Grains

Fig.2 shows the dependence of the average scalar dislocation density (ρ) as being typical of the interior of a present grain, on the subgrain size (d). The data are given for Cu, Ni and Cu-Al-O alloy. With the increase of ρ and d values, the transition takes from small almost dislocation free grains to such with chaotic substructure and further to the subgrains with cell substructure. It should be pointed out that for all three materials the UFG state was reached by different methods. Nevertheless in the range of $d = 100...500$ nm the average dislocation density decreases practically linearly with decreasing subgrain size in all materials. This fact is first of all caused by intensive interaction of dislocations with the grain boundaries and their incorporation at the grain boundaries. From the dependence $\rho(d)$ one can also see that the dislocation density ρ rushes to zero near by the grain size $d = 100$nm. Such results were already predicted theoretically

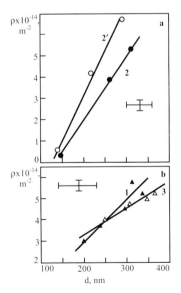

Figure 2: The dependence of scalar dislocation density (ρ) on the subgrain size (d) in UFG Cu (1), Ni (2,2') and copper alloy (3). In the case of Ni the straight line (2) corresponds to as-received state (ECAP); the straight line (2') corresponds to a state after post-deformation annealing.

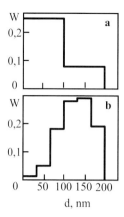

Figure 3: The size distribution of dislocation free subgrains: (a) UFG Cu, (b) UFG Cu-Al-O alloy

[12]. For the first the same data were received in our work for UFG Cu [11]. Here a similar dependence was received for a number of materials.

Similar tendencies seen be from partial histograms of the size distribution of dislocation free subgrains. As one can see in Fig.3 the interval of dislocation free subgrain sizes stretches from the smallest sizes up to $d = 200$nm independently of the material type. This indicates the universal nature of the phenomenon. For copper alloy the maximum of distribution function is shifted towards the larger grain size. Solid solution hardening in the alloy hinders forcing out dislocations from an interior of grains to grain boundaries. This effect is absent in pure copper.

Dislocation free subgrains play a special role for the mechanical properties of UFG polycrystals [9]. Because of their small size such subgrains contribute strongly to strength. A proportion of the dislocation free grains as compared with the total grain number vary from 10 to 40 %. The volume fraction of these grains, however, does not exceed the value 0.05–0.20, because of their small grain size.

3.3 Types of Subboundaries and the Disorientations

In our work [7] a combined classification of the subboundaries observed in UFG materials was suggested. Essentially, the classification unifies basically different approaches by R. Z. Valiev [3] and N. Hansen [13]. While examining a large number of TEM images of UFG Ni and Cu structures the following subboundaries types were identified out: (1) dense dislocation walls (DDW), (2) ordinary dislocation walls (ODW), (3) subboundaries with stripped contrast containing dislocations (SDSB) and dislocation free (SSB) ones. A partially similar classification was suggested in [14]. The fractions of subboundaries of different type for as-received and annealed states were measured for Ni prepared by ECAP (Table 1). One can see that in this material the SSB account for almost a half of the boundaries whereas the DDW account for 0.2–0.3. The total number of close type boundaries, i.e. the ODW and the SDSB, is closely to 0.3. The data in Table 1 clearly show that UFG metal prepared by SPD contains different types of subboundaries in comparable quantities. The annealing treatment poorly affects the fraction of subboundaries of different types.

Table 1: Fractions of the different boundary types and related disorientations in UFG nickel

Processing state	Fraction of different boundary types			Average azimuth disorientation (deg)
	SSB	SDSB+ODW	DDW	
ECAP, as-received state	0.42	0.35	0.23	6 ± 2
ECAP + annealing (125°, 3 h)	0.43	0.29	0.29	5 ± 3

Measurements of disorientations showed that after one pass of ECAP the average azimuth disorientation across a subboundary in Ni is equal to 6°. This is significantly larger than it was observed in Al [15]. Evidently this difference is explained by the larger subgrain size measured in Al [15]. Annealing of Ni after ECAP did not basically change the average disorientation across the subboundary.

3.4 Internal Stresses and Joint Disclinations

UFG materials have an non-equilibrium structure and significantly high internal energy. This energy is mainly contained in the internal stress fields. Non-equilibrium boundaries of grains, subgrains and cells, joint disclinations, second phase particles and dislocations supply this energy [11,16–18]. In the present paper the amplitudes of internal stresses were measured by three methods (see above). In the following we present the data for UFG nickel prepared by ECAP. The average value of stresses measured by X-ray technique in as-received sample is equal to 230 MPa (Table 2). After annealing this value decreases down to 190 MPa. The TEM method, used for measuring the curvature radius of the dislocation lines, allows to receive the stress values in local parts of a sample. These data are also given in Table 2. They confirm to the fact that the non-equilibrium boundaries of all types appear to be the main sources of the internal stress field.

Table 2: Internal stresses in UFG nickel

Sites of measurements	τ_i, MPa (± 50 MPa)	
	ECAP, as-received	ECAP + annealing (125°, 3 h)
Average volume value	232	192
Boundaries of grains and subgrains	255	240
Subgrain boundaries	260	250
Cell boundaries	350	410
Stripped dislocation free boundaries	540	240
Stripped dislocation boundaries	130	210
Dense dislocation walls	230	240
Inside grains of any type	170	120
Inside grains with network DSS	160	100
Inside dislocation free grains	130	130

Inside the subgrains the stresses turned out to be lower. Cell boundaries are more stressed than the grain and subgrain boundaries. Among the latter large contributions to the internal stresses come from the SSBs and DDWs. Considering the three subgrain types the internal stresses are the largest in the grains with cell substructure. The joint disclinations in triple lines were determined from the bending extinction contours seen in the TEM images. In deformed sample disclinations are found in 20 % of the triple lines. After annealing treatment the fraction of such triple lines is decreased down to 15 %. Frank (or rotation) vector ω of disclinations is considerably large. This can be seen from the average values of the local bending-torsion χ, reaching the values of $1 \cdot 10^6 \, \text{m}^{-1}$. Annealing of UFG nickel does not considerably change the average value of χ.

4 References

[1]   Investigations and Applications of Severe Plastic Deformation (Eds.: T.C. Lowe, R.Z. Valiev), NATO Science Series, 3. High Technology, Kluwer Academic Publishers, The Netherlands, 2000, Vol. 80, p. 394
[2]   Ultrafine Grained Materials 2 (Eds.: Y.T. Zhu, T.G. Langdon, R.S. Mishra, S.L. Semiatin, M.J.Saran, T.C. Lowe), TMS, Warrendale, USA, 2002, p. 685
[3]   R.S. Valiev, R.K. Islamgaliev, I.V. Alexandrov, Progr. Mater. Sci., 2000, 45, 103–189
[4]   B. Mingler, H.P. Karnthaler, M. Zehetbauer, R.Z. Valiev, Mater. Sci. Eng., 2001, A 319-321, 242–245
[5]   A.Yamashita, D. Yamaguchi, Z. Horita, T.G. Langdon, Mater. Sci. Eng.,2000, A 285, 100–106
[6]   N.A. Koneva, N.A. Popova, L.N. Idnatenko et al., Investigations and Applications of Severe Plastic Deformations (Eds.: T.C. Lowe, R.Z. Valiev), NATO Science Series, 3. High Technology, Kluwer Academic Publishers, The Netherlands, 2000, 121–126
[7]   N.A. Koneva, A.N. Zhdanov, L.N. Ignatenko et al., Ultrafine Grained Materials 2 (Eds.: Y.T. Zhu, T.G. Langdon, R.S. Mishra, S.L. Semiatin, M.J. Saran, T.C. Lowe), TMS, Warrendale, USA, 2002, 505–514
[8]   A.Vinogradov, S. Hashimoto, V. Patlan, Kitagawa, Mater. Sci. Eng., 2001, A 319-321, 862–866
[9]   E.V. Kozlov, A.N. Zhdanov, L.N. Ignatenko et al., Ultrafine Grained Materials 2 (Eds.: Y.T. Zhu, T.G. Langdon, R.S.Mishra, S.L. Semiatin, M.J. Saran, T.C. Lowe), TMS, Warrendale, USA, 2002, 419–428
[10] N.A.Koneva, E.V.Kozlov, N.A.Popova, et al., Structure, phase transformations and properties of nanocrystalline alloys (Eds.: G.G. Talyz, N.I.Noskova), RAN YO IFM, Ekaterinburg, RF, 1997, 125–140
[11] E.V. Kozlov, N.A. Popova, U.F. Ivanov et al., Ann. Chim. Fr., 1996, 21, 427–442
[12] V.I. Vladimirov, A.E. Romanov, Disclinations in Crystals, Leningrad, Nauka, RF, 1986, p. 223 (in Russian).
[13] N.Hansen, D.A.Hughes, Phys. Stat. Sol.,1995, 149, 155–172
[14] P.L.Sun, P.W.Kao, C.P.Chang, Mat. Sci. Eng., 2000, A 283, 82–85
[15] A.Chakkingal, A.B. Suriadi, P.F. Tomson, Scr. Mat., 1998, 39, 677–684
[16] A.A.Nazarov, A.E.Romanov, R.Z.Valiev, Nanostructured Materials, 1994, 4, 93–101
[17] A.A.Nazarov, A.E.Romanov, R.Z.Valiev, Scr. Mater. 1996, 34, 729–734
[18] M.Zehetbauer, T.Ungar, R.Kral et al., Acta. Mater., 1999, 47, 1053–1061

Microstructure Development of Copper Single Crystal Deformed by Equal Channel Angular Pressing

Tetsuo Koyama[1], Hiroyuki Miyamoto[1], Takuro Mimaki[1], Alexei Vinogradov[2] and Satoshi Hashimoto[2]
[1] Doshisha University, Kyotanabe, Japan
[2] Osaka City University, Osaka, Japan

1   Introduction

The equal-channel angular (ECA) pressing is most promising technique to produce ultra-fine grained (UFG) materials. Among the parameters of the ECA pressing, processing route, i.e., rotation of a billet around its axis between each pass, has been demonstrated to affect the microstructure development and grain refinement appreciably [1–4]. By the processing route, macroscopic shear plane can be controlled with respect to the texture and microstructure. Thus, it can be assumed that it is not accumulative strain (dislocation density) but an interaction between the shear plane and microstructures that governs the grain refinement [2]. In general sense, however, it is pointed out that during plastic deformation, a high angle grain boundary (HAGB) can be generated by the extension of a pre-existing boundary and/or generation by grain subdividing [5,6]. The former process is strain dependent and therefore high accumulative strain could result in grain refinement whereas the second is a consequence of the crystallographic nature of plastic deformation. ECA pressing can take advantage of the second process. In this context, examining the microstructural development in a single crystalline billet during ECA pressing provides us a deeper insight of (1) the role of pre-exist grain boundary on the grain refinement process by comparing with that of polycrystalline counterpart, (2) grain refinement mechanism from crystallographic viewpoint. In the previous study [7], we found that after eight passes via route C, the microstructure of single crystalline billet was still banded structure whereas that of a polycrystal was equiaxed UFG structure. In the current study, the effect of initial orientation on the grain refinement was examined. The role of interaction between macroscopic shear plane and shear bands was emphasized.

2   Experimental Procedures

The billets of copper single crystals (20 mm in diameter and 100 mm in length) of commercial purity (99.95 wt%) were grown by the Bridgeman method. The initial crystallographic orientation of the seven single crystalline billets were described in terms of intrusion direction (ID) and extrusion direction (ED) as shown in Table 1. The seven billets are categorized into three groups according to the previous study on microstructural characteristics after the first pass [7], that is, groups I (billets 1,2 and 3), II (billets 4 and 5) and III (billets 6 and 7) respectively. The microstructural characteristics of the three groups are briefly described in the next section. The geometry of the channel in ECA-die and coordinate system are schematically represented in Fig. 1, where $\theta = 90$ degrees, $\eta = 15$ degrees, and $\varphi = 20$ mm and thus, shear strain subjected in one pass was estimated to 1.08. The billets were then pressed for four passes via the so-called route B and C where the billets were rotated by 90 and 180 degrees around the axial direction after each pass, respectively [8]. In the route C, a billet was subjected to a shear strain in the same plane but in the opposite direction in the successive passes whereas the shear plane is crossed in an alternative pass in the route B [8]. The orientation and texture after one pass were analyzed by electron back-scattering diffraction (EBSD). The microstructure after each pass was observed by optical microscopy, scanning electron microscopy (SEM) and transmission electron microscopy (TEM). In the present study, the microstructure close to the die wall, which was generally different from the interior part, was neglected since the stress state in this region is rather complex because of friction effects.

Table 1: Initial orientation of single crystal billets

Billet number	1	2	3	4	5	6	7
I D	<001>	<111>	<111>	<001>	<011>	<011>	<011>
Group	I	I	I	II	II	III	III

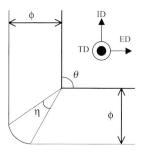

Figure 1: Schematic diagram showing the geometry of ECA pressing

3 Experimental Results and Discussion

3.1 Macroscopic Appearance After one Pass

Figure 2 shows the macroscopic appearances of cross-sections of a selected billet in each group. The surfaces for observation were mechanically polished and subsequently etched by nitric acid. The chemical etching revealed a macroscopically heterogeneous feature with banded structures in the group I as shown in figures 2(a, d). The plane of the bands was approximately parallel to the macroscopic shear plane, and did not correspond to a low index crystallographic plane. As described later, a detail analysis by EBSD revealed that these bands consist of fine bands of 4 to 10 µm in width with an orientation different from the matrix. These microscopic features of the bands are very similar to those presented by Wagner et al. [9] and Morii et al. [10], and are therefore considered to be shear bands.

In the group II, different type of bands was observed in both the vertical and horizontal directions on the entire cross-section (figure 2(b)). The macroscopic features of the bands are very similar to deformation bands observed in cold-drawn copper single crystals [11]. In contrast to the groups I and II, the surfaces of group III specimens are macroscopically smooth and featureless (figure 2(c,f)). Neither deformation bands nor shear bands were recognized by macroscopic observation, suggesting that the whole billet, except for the periphery close to die wall, was deformed by homogenous crystal slip.

The orientation image map (OIM) and {111} pole figures of the corresponding areas are also shown in figure 3. Both the pole figure and OIM clearly show the orientation splitting and generation of HAGB as a consequence of shear banding in a group I crystal. The end orientation of the matrix and the shear band are close to A {111}<112> and C {100}<011>, respectively, where {hkl} is parallel to the macroscopic shear plane, and <uvw> the shear direction. In this case, the shear direction is tilted with 45 degrees from both ID and ED directions and the shear plane normal is perpendicular to the shear direction. Both orientations have a common <110> direction parallel to TD and local crystal rotation around TD may lead to C {100}<011> in the shear bands as suggested by Hirsch et al. [12].

In the group II, the end orientation is close to the cubic one {100}<001>, but split symmetrically in terms of axial direction. In contrast to groups I and II, the billets in the group III retained their single crystalline structure with small orientation spreading and the end orientation showed B{112}<110>. These end orientations of A{111}<112>, C{001}<110> and B{112}<110>, except for the near cubic orientation in the group II have been recognized as the main components of shear texture developed during ECAP [13,14] and torsion straining [15]. The shear texture consists of two fibers, i.e., the A-fiber with a crystallographic slip plane parallel to the macroscopic shear plane and the B-fiber with a crystallographic slip direction parallel to the shear direction [13]. The A{111}<112> is the main component of the A-fiber, and B{112}<110> and C {001}<110> are the main components of the B-fiber.

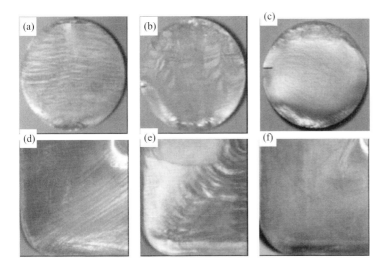

Figure 2: Macroscopic appearances after one pass, (a)-(c) group I–III, respectively from ED, (d)-(f) groups I–III from TD, respectively

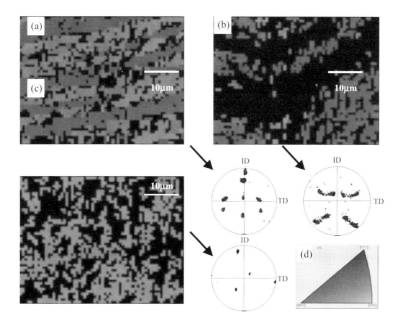

Figure 3: OIM and {111} pole figures of (a) group I, (b) group II, and (c) group III after one pass

3.2 Effect of the Route on the Microstructures

The effect of the processing route on the development of microstructures was mainly examined by TEM. TEM micrographs of three groups after one and four passes are shown in Figs. 4. After one pass, a clustered microbands of 200 to 300 nm in width were densely observed parallel to the shear direction in group I (Figs. 4(a)). The clustered microbands are microscopic features of the shear bands [11]. Selected area diffraction pattern (SADP) shows misorientation between the microbands are very small. In the group II and III, elongated cellular structures are dominant (Figs. 4(b,c)). Although some shear bands were revealed by TEM, the density of shear bands was much lower than that in group I. It is of interest to note that in group II, the microbands developed in two directions as shown in Fig 4(b).

After four passes via the route B, equiaxed structures with grain size smaller than 1 μm were generally observed in all three groups and the banded structures were not recognized (Figs. 4(d)–(f)). Furthermore, SADP shows that grains were fragmented by grain boundaries with large misorienation. It should be noted that the grain size of group III is slightly larger than those of group I and II. After four passes via the route C, however, the banded structures were still recognized in all groups as shown in figure 4(g–i). The group III exhibits the most distinct banded structures with very little misorientation. Thus, single crystalline billets were generally fragmented more effectively via route B than route C. This is consistent with the result reported on Aluminum by Iwahashi et al. [1] although otherwise was reported [14]. In summary, the grain refinement was best accomplished by a combination of shear bands formation (group I) and the route B whereas banded structure with less fragmentation was most prominent as a result of pressing of homogenous structure processed via the route C.

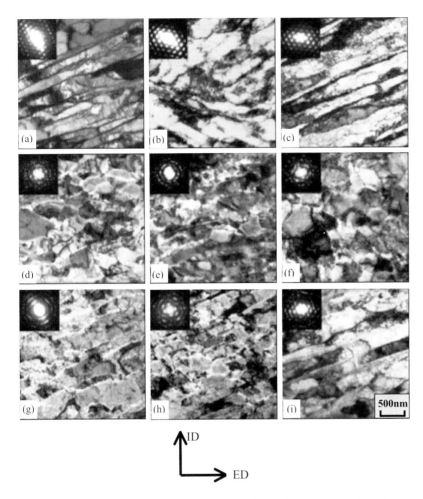

Figure. 4: TE micrographs and diffraction patterns (TD direction) after ECA pressing: (a)–(c) after one pass, ,(d)–(f) after four passes by route B, and (g)–(i) after four passes by route C, (a) (d) (g) group I, (b) (e) (h) group II and (c) (f) (i) group III

4 Summary

Copper single crystals were subjected to ECA pressing in order to investigate the grain refining mechanism from crystallographic aspects. The microstructural changes and orientations after the first pass were strongly influenced by the initial orientation and categorized into three groups, i.e., heterogeneous structure consisting of shear bands (group I), structure of deformation bands (group II) and homogeneous slip structure (group III). Each group showed a common end orientation as a result of lattice rotation, and these end orientations were recognized as the main components of shear texture of polycrystals. The billet were further pressed up to four passes, and the microstructures via the route B showed equi-axial structure whereas those via the route C showed relatively banded structure irrespective of the initial orientation. Best refined structure was obtained by a combination of shear band formation and pressing via the route B.

5 Acknowledgment

The authors are grateful to Prof. U.Erb, Depertment of Materials Science, University of Toronto for valuable comments, and to Dr T. Yamasaki for his great contribution to the study.

6 References

[1] Y. Iwahashi, Z. Horita, M.Nemoto and T.G. Langdon, Acta mater. 1998,46, 3317–3331
[2] Y.T. Zhu and T.C. Lowe, Mater. Sci. Eng. A, 2000, 291, 46–53
[3] M.Furukawa, Z. Horita, M.Nemoto, R.Z.Valiev and TG.Langdon, Mater. Charact, 1996, 37, 277–283
[4] D.H.Shin, I. Kim, J.Kim, K.T. Park, Acta Mater., 2001, 49, 1285–1292
[5] P.B.Prangnell and J.R.Bowen, Ultrafine grained materials 2, ed by Zhu YT, Langdon TG, Mishra RS, Semiatin SL, Saran MJ, Lowe TC, 2002; 89
[6] D.A.Hughes, Q.Liu, D.C.Chrzan and N.Hansen, Acta Mater. 1997; 45, 105–112
[7] H. Miyamoto, U. Erb, T. Koyama., T. Mimaki, A. Vinogradov and S. Hashimoto, to be submitted
[8] V.M.Segal, Mater. Sci. Eng. A, 1995; 197,157–164
[9] P.Wagner, O. Engler and K. Lucke, Acta Metall. Mater., 1995, 43, 3799–3812
[10] K. Morii, H. Mecking and Y. Nakayama, Acta Metall. 1985, 33, 379–386
[11] N.Inakazu, and H.Yamamoto, J. Inst. Metals, 1980, 44, 1356
[12] J. Hirsch and K. Lucke, Acta Metall., 1988, 36, 2883–2904
[13] W.H.Huang, L. Chang, P.W. Kao and C.P. Chang, Mater. Sci. Eng. A, 2001, 307, 113–118
[14] A.Gholinia, P. Bate and P.B. Prangnell, Acta Mater., 2002, 50, 2121–2136
[15] G.R.Canova, U.F. Kocks and J.J. Jonas, Acta Metall. 1984, 32, 211–226

TEM Investigations of Ti Deformed by ECAP

B. Mingler, L. Zeipper, H. P. Karnthaler and M. Zehetbauer
Institute of Materials Physics, University of Vienna, Wien, Austria.

1 Introduction

Commercially pure titanium (CP-Ti) is chemically inert and biologically more compatible than Ti alloys [1]. Therefore it is preferentially used for medical implants and devices. As compared to Ti alloys coarse grained CP-Ti lacks the necessary strength. In the present study equal-channel angular pressing (ECAP) deformation is used to achieve smaller grain sizes and producing work pieces with dimensions large enough for technical applications. The results can be compared with those from high pressure torsional (HPT) deformed samples that received after the deformation similar heat treatments [2]; but it should be pointed out that the HPT deformed samples are in most cases much too small for manufacturing technical applications. The use of a heating holder during the transmission electron microscopy (TEM) investigations allows in-situ observations of the changes in the microstructure during annealing with increasing temperatures. It is the aim of this work to correlate the microstructures deduced from the TEM results of the different samples with their hardness values.

2 Experimental Procedure and Results

Since Ti samples must have a certain ductility during ECAP to prevent crack formation they must be processed at elevated temperatures. Therefore CP-Titan grade 2 samples with a diameter of 40 mm and a length of about 120 mm were ECAP processed in Ufa (Russia) at about 450 °C at a pressing speed of 6 mm/s for 1 and 8 passes by route Bc (90° turn after each pass). 1 ECAP pass corresponds to a true strain of 1.13 [3]. After the last ECAP pass the samples were water quenched. Additional heat treatments after the ECAP deformation at 300, 400, 500 and 900 °C were done in vacuum for 10 min with subsequent cooling down to room temperature in the furnace. The influence of different numbers of ECAP passes and of different heat treatments after ECAP on the microstructure was studied by TEM. All samples were observed in longitudinal and transversal sections. The grains in longitudinal sections showed a tendency for elongation in one direction.

The TEM investigations of the ECAP processed samples yield the result that in contrast to the initial state exhibiting a grain size of several μm the microstructure after 1 ECAP pass (from a transversal section) consists of a mixture of equiaxed grains with a diameter of 300–900 nm and elongated grains with lengths up to 2 μm. An area containing elongated grains is imaged in Fig. 1.

Figure 1: TEM micrograph of an area with elongated grains in a transversal section in Ti deformed by 1 ECAP pass

Fig. 2 shows the microstructure after 8 ECAP passes (from a transversal section). It consists of equiaxed grains with diameters from 300–900 nm. An additional annealing of the samples after 8 ECAP passes at 300, 400 and 500 °C seems to have little influence on the microstructure. Fig. 3 shows a characteristic area from a sample with 8 ECAP passes that received an additional annealing at 500°C revealing that the microstructure is very similar to that imaged in Fig. 2.

Figure 2: TEM micrograph of equiaxed grains in a transversal section in Ti deformed by 8 ECAP passes

Figure 3 : TEM micrograph of Ti deformed by 8 ECAP passes and with additional annealing at 300–500 °C (transv. sect.). The structure is similar to that of Fig. 2

An annealing for 10 min at 900 °C after 8 ECAP passes yields a completely different microstructure compared to that of annealing at lower temperatures (300–500 °C). It leads to large grains with diameters of several µm containing a low density of dislocations; frequently they are straight and arranged parallel to each other (Fig. 4a).

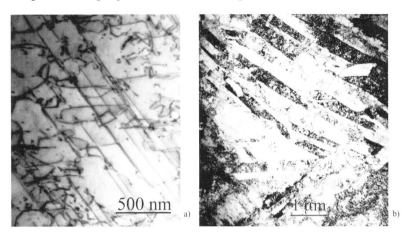

Figure 4: TEM micrographs of Ti deformed by 8 ECAP passes and with additional annealing at 900 °C (transv. sect.) a) large grains with diameters of several µm containing parallel, straight dislocations b) in some areas lamella-like structures consisting of the high temperature β-phase (bcc) are visible

In some areas lamella-like structures are observed (Fig. 4b); these structures were analyzed using diffraction patterns and determined to consist of the high temperature β-phase (bcc).

The observation of the evolution of the microstructure during in-situ heating of a sample with 8 ECAP passes revealed a dramatic increase of the grain sizes starting at temperatures between 700 and 800 °C. After 10 min heating at 800 °C grain sizes larger than 1 µm were observed as shown in Fig. 5.

Figure 5: TEM micrograph of Ti deformed by 8 ECAP passes and in-situ heated to 800°C (10 min) (transv. sect.).

In Fig. 6 diffraction patterns of different samples are compared with each other (they were all taken from areas with about 5 μm diameter). Fig. 6a shows the diffraction pattern from an area with elongated grains in the sample with 1 ECAP pass (cf. Fig. 1) yielding spots with a rather high intensity and small spreading.

Fig. 6b is from an area with equiaxed, small grains which are characteristic for both the samples with 8 ECAP passes (cf. Fig. 2) and those with additional annealing at 300–500 °C (cf. Fig. 3); many spots are arranged on rings exhibiting even intensity. The diffraction pattern of Fig. 6c stems from a single grain in the sample with 8 ECAP passes plus annealing for 10 min at 900 °C (cf. Fig. 4).

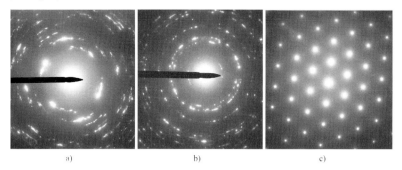

Figure 6: TEM diffraction patterns taken from areas with a diameter of about 5 μm (transv. sect.) a) from an area with elongated grains in the sample deformed by 1 ECAP pass (diffraction spots with a rather high intensity and small spreading) b) characteristic for all samples deformed by 8 ECAP passes and additional annealing at 300–500 °C (many spots arranged on rings indicating the large misorientations of the grains) c) from a single grain in the sample deformed by 8 ECAP passes and with annealing for 10 min at 900 °C (beam direction is [0001]).

3   Discussion

It was shown that a single ECAP pass leads to a strong decrease of the grain sizes from some μm to the submicrometer range (cf. Fig. 1). In some areas elongated grains were observed in transversal sections which is in agreement with observations in other works [4, 5].

The small equiaxed grains in samples with 8 ECAP passes (cf. Fig. 2) are considered to be the consequence of their rotating by 90° around their axis after each pass (in total the samples were rotated two times around their axis which means that no preferred direction for elongation is remaining). The deformation behavior after multiple passes is difficult to analyze since in hcp structures in principle 30 slip systems are possible; their activation depends on the value of c/a. In Ti ($c/a = 1.58$) the dominant operative slip occurs on basal and prismatic planes and in order to accommodate straining in the c-direction slip on the first- and second-order pyramidal planes can become possible [6].

The misorientations of the grains can be deduced from diffraction patterns as shown in Fig. 6. After one ECAP pass only a few diffraction spots have a rather high intensity indicating the presence of only a few grain orientations. The spreading of these spots corresponds to a total misorientation of about 30° (cf. Fig. 6a). For the samples with 8 ECAP passes the diffraction spots show rather similar intensities and they are arranged uniformly around the transmitted beam (cf. Fig. 6b). This means that the amount of high-angle grain boundaries is much higher than in the samples deformed by 1 ECAP pass.

The annealing of the samples with 8 ECAP passes at temperatures of 300, 400 and 500 °C has no significant influence on the microstructure compared to the sample without additional annealing (cf. Fig. 2 and 3). It is proposed that these temperatures of annealing are too low compared to dynamic recovery effects occurring already during the ECAP process done at 450 °C (in Al dynamic recovery effects were observed at room temperature RT (0.32 T_m) during cold rolling [7]). Therefore after 8 ECAP passes (true strain $\varepsilon \sim 9$) an additional annealing up to temperatures of 500 °C can be applied to increase ductility (about 10 %) without loosing hardness. Hardness measurements showed that there is a significant hardness increase of about 20% from the initial state (before ECAP) to the state with 1 ECAP pass whereas the hardness increase from 1 ECAP pass to 8 ECAP passes is less than 10 %. Heat treatments at 300 (0.3 T_m), 400 (0.35 T_m) and 500 °C (0.4 T_m) for 10 min after 8 ECAP passes have only little influence on the hardness, whereas a heat treatment at 900 °C (0.6 T_m) decreases the hardness to a value smaller than that after 1 ECAP pass. Contrary to this an annealing at 500 °C after high pressure torsion (true strain $\varepsilon \sim 7$, pressure ~ 5 GPa) decreases the hardness by about 30 % [2]. This different annealing response is explained by the different deformation temperatures (HPT: RT; ECAP: $\sim$ 450 °C).

In the case of an annealing at 900 °C for 10 min strong recrystallization occurs and the microstructure changes completely to grain sizes of some µm (cf. Fig. 4). In addition the high temperature β-phase (bcc) was observed at RT indicating that the speed of cooling was fast enough to quench-in this phase. The in-situ heating experiment showed that strong recrystallization in the thin foils started at temperatures between 700 and 800 °C (0.50 and 0.55 T_m) (cf. Fig 5).

The TEM results are in good agreement with the hardness values (this confirms the very good correlation between microstructure and hardness as observed in the case of ECAP deformed Cu [8]). Two parameters seem to be most important: grain size and grain boundary angles. Samples with grain sizes of some µm (initial state, 8 ECAP passes with annealing at 900 °C) have the lowest hardness whereas samples with grain sizes in the submicrometer regime (1 ECAP pass, 8 ECAP passes with annealing at 300–500 °C) have higher hardness values. The lower hardness of the sample with 1 ECAP pass as compared to that with 8 ECAP passes is explained by the lower amount of high angle boundaries and by the existence of elongated grains in some areas.

4   Conclusions

- Ti deformed by 1 ECAP pass shows in some areas elongated grains in the transversal sections. Only few of the grain boundaries are high-angle boundaries.
- Ti deformed by 8 ECAP passes and also with additional annealing for 10 min at 300, 400 and 500 °C shows equiaxed grains (diameter: 300–900 nm) in the transversal sections. This is a consequence of their rotating by 90° around their axis after each ECAP pass. Most of the grain boundaries are high angle boundaries. Annealing up to a temperature of 500°C after the ECAP passes has no significant influence on the grain sizes and the hardness but increases the ductility by 10 %. It is proposed that dynamic recovery occurring during the ECAP (done at 450 °C) anticipates the effect of annealing.

- After 8 ECAP passes and with annealing for 10 min at 900 °C strong recrystallization effects (grain sizes of several μm) and the appearance of the high temperature βTi phase (bcc) were observed.
- An in-situ heating of a TEM sample with 8 ECAP passes showed that recrystallization starts between 700 and 800 °C.

5 Acknowledgements

Financial support by the Austrian Science Foundation (FWF project 12945 PHY) is gratefully acknowledged.

6 References

[1] Stolyarov, V.V.; Zhu, Y.T.; Lowe, T.C.; Valiev, R.Z., Mater. Sci. Eng. A303 2001, 82–89
[2] Popov, A.A.; Pyshmintsev, I.Y.; Demakov, S.L.; Illarinov, A.G., Lowe, T.C.; Sergeyeva, A.V.; Valiev, R.Z., Scripta Mater. 1997, 37, 1089–1094
[3] Iwahashi, Y.; Wang, J.; Horita, Z.; Nemoto, M.; Langdon, T.G., Scripta mater. 35 1996, 143–146
[4] Shin, D.H.; Kim, I.; Kim, J.; Zhu, Y.T., Mater. Sci. Eng. A334 2002, 239–245
[5] Kim, I.; Kim, J.; Shin, D.H.; Liao, X.Z.; Zhu, Y.T., Scripta Mater. 2003, 48, 813–817
[6] Balasubramanian, S; Anand, L., Acta Mater. 2002, 50, 133–148
[7] Schafler, E.; Zehetbauer, M.; Hanak, P.; Ungar, T.; Hebesberger, T.; Pippan, R.; Mingler, B.; Karnthaler, H. P.; Amenitsch, H.; Bernstorff, S., in: Investigations and Applications of Severe Plastic Deformation (Ed.: T.C. Lowe and R.Z. Valiev), Kluwer Academic Publishers, Netherlands, 2000, 163
[8] Mingler, B.; Karnthaler, H. P.; Zehetbauer, M.; Valiev, R. Z., Mat. Sci. Eng. A319–321 2001, 242–245

Microstructural Evolution during Severe Deformation in Austenitic Stainless Steel with Second Phase Particles

H. Miura, H. Hamaji and T. Sakai
Department of Mechanical Engineering and Intelligent Systems, University of Electro-Communications, Chofu, Tokyo, Japan

1 Abstract

The effect of large amount of dispersed particles on ultra-fine grain (UFG) evolution by severe large deformation was investigated. For that purpose, austenitic stainless steels containing about 1 vol.% of second phase particles, which radii ranging from 10 to 140 nm, were multi-axially compressed to a strain of $\varepsilon = 6$ at maximum at 873 K. Microstructures of dislocation walls, sub-boundaries and grain boundaries were gradually developed as strain increased. UFG evolution, ranging up to 0.2 µm uniformly took place at $\varepsilon = 6$. UFGs evolved both in austenitic stainless steels with particles and without particles. The feature of grain boundaries or dislocation walls in the stainless steel with higher density of particles, however, were not so sharp nor highly contrasted compared with those of particle-free austenitic stainless steel. Recovery process seemed very important for UFG evolution in materials with dispersed fine particles.

2 Introduction

Recently special attention is being paid to grain refinement by severe large deformation process to achieve nano-meter size grains. Finer grain distribution brings higher strength of materials due to reduction of mean-free pass of dislocations by grain boundaries. By usage of severe deformation process, as multi-axial compressions (MAC) [1–3], equal channel angular extrusion (ECAE) [4] and accumulative roll-bonding (ARB) [5], ultra fine grains (UFGs) up to 0.2 µm in diameter were easily and homogeneously evolved. Such evolved UFGs were revealed to cause 400 MPa and 3.7 GPa hardness, which are comparable to 1.2 GPa and 11 GPa strength [6], even in 4N purity copper [2] and austenitic stainless steel [1], respectively. However, such UFG material formed by severe deformation process possesses fatal property of thermal instability due to stored large strain energy and large volume fraction of grain boundaries. They are both those for driving force of recrystallization and grain growth.

It is well-known that dispersion particles can contribute not only to strengthening but also to thermal stability impeding the occurrence of recrystallization and grain growth. Particles lower the mobility of dislocations by Orowan mechanism or Srolovitz mechanism [7] and impede migration of grain boundaries by Zener pinning force at elevated temperature [8]. Dispersion of the fine particles, therefore, brings two favorable nature into the UFG materials. Nevertheless, the effects of dispersion particles on the UFG evolution by severe deformation process and its thermal stability have not been systematically investigated yet, as far as the authors know.

In the present study, as the first step of the above mentioned investigation, the evolution of UFGs in an austenitic stainless steel with fine precipitates is studied by MAC method. For com-

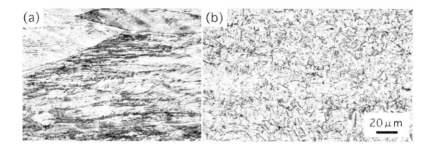

Figure 1: Microstructures of an austenitic stainless steel of (a) as-hot rolled, followed by (b) aging at 1173 K for 27 h, respectively

parison, an austenitic stainless steel with no precipitates is also prepared and MAC processed. The effect of particle dispersion on the UFG evolution is analyzed and discussed.

3   Experimental Procedure

Hot-rolled austenitic stainless steel of Fe-18Cr-10Ni-0.8Mn-0.56Nb-0.4Si-0.07C-0.03P-0.03Al (mass %) was aged to have about 1 vol.% precipitates at 1173 K for 27 h. The samples of as-hot-rolled and after aging will be referred as samples A1 and A2, respectively. In both samples, fine particles uniformly precipitated. However, in sample A1, both coherent and incoherent particles of 10 and 60 nm in diameter respectively were precipitated, while in sample A2 only incoherent particles of 140 nm in diameter were precipitated. The optical microstructures of these samples before deformation are shown in Fig. 1. The grain size of the sample A2 was 7 µm, though that of A1 could not be measured due to as-rolled structure.

These samples were multi-axially compressed at a temperature of 873K and at a strain rate of $3 \cdot 10^{-3} \, \text{s}^{-1}$ in vacuum after machining to a rectangular shape of 1:1.22:1.5 aspect ratio. The details of the MAC process are described elsewhere [1–3]. At each pass step, a pass strain of 0.4 was accumulated to the samples. The samples were multi-axially deformed to a total true strain of $\varepsilon = 6.0$. After the deformation, evolved microstructure was observed using optical microscope (OM) and transmission electron microscope (TEM). Furthermore, Vickers hardness was measured after the deformation. A Fe-32Ni austenitic stainless steel referred as sample B, which included no particles and had about 30 µm initial grain size, was also MAC processed and investigated for comparison.

4   Results and Discussion

Figure 2 shows true stress-cumulative true strain curves for the samples A1 and A2. The feature of the flow curves resemble to that of dynamic recovery (DRV) or dynamic recrystallization (DRX), showing work hardening at lower strain and, then, becomes steady-state flow or shows slight work softening at higher strain region. Because no occurrence of DRX was revealed in neither samples by OM and TEM analysis, the work-softening in the sample A2 would be presumably caused mainly by DRV process as rearrangement of dislocation and dislocation walls to evolve UFG structure. It is notable that a clear peak appears in the sample A1 at the first compression followed by small stress oscillation. This should be due to the effect of strong as-hot-rolled texture and super-saturated solute atoms.

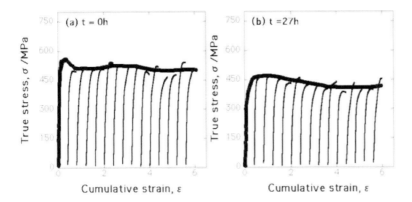

Figure 2: True stress vs. cumulative true strain curves under MAC at 873 K for the samples (a) as-hot rolled ($t = 0$ h) and (b) aged at 1173 K for 27 h, respectively

Change in room-temperature Vickers hardness depending on strain was measured. The result is summarized in Fig. 3. In both samples, the hardness abruptly increases at the beginning of straining by the existence of precipitates, and then, shows gradual decrease at higher strain region. This tendency corresponds well to that of flow curves shown in Fig. 2. What more interesting in Fig. 3 is that the much higher hardness in sample A1 compared with that of A2 at lower strain region, while the difference becomes smaller at higher strain region. The hardness of sample B processed to $\varepsilon = 6$ was 1.9 GPa. These facts would suggest the important contribution of particles, solute atoms and grain size to the mechanical strength also in UFG materials.

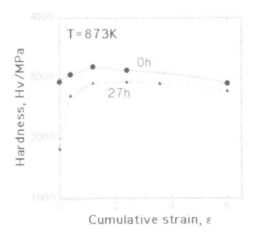

Figure 3: Vickers hardness change depending on cumulative true strain under MAC at 873 K for the samples of as-hot rolled ($t = 0$ h) and aged for 27 h ($t = 27$ h).

Microstructural evolution during MAC process was observed using OM and the photographs are exhibited in Fig. 4. In both photographs, the grains look to become finer by straining to $\varepsilon = 6$ compared with those in Fig. 1. Grain refinement starts at the initial grain boundaries and spreads into grain interiors [2]. It is interesting to note that the initial grain boundaries still remain even after MAC process to $\varepsilon = 6$. The difference between Figs. 5 (a) and (b), therefore, seems not so clear.

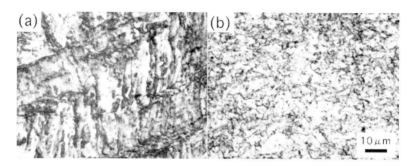

Figure 4: Optical micrographs of microstructure evolved after straining up to $\varepsilon = 6$ under MAC at 873 K for the samples of (a) as-hot rolled, followed by (b) aging at 1173 K for 27 h, respectively

TEM micrographs of the micro-structure of samples A1 and A2 deformed to a strain of $\varepsilon = 6$ are shown in Fig. 5. For comparison, a typical TEM photograph of the sample B deformed by the same process is also shown here. It can be seen that ultra-fine grains or subgrains, which size is less than about 1 μm, is uniformly developed in all the samples. The microstructures were developed under MAC process by formation of microbands and dense dislocation walls, and gradual increase in misorientation angles among the formed subgrains resulted in the full evolution of UFGs. When the samples A1 and A2 were MAC processed at lower temperature, $T = 773$ K, the boundaries looked like dense dislocation walls.

Selected-area-electron-diffraction (SAED) analysis revealed that disorientation angles among the evolved (sub)grains were smaller than those processed at 873 K. This temperature-dependent microstructural evolution indicates the important role of recovery process for UFG evolution. That is, the sufficient DRV process as dislocation climbing and boundary migration, is necessary for the evolution of grain boundaries from dense dislocation walls and, therefore, UFGs. At this strain region, the sizes of subgrains and grains become almost the same [2]. However, the developed microstructure differs much depending on the sample. The (sub)grain size in A2 is the smallest, ~0.2 μm, and B the largest, ~1 μm. The relatively larger grain size of the sample B should be due to much easier occurrence of recovery process such as rearrangement of dislocations and migration of dislocation walls. It is characteristic that the feature of grain boundaries is not so sharp nor clear in sample A1 (Fig. 5 (a)) compared with those in the other samples. The wavy stripe contrast in Fig. 5 (a) suggests the presence of large residual strain. SAED analysis also revealed the residual stress by long streaks at the diffraction spots. The misorientations appear to be emphasized by heavy distortion in the (sub) grains. The evolved microstructure in the sample A1, therefore, may be different from that of typical UFGs. These results would be because solute atoms and higher density of particles (finer precipitates), especially such as coherent particles, impede the recovery process.

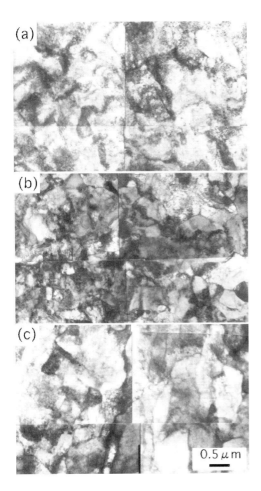

Figure 5: Microstructures in austenitic stainless steels evolved by straining to $\varepsilon = 6$ under MAC at 873 K in the samples of (a) as-hot rolled, (b) aged for 27h and (c) particle free, respectively

Such UFGs can be easily and uniformly developed by MAC even at room temperature [2]. When the present austenitic stainless steels with particles were MAC processed at lower temperatures, evolution of UFGs was evidently delayed, though the result is not shown here. This result indicates that recovery process is necessary for easier evolution of UFGs in dispersion-hardened materials. These results indicate that recovery process is essentially one of the most important factors controlling the evolution of UFGs. Original grain size also affects the UFG evolution; the UFG evolution becomes easier when the original grain size is finer [9]. This effect of original grain size should also influence the evolved microstructure in Figs. 4 and 5.

Ferry and Humphreys reported that deformation zone (DZ), which causes grain-refinement around the particles (3 μm), was developed around the particles by large straining [10]. However, no DZ around the particles in the present case was observed even after straining to $\varepsilon = 6$. This result indicates that DZ is hardly formed even by severe straining when the particle size is enough small. It can be summarized, therefore, that the dispersed fine particles seemed not to strongly stimulate nor impede the evolution of UFGs, while the particles contributed much to strengthening of the materials.

5 Conclusions

The effect of particle dispersion on ultra-fine grain (UFG) formation by multi-axial compression was investigated. UFGs were evolved both in austenitic stainless steels with particles and without particles. However, the evolution of UFGs looked to be disturbed in a sample with relatively higher density of particles (or smaller size of particles) and super-saturated-solute atoms, while not in the sample with lower density of particles (or larger particles). Such disturbance was also manifested by the occurrence of unclear boundaries in the evolved microstructure. It was concluded that ease occurrence of recovery process was one of the important factors controlling UFG evolution and caused the difference of such structural evolution.

6 Acknowledgements

The authors acknowledge the material supply from Nippon Steel Corporation, Japan, and financial support received from Grant-in-Aid for Scientific Research from Ministry of Education, Science and Culture under Grant No. 14550700.

7 References

[1] A. Belyakov, T. Sakai, H. Miura, R. Kaibyshev, K. Tsuzaki, Acta Mater., 2002, 50, 1547–1557
[2] A. Belyakov, T. Sakai, H. Miura and T.Tsuzaki, Phil. Mag. A, 2001, 81, 2629–2643
[3] A. Belyakov, T. Sakai, H. Miura, Mater. Trans., JIM, 2000, 41, 476–484
[4] Y. Iwahashi, Z. Horita, M. Nemoto and L. G. Landon, Acta Mater. 1997, 45, 4733–4741
[5] Y. Saito, H. Utsunomiya, N. Tsuji and T. Sakai, Acta Mater., 1999, 47, 579–583
[6] S. Takaki, Ultrafine grained steel, ISUGS 2001 (Ed.: S. Takaki and T. Maki), ISIJ, 2001, 42–50
[7] D. J. Srolovitz, R. Petkovic-Luton and M. J. Luton, Phil. Mag. A.1983, 48, 795–809
[8] P. A. Manohar, M. Ferry and T. Chandra, ISIJ Int. 1998, 38, 913–924
[9] A. Belyakov, K. Tsuzaki, H. Miura and T. Sakai, Acta Mater., 2003, (in press)
[10] M. Ferry and F. J. Humphreys, Acta Mater., 1996, 44, 3089–3103

Structural Models and Mechanisms for the Formation of High-Energy Nanostructures under Severe Plastic Deformation

A. N. Tyumentsev[1], A. D. Korotaev[2], Yu. P. Pinzhin[1], I. A. Ditenberg[2], I. Yu. Litovchenko[1], N. S. Surikova[2], S. V. Ovchinnikov[1], N. V. Shevchenko[2], R. Z. Valiev[3]

[1]Institute of Strength Physics and Materials Technology, RAS, Tomsk, Russia
[2]Siberian Physicotechnical Institute, Tomsk, Russia
[3]Institute of Physics of Advanced Materials Ufa State Aviation Technical University, Ufa, Russia

1 Introduction

Severe plastic deformation is characterized by the intense strain hardening and low efficiency of dislocation mechanisms of deformation (through a noncorrelated motion of dislocations). For this reason the processes of plastic flow and crystal lattice reorientation, which occur under conditions of the formation of nanostructural states [1–6], are controlled by cooperative deformation mechanisms. The cooperative character of deformation implies, on the one hand, collective behavior of elementary defects ensembles, which are deformation carriers, and, on the other hand, concurrent work of various modes of crystal deformation and reorientation, such as dislocation, disclination, and diffusion modes, mechanical twinning, and phase transformations. This paper briefly summarizes the results of a study of the role played by the mentioned cooperative mechanisms for the fragmentation of a crystal during the formation of nanostructural states in metal alloys (Cu, Ni, Ti, austenitic steels) and intermetallic compounds (Ni_3Al, TiNi) under various conditions of severe plastic deformation (SPD).

2 Mechanisms for the Deformation and Reorientation of a Crystal

Investigations of the microstructure of nanostructural (NS) materials produced by severe plastic deformation (SPD) methods show that the most important features of their defect substructure are high values (some tens of degrees per micrometer) of lattice curvature and high densities of partial disclinations at intercrystallite boundaries [1–8]. The local internal stresses in the zones where the defect density peaks approach the theoretical strength of a crystal, $\sigma_{loc.} \sim E/20$ [1–9] and their gradients reach $\partial\sigma_{loc.}/\partial r \approx E / 7 \ \mu m^{-1}$ [7–9]. In combination with the low dislocation activity, the above features are responsible for the formation of new high-energy deformation carriers and for the development of cooperative crystal reorientation mechanisms. The type of deformation carrier, and, hence, the reorientation mechanism, is determined by the conditions under which the deformation occurs, by the character of the elastic-stress state, and by the crystal structure and properties.

2.1 The Disclination Mechanism for the Lattice Reorientation

Invoking this mechanism to describe the fragmentation of a crystal, we are based on the fact that the formation of $\hat{\chi}$ misorientation boundaries is often preceded by a structural state with a high lattice curvature [1, 2, 8, 9]. Generally, this curvature is such that the values of the curvature rotor $-\nabla \times \chi^p$, whose plastic part is the continual disclination density tensor $\bar{\rho}\,\Omega \equiv -\nabla \times \chi^p$, appear to be non-zero. Such a curvature and its disclination model are presented in Fig. 1 a, b. This state is structurally unstable and its relaxation through collective rearrangements of strongly interacting chaotically distributed dislocations of like sign into dislocation walls or networks results in the formation of discrete misorientation boundaries (Fig. 1 c, d).

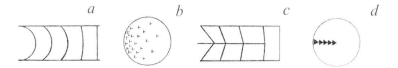

Figure 1: Sketches of lattice misorientations (a, c) and their dislocation (b) and disclination (d) models in a zone with a high continual disclination density prior to (a, b) and after the relaxation of this zone into a boundary with a variable misorientation vector (c, d).

This mechanism was observed in all materials we tested (Ni [4–6], Cu [9, 10], Ni$_3$Al [1–3, 6, 8], austenitic steels [8, 11, 12], and TiNi alloys [12]) and seems to be the most general mechanism for the reorientation of a crystal under SPD. As can be seen from Fig. 1 c, d, for nonzero values of $-\nabla \times \bar{\chi}$, the model based on this mechanism predicts an important feature of the microstructure of submicrocrystalline (SMC) materials, namely, intercrystallite boundaries with varying misorientation vectors, whose substructure is representable by pileups of continuously distributed partial disclinations or by Somigliana dislocations (for details, see [1, 4–6, 9]). In our opinion, these structural features of intercrystallite boundaries are responsible for the highly nonequilibrium state of grain boundaries which appear under SPD.

2.2 Nonequilibrium Point Defect Flows in Stress Fields

The high local stress gradients typical of the substructures shown in Fig. 1 and the high concentration of nonequilibrium point defects generated in the course of deformation determine the feasibility of the quasi-viscous (diffusion) mechanism for the crystal reorientation that is governed by local flows of these defects in the fields of the gradients of the stress tensor diagonal components. Analysis [13, 14] has shown that the rate of reorientation ($\dot{\varphi}$) of a microregion of size d, provided there exists a pressure gradient $\partial P/\partial r$ ($P \sim \sigma_{11} + \sigma_{22} + \sigma_{33}$) along the region boundaries, can be estimated by the formula

$$\dot{\varphi} \approx n \times (D/kT) \times (\Omega/d) \times (\partial P/\partial r)$$

where n is the nonequilibrium concentration of point defects, D is the coefficient of their diffusion, T is the temperature, and Ω is the volume of a vacancy or interstitial atom. For the mass transfer occurring by the vacancy mechanism with a vacancy migration activation energy of

0.6 eV, at room temperature, and with typical values of n (~10^{-4}) and d (~0.1 μm), the reorientation rate $\dot\varphi_V$ is of the order of 10^{-6} s^{-1}. If the interstitial atoms flows are taken into account, we get $\dot\varphi_\mu \sim 1 s^{-1}$, and with this reorientation rate large-angle misorientation boundaries can be formed in the course of active plastic deformation.

A good illustration of the important role played by nonequilibrium point defect flows in the formation of NS states is the modification of the surface layer in the process of ion implantation of nanostructural Ni_3Al produced by high pressure torsion (HPT). Upon HPT, an NS state was formed in this material (Fig. 2a) with crystallites of size $d \sim 0.1$–0.3 μm and a highly defective substructure with high local stresses and high stress gradients [1, 2]. Ion implantation performed at room temperature with an irradiation dose of $\approx 10^{16}$ cm^{-2} results in the high concentration of nonequilibrium vacancies and interstitial atoms. Under such conditions a nanocrystalline (NC) state (Fig. 2b) with the nanocrystal size ranging from a few nanometers to several tens of nanometers was detected in the ion-implanted layer of thickness ~0.2 μm.

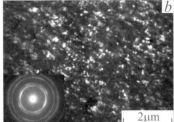

Figure 2: Microstructure of the surface layer of the Ni_3Al alloy subjected to HPT (a) followed by ion implantation at room temperature (b)

In crystals with a low defect density, the modification of the microstructure resulting from ion implantation performed in the mentioned regime shows up only in an increased density of defects (dislocations, dislocation loops, point defects, and clusters of point defects). Therefore, the formation of the NC structure (Fig. 2b) seems to be a result of the relaxation of nanometer-scale local gradients by the point defect flows generated in the course of ion implantation. In addition, the mentioned defects, promoting the dislocation creeping, may activate the collective dislocation-disclination reorientation modes (see Fig. 1).

2.3 Dynamic Phase (Martensitic) Transformations Mechanism

This mechanism is a consequence of the phase instability of a crystal in high local stress fields. We first discovered it in strain localization bands (SLB's) formed in a Ni_3Al-base alloy upon HPT [8] and in austenitic steels upon rolling [8, 11, 12]. In this case, the reorientation of the lattice occurs through forward plus reverse martensitic transformations, the latter developing via an alternative system. The reorientation of a crystal in the course of forward plus reverse martensitic transformations of different types has been analyzed theoretically and atomic mechanisms for this reorientation have been proposed in [12].

Let us illustrate these mechanisms by the example of the $\gamma \rightarrow \alpha \rightarrow \gamma$ transformation in an austenitic steel. Figure 3 a presents schematically the mutual atomic rearrangements in the plane of the forward $\gamma \rightarrow \alpha$ phase transformation for the Kurdyumov–Sachs (K–S) orientation relationship. These rearrangements include the tensile-compressive deformation in the directions indicated by the arrows and the turn of the transformation plane about the type <110> axis normal to this plane by an angle of 5.23°. According to [15], a change of the system of the reverse $\alpha \rightarrow \gamma$ transformation implies a change of the transformation plane. It has been shown [12] that the $\gamma \rightarrow \alpha \rightarrow \gamma$ transformations in austenitic steels result in the reorientation of the crystal in the transformation zones about the axis close to the <110> axis by angles of about 60°. A variant of this type of transformation is illustrated by Fig. 3b. This variant corresponds to the $\gamma \rightarrow \alpha \rightarrow \gamma$ transformation in which, in the course of the forward and reverse transformation, the general direction $[01\bar{1}]_\gamma \| [\bar{1}1\bar{1}]_\alpha$ is preserved in the K–S orientation relationship. This direction is thus the direction of the vector of reorientation by the angle equal to the angle between the planes of the forward and reverse transformations in the intermediate α phase $\theta = 120° [011] = 60° [0\,1\,1]$.

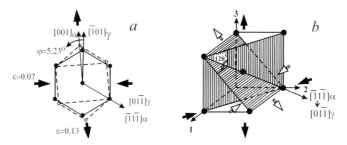

Figure 3: The 60^0<110> variant of the reorientation of a crystal upon the $\gamma \rightarrow \alpha \rightarrow \gamma$ transformation

It has been shown [12] that the B2→B19 (B19`)→B2 transformation in TiNi alloys is a mechanism for the deformation twinning in planes with complex Miller indices. Recent results obtained by the authors of [12] suggest that twins of this type actively participate in the formation of NS states in the mentioned alloys under severe plastic deformation.

The well-known effect of a decrease in elasticity moduli near the point of martensitic transformation has the result that the material softens in the zones of dynamic phase transformations. This is accompanied by the activation of almost all currently known deformation modes such as dislocation slipping (due to a decrease in critical shear stresses), diffusion deformation (due to a decrease in the activation energy of the formation and migration of point defects), and collective disclination reorientation modes (developing with the participation of the mentioned mechanisms). Owing to this and due to the complicated configuration of the fields of *high* local stresses in the zones of dynamic phase transformations, highly defective structural states with high continual densities of disclinations and discrete misorientation boundaries with variable θ vectors, similar to those presented in Fig. 1, and even fragmented nanocrystalline-scale structures are formed in these zones (Fig. 4).

2.4 Mechanical Twinning

In our opinion, mechanical twinning is a more general mechanism for the formation of NS materials than it is commonly believed nowadays. The fact that this mechanism is undervalued stems in the main from the methodical difficulties of detecting deformation twins. These difficulties arise from the highly defective character of NS states, namely from the presence in twinning zones of large continuous and discrete misorientations, which result from the occurrence of dislocation-disclination modes of lattice reorientation in the course of or after the formation of twins. This leads to a considerable (up to 10° and even larger) departure of their orientation from that of twins and to the absence of diffraction peaks characteristic of twinning configurations in electron diffraction patterns. In this case, twins can be detected only by the dark-field analysis of misorientations. A study performed with this type of analysis has shown that mechanical twinning is a leading mechanism for the formation of large-angle boundaries in NS states in practically all alloys we tested and under all deformation conditions we realized, namely, in austenitic steels and TiNi alloys upon rolling, in a Ni_3Al-base alloy subjected to HPT, and in copper went through equal channel angular pressing and HPT.

Figure 4 gives an example of a rolled high-nitrogen austenitic steel in which the formation of the NS state is the result of all above fragmentation mechanisms. For $\varepsilon \leq 30$–40 %, the principal fragmentation mechanism is the formation of deformation microtwins of thickness ~30 nm (Fig. 4 a). As the strain is increased, SLB's are formed (see Fig. 4 b) in which the crystal reorientation occurs through dynamic phase ($\gamma \rightarrow \alpha \rightarrow \gamma$) transformations. In the course of the interaction of SLB's with twins, the reorientation of the lattice with respect to the original crystal may occur in different variants with vectors of reorientation in type <110> directions by angles of 10.5, 35, 49.5, 60, and 70.5°. The result of the softening (decrease in elasticity moduli) of the crystal under the conditions of phase transformation is the activation of dislocation-disclination and diffusion mechanisms for the fragmentation inside SLB's, which leads to the formation of nanocrystalline structure in these bands and to substantial smearing of the diffraction maxima in electron diffraction patterns (see Fig. 4 b). After deformation with $\varepsilon \geq 90$ %, as a result of the combined action of all mentioned mechanisms, a nanocrystalline structural state with a high large-angle boundary density and with the nanocrystal size ranging between a few nanometers and several tens of nanometers is formed.

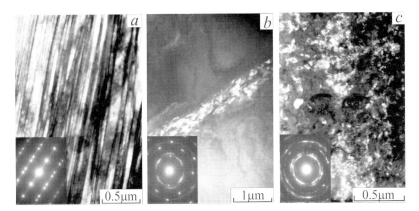

Figure 4: Deformation microtwins (a), a $\gamma \rightarrow \alpha \rightarrow \gamma$ reorientation band (b), and a nanostructural state (c) in 1718 high-nitrogen austenitic steel after deformation by rolling at room temperature. $\varepsilon \approx 40$ (a), 50 (b), and 90 % (c).

3 Acknowledgements

The research described in this publication was made possible in part by Grant T00-5.8-2918 from the Ministry of Education of Russian Federation and Grant TO-016-02 of U.S. Civilian Research and Development Foundation.

4 References

[1] Tyumentsev A.N., Tretjak M.V., Pinzhin Yu. P. et. al., Physics of Metals and Metallography, 2000, Vol. 90, 5, pp. 461–470
[2] Tretiak M. V., Tyumentsev A. N., Physical Mesomechanics, 2000, Vol. 3, No 3, 23–27
[3] Tyumentsev A. N., Pinzhin Yu. P., Tretjak M. V. et. al., Theoretical and Applied Fracture Mechanics, 2001, Vol. 35, 2, pp. 155–161
[4] Valiev R. Z., Islamgaliev R. K., Tyumentsev A. H., in Proceedings of the International Workshop on Local Lattice Rotations and Disclinations in Microstructures of Distorted Crystalline Materials, 10-14 April, 2000, Rauschenbach/Erzgebirge, Solid State Phenomena, 2002, Vol. 87, pp. 255–264
[5] Tyumentsev A. N., Pinzhin Yu. P., Korotaev A. D. et. al., Physics of Metals and Metallography, 1998, Vol. 86, No 6, pp. 604–610
[6] Tyumentsev A.N., Tretjak M.V., Korotaev A.D. et. al., Proceedings of the NATO Workshop on Investigations and Applications of Severe Plastic Deformation, 10–14 June, 1999, Moskow/Russia, NATO Science Series, 3. High Technology, Vol. 80, 2000, pp. 127–132
[7] Korotaev A. D., Tyumentsev A. N., Pinzhin Yu. P., Theoretical and Applied Fracture Mechanics, 2001, Vol. 35, 2, pp. 163–169
[8] Korotaev A. D., Tyumentsev A. N., Litovchenko I. Yu., The Physics of Metals and Metallography 2000, Vol. 90, Suppl. 1, pp. S36–S47
[9] Tyumentsev A. N., Panin V. E., Derevyagina L. S. et. al., Physical Mesomechanics, 1999, Vol. 2, No 6, 105–112
[10] Tyumentsev A. N., Panin V. E., Ditenberg I. A. et. al., Physical Mesomechanics, 2001, Vol. 4, No 6, 77–85
[11] Litovchenko I. Yu., Tyumentsev A. N. and Pinzhin Yu. P. et. al., Physical Mesomechanics, 2000, Vol. 3, No 3, 5–13
[12] Tyumentsev A. N., Litovchenko I. Yu. Pinzhin Yu. P. et. al., in Proceedings of the International Seminar "Mesostructure", 4-7 December, 2001, St-Petersburg/Russia, Voprosy Materialovedenija, 2002, Vol. 29, 1, pp. 314–335
[13] Korotaev A. D., Tyumentsev A. N. and Sukhovarov V. F., Dispersion Hardening of Refractory Metals, 1989, Nauka, Novosibirsk
[14] Korotaev A. D., Tyumentsev A. N., Gonchikov V. Ch. Olemskoy A. I., Izv. Vyssh. Uchebn. Zaved. Fizika, 1991, No 3, 81–92
[15] Kassan-Ogly F. A., Naish V. E., Sagaradze I. V., Fizika Metallov i Metallovedenie 1995, Vol. 80, 5, pp. 14–27

Grain Refinement and Microstructural Evolution in Nickel During High–Pressure Torsion

A. P. Zhilyaev[1*], G. V. Nurislamova[2], B.-K. Kim[3], M. D. Baró[1], J. A. Szpunar[3] and T. G. Langdon[4]

[1] Universitat Autònoma de Barcelona, Bellaterra, Spain
[2] Institute for Physics of Advanced Materials, Ufa State Aviation Technical University, Ufa, Rusia
[3] Department of Metals and Materials Engineering, McGill University, Montreal, Canada
[4] Departments of Aerospace & Mechanical Engineering and Materials Science, University of Southern California, Los Angeles, USA

1   Introduction

Significant interest has developed recently in investigating bulk materials by severe plastic deformation (SPD). These procedures are attractive because they are capable of producing large bulk samples free of any residual porosity [1, 2].

It is now well established that SPD processing can lead to a very significant refinement in the grain size of a wide range of materials including pure metals, metallic alloys and intermetallics [3, 4]. Furthermore, these ultrafine grain sizes, which are typically in the submicrometer or nanometer range, produce new and unusual physical properties such as decreases in the elastic moduli, decreases in the Curie and Debye temperatures, increases in the rates of diffusion and improved magnetic properties [2]. If the ultrafine grain sizes are reasonably stable at elevated temperatures, there is a potential for achieving superplastic ductility at both unusually low testing temperatures and exceptionally rapid strain rates [5]. Recently it has been shown that SPD processing is capable of producing materials that combine both high strength and high ductility [6] and this unusual combination has been attributed specifically to the development of unique nanostructures in SPD.

Two different procedures are generally used in SPD processing. The first, known as Equal-Channel Angular Pressing (ECAP), involves pressing a bar or rod through a die within a channel bent into an L-shaped configuration [7]. The second, known as High-Pressure Torsion (HPT), involves subjecting a sample, in the form of a thin disk, to a hydrostatic pressure and concurrent torsional straining [8].

Generally, it is accepted in HPT that an applied pressure of ~5 GPa and more than 5 rotations of the sample in torsion are enough to produce reasonably homogeneous microstructures all through the sample with as-processed grain sizes of ~100 nm or smaller [2, 3] but the precise relationships linking the pressure, total strain and location within the sample have not been studied in any detail. The present paper is devoted to clarifying this deficiency by conducting a series of careful experiments on pure nickel subjected to HPT processing and by examining, throughout the deformed disks, the variation of the microhardness and the characteristics of the microstructure using transmission electron microscopy and orientation imaging microscopy.

---

*On leave from Institute of Mechanics, Russian Academy of Science, Ufa, Russia

2   Experimental Material and Procedures

High purity nickel (99.99 %) was selected for this investigation [9, 10]. High-pressure torsion was applied at room temperature using the same procedure as in earlier studies [9–11]. Specifically, samples were prepared in the form of small disks with diameters of ~10 mm and thicknesses of ~0.3 mm. The influence of the applied pressure was evaluated by subjecting samples to a total of 5 complete revolutions under separate imposed pressures of 1, 3, 6 and 9 GPa. The influence of the accumulated strain was evaluated by maintaining a constant pressure of 6 GPa and torsionally straining separate samples through totals of 0.5, 1, 3, 5 and 7 revolutions.

Following HPT, precise microhardness measurements were taken on the surfaces of the samples along diameters positioned at rotational increments of 45° about the central point of each disk. These measurements were taken in incremental steps along every diameter at positions ~1.25 mm apart and at each point the average microhardness was determined from four separate measurements clustered around the selected position. Samples were examined by transmission electron microscopy (TEM) after HPT by cutting small pieces from either the centres or the peripheral areas of the disks in the immediate vicinity of the edge.

Orientation imaging microscopy (OIM) was used to obtain detailed quantitative information about the microstructures developed in HPT. The grain boundary statistics were analysed using orientation data recorded with a Philips XL-30 FEG scanning electron microscope (SEM) and a TSL orientation imaging system [11–13]. The electron backscatter diffraction (EBSD) patterns were observed on the screen with a charge coupled device (CCD) camera and they were collected from a tilted bulk sample. Automatic processing and analysis of these patterns provided quantitative information on the local texture in the sample, the distributions of grain boundary misorientations and the nature of the as-processed microstructure. In this investigation, the angular resolution of the EBSD measurements was of the order of ~1° and the average confidence index for all measurements was higher than ~0.4.

3   Experimental Results and Discussion

Figures 1 and 2 show the influence of the applied pressure on the microstructures and microhardness profiles for disks subjected to 5 whole revolutions under applied pressures of 1 and 9 GPa. All data are plotted in the form of three-dimensional meshes and the local hardness values within these meshes are then projected onto the lower plane and represented as hardness contours. Inserts depict TEM micrographs taken from the center and edge of the HPT disks.

Inspection shows that all measured values of the local microhardness are larger than for the unprocessed nickel where initially there was a hardness of ~1.4 GPa. It is also apparent that the microhardness values are generally non-uniform across the diameters of the samples and there are lower hardness values in the centres of the disk. The meshes plotted in Figs 1 and 2 show that an increase in the applied pressure from 1 to 9 GPa leads to an overall increase in the values of Hv especially in the centres of the disks. An asymmetry visible at some of the applied pressures was attributed to the development of non-homogeneity of hardening during deformation in HPT since it is apparent that the evolution of hardness with increasing pressure is essentially a non-homogeneous process. At the lowest pressure of 1 GPa the microhardness increases primarily at the outer edge of the disk as anticipated from the higher strains evident in Fig. 1. However, this non-uniformity is gradually removed so that, at a pressure of 9 GPa, there is a reasonably homogeneous distribution of hardness values all lying at a level of > 3 GPa. The increase in the average value of the microhardness with increasing applied pressure appears to be

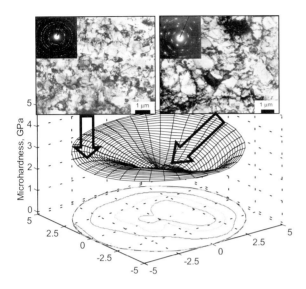

Figure 1: Three-dimensional representations of the local microhardness for disks subjected to 5 whole revolutions under applied pressures of 1 GPa. Inserts show microstructures together with the associated SAED patterns taken from the indicated areas

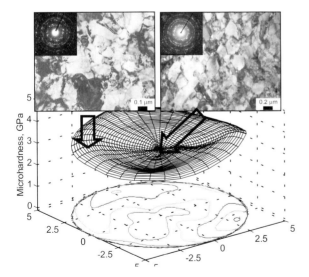

Figure 2: Three-dimensional representations of the local microhardness for disks subjected to 5 whole revolutions under applied pressures of 9 GPa. Inserts show microstructures together with the associated SAED patterns taken from the indicated areas.

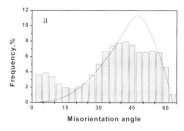 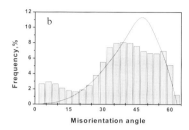

Figure 3: Distributions of grain boundary angles in (a) the centre and (b) the periphery after HPT processing through $N = 5$ revolutions under an applied pressure of $P = 6$ GPa: the solid lines denote a random distribution of misorientation angles

due to the development of higher microhardness values around the periphery of the disk and the subsequent sweeping of these areas of higher hardness across the disk

Examination of the microstructures obtained by TEM reveals important characteristics. First, the mean grain size at the centre of the disk tends to be larger than at the periphery. In practice, the largest difference between the centre and the periphery occurs at the smallest applied pressure of $P = 1$ GPa (Fig. 1) where the cell or subgrain size in the central region is ~2–3 times larger than at the edge: in both locations, the azimuthal spreading of spots in the SAED patterns may indicate high internal stresses and it suggests the presence of subgrains. By contrast, the microstructures are reasonably similar at the periphery and the centre of the disk for the condition where $P = 9$ GPa (Fig. 2) for $N = 5$ whole revolutions and in both locations the grains are essentially equiaxed with an average size of ~0.17 mm. These results are consistent with earlier TEM observations [14].

The grain boundary (GB) statistics were measured for totals of 1145 and 1155 grains from the edge and the centre of the HPT sample, respectively. These GB statistics are presented both in terms of the GB character distributions as measured by the misorientation angles and by the reciprocal density of coincidence sites where four distinct sets of boundaries were defined: low-angle boundaries ($\Sigma 1$), twins or $\Sigma 3$ boundaries, other special GBs ($\Sigma 5$–30) and high-angle random grain boundaries. Following conventional practice, the various types of GB were classified according to CSL-theory using the nearness (or Brandon) condition of $\Delta\theta = 15° \cdot \Sigma^{-1/2}$ [15] with all low-angle boundaries designated as $\Sigma 1$ boundaries and having misorientations up to a maximum of $15°$.

Figure 3 shows the GB distributions for the misorientation angles from measurements taken at the centre and at the periphery after HPT for $P = 6$ GPa with $N = 5$ revolutions and Table 1 shows the statistical data in terms of these four sets of GB: included in Fig. 3 are solid lines denoting a random distribution of misorientation angles and the statistical requirements for a random distribution are also given in the last line of Table 1. The two histograms shown in Fig. 3 have essentially the same appearance thereby demonstrating that the microstructures are similar between the centre and the edge of the disk under these experimental conditions. However, both distributions have a bimodal character with a peak at low angles which is higher than anticipated for a true random distribution. Furthermore, it is shown in Table 1 that the fractions of low-angle boundaries are ~15.0 ± 0.3 % near the periphery and ~18.3 ± 0.2 % near the centre and both of these values are significantly higher than the value of 2.1% anticipated for a random distribution. In Table 1, the fractions of twins ($\Sigma 3$) and special $\Sigma 5$–30 boundaries in the centre and periphery regions of the sample are essentially identical to within experimental error but both fractions are significantly higher than those anticipated for a random distribution. By contrast, the fractions of high-angle random boundaries are also fairly similar in the centre and the periphery but in this case the fractions (~ 65–67 %) are lower than the fraction of 89.3 % required for a true random distribution. Similar results, including a lower fraction of high-angle random boundaries, were also reported earlier for pure Ni after processing by ECAP [10] and HPT [16].

Table 1: Grain boundary character distributions () as a percentage in HPT nickel where $P = 6$ GPa and $N = 5$ revolutions.

Samples \ Boundaries	$\Sigma 1$	$\Sigma 3$	Other, $\Sigma 5$–30	High-angle random
Centre	18.3 ±0.2	3.0 ±0.5	13.5 ±0.3	65.0 ±0.1
Periphery	15.0 ±0.3	4.1 ±0.5	13.5 ±0.3	67.4 ±0.1
Random distribution	2.1	1.6	7.0	89.3

4 Concluding Remarks

1. Using the three separate procedures of microhardness measurements, transmission electron microscopy and orientation imaging microscopy, it is shown that HPT may be used effectively to develop an essentially homogeneous microstructure throughout the sample provided the applied pressure and the torsional strain (as measured in terms of the number of revolutions) are sufficiently high. For pure Ni, a homogeneous microstructure of equiaxed grains may be achieved if the applied pressure is at or above ~6 GPa and the sample is subjected to at least ~5 whole revolutions.
1. From quantitative measurements using orientation imaging microscopy, it is shown that the homogeneous microstructures produced by HPT contain larger fractions of low-angle, twin and special boundaries but a smaller fraction of high-angle random boundaries than anticipated in a random distribution. However, the distributions of misorientation angles in these homogeneous microstructures are essentially identical in the centre and at the periphery of the deformed disks.

5 Acknowledgements

This work was partially supported by the U.S. Army Research Office under Grant No. DAAD19-00-1-0488. One of the authors (APZ) thanks the DGR of Generalitat of Catalonia and the CICYT (MAT2001-2555) for financial support.

6 References

[1] T.C. Lowe, R.Z. Valiev (eds,), Investigations and Applications of Severe Plastic Deformation, Kluwer Academic Publishing, Dordrecht, Netherlands, 2000, pp. 395
[2] R. Z. Valiev, R. K. Islamgaliev, I.V. Alexandrov, Prog. Mater. Sci. 2000, 45, 103–189
[3] R. Z. Valiev, N. A. Krasilnikov, N. K. Tsenev, Mater. Sci. Eng. 1991, A137, 35–40
[4] R. Z.Valiev, A. V. Korznikov, M. M. Mulyukov, Mater. Sci. Eng. 1993, A168, 141–148
[5] S. X. McFadden, R. S. Mishra, R. Z. Valiev, A. P. Zhilyaev, A. K. Mukherjee, Nature, 1999, 398, 684–686
[6] R. Z. Valiev, I. V. Alexandrov, Y. T. Zhu, T. C. Lowe, J. Mater. Res. 2002, 17, 5–8
[7] V. M. Segal, V. I. Reznikov, A. E. Drobyshevskiy, V. I. Kopylov, Russ. Metall., 1981, 1, 99–105
[8] N. A. Smirnova, V. I. Levit, V. I. Pilyugin, R. I. Kuznetsov, L. S. Davydova, V. A. Sazonova, Fiz. Metal. Metalloved. 1986, 61, 1170–1177
[9] A. P. Zhilyaev, S. Lee, G. V. Nurislamova, R. Z. Valiev, T. G. Langdon, Scripta Mater. 2001, 44, 2753–1758
[10] A. P. Zhilyaev, G. V. Nurislamova, M. D. Baró, R. Z. Valiev, T.G. Langdon, Metal. Mater. Trans. A, 2002, 33A, 1865–1868
[11] A. P. Zhilyaev, B.-K. Kim, G. V. Nurislamova, M. D. Baró, J. A. Szpunar, T. G. Langdon, Scripta Mater. 2002, 48, 575–580
[12] D. J. Dingley, D. P. Field, Mater. Sci. Technol. 1996, 12, 1–9
[13] B.-K. Kim, J. A. Szpunar, Scripta Mater. 2001, 44, 2605–2610
[14] R. K. Islamgaliev, F. Chmelik, R. Kuzel, Mater. Sci. Eng. 1997, A237, 43–51
[15] D. G. Brandon, Acta Metall. 1966, 30, 1479–1484
[16] [16] P. Zhilyaev, G. V. Nurislamova, Bae-Kyun Kim, M. D. Baró, J. A. Szpunar and T. G.Langdon. Acta Mater. 2003, 53(3), 753–765

VII Analyses of SPD Materials by Selected Physical Methods

The Meaning of Size Obtained from Broadened X-ray Diffraction Peaks

Tamás Ungár
Department of General Physics, Eötvös University, Budapest, Hungary

1   Abstract

X-ray diffraction peak profile analysis (DPPA) is a powerful tool for the characterisation of microstructures either in the bulk or in loose powder materials. The evaluation and modelling procedures have been developed together with the experimental techniques. The dislocation density and structure, as obtained from peak profile analysis, is in good correlation with TEM observations. As for the crystallite size, in some cases a good correlation, however, in other cases definite discrepancies with TEM results can be observed. In the present work those literature data are critically reviewed where crystallite size or grain size have been determined by DPPA and TEM simultaneously. The correlation and discrepancies between DPPA and TEM results are discussed in terms of the microstructures in different types of specimens.

2   Introduction

X-ray diffraction peak profile analysis is a powerful tool for the characterisation of microstructures either in the bulk or in loose powder materials. The evaluation and modeling procedures have been developed together with the experimental techniques. Special double-crystal high resolution diffractometers [1] and high resolution powder diffractometers at synchrotrons [2,3] have small instrumental peak breadths compared to the physical broadening of diffraction peaks. [4] Diffraction patterns measured by these instruments require "almost no instrumental corrections and thus provide the physical peak profiles as directly measured data. High quality standard materials, especially well annealed LaB^6 with crystallite size larger than several micrometers, can be used for instrumental corrections straightforward. [5,6] As a result, good correlation can be observed between data obtained in different conventional or synchrotron laboratories. The ideal powder diffraction pattern consists of narrow, symmetrical, delta-function like peaks, positioned according to a well defined unit cell. A number of deviations from the ideal powder pattern are related to the microstructure of materials and are the subject of peak profile analysis. (i) Peak shift is related to internal stresses or planar faults, especially stacking faults or twinning. (ii) Peak broadening indicates crystallite smallness and microstresses, however, stress gradients and/or chemical heterogeneities can also cause peak broadening. (iii) Peak asymmetries can be caused by long-range internal stresses, planar faults or chemical heterogeneities. (iv) Anisotropic peak broadening can result from anisotropic crystallite shape or anisotropic strain. Peak profiles reveal at least four fundamental features: (*a*) shifts, (*b*) broadening, (*c*) asymmetries and (*d*) anisotropic broadening. Microstructural properties can be summarised at least into the following nine different categories: (*1*) internal stresses, (*2*) stacking faults, (*3*) twinning, (*4*) crystallite smallness, (*5*) microstresses, (*6*) long-range internal stresses,

(7) chemical heterogeneities, (8) anisotropic crystallite shape or (9) anisotropic strain. There is no one-to-one correlation between the different peak profile features and the different microstructural properties. For this reason the interpretation of peak profiles in terms of microstructural properties becomes more reliable if the results of other methods, e.g. transmission or scanning electron microscopy (TEM or SEM), are also used. On the other hand, the results of other methods, especially TEM or SEM, can be refined and/or Peaks

All the different microstructural properties from (1) to (9) have been treated on well established physical basis. Internal stresses are described by the elastic properties of crystals. The determination of stacking faults and twinning from diffraction profiles is worked out for cubic and hexagonal crystals. [7,8] Size broadening is treated in detail for spherical or non-spherical crystallites. [9,10] The evaluation of microstresses is worked out in detail for dislocations [11–17] Long-range internal stresses are related to the dipole polarization of dislocations. [18,19] Anisotropic crystallite shape are a delicate phenomenon which has been dealt with by special elegance in the case of ZrO nanoparticles. [20] Finally, strain anisotropy is a general feature of peak profiles which is closely related to the elastic properties of crystals. [21–26]

Both, the different peak profile features and the different microstructural properties are complex, and the two sides, e.g. the experimental features and the physical properties can combine in many different ways. This fact makes it practically impossible (at least at this moment) to have a general description of DPPA which would be able to treat all the microstructural properties by a unified general theory. It is the experimentators\9 task to select those microstructural properties which are most relevant to be considered in a particular experiment. At the same time, just because of the complexity of both, the experiment and the microstructural properties, it is probably not possible to produce a model-independent general theory for diffraction peak profile analysis which could tell us something more specific than the obvious about materials properties. For the same reason, though there have been many attempts to develop model independent descriptions of diffraction peak profiles, these have not yet been too successful. In the present work we follow the type of peak profile analysis (often called line profile analysis) in which the theory is based on a well defined specific microstructural model, in particular: small crystallites and dislocations.

In one-component pure materials or single phase compounds the effects of microstrain, strain-anisotropy and crystallite size have been treated together successfully. [23,26–32] The microstrain is related to dislocations, strain-anisotropy is treated in terms of the anisotropic contrast of dislocations and crystallite size is described by appropriate size distribution functions and proper assumptions about shape. With these ingredients we take it as granted that the evaluation of diffraction peak profiles in terms of microstructural properties is correct. The dislocation density and structure, as obtained from peak profile analysis, is in good correlation with TEM observations, cf. [4] As for the crystallite size, in some cases a good correlation, however, in other cases definite discrepancies with TEM results can be observed. In the present work those literature data are critically reviewed where crystallite size or grain size have been determined by DPPA and TEM simultaneously. The correlations and discrepancies between DPPA and TEM results are discussed in terms of the microstructures in different types of specimens.

3 Principles of X-ray Diffraction Peak Profile Analysis

3.1 Strain Broadening

According to the kinematical theory of scattering diffraction profiles are the convolution of the size and distortion profiles, I^S and I^D, respectively: $I^F = I^S*I^D$, where the superscript F indicates physical profiles free of instrumental effects. [6] The Fourier transform of this is the Warren-Averbach equation, cf. in. [7] This equation has a wide generality as proved theoretically by several authors. [8–10] One of the main challenge related to this equation is the way in which the size Fourier coefficients, A_L^S, and the mean square strain, $<\varepsilon_{S;L}^2>$, are interpreted.

The interpretation of the mean square strain can either be done phenomenologically or on the basis of more specific microscopical models. First was made by Warren suggesting that if the displacements of atoms follow a Gaussian distribution then the mean square strain is a constant, i.e $<\varepsilon_{S;L}^2> = <\varepsilon^2>$, independent of either g or L. In this case plotting $\ln A_{S;L}$ versus g^2 for given values of L the descending slops yield the values of $<\varepsilon^2>$. Experiment has shown that the mean square strain is never independent neither of g nor of L. [7,11–17,23–32] The L dependence reveals a double singularity, at small values $<\varepsilon_{S;L}^2>$ diverges logarithmically and at large L values it decreases like $1/L$, latter indicating the longe range character of the mean square strain. Krivoglaz, [11] Wilkens, [12] Groma et al., [14] Gaál, [13] vanBerkum et al., [34] Kamminga and Delhez, [17] Levine and Thomson, [15] and Groma, [16] all have shown that this particular behavior of the mean square strain indicates that the major contribution of strain to diffraction peak broadening comes from dislocations.

The g dependence of the mean square strain is known in powder diffraction as strain anisotropy. [35] This means that neither the breadth nor the Fourier coefficients of the diffraction profiles are monotonous functions of the diffraction angle. Two different models have been developed so far for the interpretation of strain anisotropy: (i) a phenomenological model based on the anisotropy of the elastic properties of crystals [27] and (ii) the dislocation model based on the mean square strain of dislocated crystals. [29] The dislocation model of $<\varepsilon_{S;L}^2>$ takes into account that the contribution of a dislocation to strain broadening depends on the relative orientations between the diffraction vector, g, and the line and Burgers vectors of dislocations, l and b, respectively, in a similar way as the contrast effect of dislocations in electron microscopy. Anisotropic contrast can be summarised in contrast factors, C, which can be calculated numerically on the basis of the crystallography of dislocations and the elastic constants of the crystal. [12,14,24,28,36–42] In the case of untextured polycrystals, or if all the possible Burgers vectors are randomly populated, or if the specimen is a loose powder, the individual contrast factors can be averaged over the permutations of the hkl indices. It can be shown that the average contrast factor, $\bar{C}$, is a linear function of the fourth order invariants of the hkl indices. [24] In the case of cubic and hexagonal crystals this can be witten as:

$$\bar{C} = \bar{C}_{h00}(1-qH^2) \quad \text{and} \quad \bar{C}_{hkl} = \bar{C}_{hk0}\left[1+q_1x+q_2x^2\right] \tag{1}$$

respectively, where $\bar{C}_{h00}$ and $\bar{C}_{hk0}$ are the average dislocation contrast factors for the h00 and hk0 reflections, respectively, $H^2=(h^2k^2+h^2l^2+k^2l^2)/(h^2+k^2+l^2)^2$; q, q_1 and q_2 are parameters depending on the elastic constants and on the character of dislocations in the crystal (e.g. edge or screw, or basal, "prismatic or pyramidal, respectively). In Equation 1 $x = (2/3)(l/ga)^2$, where l and a are the prysmatic index and the "basal lattice constant, respectively, in the case of hexagonal crystals. Detailed accounts and compilations of the q, q_1 and q_2 parameters can be found in. [39,41,42] Here, we note that the fourth order invariants of the hkl indices appear also in the phenomenological interpretation of anisotropic strain broadening given by Stokes and Wilson [23] and Stephens. [27]

3.2 Size Broadening

Size broadened profiles can be described by assuming (i) a size distribution function and (ii) the shape of crystallites, or coherently scattering domains. There is experimental evidence that the log-normal size distribution function, $f(x)$, "given by the median m and the variance ρ, can describe crystallite size distribution in a wide range of bulk or loose powder materials. [9,43–47] Hinds [48] has shown that with m and ρ the arithmetic-, the area- and the volume weighted average crystallite diameters are:

$$\langle x \rangle_j = m \exp(k\sigma^2) \qquad (2)$$

where k = 0.5, 2.5, and 3.5 in the case of arithmetic-, area- and volume weighted average and j stands for these different averages, respectively.

3.3 The Evaluation Procedure

A numerical evaluation procedure has been worked out which provides five or six physical parameters in the case of cubic or hexagonal crystals, respectively. [49] In cubic crystals these parameters are: (i) the mean, m, and (ii) the variance, ρ, of the log-normal size distribution function, (iii) the dislocation density, ρ, (iv) the dislocation arrangement parameter, $M = R \sqrt{\rho}$, and (v) the q parameter in the dislocation contrast factor. In hexagonal crystals (i) to (iv) are the same, but instead of a single q parameter in the dislocation contrast factors, two parameters, q_1 and q_2 are provided in accordance with Equation 1.

The numerical procedure has the following steps. (*a*) The Fourier coefficients of the measured physical profiles are calculated by a non-equidistant sampling Fourier transformation, (*b*) the Fourier coefficients of the size and strain profiles are calculated, (*c*) the experimental and the calculated Fourier coefficients are compared by the Marquardt-Levenberg [50,51] least squares procedure using the Gnuplot program package under the unix systems. The Gnuplot programme package has been augmented by the non-equidistant sampling Fourier transformation and the fast Fourier transformation procedures. [49] The procedure is freely available for non-commercial, scientific purposes through the internet as a frontend: http://www.renyi.hu/mwp. A username and password for access to the frontend is instantly provided by the authors on request by email at either ribarik@renyi.hu or ungar@ludens.elte.hu.

4 Crystallite Size and Size-distributions Determined by DPPA and TEM

There is a limited number of experimental work available in the literature where X-ray diffraction peak profile analysis (DPPA) and transmission electron microscopy (TEM) have been carried out simultaneously to study the microstructure. In the following we present a concise review in three categories: (1) when X-ray and TEM sizes are identical, (2) when X-ray and TEM sizes are in good correlation and (3) when X-"ray and TEM sizes are in discrepancy and therefore need discussion.

4.1 When X-ray and TEM Sizes are Identical

In the sintering process of bulk silicon-nitride (Si_3N_4) ceramics the grain size of the starting powder is an important factor. [52,53] Si_3N_4 loose powders or bulk specimens can be produced in many different ways. [54–56] Figure 1 (by courtesy of Gubicza et al. [55]) shows the TEM image of a commercial (LC12, Starck, Germany) Si_3N_4 powder produced by nitridation of silicon and subsequent milling. A large number of images were evaluated for particle size. About 300 particles were chosen at random in different areas in different images. The frequency of the measured diameters, scaled at the left ordinate of the plot, is shown in Fig. 2 (by courtesy of Gubicza et al. [55]) as bar graphs. The solid line in the figure is the size distribution function obtained by DPPA. The $f(x)$ function is scaled at the right ordinate of the plot. The agreement between the size distributions obtained by TEM and X-rays is good. Similar comparison of the crystallite size distributions has been made for a nanodisperse Si_3N_4 powder which has been synthesized by the vapor phase reaction of silicon-tetrachloride and ammonia in a thermal plasma reactor and crystallized at 1500∘C. [56] This matter represents a rare case when the TEM size distribution indicates smaller particle size than DPPA. The reason of this is that in the initial state the plasma-thermally synthesized specimen contained an amorphous fraction of about 80 % (see in Table 1, in ref. [56]). After crystallisation at 1500 ∘C an amorphous fraction of about 20 % is still retained. In the TEM images the smallest amorphous particles cannot be distinguished from the particles of the crystalline phase which suppresses the average crystallite

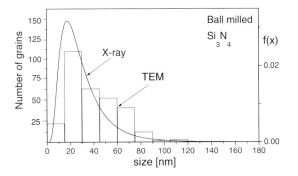

Figure 1: TEM image of a commercial (LC12, Starck, Germany) Si_3N_4 powder produced by nitridation of silicon and subsequent ball milling.

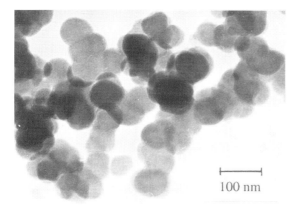

Figure 2: The size-distribution of the Si_3N_4 particles measured by TEM intersection counting (bar graph) and by X-ray DPPA (solid line).

size measured by TEM. The number of particles sampled in the present X-"ray diffraction experiments is about 10^9 which is many orders of magnitude larger than the number evaluated in the TEM experiments. With this in mind we consider that the agreement between the X-ray and TEM size distributions is good.

Langford and coworkers have studied the effect of size distribution on X-ray powder diffraction peak profiles. [9] "Nanocrystalline powder specimens of CeO_2 were prepared from the decompositon of an oxide-nitrate, $Ce_2O(NO_3) \cdot H_2O$, under N_2 atmosphere with a heating rate of 3 K/h up to 503 K. The specimen was additionally annealed for 24 h at the same temperature. The high resolution TEM (HREM) image (see Fig. 3, in [9]) has indicated crystallites of the diameter of about 4±1 nm. Assuming log-normal size distribution the diffraction patterns were fitted by a Rietveld type whole-powder-pattern fitting procedure. The size distribution provided by the X-ray fitting procedure is in perfect agreement with the bar diagram obtained by evaluating the TEM images. [9]

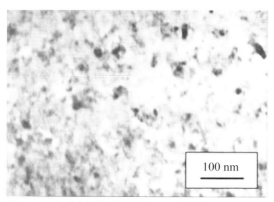

Figure 3: The bright field TEM image of the fast cooled ribbon of Hf11Ni89

Loose powder of ZnO was prepared from the thermal "decomposition of zinc-hydroxinitrate, $Zn_3(OH)_4(NO_3)_2$. [20] In order to avoid secondary growth after the first decomposition from the precursors the ambient conditions were carfully controlled. [57] The HREM image in Fig. 10b in Louër et al. [20] shows well defined cylindrical nanoparticles of ZnO with aspect ratios of about 3. The Williamson-Hall plot in Fig. 4 in Louër et al. [20] revealed a strong anisotropy with hkl, however, with clear indication for the absence of strain, e.g. the full width at half maximum (FWHM) of the 002/004 and 100/300 pairs of reflections are identical within the experimental error. Based on this behaviour of the Williamson-Hall plot, authors evaluated the shape anisotropy in terms of a cylindrical particle model. The average cylinder diameter and height was obtained as 8±2 nm and 20±5 nm, respectively, in excellent agreement with the HREM results.

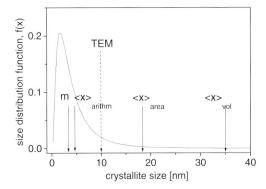

Figure 4: The crystallite size distribution function determined by X-ray DPPA of Hf11Ni89, the different averages according to Equation 2 and the average size estimated from the TEM image in Figure 3 together with the median of the size distribution function are also indicated in the figure.

The above examples show that in the case of loose powder particles, as long as the particle diameter is within the sensitivity range of X-ray diffraction peak broadening, i.e. smaller than about one micrometer, TEM or HREM particle sizes are in perfect agreement with size and size-distribution results provided by X-ray DPPA.

4.2 When X-ray and TEM Sizes are in Good Correlation

Electrodeposited nanocrystalline Ni foils were produced by pulse plating onto titanium substrate. [58,59] The X-ray DPPA provided: $<x>_{area} = 12$ nm and $<x>_{vol} = 38$ nm for the area- and volume-weighted mean crystallite diameters, respectively. [60] Although the TEM image obtained on the same specimen has not been evaluated quantitatively for crystallite or grain size, a good qualitative correlation between the X-ray and TEM results can be observed. [58,59]

High purity Ni powder was ball milled for 96 h in a planetary mill. [47] Dark field and HREM images were evaluated quantitatively. The size distributions obtained by X-ray DPPA were found to be in good correlation with the TEM evaluations.

Nanocrystalline Pd pallets were prepared by inert-gas condensation and subsequent compaction. [61,62] Dark field TEM images were evaluated quantitatively. The 111/222 pair of X-"ray diffraction peaks were evaluated by the classical Warren-Averbach method. [7] It was shown that from the size Fourier coefficients both, the area- and the volume-weighted mean crystallite sizes can be evaluated. From these two mean size values the size distribution function has been determined. A good correlation between the bar-diagram obtained from the TEM images and the size-distribution function provided by the X-ray DPPA was observed.

Nanocrystalline Ni_5Hf particles were produced by careful crystallization from the amorphous state. [63] The amorphous ribbons of 3 mm width and 11 λm thickness were produced from a master alloy of Hf11Ni89 melted and fast quenched by the melt-spinning technique. The bright field TEM image of the ribbon with the small particles is shown in Fig. 3. The diffraction patterns show a strong amorphous halo and well "defined Debye-Sherrer rings indicating the coexistence of the amorphous and crystalline phases. The black contrasts in the image are the Ni_5Hf particles. [64] The crystallite size distribution function determined by X-ray DPPA is shown in Fig. 4. The measured values of the meadian and variance are: $m = 3.3\pm0.2$ nm and $\rho = 0.82\pm0.02$. The arithmetic-, the area- and the volume-weighted mean crystallite size values calculated from m and ρ according to equation (2) are also shown in the figure. They indicate a wide size distribution in good correlation with the TEM image where very small and somewhat larger particles can be seen together. From the TEM image the particle size has only been determined qualitatively and is indicated in Fig. 4 by a dashed line.

Nanocrystalline Cu was prepared by inert-gas condensation and subsequent hot compaction at Argonne National Laboratory [65]. The crystallite size distribution functions of specimens deformed up to about 10 % strain was determined shortly after sample preparation and after six months natural ageing at room temperature. [65] The log-normal size distribution functions were determined both by TEM and X-ray DPPA. The comparison showed on the one hand that there was a slight grain growth during natural ageing at room temperature, on the other hand, there was a good correlation between the TEM and X-ray size distribution functions.

4.3 When X-ray and TEM Sizes are in Discrepancy

The microstructure of the 1.4914 ferritic/martensitic steel from the European Fusion Materials Technology Programme (EFMTP) was investigated by TEM by Marmy et al. [66] and Marmy and Victoria. [67] It was found that the strength of the tempered martensite is caused by the carbide precipitates and by the lath structure which is strengthened by a high density of dislocations, after tensile deformation or low cycle fatigue a well developed dislocation cell structure is formed leaving a small amount of the original dislocation density in the martensite. The TEM images revealed strongly elongated laths of the ferritic/martensitic material in which the laths are subdivided by dislocation cell walls. The dislocation densities and arrangement parameters, ρ and M, the parameters of the log-normal size distribution function, ρ and m, and the area-weighted mean crystallite diameters, <x> area, obtained by eq. (2) for the normal state and after fatigue at RT and 250 ∘C are shown in Table 2 of Ref. [67]. It can be seen that while the dislocation density increases during fatigue, especially at 250 ∘C, the area-weighted mean crystallite diameter increases also considerably. This indicates that during fatigue, especially at higher temperatures, the dislocation density in the cell walls increases strongly, whereas the dislocation cell size or the subgrain size increases too. In fact, the two parameters, i.e. the dislocation density and the area-weighted mean crystallite diameters, $\langle\pi^2 r\rangle_E$ and $\langle x\rangle_{area}$, are in good correlation with TEM observations and other experiments on fatigued specimens.

A 99.98 % copper specimen was extruded and subsequently deformed by equal channel angular pressing (ECAP) in a single pass. [43] For the purpose of the X-ray diffraction measurements, in order to avoid machining effects, an approximately 100 λm surface layer was removed from the specimen surface by chemical etching. The crystallite size distribution function was determined by DPPA. TEM images were evaluated in two steps. [30] In a first step the well defined, large angle grain boundaries were drawn into a contour map. This procedure was carried out on about ten similar images. The linear intersection method was used to produce the bar graph of the distribution of crystallite-diameters. It was established that the TEM size distribution gives much larger size values than the X-ray method. In a second step typical grains in good contrast orientation were selected and finer contour maps were produced. These contour maps were also evaluated by the linear intersection method. The size distribution obtained from the finer contours corresponds to subgrains or dislocation cells and this size distribution is close to that determined from X-rays. [30] This indicates that in a bulk specimen the crystallite size and size-distribution obtained by the X-ray method is closer or almost equal to the subrain or dislocation cell size or size-distribution.

Submicron grain-size titanium was produced by ECAP and cold rolled subsequently. [68] The crystallite size distribution and the dislocation density and the type of active slip systems were studied by X-ray diffraction. [69] The median and the variance of the crystallite size-distribution were obtained as: $m = 39$ nm and $\rho = 0.14$, respectively. Due to the small value of ρ these values provide the area-weighted mean crystallite size close to the value of m: $\langle x\rangle_{area} = 41 \pm 3$ nm. The mean size of the dislocation cells obtained from TEM images was about 45 ± 5 nm, in a relatively good correlation with the X-ray results. [70] It is noted however, that the grain size observed in TEM is much larger, about 270 ± 20 nm (by counting over some 150 grains in the TEM images). [70]

5 Conclusions

The examples discussed above show that:

1. the interpretation of crystallite size obtained from X-ray diffraction peak broadaning needs thorough discussion in terms of microstructure,
2. a single parameter obtained either from the reciprocal of the FWHM, or the reciprocal of the integral peak breadth, or the size Fourier coefficients alone cannot tell anything about a unit as an element of the microstructure,
3. the crystallites or coherently scattering domains always have a distribution, in other words they are usually polydisperse,
4. the size distribution of crystallites can be determined from X-ray diffraction peak broadaning if assumptions are made about (a) the type of distribution and (b) the shape of crystallites,
5. the knowledge of the size distribution enables to determine different averages, especially, the arithmetic-, area- and volume-weighted averages, respectively,
6. only the appropriately weighted averages can be compared directly with size values obtained from other measurements, especially TEM investigations,

7. TEM counting of average sizes is most appropriately correlated with the area weighted mean crystallite size from X-ray diffraction,
8. the examples discussed here have shown that the X-"ray determined crystallite size can be (1) either identical with the TEM particle- or grain size, (2) or can give the average size of subgrains or dislocation cells,
9. (1) is observed either in loose powders of submicron grain size or nanocrystalline bulk materials, whereas (2) is more often the case in bulk materials containing strain or dislocations,
10. the general conlusion is, that a shear number of size provided by X-ray diffraction peak broadaning does not mean much without a thorough discussion of the microstructure, possibly supported by auxiliary observations, best by TEM.

6 Acknowledgement

This work was supported by the Hungarian Scientific Research Fund, OTKA, Grant Nos. T031786, T034666 and T029701.

7 References

[11] M. Wilkens, H. Z. Eckert, Naturforschung 1964, 19a, 459.
[12] A. N. Fitch, Mater. Sci. Forum 1996, 219, 228.
[13] R. E. Dinnebier, P. W. Stephens, M. Pink, J. Sieler, "Abstract in Nat. Synchrotron Light Source Activity Rep.' 1995, Publ. Brookhaven Nat. Lab., BNL 52496, 1995, p. B47.
[14] M. Wilkens, in Proc. 8th Int. Conf. Strength Met. Alloys (ICSMA 8) (Eds.: P. O. Kettunen, T. K. Lepistö, M. E. Lehtonen), Tampere, Finland, Pergamon Press, 1988, p.47–152.
[15] D. Louër, J. I. Langford, J. Appl. Cryst. 1988, 21, 430.
[16] J. I. Langford, R. Delhez, Th. H. De Keijser, E. J. Mittemeijer, Austr. J. Phys. 1988, *4*, 173.
[17] B. E. Warren, Progr. Metal Phys. 1959, *8*, 147.
[18] M. M. J. Treacy, J. M. Newsam, M. W. Deem, Proc. Roy. Soc. London A 1991, 433, 499.
[19] J. I. Langford, D. Louër, J. Appl. Cryst. 2000, 33, 964.
[20] P. Scardi, M. Leoni, J. Appl. Cryst. 2000, 33, 184.
[21] M. A.Krivoglaz, in X-ray and Neutron Diffraction in Nonideal Crystals, Springer-Verlag, Berlin Heidelberg New York, 1996.
[22] M. Wilkens, in Fundamental Aspects of Dislocation Theory (Eds.: J. A. Simmons, R. de Wit, R. Bullough), Vol. II. Nat. Bur. Stand. (US) Spec. Publ. No. 317, Washington, DC. USA, 1970, p. 1195–1221.
[23] I. Gaál, in Proc. 5th Risø Int. Symp. on Metallurgy and "Material Science (Eds.: N.H. Andersen, et al.), Riso Nat. Lab., Roskilde, Denmark, 1984, p. 249–254.
[24] I. Groma, T. Ungár, M. Wilkens, J. Appl. Cryst. 1988, 21, 47.
[25] L. E. Levine, R. Thomson, Acta Cryst. A 1997, 53, 590.
[26] I. Groma, Phys. Rev. B 1998, 57, 7534.
[27] J. D. Kamminga, R. Delhez, J. Appl. Cryst. 2000, 33, 1122.

[28] H. Mughrabi, Acta metall. 1983, 31, 1367.
[29] H. Mughrabi, T. Ungár, W. Kienle, M. Wilkens, Phil. Mag. 1986, 53, 793.
[30] D. Louër, J. P. Auffrdic, J. I. Langford, D. Ciosmak, J. C. Niepce, J. Appl. Cryst. 1983, 16, 183.
[31] Stokes, A. R.; Wilson, A. J. C., Proc. Phy. Soc. London, 1944, 56, 174–181.
[32] Ungár, T.; Tichy, G., Phys. Stat. Sol. (a) 1999, 171, 425.
[33] Scardi, P.; Leoni, M., J. Appl. Cryst., 1999, 32, 671–682.
[34] Dinnebier, R. E., Von Dreele, R., Stephens, P. W., Jelonek, S.; Sieler, J., J. Appl. Cryst., 1999, 32, 761–769.
[35] P. W. Stephens, J. Appl. Cryst. 1999, 32, 281.
[36] R. W. Cheary, E. Dooryhee, P. Lynch, N. Armstrong, S. Dligatch, J. Appl. Cryst. 2000, 33,1271.
[37] T. Ungár, A. Borbëly, Appl. Phys. Lett. 1996, 69, 3173.
[38] A. Rëvësz, T. Ungár, A. Borbëly, J. Lendvai, Nanostructured Materials 1996, 7, 779.
[39] T. Ungár, S. Ott, P. G. Sanders, A. Borbëly, J. R. Weertman, Acta Mater. 1998, 10, 3693.
[40] T. Ungár, J. Gubicza, G. Ribárik, A. Borbëly, J. Appl. Cryst. 2001, 34, 298.
[41] D. Kaptás, L. F. Kiss, J. Balogh, J. Gubicza, T. Kemëny, I.'Vincze, Hyperfine Interactions 2002, 141/142, 175.
[42] T. Kemëny, L. F. Kiss, D. Kaptás, J. Balogh, L. Bujdosó, J. Gubicza, I. Vincze, Mat. Sci. Eng. A 2002, in press.
[43] H.-J. Bunge, J. Fiala, R. L. Snyder, (Eds.), X-ray Powder Diffraction Analysis of Real Structure of Materials, IUCr series, Oxford Univ. Press, 2000.
[44] J. G. M. Van Berkum, A. C. Vermuelen, R. Delhez, T. H. de Keijser, E.J. Mittemeijer, J. Appl. Cryst. 1994, 27, 345.
[45] G. Caglioti, A. Paoletti, F. P. Ricci, Nucl. Instrum. 1958, *3*, 223.
[46] P. Klimanek, R. Kuzel Jr., J. Appl. Cryst. 1988, 21, 59.
[47] R. Kuzel Jr., P. Klimanek, J. Appl. Cryst. 1988, 21, 363.
[48] R. Kuzel Jr., P. Klimanek, J. Appl. Cryst. 1989, 22, 299.
[49] T. Ungár, I. Dragomir, A. Rëvësz, A. Borbëly, J. Appl. Cryst. 1999, 32, 992.
[50] A. Borbëly, J. H. Driver, T. Ungár, Acta Mater. 2000, 48, 2005.
[51] I. Dragomir, T. Ungár, J. Powder Diffraction 2002, 17, 104.
[52] I. Dragomir, T. Ungár, J. Appl. Cryst. 2002, 35, 556.
[53] R. Z. Valiev, E. Kozlov, V. Ivanov, F. Yu. J. Lian, A. A. Nazarov, B. Baudelet, Acta metall. mater. 1994, 42, 2467.
[54] C. D. Terwilliger, Y. M. Chiang, Acta Met. Mater. 1995, 43, 319.
[55] C. E. Krill, R. Birringer, Phil. Mag. A 1998, 77, 621.
[56] T. Ungár, A. Borbëly, G. R. Goren-Muginstein, S. Berger, A. R. Rosen, Nanostructured Materials 1999, 11, 103.
[57] P. Scardi, M. Leoni, J. Acta. Cryst. A 2002, 58, 190.
[58] W. C. Hinds, Aerosol Technology: Properties, Behavior and Measurement of Airbone Particles, Wiley, New York, 1982.
[59] G. Ribárik, T. Ungár, J. Gubicza, J. Appl. Cryst. 2001, 34, 669.
[60] K. Levenberg, Quart. Appl. Math. 1944, *2*, 164.
[61] D. W. Marquardt, J. Appl. Math. 1963, 11, 431.
[62] J. Gubicza, P. Arató, F. Wëber A. Juhász, J. Mat. Sci. Lett. 1998, 17, 615.
[63] J. Gubicza, F. Wëber, Mater. Sci. Eng. A 1999, 263, 101.

[64] J. Szëpvölgyi, I. Mohai, J. Gubicza, J. Mater. Chem. 2001, 11, 859.
[65] J. Gubicza, J. Szëpvölgyi, I. Mohai, L. Zsoldos, T. Ungár, Mat. Sci. Eng. A 2000, 280, 263.
[66] J. Gubicza, J. Szëpvölgyi, I. Mohai, G. Ribárik, T. Ungár, J. Mater. Sci. 2000, 35, 3711.
[67] M. Loüer, M. Loüer, D. Weigel, C.R. Acad. Sci. 1970, 270, 881.
[68] E. Tóth-Kádár, I. Bakonyi, A. Sólyom, J. Hering, G.'Konczos, F. Pavlyák, Surf. Coat. Technol. 1987, 31, 31.
[69] I. Bakonyi, E. Tóth-Kádár, L. Pogány, A. Cziráki, L. Gerocs, K. Varga-Josepovits, B. Arnold, K. Wetzig, Surf. Coat. Technol. 1996, 78, 124.
[70] T. Ungár, A. Rëvësz, A. Borbëly, J. Appl. Cryst. 1998, 31, 554.
[71] R. Birringer, H. Gleiter, H.-P. Klein, P. Marquartdt, Phys. Lett. 1984, A102, 365.
[72] C. E. Krill, R. Birringer, R., Phil. Mag. A 1998, 77, 621.
[73] I. Bakonyi, F. Mehner, M. Rapp, A. Cziráki, H. Kronmüller, R. Kirchheim, Z., Metallkd. 1995, 86, 619.
[74] J. Gubicza, G. Ribárik, I. Bakonyi, T. Ungár, J. Nanosci. and Nanotechn. 2001, *1*, 343.
[75] P. G. Sanders, G. E. G.; Fougere, L. J. Thompson, J. A. Eastman, J. R. Weertman, Nanostruct. Mater. 1997, *8*, 243.
[76] P. Marmy, R. Yuzhen, M. Victoria, J. Nuclear Mater. 1991, 179, 697.
[77] P. Marmy, M. Victoria, M., in Proc. Int. Conf. Strength of Metals and Alloys ICSMA 9 (Eds.: D. G. Brandon, R.'Chaim, R. A. Rosen), Vol. 2, Haifa, 1991, p. 841.
[78] V. V. Stolyarov, Y. T. Zhu, I. V. Alexandrov, T. C. Lowe, R. Z. Valiev, Mater. Sci. Eng. A 2001, 299, 59.
[79] J. Gubicza, I. C. Dragomir, G. Ribárik, Y. T. Zhu, R. Z. Valiev, T. Ungár, Mater. Sci. Forum 2003, 414-415, 229.
[80] Y. T. Zhu, J. Y. Huang, J. Gubicza, T. Ungár, Y. M. Wang, E. Ma, R. Z. Valiev, J. Mater. Res., submitted.
[81] T. Ungár, A. Revësz, A. Borbëly, J. Appl. Cryst. 1998, 31, 554.

Ultra Fine Grained Copper Prepared by High Pressure Torsion: Spatial Distribution of Defects from Positron Annihilation Spectroscopy

J. Cizek [1], I. Prochazka [1], G. Brauer [2], W. Anwand [2], R. Kuzel [1], M. Cieslar [1], R. K. Islamgaliev [3]
[1] Charles University in Prague, Faculty of Mathematics and Physics, Prague, Czech Republic
[2] Institut für Ionenstrahlphysik und Materialforschung, Forschungszentrum Rossendorf, Dresden, Germany
[3] Institute of Physics of Advanced Materials, Ufa State Aviation Technical University, Ufa, Russia

1 Introduction

High-pressure torsion (HPT) and equal channel angular pressing (ECAP) are techniques based on severe plastic deformation (SPD). They are used to produce relatively large amounts of ultra fine-grained (UFG) materials without any porosity [1]. HPT-made metallic UFG specimens are typically disk shaped with a diameter of $\approx$ 10 mm and a thickness of $\approx$ 0.3 mm. In an ideal case, the specimens would exhibit axial symmetry of UFG structure. In real case, however, some deviations from axial symmetry of UFG structure may be expected. In addition, radial as well as depth variations of grain size and/or defect density may result from unequal degrees of plastic deformation during the preparation of specimens. In the ECAP technique, ingots of a diameter given by channel cross-section are produced. The shear deformation during ECAP pressing is the same along the whole ingot. It means that, contrary to the HPT-made specimens, variations of defect density and grain size with depth are not expected.

The investigation of lateral and depth distributions of grain size and defects in UFG materials prepared by SPD is very important for an understanding of processes which take place during the preparation procedure and underlie the formation of an UFG structure. Previous transmission electron microscopy (TEM) studies of UFG metals prepared by HPT [2,3] revealed a fragmented structure with high angle misorientation of neighbouring grains. Grain interiors are almost free of dislocations and separated by distorted regions (DRs) with a high-dislocation density situated along grain boundaries (GBs) [2,3]. Strong inhomogeneities in the dislocation distribution have also been confirmed by X-ray diffraction (XRD) [4], synchrotron radiation [5] and positron lifetime (PL) spectroscopy [3,6]. In our recent PL investigations of HPT-made UFG copper [3,6], two kinds of defects could be found in the material: (i) dislocations in DRs along GBs and (ii) microvoids of a size comparable to a few vacancies distributed homogeneously throughout grains.

The present contribution is concerned with the spatial distribution of defects in SPD-made UFG copper. For specimens produced by HPT, detailed data were collected on defect densities as functions of radial distance from specimen axis as well as the depth from the specimen surface. Two mutually complementary positron-annihilation spectroscopy (PAS) techniques were employed as a principal experimental tool in the present work: (a) slow positron implantation spectroscopy (SPIS) by measurement of Doppler broadening (DB) of annihilation radiation, and (b) conventional PL and DB measurements with a ^{22}Na positron source. Results obtained by these techniques were correlated with TEM, XRD and microhardness measurements. Since the results obtained for HPT-made specimens will be given in details elsewhere [7] they are only briefly reproduced here. In addition, first PAS measurements on ECAP produced specimens are reported here and compared with those for the HPT technique.

2   Experimental

PAS makes use of the annihilation of electron–positron pairs in matter into which energetic positrons were implanted. The annihilation process results in an emission of annihilation photons, which convey desired information about the material studied. Basic PAS observables are the positron lifetime and the Doppler broadening of annihilation radiation. The former quantity reflects the local electron density scanned by a positron while the latter one depends on the electron momentum distribution near the annihilation site. Both observables can be used for a characterisation of open-space defects capable of capturing positrons into a localised state: vacancies, dislocations, a few-vacancy clusters, etc. PAS is nowadays widely recognised as an effective non-destructive tool for investigations of materials structure on an atomic level [8]. The potential of PAS for such studies arises from the facts that thermalised positrons penetrating through matter behave like a probe of the size of a few nanometers. Since positrons diffuse in metals to distances typically amounting to several hundreds of nanometers, an information from statistically significant volume is obtained.

SPIS experiments were performed on the magnetically guided positron beam system "SPONSOR" [9] at Rossendorf. Positron energies E varied between 30 eV and 35 keV. Within this energy range, a region from sample *surface* up to $\approx$ 1.5 μm depth is probed by the positron beam. The diameter of the beam spot was $\approx$ 4 mm within the whole positron energy range. Energy spectra of annihilation photons were measured by means of a HPGe spectrometer with an energy resolution of 1.25 keV at 0.511 MeV. The shapes of Doppler broadened annihilation peaks were characterised in terms of ordinary S and W parameters proportional to the central (low momentum) and wing (high momentum) parts of peaks, respectively. Hence, the larger S-values mean an enhanced role of positron trapping in open-volume defects and variations of S with energy E reflect the depth dependence of defect concentration and of positron back diffusion to the surface. The S(E) dependencies were analysed using the VEPFIT code [10].

Due to higher initial energies of positrons, conventional PL and DB measurements probe *volume* properties of specimens. These measurements were conducted at Prague. A 1.3 MBq ^{22}NaCl source sealed between 2 μm thick mylar foils was used (the diameter of the activity spot was $\approx$ 3 mm). A high-resolution BaF$_2$ spectrometer [11] was utilised in PL measurements (time resolution of $\approx$ 150 ps FWHM at $\approx$ 75 s^{-1} coincidence count rate). At least 10^7 counts were collected in each spectrum. The spectra were decomposed into exponential components by means of a maximum likelihood procedure [11]. Conventional DB measurements were performed using a standard HPGe spectrometer with an energy resolution of 1.2 keV FWHM at 0.511 MeV.

XRD measurements were carried out with the aid of powder diffractometers using Cu K$_\alpha$ radiation, for which penetration depth in Cu amounts to $\approx$ 6–7 μm. Thus a layer, still relatively close to the specimen surface, was probed. A full description of XRD studies as well as TEM and microhardness measurements performed in this study will be given elsewhere [7].

HPT specimens with UFG structure were made of high purity copper (99.99 %) subjected to torsion under a pressure of 6 GPa at room temperature (true logarithmic strain e = 7). In order to increase the depth range of SPIS and XRD, some of the UFG Cu specimens were, after characterisation of the as-prepared state, subjected to a controlled chemical etching in a solute of HNO$_3$ (25 %), H$_3$PO$_4$ (12.5 %), CH$_3$COOH (12.5 %) and H$_2$O (50 %) at room temperature.

ECAP specimens for PAS measurements were obtained by cutting discs of ≈ 1 mm thickness from the virgin ingot. In order to remove defects introduced by cutting, a similar etching procedure as described above was applied.

3 Results

Conventional PL and SPIS measurements of S parameter vs. positron energy E, S(E), were performed with a well annealed (850 °C/30 min) pure Cu (99.999 %) *reference* specimen. A single (bulk) lifetime of 114.5 ± 0.1 ps and the diffusion coefficient of positrons in Cu, D_+ = 1.86 ± 0.08 cm^2 s^{-1}, were obtained in accordance with the literature data [12,13].

3.1 HPT-made UFG Cu

As-received specimen. The S(E) dependence measured by SPIS with the beam hitting the centre of the disk is shown in Figure 1 and compared with the reference Cu. The S values of Figure 1 are higher for UFG Cu than for the reference Cu above 5 keV positron energy. This is a clear evidence of positron trapping at open volume defects in the HPT-made UFG Cu. The absence of any plateau of the observed S(E) curve for the UFG Cu, which remained slightly decreasing up to the highest positron energy measured (see Figure 1), implies a decrease of defect density with depth. The two kinds of defects in HPT-made UFG copper were identified in our recent study [6]: dislocations inside the DRs along GBs, and microvoids inside grains. Thus the observed decrease in S(E) may occur due to a decrease in the concentration of microvoids or the dislocation density, and also due to increase in the mean crystallite size (decrease in the volume fraction of DRs).

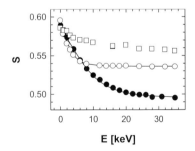

Figure 1: S(E) dependencies measured for as received (□), etched (◊) UFG Cu and annealed Cu (●). VEPFIT results are shown by the solid line.

To study the local variations of defect density, the S(E) dependencies were measured with the beam centred at several positions in radial distances of r = 3 mm from the specimen axis. Local changes in S values were observed. Generally the S parameters appeared to be lower at "outer" positions than at the "centre" one. XRD has indicated [7] no significant changes of the dislocation density or the mean crystallite size with r. Therefore, the lowering of S parameter observed in the subsurface layer indicates a decrease in concentration of microvoids with increasing r.

Conventional PL and DB measurements were carried out for several values of r with the same specimens. Two components, arising from positron trapping in dislocations and at microvoids, were identified in the PL spectra as shown in Fig 2. The lifetime of the microvoid component, τ_2, which exhibited a slightly non-monotonic variations with r, reflects the size of microvoids. From theoretical results of Ref. [6], the mean volume of microvoids was estimated to be 3-4 agglomerated vacancies. A slight increase of the size of microvoids occurs for $r > 3.5$ mm, see Figure 2. The intensity of microvoid component, I_2, shows a remarkable increase in the interval $0 < r < 2$ mm. No variations of dislocation density with r were indicated by TEM and XRD [10]. Thus a radial increase of the concentration of microvoids towards higher r is implied by observed I_2's. It is further supported by: (i) an increase of S parameters with r found by means of conventional DB, and (ii) the measured increase of microhardness with r (see Ref. [10] for details). On the other hand, SPIS results for a thin layer close to surface indicate slightly lower S values at $r = 3$ mm than at the sample centre. It means that spatial distribution of defects very close to the surface can be different from that at larger depth seen by the conventional PAS.

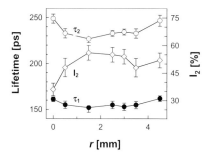

Figure 2: Positron lifetimes τ_1 and τ_2 and intensity I_2 as a function of radial distance r

An *etched specimen* was then used in SPIS and XRD where layers of ≈ 35 and ≈ 50 μm thickness were removed, respectively. The measured $S(E)$ dependence was included in Figure 1. The S values intermediate between the as prepared and reference states were observed for high positron energies. This is a manifest of a decreased but still existing positron trapping at ≈ 35 μm depth. A clearly visible plateau for $E > 15$ keV means that, contrary to the subsurface layer, no significant changes of the defect density, or the grain size, with depth occur at ≈ 35 μm depth. VEPFIT analysis provided an effective positron diffusion length of $L_+ = 35 \pm 1$ nm. This value does allow to estimate the mean dislocation densities to be $(1.1 \pm 0.3) \cdot 10^{15}$ and $(7.0 \pm 0.5) \cdot 10^{14}$ m^{-2} for the subsurface layer and $t \approx 35$ μm, respectively. XRD results for the etched specimen suggest an increased coherent domain size (i. e. a decreased volume fraction of DRs and hence the mean dislocation density) compared to the as received state (see Ref. [10] for details). One can also note a reasonable agreement of XRD data on coherent domain size with TEM results regarding the estimated mean grain size [10]. No more changes of S with r were detected at $t \approx 35$ μm.

3.2 ECAP made UFG Cu

SPIS measurements were carried out with the beam hitting the centre of the specimen and the two opposite edge positions (r 3 mm). In all cases, exactly the same S(E) curves were obtained. This suggests that there are no significant radial changes of the defect density. In Figure 3, $S(E)$ curves measured for etched HPT and ECAP specimens are shown. A plateau of the $S(E)$ curve is clearly visible for E > 15 keV for both specimens. The bulk S value is lower for the ECAP specimen, indicating a lower degree of positron trapping at open volume type defects compared to the HPT specimen. The same conclusion follows also from effective positron diffusion lengths obtained by VEPFIT: $L_+ = 42 \pm 2$ and 35 ± 1 nm respectively, for ECAP and HPT samples. This seems to indicate that a larger mean grain size may result from the ECAP compared to the HPT procedure.

Conventional PL measurements performed for the ECAP specimen (central position of the positron source) revealed two components with lifetimes similar to those observed in the HPT case. This suggests that positron trapping at dislocations and in microvoids in the ECAP sample takes place. However, the intensity I_2 of the microvoid component is lowered to about 20 %. This is less than half of the value observed in the HPT specimen and may be regarded as a result of a higher concentration of microvoids in HPT specimen.

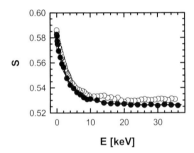

Figure 3: $S(E)$ dependencies measured for etched HPT (○) and ECAP (●) UFG Cu. VEPFIT results are shown by the solid line

4 Conclusions

Details of lateral and depth distributions of defects in HPT-made UFG Cu could be investigated using PAS techniques (SPIS, conventional PL and DB) in combination with other methods, like TEM, XRD and microhardness. It is known that UFG Cu prepared by HPT contains two types of lattice defects: dislocations in DRs along GBs, and microvoids of a size of a few vacancies inside grains [3,6]. These findings were confirmed by PAS in the present work. Moreover, mean dislocation densities derived from the present PAS data agree well with those given by XRD. Furthermore, PAS has provided a unique information about the size and concentration of microvoids formed, and, differences between HPT and ECAP-made specimens could be reflected.

5   Acknowledgements

This work was partially supported in the frame of Bilateral Co-operation in Science and Technology between The Ministry of Education, Youth and Sports of Czech Republic and The International Bureau of The BMBF (Germany), contract number CZE 00/035. Financial supports granted by The Ministry of Education, Youth and Sports of Czech Republic (projects KONTAKT ME 465 and COST OC523.50), The Grant Agency of Czech Republic (project 106/01/D049) and The Grant Agency of Charles University (project 187/2001) are highly acknowledged.

6   References

[1]   Valiev, R.Z.; Islamgaliev, R.K.; Aleksandrov, I.V., Prog. Mater. Sci. 2000, 45, 103
[2]   Islamgaliev, R.K.; Chmelik, F; Kuzel, R., Mater. Sci. Engineering A, 1997, 237, 43
[3]   Cizek, J.; Prochazka, I.; Vostry, P.; Chmelik, F; Islamgaliev, R.K, Acta Phys. Polonica A, 1995, 95, 487
[4]   Aleksandrov, I.V.; Zhang, K.; Kilmametov, A.R.; Lu, K.; Valiev R.Z,: Mater. Sci. Engineering A, 1997, 234-236, 321
[5]   Zehetbauer, M.; Ungar, T.; Kral, R.; Borbely, A.; Schafler, E.; Ortner, B.; Artenitsch, H.; Bernstorff, S., Acta Mater., 1999, 47, 1053
[6]   Cizek, J.; Prochazka, I.; Cieslar, M.; Kuzel, R.; Kuriplach, J.; Chmelik, F.; Stulikova, I.; Becvar, F.; Melikhova, O.; Islamgaliev, R.K., Phys. Rev. B 2002, 65, 094106
[7]   Cizek, J.; Prochazka, I.; Brauer, G.; Anwand, W.; Kuzel, R.; Cieslar, M.; Islamgaliev, R.K., phys. stat. sol. (a) 2003, 195, 335
[8]   Hautojärvi, P.; Corbel, C., in Positron Spectroscopy of Solids, Proceedings of the International School of Physics «Enrico Fermi» Course CXXV (edited by A. Dupasquier and A.P. Mills, jr.), Varenna 1993, IOS Press, 1995, p.491
[9]   Anwand, W.; Kissener, H.-R.; Brauer, G., Acta Phys. Polonica A 1995, 88, 7
[10] van Veen, A.; Schut,H.; Clement, M.; de Nijs, J.M.M.; Kruseman, A.; Ijpma, M.R., Appl. Surf. Sci. 1995, 85, 216
[11] Becvar, F.; Cizek, J.; Lestak, L.; Novotny, I.; Prochazka, I; Sebesta, F., Nucl. Instr. Meth. A 2000, 443, 557
[12] Seeger, A.; Major, J.; Banhart, F., phys. stat. sol. (a) 1987, 102,171
[13] Soininen, E.; Huomo, H.; Huttunen, P.A.; Mäkkinen, J.; Vehanen, A.; Hautojärvi, P., Phys. Rev. B 1990, 41, 6227

Anelastic Properties of Nanocrystalline Magnesium

Zuzanka Trojanová[1], Bernd Weidenfeller[2], Pavel Lukáč[1], Werner Riehemann[2], Miroslav Stank[1]
[1]Department of Metal Physics, Charles University, Praha, Czech Republic
[2] Department of Materials Engineering and Technology, Technical University Clausthal, Clausthal-Zellerfeld, Germany

1 Abstract

Internal friction in nanocrystalline Mg was measured by force vibration method at three frequencies. The specimens were prepared by milling procedure in an inert atmosphere and were subsequently compacted and hot extruded. The linear grain size of specimens used was estimated by X-ray line profile analysis to be about 100 nm. Two developed peaks in the internal friction spectrum were obtained at temperatures $\approx$ 370 K and $\approx$ 620 K. Influence of prestraining and annealing was observed. The activation energy of the low temperature peak was estimated to be 1.2 eV and its activation volume $V_D = 108\ b^3$. The experimental results are analysed and discussed.

2 Introduction

During dynamic testing, an oscillatory strain is applied to the material and resulting strain developed in material is measured. While for an ideal solids the response is instantaneous (the Hooke's law takes place) in an anelastic solid the stress and strain waves are not in phase, the phase shift (phase angle φ) between the stress and strain appears. Internal friction spectrum (the temperature or frequency dependence of tg φ) is measured. Internal friction in a material is the result of internal processes occurring during alternating stress cycles imposed on it and these processes originate in the interactions between the structural components in material. The presence of point, line, and planar defects within a stressed material often causes internal friction to occur because of the atomic movement, rearrangement, or realignment of the defects under application of the stress. These techniques can be used to characterise types of structure defects. Nanocrystalline materials are characterised by small grain size, typically in the range 1–100 nm. They have prominent properties owing to their high volume fraction of grain boundaries. There was a possibility to study role of dislocations and grain boundaries in a nanocrystalline material and to contribute to explanation processes occurring during plastic deformation of nanocrystalline magnesium. Internal friction of nanocrystalline materials was reported only in a few papers [1–6]. In the internal friction spectrum of high purity magnesium (99.9999 %) in medium temperature range three peaks were observed in papers [7,8]. High temperature peak in 99.9 % Mg was found by Kê [9] at $\approx$ 470 K and 0.5 Hz. Both materials were regularly polycrystalline.

3 Experimental Procedure

Nanocrystalline Mg samples were prepared by milling procedure in an inert atmosphere and were subsequently compacted and hot extruded. The linear grain size of specimens used was estimated by X-ray line profile analysis to be about 100 nm. The internal friction spectra were measured in the DMA 2980 apparatus in a single heating and cooling process. Three frequencies 0.5, 5 and 50 Hz were used. Throughout the measurements the strain amplitude was $1.2 \cdot 10^{-4}$. No grain growth during heating up to 450 °C has been observed [10]. Optical spark analysis revealed that the material after ball milling was not contaminated by impurities as iron from the milling balls or from the extruder tools [10]. No porosity was detected.

4 Results and Discussion

Figure 1 shows the temperature dependence of the internal friction tg φ measured at 0.5 Hz during heating (heating rate 1K/min.). Two peaks have been found: the weak low temperature peak $P_1 \approx 370$ K and $P_2 \approx 620$ K. The heating and cooling routes were not identical as demonstrated in Figure 2. The higher values of the internal friction reveal about some relaxation processes occurring in the sample at temperatures higher than 500 K. Figure 3 shows the internal friction of nanocrystalline Mg at different frequencies during heating cycle. It can be seen that the peaks are shifted to higher temperature with increasing frequency, indicating that they are the relaxations peaks. The high temperature peak was shifted at 5 and 50 Hz to the temperatures out of measured interval. The internal friction peaks are assumed to be imposed by a background IF_b expressed by

$$IF_b = A + B \exp(-C/kT), \qquad (1)$$

where k is the Boltzmann constant, T is the absolute temperature, A, B, and C are constants.

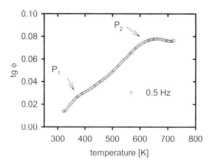 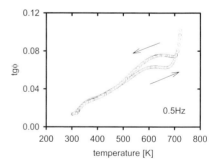

Figure 1: Internal friction spectrum measured at 0.5 Hz with two peaks

Figure 2: Internal friction spectrum measurend in heating and cooling routes at 0.5 Hz

After subtracting the background by using a fitting program PeakFit, the maximum temperature T_P was estimated for the low temperature peaks. The peak widths are broader than that for a

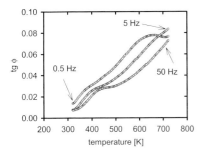

Figure 3: Frequency dependence of the internal friction spectrum

Debye peak, characterised by single relaxation time. The internal friction peak appears at the condition $\omega\tau = 1$ [11] i. e.

$$\omega\tau = \omega\tau_0 \exp(\Delta H / kT), \qquad (2)$$

where ω is the angular frequency $\omega = 2\pi f$, f is the measuring frequency), τ is the mean relaxation time, τ_0 pre-exponential factor, and ΔH is the mean activation enthalpy. Figure 4 shows the semilogarithmic plot of the frequency versus reciprocal value of the peak temperature T_P, so called Arrhenius plot. From the slope and intercept of the straight line the mean activation enthalpy ΔH for the low temperature peak has been obtained to be 1.2 eV, and the mean pre-exponential factor $\tau_0 \approx 8 \cdot 10^{-18}$ s. Relaxation peaks at medium temperatures can be attributed either to mechanisms related to dislocation mobility or mechanisms involving grain boundaries. Thus, it is important to study, on the one hand, the similarity existing between single crystals and polycrystals spectra, and, on the other hand, the influence of the nature and distribution of dislocations on the internal friction spectra. As far as we know, the internal friction spectrum of Mg single crystals has not been measured in the medium temperature range. The main slip mode in Mg is the basal one. Dislocations glide through the kink-pair formation according the model originally proposed by Seeger [12,13] and afterwards developed by the same author [14]. This mechanism gives rise to the Bordoni peak in the internal friction spectrum and it appears at frequencies about 1 Hz at approximately $1/3\ T_D$ (Debye temperature). For the case of magnesium it appears below 100 K [15]. Internal friction in a high purity magnesium (99,9999 %) has been studied in [7] at 1 Hz using torsion pendulum. The performed experiences showed in temperature interval above the room temperature an internal friction peak depicted by authors P_D placed at ≈ 340 K and peak P_1 at ≈ 430 K. The position of the P_D peak is very close to our P_1 peak. Authors of [7] attributed this peak to dislocation effects. As far as our peak P_1 has a dislocation nature, the height of the peak should be influenced by the sample kept at a low temperature. Deformation by rolling at room temperature ($\varepsilon \approx 2.55$ %) gives rise to a spectrum presented in Figure 5 in which, on one hand, P_1 peak is stronger in the first run after prestraining and, on the other hand, the P_2 peak was not affected. In the second and following runs the height of the P_1 peak is stable, approximately the same as before rolling. The result of this experiment confirmed the hypothesis about the dislocation nature of this peak. In our previous study, the amplitude dependence of decrement in nanocrystalline magnesium has been measured [16]. The existence of the amplitude dependent component of decrement indicated free dislocation loops

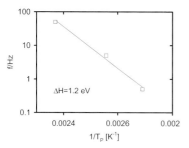

Figure 4: Semilogarithmic plot of frequency versus reciprocal peak temperature $1/T_P$ for the low temperature peak P_1

in the material. Newly created dislocations which were not anchored by pins (very probably foreign atoms or their small clusters) annihilated during measurement at higher temperatures. Creating of new dislocations is very probably a difficult process owing to very short length of the Frank-Read sources. On the other hand, we can consider increased diffusivity at elevated temperatures [17]. Dislocation annihilation supported by enhanced diffusivity decreased the total dislocation density.

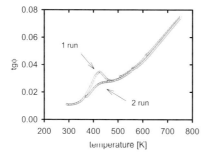

Figure 5: Influence of prestraining of $\varepsilon = 2.55$ % on P_1 peak height

In the case of relaxation processes linked to dislocation motion, the activation area $A = V/b$ (b is the Burgers vector) is the area that the dislocation has to move in order to activate the process, this means to cross over the barrier. Therefore, the product $V \cdot \sigma$ represents the work done by the applied stress to promote the motion over the local barrier. Neglecting the entropy term, the activation volume can be expressed as

$$V = -\frac{\partial \Delta H}{\partial \sigma} = -k \ln(\omega \tau_0) \frac{\partial T_p}{\partial \sigma}, \qquad (3)$$

using eq. (2) and the condition $\omega \tau = 1$ for the peak appearance. Figure 6 shows the internal friction spectrum measured at different strain amplitudes ε (maximum strain, $2 \cdot 10^{-4}$ or less). As the strain amplitude increases the peak shifts towards to lower temperatures and its height increases while the background increases moderately.

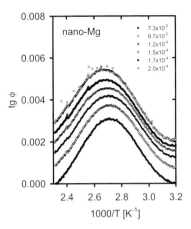

Figure 6: Effect of strain amplitude on the peak temperature

Taking for $\ln(\omega\tau_0) = -37.53$ ($\tau_0 = 8 \cdot 10^{-18}$ s, $\omega = 2\pi$ Hz), and for $\partial T_p/\partial\sigma = (1/E)\, 4.8 \cdot 10^4$ (linear regression from Fig. 7, $\sigma = E\varepsilon$, E is the Young's modulus), we obtain for the activation volume $V(\exp) = 5.9 \cdot 10^{-28}$ m^3 and the dislocation activation volume $V_d = \psi V_{\exp} = 3.5 \cdot 10^{-27}$ m^3 = 108 b^3 (E = 42 GPa, and the Taylor factor $\psi = 6$). The activation parameters i.e. the activation energy and the activation volume estimated for magnesium of the high purity in [8] amount to $\Delta H = 1$ eV and $V = 1100\ b^3$. The higher value of the activation volume is a consequence of the lower purity of our material. Shorter dislocation segments are pinned by solute atoms.

The high temperature peak P_2 was estimated only at the lowest frequency. Peaks for higher

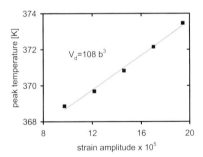

Figure 7: A linear regression of the peak temperature versus strain amplitude yields a value of the activation volume

frequencies were very probably shifted to temperatures out of the interval investigated. Thus, we are not able to estimate the activation enthalpy of this peak. This peak can be very probably attributed to a grain boundary mechanism. The estimated difference between the spectrum measured in heating half cycle and cooling one (Figure 2) is very probably caused by relaxation processes in grain boundaries. Such behaviour was observed only in the first cycle, then a stable peak during re-heating and cooling was found. The position as well as the height of this peak was not influenced by prestraining of the sample. The high temperature peak connected with grain boundary relaxation of polycrystalline magnesium has been described in [9] and [18]. There exists a wide controversy about the mechanisms, which control the grain boundary peak in metals and alloys [19]. In fact the characteristic grain boundary relaxation was also related to dislocation slip controlled by jog climb and vacancies diffusion, the activation energy being close to $\approx$ 1 eV [7,8]. Nevertheless, this point is an open discussion. A damping peak at around 425 K for frequency 1 Hz has been estimated in the internal friction spectrum of magnesium alloy AZ91 [20]. A higher peak temperature of the nanocrystalline magnesium estimated in this work is very probably done by the presence of magnesium oxide at the grain boundaries coming from the manufacturing process. The grain surface has been passivated by the content of 1–2 % of O_2 in the inert atmosphere to prevent the danger of fire. Oxide particles at the grain boundaries stabilised the grains and were the reason why we did not observe grain growth.

5 Conclusions

The internal friction spectrum of nanocrystalline magnesium has been measured from room temperature up to 450 °C. Two relaxation peaks were found. The position of the peaks depends on measuring frequency. The activation enthaly of the low temperature peak ΔH = 1.2 eV was estimated by the frequency shift in the temperature scale. The pre-exponential factor was found to be $t_0 = 8 \cdot 10^{-18}$ s. The sensitivity of the peak height to the sample predeformation confirmed the assumption about the dislocation nature of this peak. It indicates the fact that our material has a definite density of free dislocation loops. Nanocrystalline magnesium is a very stable material due to oxide particles situated at grain boundaries. This is very probably also the reason why a high temperature relaxation peak, which was attributed to the grain boundary processes, has been found at higher temperature comparing with polycrystalline magnesium.

6 Acknowledgements

This work was supported by the Grant Agency of the Academy of Sciences of the Czech Republic (Grant No. A2041203). The authors are grateful for the support offered by the Czech and German authorities under the Exchange Programme CZE 01/029.

7 References

[1] Weller, M., Diehl, J., Schaefer, H.E., Phil. Mag. A 36, 1991, 527–533
[2] Bonetti, E., Pasquini, L., Campari, E.G., Nanostructured Mater. 10, 1998, 457
[3] Weins, W.N., Makinson, J.D., de Angelis, R.J., Axtell, S.C., Nanostructured Mater. 9, 1997, 509–12
[4] Okuda, S., Tang, F., Tamimoto, H., Iwamoto, Y., J. Alloys Comp. 211/212, 1994, 494
[5] Bonetti, E., Campari, E.G., del Bianco, L., Nanostructured Mater. 11, 1999, 709–720
[6] Wang, Y.Z., Cui, P., Wu, X.J., Huang, J.B., Cai, B., phys. stat. sol. (a) 186, 2001, 99–104
[7] Nó, M.L., Oleaga, A., Esnouf, C., San Juan, J., phys. stat. sol. (a) 120, 1990, 419–427

[8] Nó, M.L.,in Mechanical Spectroscopy Q^{-1} 2001 (Eds. R. Schaller, G. Fantozzi, G. Gremaud) Trans. Tech. Publ., Switzerland 2001, 247–267
[9] Kê, T.S., Phys. Rev. 71, 1947, 533.
[10] Ferkel, H., Mordike, B.L., Mater. Sci. Eng. A298, 2001, 193–199
[11] Nowick, A.S., Berry, B.S., in Anelastic Relaxations in Crystalline Solids. Academic Press, New York/London 1972
[12] Seeger, A., Phil. Mag. 1, 1956, 651
[13] Seeger, A., Donth, H., Pfaff, F., Disc. Faraday Soc. 23, 1957, 19
[14] Seeger, A., Journal de Physique, C5 42, 1982, 311
[15] Seyed Reihani, S.M., Esnouf, C., Fantozzi, G., Revel, G., J. Physique Letters 39, 1978, L 429
[16] Trojanová, Z., Lukác, P., Ferkel, H., Riehemann, W., Mater Sci. Eng. A, in press
[17] Birringer, R., Hahn, H., Höfler, H., Karch, J., Gleiter, H., Diffusion and Defect Data, Trans. Tech. Publ., Aemannsdorf, 1988, 17
[18] Esnouf, C., Fantozzi, G., J. Phys. 42, 1981, C5–445
[19] Benoit, W., in Mechanical Spectroscopy Q^{-1} 2001 (Eds. R. Schaller, G. Fantozzi, G. Gremaud) Trans. Tech. Publ., Switzerland 2001, 291–305
[20] Lambri, O.A., Riehemann, W., Trojanová, Z., Scripta Mater. 45, 2001, 1365–1371

X-ray Peak Profile Analysis on the Microstructure of Al-5.9%Mg-0.3%Sc-0.18%Zr Alloy Deformed by High Pressure Torsion Straining

Jen Gubicza[1,2], Dániel Fátay[1], Krisztián Nyilas[1], Elena Bastarash[3], Sergey Dobatkin[3] and Tamás Ungár[1]

[1] Department of General Physics, Eötvös University, Budapest, Hungary
[2] Department of Solid State Physics, Eötvös University, Budapest, Hungary
[3] Moscow State Steel and Alloys Institute (Technological University), Moscow, Russia

1 Abstract

The microstructure of plastically deformed Al-5.9at%Mg-0.3%Sc-0.18%Zr alloy has been investigated. The severe plastic deformation has been performed by high pressure torsion straining (HPT) up to 15 revolutions at room temperature. The microstructure as a function of the number of revolutions is studied by X-ray diffraction peak profile analysis. It is concluded that the HPT technique results in nanostructure even after 0.5 turn with very high dislocation density. The crystallite size decreases and the dislocation density increases with the number of revolutions, however, after five turns they go into saturation. The edge and the dipole character of the dislocation structure becomes stronger with the increase of the number of revolutions. The value of the crystallite size determined by X-ray peak profile analysis is a bit smaller than the grain size obtained by transmission electron microscopy (TEM).

2 Introduction

It is well known that alloying of aluminium with scandium with appropriate selection of the composition results in the increase of the recrystallisation temperature above the solidus temperature [1]. As a consequence of this, the heat-treatment of the alloy does not result in grain-growth preserving the high value of the strength. The increase of the recrystallisation temperature is caused by the formation of coherent Al_3Sc phase in the grain boundaries which hinders the grain boundary motion in the Al alloy [1,2]. During the long term heat-treatments of the Al alloys, the grain boundary pinning capability of Al_3Sc particles tends to decrease due to the dissolution and the coarsening of these precipitates [1,2]. For increasing the thermal stability of Al_3Sc particles, Zr is added to Al alloys together with Sc [1]. The $Al_3(Sc_{1-x}Zr_x)$ phase formed in these alloys preserves all of the positive effects of Al_3Sc mentioned above [1,2].

The need of producing bulk nanostructured Al alloys leads to the elaboration of the severe plastic deformation (SPD) techniques. In the recent years the SPD methods, e.g. high pressure torsion (HPT), have been widely and successfully used for producing ultrafine grained structure in Al alloys [3,4]. This nanostructure can be retained even at high temperatures with the addition of Sc [5]. The grain size of Al-Sc alloys produced by SPD can be further reduced by adding Mg [5,6]. It was found that the Mg concentration has a strong effect on the superplastic behaviour of these alloys [6].

The analysis of the X-ray diffraction profiles provides information about the microstructure of materials with good statistics. The development in the computer technique permits the fitting of the whole diffraction profiles by theoretical functions calculated on the basis of the model of the microstructure [7–10]. These procedures provide the shape and the size distribution of the crystallites and parameters characterising the defect structure of the crystal lattice. Using the peak profile functions calculated for dislocations, the density and the dipole character of dislocations as well as the dominant dislocation slip system can be determined [11].

Although the effect of the Sc, Zr and Mg additions on the recrystallisation resistance and the strength of the aluminium alloys has been extensively studied [e.g. 1,6], only little information exist on the effect of the extent of SPD on the microstructure of Al-Mg-Sc-Zr alloys. In this paper the forming of the nanostructure in Al-5.9at%Mg-0.3%Sc-0.18%Zr alloy during HPT is studied by X-ray peak profile analysis. The changes of the crystallite size and the dislocation structure as a function of the number of revolutions are discussed and compared with the results of transmission electron microscopy (TEM).

3 Experimentals

An aluminium alloy with the composition of Al-5.9at%Mg-0.3%Sc-0.18%Zr was deformed by HPT. The deformation was performed under 4 GPa pressure on Bridgman anvil at room temperature. The samples of 15 mm in diameter and 0.6 mm in thickness were deformed by torsion of 0.5, 1, 3, 5, 10 and 15 rotations.

The microstructure of the high pressure torsion strained specimens was studied by X-ray diffraction peak profile analysis. The diffraction profiles were measured by a Philips X'pert powder diffractometer using Cu anode and pyrolithic graphite secondary monochromator. The instrumental corrections were made by using the powder pattern of a Si standard (NBS 640a) and the usual Stokes correction procedure. The first 8 reflections were evaluated by the method of Multiple Whole Profile fitting as it is described in Section 3 of this paper.

The microstructure was also investigated by TEM using JEM–100CX microscope. The average grain size was measured on the dark-field micrographs by the linear intercept method.

4 Evaluation Procedure of the X-ray Diffraction Profiles

The X-ray diffraction peak profiles were evaluated by the Multiple Whole Profile (MWP) fitting procedure [9,10]. In this method after the background subtraction and the instrumental corrections, the Fourier transforms of the measured profiles are fitted by the product of the theoretical functions of size and strain broadening. The calculation of the theoretical funtions is based on a model of the microstructure. In this model it is assumed that the crystallites have spherical shape with log-normal size distribution and the lattice strain is caused by dislocations. The procedure is based on five fitting parameters in the case of cubic crystals: (i) the median (m) and (ii) the variance (σ) of the log-normal size distribution function, (iii) the dislocation density (ρ), (iv) the effective outer cut-off radius of dislocations, R_e, and (v) the parameter q describing the average dislocation contrast factors and the edge or screw character of dislocations. The higher the value of R_e relative to the average distance between the dislocations ($\rho^{-0.5}$), the weaker the dipole character and the screening of the displacement fields of dislocations [12]. The area-weighted mean crystallite size is calculated from σ and σ as [7,9]:

$$<x>_{area} = m \exp(2.5\,\sigma^2). \tag{1}$$

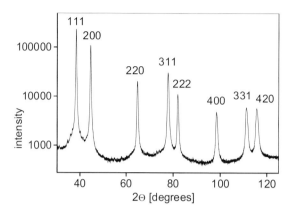

Figure 1: The X-ray diffraction pattern of the alloy deformed by HPT straining for 0.5 revolution

5 Results and Discussion

The X-ray diffraction pattern of the alloy deformed by HPT straining for 0.5 revolution is shown in logarithmic intensity scale in Fig. 1. As it can be seen from the figure the main phase is the Al solid solution. The lattice parameter of the aluminium solid solution is 0.40721 ± 0.00002 nm which is higher than the value for pure Al, 0.40494 nm. This is probably resulted from the solution of the larger Mg atoms into the Al lattice. Assuming that the increase of the lattice constant is caused only by the Mg solute atoms, the Mg concentration in the solid solution is obtained to be 5.38 ± 0.03 at.% which is less than the nominal value of 5.9 at.%. In the tail parts of 111 and 200 reflections of the Al solid solution, the peaks of minor phases are observed. For the sake of better visibility, a part of the diffractogram in Fig. 1 is shown in a wider 2Θ scale in Fig. 2a. The open squares, the open circles and the solid squares in this figure represent the Bragg peaks of $Al_3(Sc_{1-x}Zr_x)$, Al_3Mg_2 and Al_2Zr phases, respectively. The formation of the Mg-rich Al_3Mg_2 phase may be the reason of that the measured Mg concentration in Al solid solution is lower than the nominal value. The volume fraction of the minor phases estimated from the peak intensities is under 0.3 %. Consequently, these phases are not included in the evaluation of the microstructure and their peaks are subtracted from the profiles of Al solid solution as a background. The lattice parameter of the Al solid solution does not change significantly with the increase of the number of revolutions while the peaks of the minor phases gradually disappear. Fig. 2b shows the X-ray diffraction pattern of the specimen deformed for 15 revolutions in logarithmic intensity scale. It can be seen that the peaks of the minor phases almost completely disappeared.

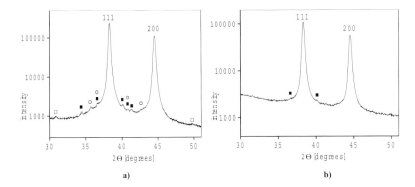

Figure 2: The parts of the diffractograms corresponding to the 2Θ between 30 and 51° for (a) 0.5 and (b) 15 revolutions. The open squares, the open circles and the solid squares represent the Bragg peaks corresponding to $Al_3(Sc_{1-x}Zr_x)$, Al_3Mg_2 and Al_2Zr phases, respectively

The measured and the fitted Fourier transforms are shown in Fig. 3 for the specimen subjected to 15 revolutions. The difference between the measured and the fitted values is also plotted in the lower part of the figure with the same scaling as in the main part of the figure. The agreement between the measured and the fitted Fourier transforms is good. The quality of the fitting is characterized quantitatively by the weighted least-squares error (R_{wp}) where the weights are the reciprocal of the observed Fourier transforms. The microstructural parameters obtained from the fitting procedure and the values of R_{wp} are listed in Table 1.

Table 1: The median, m, the variance, σ, of the crystallite size distribution, the area-weighted mean crystallite size, $<x>_{area}$, the dislocation density, ρ, the outer cut off radius of dislocations, R_e, and the weighted least-squares error, R_{wp}, for different revolutions of HPT straining

Revolutions	m [nm]	σ	$<x>_{area}$ [nm]	q	ρ [10^{14}m^{-2}]	R_e [nm]	R_{wp} [%]
0.5	39 ± 4	0.36 ± 0.03	54 ± 5	0.64 ± 0.04	16 ± 2	25 ± 2	0.6
1	37 ± 3	0.32 ± 0.02	48 ± 4	0.59 ± 0.04	16 ± 2	26 ± 3	0.8
3	29 ± 3	0.29 ± 0.02	36 ± 4	0.44 ± 0.03	14 ± 2	16 ± 2	0.8
5	26 ± 3	0.37 ± 0.03	37 ± 4	0.46 ± 0.04	24 ± 2	16 ± 2	1.0
10	30 ± 3	0.30 ± 0.02	37 ± 4	0.42 ± 0.03	23 ± 2	13 ± 2	1.0
15	31 ± 3	0.33 ± 0.03	40 ± 4	0.49 ± 0.04	24 ± 2	12 ± 1	1.1

The area-weighted mean crystallite size as a function of the number of revolutions is plotted in Fig. 4. It can be established that the HPT straining technique resulted in nanocrystalline state with crystallite size of 54 nm even after 0.5 revolution. A slight decrease of the size of the crystallites can be observed up to 5 revolutions but afterwards it goes into saturation. The dislocation density has a high value even after 0.5 revolution, $16 \cdot 10^{14}$ m^{-2}, and it increases slightly with the increase of the number of turns. The character of dislocations can be obtained from the value of the q parameter. The q parameter values in aluminium for pure screw and edge disloca-

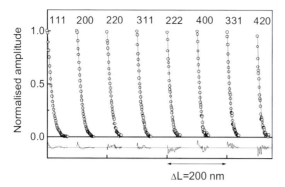

Figure 3: The measured (open circles) and the fitted (solid line) Fourier transforms for the specimen subjected to 15 revolutions. The difference between the measured and the fitted values is also plotted in the lower part of the figure with the same scaling as in the main part of the figure

tions of the <110>{111} slip system are 1.33 or 0.36, respectively [11]. The experimental values of the q parameters obtained for the present HPT deformed alloys are listed in Table 1. For all the specimens the value of q is lower than the arithmetic mean of the q parameter values corresponding to the pure edge and screw cases. This indicates that the dislocations have more edge than screw character. A definite tendency of the decrease of the q parameter with the number of rotations can be observed, indicating that the edge character becomes stronger with increasing deformation. The outer cut off radius of dislocations, R_e, decreases with the number of revolutions (see Table 1) which means that the dipole character of the dislocations increases, i.e. the screening of the strain field of adjacent dislocations becomes stronger with the deformation. The grain size was determined from TEM micrographs for all the specimens and plotted in Fig. 4. It can be established that the values of the crystallite size obtained by X-ray peak profile analysis is somewhat smaller than the grain size determined by TEM.

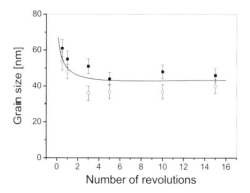

figure 4: The area-weighted mean crystallite size obtained from X-ray peak profile analysis (open circles) and the grain size determined by TEM (solid squares) as a function of the number of revolutions

6 Conclusions

The microstructure produced by HPT straining in Al-5.9%Mg-0.3%Sc-0.18%Zr alloy was studied by X-ray peak profile analysis. It was found that nanostructure was formed even after 0.5 revolution with a crystallite size of 54 nm and with very high dislocation density of $16 \cdot 10^{14}\,m^{-2}$. The size of the crystallites decreased and the dislocation density increased slightly when the number of turns increased. The edge and dipole character of the dislocation structure became stronger during deformation. After 5 revolutions the microstructure did not change with the increase of turns.

7 Acknowledgements

This work was supported by the Hungarian Scientific Research Fund, OTKA, Grant Nos. T031786, T034666, T029701 and by the Magyary Zoltán postdoctoral program of Foundation for Hungarian Higher Education and Research (AMFK). The authors greatly thank Prof. V.V. Zakharov and Dr. T.D. Rostova who kindly presented Al–Mg-Sc-Zr alloy 01570.

8 Reference

[1] V. G. Davydov, T. D. Rostova, V. V. Zakharov, Yu. A. Filatov, V. I. Yelagin, Mat. Sci.Eng. A 2000, 280, 30–36
[2] E. A. Marquis, D. N. Seidman, Acta mater. 2001, 49, 1909–1919
[3] V. V. Stolyarov, V. V. Latysh, V. A. Shundalov, D. A. Salimonenko, R. K. Islamgaliev, R. Z. Valiev, Mat. Sci. and Eng. A 1997, 234–236, 339–342
[4] I. V. Alexandrov, R. Z. Valiev, Scripta mater. 2001, 44, 1605–1608
[5] S. V. Dobatkin, V. V. Zakharov, R. Z. Valiev, in Proc. of the First Joint Inter. Conf. "Recrystallization and Grain Growth"(Eds.: G. Gottstein and D. A. Molodov), Springer-Verlag, 2001, 509–514
[6] M. Furukawa, A. Utsunomiya, K. Matsubara, Z. Horita, T. G. Langdon, Acta Mater. 2001, 49, 3829–3838
[7] J. I. Langford, D. Louer, P. Scardi, J. Appl. Cryst. 2000, 33, 964–974
[8] P. Scardi, M. Leoni, Acta Cryst. A 2002, 58, 190–200
[9] T. Ungár, J. Gubicza, G. Ribárik, A. Borbély, J. Appl. Cryst. 2001, 34, 298–310
[10] G. Ribárik, T. Ungár, J. Gubicza, J. Appl. Cryst. 2001, 34, 669–676
[11] T. Ungár, I. Dragomir, Á. Révész, A. Borbély, J. Appl. Cryst. 1999, 32, 992–1002
[12] M. Wilkens, in Fundamental Aspects of Dislocation Theory (Ed.: J. A. Simmons, R. de Wit, R. Bullough), Nat. Bur. Stand., Washington DC, USA, 1970, Vol. II. Spec. Publ. No.317, p.1195

Evolution of Microstructure during Thermal Treatment in SPD Titanium

E. Schafler[1,2], L. Zeipper[1,3], M. J. Zehetbauer[1]

[1] Institute of Materials Physics, University of Vienna, Austria
[2] Erich Schmid Institute of Material Science, Leoben, Austria
[3] ARC Seibersdorf research GMBH, Seibersdorf, Austria

1 Introduction

Severe Plastic Deformation (SPD) techniques (e.g. equal channel angular pressing (ECAP), high pressure torsion (HPT)) are methods for producing highly dense, bulk ultrafine-grained (submicron-grain-sized or nanostructured) materials [1]. The materials produced by SPD techniques reveal an attractive combination of high strength and good ductility [2–4].

As the heaviest and strongest representative of the light metals, Ti is of high technical interest, especially for medical applications due to its very good biocompatibility. For good mechanical resistance the strength and ductility should be higher. Currently alloys of Ti with Al and V are used to accomplish the mechanical requirements. Applying SPD techniques to pure Ti or even its alloys the mechanical properties should be improved in order to extend lifetime of e.g. a medical implant and at the same time can enhance the biocompatibility by using more pure Ti.

Since Ti lacks sufficient ductility at lower temperatures the ECAP was performed at elevated temperature of 450 °C at the beginning of the deformation process [1,5]. The homologous deformation temperature of about 0.4 T_m and the high strains imply dynamic recovery effects even during deformation. This recovery leads to a more intrinsic equilibrium state of the material which is achieved by the deformation induced defects due to the high strain. Also more microstructural homogenization can be expected which increases the stability as well as the isotropy of mechanical parameters. Now the question arises whether this recovery process has led to an optimal material state. For clarification an annealing experiment has been made with subsequent investigation by different methods in order to correlate macroscopic (mechanical) properties and microstructural features: i) microhardness measurements (HV), ii) scanning electron microscopy (SEM) and iii) Multi-Reflection X-Ray Profile Analysis (MXPA). X-ray diffraction peak profiles are broadened due to small crystallite sizes and lattice distortions. The two effects can be separated on the basis of the different diffraction order dependence of peak broadening. The standard methods of peak profile analysis based on the full widths at half maximum (FWHM), the integral breadths and the Fourier coefficients of the profiles provide the coherently scattering domain size and the mean square strain [6–8].

2 Experimental

2.1 Material, Deformation and Thermal Treatment

Commercial pure Ti has been formed to billets of 40 mm in diameter and 210 mm length. With a starting temperature of 450°C the material has been deformed by 8 passes of Equal Channel Angular Pressing with a die angle of 90° and deformation route B_C that means rotation of the billet by 90° in the same direction between each pass. For the investigation the samples have been cut by spark erosion from the centre of the billet, grinded, mechanically polished and afterwards electro-polished.

The thermal treatment was done under high vacuum conditions at 300 °C, 400 °C and 500 °C for each 10 minutes. Individual samples have been taken for each temperature.

2.2 Microhardness Measurements

For the microhardness measurements a Zeiss-Axioplan microscope with computer imaging system and evaluation software was used. The indentation apparatus was of type MHT-4 (A. Paar, Graz). The indentation force was 1 N and the rate 0.1 N/s. The loading time of indentation was 10s.

2.3 Scanning Electron Microscopy

A LEO Stereoscan 1525 Scanning electron microscope was used to take micrographs with a detector for Back Scattered Electrons (BSE). In a single-phase material, the energy of the detected back scattered electrons depends on both, the orientation of the crystallites with respect to the direction of the insistent electron beam, and the density of dislocations. Thus, different orientated structural elements appear in these micrographs as regions of different grey-values and allow an estimation of the typical microstructural sizes.

2.4 Multi-Reflection X-Ray Profile Analysis

We used the 0.3 × 3 mm line focus of a Siemens rotating Cu anode operated at 45kV and up to 100 mA. The primary X-ray beam was monochromatized by asymmetrically cut plane Ge crystal using the 220 reflection and tuned to CuKα_1 line (λ = 1.54 Å). The dimension of the footprint of the beam on the sample was about 0.1 × 1 mm^2. The rotation of the sample provided an illumination of 1mm in diameter. The scattered radiation was registered by a linear position sensitive X-ray detector type OED-50 (Braun, Munich) set up at a distance of 450 mm from the sample. The following reflections have been measured and considered for evaluation: {010}, {011}, {110}, {112}, {021}, {210} and {121}.

2.4.1 Evaluation Procedure

According to the kinematical theory of X-ray diffraction, the physical profile of a Bragg reflection is given by the convolution of the size and the distortion profiles [7]

$$I^P = I^S * I^D, \quad (1)$$

where the superscripts S and D stand for size and distortion, respectively. The Fourier transform of this equation gives [7]

$$A(L) = A^S(L) A^D(L), \quad (2)$$

where $A(L)$ are the absolute values of the Fourier coefficients of the physical profiles, A^S and A^D are the size and the distortion Fourier coefficients and L is the Fourier variable. Assuming that the lattice distortions are caused by dislocations A^D can be given in the form [9-11]

$$A^D(L) = \exp[-\rho B L^2 f(\eta) g^2 \bar{C}], \quad (3)$$

where L is the Fourier variable, $B = \pi b^2/2$, b is the absolute value of the Burgers vector, ρ is the dislocation density, $\eta \sim L/R_e$, R_e is the effective outer cut-off radius of dislocations, g is the absolute value of the diffraction vector, $\bar{C}$ is the average dislocation contrast factor and $f(\eta)$ is a function derived explicitly by Wilkens [11]. It can been shown that for a hexagonal polycrystalline specimen $\bar{C}$ can be given in the following form [14]:

$$\bar{C} = \bar{C}_{hk0} \left[1 + q_1 x + q_2 x^2 \right], \quad (4)$$

where $x = (2l^2)/(3a^2 g^2)$, a is the lattice parameter in the hexagonal basal plane, l is the fourth index of reflection, q_1 and q_2 are parameters depending on the elastic constants and on the dislocation slip systems active in the crystal [14].

In bulk nanostructured materials, the crystallite-size distribution can be described by log normal function [15–18] assuming spherical crystallite shape with log-normal size distribution [12,13]. If s is the variance and m is the median of the log-normal size distribution function, the arithmetic, the area- and the volume-weighted mean crystallite sizes are obtained as [19]:

$$<x>_{arit} = m \exp(0.5\, \sigma^2), \quad (5)$$

$$<x>_{area} = m \exp(2.5\, \sigma^2), \quad (6)$$

$$<x>_{vol} = m \exp(3.5\, \sigma^2). \quad (7)$$

A numerical procedure has been worked out for fitting the Fourier transform of the experimental profiles by the product of the theoretical functions of size and strain Fourier transforms given in Eqs. (3) and (5) [12,13]. The procedure has six fitting parameters for hexagonal crystals: m and σ corresponding to size broadening, ρ and M corresponding to strain broadening, q_1 and q_2 corresponding to the strain anisotropy [12,13].

3 Results

The results of the microhardness investigations are shown in Figure 1. The material reveals a microhardness of 2344 MPa in the as deformed state. This is ~ 30 % higher than the annealed material has. After annealing at temperatures between 350 °C to 450 °C there occurs an increase of microhardness by about 5 % compared to the as deformed state. At a temperature of 500 °C the hardness decreases about to the level of the as processed material. This corresponds to the results of tensile tests with the same material [20], where the ultimate tensile strength does not change, whereas the ductility can be increased by 30 %.

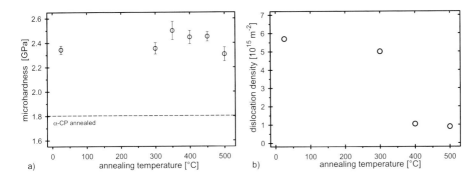

Figure 1: a) Microhardness as a function of annealing temperature. The dashed line indicates the values of the annealed state. b) Dislocation density obtained by MXPA as a function of annealing temperature.

The X-ray diffraction analysis yields microstructural parameters like dislocation density and the coherently scattering domain size. In Fig.1b the evolution of the dislocation density with increasing annealing temperature is shown. The initial value of the dislocation density measured with the as processed material is about $5 \cdot 10^{15} m^{-2}$ and first decreases very slowly to 300°C and with a strong drop at 400 °C reaching a value of about $1 \cdot 10^{15} m^{-2}$. This corresponds to a homologous temperature of 0.35 T_m.

In Fig.2a and Fig.2c SEM micrographs of the as deformed state and the material annealed at 400° C are presented as they have been mapped by the SEM in orientation contrast mode. Qualitatively the annealed state seems to have a smaller and sharper microstructure. Fig.2b and Fig.2d show the results of a digital image analysis of the SEM micrographs. The original images were processed in several steps in order to get a fragmented image. Afterwards, using the area method and assuming circular elements the grain/crystallite sizes have been evaluated using MATLAB. From the fragmented micrographs log-normal size distributions can be evaluated (Fig. 3b) providing the arithmetic mean diameters m_a as a characteristic value for the grain size. m_a is decreasing from 940 nm in the as processed state to about 700 nm after annealing at 300 °C and 400 °C (Fig.3a). The results for the coherently scattering domain size are of the order of 50 nm to 100 nm corresponding to other investigations from transmission electron microscopy [21] and MXPA [22].

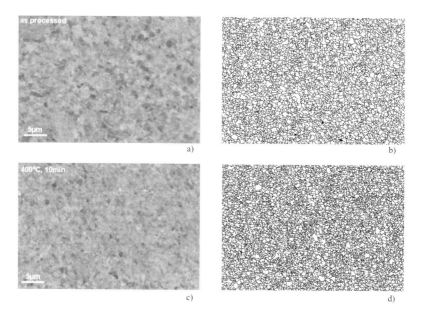

Figure 2: SEM Micrographs showing the microstructure and the related fragmentation sketches evaluated by MATLAB of the initial and an annealed state: a) and b) as deformed; c) and d) annealed at 400 °C for 10 min

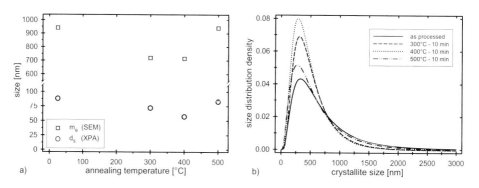

Figure 3: a) Structural sizes as determined by SEM (m_a) and MXPA (d_s) as a function of the annealing temperature. b) Log-normal size distributions determined from the SEM-micrographs.

4 Discussion

After processing the α-CP Ti material by equal channel angular pressing with a starting temperature of 450 °C, the strength (represented by the microhardness in Fig. 1(a)) cannot be increased significantly by temperature treatment, as it is the case for Ti deformed by high pressure torsion at room temperature [23]. The reason may be found in the higher defect densities remaining after cold working causing also a higher fragmentation of the material. The error bars in Fig. 1(a) indicate the standard deviation, which is very low (about 2–3 %) at all investigated sample sets, indicating a good homogeneity of the materials. The dynamic recovery processes during the deformation process are enough in order to receive an ultra fine structured material with good homogeneity of strengthening improvement.

The evolution of the microstructural sizes both observed by scanning electron microscopy and derived by MXPA (Fig. 3a) exhibits an inverse change compared to the microhardness (Fig. 1a). The evolution of the coherently scattering domain size d_s with temperature shows a very similar behavior to m_a, although the values are of different orders of magnitude. According to the definition the "coherently scattering domains" represent regions of the crystal lattice in which X-ray diffraction is coherent. This coherent scattering is very sensitive to any distortions caused by dislocation, stacking faults or other defects bounding "absolute" single crystals, which cannot be separated by SEM.

The residual internal stresses caused by the lattice distortions from warm deformation process are sufficiently high to initiate the motion of free dislocations at moderate temperatures (0.3 T_m). Under the assumption that they arrange to new cell walls the scattering domain size will shrink because of increased orientation tilt. Other dislocations may also be incorporated in existing subgrain boundaries increasing the tilt between the subgrains but most of them will annihilate at higher temperatures and cause the drop in dislocation density (Figure 1b). The same explanation may be applied to the structural size determined by SEM.

So two strengthening mechanisms may compete during this temperature treatment: the Hall-Petch effect and dislocation hardening. While the grain size decreases also the dislocation density decreases yielding a less significant increase of the microhardness. At 500 °C the grain size again increases and the microhardness decreases.

When drawing Hall-Petch relationships for the grain size determined by SEM and the coherently scattering domain size, either of type $d^{-1/2}$ or type d^{-1} the much better regression coefficient results from the coherently scattering domain size. The reason probably is that the domain size somewhat reflects also the dislocation structure being present in grain interior. A close discussion of this feature is given in [24].

4 References

[1] R. Z. Valiev, R. K. Islamgaliev and I. V. Alexandrov, Progr. Mater. Sci. 2000, 45, 103
[2] V. V. Stolyarov, Y. T. Zhu, I. V. Alexandrov, T. C. Lowe and R. Z. Valiev, Mater. Sci. Eng. A 2000, 299, 59
[3] V. V. Stolyarov, Y. T. Zhu, I. V. Alexandrov, T. C. Lowe and R. Z. Valiev, Mater. Sci. Eng. A 2001, 303, 82
[4] D. Jia, Y. M. Wang, K. T. Ramesh, E. Ma›, Y. T. Zhu, and R. Z. Valiev, Applied Physics Letter 2001, 79, 611
[5] V. V. Stolyarov, Y. T. Zhu, I. V. Alexandrov, T. C. Lowe and R. Z. Valiev, J. Nanoscience and Nanotechnology 2001, 1, 237
[6] G. K. Williamson and W. H. Hall, Acta Metall. 1953, 1, 22
[7] B. E. Warren and B. L. Averbach, J. Appl. Phys. 1950, 21, 595
[8] B. E. Warren, Progr. Metal Phys. 1959, 8, 147
[9] M. A. Krivoglaz, Theory of X-ray and Thermal Neutron Scattering by Real Crystals 1996, Plenum Press, New York.
[10] M. Wilkens, phys. stat. sol. (a) 1970, 2, 359

[11] M. Wilkens, Fundamental Aspects of Dislocation Theory 1970, ed. J. A. Simmons, R. de Wit, R. Bullough, Vol. II. Nat. Bur. Stand. (US) Spec. Publ. No. 317, Washington, DC. USA, p. 1195
[12] T. Ungár, J. Gubicza, G. Ribárik and A. Borbély, J. Appl. Cryst. 2001, 34, 298
[13] G. Ribárik, T. Ungár and J. Gubicza, J. Appl. Cryst. 2001, 34, 669
[14] I. C. Dragomir and T. Ungár, J. Appl. Cryst. 2002, 35, 556
[15] Ch. D. Terwilliger and Y. M. Chiang, Acta Met. Mater. 1995, 43, 319
[16] T. Ungár, A. Borbély, G. R. Goren-Muginstein, S.Berger and A. R. Rosen, Nanostructured Materials 1999, 11, 103
[17] J. I. Langford, D. Louër and P. Scardi, J. Appl. Cryst. 2000, 33, 964.
[18] J. Gubicza, J. Szépvölgyi, I. Mohai, G. Ribárik and T. Ungár, J. Mat. Sci. 2000, 35, 3711.
[19] W. C. Hinds, Aerosol Technology: Properties, Behavior and Measurement of Airbone Particles 1982, Wiley, New York.
[20] L. Zeipper, M. Zehetbauer, E, Schafler, B. Mingler, G. Korb, this issue
[21] Y.T. Zhu, J.Y. Huang, J. Gubicza, T. Ungár, Y.M. Wang, E. Ma and R.Z. Valiev, submitted to Acta Mater. 2002
[22] J. Gubicza, I. C. Dragomir, G. Ribárik, Y. T. Zhu, R. Z. Valiev, T. Ungár, Mater.Sci.For. 414–415, 2002, 229
[23] A.V. Sergueeva, R.Z. Valiev, A. K. Mukherjee, Proc. "Ultrafine Grained Materials II" 2002, ed. Y. Zhu et. al., TMS Annual Meeting, Seattle
[24] A. Dubravina, M. J. Zehetbauer, E. Schafler, I.V. Alexandrov, Proc. 13th ICSMA, to be published in Mater. Sci. Eng. A

# VIII Influence of Deformation Parameters to SPD Nanostructures

The Role of Hydrostatic Pressure in Severe Plastic Deformation

M. J. Zehetbauer[1], H. P. Stüwe[2], A. Vorhauer[2], E. Schafler[1] and J. Kohout[3]
[1] Institute of Materials Physics, University of Vienna, Vienna, Austria
[2] Erich Schmid Institute of Materials Science, Austrian Academy of Sciences, Leoben, Austria
[3] Department of Physics, Military Academy Brno, Brno Czech Republic

1 Abstract

The contribution presents several experimental examples which show that the presence of an enhanced hydrostatic pressure—as compared to conventional large deformation modes—is one of the main features of severe plastic deformation (SPD). At the example of systematic high pressure torsion experiments with Cu at room temperature, strength measurements after deformation showed that the onset strains of deformation stages III, IV, and V are not affected by the pressure applied; however, the "related onset flow stresses increase by at least 10 % of the values of low pressure torsion, per GPa of pressure "increase. During deformation, increases of flow stresses by at least 40 % of the values of low pressure torsion, per GPa of pressure increase, have been found. From comparisons with tests on Ni, the increases appear to grow with the materials melting temperature. For a theoretical explanation of flow stress increases the pressure induced changes of i) the elastic moduli, and ii) the formation energy of lattice defects. While contribution i) is almost negligible, contribution ii) accounts for an increase of flow stress during deformation by about 15 % per GPa of pressure increase. The difference left to "experiment has to be attributed to a third contribution, i.e., the pressure specific evolution of the structure. For this contribution, a modification of the model of "Zehetbauer and Les [1–3] is introduced which is based on the pressure caused decrease of lattice diffusion. The latter is thought to restrict the diffusion controlled annihilation of "dislocations, thus leading to a higher density of vacancies, dislocations and/or grain boundaries causing the higher stress level observed.

2 Conventional and Severe Plastic Deformation

From the first attempts of using techniques of the so-called severe plastic deformation (SPD) it was not clear which common feature makes them capable of achieving ultrafine grained and nanocrystalline structures, in contrast to conventional large strain cold work techniques which in general lead to such scales in one dimension only. Considering the different modes recently designated as SPD methods, they all have in common that the material experiences certain constraints preventing free plastic flow: While in Equal Channel Angular Extrusion (ECAE, [4–6] Fig. 1a) and Twist Extrusion (TE, [7,8] Fig. 1b) this constraint rules across the deformation axis, it is the constraint in all sample dimensions with the high pressure torsion (HPT, [5,9–11], Figs. 1c,2a), and also with the extrusion—compression process in cyclic extrusion compression (CEC, [12,13] Fig. 1d), to mention only a few of SPD modes. In case of HPT, this constraint can be quantified in terms of the hydrostatic pressure p applied, and from a collection of literature data it becomes clear that the extent of p during plastic deformation determines not only the resulting grain size but also the strength of material related (see Table 1 with the UTS as a measure of the strength).

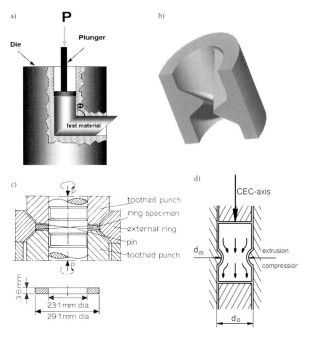

Figure 1: Different modes of Severe Plastic Deformation. a) Equal channel angular pressing (extrusion) (ECAP, ECAE) [4–6]. b) Twist extrusion (TE) [7,8]. c) High pressure torsion (HPT) (\(ring\) test, [11]). d)'Cyclic extrusion compression (CEC) [12,13]

Table 1: Different modes of Severe Plastic Deformation, yielding different final cell/subgrain sizes, shapes and Ultimate Tensile Strengths (UTS), by the example of pure Cu. The amount of hydrostatic pressure p is given in parentheses

SPD mode	Cell / subgrain size [nm]	UTS [MPa]
Conventional large strain deformation Zehetbauer and Seumer [14]	300×650 ($p = 0.1$ MPa)	400
ECAE – equal channel angular extrusion routes "A" & "B_c" Pilhatsch [15], Mingler [16]	300	450–500
CEC – Cyclic extrusion – compression Zughaer and Nutting [17]	400	450
HPT – high (hydrostatic) pressure-torsion		
Sturges et al. [18]	($p = 0.35$ GPa)	500
Saunders [19]	250 ($p = 0.5$ GPa)	570
Stegelmann et al. [20] Hebesberger et al. [10,22]	200 ($p = 6.0$ GPa)	600
Dubravina and Valiev [21]	100 ($p = 10$ GPa)	< 700

Therefore, this situation firmly suggested to perform a systematic study of the influence of the hydrostatic pressure acting during plastic deformation, and its effects to the resulting grain size and strength. Since HPT is the only method of SPD which allows for such a well controlled investigation, a standard compression machine at the Erich Schmid institute in Leoben, has been equipped with a rotating lower plunger and a load cell for the torque measurement. Up to now, several tools have been developed in order to account for a very high applicable hydrostatic pressure.

2.1 Stress Measurements after High Pressure Torsion

A sketch of the used HPT tool is given in Fig. 2a. [10,22] It consists of two plungers, a fixed and a rotating one, each with a cylindrical recipient to "accept less than half the sample to ensure hydrostatic pressure conditions. After insertion of the sample, it is compressed up to the desired value of p and the rotation of plunger is started. Due to the high pressure, the friction at the surface of the sample is sufficiently high that it undergoes a deformation by torsion. Fig. 2b shows the results from our first experiments done with a different tool at $p = 0.8$ GPa in polycrystalline Cu. [20,23] Interruptions of deformation were carried out in order to check the real rotation of the sample. The strength (for measurements of strength ρ, Vickers microhardness (HV) tests were applied using the relation $\rho = HV/3$) after deformation at this pressure is at least 10 % higher than after conventional torsion under atmospheric pressure, and for $p = 2.0$ GPa this difference amounts to about 20 % (see Fig. 3a). This Figure also gives a closer answer to the question of pressure induced strength increase. In the \(Mecking-plots\) of hardening coefficient vs. flow stress, all onset stresses τ_{III}, τ_{IV} and τ_V of deformation stages III, IV, and V are shifted by about the same amount so that one can conclude that the "effect of hydrostatic pressure to the resulting strength is "independent of deformation strain. [23] Such an increase of "final stress was also observed in CEC deformed Al5Mg [13] but by a quite smaller amount (Fig. 3b) which indicates that the hydrostatic pressure during CEC must be, although present, much smaller than 1 GPa.

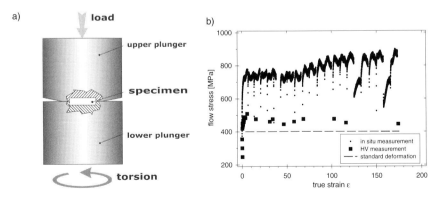

Figure 2: a) Sketch of the HPT-device used (for details see also [8,20]). b) Strength-strain characteristics at HPT deformation of Cu performed at $p = 0.8$ GPa. The highest line represent, the flow stress during deformation evaluated from torsion momentum by Equation 1a, the squares represent flow stresses $\rho = HV/3$ derived from Vikkers microhardness HV data

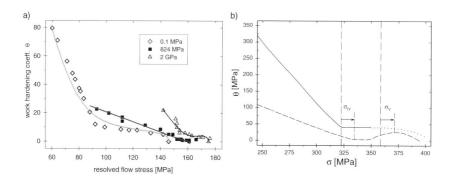

Figure 3: Shift of deformation stages III, IV, and V at different SPD modes: a) Cu, HPT deformation at different hydrostatic pressures p = 0.1 MPa, 0.8 and 2 GPa [20,23]. b) Al-5 wt.-% Mg, CEC deformation (dashed line) in comparison with standard deformation (from [13])

2.2 Measurements of Lattice Defect Densities after High Pressure Torsion

By means of the multiple reflection X-ray peak profile analysis (MXPA), [24] dislocation densities were obtained in Cu after deformation by HPT with different hydrostatic pressures. [20,23,25] In Figures 4, 6b, dislocation densities measured for 4 different hydrostatic pressures being p = 0.8, 1.3, 5, 8 GPa are given, and it is evident that the dislocation density ρ increases by up to a factor 3 if one compares the ρ values for these pressures at the same strain (for large strains see Fig. 4, Table 2). The continuous increase of dislocation density is remarkable when considering the absolute decrease of flow stress measured after HPT, too; this means that there "occurs some rearrangement but no loss of dislocations after deformation so that one can assume similar densities to exist during HPT deformation, so that these density data are justified to be used for the fitting of in-situ stress-strain relationsship in section 4.3. Another increase seems to occur with the concentration of deformation induced vacancies (see Schafler et al. [26], Table 2). In these effects the keys may lie for the "understanding of the development of ultrafine-grained and/or nanocrystalline structure: The more dislocations are produced and stored, the more grain boundaries can be built from them. The enhanced concentration of vacancies may contri-

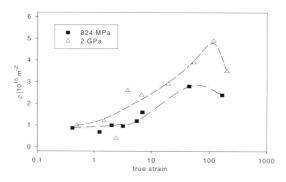

Figure 4: Evolution of dislocation density in HPT deformed Cu, as a function of deformation and hydrostatic pressure [20,23]

bute to more grain boundary sliding which has been "often observed with SPD materials when they exhibit superplasticity at far lower temperatures than with conventional materials.

Table 2: Results of fits to HPT in-situ stress-strain data of Cu by the model of Zehetbauer and Kohout [29]. The input data were taken from MXPA measurements dislocation density ρ, screw fraction f_1, and from the literature [33] combined with Equation 16 (vacancy migration enthalpy δH_{eff})

INPUT			OUTPUT				
p [GPa]	ρ [10^{15}m^{-2}] (at $\gamma = 90$)	δH_{eff} [eV]	α_1	α_2	δG [eV]	c (at $\gamma = 90$)	c_{th} [a]
0.0001	2.0	0.2	0.21	0.4	0.79	1.2×10^{-12}	1.02×10^{-20}
0.824	2.86	0.26	0.35	0.5	0.71	1.3×10^{-10}	
2.0	3.5	0.34	0.36	1.1	0.74	2.31×10^{-8}	
5.0	4.16	0.54	0.8	1.25	0.64	1.31×10^{-5}	
8.0	7.09	0.75	1.02	1.4	0.64	1.04×10^{-1}	
$f_1 = 0.7$, $d\gamma/dt = 10^{-2}$ s^{-1}							

[a] By vacancy formation enthalpy $H_{form} = 1.28$ eV [31].

2.3 Stress Measurements during High Pressure Torsion

For the measurements of stress–strain characteristics during torsion, the torque was monitored as a function of rotations applied. Resolved shear stresses and resolved shear strains have been calculated by the equations

$$\tau = (1/M_\tau)\frac{3Q}{2\pi R^3} \quad (1a), \quad \gamma = M_\tau \frac{R\varphi}{d} \quad (1b) \tag{1ab}$$

where Q is the measured torque, R is the effective radius of the sample recipient, φ is the torsion angle, and d is the sample thickness. $M\tau$ is the Taylor factor for torsion which has been taken as $M\tau = 1.65$ for face-centered cubic (fcc) metals presented in this paper. It is seen from Figure 2b that the flow stress under hydrostatic pressure during deformation ($p = 0.8$ GPa here) may get up to 40 % higher than that after deformation. As for a more systematic investigation, HPT tests were carried out in polycrystalline Cu and Ni at hydrostatic pressures $p = 1, 3, 5,$ and 8 GPa. For the case of Cu (Fig. 5a), the in-situ measured resolved shear stress seems to be about a factor 2, 5, and 7.7 times higher than the steady-state stress reached at conventional torsion with normal pressure conditions $p = 0.1$ MPa. In Ni (Fig. 5b) the enhancement of shear stresses with the same numbers of pressure even reaches factors 2.7, 8.3, and 12.5. Although these flow stresses appear as very high, they are still far from theoretical critical flow stress which amounts to about 9 GPa in Cu and 16 GPa in Ni. If one agrees to compare the stresses of the significant maxima of different dependences within a range of shear strains $\gamma = 10...20$, the observed pressure induced increase of flow stress appears to be roughly linear in hydrostatic pressure p, being $\Delta\tau/\tau = 0.8$ p/GPa for Cu, and $\Delta\tau/\tau = 1.5$ p/GPa for Ni.

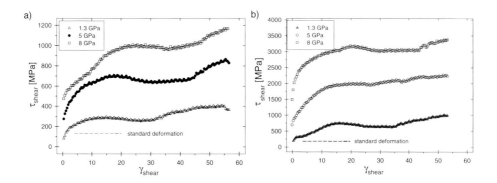

Figure 5: Data of resolved shear stress vs. resolved shear strain measured during deformation by High Pressure Torsion, at hydrostatic pressures p = 1.3, 5, and 8 GPa, in polycrystalline a) Cu; b) Ni

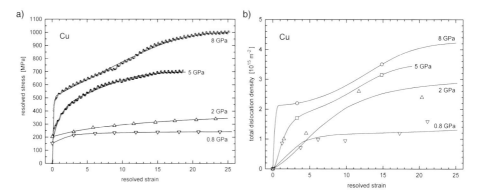

Figure 6: Measured data (points), and fits (lines) calculated by the model of Zehetbauer and Kohout [29] for different hydrostatic pressures p as indicated, for HPT deformed Cu. a) In-situ stress strain relationships; b) dislocation density evolution

One should compare these values with existing literature. There have been published several data by Haasen and Lawson [27] which performed tension experiments under $p = 500$ MPa mainly in single crystalline materials. However, the results reach from $\Delta\tau/\tau = 0.2$ p/GPa for Cu and Ni, $\Delta\tau/\tau = 0.3$ p/GPa for Al, to $\Delta\tau/\tau = 0.6$ p/GPa for α-CuZn single crystals, with a tendency to smaller values for polycrystals. More recently, there have been carried out torsion tests under p = 0.1 and 350 MPa by Sturges et al. [18] on polycrystalline Cu who, resulted in $\Delta\tau/\tau = 0.42$ p/GPa. To sum up, the situation seems to be not very uniform, and some estimations are presented in the next section which magnitude of the effect at all is to be expected from theoretical considerations.

3 Theoretical Estimates and Modeling

Under hydrostatic pressure the flow stress of metals τ is raised by four increments: $\Delta\tau_1$ is caused by the increase of the shear modulus μ; $\Delta\tau_2$ and $\Delta\tau_3$ are caused by the work done to create the excess volumes of dislocations and vacancies, and $\Delta\tau_4$ is caused by a change in microstructure. After release of pressure, only $\Delta\tau_4$ remains as increased strength of the "material.

3.1 Elastic Modulus Effect

Under a hydrostatic pressure p the flow stress will increase like the shear modulus μ as

$$\frac{\Delta\tau_1}{\tau} = \frac{1}{\mu}\frac{d\mu}{dp} p \qquad (2)$$

and assuming that Young's modulus E changes with pressure like μ we obtain

$$\frac{\Delta\tau_1}{\tau} = \frac{1}{E}\frac{dE}{dp} p \approx \frac{1}{E}\frac{m+4}{3} p \qquad (3)$$

where m is the exponent in the repulsive part of atomic potential $U = \frac{A}{x^m} - \frac{B}{x}$.

Applied to the case of Cu (E = 125 GPa, and m = 3.75) one finds $\Delta\tau_1/\tau = 0.02\ p/\text{GPa}$ for the "modulus" – effect of p.

3.2 Effect of Excess Volumina of Deformation Induced Lattice Defects

During deformation, lattice defects like dislocations and vacancies are generated. Their excess volumes dVdisloc and dVvacanc must be created against the hydrostatic pressure p. For a deformation step $d\gamma$, this leads to stress increments $\Delta\tau_{2,3}$ which can be estimated from

$$\Delta\tau_2\, d\gamma = \frac{dV_{disloc}}{V} p \approx 1.5\, b^2\, d\rho\, p \qquad (4)$$

for dislocations with volume density ρ, and

$$\Delta\tau_3\, d\gamma = \frac{dV_{vacanc}}{V} p \approx b^3\, dc\, p \qquad (5)$$

for vacancies with volume concentration c.

Assume that the strain $d\gamma$ is achieved by the formation of N square dislocation loops with area $A = 2K \times 2K$ in unit volume, where K is the mean path of both screw and edge dislocations. Then we have

$$d\gamma = b\frac{A}{V}N = b\frac{4\Lambda^2 N}{V} \quad (6a) \quad \text{and} \quad d\rho = \frac{8\Lambda N}{V} \quad (6b) \tag{6ab}$$

Thus, one receives for the rate of dislocation production

$$\frac{d\rho}{d\gamma} = \frac{2}{\Lambda b} \tag{7}$$

Inserting (7) into (4), the stress increase $\Delta\tau_2$ due to dislocation generation at pressure p reads as

$$\Delta\tau_2 = 3\frac{b}{\Lambda}p \tag{8}$$

Calculating now the stress increase $\Delta\tau_3$ due to vacancy generation at pressure p, we need an expression for dc, the number of vacancies which is generated during deformation step $d\gamma$. We make use of the "moving jog" mechanism suggested by van Bueren. [28] The number of produced jogs is the total slip area which the loops have to overcome, times the forest dislocation density, then $(4\ \Lambda^2\ N/\ V)\ \rho$. To get dc, we have to multiply this expression with the number of point defects generated during the movement of the jogs for a mean free path of Λ, being $1/8\ \kappa/b$. The factor $1/8$ accounts for 3 facts: i) that jogs are generated in screw dislocations only, ii) that only screw forest dislocations generate jogs in the moving screws, and iii) that, in average, only 50 % of generated point defects will be vacancies. With $d\gamma$ from (6a), the vacancy production rate is then

$$\frac{dc}{d\gamma} = \frac{\rho\Lambda}{8b^2} \tag{9}$$

and Equation 5 reads as

$$\Delta\tau_3 = \frac{\rho\Lambda}{8}bp \tag{10}$$

Let us reasonably assume now that the mean free path of dislocations Λ, takes a value so that the sum $(\Delta\tau_2 + \Delta\tau_3)$ gets a minimum. Then $\Lambda = \sqrt{(24/\rho)}$, and the sum of both types of "intrinsic hardening terms results as

$$(\Delta\tau_2 + \Delta\tau_3) = \sqrt{\frac{3}{2}\rho}\ bp \tag{11}$$

Inserting $(\sqrt{\rho})\ b = \tau/\alpha\mu$, using $\alpha = 0.2$, one receives $(\Delta\tau_2 + \Delta\tau_3)/\tau = 0.13\ p/\text{GPa}$ for the "defect volumina" – effect of p.

Summing up the \(modulus\) effect and the "defect volumina" effect, we arrive at

$$\frac{\sum\Delta\tau_i}{\tau} = 0.15\ p/\text{GPa} \tag{12}$$

According to our own experiments and those reported by Sturges et al., [18] the experimental value in Cu amounts to $\Sigma D\tau_i/\tau \geq 0.4$ p/GPa. Obviously, the difference left to (12) corresponds to $\Delta\tau_d/\tau$, the stress increase from the permanent changes in microsucture due to the influence of hydrostatic pressure p. Such changes have, indeed, been identified by MXPA as the increasing dislocation density (Figs. 4,6b), [20,23,25] and by EBSP measurements as the decreasing grain size. [22] The next section gives an idea how this effect can be understood and described by modeling.

3.3 Modeling of Defect Density Evolution with Deformation under Hydrostatic Pressure p

Recently, Zehetbauer and Kohout [29] have adapted the large strain work hardening model of Zehetbauer [1] and "Zehetbauer and Les [2,3] to cases of extended hydrostatic pressure being present during deformation. In order to understand the reasons and details of the adaption, the original model is shortly introduced. The model is based on the assumption that the screw and edge dislocations do not interact and thus are placed in separate regions, screws in cell interiors, edges in cell walls. Then the macroscopic hardening $\Theta = d\tau/d\gamma$ is described in terms of hardening contributions $\eta_1 = d\tau_1/d\gamma$ of screw regions with size L_1 (corresponding to cell interiors) and $\Theta_2 = d\tau_2/d\gamma$ of edge regions with size L_2 (corresponding to cell walls). τ_1 and τ_2 are the plastic resistances of screw and edge regions, respectively. By means of volume fractions $f_1 = [L_1/(L_1 + L_2)]^3$ for screws, and $f_2 = 1 - f_1$ for edges, one can write the macroscopic hardening as

$$\Theta = f_1 \Theta_1 + f_2 \Theta_2 \tag{13}$$

and derive for the respective hardening contributions the equations:

$$\Theta_1 = C_1 - C_2 \cdot \tau_1 \tag{14a},$$
$$\text{and} \quad \Theta_2 = C_3 - C_4 [\tau_2 - \tau_2(\gamma=0)] \cdot \tau_2^5 \tag{14b}$$

The specific stresses τ_i are related to dislocation densities ρ_i via $\tau_i = \alpha_i \mu b \sqrt{\rho_i}$ where μ is the shear modulus and b the Burgers vector. The constants $\{C_i\}$ which describe storage and annihilation of dislocations are obtained from fits to the experimental τ/γ relationships while the specific interaction "parameters α_i are obtained from fits to the experimental data $\rho(\gamma)$, using the relation $\rho = f_1 \rho_1 + f_2 \rho_2 = (f_1 \tau_1^2/\alpha_1^2 + f_2 \tau_2^2/\alpha_2^2)/b^2\mu^2$. The f_i are accessible from XPA measurements, [18] yielding $f1^* = L_1/(L_1 + L_2)$. The $\{C_i\}$ cannot take arbitrary values since they are connected to important physical parameters (indicated below by cursive letters) which have to agree with reality:

- $C_1 = \alpha_1 \mu /2 \beta_1$, $C_3 = \alpha_2 \mu/2 \beta_2$; with β_1, β_2 as the dislocation storage rates for screws and edges, respectively, where $\beta_i \equiv \Lambda_i/(1/\sqrt{\rho_i})$ reflects the relation between mean free path K_i of screws (edges) and their distance $(1/\sqrt{\rho_i})$ (see Zehetbauer and co-workers [1–3]);
- $C_2 = 1/(d\gamma/dt)\omega_D \exp(-\delta G/kT)$; with δG as the enthalpy of screw annihilation
- $C_4 = c/\{\sqrt{\pi}/2 \ (1-\nu) \ \exp(\delta H/kT) \times [\tau_2 - \tau_2(\gamma=0)] \ d\gamma/dt \ kT$
  $\alpha_2^4 b^2 \mu^{3/2}/(\Omega D_{c0})\}$ (15)

with δH as the enthalpy of vacancy migration (edge annihilation), ω_D as Debye's freuency, ν as Poisson's ratio, Ω as the atomic volume, and D_{c0} as the core diffusion coefficient.

One can ask now to which quantities shown above the "hydrostatic pressure has an influence. As we learned in the previous section 4.1, the influence to the elastic moduli μ, E is very small, and it will be therefore neglected in what follows. Moreover, no influence of p is expected to annihilation processes which operate by thermal activation in terms of ‚conservative' dislocation slip by cross slip of screws, in contrast to "non-conservative" dislocation movement as through climbing of edges where the volume is locally changing due to continuous mass transfer. Because of the latter, it is mainly the constant C_4 which seems to be strongly affected by p: An extended value of p induces an extra work $p\Omega$ to be achieved when a vacancy is to migrate through the lattice. This leads to a lower diffusion coefficient due to an enhancement in the "effective vacancy migration enthalpy δH_{eff} (compare e.g., "Philibert [30])

$$\delta H_{eff} = \delta H + p\Omega \tag{16}$$

Therefore, by replacing δH by δH_{eff} in Equation 15, one immediately sees that C_4 is decreased which leads to a restriction of edge annihilation term in Equation 14b and thus, to an enhanced edge and total dislocation density (A similar idea has been independently reported by Valiev et al. [32] who, however, did not follow it in a quantitative way). On the other hand, vacancies will be less consumed by annihilating edge dislocations and therefore the vacancy concentration c is expected to increase, too. This indeed turned out when we successfully fitted the first parts of experimental strength data (Fig. 5a) as well as of dislocation densities (Fig. 5b) with the modified model.

Table 2 shows the input data for the hydrostatic pressure applied, and the related dislocation density near to saturation (measured by MXPA at resolved strains $\gamma = 90$, see also Fig. 4). For comparison, data for conventional deformation at'atmospheric pressure according to [14] have been added. $\delta H_{eff}(p)$ has been calculated by means of (14) with $\Omega = 1.1 \times 10^{-29}$ m^3 as the atomic volume in Cu, and $\delta H = 0.2$ eV for the vacancy migration enthalpy measured by Sassa et al. [33] for conventionally deformed pure fcc materials. As concerns the output values, we note that the dislocation configuration "parameters α_1, α_2 continuously increase with increasing "hydrostatic pressure p, which appears to be a consequence of the defect volume effect discussed in section 4.1. The activation enthalpy for screw dislocation indeed is not markedly "affected by p, whereas the concentration of deformation vacancies (similar to ρ, the data given in Table 2 are related to a resolved strain of $\gamma = 90$) shows a dramatic increase with "increasing pressure. The values not only exceed that of thermal equilibrium by at least 10 orders of magnitude, they are higher than that of conventional large strain deformation of Cu done at $p = 0.1$ MPa by at least a factor 100. But this "appears still as too low in view of recent experimental results by Schafler et al. [26] who found atomic vacancy concentrations of order c $\approx 10^{-4}$ in ECAP deformed Cu. It is no doubt that there is still a need for further experimental investigations.

4 Conclusions

Systematic investigations of the influence of the hydrostatic pressure to the strength of pure Cu have been undertaken by means of a high pressure torsion facility which has been equipped with suitable torque and pressure cells. The results showed that

1. the increase of flow stress due to enhancement of hydrostatic pressure by 1 GPa during deformation (\(dynamic\) increase) amounts to at least 40 % of the flow stress known from standard deformation at atmospheric pressure; the increase of flow stress after deformation at enhanced hydrostatic pressure by 1 GPa ("static" increase) still is at least 10 %.
2. both stress increases seem to be larger with lower homologous deformation temperature of the material in question;
3. estimations of both the pressure specific modulus change and the pressure specific effect from finite lattice "defect volume show that these in sum account for less than one halfth of observed total dynamic flow stress increase. Thus, the part left must be ascribed to the pressure specific microstructural evolution. This conclusion is illustrated by " X-ray Bragg Profile measurements which showed a pressure induced increase of dislocation density by a factor of 1.8 for 2 GPa increase of pressure, up to a factor of 4 for 8 GPa increase of pressure, and by EBSP investigations which exhibited a pressure induced decrease of grain size by a factor 1.5 for 7 GPa increase.
4. the pressure specific increase of dislocation density can be quantitatively described by a model which uses the idea that the extent of p affects the diffusion and thus the annihilation of deformation induced lattice defects.

5 References

[1] M. Zehetbauer, Acta Metall. Mater. 1993, 41, 589
[2] P. Les, M. Zehetbauer, Key Eng. Mater. 1994, 97–98, 335
[3] M. Zehetbauer, P. Les, Proc. of the 2nd Symp. Mater. Struct. Micromech. Fract. (MSMF–2), June 1998, Brno (Czech Republic) Kovove Mater. 1998, 36(3), 153
[4] V. M. Segal, V. I. Reznikov, A. E. Dobryshevshiy, V. I. Kopylov, Russ. Metall. (Engl. Trans) 1981, *1*, 99
[5] R. Z. Valiev, I. V. Islamgaliev, I. V. Alexandrov, Prog. Mater. Sci. 2000, 45, 103
[6] Y. Iwahashi, J. Wang, Z. Horita, M. Nemoto, G. Langdon, Scripta Mater. 1996, 35, 143
[7] Y. Beygelzimmer, V. Varyukhin, D. Orlov, B. Efros, V.'Stolyarov, H. Salimgareyev, in Proc. of the 2nd Int. Symposium on Ultrafine Grained Materials, 2002 TMS Annual Meeting, Seattle, USA, Feb. 17-21, 2002, The Minerals, Metals and Materials Society, Warrendale 2002, pp. 43–46
[8] D. Glukhov, Russian Patent 21 916 552, 2001
[9] P. W. Bridgman, Studies in Large Plastic Flow and Fracture, McGraw-Hill, New York 1952
[10] T. Hebesberger, R. Pippan, H. P. Stüwe, in Proc. of the 2nd Int. Symposium on Ultrafine Grained Materials, 2002 TMS Annual Meeting, Seattle, USA, Feb. 17–21, 2002, The Minerals, Metals and Materials Society, Warrendale 2002, pp. 133–140
[11] S. Erbel, Met. Technol. 1979, *6*, 482
[12] J. Richert, M. Richert, Aluminium 1986, 62, 604
[13] M. Richert, H. P. Stüwe, M. J. Zehetbauer, J. Richert, R.'Pippan, C. Motz, E. Schafler, Mater. Sci. Eng. A 2003, in press
[14] M. Zehetbauer, V. Seumer, Acta Metall. Mater. 1993, 41, 577
[15] C. Pilhatsch, Diploma Thesis, University of Vienna, Austria 2003, to be published

[16] B. Mingler, H. P. Karnthaler, M. Zehetbauer, R. Valiev, Mater. Sci. Eng. A 2001, 319–321, 242
[17] H. J. Zughaer, J. Nutting, Mater. Sci. Technol. 1992, 8, 1104
[18] J. L. Sturges, B. Parsons, B. N. Cole, in Mechanical Properties at High Rates of Strain (Ed: J. Harding), Conf. series 47, The Institute of Physics, London 1980, pp. 35–48
[19] I. Saunders, J. Nutting, Metals Sci. 1984, 12, 571
[20] L. Stegelmann, Diploma Thesis, RWTH Aachen 2001
[21] A. Dubravina, R. Valiev, unpublished
[22] T. Hebesberger, R. Pippan, H. P. Stüwe, in Proc. of the 2nd Int. Conf. on Nanomaterials by Severe Plastic Deformation: Fundamentals, Processing, Applications, Wien, Austria, Dec. 9–13, 2002, Wiley VCH, Weinheim 2003, to be published
[23] M. Zehetbauer, in Proc. of the 2nd Int. Symposium on "Ultrafine Grained Materials, 2002 TMS Annual Meeting, Seattle, USA, Feb. 17–21, 2002, The Minerals, Metals and Materials Society, Warrendale 2002, pp. 669–678
[24] T. Ungár, J. Gubicza, G. Ribárik, A. Borbély, J. Appl. Cryst. 2001, 34, 298
[25] A. Dubravina, M. Zehetbauer, E. Schafler, I. Alexandrov, in Proc. of the 13th Int. Conf. on Strength of Materials (ICSMA 13), Budapest, Hungary, Aug. 2003, Mater. Sci. Eng. A, 2003, to be published
[26] E. Schafler, A. Dubravina, Z. Kovacs, in Proc. of the 2nd Int. Symposium on Ultrafine Grained Materials, 2002 TMS Annual Meeting, Seattle, USA, Feb. 17–21, 2002, The Minerals, Metals & Materials Society, Warrendale, 2002, pp. 605–613
[27] P. Haasen, A. W. Lawson, Jr, Z. Metallkd. 1958, 6, 280
[28] H. G. Van Bueren, Acta Metall. 1955, 3, 519
[29] M. Zehetbauer, J. Kohout, in Proc. of the 9th Int. Symp. Plasticity and Current Appl. "Plasticity'02", Aruba, Netherlands, Jan. 3–8, 2002 (Eds: A. S. Khan, O. Lopez-Pamies), NEAT Press, Maryland 2002, pp. 543–545
[30] J. Philibert, Atom Movements Diffusion and Mass Transport in Solids, Les Editions de Physique, Les Ulis 1991, p. 110
[31] H. J. Wollenberger, in Physical Metallurgy (Eds: R. W. Cahn, P. Haasen), Chap. 18, North Holland, Elsevier Science BV., Amsterdam 1996, pp. 1621–1721
[32] R. Z. Valiev, Yu. V. Ivanisenko, E. F. Rauch, B Baudelet, Acta Mater. 1996, 44, 4705
[33] K. Sassa, W. Petry, G. Vogl, Philos. Mag. A 1983, 48, 41

Influence of the Processing parameters at High Pressure Torsion

T. Hebesberger[1,3], A. Vorhauer[1], H.P. Stüwe and R. Pippan[1,2]
[1]Erich Schmid Institute of Materials Science Austrian Academy of Sciences, Leoben (A)
[2]Christian Doppler Laboratory for Local Analysis of Deformation and Fracture, Leoben (A)
[3]now VOEST-ALPINE STAHL, GmbH

1 Introduction

Severe plastic deformation techniques induce large plastic strains in materials at relatively low temperatures. These techniques have the potential of refining the microstructure of metals and alloys to the submicrometer or even to the nanometer range [1–4]. The microstructure of a severe plastic deformed material should depend on the applied strain, the strain path – which is given by the applied technique – the strain rate, the temperature and the pressure. In the present paper we focus our attention to the effect of strain, temperature and pressure. High pressure torsion, HPT, is a proper tool to investigate these parameters in an extreme wide range.

The paper could be interpreted as a fundamental investigation of strain, pressure and temperature on the microstructural evolution during severe plastic deformation or as a study of the most important processing parameters of the HPT-technique. Recrystallized pure copper is used. The developed microstructures were analyzed by back scatter electron (BSE) imaging and by an electron back scatter diffraction technique.

2 Materials and Procedure

High purity copper (99.99 %) was selected as the material for investigation. An annealing at 600 °C for two hours led to a fully recrystallized microstructure with a mean grain size of about 80μm.

Figure 1 schematically shows the HPT tool. A disk shaped sample is placed between two stamps, with a cavity which is somewhat smaller than the sample. After the pressure is imposed, one stamp is rotated. The friction stress deform the sample by shear under hydrostatic pressure. In order to avoid a sliding between the sample and the stamp, the surface of the stamp is cleaned by sand blasting. Further details about HPT are given in [4–7].

The diameter of the samples was 8mm and the thickness 0.4 mm (in some cases 0.6 mm). The applied pressure was varied in the range between 850 MPa and 8 GPa. 850 MPa is the minimum pressure to provide sufficient friction for applying the torsion momentum. At pressures smaller than 850 MPa the sample can slip over the surface of the stamp. The amount of deformation was varied in a wide range, from 0.1 to 40 revolutions. The lower stamp rotated with 0.2 revolutions per minute yielding a strain rate of about 0.2 s^{-1} for the edge of the samples. The experiments were performed at room temperature, at 120 °C and at 200 °C.

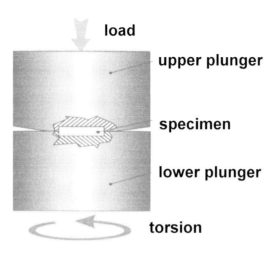

Figure 1: Schematic illustration of the used HPT-device

3 Microstructural Evolution

3.1 The Effect of Strain at Room Temperature

In order to get an impression of the effect of strain on the evolution of microstructure in Fig. 2 the BSE images at an equivalent strain (von Mises) of 1,2,4,8,16,32,64,128,256 and 512 is depicted ($\varepsilon = \gamma \cdot 3^{-0.5}$, $\gamma = (2\pi\ r\ n) / t$, r: radius, t: thickness of the sample, n: number of revolutions). From one image to the other the strain is increased by a factor of 2. In all cases the micrographs are taken from the topside of the HPT samples, i.e., the observation direction is parallel to the torsion axis. At a strain of 1 a coarse irregular substructure develops with a typical size of few microns. With increasing strain the size of this irregular structure decreases and at a strain between 4 and 8 a more regular, nearly equiaxed structure develops. A further increase in the strain induces an increase in the regularity of the microstructure and causes a further refinement of the microstructure. At strains larger than 32 the size of the structural elements and the distribution of the size of the structural elements remain nearly constant. In this saturation stage the size of the largest structural elements is about 300 nm whereas the size of the smallest structural elements is about 100 nm. In order to quantify the evolution of disorientation between structural elements, the deformed samples were subjected to local orientation measurements by electron back scatter diffraction. The electron beam of the SEM, working in the spot mode, scans a selected area with a certain step size (in this investigation 100 nm). Point by point a diffraction pattern is obtained, captured, automatically indexed by the system software (Tex Sem) and is usually represent in orientation maps. The used system has a accuracy in the determination of the crystallographic orientation of ± 1° and a spatial resolution somewhat smaller than 100 nm. In order to quantify the disorientation between structural elements, without a data clean up, a new developed procedure [8] was applied which analyzes the orientation data of a map with respect to disorientation between data points and their distance. Such analysis gives a distribution of disorientation angle as a function of data point distance which can be used to cha-

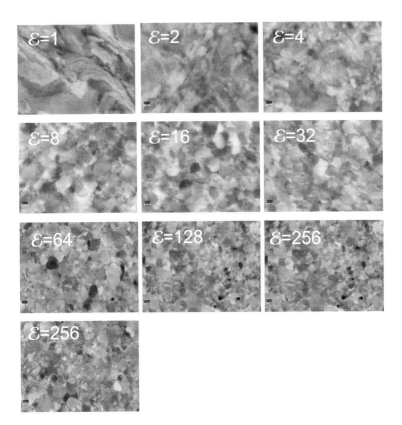

Figure 2: BSE images of HPT-deformed copper at a strain 1, 2, 4, 8, 16, 32, 64, 128, 256, 512

racterize disorientations between the first, the second and third nearest neighboring structural elements.

The Figs. 3 and 4 show the effect at strain on orientation between structural elements. It is important to note that the size of the scanned region is significantly smaller than original grain size. From the distribution of disorientation between data points it is visible that most nearest neighboring structural elements have at strain of 2.7 a disorientation of 3 or 4°, second nearest neighbors have a disorientation of about 5° and the 10th nearest neighbors have a disorientation of about 9°. This indicates clearly that a certain continuous change of orientation is superimposed by a fluctuation by about 4° from structural element to structural element. With increasing strain this disorientation between structural elements increases [5] and reaches about a random distribution of neighboring structural elements at a strain of about 10 (Fig. 3).

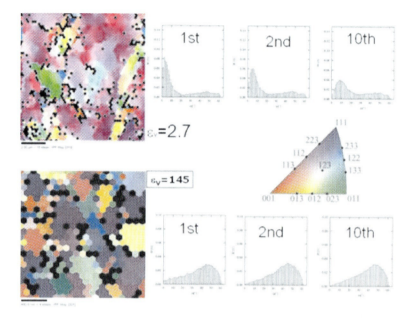

Figure 3: Orientation map, giving the crystal orientation in respect to a direction perpendicular to the surface (parallel to the torsion axis) and the determined frequency values of disorientation angles between data points, for data point distances, which correspond about to data points, of 1st, 2nd and 10th nearest neighboring structural elements at a strain of 2.7 and 147, respectively

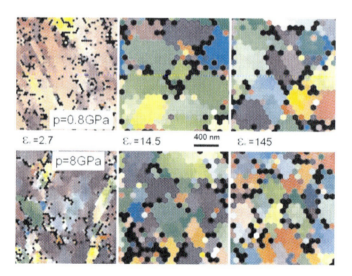

Figure 4: Comparison of the orientation maps of HPT-deformed copper at a strain of 2.7, 14.5 and 145 at pressures of 0.8 and 8 GPa

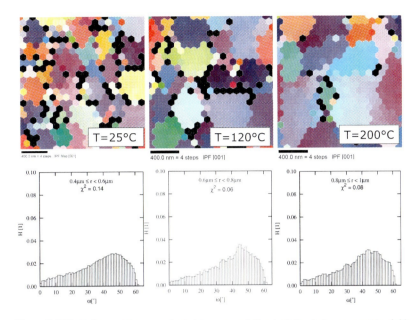

Figure 5: Effect of temperature on orientation maps and disorientation between nearest neighboring structural elements at a strain of 145

3.2 The Effect of Pressure

In Fig. 4 the orientation maps of samples deformed at a pressure of 0.8 and 8 GPa to different strains of 2.7, 14.5 and 145 are compared. It is obvious from the colour code that orientation distribution at different strains are very similar. A more quantitative comparison of the disorientation distribution in [6] comes to the some conclusion. From this investigation only one difference is evident, the size of the structural elements at larger pressures is somewhat smaller. The size of the largest structural elements at 0.8 GPa and 8 GPa is about 300 nm and 200 nm, respectively.

3.3 The Effect of Temperature

In Fig. 5 the orientation maps as well as the disorientation of the nearest neighboring structural elements of HPT samples deformed to a strain of 145 at a pressure of 2 GPa at room temperature, 120 °C and 200 °C are compared. It is worth to note that the depicted microstructures are the "saturation" microstructures at the corresponding temperatures. The distribution of disorientation angle of the nearest neighboring structural elements have about random distribution which is not affected by the temperature. However, the size of structural elements decreases significantly with increasing temperature. The size of the largest structural elements at room temperature, at 120 and 200 °C is about 300 nm, 450 nm and 600 nm, respectively. Furthermore it seem that the size distribution is also affected by the temperature, at small temperatures a larger variation of the size of the structural elements than at high temperatures occurs.

4 Conclusion

High purity copper has been subjected to severe plastic deformation by HPT to different strains at different pressures and different temperatures.

- The size of elements of the developed substructure decreases with increasing strain and reaches a saturation in size of structural elements and distribution of sizes of the structural elements at a strain of about 50.
- The disorientation between the structural elements increases with increasing strain, till it reaches a nearly random distribution of disorientation at a strain of about 10.
- An increase in pressure decreases the resulting size of the structural elements and an increase in temperature leads to coarser microstructure, whereby increasing the temperature shows a much larger effect.

5 Acknowledgement

The partly financial support by the Austrian Fonds zur Förderung der wissenschaftlichen Forschung (Project P 12944-PHY) is gratefully acknowledged.

6 References

[1] M. Furukawa, Y. Iwahashi, Z. Horita, M. Nemeto, T.G. Langdon, Materials Science and Engineering A 257, (1998) 328
[2] K. Oh-ishi, Z. Horita, M. Furukawa, M. Nemeto, T.G. Langdon, Metallurgical and Materials Transactions 29A (1998) 2011
[3] T.C. Lowe, R.Z. Valiev, Investigations and Applications of Severe Plastic Deformation, Kluwer Academic Publishers, Dordrecht (2000)
[4] R.Z. Valiev, R.V. Islamgaliev, I.V. Alexandrov, Progress in Materials Science 45 (2000) 104
[5] T. Hebesberger, Doctoral Thesis, University of Leoben, Austria (2001)
[6] T. Hebesberger, R. Pippan, H.P. Stüwe, Effect of pressure on the final grain size after high pressure torsion, in: Ultrafine Grained Materials II, Eds. T. Zhu et al., TMS, 2002, 133–140
[7] M. Zehetbauer, Features of severe plastic deformation as compared to conventional deformation modes, in: Ultrafine Grained Materials II, Eds. T. Zhu et al., TMS, 2002, 669–678
[8] A. Vorhauer, T. Hebesberger, R. Pippan, Disorientations as a function of distances: A new procedure to analyse local orientation data, Acta Mater. (2003) in press

Mechanical Properties and Thermal Stability of Nano-Structured Armco Iron Produced by High Pressure Torsion

Yu. Ivanisenko[1], A.V. Sergueeva[2], A. Minkow[1], R.Z. Valiev[3], and H.-J. Fecht[1,4]
[1] Division of Materials, Ulm University, Ulm, Germany
[2] Chemical Engineering and Material Science Department, University of California, Davis, USA
[3] Institute of Physics of Advanced Materials, Ufa State Aviation Technical University, Ufa, Russia
[4] Institute of Nanotechnology, Research Center Karlsruhe, Karlsruhe, Germany

1 Introduction

Features of mechanical properties revealed recently in some bulk nanostructured (NS) materials such as very high strength and ductility [1], low temperature of brittle-ductile transition [2], low temperature superplasticity [3] or inverse Hall-Petch relation [4] call for more attentive investigations of such materials. In recent years, a big progress was made in investigation of NS materials prepared by means of compaction of ball milled powders [5, 6], however, such materials usually contain contamination and residual porosity. It is known, that NS materials are very sensitive to their initial microstructures. Two grades of a nanocrystalline material (same composition and structure) that are prepared by different procedures and that have a similar grain size may exhibit quite different mechanical properties [7, 8]. As a result, the existing experimental data on mechanical behaviour of the NS materials are at times ambiguous and contradictory. In that respect, the use of the methods of severe plastic deformation, such as High Pressure Torsion (HPT) and Equal Channel Angular Pressing (ECAP) is attractive, because there is a possibility to produce bulk nanostructured fully dense metals and alloys [1].

The aim of this work was to study the evolution of structure and tensile properties of nanostructured Armco iron upon heat treatment.

2 Experimental

Commercial Armco iron (in wt. %: Fe-0.035% C-0.008%S-0.0012%P-0.2%Mn-0.2%Si) was used for this investigation. The nanostructure was produced by high pressure torsion (HPT) deformation under a pressure of 6 GPa as described in [9]. The as-processed samples were annealed for 1 hour at temperatures of 100-600 °C in Ar atmosphere.

The structure of the samples was studied on a JEM-2000EX electron microscope at an accelerating voltage of 200 kV. The size of structural components was determined from dark (as-processed state, annealing at 200, 300 and 400 °C) and bright field (annealing at 500 and 600 °C) images obtained with 50000-fold magnification. The measurements were carried out in at least four fields of vision each of which contained minimum of 70 grains. The measurement error did not exceed 15 % at a confidence level of 50 %. The structure was studied from sample areas being distanced by $R = 3$ mm from the sample centre, at least three thin foils were studied for each state.

Samples were investigated by orientation imaging microscopy (OIM) in a scanning electron microscope (FEG-SEM) LEO 1550 operating at a voltage of 20 kV, INCA Crystal software (Oxford Instruments) was used for indexing electron back scattering diffraction (EBSD) patterns. The analysis was performed for approximately 500 grains mapped in 3–5 OIM scans for each sample. The mean grain size was estimated from the obtained grain maps. The fracture surface morphology of NS Armco iron after tensile tests was studied on the same SEM operating at a voltage of 5 kV. The internal lattice strains were determined from the broadening of the $(110)_\alpha$, and $(220)_\alpha$ lines using the Warren-Averbach method (using a DRON-4 diffractometer and Co K_α radiation) from the broadening of the $(110)_\alpha$, and $(220)_\alpha$ peaks.

The mechanical properties of Armco iron were studied by means of tensile tests. Tensile specimens of 1-mm gauge length × 1 mm width were cut by spark erosion from disc-shaped HPT samples, by avoiding the central area with lowest strain. The gauge part of the sample was situated at a distance of approximately 3 mm from the disc center. The surfaces of the samples were ground and polished before tests using a diamond paste. The tensile tests were performed using a custom-built computer-controlled constant-strain-rate tensile machine with a displacement resolution of 5 μm and a load resolution of 0.01 N. The displacement rate is controlled by a stepper motor that moves the crosshead at a pre-defined velocity profile as dictated by a microprocessor. The tensile tests were performed at room temperature and at strain rate of $10^{-3}\,s^{-1}$.

3   Results

3.1   Microstructure of NS Armco iron

Figure 1 shows the typical microstructure of Armco iron in the HPT-processed state and after subsequent annealing at different temperatures. It is seen that the as-processed structure is fine-grained and quite uniform. The mean grain size determined from the dark field images is 140 ± 20 nm, however some larger grains with the size of 200 nm can be seen as well. The selected area electron diffraction patterns were taken from 0.5 μm² areas. For the as-processed sample (Fig. 1a) the rings formed by many individual spots were observed, that is attribute to the finely dispersed and highly misoriented grain structures [1, 10].

After annealing of the as-processed samples in the temperature interval 200–400 °C the diffraction contrast on the TEM images of the structure becomes more homogenous (Fig. 1b,c). The grains with banded contrast at their boundaries typical for equilibrium grain boundaries (GBs) [10, 11] appear in the structure. After annealing at 400 °C the fraction of such grains is approximately 30 % (Fig. 1c), and after annealing at 500 °C almost all GBs are in equilibrium (Fig. 1d).

The attempts to obtain EBSD patterns for the as-processed sample failed probably because of high internal stresses. Strain at the grain boundary together with the small grain size limits the resolution. After decreasing the internal stresses by annealing at 200 °C (see below) the performing of OIM analysis became possible. This analysis revealed that 80 % of GBs are high angle, however there was a visible peak at the misorientation angle of 5° as well (Fig. 2). Annealing at 400 °C leads to slight changes in the character of misorientations distribution: the amount of low angle misorientations decreases, and high angle - increases (Fig. 2).

The data on the mean grain size estimated from the dark and bright TEM images and OIM maps (for the samples annealed at 200 and 400 °C) and the elastic lattice strains for the samples in the as-processed state and after different annealing are presented on the Figure 3. The small grain size (<200 nm) is retained up to the annealing temperature of 400 °C, but the high level of elastic lattice strains in the as-processed sample (0.3 %) gradually decreases with increase in annealing temperature and at the temperature of 400 °C it corresponds to that of coarse-grained material (Fig. 3).

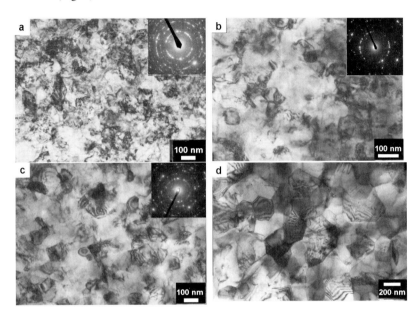

Figure 1: TEM micrographs of Armco iron after HPT and annealing: (a) - as-processed state; (b) –annealed at 200 °C for 1 hour; (c) - annealed at 400 °C for 1 hour; (d) - annealed at 500 °C for 1 hour: bright field image and diffraction pattern.

3.2 Mechanical Behaviour

Figure 4 shows the typical stress-strain curves in tensile tests of Armco iron in as-processed state and after annealing at 200, 400 and 500 °C for 1 hour. In the current investigation the yield stress was determined as the stress of the proportionality limit at the beginning of yielding. One can see that the as-processed samples demonstrate the high yield stress of 1000 MPa and rapid strain hardening after yielding with limited ductility (~3 %). Annealing at 200 °C for 1 hour leads to brittle fracture of the sample just after yielding with slight increase in yield stress. After annealing at temperatures of 400 °C yield stress of the samples decreases to 840 MPa. The elongation of the sample was about 1.5 %. The grain growth at the temperature of 500 °C leads to decreasing of the yield stress to 500 MPa. This sample has demonstrated the elongation of 10%.

Though the samples have displayed a very limited ductility, SEM observations of their surface have revealed a very narrow (about 100 µm) necking area near the fractured end with well developed deformation relief (Fig. 5 a,b). That shows that material has some ductility, however rapid strain localization leads to fracture. On the surface of the as-processed sample one can observe wavy lines of localized plasticity [5] and it is seen that cracks nucleate along them (Fig.

5a). In the annealed at 400 °C sample deformation relief displays the extended parallel to the tensile axis lines (Fig. 5b).

4 Discussion

The results obtained show that severe plastic deformation by HPT leads to the formation of uniform ultra-fine structure with a mean grain size of 140 ± 20 nm in Armco iron. After such a treatment more than 80 % of grains have high angle boundaries, however there is a visible peak at the misorientation angle of 5° (Fig. 2). Very similar results of grain boundary statistic were obtained for HPT deformed Ni by Zhilyaev et. al [12]. Unlike the usual high-angle grain boundaries, GBs in severely deformed materials are in specific non-equilibrium state with the high density of defects [11]. Possibly, the non-equilibrium grain boundaries are responsible for the high level of internal stresses in the as-processed sample (Fig. 3).

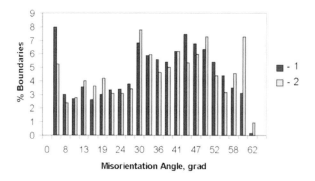

Figure 2: Grain boundary distribution by misorientation angles in Armco iron after HPT deformation and subsequent annealing at: 200 °C - 1 and 400 °C - 2

Annealing in the temperature range of 200–400 °C does not lead to a notable grain growth, but causes gradual recovery process in structure including the non-equilibrium grain boundaries (Fig. 1 (b,c), Fig. 3). Some increase of mean grain size and the amount of big grains at that temperature range accompanied by decrease of the amount of low angle GBs (Fig. 2) can be explained by the occurring of coalescence of neighbouring low-misoriented grains. It is worthwhile to notice that even after annealing of the as-processed sample at the temperature of 500 °C the mean grain size is quite small (320 nm) being e.g. much smaller than the grain size obtained after recrystallization at the same temperature in cold rolled low carbon steel [13].

The values of yield stress of 1000 MPa and ultimate stress of 2200 MPa for the as-processed Armco iron correlate well with the small grain size and are consistent with results of Jia et al. [6] obtained on NS iron with a mean grain size of 80 nm prepared by compacting of ball milled powders. Along with the high strength the as-processed sample as well as annealed ones have demonstrated very limited ductility. Only after annealing at 500 °C increasing the mean grain size to 320 nm, a true strain of 10 % was achieved that was accompanied by decreases in both the yield stress and ultimate stress.

457

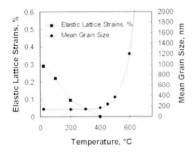 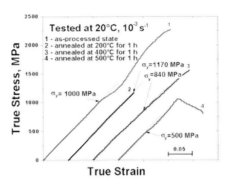

Figure 3: Mean grain size and Elastic Lattice Strains as a function of annealing temperature for NS Armco iron

Figure 4: Stress-strain curves of Armco iron in the as-processed state and after annealing at different temperatures

Expectations of high ductility in metals and alloys with decreasing mean grain size down to nanoscale are based on the possibility of grain boundary sliding at the temperatures much lower than that typical of conventional superplasticity. Actually such a process was reported in several studies in the past [3,14,15]. However, the necessary requirement for the occurrence of grain boundary sliding is the accommodation processes, such as movement of grain boundary dislocations and vacancies [16], requiring a high diffusion activity. The self-diffusion coefficient of iron is very low at 20 °C [13]; thus it makes accommodation of grain boundary sliding difficult. As a result the strong strain localization in shear bands (Fig. 5 a,b) leads to a rapid fracture. Mechanisms of strain localization still remain unclear for the material under investigation.

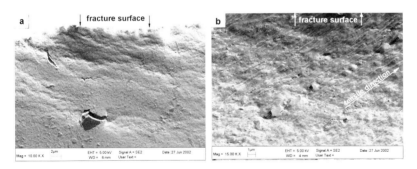

Figure 5: Deformation relief after tensile tests on the surface of the samples close to the fracture area: as-processed (a), and after annealing at 400 °C, 1 h (b)

5 Conclusions

1. It was found that more that 80 % of grains in HPT-processed Armco iron with a mean grain size of 140 nm have high angle misorientations, and that the grain boundaries are in a non-equilibrium state associated with high internal stresses.

2. The small grain size retained up to an annealing temperature of 400 °C, and the transformation of non-equilibrium grain boundaries into equilibrium ones was observed. Further increase in the annealing temperature led to a visible grain growth.
3. The as-processed Armco iron exhibited very high levels of yield stress of 1000 MPa and ultimate stress of 2200 MPa while the ductility has been rather limited. The ductility of the as-processed samples could not even been enhanced by annealing treatment at temperatures of about 400 °C. Only an increase of the mean grain size to more than 320 nm after annealing at 500 °C enabled elongations of about 10 %, but then, however, the yield stress decreased.

6 Acknowledgement

The authors would like to thank I.M. Safarov from IMSP RAS for the help with TEM investigations. This research was supported in part by Alexander von Humboldt Foundation (Yu. Ivanisenko) and by Deutsche Forschungsgemeinschaft (G.W. Leibniz program).

7 References

[1] Valiev R.Z.; Islaimgaliev R.K.; Alexandrov I.V.; Progr. in Mat. Sci., 2000, 2, 103–189
[2] Safarov I.M. et al; Phys. Met. Metallogr. 1992, 3, 303–308
[3] McFadden S.X.; Mishra R.S.; Valiev R.Z.; Zhilyaev A.P.; Mukherjee A.K.; Nature, 1999, 398, 684–686
[4] Song H.W.; Guo S.R.;. Hu Z.Q.; NanoStruct. Mater 1999, 2, 203–210
[5] Malow T.R.; Koch C.C.; Acta mater. 1998, 18, 6459–6473
[6] Jia D.; Ramesh K.T.; Ma E.; Scripta Mater. 2000, 42, 73–81
[7] Weertman J.R.; Sanders P.G.; Solid State Phenomena, 1994, 35-36, 249–262
[8] Suryanarayana C.; International Materials Revew, 1995, 40, 41
[9] Valiev R.Z.; Ivanisenko Yu. V.; Rauch E.E.; Baudelet B.; Acta Mater., 1996, 12, 4705–4712
[10] Valiev R.Z.; Korznikov A.V.; Mulyukov R.R.; Phys. Met. Metallogr,. 1992, 4, 373
[11] Valiev R.Z.; Gertsman V.Yu.; Kaibyshev O.A.; Phys.Stat.Sol. (a) 1980, 61, 95–102
[12] Zhilyaev A.P. et al. Scripta Mater., 2002, 8, 757–580
[13] Steel. A Handbook for Materials Research and Engineering. Ed. By Verein Deutcher Eisenhüttenleute, Vol.1, Berlin: Springer-Verlag, 1992, p. 737
[14] Betz U.; Padmanabhan K.A.; Hann H.; Journal of Mat. Sci., 2001, 36, 5811–5822
[15] Valiev R. Z., Kozlov E. V.; Ivanov Yu. F.; Lian J.; Nazarov A. A.; and Baudelet B.; Acta Metall. Mater. 1994, 42, 2467–2475
[16] Kaibishev O.A.; Superplasticity of Alloys, Intemetallides and Ceramics. Berlin: Springer-Verlag, 1992, p. 237

Properties of Aluminum Alloys Processed by Equal Channel Angular Pressing Using a 60 Degrees Die

Minoru Furukawa[1], Hiroki Akamatsu[2], Zenji Horita[2] and Terence G. Langdon[3]
[1]Fukuoka University of Education, Munakata, Japan
[2]Kyushu University, Fukuoka, Japan
[3]University of Southern California, Los Angeles, U.S.A.

1   Abstract

Tests were undertaken to evaluate the significance of performing equal-channel angular pressing using a die having an angle of 60° between the two channels. The tests were conducted using samples of pure aluminum and an Al-1% Mg-0.2% Sc alloy. The results lead to the conclusion that the mechanical properties of samples in the as-pressed condition are similar for both 60° and 90° dies provided the data are compared at the same equivalent strains.

2   Introduction

Equal-channel angular pressing (ECAP) is a processing procedure in which severe plastic deformation is imposed on a sample by pressing it through a die contained within a channel that is bent through an abrupt angle [1]. It is now well-established that processing by ECAP has the potential for achieving very significant grain refinement in polycrystalline metals [2,3], generally to the submicrometer level and possibly even to the nanometer level. In practice, the strain imposed in a single passage through the die in ECAP depends upon two internal angles: the angle Θ between the two parts of the channel and the angle Ψ representing the outer arc of curvature where the two channels intersect [4]. It can be shown that the angle Θ is especially significant in determining the strain and in general the imposed strain is close to ~1 for all values of Ψ when $\Theta = 90°$ [5].

Numerous experiments have been reported to date using dies having different values of Θ. There are many reports where the value of Θ is either 90° or 120° and there are also some reports of experiments using dies having larger angles up to 135°. To place these various results in perspective, a detailed evaluation of the as-pressed microstructures was conducted using ECAP dies having four different channel angles: specifically, the values of Θ were 90°, 112.5°, 135° or 157.5°, respectively [6]. From these experiments it was concluded that an ultrafine microstructure of reasonably equiaxed grains, separated by boundaries having high angles of misorientation, is attained most readily when using a die having a channel angle that is close to, or equal to, 90° so that a very intense strain is imposed on each separate pass through the die. The implication from these results is that equiaxed microstructures may be achieved even more readily when the channel angle Θ is reduced below 90° since this will give an even larger strain on each pass. However, there have been no reports to date of any experiments in which ECAP was conducted using a die having a value of $\Theta < 90°$. Accordingly, the present experiments were initiated to evaluate the microstructures and mechanical properties which may be achieved in pure aluminum and an Al-based alloy when using a die having a channel angle of $\Theta = 60°$.

3   Experimental Materials and Procedures

The experiments were conducted using aluminum of 99.99 % purity and an Al-1 wt % Mg-0.2 wt % Sc alloy. The pure Al was supplied in a hot-rolled condition and it was homogenized for 24 hours at 753 K, swaged into a rod with a diameter of 10 mm and then the rod was cut into lengths of ~60 mm. Prior to ECAP, the pure Al was annealed for 1 hour at 773 K and air cooled to give an initial grain size of ~1 mm. The Al-Mg-Sc alloy was prepared by casting into an ingot with dimensions of $18 \times 60 \times 160$ mm^3 and it was also homogenized for 24 hours at 753 K, cut into bars with dimensions of $15 \times 15 \times 120$ mm^3 and then swaged and cut to the same dimensions as for the pure Al. The alloy was given a solution heat treatment for 1 hour at 883 K followed by rapid quenching. The grain size of the alloy prior to ECAP was ~500 μm. Further details on the ECAP processing of these materials when using a die with $\Theta = 90°$ were given earlier for pure Al [5,7] and the Al-1% Mg-0.2% Sc alloy [8], respectively. Information is also available on ECAP of an Al-3% Mg-0.2% Sc alloy [9-12].

All of the pressings were conducted using a solid ECAP die constructed from SKD11 tool steel (Fe-1.2~1.4wt% C-11~13% Cr-0.8~1.2% Mo-0.5% V). The die was fabricated with an internal angle of $\Theta = 60°$ and special steps were taken to determine the value of the internal angle representing the arc of curvature, Ψ. When the sample passes through the shearing plane at the intersection of the two channels, it is sheared into a rhombohedral shape [13] such that the outer dimensions at the sheared end of the sample provide a direct measure of the shear strain, γ. Thus, using the conventional relationship for the imposed strain in ECAP [4] and taking $\Theta = 60°$, the imposed strain, ε (= $\gamma/\sqrt{3}$), was estimated as ~1.6 corresponding to an internal arc of curvature of $\Psi \approx 30°$. Thus, each pass through the die imposes a strain of ~1.6 and the total strain is therefore 1.6 N where N is the number of passes. All of the pressings were conducted at room temperature with pure Al taken to a maximum of 4 passes and the Al-Mg-Sc alloy to a maximum of 8 passes. All specimens were pressed using route B_C where the samples are rotated by 90° in the same sense between each pass [14] and it has been established that this route is preferable for attaining an equiaxed microstructure [15].

Following ECAP, samples were examined using optical microscopy (OM) and transmission electron microscopy (TEM). Selected area electron diffraction patterns were recorded using an aperture size of ~12.3 μm. Tensile specimens were machined from the as-pressed rods with gauge lengths of 5 mm and cross-sectional areas of 3×2 mm^2. Tensile tests were performed at 673 K and the specimens were heated to the testing temperature over a period of ~30 minutes and then held at temperature for ~10 minutes prior to the start of each test. Specimens were pulled to failure at strain rates from $1.0 \cdot 10^{-3}$ to $3.3 \cdot 10^{-2}$ s^{-1} and air cooled.

4   Experimental Results and Discussion

4.1   Microstructures after ECAP Using a 60° Die

In order to understand the nature of the deformation occurring when a sample passes through an ECAP die having a channel angle of $\Theta = 60°$, it is first necessary to construct the shearing patterns associated with this type of die. An earlier report documented the shearing patterns associated with dies having $\Theta = 90°$ and 120° [16] and the equivalent patterns for $\Theta = 60°$, including

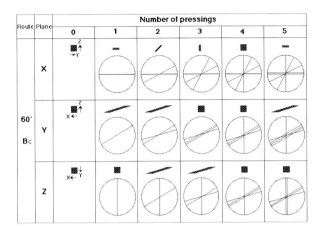

Figure 1: Schematic illustrations of grain shapes and shearing patterns on the X, Y and Z planes using a 60° die in route B_C: results are shown for a total of 5 passes

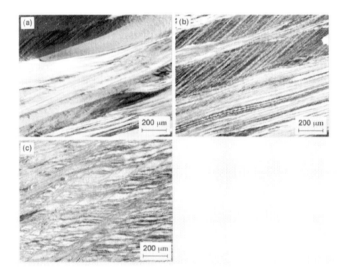

Figure 2: Optical microstructures of pure Al on the Y plane after ECAP using a 60° die and route B_C for totals of (a) 1 pass, (b) 2 passes and (c) 4 passes

the grain distortions, are shown in Fig. 1 when using route B_C: as previously, the X, Y and Z planes correspond to the transverse, flow and longitudinal planes, respectively.

Figures 2 and 3 provide microstructural information on the Y plane after ECAP for the samples of pure Al using OM and TEM, respectively: these specimens were pressed through the 60° die using route B_C for totals of (a) 1 pass with a strain of ~1.6, (b) 2 passes with a strain of ~3.2 and (c) 4 passes with a strain of ~6.4.

Inspection of Figs 2(a) and (b) shows that the original grains are now inclined by ~16° to the X axis after 1 and 2 passes, and this is in agreement with the prediction given for the Y plane in

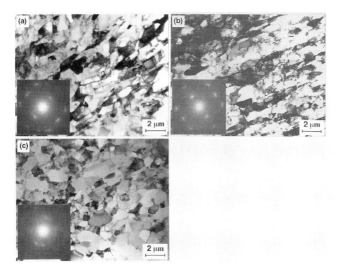

Figure 3: Microstructures of pure Al on the Y plane using TEM after ECAP with a 60° die and route B_C for totals of (a) 1 pass, (b) 2 passes and (c) 4 passes

the central row of Fig. 1. After 4 passes shown in Fig. 2(c), there is an essentially equiaxed array of grains as also predicted for $N = 4$ in Fig. 1. Thus, on the macroscopic scale, there is excellent agreement between the experimental observations and the theoretical predictions for a die with a channel angle of 60°. An earlier investigation demonstrated similar agreement between experiment and theory when using a conventional die with $\Theta = 90°$ [17]. For the TEM photomicrographs, Fig. 3 (a) shows bands of subgrains inclined at ~30° to the X-axis and with a width of ~0.8 μm whereas in Fig. 3(b) the subgrains are inclined at ~15°–30° from the X-axis: the SAED patterns after 1 and 2 passes are net patterns demonstrating the presence of low-angle boundaries although the diffraction spots tend to be slightly broader in Fig. 3(b). In Fig. 3(c) after 4 passes, the diffraction spots are distributed randomly and careful observations suggested ~70 % of the total area contained high-angle boundaries and ~30% of the total area contained low-angle boundaries. In this condition, there is an array of essentially equiaxed grains with an average grain size of ~1.1 μm which is only marginally smaller than the grain size of ~1.2 μm recorded in pure Al with $\Theta = 90°$ [5,7]. These TEM observations are generally consistent with the predictions in Fig. 1.

4.2 Mechanical Properties at Elevated Temperatures after ECAP Using a 60° Die

The preceding results show that the grain size attained in pure Al with a 60° die (~1.1 μm) is only slightly smaller than the grain size achieved in the same material with a 90° die (~1.2 μm). A similar result was found also for the Al-1% Mg-0.2% Sc alloy where the measured grain size on the Y plane after a total of 8 passes was ~0.31 μm which compares with ~0.36 μm when using a 90° die for the same number of passes [8,18]. It was shown earlier that these grains are reasonably stable at high temperatures [8,18] but nevertheless there was some grain growth in the period of ~40 minutes required for heating to the test temperature and temperature stabilization. For example, for the material pressed through a total of 8 passes with a 60° die the grain size was ~1.3 µm at the start of tensile testing whereas the grain size was ~1.2 µm in the material pressed through a total of 8 passes with a 90° die.

Figure 4(a) shows the variation of the elongation to failure with the imposed strain rate for specimens tested at 673 K using a range of strain rates. The results in Fig. 4(a) represent three different processing conditions: with a 90° die for 8 passes where the elongations are the lowest, with a 60° die for 6 passes where the elongations are intermediate and with a 60° die for 8 passes where the elongations are a maximum. At a strain rate of $3.3 \cdot 10^{-3}$ s^{-1}, the elongations to failure are 560 %, 800 % and 1020 % for these three pressing conditions, respectively. Thus, although the grain sizes in the two materials pressed through 8 passes are reasonably similar for the 60° and 90° dies, there is a significantly lower elongation to failure in the material pressed through the conventional 90° die. The reason for this apparent disparity can be seen by inspection of Fig. 4(b) where the elongation to failure with a strain rate of $3.3 \cdot 10^{-3}$ s^{-1} is plotted as a function of the equivalent strain for an unpressed sample, for samples pressed through a 60° die and for the sample pressed for 8 passes through a 90° die. Thus, the experimental point for the sample pressed through the 90° die lies at a similar equivalent strain, and shows a similar elongation to failure, as the sample pressed through only 5 passes with a 60° die. These results suggest that the fraction of high-angle boundaries probably increases with increasing equivalent strain in ECAP.

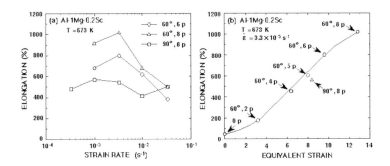

Figure 4: Elongation versus (a) strain rate and (b) equivalent strain for specimens of the Al-Mg-Sc alloy tested at 673 K after pressing with 60° or 90° dies

5 Summary and Conclusions

1. Samples of pure aluminum and an Al-1% Mg-0.2% Sc alloy were processed by ECAP using a 60° die. The resultant grain sizes were only marginally smaller than those attained using a conventional 90° die.
2. Experiments show the elongations to failure are significantly larger when processing with the 60° die but the elongations are essentially identical when specimens processed using 60° and 90° dies are compared at similar equivalent strains.

6 Acknowledgements

This work was supported in part by the Light Metals Educational Foundation of Japan and in part by the U.S. Army Research Office under Grant No. DAAD19-00-1-0488.

7 References

[1] V.M. Segal, V.I. Reznikov, A.E. Drobyshevskiy, V.I. Kopylov, Russian Metall., 1981, 1, 99–105
[2] R.Z. Valiev, N.A. Krasilnikov, N.K. Tsenev, Mater. Sci. Eng., 1991, A137, 35–40
[3] R.Z. Valiev, A.V. Korznikov, R.R. Mulyukov, Mater. Sci. Eng., 1993, A168, 141–148
[4] Y. Iwahashi, J. Wang, Z. Horita, M. Nemoto, T.G. Langdon, Scripta Mater., 1996, 35, 143–146
[5] Y. Iwahashi, Z. Horita, M. Nemoto, T.G. Langdon, Acta Mater., 1997, 45, 4733–4741
[6] K. Nakashima, Z. Horita, M. Nemoto, T.G. Langdon, Acta Mater., 1998, 46, 1589–1599
[7] Y. Iwahashi, Z. Horita, M. Nemoto, T.G. Langdon, Acta Mater., 1998, 46, 3317–3331
[8] M. Furukawa, A. Utsunomiya, K. Matsubara, Z. Horita, T.G. Langdon, Acta Mater., 2001, 49, 3829
[9] Z. Horita, M. Furukawa, M. Nemoto, A.J. Barnes, T.G. Langdon, Acta Mater., 2000, 48, 3633–3640
[10] S. Komura, Z. Horita, M. Furukawa, M. Nemoto, T.G. Langdon, J. Mater. Res., 2000, 15, 2571–2575
[11] S. Komura, M. Furukawa, Z. Horita, M. Nemoto, T.G. Langdon, Mater. Sci. Eng., 2001, A297, 111–118
[12] S. Komura, Z. Horita, M. Furukawa, M. Nemoto, T.G. Langdon, Metall. Mater. Trans. A, 2001, 32A, 707–716
[13] M. Furukawa, Z. Horita, M. Nemoto, T.G. Langdon, J. Mater. Sci., 2001, 36, 2835–2843
[14] M. Furukawa, Y. Iwahashi, Z. Horita, M. Nemoto, T.G. Langdon, Mater. Sci. Eng., 1998, A257, 328–332
[15] K. Oh-ishi, Z. Horita, M. Furukawa, M. Nemoto, T.G. Langdon, Metall. Mater. Trans. A, 1998, 29A, 2011–2013
[16] M. Furukawa, Z. Horita, T.G. Langdon, Mater. Sci. Eng., 2002, A332, 97–109
[17] Y. Iwahashi, M. Furukawa, Z. Horita, M. Nemoto, T.G. Langdon, Metall. Mater. Trans. A, 1998, 29A, 2245–2252
[18] K. Matsubara, A. Utsunomiya, M. Furukawa, Z. Horita, T.G. Langdon, in *The Fourth Pacific Rim International Conference on Advanced Materials and Processing (PRICM4)* (Ed. S. Hanada, Z. Zhong, S.W. Nam, R.N. Wright), The Japan Institute of Metals, Sendai, Japan, 2001, pp. 2003–2006

Deformation Behaviour of Copper Subjected to High Pressure Torsion

A. A. Dubravina, I. V. Alexandrov, R. Z. Valiev
Institute of Physics of Advanced Materials, Ufa State Aviation Technical University, Ufa, Russia
A. V. Sergueeva
Department of Chemical Engineering & Materials Science, University of California, Davis, USA

1 Introduction

Nanostructured materials processed by severe plastic deformation (SPD) can exhibit new and extraordinary mechanical behaviour [1, 2]. For example, these materials display high strength properties, low temperature and high strain rate superplasticity, and so on. Recently it has been demonstrated that the formation of nanostructures by SPD in metals and alloys leads to a significant increase in strength while the high level of ductility preserves [3, 4]. However, many researches failed to achieve both enhanced strength and enhanced ductility in materials using SPD. Contradictory results in the works by different scientists can be explained by the fact that mechanical properties of SPD nanomaterials are very sensitive to specific features of forming nanostructures. Such peculiarities of the microstructures depend on the applied SPD scheme, deformation routes and regimes, geometry of samples, etc. [1].

Our recent investigations have demonstrated that microstructures of various types can be formed in Cu by SPD using high pressure torsion (HPT) deformation to various strain level [5]. On the other hand, it is well known, that nanostructures processed by SPD are non-equilibrium and have low thermal stability due to specific defect structure of grain boundaries [1]. Previously it was demonstrated [6] that annealing of nanostructured Cu processed by HPT at temperatures lower than 150 °C does not lead to the grain growth but to recovery of the grain boundary structure and to relaxation of high internal stresses. Microstructures formed in the as processed by HPT and subsequently annealed material possess different densities of crystal lattice defects, different size of structural units as well as different types of boundaries [5,6]. It was shown that a change of the microstructure type influences significantly the microhardness value of Cu [1, 5, 7].

Based on previous data, the general aim of the current work is detailed investigation of the deformation behavior of Cu produced by HPT with and without subsequent annealing in terms of dependence on the type of microstructure formed.

2 Experimental Techniques

HPT deformation can be presented as a process, which includes compression and torsion under high imposed pressure (Fig. 1). The sample is deformed by shear owing to friction forces arising between the rotated anvils and the sample.

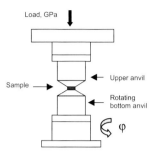

Figure 1: Scheme of high pressure torsion deformation

The disk type samples (∅10×0.7 mm) of annealed Cu (99.9 %) possessing an equiaxed microstructure with an average grain size 120 μm were subjected to HPT under the applied pressure of 5 GPa with a different number of rotations (Table 1) at ambient temperature. Strain values at HPT (Table 1) were defined as a total of strains for cases of pure compression and pure torsion straining.

Table 1: Strain values of Cu samples deformed by HPT (R = 2.5 mm – a half-radius distance)

Rotation number	0 (compression at P = 5GPa)	0.1	0.2	0.4	0.5	1	5
Strain	0.8		4	7	16	20	44	325

Transmission electron microscopy investigations were carried out for the areas close to the radius center of the HPT samples. Flat tensile specimens with a 1-mm gage length × 1 mm width were electro discharge machined from the discs. The tensile tests were performed using a custom-built computer controlled constant strain rate tensile test machine. The strain is measured through crosshead displacement measurements via LVDT with a resolution of 5 microns. A load resolution is 0.01 N. The data are plugged in to the true stress and true strain. The gauge of the tensile specimens corresponded to the half-radius distance from the center of the HPT samples. Tensile tests were conducted at strain rate 10^{-3} s^{-1} and at temperatures 20 °C and 130 °C. Before mechanical tests at elevated temperature the HPT samples were annealed at 130 °C during 10 minutes. Such a treatment provided occurring of the recovery processes in the HPT Cu without changing the grain size [6].

3   Results and Discussion

Transmission electron microscopy investigations showed that different types of microstructure are formed in Cu at HPT depending on the accumulated strain [5]. One of the types of microstructure is cellular structure (CS) with high dislocation density ($10^{10} - 10^{11}$ cm^{-2}) that is typical for the strain at HPT e = 0.8 / 4 (Fig 2, a) [5]. The average cell size at e = 0.8 is 970 ± 110 nm and at e = 4 is 400 ± 60 nm. The wide boundaries of cells consist of individual lattice dislocations. The second type of microstructure is the subgrain one (SGS) (e = 7 – 16) with low-angle

boundaries and low dislocation density in subgrain interiors (Fig. 2, b) [5]. The third type is nanostructure (NS) ($e = 20 / 325$) characterized by high-angle misorientations
between neighboring grains and the high level of internal elastic microdistortions in vicinity of grain boundaries (Fig. 2, c) [5]. At the same time the average sizes of structural elements in SGS and NS are the close to each other and equal to 225 ± 15 nm at $e = 7$ and 220 ± 10 nm at $e = 44$ correspondingly.

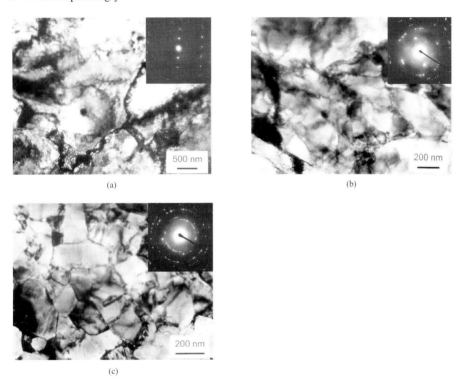

Figure 2: Microstructures formed in Cu at HPT with different strains: $e = 4$ (a); $e = 7$ (b); $e = 44$ (c)

The flow stress-strain curves of room temperature tensile tests for coarse-grained Cu and Cu subjected to HPT are presented in Figure 3. It appears that in case of HPT Cu the stress-strain curves differ significantly from that for coarse-grained Cu. Typical features of deformation behavior of the HPT Cu are high yield stresses which more than two times higher than that for coarse-grained Cu; remarkable strain hardening effect; short period of uniform deformation; high level of ultimate strength; reduced ductility.

The values of yield strength (YS) and ultimate strength (US) of HPT Cu samples gradually rise up with the increase in strain at HPT and become constant for the samples deformed at $e \geq 20$ (Fig. 4). At the same time, the value of elongation to failure (δ) decreases and also becomes constant (Fig. 4) but at smaller HPT strains ($e = 7$).

Room temperature mechanical test data analysis along with microstructural investigation has revealed a significant difference in the level of mechanical characteristics of HPT Cu depending on the initially formed type of microstructure. The formation of cellular microstructure in coarse-grained Cu leads to the increase in yield strength by 55 % and in ultimate strength by 3 % due to dislocation strengthening. The value of elongation to failure decreases by 45 % (Fig. 4).

The subsequent transformation of cellular structure by formation of subgrains results in the further increase in strength characteristics of HPT Cu (YS by 30% and US by 20 % as compared to the same characteristics for Cu with cellular microstructure ($e = 0.8$)) and in the decrease in the ductility characteristics (δ by 40 %) (Fig. 4). This change in the level of mechanical properties can be explained by a refinement of structural elements as well as by qualitative change of the boundaries type between structural elements (substructure strengthening).

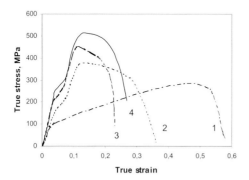

Figure 3: Stress-strain curves of room temperature tensile tests for coarse grained Cu (1) and Cu processed by HPT at different strain: $e = 0.8$ (2); $e = 7$ (3); $e = 20$ (4)

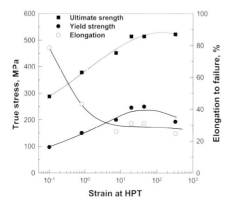

Figure 4: Changing mechanical properties of Cu during HPT; tensile tests were conducted at 20°C

The nanostructured Cu processed by HPT is characterized by the maximum strength parameters in comparison with CS and SGS Cu. The values of yield strength and ultimate strength in nanostructured Cu exceed the corresponding values for Cu with subgrained microstructure by 25 % and 15 % respectively (Fig. 4). According to the transmission electron microscopy data average sizes of structure elements in subgrained and nanostuctured Cu are equal. The difference in the strength level of NS and SGS Cu can be caused both by increasing boundaries misorientations (grain boundaries strengthening) and appearance of long-range fields of elastic stresses near grain boundaries [1]. At the same time it seem that the change of the formed grain boundary type does not influence significantly the material ductility (Fig. 4).

The true flow stress-strain curves of tensile tests at elevated temperatures for HPT Cu annealed at 130 °C are represented in Fig. 5. One can observe that low-temperature annealing and enhanced the temperature of mechanical testing have led to a change of deformation behavior of nanostructured Cu. In particular, values of elongation to necking (uniform deformation) and elongation to failure increased significantly (Fig. 5). For the annealed nanostructured Cu the value of elongation to failure is about of 75 % of that for coarse-grained Cu (Fig. 5) while for as-deformed nanostructured Cu tested at room temperature it is of only 40 % of elongation to failure for coarse-grained Cu (Fig. 3). The values ratio of ultimate strength for nanostructured and coarse-grained Cu tested at 20 °C and at 130 °C is 1.8 and 1.7, respectively.

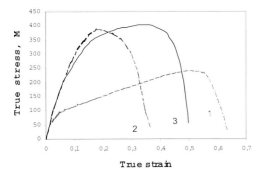

Figure 5: Stress-strain curves of tensile tests at 130 °C for (1) coarse-grained Cu; (2) HPT Cu with subgrain structure ($e = 7$); (3) HPT Cu with nanostructure ($e = 20$)

Also, it can be seen in Fig. 5 that the deformation behaviour of Cu with substructure (subgrains) and nanostructure differs greatly at 130 °C. The samples of nanostructured Cu are characterized by higher ductility characteristics. Thus, the value of elongation to necking (uniform deformation) more than two times higher than that for SGS Cu. However, the yield strength and ultimate strength for Cu with both types of microstructure are almost the same (Fig. 5).

SGS Cu tested at 130°C has demonstrated a mechanical behaviour typical for that of metallic materials at low temperatures when the dislocation slip is a main mechanism of deformation. At the same time, a significant difference in ductility of SGS and NS Cu at the similar level of strength characteristics suggests the existence of some additional mechanism of deformation in nanostructured Cu. As it was shown earlier [3,8] for nanostructured Cu processed by SPD using equal-channel angular pressing, grain boundary sliding could be involved in deformation process in such materials even at low temperatures. Thus probably it is the activation of grain boundary sliding that is responsible for the enhanced ductility in HPT nanostructured Cu at elevated temperatures. This issue will be investigated in more detail in our future works.

4 Conclusions

The effect of microstructure type on the mechanical properties of pure Cu subjected to HPT has been investigated. It was revealed that:

1. Mechanical behavior of Cu at room temperature tensile tests is characterized by the increase in strength properties and the decrease in ductility with microstructures changing in the following sequence: coarse-grained → cellular → subgrained → nanostructured.
2. Low-temperature annealing and a rise in temperature of mechanical tests (till 130°C) result in a significant change of the mechanical behavior of nanostructured Cu. Under these conditions nanostructured Cu displays enhanced ductility close to that of coarse-grained Cu. At the same time, the strength characteristics remain at a rather high level. Apparently, such unique combination of mechanical properties is stipulated by the activation of grain boundary sliding at tensile tests of nanostructured Cu..

5 Acknowledgements

This work was supported partly by the INTAS project 01-0320 and by a grant from the U.S. National Science Foundation (NSF-DMR-9903321 and NSF-DMR-0240144).

6 References

[1] R. Z. Valiev, R. K. Islamgaliev, I. V. Alexandrov, Progr. Mater. Sci. 2000, 45, p. 103
[2] R. Z. Valiev, in *Ultrafine Grained Materials II* (Ed.: Y. T. Zhu, T. G. Langdon, R. S. Mishra, S. L. Semiatin, M. J. Saran, T. C. Lowe), TMS, Warrendale, PA, 2002, 313–322
[3] R. Z. Valiev, I. V. Alexandrov, Y. T Zhu, T. C. Lowe, JMR, 2002, 17, 1, 5–8
[4] R. Z. Valiev, NATURE 2002, 419, 887–889
[5] I. V. Alexandrov, A. A.Dubravina, H. S. Kim, Defect and Diffusion Forum 2002, 208–209, 229–232
[6] V. Y. Gertsman, R. Birringer, R. Z. Valiev, H. Gleiter, Scripta Metal. & Mat. 1994, 30, 229–234
[7] N. M. Amirkhanov, J. J. Bucki, R. K. Islamgaliev, K. J. Kurzydlowski, R. Z. Valiev, J.Metast.and Nanocryst.Mat. 2001, 9, 21–28
[8] R. Z. Valiev, E. V. Kozlov, Yu. F. Ivanov, J. Lian, A. A. Nazarov, B. Baudelet, Acta metall. mater. 1994, 42, 7, 2467–2475

Features of Equal Channel Angular Pressing of Hard-to-Deform Materials

G.I. Raab, E.P. Soshnikova
Ufa State Aviation Technical University, Ufa, Russia

1 Introduction

Ultra fine-grained (UFG) materials possess unique physical and mechanical properties and can be successfully used in industry [1,2]. Traditional methods of hot plastic working (extrusion, rolling) are not efficient enough for fine-grain processing of bulk hard-to-deform materials. At present, for achievement of UFG structure in such materials a special method of severe plastic deformation (SPD) is used. In particular, a multi-pass procedure of equal channel angular pressing (ECAP) is an efficient method for processing bulk UFG materials [1,2,3]. However, application of ECAP for processing such hard-to-deform materials as steels, titanium alloys, high-melting point materials reveals some difficulties due to the materials low workability [4,5]. Evidently, both the properties of the material and the parameters of ECAP influence the workability.

The goal of the present work was to investigate the influence of ECAP regimes and routes on workability of UFG i during multi-pass pressing.

2 Materials and Experimental Procedure

Initial billets out of the CP Ti with the dimensions ∅ 20×80 mm were used for investigations. ECAP was performed using a hydraulic press having a force of 1.6 MN. The temperature of pressing was 300 and 400 °C; the velocity of traverse movement was 6 mm/sec. The diameter of channels ECAP die was 20.2 mm (inlet) and 19.8 mm (outlet). The angle of channel intersection was 120 degrees. The routes of pressing are as follows. Route A – without a turn of a billet, route B – a turn through 90°, route – a turn through 180°. The lubricant ROSoil "A" was applied [6]. The accumulated true strain was determined by the equation $e = 2N(\cot\varphi/2)/\sqrt{3}$ where φ is the angle of channel intersection, N is number of pass [3]. Hydrostatic pressure σ in the center of deformation was taken equal to the specific load on the punch at the end of pressing. Data on flow stress and strain of the CP Ti at temperatures 300 and 400 °C were obtained by compressive tests of standard samples at a test-machine «Instron». Investigations of structure were made by TEM on a JEM-100B microscope.

For evaluation of fracture the probability function of crack initiation $F = F[\varepsilon, \sigma]$ was applied [7], where ε is the strain, σ is the stress.

3 Damage Model at ECAP

At severe plastic deformation the damage should be considered as a relationships of an aggregate of current levels of structure defects density (dislocations, disclinations, etc) and pores, as well as internal stresses to critical ones. It is known that parameters characterizing damage are straightly connected with accumulated strain. The basis concept of damage mechanics is that the damage ω follows a power law in terms of the accumulated plastic shear strain, ε [8,9].

$$\omega = (\varepsilon/\varepsilon_c)^a \qquad (1)$$

where ε_c is the critical level of accumulated shear strain at fracture, a is the coefficient of the power function.

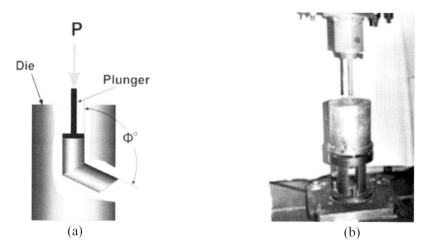

Figure 1: Scheme (a) and die set (b) of ECAP

The critical level of accumulated shear strain ε_c firstly depends on the character of accumulated strain at the moment of fracture, and on the parameter of the stress state that is described by the known equation [10]

$$\varepsilon_c = \varepsilon_c (H, T, \mu_\sigma, \sigma/\tau) \qquad (2)$$

where H is the intensity of shear strain rate, T is the temperature, is Loder coefficient, and σ/τ is the stress index.

In case when ECAP deformation occurs under isothermal conditions at constant strain rate and angle of channel intersection, the parameters (H, T, μ_σ) can be taken as constant. Then

$$\varepsilon_c = \varepsilon_c (\sigma/\tau) \qquad (3)$$

In ε_c–σ/τ coordinates the dependence (3) represents a limit diagram of ductility for fixed temperature-strain rate parameters and can be used to determine conditions of fracture [10]. However, while plotting diagrams of ductility one should take into account the fact that both the change in the sign of deformation and the increase in steps of deformation, increase the total value of accumulated strain [10]. That is why ε can be presented as a function depending on the stress index and the path (way) which a material particle undergoes during deformation. Taking into account the fact that the path (way) of the material particle depends on the route of deformation, i.e. the direction and value of the vector of ε at each pass of loading and the number of passes N, equation (3) can be written as

$$\varepsilon_c = \varepsilon_c (N, \varepsilon, \sigma/\tau) \qquad (4)$$

Basing on the analysis made one can assume that the stress index parameter, the route of pressing and the number of passes, i.e. the accumulated strain $\varepsilon = \varepsilon (N)$ may influence the damage of a billet at ECAP under isothermal conditions. This work has intended a check of damage mechanics concept for the case of ECAP deformation.

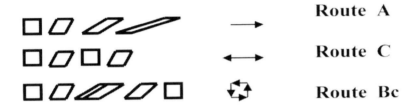

Figure 2: Scheme of deformation of the material particle depending on the route of ECAP

4 Results and Discussion

The experiments on ECAP of titanium were performed at pressing temperatures 300 and 400 °C using the die set of ECAP shown in Fig.1b. Pressing was performed until occurrence of the first shear crack on the portion of the billet corresponding to the steady flow stage. The appearance of the typical crack is shown in Fig.3.

As a result from the measurements of ECAP parameters, curves of σ/τ, σ as a function of Σe were plotted (Fig. 4).

During processing at the temperature 300 °C by routes , and the first surface cracks with the depth about 2 mm in the direction of shear plane occurred after 12, 13 and 17 passes and the accumulated strain was $\Sigma e = 8{,}4$; $\Sigma e = 9{,}1$ and $\Sigma e = 11{,}9$, respectively. Evaluation of efficiency of accumulated strain of pressing routes has shown that route C is most efficient.

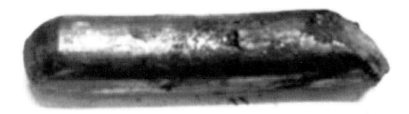

Figure 3: Image of the samples and character of cracks after ECAP

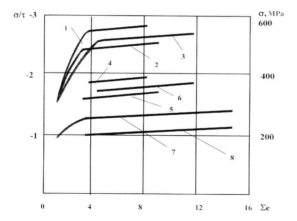

Figure 4: Dependencies of hydrostatic pressure σ and stress index σ/τ (for the established stage) on the accumulated strain at ECAP for CP Ti. 1, 2, 3 - σ for route A, Bc, C, respectively, 4, 5, 6 - σ/τ for route A, Bc, C, respectively, for the temperature of pressing 300 °C; 7 - σ, 8 - σ/τ for route A, for the temperature of pressing 400 °C

Investigation of the influence of the temperature on workability during ECAP was performed for route A. The temperature of pressing was increased up to $T = 400$ °C. The first cracks in the direction of shear plane occurred after the 21st pass when the accumulated strain was $\Sigma e = 14{,}7$.

The analysis of dependencies has shown that at 300 °C there occurs stabilization of parameters σ/τ and σ by route after $\varepsilon = 3.5$ ($N = 5$), by route – after $\varepsilon = 2.8$ ($N=4$), by – after $\varepsilon = 4{,}2$ ($N=6$). A more 'soft" scheme of the stress state ($\sigma/\tau = -1{,}9$) is realized by route A. The specific feature of the dependencies obtained is that after 4-6 passes of pressing ($\Sigma e = 2.8$–4.2) in case of a small decrease in σ/τ and a small growth in σ the super high strains ($\Sigma e = 4$–8) are accumulated until fracture of the billet. One can assume that it may occur due to the change of the mechanism of deformation when recovery processes become dominating.

One should note that the increase in the temperature of pressing from 300 °C to 400 °C does not exert an essential influence on the character of changes of σ/τ and σ. The numerical value of σ decreases by a factor of twice and σ/τ increases from $-1{,}9$ to -1. The scheme of the stress state at 400 °C becomes more "rigid".

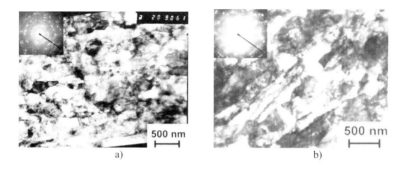

Figure 5: Structure of CP Ti billets after ECAP by route A at temperatures 300 °C (a) and 400 °C (b)

Figure 6: General images of billets, being (a) the initial billet, (b) the UFG billet processed at $T = 300$ °C by route Bc with $\varepsilon = 7$ ($N = 10$), (c) UFG billet processed at 300C by route C with $\varepsilon = 9,8$ ($N = 14$)

Analyzing the dependencies in terms of damage, one can note that the character of growth of damage for all routes corresponds to dependence (1), i.e. with increasing accumulated strain the damage increases. The results of investigations confirm dependence (4) that damage depends on the route (way), which a material particle undergoes during the process of deformation (Fig.2). It has been revealed that the minimum damage for one pass is accumulated in case of alternating character of the shear strain applied (route). The value of accumulated strain ($\Sigma e = 11,9$) being the largest one as compared to the other routes, confirms this. When the vector of deformation concentrates in one direction (route A) and when the vector gradually changes to the opposite one (route), the damage grows more intensively.

The influence of the deformation temperature with ECAP by route A on the microstructure is shown in Fig. 5. After applying about the same accumulated strain, the microstructures observed from vertically longitudinal sections of billets testify that, independently of the deformation temperature, a structure with elongated grains evolves during route A-pressing. At the temperature 300 °C the elongation of grains is less marked. Moreover, the decrease in the deformation temperature leads to the formation of more fine grains.

The results of studies performed provided optimizing regimes of ECA pressing of billets out of Ti. Billets without defects were processed at the temperature 300 °C by routes and at the accumulated strain $\Sigma e = 7$ and $\Sigma e = 9,8$, respectively (Fig. 6).

5 Conclusions

1. Investigations of the influence of the pressing route on the accumulated strain until occurrence of damage in CP Ti samples during multi-pass ECAP at a channel angle 120° and

pressing temperatures 300 and 400 °C have shown that ECAP by route C provides the lowest damage per ECAP pass.
2. Investigations of the influence of the pressing route on the stress index σ/τ of the ECAP P Ti have revealed that maximum absolute values of σ/τ are obtained by using route A, and minimum ones by using route A. more 'soft" scheme of the stress state ($\sigma/\tau = -1,9$) is realized by route A.
3. After applying $N = 4$–6 passes of pressing, stabilization of the hydrostatic pressure σ occurs. For route Bc it occurs after $N = 4$, for route C after $N = 6$, and for route A after $N = 5$.
4. Optimization of ECAP regimes and - routes provides the possibility to process bulk billets of UFG CP Ti. Billets without defects were processed at the temperature 300 °C by routes and at accumulated strains $\varepsilon = 7$ ($N = 10$) and $\varepsilon = 9,8$ ($N = 14$), respectively.

6   Acknowlegments

The investigations were made under support of International Projects ISTC 2398p and INTAS 01-0320.

7   References

[1] Valiev, R.Z., Alexandrov, I.V. Nanostructured materials from severe plastic deformation. – M.: LOGOS, 2000. – 272 p
[2] Valiev, R.Z., Stolyarov, V.V., Latysh, V.V., Raab, G.I., Zhu, Y.T., Lowe, T.C. Proceedings of Int. Conf. Titanium-99, eds.:acad. Gorynin I.V. und prof. Ushkov S.S., Nauka. – V.3. – 2000. – . 1569–1572
[3] Processes of plastic structure formation of metals. V.M. Segal, V.I. Reznikiov, V.I. Kopylov , Minsk: Nauka i Tekhnika, 1994. p. 232
[4] DeLo, D.P.; Semiatin, S.L., Metall. Mater. Trans A, 30A, 1999, p. 2473–2481
[5] DeLo, D.P.; Semiatin, S.L., Ultrafine Grained Materials II, TMS, Warrendale, PA, 2002, p. 539–546
[6] ROSoil ""- technological lubricant TU No. 0258-017-06377289-99
[7] Physical nature of failure of metals. Vladimorov V.I. M.: Metallurgia, 1984. P. 280
[8] Kashanov, L.M. Introduction to Continuum Damage Mechanics, Kluver Academic Publishers, Dordrecht, 1986
[9] Lapovok, R.Ye.; Cottam, R.E., Ultrafine Grained Materials II, TMS, Warrendale, PA, 2002, p. 547–555
[10] Theory of plastic deformation of metals. E.P. Winksov, Y.Jonson, V.L. Kolmogorov .; Ed. E.P. Winkson, A.G.Ovchinnikov-M.: Mashinostroenie, 1983. p. 598, il

IX   New Methods of SPD

ARB (Accumulative Roll-Bonding) and Other New Techniques to Produce Bulk Ultrafine Grained Materials

Nobuhiro Tsuji[1], Yoshihiro Saito[1], Seong-Hee Lee[2] and Yoritoshi Minamino[1]
[1] Osaka University, Suita, Japan
[2] Mokpo National University, Mokpo, Korea

1 Abstract

Accumulative roll-Bonding (ARB) is a severe plastic deformation (SPD) process invented by the authors in order to fabricate ultrafine grained metallic materials. ARB is the only SPD process applicable to continuous production of bulky materials. In the process, 50 % rolled material is cut into two, stacked to be the initial dimension and then rolled again. In order to obtain one-body solid material, the rolling in ARB is not only a deformation process but also a bonding process (roll-bonding). By "repeating this procedure, SPD of bulky materials can be realized. In this review paper, various kinds of new SPD mechanical properties of the ARB processed materials are indicated.

2 Introduction

Ultrafine grained (UFG) metallic materials whose mean grain size is smaller than 1 µm are expected to perform prominent mechanical properties. In order to put the UFG materials to structural use, they must have bulky dimensions. "Severe plastic deformation (SPD) is a hopeful process to fabricate bulky materials with submicrometer grain sizes. Before SPD, grain refinement of metallic materials has been mainly achieved by conventional plastic working and subsequent annealing which results in recrystallization by nucleation and growth (discontinuous recrystallization). The minimum grain size achieved in this route has been about 10 µm. The total "reduction in conventional cold-rolling in industries, for example, is 60 ~ 80 %, which corresponds with von Mises true strain (ε) of 1.06 ~ 1.86, while very large plastic strain over 4.0 is applied to the materials in the SPD processes, so that the UFG microstructures form in the heavily deformed materials. In recent years, various kinds of SPD processes have been proposed and the research works on the SPD and the UFG materials have been energetically conducted. In the present paper, the authors try to overview the various SPD processes for ultragrain refinement, especially putting stress on the "Accumulative Roll-Bonding (ARB) process [1–3] which is a promising SPD process invented by the present authors for bulky materials.

3 Various SPD Processes

3.1 Two Successful SPD Processes

The most famous and successful SPD processes are the equal channel angular extrusion (or pressing) (ECAE or ECAP) [4] and the high pressure torsion (HPT) [5] whose outlines are illustrated in Figure 1. In ECAE, the material is put into an angular channel-die and have a simple shear deformation. The plastic strain per pass depends on the channel-angle as well as the corner radius of the die, and typical equivalent strain (ε) in 90° die is about 1.0. [6] Because cross-sectional area is constant after the pass, the procedures can be repeated limitlessly so that large plastic strain is imposed on the materials. Numerous numbers of researchers are now studying the ECAE process all over the world, since it is easy to install the process, which does not require special equipments except for the dies and a press-machine, in the laboratories. The ECAE process has been applied to various kinds of metals and alloys and succeeded in producing the UFG microstructures after several passes in most cases. However, the major materials used are light metals like Al-alloys and Mg-alloys and the trials for steels are limited. [7,8] This is presumably because large force is required to put the materials through the channel-die, overcoming large flow stress and frictional stress. The hydrostatic compressive stress field in the ECAE process is an advantageous point to prevent fracture of the materials. However, cracking due to shear localization has been reported in less-workable materials such as Ti-Al-V alloy. [9] Though the ECAE can certainly fabricate bulk materials, typical size of the samples is still small. Further, the ECAE is principally not a continuous process but a batch process. Several trials for the continuous ECAE processing have been done, as will be shown later. Disc sample is torsionally deformed under high pressure of several GPa in the HPT process. [5] Because of no change in the sample dimension, very large shear strain can be achieved along the periphery. It should be noted, however, that the amount of shear strain (so that the strain rate as well) differs depending on the radial position. There is no hope for bulky applications of this process, since the HPT can be applied only to small and thin disc samples. However, the process might be scientifically useful. For example, solid-state amorphization and nanocrystallization of NiTi by this process has been recently reported. [10] Because both of these successful SPD processes use simple shear deformation principally, the role of shear for the ultra-grain refinement has been argued. Although there are still many unclear issues on the role of shear strain, the authors think that the shear deformation is not necessarily required for the UFG formation. This is because the UFG microstructures can be produced by the ARB and the conventional rolling, as will be discussed later. It is obvious, however, that the primary important point in SPD is the de-

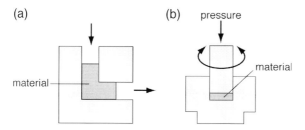

Figure 1: Outlines of a) ECAE and b) HPT processes

formation without dimensional changes of the materials. Simple shear is a deformation without change in height (thickness) of the materials, so that it is certainly advantageous for SPD.

3.2 Continuous ECAE Processes

From a viewpoint of practical application, one of the most disadvantageous points in the ECAE process is that it is not a continuous process but a batch process, as was pointed out above. A few experiments to make the ECAE continuous have been attempted (Fig. 2). Saito et al. [11,12] developed the Conshearing process for continuous ECAE of sheet metals (Fig. 2a). They equipped the ECAE die at the end of the Satellite–Mill they previously invented. [13] The Satellite–Mill is the special rolling mill which can make the rotating speed of all the satellite rolls constant. As a result, compression force along the rolling direction (RD) appears in the materials between adjacent satellite rolls. Folding of the materials is prevented by the guideshoe equipped between the satellite rolls. They used the compression force in the Satellite–Mill to put the sheet into the ECAE die. The Conshearing was applied to a commercial purity aluminum up to 4 cycles and succeeded in fabricating sound sheets, but UFG microstructures have not been obtained. [12] Because they used sheet materials, it was probably not effective to impose ideal shear strain owing to the bending and bending-back deformation, and friction. [12] Further, it is difficult to make the channel-angle 90° in this configuration (they used 125°) [11,12] On the other hand, the processed materials showed unique textures. [11] Lee et al. [14] developed another continuous ECAE process, named the continuous confined strip shearing (C2S2) process, which is principally the same as the conshearing. They seem to use conventional two-high mill, but the surface of the lower roll was mechanically roughened in order to feed the material into the ECAE die. As a result, the surface quality of the specimens would be worse than the conventionally rolled materials. The strips with a dimension of $1.55^T \times 20^W \times 1000^L$ mm^3 were processed by the C2S2 process and the UFGs similar to those "obtained in the conventional ECAE have been reported in 1050-Al, though the difficulties and the disadvantages are not obvious from the limited publications at the present moment. The channel-angle was varied from 100° to 140°, and it was found that the critical strain to form the UFGs increases with increasing the angle. [14]

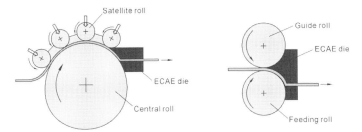

Figure 2: Outlines of a) Conshearing and b) C2S2 processes

3.3 Other Processes

Other unique SPD processes are summarized in Figure 3. Their main concept is the repetition of plastic deformation without change in the dimension of the materials, which is common in all the special SPD processes. The cyclic extrusion compression (CEC; Fig. 3a) is a repetition of the extrusion to decrease the sample diameter and the compression to increase the diameter to the initial dimension. [15,16] The hydrostatic compressive stress field in the process would be preferable to avoid fracture of the materials, although it must be a batch process for the materials with limited sizes. There are only limited number of reports about ultra-grain refinement by the CEC process, [17] but bulk mechanical alloying of Ag-70at.%Cu has been achieved. [16] Figure 3b shows the outline of the continuous cyclic bending (CCB) process proposed by Takayama et al.. [18] This is a repetition of bending and bending-back. Though the maximum strain at the surface achieved in the single bending pass is small ($\varepsilon = 0.1056$), [18] it is very easy to repeat the process and then large total strain can be summed after many passes. However, UFG microstructures have not been developed by this process. The results of this attempt can suggest us two significant points: i) Even if large amount of total strain is imposed as a summation of small strain, UFGs do not necessarily form in the materials. ii) Returning deformation, like bending and bending back in the CCB, seems disadvantageous for ultra-grain refinement. Recently, a new SPD process which seems hopeful for continuous production of bulky UFG materials has been proposed. [19,20] It is the repetitive corrugation and straightening (RCS) process illustrated in Figure 3c. The process illustrated in Figure 3c is, however, still in imagination, and the actual experiments are simulated in simple bending by use of a pressing machine equipped with dies. [19,20] Although the principle of the RCS is bending and bending-back, which is the same as the CCB, the UFG microstructure was reported in RCS processed Cu. [20] Probably the difference between the CCB and the RCS is the amount of strain per pass. However, when the strain per pass is increased by reducing the corner radius of the tools, the strain must inevitably localize in limited parts of the materials depending on the radius of the tools (dies). It is also the case even if the continuous process illustrated in Figure 3c could be realized.

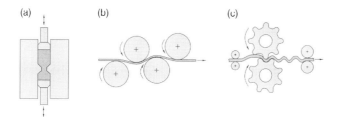

Figure 3: Outlines of a) CEC, b) CCB, and c) RCS processes

4 Accumulative Roll-Bonding (ARB)

4.1 Principle of the ARB

The accumulative roll-bonding (ARB) [1–3] is an only SPD process using rolling deformation itself. The ARB was invented by the present authors in 1998. [1] The principle of the ARB is illustrated in Figure 4. Rolling is the most advantageous metalworking process for continuous production of plates, sheets and bars. However, the total reduction applied to the materials is substantially limited because of the decrease in the cross-sectional dimension of the materials with increasing the reduction. In the ARB process, 50 % rolled sheet, for example, is cut into two, stacked to be the initial dimension, and then rolled again. In order to obtain one-body solid materials, the rolling in the ARB is not only a deformation process but also a bonding process (roll-bonding). To achieve good bonding, the surface of the materials is degreased and wire-brushed before stacking, and the roll-bonding is sometimes carried out at elevated temperatures

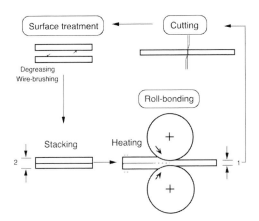

Figure. 4: Schematic illustration showing the principle of the accumulative roll-bonding (ARB) process

below the recrystallization temperature of the materials. Since the above mentioned procedures can be repeated limitlessly, it is possible to impose very large plastic strain on the materials in the ARB process. Table 1 summarizes the geometrical changes of the samples during the ARB, provided two pieces of the sheets 1 mm thick are stacked and roll-bonded by 50 % reduction per cycle. It is interesting that the number of the initial sheets included in the sheet ARB processed by n cycles becomes 2&hoks;n&hoksE;. For example, 1024 initial sheets are included in the 10-cycles ARB processed materials, so that the mean thickness of the initial sheet is smaller than 1 µm. Because von Mises equivalent strain of the 50 % rolling is 0.80, the total equivalent strain after n cycles is 0.8 n. It has been found by our studies that bonding in the ARB process is not difficult generally. For example, good bonding of low-carbon steel can be achieved even at ambient temperatures. [3] However, surface treatment is indispensable for bonding. Furthermore, there is a critical rolling reduction in one pass roll-bonding, below which it is difficult to achieve sufficient bonding. Though the critical reduction depends on the materials and the processing temperatures, more than 35 % reduction by one pass is necessary in general, so that the rolling-force becomes big compared with the conventional rolling. Except for the enough capacity of the rolling mill which can realize one-pass heavy roll-bonding, there are no special requirements in the equipments for the ARB. The serious problem in the ARB process is fracture of the materials. [3] Because large amount of total plastic strain is accumulated in the materials and the rolling is not a hydrostatic process, edge-cracks sometimes occur in the sheets especially at higher cycles. In certain kinds of the materials, such as Al–Mg alloy, the edge cracks greatly propagate into the centre of the sheets. In that case, it becomes impossible to proceed to the subsequent cycles. However, there are several small techniques to avoid such cracking, and the sound bulky sheets can be fabricated by the ARB process in most of the metallic materials. [1–3,21–34] In cases of quite ductile materials, for example, pure aluminum and iron, the UFG sheets having a dimension of $1^T \times 50^W \times 300^L$ mm^3 can be fabricated without cracking by the ARB process even in the university laboratories. [3]

Table 1: Geometrical changes of the materials during the ARB where two pieces of the sheets 1mm thick are roll-bonded by 50 % reduction per cycle

No. of cycles	1	2	3	4	5	6	7	8	9	10	n
No. of layers	2	4	8	16	32	64	128	256	512	1024	2^n
No of bonded boundaries	1	3	7	15	31	63	127	255	511	1023	2^n-1
Layer interval (μm)	500	250	125	62.5	31.2	15.6	7.8	3.9	1.9	0.96	$1000/2^n$
Total reduction (%)	50	75	87.5	93.8	96.9	98.4	99.2	99.6	99.8	99.9	$(1-1/2^n)\times 100$
Equivalent strain	0.8	1.6	2.4	3.2	4.0	4.8	5.6	6.4	7.2	8.0	$\left(\frac{2}{\sqrt{3}}\ln 2\right)n = 0.8\,n$

4.2 ARB Processed Materials

The materials ARB processed by several cycles are filled with the UFGs. Figure 5 shows the TEM microstructures indicating the typical UFGs in various kinds of the ARB processed Al-alloys and steels. [3] Independent of the kind of the materials, the clear UFGs whose diameters are smaller than 1 μm are observed. The UFGs are surrounded by clear but irregular shaped boundaries and the number of dislocations inside the grains seems small. These features are si-

Table 2: Microstructure, grain size and tensile strength of the various kinds of metals and alloys ARB processed in Osaka University.

Materials [mass %]	ARB Process	Microstructure	Grain size [μm]	Tensile strength [MPa]
4N-Al	7 cycles at RT	pancake UFG	0.67	125
100-Al (99 % Al)	8 cycles at RT	pancake UFG	0.21	310
5052-Al (Al-2.4Mg)	4 cycles at RT	ultrafine lamellae	0.26	388
5083-Al (al-4.5Mg+0.57Mn)	7 cycles at 100°C	ultrafine lamellae	0.08	530
6061-Al (Al-1.1Mg-0.63Si)	8 cycles at RT	ultrafine lamellae	0.10	357
7075-Al (Al-5.6Zn-2.6Mg-1.7Cu)	5 cycles at 250°C	pancake UFG	0.30	376
OFHC-Cu	6 cycles at RT	ultrafine lamellae	0.26	520
Cu-0.27Co-0.09P	8 cycles at 200°C	ultrafine lamellae	0.15	470
Ni	5 cycles at RT	ultrafine lamellae	0.14	885
IF steel	7 cycles at 500°C	pancake UFG	0.21	870
0.041P-IF	5 cycles at 400°C	pancake UFG	0.18	820
SS400 steel (Fe-0.13C-0.37Mn)	5 cycles at RT	ultrafine lamellae	0.11	1030
Fe-36Ni	7 cycles at 500°C	ultrafine lamellae	0.087	780

milar to those observed in the materials heavily deformed by other SPD processes. It should be noted, however, the microstructures in Figure 5 were observed from the normal direction (ND) of the sheets. The most characteristic feature of the UFGs in the ARB processed materials is the elongated morphology. Figure 6a is a typical microstructure of the elongated ultrafine grains in the ultra-low-carbon IF (interstitial free) steel ARB processed by seven cycles ($\varepsilon = 5.6$) at 500 °C. [3,26,30] The grains are elongated along RD. Quite similar microstructures appear in the ARB processed aluminum alloys. [3,25,29] These microstructures resemble the lamellar boundary structures observed in heavily deformed materials. [35] Figure 6b is the misorientation map of the identical regions obtained by TEM/Kikuchi-line analysis. [36] Such orientation measurements clearly confirm that the elongated UFGs are not subgrains but grains surrounded by high-angle grain boundaries. The high density of the high-angle grain boundaries in the ARB processed materials has been also confirmed in macroscopic regions by means of SEM/EBSD analysis. [30] At the same time, however, the UFGs have characteristics as deformation microstructures, such as the elongated grain shape and the dislocation substructures inside the grains. The formation mechanism of the UFGs during SPD is still an issue under discussion. However, recent investigations suggest that the formation process of the UFGs is not conventional discontinuous recrystallization but continuous recrystallization (or in-situ recrystallization) characterized by ultrafine grain subdivision, recovery to form clear UFGs, and short range grain

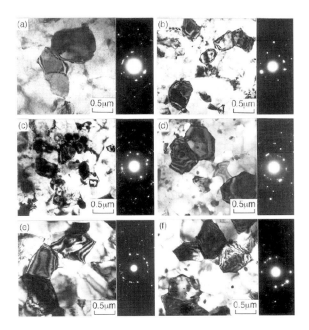

Figure 5: TEM microstructures of the typical UFGs in various kinds of ARB processed materials. Observed from ND. a) 1100-Al (99 %Al) ARB processed by 8 cycles (e = 6.4) at 473 K. b) 5083-Al (Al-4.5 %Mg) ARB processed by 7 cycles (e = 5.6) at 573 K. c) 6061-Al (Al-1.1 %Mg–0.4 %Si) ARB processed by 6 cycles (e = 4.8) at RT. d) 7N01-Al (Al-4.4 %Zn-1.8 %Mg) ARB processed by 5 cycles (e = 4.0) at 523 K and then annealed at 473 &#128>;K for 900 s. e) IF steel ARB processed by 5 cycles (e = 4.0) at 773 K, and then annealed at 773 K for 600 s. f) Plain low-carbon steel (SS400; Fe-0.13 %C-0.37 %Mn) ARB processed to a total strain of 4.0 at RT., and annealed at 833 K for 1.8 ks.

(a)

(b)

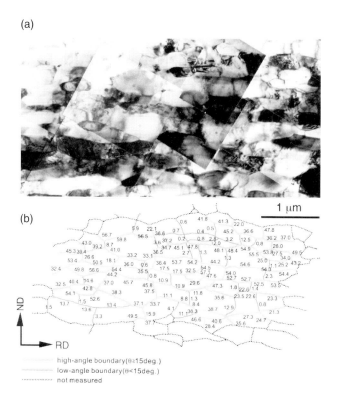

Figure 6: TEM image a) and corresponding misorientation map b) of the IF steel ARB processed by 7 cycles (ε = 5.6) at 773 K. Observed from TD. The misorientations (deg.) indicated in b) were calculated from the accurate orientations of the elongated grains measured by TEM/Kikuchi-line analysis.

boundary migration. [37,38] The ARB processed materials having the elongated UFG structures perform very high strength. [1,3,21,24,25,27,31,32] The grain size and the tensile strength of the various UFG materials fabricated by the ARB are summarized in Table 2. In most cases, the mean grain thickness of the pancake-shaped UFGs or the ultrafine lamellar structures are 100 ∼ 200 nm. The materials with higher purity tend to show larger grain size. The ARB at lower temperature results in smaller grain size within the similar materials. The UFG materials perform the tensile strength two to four times higher than those of the starting materials having conventional grain sizes. On the other hand, the ARB processed materials have limited tensile elongation owing to early plastic instability. [31] It has been also clarified that the ARB processed 5083-Al alloy with UFG microstructure performs low-temperature superplasticity at 200 °C. [22,23]

4.3 Role of Redundant Shear in the ARB

The UFG microstructures are likely to be formed by the ultrafine grain subdivision during SPD principally. [37] As a result, the elongated UFG structures in the ARB processed materials are quite similar to the lamellar boundary structures which have been observed in the materials heavily deformed by conventional rolling. [35] This indicates that the UFG structures are formed not only by shear deformation like ECAE and HPT but also by another mode of plastic deformation, such as rolling (ARB). However, Huang et al. [29] recently clarified that the formation of the ultrafine structure is much faster in the ARB processed 99 % Al than that in the conventionally rolled 99 % Al. For example, the mean misorientations in the ARB processed specimen is significantly larger than that in the conventionally cold-rolled specimen at the identical strains. [29] This means that the ARB process is more advantageous for ultra-grain refinement than the conventional rolling. A probable reason for this difference between the ARB and the conventional rolling is the redundant shear strain. The roll-bonding in the ARB process has been usually carried out without lubrication. It has been known that large amount of redundant shear strain is applied at subsurface regions of the sheets in the rolling under the less-lubricated conditions. [39] In the ARB of 1100-Al, the redundant shear strain has been quantitatively evaluated by Lee et al. [28] Figure 7 shows the flection of the embedded pin after the 1st cycle of the ARB. The straight pin embedded vertically to the sheet before roll-bonding has been greatly bent especially near the surface, which indicates large redundant shear strain was imposed due to large friction between the rolls and the specimen. The shear strain (ε) distribution through thickness, which was calculated from the flection in Figure 7, is shown in Figure 8a. [28] The shear strain (χ) just below the surface reaches up to 8 ($\varepsilon = 4.6$). It has been also clarified that the subsurface regions of the ARB processed sheets have a kind of shear texture completely different from the center regions. [1,30] The characteristic in the ARB processed sheet is the complicated distributions of the redundant shear strain after several cycles, as is shown in Figure 8. Because the 50 % rolled sheet is cut and stacked between cycles, half of the surface which had undergone the severe shear deformation comes into the center. As a result, different from conventional rolling, the sheared regions do not localize only at subsurface layers but complicatedly distribute through thickness of the sheets as the ARB cycle proceeds (Fig. 8). Because shear deformation does not change the thickness of the materials, the integrated area in Figure 8 (integrated shear strain) should be added to the equivalent strain of the sheet. That is, the substantial strain applied during the n-cycles of the ARB without lubrication is much larger than 0.8 n, which was calculated only from the reduction in thickness. The substantially larger strains accumulated must be one of the reasons for the faster development of the UFG structure in the ARB materials. Furthermore, the effect of changing strain path should be taken into account as well. The ARB processed materials have had quite complicated combination of plane-strain deforma-

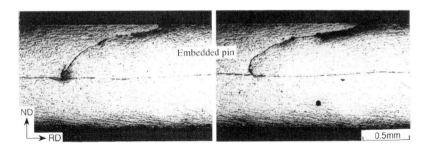

Figure 7: Optical microstructures showing the flection of the embedded pin in the 1100 aluminum sheet ARB processed by one cycle at ambient temperature without lubrication. Observed on a longitudinal cross-section

tion and shear deformation, depending on the thickness location and the number of cycles. This means that the strain path at each region greatly changes between shear and plane-strain compression in every cycle. As a result, the centre part of the ARB processed sheets show relatively weak texture even after large rolling reduction. Although the role of the strain path for ultra-grain refinement is still under discussion, the present results in the ARB suggests that changing the strain path is effective for the formation of the UFGs, in other words, for the faster ultrafine grain subdivision.

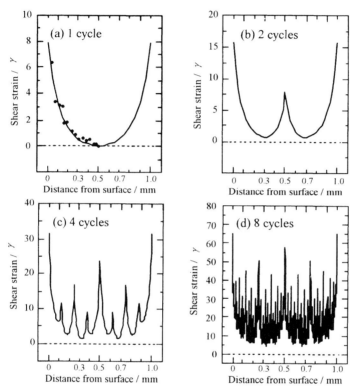

Figure 8: Distribution of shear strain through thickness of the 1100-Al ARB processed by a) 1, b) 2, c) 4, and d) 8 cycles at RT

5 Summary

Various kinds of SPD processes were overviewed and the characteristics of the ARB process was summarized. The formation of the UFG structure was discussed, putting stress on the role of shear deformation. It was shown that the ARB is the promising SPD process which can produce the bulky UFG materials continuously. By use of the advantages in the ARB process, formation mechanism of the UFGs as well as the properties of the UFG materials should be further studied in the future.

6 Acknowledgement

The present work was financially supported by Industrial "Technology Research Grant Program' 01 from NEDO of Japan under project ID 01A23025d, Grant-in-Aid for Scientific "Research (No.14702052), and the 21st Century COE Program "Center of Excellence for Advanced Structural and Functional Materials Design" in Osaka University from the Ministry of Education, Sports, Culture, Science and Technology of Japan.

7 References

[1] S. Saito, N. Tsuji, H. Utsunomiya, T. Sakai, R. G. Hong, Scripta Mater. 1998, 39, 1221
[2] Y. Saito, H. Utsunomiya, N. Tsuji, T. Sakai, Acta Mater. 1999, 47, 579
[3] N. Tsuji, Y. Minamino, Y. Koizumi, Y. Saito, in Proc. of the 11th Int. Symp. on Processing and Fabrication of Adv. Mater. (PFAM XI), ASM, Materials Park, OH 2003, in press.
[4] M. Nemoto, Z. Horita, M. Furukawa, T. G. Langdon, Met. Mater. 1998, *4*, 1181
[5] Z. Horita, D. J. Smith, M. Furukawa, M. Nemoto, R. Z. Valiev, T. G. Langdon, Mater. Res. 1996, 11, 1880
[6] Y. Iwahashi, J. Wang, Z. Horita, M. Nemoto, T. G. Langdon, Scripta Mater. 1996, 35, 143
[7] D. H. Shin, B. C. Kim, Y. S. Kim, K. T. Park, Acta Mater. 2000, 48, 2247
[8] A. Azushima, K. Aoki, Mater. Sci. Eng. A 2002, 337, 45
[9] S. L. Semiatin, J. O. Brown, T. M. Brown, D. P. DeLo, T. R. Bieler, J. H. Beynon, Metall. Mater. Trans. A. 2001, 32A, 1556
[10] A. V. Sergueeva, C. Song, R. Z. Valiev, A. K. Mukherjee, Mater. Sci. Eng. A. 2003, 339, 159
[11] Y. Saito, H. Utsunomiya, H. Suzuki, T. Sakai, Scripta Mater. 2000, 42, 1139
[12] H. Utsunomiya, Y. Saito, H. Suzuki, T. Sakai, Proc. Inst. Mech. Eng. B. 2001, 215, 947
[13] Y. Saito, T. Watanabe, H. J. Utsunomiya, Mater. Eng. Perform. 1992, *1*, 789
[14] J. C. Lee, H. K. Seok, J. Y. Suh, Acta Mater. 2002, 50, 4005
[15] J. Richert, M. Richert, Aluminium 1986, 62, 604
[16] T. Aizawa, K. Tatsuzawa, J. J. Kihara, J. Fac. Eng., Univ Tokyo, Ser. B 1993, XLII, 261
[17] M. Richert, Q. Liu, N. Hansen, Mater. Sci. Eng. A 1999, A260, 275
[18] Y. Takayama, M. Yamaguchi, T. Tozawa, H. Kato, H. Watanabe, T. Izawa, in Proc. of the 4th Int. Conf. on "Recrystallization and Related Phenomena (Rex\999), The Jpn. Inst. Metals, Sendai, Japan 1999, pp. 321–326
[19] Y. T. Zhu, H Jiang, J. Huang, T. C. Lowe, Metall. Mater. Trans. A. 2001, 32A, 1559
[20] J. Y. Huang, Y. T. Zhu, H. Jiang, T. C. Lowe, Acta Mater. 2001, 49, 1497
[21] N. Tsuji, Y. Saito, H. Utsunomiya, S. Tanigawa, Scripta Mater. 1999, 40, 795
[22] N. Tsuji, K. Shiotsuki, Y. Saito, Mater. Trans. JIM 1999, 40, 765
[23] N. Tsuji, K. Shiotsuki, H. Utsunomiya, Y. Saito, Mater. Sci. Forum. 1999, 304–306, 73
[24] S. H. Lee, T. Sakai, Y. Saito, H. Utsunomiya, N. Tsuji, Mater. Trans. JIM 1999, 40, 1422
[25] Y. Ito, N. Tsuji, Y. Saito, H. Utsunomiya, T. J. Sakai, Jpn. Inst. Metals. 2000, 64, 429
[26] N. Tsuji, R. Ueji, Y. Saito, Materia Japan 2000, 39, 961
[27] T. Sakai, Y. Saito, T. Kanzaki, N. Tamaki, N. J. Tsuji, JCBRA 2001, 40, 213

[28] S. H. Lee, Y. Saito, N. Tsuji, H. Utsunomiya, T. Sakai, Scripta Mater. 2002, 46, 281
[29] X. Huang, N. Tsuji, N. Hansen, Y. Minamino, Mater. Sci. Eng. A. 2003, 340, 265
[30] N. Tsuji, R. Ueji, Y. Minamino, Scripta Mater. 2002, 47, 69
[31] N. Tsuji, Y. Ito, Y. Saito, Y. Minamino, Scripta Mater. 2002, 47, 893
[32] N. Kamikawa, N. Tsuji, Y. Saito, Tetsu-to-Haganè 2003, 89, 273
[33] N. Tsuji, Y. Ito, H. Nakashima, F. Yoshida, Y. Minamino, Mater. Sci. Forum. 2002, 396–402, 423
[34] H. W. Kim, S. B. Kang, Z. P. Xing, N. Tsuji, Y. Minamino, Mater. Sci. Forum 2002, 408–412, 727
[35] N. Hansen, D. J. Jensen, Phil. Trans. R. Soc. Lond. A 1999, 357, 1447
[36] S. J. Zaefferer, J. Appl. Cryst. 1999, 33, 10
[37] N. Tsuji, R. Ueji, Y. Ito, Y. Saito, in Proc. of the 21st Risø Int. Symp. on Mater. Sci., Risø National Laboratory, Denmark, 2000, pp. 607–616
[38] F. J. Humphreys, P. B. Prangnell, R. Priestner, Curr. Opin. Solid State Mater. Sci. 2001, 5, 15
[39] T. Sakai, Y. Saito, M. Matsuo, K. Hirano, K. Kato, ISIJ Int. 1988, 28, 1028

Optimal SPD processing of plates by Constrained Groove Pressing (CGP)

Jon Alkorta, Javier Gil Sevillano
Centro de Estudios e Investigaciones Técnicas de Guipúzcoa (CEIT) y TECNUN (University of Navarra). San Sebastian. Basque Country (Spain)

1 Introduction

In the recent years considerable efforts have made in order to produce ultrafine grained materials by imposing large plastic straining by many different techniques such as equal-channel angular pressing (ECAP) [1–4], severe plastic torsion straining (SPTS) [1, 5–7] or accumulative roll bonding (ARB) [8–10]. However, ECAP and SPTS seem to be of little use when manufacturing plate-shaped materials efficiently and only ARB is appropriate for producing SPD plates or sheets. Efforts, therefore, have been focused not only on analysing the properties of the materials subjected to severe plastic deformation (SPD) but also on the design and development of new techniques that could compete with the processes cited before. One of the most recent contributions from that point of view is the constrained groove pressing (CGP) developed by Shin et al. in 2002 [11].

CGP consists in pressing a sample by a set of grooved dies tightly constrained by a wall (Figure 1). The sample is, therefore, forced to deform ideally by simple shear in the inclined region whereas the flat regions keep undeformed. In the introductory paper on CGP, Shin et al. proposed a deformation path (called Route I from now on) consisted in successive CGP + flattenings with 180° rotations between them in order to obtain an homogeneous deformation field along the whole sample. But this is not the only possible path; in fact, this technique provides us with a large variety of possible deformation routes. So, which of them is the optimal for obtaining ultra-fine grained materials?

Concerning this, recently Dupuy et al. [12, 13] based on the work carried out by Schmitt et al. [14] have proposed a simple and effective way to relate the deformation path to its capacity for grain refinement. They assume the fact that an orthogonal deformation path activates new slip systems which could promote cell misorientation and activate more efficiently grain refinement mechanisms; thus, they introduced an α parameter which determines the strain path change between two successive deformation steps:

$$\alpha = \frac{\varepsilon_p : \varepsilon}{\sqrt{\varepsilon_p : \varepsilon_p} \sqrt{\varepsilon : \varepsilon}} \quad (1)$$

where ε_p and ε are the plastic strain tensors corresponding respectively to the first deformation mode and the subsequent deformation mode. The value of α varies from 1 (no change in the strain path) to –1 (Bauschinger test); the intermediate situation with $\alpha \approx 0$ corresponds to the situation where the slip systems pertaining to the first loading mode are latent during the restraining and vice versa. The α parameter, therefore, provides us with an efficient geometrical tool for designing the best route to be followed during a CGP process.

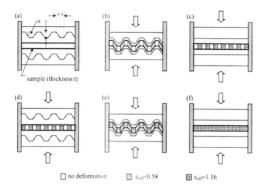

Figure 1: Schematic sequence of the Route I. From Shin et al. [11].

The present work tries to optimise, using finite element analysis, the CGP process and the deformation route by the design of a new route (called route II from now on) that improves the α parameter. This new processing route includes 90° rotations between CGPs.

2 CGP Description

As cited before, Constrained Groove Pressing consists in pressing a sample with a set of grooved dies tightly constrained by a wall (Figure 1 from Shin et al. [11]). The deformation induced is inhomogeneously distributed along the sample width in such a way that the inclined region is deformed ideally by simple shear ($\gamma = \tan \theta$) and the flat region keeps undeformed. This distribution allows us to design a deformation route in order to homogenise the deformation induced and optimise the α parameter.

2.1 Route I

Route I was designed by Shin et al. [11] in their introductory paper on CGP. It consists in a set of consecutive CGPs + flattenings (Table 1 and Figure 1) plus sample rotations that locate the sample in such a way that the previously undeformed regions can be deformed and vice versa. This route is based, therefore, on a systematic strain reversal that can lead to a low cycle fatigue (LCF) which most probably leads to an inefficient grain refinement.

Table 1: Deformation routes for CGP

Step number	Route I	Route II
1	CGP + flattening	4 x (CGP + 90° rotation) + flattening
2	180° rotation	Change of tools
3	CGP + flattening	4 x (CGP + 90° rotation) + flattening

2.2　Route II

Route II is the route proposed in the present work in order to avoid some of the inconvenients from the Route I. The characteristics of this route are shown in Table 1. The route starts with a set of 4 CGPs followed each of them by a 90° rotation (the rotation axis is parallel to the pressing direction and stands on the middle of a flat region); after the fourth pressing we recover the plate-like shape by flattening the sample with a set of flat dies (recommended but only required when completing the whole process). After this first step, the sample is deformed inhomogeneously along the sample and reaches a similar structure to the obtained by the first step of the route I (see Fig. 1c). The process continues with a change of the set of tools in order to change the location of the grooves in such a way that the undeformed region can be deformed by the next step and vice versa (the rotation axis stands now on the middle of an inclined region of the sample). Afterwards, a new set of four CGPs + 90° rotations is performed and the final flattening recovers the plate-like shape.

3　Finite Element Analysis Results and Discussion

All the elastic-plastic FEM analysis have been carried out using Abaqus 6.1 Explicit[15].

For the Route I two-dimensional elasto-plastic FEM simulations under plane-strain conditions are considered, i.e. strain along normal direction is assumed to be zero. Therefore, 4-node bilinear plane-strain elements are used with hourglass control and reduced integration. For the Route II three-dimensional elasto-plastic FEM simulations are considered using 8-node brick elements with hourglass control and reduced integration.

All the simulations were carried out for half a groove period using symmetric boundary conditions in order to simplify the problem. The material properties used correspond to an Al-Mg alloy (Al 5083 O-Temper) [16] (Table 2) with a Voce-type strain hardening behaviour [17]. Hard contact interactions have been assumed between the sample and the die with no friction.

Table 2: Characteristics of the material used for FE simulations

	Value
Young's modulus	69 Gpa
Poisson's ratio	0.3
Yield stress	145 Mpa
Saturation stress	290 Mpa

3.1　CGP Optimization

Firstly, a set of FE analysis were carried out in order to optimize the different parameters that influences over the deformation obtained after a constrained groove pressing. The most important one is the ratio between the sample thickness and the groove period (r parameter from now on). Four different values of r were analized: 1/8, ¼, ½ and 1. For the lowest value of r necking was observed and for the highest value of r large strain inhomogeneities were observed that led to folds at the sample surface (Figure 2). Therefore, we conclude that the optimum value of r

lies between 0.25 and 0.5 (Shin et al. used $r = 0.25$). It is remarkable the fact that if the groove angle is reduced (Shin et al. used $\theta = 45°$) the window for the optimal r value is wider although the shear strain per pressing diminishes ($\gamma = \tan \theta$) at the same time. For Route II $r = 0.5$ and $\theta = 45°$ were used.

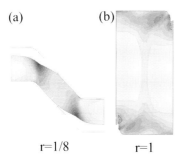

Figure 2: Necking and folds for, respectively, low and high values of r parameter

3.2 Route I

Figure 3 shows the deformation field obtained by FE analysis for Route I after the first and the third step and the strain profile along the cross section of the sample (Table 1). After the third step, a peak-valley deformation structure can be seen along the sample width, in contrast with the homogeneous ideal deformation predicted by simpler analysis (Shin et al., [11]). Besides, it is remarkable the fact that the second set of CGP + flattening induces less deformation than the first one; since simulations for non-hardening materials shown similar effects, we can conclude

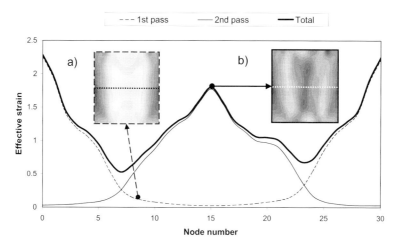

Figure 3: FEM results for CGP route I. Pictures correspond to the strain field after the first pass (a) (Step 1) and the second pass (b) (step 3). The thin lines correspond to the deformation profile along the cross section path (dotted line in figures) for the first and the second pass separately. The thick line corresponds to the total deformation profile after the second pass.

that this is not due to the strain-hardening behaviour of the material but to elastic recovery effects. After the first step the material maintains a residual deformation due to the elastic recovery that reduces the induced deformation effectivity for the subsequent deformation steps.

3.3 Route II

Figure 4 shows the deformation field obtained by FE analysis for Route II after the first and third step (Table 1) and the strain profile through a path which connects the highly deformed peaks. After the first step, a chessboard-like deformation field can be seen, with contiguous non-deformed and deformed regions. After the grooved tool change non-deformed regions stand on the inclined regions of the new grooved dies and vice versa. This leads to the structure shown in Figure 4. Again, a slight peak-valley structure is observed with effective strains ranged from $\varepsilon \approx 2$ to $\varepsilon \approx 4$. A loss of induced deformation efficiency in the second set of deformation (step 3) due to elastic recovery effects is observed as well.

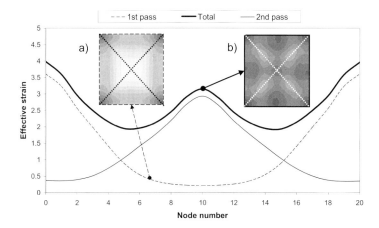

Figure 4: FEM results for CGP route II. Pictures correspond to the strain field for the plane normal to the pressing direction after the first pass (a) (Step 1) and the second pass (b) (step 3). The thin lines correspond to the profile of the average deformation contribution of the first and the second passes along the diagonal dotted paths. The thick line corresponds to the total deformation profile after the second pass.

3.4 α Parameter

Figure 5 shows the average of the α parameter calculated for all the elements that underwent an effective strain higher than 0.15 on each of two consecutive deformation passes. The values obtained are compared with the theoretical values obtained by Dupuy et al. [12] for different ECAP routes with a die angle equal to 90°. As it can be seen in figure 5, Route I is nearly pure strain reversal (as expected) whereas Route II presents an enhanced α parameter close to the predicted for ECAP (route B).

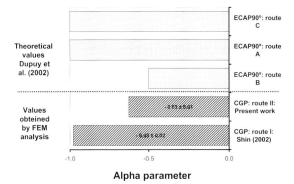

Figure 5: FEM estimations for the α parameter for CGP (Routes I and II). Comparison between CGP 90° (from Dupuy et al. [12]) and CGP

4 Conclusions

FE analyses have been used in order to optimise the CGP process. It has been concluded that the optimal value of the sample thickness to groove period ratio lies between 0.25 and 0.5 for a groove angle $\theta = 45°$. Besides, it has been seen that a reduction of the groove angle to 30° leads to an enhancement of the window for optimal r parameter to the detriment of the induced shear strain.

Route I proposed by Shin et al. [11] was analysed by FE analysis. In contrast with the ideally predicted deformation field, FE analyses show a marked peak-valley deformation structure along the sample width. Besides, a loss of induced strain effectivity has been detected for the second set of deformations (step 3) due to elastic recovery effects. Concerning the α parameter, Route I being a mere strain reversal behaviour, it is probably rather inefficient for grain refinement.

In order to improve the process a new deformation route (Route II) was proposed; this new deformation route significantly improves the α parameter and reduces the equivalent strain inhomogeneity.

5 Acknowledgements

This work has been performed as part of the project MAT2002-04343-C03-03 under the sponsorship of the General Direction of Research of the Spanish Ministry of Science and Technology (MCYT).

6 References

[1] R.Z. Valiev, R.K. Islamgaliev, I.V. Alexandrov. Prog. Mat. Sci., 2000, 45, 103–189
[2] V.M. Segal. Mater. Sci Eng., 1995, A197, 157–164

[3] Y. Iwahashi, Z. Horita, M. Nemoto, T.G. Langdon. Acta Mater., 1998, 46, 3317–3331
[4] A. Gholinia, P.B. Prangnell, M.V. Markushev. Acta Mater., 2000, 1115–1130
[5] R.Z. Valiev, O.A. Kaibyshev, R. Kuznetsov, R. Sh. Musalimov, N.K. Tsenev. Proc. Acad. Sci. U.S.S.R., 1988, 301, 864
[6] R.Z. Valiev, A.V. Korznikov, R.R. Mulyokov. Mat. Sci. Eng., 1993, A168, 141–148
[7] V. Y. Gertsman, R. Birringer, R.Z. Valiev, H. Gleiter. Scr. Metall. Mater., 1994, 30, 229–234
[8] Y. Saito, N. Tsuji, H. Utsunomiya, T. Sakai, R.G. Hong. Scripta Mater., 1998, 39, 1221–1227
[9] N. Tsuji, Y. Saito, H. Utsunomiya, S. Tanigawa, Scripta Mater., 1999, 40, 795–800
[10] Y. Saito, H. Utsunomiya, N. Tsuji, T. Sakai. Acta Mater., 1999, 47, 579–583
[11] D. H. Shin, J-J. Park, Y-S. Kim, K-T Park. Mater. Sci. Eng., 2002, A328, 98–103
[12] L. Dupuy, E.F. Rauch. Mater. Sci. Eng., 2002, A337, 241–247
[13] L. Dupuy. Comportement mécanique d'un alliage d'aluminium hyper-déformé. Ph.D. Thesis. Institut National Polytechnique de Grenoble. 2000
J. H. Schmitt, E. L. Shen, J. L. Raphanel, Int. J. Plast. 1994, 10, 535–551
[14] ABAQUS is a registered trademark of Hibbitt, Karlsson & Sorensen, Inc.
[15] ASM Handbook, Formerly 10th edition, Metals Handbook, 1990, Vol.2, p. 93.
[16] J. Gil Sevillano, P. Van Houtte, E. Aernoudt. Prog. Mater. Sci., 1980, 25, 69–412

Comparative Study and Texture Modeling of Accumulative Roll Bonding (ARB) processed AA8079 and CP-Al

C. P. Heason and P. B. Prangnell
UMIST, Manchester, UK

1 Abstract

Grain refinement of aluminum by ARB, has been found to be improved and stabilized by the presence of coarse second phase particles. It is thought that second phase particles help prevent a strong Shear and Copper texture developing during ARB processing, by texture ratcheting, which leads to unrefined bands in single phase alloys. Predictions from a Taylor model, based on actual deformation histories for the ARB processed sheet, go some way to confirm this hypothesis.

2 Introduction

Severe deformation processing has attracted increasing interest over recent years, due to its potential for producing ultra-fine grain (UFG) structures in bulk materials at a relatively low cost [1]. However, most severe deformation techniques are impractical for producing UFG sheet materials on a commercial scale. A relatively new method, which lends itself to large scale production of UFG sheet, is Accumulative Roll Bonding (ARB) developed by Saito et al. [2]. The ARB process involves roll-bonding two sheets using a 50 % reduction. The roll-bonded sheet is then cut in half, the two halves stacked and the process repeated until the desired strain is achieved. Theoretically, the technique can be used to build up unlimited strains in a material, because there is no change in the sheet dimensions. To date, ARB processing has been less extensively investigated than other severe deformation techniques, like ECAE. Tsuji et al. [3] have presented evidence of UFG structures in ARB processed sheet. TEM measurements of the boundary misorientations have shown that a high fraction of high angle grain boundaries (HAGBs) are present, at least in the sheet normal direction [4]. However, as the misorientations between the grains/subgrains have only been determined in a few cases, there is still insufficient evidence to tell whether ARB processed materials always contain uniform submicron grain structures through the full sheet thickness. More recent work by the current authors [5, 6], using high resolution EBSD analysis, has shown that coarse unrefined bands can be retained at very high strains. The cause of this inhomogeneity has been proposed to be related to the strong texture that can develop during ARB processing.

The work described in this paper follows on from previous research [5,6] and investigates the relationship between microstructure and texture in more detail, by using a Taylor model to predict the texture changes that occur during cyclic ARB processing. Comparisons are also made between the microstructures and textures produced in a largely single-phase alloy (AA1100) to those found in a particle containing material (AA8079).

3 Experimental

Cast blocks of 97.5% commercial purity (CP) aluminium (AA1100) and AA8079 (Al-1.3Fe-0.09Si) were conventionally warm rolled to 2 mm thick (true strain of –3.5) without recrystallisation. The CP-Al is largely a single phase material, whereas the AA8079 alloy contained a large volume fraction (~ 2.5 vol.%) of coarse 1-5 µm $Al_{13}Fe_4$ particles, with an interparticle spacing of ~ 10 µm. Sheets of both alloys were processed by ARB at 80 °C using a mill with 255 mm diameter rolls and no lubrication, to a total true strain of ~ 11 (10 ARB cycles). The sheet surfaces were degreased with acetone and wire brushed prior to each ARB cycle and preheated to a temperature of 80 °C (~ 10 min.), before roll bonding. ND-RD sections were taken from the centre of the sheets and analysed, as a function of strain, using Electron Back Scattered Diffraction (EBSD). High-resolution EBSD maps were obtained using a Philips XL30 FEG-SEM interfaced to an HKL Channel 5 EBSD system. Both fine (0.1 µm) and coarse (0.5 µm) step size maps were used to study the deformation structures at specific locations, and the texture distribution through the sheet thickness, respectively. In the EBSD data, a misorientation cut off of 1.5° was used to eliminate misorientation noise and high angle boundaries (HAGBs) were assumed to have misorientations of greater than 15°.

4 Results and Discussion

The deformation structures of the ARB processed sheets were characterized using high resolution EBSD maps (0.1 µm step size). Example maps are shown in Fig. 1, for both alloys, taken in the RD-ND plane just above the center bond line. It can be observed that after 4 ARB cycles (ε_{true} = 6.4) the microstructure of the particle containing AA8079 alloy (Fig. 1b) is more refined, compared to the CP-Al (Fig. 1a), and already contains mainly submicron 'grains'. At the same strain, the CP-Al alloy still contained high aspect ratio fibrous grains, although they were generally ~ 1 µm thick. During deformation, large lattice rotations can occur around second phase particles, which increases local misorientations and thus produces new HAGBs at relatively low strains. Coarse second phase particles also randomize the matrix flow behavior, breaking up fibrous lamellae deformation structures and reducing the textural strength, leading to more rapid rates of grain refinement. Surprisingly, it was found that there was a more dramatic difference between the microstructures produced in the two alloys after a larger number of ARB cycles. After 7 ARB cycles (ε_{true} = 8.7) (Fig.1c and 1d), unrefined bands were seen near the center of the CP-Al alloy, whereas a reasonably homogenous submicron grain structure was produced in the AA8079 alloy throughout the entire sheet thickness, which did not change significantly with subsequent passes. The coarse unrefined regions in the CP-Al material first appeared near the center bond line of the sheet at strains of ~ 6. As the center region moves out towards the surface on subsequent ARB cycles (first to the 1/4 position) and is reduced in thickness, after a large number of ARB cycles the microstructure through the sheet became more heterogeneous, with a mixture of layers of UFGs and unrefined bands, although the unrefined regions were concentrated at the center and became more pronounced as the strain level increased. Using static annealing treatments up to 100 °C no evidence could be found that these unrefined layers were formed by recrystallisation.

By analyzing EBSD maps after each ARB cycle for both alloys, quantitative measurements were made of the fraction of HAGB area as a function of strain (shown in Fig. 2). For a material to be considered as ultra-fine grained, it has been proposed that more than 70 % of the boundaries must be classed as high angle (i.e. have misorientations > 15°) and the average HAGB spacing must be less than 1 µm in all dimensions. It can be observed that the percentage of HAGB area in the AA8079 alloy (Fig. 2b) initially rapidly increases with ARB processing, reaching over 70 % HAGBs everywhere throughout the sheet thickness above a strain of 6 (4 ARB), whereupon little further change occurs, although there is a slight reduction at very high strains; i.e. a homogeneous UFG material has been produced through the sheet by a strain of 6. Contrastingly, the percentage of HAGB area in the CP-Al alloy increases more slowly, fails to reach 70 %, and in agreement with the observations described above, starts to decrease again at high strains (> 6). This decrease in HAGB fraction at high strains is more pronounced at the sheet center where it drops to ~55 % by a strain of ~ 11.

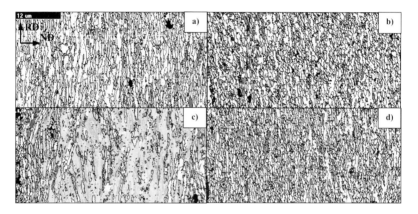

Figure 1: EBSD maps, with 0.1 µm step size, showing typical deformation microstructures near the midplane after 4ARB cycles for (a) CP-Al and (b) AA8079 and after 7 ARB cycles (c) CP-Al, and (d) AA8079. HAGBs are shown as black lines

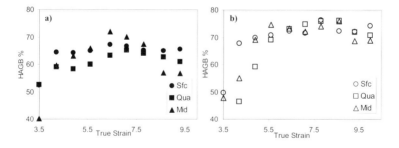

Figure 2: Graphs showing the relative percentage of HAGB area, measured near the sheet surface (Sfc), at the quarter thickness (Qua) and center (Mid) during ARB processing of (a) CP-Al and (b) AA8079

Previous work [5, 6] has shown that the unrefined bands produced in the CP-Al sample are probably caused by the development of a strong texture during ARB processing. This results from a decrease of the S {123}<634> and Brass {011}<112> rolling components and a large increase in the surface Shear {001}<110> and {112}<111> Copper components at high ARB strains. More importantly, these components are not uniformly distributed with the Shear component dominating near the surface and Copper at the sheet center. The absence of a texture spread in these regions, unlike the layers of different rolling components seen in conventionally heavily rolled lamella grain structures, means that new HAGBs can not be readily formed. Furthermore, the data above suggests that the boundary misorientations in these strong textured regions are actually reduced as the material rotates towards a single texture component with increasing ARB strain. Previously it was suggested that this effect arises because of 'texture ratcheting', which is a consequence of the cyclic nature of the ARB process [5, 6]. This occurs because when the middle of the sheet, initially deformed in plane strain, moves towards the surface on subsequent ARB cycles, all the rolling components rotate towards the same {001}<110> surface Shear texture component. When the surface is moved back to the center and deformed under plain strain conditions (as the sheets are re-stacked) the Shear texture that dominated at the surface rotates towards a *single* Copper component, resulting in an unrefined band consisting of a narrow Copper texture spread near the center plane. As ARB processing progresses, the Shear component at the surface increases in intensity and when this is rolled back into the center a more intense Copper texture is produce, and so on. In contrast, production of an ultra-fine grained microstructure requires a weak, or large spread in texture, because if a material develops a very strong texture it may be impossible for sufficiently high misorientations to evolve between all the subgrains for them to be classed as grain boundaries (this is one of the main advantages of the ECAE process where weak textures are normally seen).

If the unrefined regions present in the CP-Al sheet are due to the evolution of bands of low texture spread occurring during ARB processing, then in the more refined particle containing AA8079 alloy the texture must develop differently. Overall, the texture components tended have significantly lower intensities in the AA8079 alloy. Fig. 3 illustrates the volume percentages of specific texture components, averaged through the whole sheet thickness, for the ARB

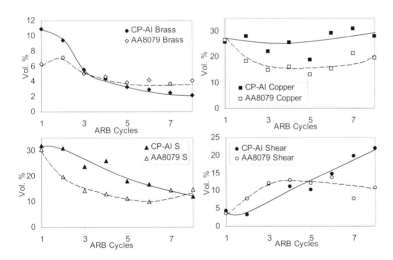

Figure 3: Graphs showing the variation in volume fraction of specific texture components during ARB processing of the CP-Al and AA8079 materials, averaged through the sheet thickness by EBSD analysis

processed CP-Al and AA8079 materials. It can be seen that the Brass and S rolling components follow similar trends for both alloys and decrease in intensity with the number of ARB passes. In contrast, the Copper component stays at a nearly constant high level (30 %) in the CP-Al material and increases slightly at high strains, while in the particle-containing alloy, it initially reduces before becoming stable at a much lower level (~20 %). Furthermore, the shear component volume fraction increases almost linearly with strain in the low particle content CP-Al material, whereas in the high particle-content AA8079 alloy it is much lower and appears to decline slowly at high strains, reaching a maximum level of ~ 13 %. This difference is clearly due to the presence of the second phase particles, which can affect the way in which texture develops during rolling [7]. During deformation, large lattice rotations around the particles can cause different textures to develop compared with the rest of the matrix and so the overall texture of the material as a whole will be weaker than for single phase material. The presence of second phase particles will also inhibit the formation of a strong single surface Shear component, as can be seen in Fig. 3. This means that the unrefined mid plane Copper texture band, which is formed by the Shear texture rotating when the surface is moved to the center on stacking the sheets, is not as strong as that produced in CP-Al. Consequently, the texture in AA8079 does not become dominated by bands of two, very strong textures (Copper at the center and Shear at the surface) and so, unrefined bands are not produced in the center of the sheet after a large number of ARB cycles.

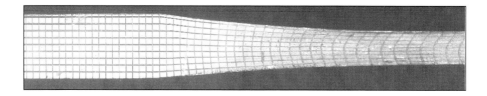

Figure 4: Image showing the deformation of a scribed grid during rolling with 50 % reduction (CP-AL)

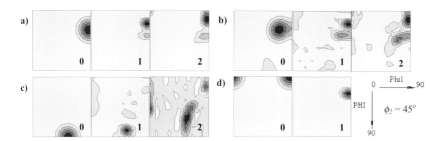

Figure 5: $\phi_2 = 45°$ sections from predicted ODFs, showing the progressive rotation of the (a) Copper {112}<111>, (b) S {123}<634> and (c) Brass {011}<112> rolling components towards the Shear {001}<110>, after each ARB pass (0,1,2) when subjected to a strain path typical of that of material near the surface of the sheet; in (d) the rapid rotation of the surface shear texture to Cu during one rolling pass under plane strain is shown

In previous work [6], a computer model, based on the Taylor approach, was used to try to understand the texture changes occurring during ARB processing. However, originally the deformation histories were only estimated. In the current work this has been remedied by measuring the actual deformation histories at different depths in the sheet during ARB processing. To determine the local strain histories, a scribed grid was inserted along the centerline of a sheet through its cross section (Fig. 4). The sample was then rolled to a 50 % reduction and stopped midway through the mill. The grid elements were assumed to follow streamlines and were used to determine the development of plastic deformation gradients through the roll gap, which were inputted into the Taylor approach model to predict how particular texture components would behave at different sheet depths during ARB.

The model calculations showed that when all the rolling components are deformed near to the surface, they do rotate towards a {001}<110> Shear texture (Figure 4a-c), which becomes stronger after repeated deformations at the surface and when the surface Shear dominated texture is then deformed at the center of the sheet it rapidly rotates to a very strong Copper orientation (Figure 4d), confirming the above described texture ratcheting hypothesis. However, the model did not predict as strong a Shear component as measured in practice and the surface shear texture was predicted to be only found in a thin surface region. In contrast, experimentally, it was found that after a large number of ARB cycles the surface Shear texture could penetrate up to a quarter of the sheet thickness. This may be because different strain paths are produced during ARB, compared to when rolling a single sheet (as used in the model), which results in more shear at the surface. Further work is needed to ascertain this.

5   Conclusions

It has been found that grain refinement of aluminum by ARB, is improved and stabilized by the presence of coarse second phase particles. It is thought that second phase particles help prevent a strong Shear and Copper texture developing during ARB processing, by texture ratcheting, which leads to unrefined regions in single phase alloys. Predictions from a Taylor approach model, based on actual deformation histories for the ARB processed sheet, go some way to confirm this hypothesis. However, the intensity and depth of shear predicted by the model is not as much as observed experimentally.

6   Acknowledgements

The authors would like to thank Dr. P. S. Bate for assistance with modeling and the EPSRC and Alcan Inc. for financial support.

7   References

[1]   Valiev, R.Z., Islamgaliev R. K., and Alexandrov I. V.; Progress in Materials Science, 2000. 45(2), p. 103–189
[2]   Saito, Y., Tsuji N., Utsunomiya H., Sakai T. and Hong R. G.; Scripta Materialia, 1998. 39(9), p. 1221–1227

[3] Tsuji, N., Saito Y., Utsunomiya H. and Tanigawa S.; Scripta Materialia, 1999. 40(7), p. 795–800
[4] Huang, X., Tsuji N., Minamino Y. and Hansen N.; Riso International Symposium on Materials Science, 2001, p. 255–262
[5] Heason, C.P. and Prangnell P. B.; 8th International Conference on Aluminium Alloys (ICAA8), Cambridge, UK, 2002, p. 429–434
[6] Heason, C.P. and Prangnell P. B.; 13th International Conference on the Textures of Materials (ICOTOM13), Seoul, Korea, 2002, p. 733–738
[7] Jensen, D.J., Hansen N., and Humphreys F. J.; 8th International Conference on the Textures of Materials (ICOTOM8), New Mexico, USA, 1987, p. 431–444

Nanocrystallization in Carbon Steels by Various Severe Plastic Deformation Processes

Y. Todaka, M. Umemoto, K. Tsuchiya,
Department of Production System Engineering, Toyohashi University of Technology,
Toyohashi, Aichi 441-8580, Japan

1 Introduction

Nanocrystalline materials have attracted considerable scientific interests in the past decade. Various severe plastic deformation methods have been proposed to produce nanocrystalline materials, such as ball milling [1,2], severe plastic torsion straining [3] and surface mechanical attrition [4,5]. Among these, extensive works have been performed on ball milling due to its simplicity, low cost and applicability to essentially all classes of materials. From our previous ball milling experiments in steels [6–9], it was found that the nanocrystalline regions have the following characteristics: 1) homogeneous structure with sharp boundary with work-hardened region, 2) ultrafine grains of less than 100 nm with almost no dislocations, 3) extremely high hardness (8 ~ 14 GPa), 4) dissolution of cementite when it exist and 5) no recrystallization and slow grain growth by annealing. Although ball milling is a useful method to produce nanocrystalline materials, it is not suitable to study the nanocrystallization mechanism since the deformation mode is quite complex and contamination is hard to avoid. To study nanocrystallization mechanism by severe plastic deformation, methods which produces simple deformation on specimens without contamination are desired.

The purpose of the present study is to demonstrate new severe plastic deformation techniques, i. e. ball drop [10,11], particle impact and shot peening processes, to produce nanocrystalline regions on the surface of bulk steel samples. The nanocrystalline regions formed by these techniques were compared with those in ball milled powder. Finally, shot peening is proposed to be the most practical process for producing nanocrystalline regions.

2 Experimental Procedures

The materials used in this study were eutectoid carbon steels of Fe-0.80C (Fe-0.80C-0.20Si-1.33Mn in mass%) and Fe-0.89C (Fe-0.89C-0.25Si-0.50Mn in mass%) with either pearlite or spheroidite structure. Silicon steel of Fe-3.29Si (Fe-3.29Si-0.01Mn in mass%) and 590 MPa class high tensile steel (Fe-0.05C-1.29Mn in mass%) were also used. Annealing of nanocrystallized specimens was carried out at 873 K for 3.6 ks by sealing in a quartz tube under a pure Ar protective atmosphere. Specimens were characterized by SEM, TEM and Vickers microhardness tester (load of 0.98 N for 10 s). Specimens for SEM observations were etched by 5 % Nital.

In a ball drop experiment, a weight with a ball attached on its bottom was dropped from a height of 1 m onto a bulk specimen with flat surface. Bearing steel ball (Fe-1Cr-1.5Cr in mass%) of φ 6 mm in diameter and weight of either 4 or 5 kg were used. The specimens were φ 15 mm in diameter and 2 to 4 mm in thickness. All experiments were carried out in air at either room temperature (R. T.) or liquid nitrogen (LN$_2$) temperature. The details of the ball drop experiment were described in our previous paper [10]. A particle impact experiment was done by a high-pressure He gas gun which can accelerate particles in a desired speed. The bore was ϕ 4.2 mm in inner diameter and 4 m in length. Bearing steel ball (Fe-1Cr-1.5Cr in mass%) with ϕ 4 mm in diameter was chosen as projectile accelerated to a speed of 120 m/s. Specimens of 30 × 30 × 3 mm were mounted at the end of bore. All the experiments were done in air at LN$_2$ temperature. A shot peening experiment was done using cast steel shot (Fe-1.0C-1.3Si-1.0Mn in mass%, HV 8 GPa) of less than ϕ 50 μm. Shot speed was 190 m/s, and shot period was 10 to 60 s (here one second of shot peening corresponds to 100 % in coverage).

3 Results and Disscussion

3.1 Ball Drop Experiment

The nanocrystalline regions formed by the ball drop technique appears usually at surface of specimens [11]. Figure 1 (a) shows a typical nanocrystalline region formed in Fe-0.89C specimen with pearlite structure by ball drop experiment (8 times of ball drops with a weight of 4 kg from a height of 1 m). The microhardness of the nanocrystalline region is as high as 11.7 GPa, which is much higher than that of the adjacent work-hardened region (4.3 GPa). It is noted that the observed microhardness and microstructure after the ball drop technique are similar to those observed in ball milled Fe-0.89C powder [12]. Figure 1 (b) and (c) show dark field (DF) image and selected area diffraction (SAD) rings of Fe-0.80C specimen with pearlite structure after ball dropped (8 times, 5 kg, 1 m). TEM samples were prepared from the cross section perpendicular to the specimen surface. The DF image shows that the ferrite grain size is the order of 100 nm. All the diffraction rings correspond to bcc ferrite, and rings corresponding to cementite are hardly detected. This indicates that cementite is mostly dissolved into ferrite. The SAD pattern taken from the area of ϕ 1.2 μm shows nearly continuous rings, indicating the random orientations of the ferrite grains. The number of ball drops necessary to produce nanocrystalline layer depends on the composition, microstructure and temperature of specimens, and ball drop conditions (weight and height). The number of drops is less for harder samples and higher energy drop conditions (larger weight and height). Pre-strain of specimens also reduces the number of ball drops. In the case of the pearlitic specimens, it was possible to produce the nanocrystalline region by one time of ball drop after cold rolling of 80 % [11,12]. Low processing temperature also reduces the number of drops. In the pearlitic Fe-0.80C specimens after ball dropped (8 times, 5 kg, 1 m) at R.T. and LN$_2$ temperature, the thicker nanocrystalline layer formed at LN$_2$ temperature than at R.T. The thickness of layer formed at LN$_2$ temperature was about 30 μm and almost twice the thickness formed at R.T. This supports that nanocrystallization by the ball drop technique is purely due to severe plastic deformation and not concerned with thermally produced martensite as a consequence of adiabatic deformation. The annealing experiment of ball dropped specimens showed that the behavior of nanocrystalline and work-hardened regions was quite different. After annealing at 873 K for 3.6 K, a much fine structure was seen in the prior nanocrystalline region, in which fine cementite particles re-precipitated. While, the recrystallization and grain growth of ferrite took place in the work-hardened region.

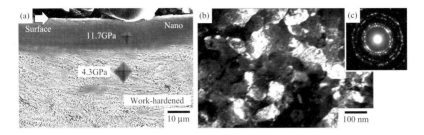

Figure 1: (a) SEM micrograph of pearlitic Fe-0.89C specimen after ball dropping (8 times, 4 kg, 1 m), and (b) TEM DF image and (c) SAD pattern of nanocrystalline region in pearlitic Fe-0.80C specimen after ball dropping (8 times, 5 kg, 1 m)

3.2 Particle Impact Experiment

Figure 2 (a) shows a typical nanocrystalline layers formed in pre-strained (82 % cold rolling) pearlitic Fe-0.80C specimen after 8 times of particle impacts at LN_2 temperature. The nanocrystalline layers formed by the particle impact technique usually appear at surface and interior of specimens. In Fig. 2 (b), the microhardness of the nanocrystalline region is as high as 9.5 GPa, which is much higher than that of the work-hardened region (4.3 GPa). The observed microhardness and microstructure after the particle impact technique are similar to those observed after ball milling and ball dropping. Figure 3 (a) shows a boundary between nanocrystalline and work-hardened regions in pre-strained (82 % cold rolling) pearlitic Fe-0.80C specimen after 8 times of particle impacts at LN_2 temperature. In the nanocrystalline region (upper part), the lamellar structure of pearlite is undetectable. Below this region, deformed pearlite structure is clearly observed. Figure 3 (b) is the enlarged micrograph of Fig. 3 (a). The thickness of cementite lamellae above the dotted curve is larger than those of below. The volume fraction of cementite above the curve is much higher than that of conventional pearlite. The observed apparent thickening of cementite lamellae may occur by the transformation of cementite to carbon over saturated ferrite and nanocrystallization of original cementite region together with the surrounding ferrite. In TEM observations, the structure of anocrystalline region produced by the particle impact deformation was similar to that of ball drop deformation. The grains with the order of 100 nm in diameter were the random orientations, and cementite was mostly dissolved into the ferrite.

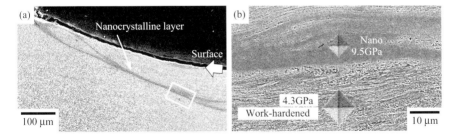

Figure 2: SEM micrographs of a typical nanocrystalline layers formed in pre-strained (82 % cold rolling) Fe-0.80C specimen with pearlite structure after 8 times of particle impacts at LN_2 temperature

508

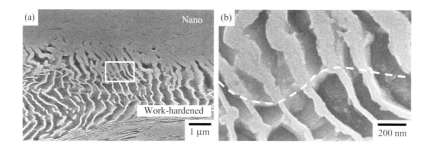

Figure 3: SEM micrographs of a boundary between nanocrystalline and work-hardened regions in pre-strained (82 % cold rolling) Fe-0.80C specimen with pearlite structure after 8 times of particle impacts at LN$_2$ temperature

3.3 Shot Peening Experiment

Figure 4 (a) shows a typical nanocrystalline region formed in Fe-0.05C-1.29Mn high tensile steel by shot peened for 10 s. The microhardness of the nanocrystalline region is as high as 6.8 GPa, which is much higher than that of adjacent work-hardened region (2.6 GPa). After annealing at 873 K for 3.6 ks, these microstructures of nanocrystalline and work-hardened regions are quite different, as shown in Fig. 4 (b). Figure 5 (a) shows a typical nanocrystalline region formed in pre-strained (84 % cold rolling) Fe-0.80C specimen with spheroidite structure by shot peened for 10 s. In the nanocrystalline region (upper part), the cementite particles are invisible. Figure 5 (b) is SEM macrograph of the shot peened specimen after annealing at 873 K for 3.6 K. The annealing behavior of the nanocrystalline and work-hardened regions is similar to those observed in ball milled, ball dropped and particle impacted Fe-0.80C specimens with spheroidite structure [6,9,11,12]. The development of nanocrystalline surface layers with peening time was studied using pre-strained (81 % cold rolling) Fe-0.80C specimens with spheroidite structure. Figure 6 shows the SEM micrographs after 10 and 60 s of shot peening. The thickness of nanocrystalline layer increased with peening time. Figure 7 (a) shows nanocrystalline region formed in Fe-3.29Si specimen by shot peened for 10 s. The sharp boundary between nanocrystalline and work-hardened regions is seen, which is similar to that observed in ball milled Fe-3.29Si powder. Figure 7 (b) and (c) show DF image and SAD rings of Fe-3.29Si specimen formed by shot peened for 60 s. TEM samples were prepared parallel to the specimen surface. The DF

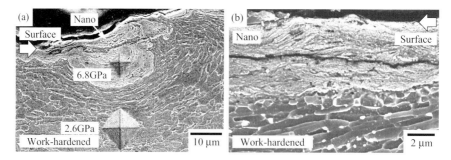

Figure 4: SEM micrographs of the nanocrystalline region formed in high tensile steel (Fe-0.05C-1.29Mn) by shot peened for 10 s (1000 % in coverage). (a) As shot peened and (b) annealed at 873 K for 3.6 ks after shot peening

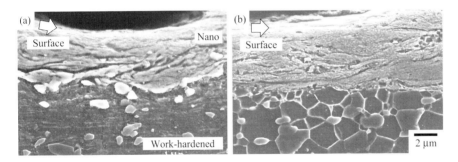

Figure 5: SEM micrographs of the nanocrystalline regions formed in pre-strained (82 % cold rolling) Fe-0.80C specimen with spheroidite structure by shot peened for 10 s (1000 % in coverage). (a) As shot peened and (b) annealed at 873 K for 3.6 ks after shot peening

image shows that the ferrite grain size is less than 20 nm, and the SAD pattern indicates that the ferrite grains are random orientations. Since bcc structure in Fe-3.29Si alloy is stable up to its melting point, nanocrystallization by shot peening is purely due to severe plastic deformation and not concerned with thermally produced martensite.

Apart from the above air blast shot peening, the surface nanocrystallization of steel induced by ultrasonic shot peening has been reported [4,5]. Liu *et al.* applied ultrasonic shot peening (3 kHz, φ 8 mm of shot size) to Fe-0.11mass%C steel. They reported the formation of a nanostruc-

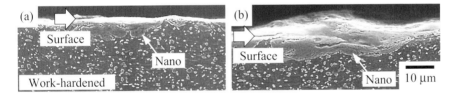

Figure 6: Peening time effect on the thickness of nanocrystalline surface layer formed by shot peening. Pre-strained (81 % cold rolling) Fe-0.80C specimens with spheroidite structure were shot peened for (a) 10 s and (b) 60 s

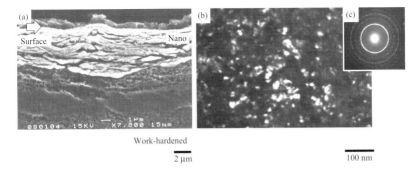

Figure 7: (a) SEM micrograph of Fe-3.29Si specimen after shot peening for 10 s (1000 % in coverage), and (b) TEM DF image and (c) SAD pattern (φ 1.2 μm in aperture size) of nanocrystalline region in Fe-3.29Si specimen after shot peening for 60 s (6000 % in coverage)

tured surface layer (average grain size of 33 nm in the top surface layer) after 1.8 ks shot peening treatment. Their result is similar with the present study. However, it is not clear whether or not these structures are same with ours, since they did not examined the hardness and the annealing behavior of nanocrystalline layer.

4 Conclusions

Nanocrystalline regions can be successively fabricated in various steels by ball mill, ball drop, particle impact and shot peening techniques. The formation of nanocrystalline regions was confirmed by TEM observations, microhardness measurements and annealing experiments. All nanocrystalline regions formed by these techniques were similar. The microhardness of nanocrystalline region was extremely higher than that of work-hardened region. Sharp boundaries between the nanocrystalline and work-hardened regions were observed. By annealing, recrystallization did not take place and slow grain growth was observed in the nanocrystalline regions. When the specimen contains cementite, it dissolves completely when the matrix is nanocrystallized.

5 Acknowledgement

This study is financially supported in part by the Grant-in-Aid by the Japan Society for the Promotion of Science (No. 14205103).

6 References

[1] C. H. Moelle, H. J. Fecht, Nanostruct. Mater. 1995, 6, 421–424
[2] S. Takaki, Y. Kimura, J. Jpn. Soc. Powder and Powder Metall. 1999, 46, 1235–1240
[3] R. Z. Valiev, Y. V. Ivanisenko, E. F. Rauch, B. Baudelet, Acta Mater. 1996, 44, 4705–4712
[4] G. Liu, S. C. Wang, X. F. Lou, J. Lu, K. Lu, Scripta Mater. 2001, 44, 1791–1795
[5] N. R. Tao, Z. B. Wang, W. P. Tong, M. L. Sui, J. Lu, K. Lu, Acta Mater. 2002, 50, 4603–4616
[6] M. Umemoto, Z. G. Liu, K. Masuyama, X. J. Hao, K. Tsuchiya, Scripta Mater. 2001, 44, 1741–1745
[7] J. Yin, M. Umemoto, Z. G. Liu, K. Tsuchiya, ISIJ Int. 2001, 41, 1389–1396
[8] M. Umemoto, Z. G. Liu, X. J. Hao, K. Masuyama, K. Tsuchiya, Mater. Sci. Forum 2001, 360-362, 167–174
[9] Y. Xu, M. Umemoto, K. Tsuchiya, Mater. Trans. 2002, 43, 2205–2212
[10] M. Umemoto, B. Huang, K. Tsuchiya, N. Suzuki, Scripta Mater. 2002, 46, 383–388
[11] M. Umemoto, X. J. Hao, T. Yasuda, K. Tsuchiya, Mater. Trans. 2002, 43, 2536–2542
[12] Y. Todaka, M. Umemoto, K. Tsuchiya, ISIJ Int. 2002, 42, 1429–1436

Severe Plastic Deformation by Twist Extrusion

Y. Beygelzimer, V. Varyukhin, D. Orlov, S. Synkov, A. Spuskanyuk, Y. Pashinska,
Donetsk Phys&Tech. Institute, 72 R.Luxembourg St., Donetsk, Ukraine

1 Introduction

Severe plastic deformation (SPD) has been shown to be one of the most effective methods for obtaining bulk ultrafine-grained materials. It is typically performed using equal channel angular pressing (ECAP) and multiple forging. In this work we describe and analyze a new method for severe plastic deformations that is based on the direct extrusion of a bulk through a twist channel, and thus it is called "twist extrusion" (TE). The idea of this method and some first experimental results were published in [1].

In this paper we describe the method, the structure and the properties of the resulting workpieces. A new approach to the investigation of the grain fragmentation at SPD is proposed and discussed. The fragmentation of grains is numerically simulated using a Cellular Model.

2 Mechanics

During twist extrusion a workpiece of commercially pure Cu (99.9 %) is extruded through the equal channel die of the shape shown on Fig.1. The impurities of Cu used were O (0.08 %), Fe, Pb, S, Zn (0.005 % each), and As, Ni, Sn, Sb, Bi (0.002 % each). The billet (25 × 15 mm in section) is extruded and stopped in the middle of processing, the material flow in the die is shown in Fig.2. The workpiece shape remains constant (Fig.3) allowing multiple processing of the specimen thus providing the desired strain accumulation.

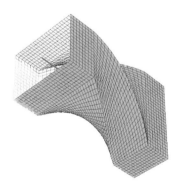

Figure 1: Shape of the twist die channel   **Figure 2:** Workpiece inside the TE die

Figure 3: Workpiece after twist extrusion

Theoretical investigation showed that the deformation is localized in the incoming and outgoing areas where the material is 'twisted' and 'untwisted'. The distribution of the equivalent strain within the cross-section of the workpiece (15×25 mm, twist die slope 60°) is shown on Fig. 4., the model used for the numerical experiments is described in [1]. This figure shows that the maximum equivalent strain of one pass is up to 1.5, with the minimum strain in the center of the billet being about 0.5.

It should be noted that TE may have either a right-hand or a left-hand twist direction which allows for deformation by different paths. Consecutive use of the right-hand and left-hand twist directions (scheme I on Fig.5) allows doubling the monotonous pass length. Here, the right-hand die twists the metal clockwise (+ at Fig.5) and then counter-clockwise (– at Fig.5), immediately followed again by the left-hand die counter-clockwise twist and then clockwise twist.

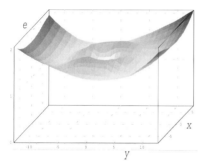

Figure 4:. Distribution of the equivalent strain within cross-section of a workpiece

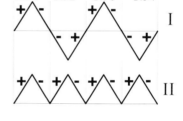

Figure 5: Effect of the twist direction change on the length of the monotonous deformation for two schemes of consecutive TE deformations. '+' and '-' denote twisting and untwisting of the billet inside each die

3   Experimental

The experimental installation shown on Figure 6 was used to process copper billets. Note that the preceding billet 6 creates the counter-pressure for the currently deformed billet 3.

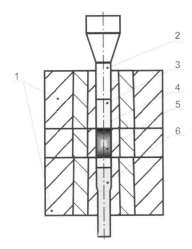

Figure 6: Scheme of installation for the direct twist extrusion: 1 – triple-layered containers; 2 – plunger; 3 – billet; 4 – thick lubrication layer; 5 – supported twist die; 6 – preceding billet

Insignificant reduction of the out-leading calibrating die cross-section in comparison with the incoming part of the die allows to increase the counter-pressure and, at the same time, compensates influence of the residual elastic deformation. The residual elastic stresses might lead to the increase of the billet cross-section during each deformation cycle due to the elastic relaxation deformation. It would require the additional mechanical adjustment of the billet shape for the next deformation step, which inevitably leads to the waste of the expensive billet material. Thus, the S_I (incoming cross-section of the die) to S_{III} (outcoming cross-section of the die) ratio should be within 1.05–1.15 range. Exceeding reduction of the billet is compensated during the next deformation cycle by means of its initial setting in the incoming die zone.

Installation allows deforming the 15×25 mm rods up to 100 mm long. Maximum work pressure is 2000 MPa, and the calculated mean equivalent strain is 1.2 for one deformation pass.

To study the copper structure arrangement during the deformation, the annealed Cu specimens ($T = 500$ °C, 4 hours) were investigated. They underwent several subsequent deformation cycles. The deformed specimens have been characterized by measurements of hardness, density and durometry. Areas from the surfaces of the deformed copper specimens were investigated by transmission electron microscopy (TEM) of foils and carbon replicas.

After the first stages of TE deformation there was a considerable, by 60 %, increase in hardness and microhardness. At the later deformation stages the increment is negligible, dependencies are in the form of saturation curves, and the metal is uniformly deformed throughout the whole volume, resulting in the formation of a uniform isotropic structure.

TEM studies of foils have shown that at the early deformation stages (two TE passes) a cellular structure is formed from previously polygonized microstructure with oriented subgrains of low-angle misorientation [2,3]. In the investigated specimens, a considerable quantity of shear band families (Fig. 7[1]) is also observed with 0.3–0.5 μm distance between high-angular disloca-

---

1. Measurements and TEM studies were performed by Prof. S.Dobatkin, Moscow Institute of Steels and Alloys, Russia.

tion walls. Normally, the high-angular dislocation walls are paired while sometimes two types of paired high-angular dislocation walls are observed.

After three TE passes the deformation structure becomes more uniform while single cells are still being observed. On the whole, the structure consists of shear bands oriented on two planes, and of subgrains. There are also well-shaped grains. The average size of the structure elements is 0.3 µm. Investigation of the surface of deformed copper specimens by means of high-resolution carbon replicas has shown that the grain size of deformed specimens (one TE pass) was in the range of 0.11–0.75 µm. Onset of recrystallization accompanied by grain growth was observed in specimens that underwent two TE passes. While after three TE passes the recrystallization process still continues, ultra-fine grained structure is formed. As a result, the boundaries become irregular and consist of small 'teeth', some porosity develops at this stage of deformation (pore size 0.02–0.15 µm). Four TE-pass deformation is accompanied by further recrystallization but the grains remain small in size.

Resulting microstructure affects the mechanical properties of the processed material: copper treated by TE has high values of the strength (σ_b = 427 MPa; σ_T = 394 MPa) and ductility (elongation δ = 17 %). For comparison, the mechanical properties of copper cold-drawn wire: σ_b = 430 MPa, σ_T = 380 MPa, δ = 1.2 %.

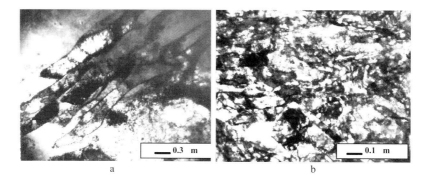

Figure 7: Microstructure of Cu after TE: a – two TE passes; b – three TE passes

4   Computer Simulation of Grain Refinement

Severe plastic deformation processes, in particular TE, still require the grain refinement mechanisms to be explained. An attempt to model the grain fragmentation during the SPD of polycrystalline materials was done through the computer simulation based on the cellular automata apparatus ("Cellular Model", [2]). Polycrystalline representative volume was modeled as a population of interconnected units which, in turn, could consist of lower scale level units. Simple units not having an internal structure were deformed by sliding along the various allowed sliding systems. For consideration of stress distribution within the limits of components, the approach of self-consistent field was used. Rotation of units and moment stresses connected with it were taken into account.

Within the framework of the Cellular Model, the following grain refinement criterion is proposed. As the moment stresses on a grain reach a critical value, that grain is fractured on the fragments arbitrary oriented within a certain range of allowed angles (Fig. 8).

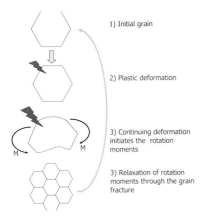
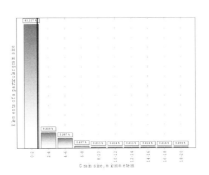

Figure 8: Grain refinement scheme

Figure 9: Grain size distribution after deformation

The obtained grain size distribution histograms are shown on Fig. 9, where it is seen that at any deformation stage, not fractured large grains are still present in the sample, which is in a good qualitative correspondence with the experimental data of other authors.

It is shown that due to the grain refinement the level of internal stresses is periodically relaxed in all parts of the billet (Fig.10). The obtained dependence of the average grain size on the strain shows again the periodic character of the grain refinement with the periodic relaxation of the internal stresses due to this refinement (Fig. 11).

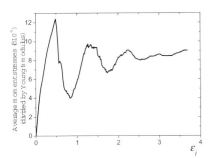
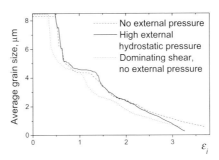

Figure 10: Moment stresses at the grains during deformation

Figure 11: Dependence of the average grain size on the the strain

5   Conclusions

Twist Extrusion produces a fine-grained structure with desirable mechanical properties comparable to those obtained by other severe plastic deformation methods. Furthermore, TE extends the capabilities of other severe plastic deformation schemes in controlling the material structure and the shape of the resulting products. In particular, using certain TE schemes it might be possible to obtain tubes and hollow sections as well as elongated profiles such as wires.

In order to avoid anisotropy of properties it is advised to combine TE with ECAP or traditional metal forming methods (e.g. rolling, drawing).

We presented a Twist Extrusion installation for deforming a work piece with a cross-section about 30 mm and a length about 100 mm. The equivalent strain during one pass equals to about 1. The temperature of the die may vary in between −170 °C and +500 °C. A hydrostatic pressure up to about 1GPa may be applied in the deformation zone.

6 References

[1] Y.Beygelzimer, D.Orlov, V.Varyukhin, Proceed. Of TMS Annual Meeting in Seattle, WA. Feb.17-21, 2002, 297–304
[2] R.A.Andriyevsky, A.M. Glezer, Fiz. Met. Metalloved, 1999, 88, No.1, 50–73
[3] A.I.Gusev, Usp. Fiz. Nauk, 1997, 168, No.1, 55–83
[4] Y.E.Beygelzimer, A.V.Spuskanyuk, Philosophical Magazine A, 1999, 79, No. 10, 2437–2459

SPD Structures Associated With Shear Bands in Cold-Rolled Low SFE Metals

K. Higashida, T. Morikawa
Kyushu University, Fukuoka, Japan

1 Introduction

Shear bands are one of the most characteristic inhomogeneity observed in deformation microstructures of cold-rolled metals. In those bands, intense shear strain such as a few hundreds percent can be localized, and a very fine-grained structure is generally formed by the severe plastic deformation (SPD). Shear bands are, therefore, understood as a kind of SPD structure, and closely related with the formation mechanism of ultra-fine grains produced by SPD.

As is well known, deformation microstructure of fcc metals is much influenced by their stacking fault energy (SFE). In cold-rolled fcc pure metals with high SFE such as Al, the microstructure is characterized by dense dislocation walls or micro bands which are thought to be geometrically necessary boundaries bearing the misorientations between adjacent areas formed by subdivision of an initial grain. [1,2] On the other hand, in low SFE fcc alloys such as α-brass, deformation twinning and shear bands are the most prominent features. [3,4]

We have been studying about the microstructural evolution due to cold rolling in austenitic stainless steels 310S whose SFE is considered to be as low as that of α-brass. In those works[5], it was found that the evolution process was separated into two steps. The first step is the deformation twinning which develops the dense twin-matrix (T-M) lamellae along the rolling direction (RD). The second step is the growth of a fine-grained structure, where the T-M lamellar structure is destroyed by the multiplication of shear bands. Here, the key process for the grain refinement is in the occurrence of shear banding in the T-M lamellae. However, the microstructural characteristics of shear bands such as the scattering of crystal orientations have not been clarified yet.

In this paper, we will report those characteristics found in shear bands, and discuss about the mechanism of transmutation from T-M lamellar structure into fine-grained structure. Special emphasis is laid on the influence of deformation twinning on the grain refinement in low SFE metals.

2 Experimental

Polycrystalline plates (initial thickness:12mm, average grain size : 100 μm) of 310S steel were rolled by reductions of 50–90 % in thickness at room temperature. Deformation structures of the rolled specimens were observed by both optical microscope and transmission electron microscope(TEM). All the observations were made from the transverse direction (TD). For preparing TEM samples, the cylindrical rod (3mmφ) was trepanned along TD from rolled sheets by spark cutting, and they were sliced into disks with 1mm thickness. Those disks were mechanically polished and electron-polished into thin foils by the twin-jet technique. The foil normal is, therefore, parallel to TD. The TEM observations were carried out with JEM-200CX operated at 200 kV in the HVEM laboratory at Kyushu University.

To examine the mechanical property of the cold-rolled specimens, tensile tests were conducted using Instron-type machine at room temperature, where the initial strain rate was $2 \cdot 10^{-2}\,\mathrm{s}^{-1}$.

3   Results and Discussion

Figure 1 shows an optical micrograph of shear bands observed in 310S cold-rolled by 50 % in thickness reduction, where a lamellar structure develops along the rolling direction. It is to be noted here that some of the lamellae are clearly sheared along the direction inclined about 40° to RD. Those bands shearing the lamellae are corresponding to the so-called shear bands. This demonstrates that severe plastic shear deformation is localized in shear bands.

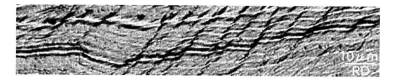

Figure 1: Optical micrographs of cold rolled 310S steel at 50 % reduction(longitudinal section)

Figure 2 shows a bright field (BF) image of a typical microstructure at 70 % thickness reduction. In this figure, bright bands inclined by 30–40° to RD are clearly seen on the background lamellae almost parallel to RD. Those bright bands correspond to shear bands observed in Figure 1. On the other hand, the lamellae observed in the background relate with deformation twinning which appears from the early stage of cold-rolling in low SFE metals such as 310S.

Figure 2: TEM image of shear bands (70 % thickness reduction)

Figure 3(a) exhibits a diffraction pattern obtained from the lamellar structure in Figure 2. This figure shows a superposition of a couple of the <011> diffraction patterns which are symmetrical to each other with respect to the {111} plane, i.e., twin and matrix double spot pattern with the <011> incidence. Figures 3(b) and 3(c) indicate dark field (DF) images obtained from the spots pointed by (b) and (c) in Figure 3(a), respectively. In these DF images, very fine lamellae less than 100nm are observed, and dark and bright areas in Figure 3(b) are reversed in Figure 3(c). The volume fraction of twin layers seems to be roughly equal to those of matrix, which can be understood from the equivalency of their spot intensities. This indicates that the lamellar structure consists of alternate stacks of twin and matrix thin layer, and shear bands are formed in those twin-matrix fine lamellae.

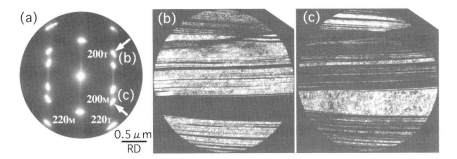

Figure 3: Diffraction pattern and dark field images of twin-matrix lamellar structure

Figure 4 exhibits an enlarged BF image of a shear band. In this figure, the shear band is running from the bottom left to the right top. Lamellar structure is seen in the top left and right bottom in the figure. In the inner part of the shear band, substructures like an aggregation of fine grains are observable, and the shape of each fine grain seems to be somewhat elongated along the extending direction of the shear band.

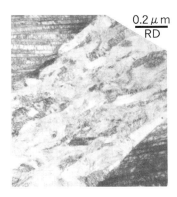
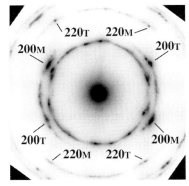

Figure 4: BF image of a shear band

Figure 5: SAD pattern obtained from a shear band

Figure 5 shows the typical image of a selected area diffraction (SAD) pattern obtained from a shear band, where the negative image is exhibited to show the details clearly. It is to be noted that arcs of Debye rings are observed, although the pattern basically coincides with the twin-matrix double spot pattern with the <011> incidence which was shown in Figure 3(a). This indicates that the shear band consists of highly misoriented small regions, and the misorientation occurs around the <011> axis. Such fine-grained structure is seen not only in the BF image in Figure 4 but also in our previous observation using dark field images. [6] In addition, note here the characteristic pattern clearly exhibited in the SAD. When comparing the {200} and {220} arcs in Figure 5 with the corresponding {200} and {220} spots in Figure 3(a), it is found that the Debye arcs appear mainly in the range of 70° between the matrix spots and the twin spots, e.g., the 70° range between the (200) spot of the matrix ($(200)_M$) and that of the twin($(200)_T$). This indicates that the crystal orientation around the <011> axis in the twinning region rotates clockwise, while in the matrix anti-clockwise.

This characteristic found in the crystal rotation is explained by the model shown in Figure 6, illustrating the crystal rotation in rolled T-M lamellae. With increasing strain due to rolling, T-M lamellae become nearly parallel to the RD as shown in 6(a). This activates the slip on the {111} planes intersecting with the lamella boundary in both the matrix and the twinned regions. Such slip deformation causes crystal rotation as shown Figure 6(b): the crystal rotation in the twinning region occurs clockwise, while in the matrix anti-clockwise in this figure. As a result, the orientation of each layer scatters between the initial orientations of the twin and matrix. In addition, the slip activation should cause the accumulation of high density of dislocations around the twin boundaries, and the dislocations accumulated at the neighboring boundaries must have opposite sense to each other, the detail of which will be described later. In order to release the highly accumulated strain energy due to the dislocations, mutual annihilation of the neighboring boundaries with opposite sense may occur locally, resulting in the formation of a kind of new grains as shown in Figure 6(c). The annihilation of those boundaries can enhance

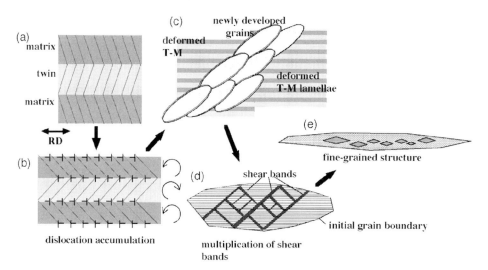

Figure 6: (a),(b) and(c) Formation process of the nucleus of a shear band, (d) multiplication of shear bands, (e) fine-grained structure

additional slip, which causes continuous nucleation of new grains. This process induces a shear strain concentration to form shear bands, where the temperature may be increased by the plastic works due to the shear localization. As was actually shown in Figure 4, we can see a kind of newly developed grains whose grain sizes are larger than the thickness of each lamella of the twin or the matrix, although it is needless to say that those grains are different from general fully recrystallized grains. Those shear bands are multiplied with increasing strain due to cold-rolling as illustrated in Figure 6(d), the T-M lamellar structure must be destroyed to form a fine-grained structure as shown in Figure 6(e).

Finally, we emphasize the effect of twinning deformation on grain-refinement in metals with low stacking fault energy. In fcc crystals with low SFE metals, twinning deformation easily occurs from the early stage of plastic deformation, which causes a very fine lamellar structure with the lamella thickness less than 100 nm as shown in Figure 3. Since a twin boundary is also regarded as a kind of grain boundary, those fine lamellae due to twinning are essential to contribute to the subdivision of an initial grain. Particularly, as illustrated in Figure 6(a), when the sense of crystal rotation in the twinned region is opposite to that in the matrix, geometrically necessary dislocations must be accumulated as a net at the boundaries. Figure 7 illustrates such dislocation behavior in the twin-matrix layered structure which is subjected to a compressive stress, where b_1 and b_2 are dislocations with the burgers vectors of a/2<110>twin and a/2<110>matrix, respectively. In this illustration, the dislocation with the Burgers vector b_1+b_2 is the boundary dislocation which should be formed as a net by usual slip dislocations in the twinned and matrix region. With the increase of compressive strain, the density of those boundary dislocations will be increased to make the tilt angle larger, which transmutes a stable twin boundary into a high energy boundary. At the same time, dislocations should be piled up against the boundaries, so that a high density of dislocations must be accumulated between the neighboring boundaries. Thus, the T-M fine lamellar structure should have a high ability of dislocation accumulation, i.e., high work-hardening rate, which can be an essential background for the formation of shear bands.

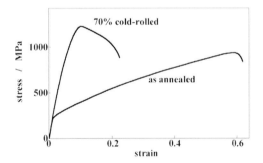

Figure 7: Dislocation accumulation in the T-M lamellae

Figure 8 shows the stress-strain curves obtained by the tensile tests of as-annealed and 70 % cold-rolled specimens. Comparing these two s-s curves, it is clearly seen that the 70 % cold-rolled specimen has the work-hardening rate much higher than as annealed. In addition, its tensile strength reaches more than 1.2 GPa. Considering that T-M lamellar structure is evolved enough in the specimen cold-rolled by 70 %, this result substantiates that the T-M lamellar

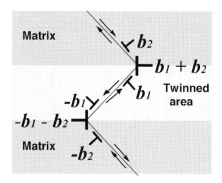

Figure 8: Stress-strain curves obtained by the tensile tests of as-annealed and 70 % cold-rolled specimens of 310S at room temperature

structure observed in the present study has a high ability of dislocation accumulation. Since shear bands are originated from these fine lamellae, the evolution of shear bands destroys the lamellae, which develops a fine-grained structure.

4   Conclusions

Formation process of a fine-grained structure by cold-rolling in 310S steels was investigated using mainly TEM. The key process is the transmutation of twin-matrix lamellae into a fine-grained structure due to shear band formation. In shear bands, the T-M lamellae are completely destroyed and a kind of new grains with the size of submicron are formed. The T-M lamellar structure has an essential role for grain refinement, since each lamella thickness is very small (less than 100nm) and the lamellae markedly enhance dislocation accumulation.

5   References

[1] B.Bay, N.Hansen, D.A.Hughes, D.Kuhlmann-Wilsdorf : Acta Metall. Mater., 40(1992), 205
[2] D. A. Hughes, N. Hansen : Metal.Trans., 24A(1993), 2021
[3] K. Morii and Y. Nakayama : Trans. JIM, 22(1981) , 857
[4] W. B. Hutchinson, B. J. Duggan and M. Hatherly : Metals Tech., 6(1979), 398
[5]   T. Morikawa and K. Higashida, Proc. of the 21st Riso Int. Symp. on Material Science, RISO National Laboratory, Denmark (2000), 467
[6] T. Morikawa, D. Senba, K. Higashida and R. Onodera, Metal Trans. JIM 40(1999), 891

Features of Mechanical Behaviour and Structure Evolution of Submicrocrystalline Titanium Under Cold Deformation

S.Yu. Mironov[1], M.M. Myshlyaev[2], G.A. Salishchev[1]

[1]Institute of Metals Superplasticity Problems, Russian Academy of Sciences, Ufa, Russia
[2]Baikov Institute of Metallurgy and Materials Science, Russian Academy of Sciences, Moscow, Russia

1 Introduction

Essential technology progress in the methods of severe plastic deformation (SPD) that was reached at last time let to receive structures with submicrocrystalline (SMC) grain sizes (0.1–1 m) in massive billets. Experimental study of their deformation behaviour showed, that SMC materials possess not ordinary combination of physics-mechanical properties. In one aspect it was noted changes of density, modulus of elasticity, diffusion activity and deviation from Hall-Petch law [1]. These cases (in combination with considerable boundaries extent increasing and their «unequilibrium» state) let authors [2] to assume about qualitative change: together with dislocation gliding in grains interior the activation of grain boundary sliding (GBS) took place, even during cold deformation. At the same time, it was found considerable strength increasing [1], presence of high dislocation density [3] and essential internal stress [1, 4], forming of crystallographic and metallographic texture [3]. All of these facts together evidence of intensive dislocation gliding in grains interior and it can be explained in frame of ordinary conceptions about grain size effect on mechanical behaviour.

Dimensional SMC interval is intermediate between «ordinary» polycrystals (above 1 µm) and nanocrystals (0.001–0.1 m). It is assumed that their behaviour would be also intermediate. So, SMC is a convenient model for nanocrystalline (NC) materials properties prognosis, for its last experimental research was very hard, in result of their receiving difficulties. According to theoretical estimations and modelling results in SMC materials the GBS and Cobble diffusion creep mast be prevailed [5, 6].

So, is it transition to SMC grain sizes accompanied by basic deformation mode changes? Or one could describe all the observed effects in frame of standard conceptions of plastic deformation (including its SPD stage) and of grain size effect? The present article is devoted to these questions for the example of SMC titanium deformation behaviour.

2 Experimental

The commercial pure titanium (impurities content, wt%: 0.443Al; 0.0965Fe; 0.0385C; 0.0073N; 0.068O) was used. SMC structure was got by means of «multiple forging» [7]. After annealing at various conditions [8] the states were got with mean grain sizes from 0.4 to 57 m. Uniaxial tensile tests were carried out at a constant crosshead speed 1 mm/min and room temperature with a test machine Shenck. There were used sheet specimens with gauge length of 20 mm, a width of 3.5 mm and a thickness of 1.5 mm. Extensometer was used in all tests.

The determination of apparent activation volume was made by method of stress relaxation [9]. Structure investigation was carried out on REM JSM-840 and TEM JEM-2000EX.

The secant method for statistical processing of structure investigations was used. The excerption was not less than 500 measurements.

3 Results

The typical microstructure of annealed SMC titanium is presented on the Figure 1a. It is seen that structure is defined by approximately equiaxial grains with sharp and straight boundaries, on which the thickness contours of extinction took place. Initial dislocation density is ~$5 \cdot 10^9$ cm^{-2}. In some relative «coarse» grains (over 0.5 m) the subboundaries are observed. Grain boundaries are almost free from dislocations, with seldom exemption. Typical distribution of grains by sizes (Figure 1b) is asymmetrical and it is defined by presence of «tail» in interval of relative coarse (over 1 m) grains. Though this grains fraction is not great, specific volume, possessed of them, is considerable.

(a)
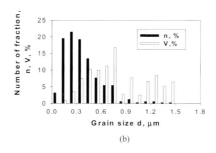
(b)

Figure 1: Initial structure: a – microstructure of annealed SMC titanium; b – distributions of grains by sizes and specific volumes of grains fractions

The results of mechanical behaviour of titanium with different mean grain sizes are presented on Figure 2. As it is seen from Figure 2a, refinement of microstructure leads to considerable (~2.5 times) increasing strength properties, which can be described by Hall-Petch dependence (Figure 2b). The change of type engineering stress-strain curves for SMC titanium was noted: Luders site and, sometimes, yield point took place (Fig 2a).

Similarly to "ordinary" coarse-grained state, the plastic flow of SMC titanium is accompanied by strain hardening. According to Figure 2c, it is characterized by non monotonous dependence on stress: With its increase the strain hardening decreases and levels out. It is important to note that there occurred a strain hardening reduction up to zero although refinement of microstructure up to the SMC range took place. As it follows from Figure 2d, the transition to SMC range of grain sizes was accompanied by a sharp decrease of uniform elongation: at d ~ 0.4 m it disappeared at all. At the same time the total elongation and true fracture strain did not show remarkable sensibility to grain size.

As it is seen from Figure 3, the apparent activation volume is little sensitive to considerable changes of mean grain sizes and strain. The effect of strain to the scalar dislocation density is presented on Figure 4. As it follows from plot, it increased about 20 times in comparison with initial state, by strain to ~25 %. Structure evolution of SMC titanium during cold deformation is presented on Figure 5a–d. Initial stage of plastic flow was defined by sharp increasing of dislocation density, as in grains interior, as in their boundaries (Figure 5a). Dislocation distribution was not uniform. Its higherdensity was observed near grain boundaries, where the formation of subboundaries took place (Figure 2b). The dislocation density in them was about $5 \cdot 10^{11}$ cm^{-2}. There were no pile-ups, tangles and cell structure. Further strain increasing leads to great structure changes (Figure 2c). Relatively sudden series of new boundaries of deformation origin appeared, which growth was approximately rectilinear and parallel to tension axis at a considerable distance. A rapid transformation of the initial equiaxial structure into band type one occurred. This morphological structure type was stable up to prolonged strain, and during its increasing only gradual reducing of transverse band dimensions took place. Shortly before destruction a new band system was found at the angle of about 55° to tension axis.

Fracture studies showed that at any case there was evidence for cup–cone and dimples. The dimples were of equiaxial form, which occupied the central part of fracture surface (Figure 2d). Its mean value was less about order than at initial grain size 57 m and was 1 m. On the periphery fracture surface elongated dimples were observed which had parabolic or half-ellipsoid form (fractures of peeling), and on "cup" walls the fractures of quasichip took place.

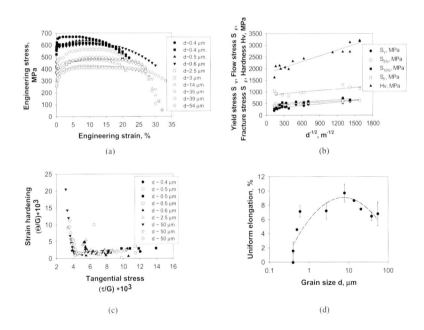

Figure 2: Effect of mean grain size on mechanical behaviour of titanium: a - engineering stress-strain curves; b – effect of grain size on the yield stress, flow stress, true fracture stress and hardness; c – strain hardening; d – uniform elongation

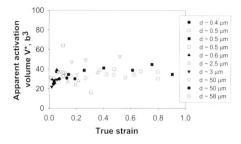 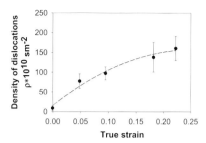

Figure 3: The effect of mean grain sizes and strains on apparent activation volume

Figure 4: The effect of true strain on dislocations density in SMC titanium

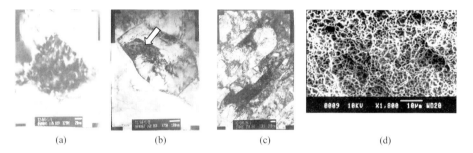

Figure 5: Structure evolution during cold deformation of SMC titanium: a – great dislocation density inside of nanosized grain; b – dislocation subboundaries formation (shown by arrow); c – band structure formation; d – dimpled fracture surface

4 Discussion

Asymmetry of histogram of grain distribution by sizes («tail» presence in the interval of their relatively «coarse» sizes, Figure 1b) illustrates the microstructure heterogeneity, resulting from severe plastic deformation. Relatively small fraction of such grains demands comparatively great specimen's volume (Figure 1b) and even small deviations of its content lead to considerable scattering of some mechanical property [8]. So, feature of initial SMC microstructure – it is necessary to know not only the mean grain size, but it is necessary to consider all features of distribution histogram for complete disparity attestation of microstructure and on this basis the mechanical behaviour prognosis.

According to direct observations of structure evolution (Figure 5a, b), speed increasing of dislocations density (Figure 4), mechanical behaviour features (high strength level, yield point, Luders site, strain hardening presence, Figure 2a, c) and apparent activation volume value (Figure 3), dislocation gliding in grains interior was one of the main deformation mode. The indirect sign of deformation mode independence from grain size is Hall-Petch law fulfilment for stress of yield, flow, fracture and microhardness (Figure 2b). So, deformation mode change, resulting from very small grain size, supposing in some works (e.g. [1, 2]), didn't take place in researched case and elementary strain events was not «particular» in comparison with more coarse grained structures. As will be shown further, differences took place at higher structure levels.

As mechanical behaviour features of SMC titanium the small strain hardening at deformation beginning (Figure 2c) and comparatively fast strain macrolocalization (Figure 2d) took place. As particularities of dislocation structure it is a relative high dislocations density (Figure 4, 5a), absence of pile-ups, tangles and cellular structure. The features of SMC titanium microstructure evolution also includes intensive forming of strain induced boundaries and accelerated transforming of initial equiaxial grain structure into band structure. The above described features of mechanical behaviour and of structure evolution of SMC titanium correspond to stage of severe plastic deformation of polycrystals. The case that they take place in SMC titanium practically from the beginning of plastic flow lets to conclusion about considerable *accelerating* of deformation events in it.

Really, high level of external stress of yield and flow (Figure 2a, b) is favourable for dislocation sources activation and for cross sliding of screw dislocation segments. Orientation arbitrariness of conjugated crystals, which implies their strain rotation to different crystallographic poles, is the cause of internal stress appearance [10], which supplemented the external stress. Their summary result – from the beginning of plastic flow the high density of mobile dislocations take place (Figure 4), which provides high rate of structure evolution.

One of the sequences of sharp increasing of mobile dislocation density is the Luders site and, sometimes, yield point presence on the initial strain stage of SMC titanium (Figure 2a). Supplementary cause, promoting this effect, is the "splitting" of dislocation low-angle boundaries, conserved from stage SMC structure processing (Bauschinger effect).

High level of stress and great density of mobile dislocations lead to intensive formation of deformation induced boundaries and disorientation increasing on it – the third stage of strain hardening is changed by the fourth stage (Figure 2b). At structural case this transition corresponds to fast transformation of initial equiaxial grains into band structure (Figure 5c). From macroscopically point of view it corresponds at accelerated strain macrolocalization to neck formation (Figure 2d).

Driving force of band structure formation is providing of deformation compatibility of conjugated crystals. Process of band structure formation in SMC titanium is most intensive on initial stage. By its further developing the transversal band size decreases rapidly to its saturating. It means, that strain passes on quality new level of appearance, which is shown by deformation induced boundary disorientation increasing [10]. Of course, transition to this strain stage is accelerated during microstructure refinement.

Experimental result comparison with date of [11] shows that neck formation process in sheet specimens of SMC titanium is not peculiar. As a feature of SMC state neck formation took place from the beginning of plastic deformation at the finest average grain size, accordingly high level of yield stress and relatively low strain hardening. In this case an uniform elongation is no longer present (Figure 2d) and plastic flow is unstable. It must be noted that in this case plane-strain condition, inherent necking in sheet specimens [10], took place from the beginning of plastic flow. So, this condition of SMC titanium macroscopic deformation is different from those for initial stage of deformation in "ordinary" coarse-grained titanium. Therefore, it must be one of the possible causes of mechanical behaviour and structure evolution features in SMC state.

Macroscopically, conditions change and sharp increasing of strain rate (about 30 times) during neck formation rapidly exhausts accommodation ability of structure and accelerates transition to specimen's destruction. Taking into account the presence of considerable plastic deformation and dimpled fracture surface (Figure 5d) it could be concluded that SMC titanium fracture was ductile as concerns microcrack initiation as well as its propagation. So, it wasn't different qualitatively from fracture of "ordinary" coarse-grained titanium [11] at same deformation conditions. As a feature of SMC titanium fracture, the increase of true fracture stress (Figure 2b) and a decrease of average dimple size on fracture surface could be observed.

Although true fracture stress was relatively high, it only accounts for about 7 % of "theoretical strength". Hence, internal stresses rule the fracture process in SMC titanium as in case of "ordinary" polycrystals. So, the refinement of microstructure up to SMC range lets "resist" the material to higher external stress by means of a more effective relaxation of internal stress. It is possible that the higher density of mobile dislocations (and, as a result, higher accommodation ability) in SMC titanium is retained to the latest stage of deformation and during fracture. Probably this provides higher density of stable microcracks in it and as a result – more dispersive dimple's size on fracture surface.

5 Conclusions

1. Initial microstructure of SMC titanium, processed by severe plastic deformation, is heterogeneous – a small fraction of relatively "coarse" grains takes considerable specific volume of specimen.
2. The most probable microscopic deformation process of SMC titanium under experimental conditions is the operation of dislocation caused glide in grain interiors.
3. The basic feature of mechanical behaviour and structure evolution of SMC titanium is considerable acceleration of deformation as a result of higher mobile dislocation density.

6 Acknowledgments

This work was supported by Russian Foundation for Basic Research, Projects N 01-02-16505 and 02-02-81021.

7 References

[1] R.Z. Valiev, R.K. Islamgaliev, I.V. Alexandrov, Bulk Nanostructured Materials from Severe Plastic
[2] Deformation, Progr. Mater. Sci. 2000, 45, 102–189
[3] R.Z. Valiev, E.V. Kozlov, Yu.F. Ivanov, A.A. Nazarov, B. Boudelet, Acta Metall. Mater, 1994, 42, 2467–2473
[4] S.Yu. Mironov, G.A. Salishchev, M.M. Myshlyaev, The Physics of Metals and Metallography 2002, 93, 362–373
[5] A.N. Tyumentsev, M.V. Tretiak, A.D. Korotaev, Yu.P. Pishin, R.Z. Valiev, R.K. Islamgaliev, A.V. Korznikov, NATO Advanced Recearch Workshop on Investigations and Applications of Severe Plastic Deformation-Moscow, Russia – 2-7 August 1999, 93–102

[6] H.S. Kim, Y. Estrin and M.B. Bush, Acta mater. 2000, 48, 493–504
[7] H.-H. Fu, D.J. Benson and M.A. Meyer, Acta Mater. 2001, 49, 2567–2582
[8] Kaibyshev .., Salishchev G.., Galeyev R.., Lutfullin R.Ya. and Valiakhmetov .R. Patent / US97/18642, WO 9817836, 1998
[9] G.A. Salishchev, S.Yu. Mironov, Izvestia VUZov. Fizika. 2001, 28–32. (in Russian).
[10] V.I. Dotsenko, Phys. Stat. Sol. (b) 1979, 11–43
[11] V.V. Rybin, Heavy plastic deformations and fracture of metals, Moscow, Metallurgy, 1986, p. 224 (in Russian)
[12] H.Conrad, Progress in Materials Science 1981, 123–403
[13] F.A. McClintock and A.S. Argon, Mechanical behavior of materials, Addison-Wesley Publishing Company, Inc., Massachusetts, USA, 1963

Ultra Grain Refinement of Fe-Based Alloys by Accumulated Roll Bonding

A.C.C Reis, I. Tolleneer, L.Barbé, L. Kestens and Y. Houbaert

Ghent University, Department of Metallurgy and Materials Science, Technologiepark 903, B-9052, Ghent, Belgium

1   Abstract

Ultrafine grain structures are being studied worldwide using different types of processes on various materials including low carbon steels. According to state-of-the-art literature equal channel angular pressing (ECAP) and powder metallurgy techniques can truly produce ultrafine to nanocrystalline materials. Unfortunately these techniques do not seem very promising with regard to industrial implementation, because it is not possible to apply these methods in a continuous process. A technique which is based on thermo-mechanical processes of a continuous nature would be much more attractive in terms of industrial upscaling. In the present approach a severe plastic deformation is applied to an interstitial free steel by submitting the material to a severe rolling reduction. In a previous work by the present authors it was already shown that conventional severe rolling reduction, of 95 % ($\varepsilon = 3.0$), does not produce a random high angle grain boundary distribution in the deformed sheet. In order to obtain rolling reductions of more than 99 % the Accumulated Roll Bonding (ARB) technique was employed. The microstructure, texture and misorientation distribution of ARB samples which were warm rolled between 500 and 680 °C to a total reduction of 99.9 % (true strain $\varepsilon = 6.9$) is studied.

2   Introduction

Obtaining ultrafine grain structures for different types of materials has been the objective of many studies during the last decade. Without the necessity to add expensive alloying elements, ultrafine structured materials offer the interesting perspective of producing improved mechanical properties such as high strength at room temperature, high speed superplastic deformation at elevated temperature and high corrosion resistance [1].

Several methods such as mechanical milling of powders [2], equal channel angular pressing (ECAP) [3], deformation induced ferrite transformation (DIFT) [4] or high pressure torsion (HPT) [5], were developed and successfully applied on a laboratory scale. All of these methods can truly produce ultrafine grained materials with an average grain size of the order of 1μm or less. The majority of these processes also display severe limitations to the extent that they are generally restricted to small sample gauges and are difficult to implement on an industrial scale. In many instances also the mechanical properties require further evaluation (e.g. the fatigue properties). In addition to these novel processing routes also the more conventional thermo-mechanical strategies were further optimized with the aim of ultra-grain refinement. The present authors have shown that ultra-fast reheating of severely cold rolled sheets to a reduction of 95 % gave rise to a considerable refinement but failed to produce a truly sub-micron crystalline structure [6]. Therefore, in the present paper, the potential is investigated of an alternative thermomechanical process which is known as *Accumulative Roll Bonding* (ARB). This technique was developed by Saito et al. [1] and was proven successful in obtaining ultrafine single phase structures with a grain size of less than 1μm [7]. The aim of this study is to reproduce an ultrafine structured titanium interstitial free (TiIF) steel by applying the ARB technique and to evaluate the microstructures and textures obtained by this process.

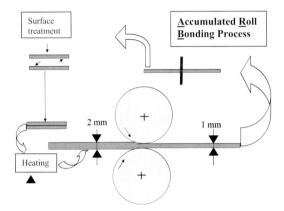

Figure 1: Schematic drawing the of ARB process [1]

2   Experimental

A TiIF steel (24 ppm C, 21 ppm N, 42 ppm S, 115 ppm P, 521 ppm Al, 443 ppm Ti and 0.1856 mass % Mn), was commercially rough rolled and laboratory finish rolled to a thickness of 4 mm with a finishing rolling temperature of 930°C. After the hot rolling the sheets were cold rolled to a reduction of 75 % in multiple passes on a two-high reversible laboratory mill in order to obtain a final thickness of 1mm. From this sheet two samples (250 mmRD × 25 mmTD × 1 mmND) were cut and stacked on top of each other. Afterwards they were warm rolled together at 680 °C with a reduction of 50 %, which produced a sample with a thickness of 1mm as a result of the diffusive bonding process occurring during rolling. The sample was subsequently cut into two pieces again, which were thoroughly brushed and cleaned with acetone in order to remove the oxide scale. These two pieces were again stacked on each other, reheated to 680 °C and warm rolled with a single pass reduction of 50 %, cf. Fig.1. This process was repeated 10 times, which produced a sample with a total reduction of 99.9 % ($\varepsilon = 6.9$) and a thickness of 1mm. In a series of subsequent experiments reheating temperatures of 610 °C, 590 °C and 500 °C were applied.

The samples were observed by optical microscopy and after that, selected samples were analyzed by scanning electron microscopy (SEM) on a XL30 Philips instrument with a LaB_6 filament and equipped with an orientation imaging microscopy (OIM) system of TSL®. These samples were prepared by standard metallographic procedures, including an electropolishing stage for the samples that would be analysed by OIM. The mechanical characterization was carried out by Vickers hardness measurements with an indentation load of 30 N.

3   Results and Discussion

Fig. 2 shows the optical micrographs of the sample after 10 rolling passes corresponding to a rolling reduction of 99.9 %. In these samples there are $2^{10} = 1024$ original sheets confined to one single sheet of 1 mm thickness. This means that a layer of approximately 1 µm thickness contains one original sheet. The metallographical study shows that although a number of the more recent bonds (particularly the one in the middle of the transverse section) is still visible, the vast majority of the $(2^{10}-1)$ bonds has completely dissolved in the structure. Apart from the long strains of quasi-spherical inclusions there is no trace left from the bonds. The EDX analysis of these particle inclusions allows to identify them as iron oxides and iron particles. Fig. 2 also shows that the oxide particles are much larger for the higher reheating temperature because the oxide surface layer which is formed during interpass furnace reheating is presumably much thicker at a higher temperature and an identical annealing time. It also shows that the aceton and brush cleaning procedure which was applied here was not sufficient to remove the entire oxide layers. Therefore, in future experiments, a controlled furnace atmosphere will be applied.

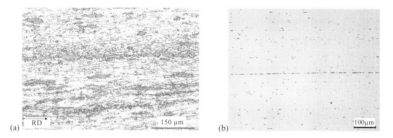

Figure 2: Microstructure of Ti-IF steel after warm reduction of 99.9 % with a furnace reheating temperature of (a) 680 °C and (b) 500 °C, respectively, and an annealing time of 10 min

Fig. 3 shows the hardness evolution as a function of the true rolling strain. In this analysis the true strain was calculated as $\varepsilon = -\ln(h_0/h)$, with h_0 and h the initial and the final thickness of the sheet, respectively. The curves show an accumulative hardening profile only for the material which was interpass annealed at 500 °C, whereas for the samples annealed at higher temperatures an important steady state regime is observed. For the highest reheating temperature of 680 °C the steady state continues to the very last pass, whereas for the two intermediate temperatures of 610 °C and 590 °C a hardening stage is observed towards the end of the process, which is followed immediately by a softening in the latter case. As in any hot deformation process, also during ARB the metal will be subjected to the interactive occurrence of softening and hardening mechanisms. The data strongly suggest that at 680 °C an equilibrium is achieved between plastic work hardening and thermally activated softening phenomena.

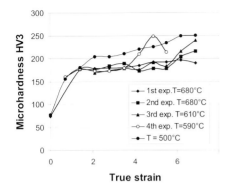

Figure 3: Hardness evolution as a function of true strain during the ARB process

In order to identify these softening processes, orientation measurements on selected samples were carried out. Fig. 4 shows an orientation map which was measured after the 10th cycle ($\varepsilon = 6.9$). The different colors of the map correspond to different crystallographic directions parallel to the sheet normal direction ND, according to the attached color key. This sample was submitted to an additional recovery annealing treatment at 500 °C during 10min in order to improve the band contrast of the Kikuchi patterns. Bold black lines represent high angle grain boundaries (HAGBs) with a misorientation angle exceeding 15 deg, whereas the fine lines delineate low angle grain boundaries (LAGBs) carrying less than 15 deg of misorientation. The OIM map of Fig. 4 displays the following interesting features: elongated deformed grains with an internal orientation spread revealing a certain cellular structure and the presence of scattered small recrystallization nuclei. The observed aspect ratio of the deformed grains (approximately 4 to 6) clearly demonstrates that an interpass recrystallization has occurred. This is further supported by the rather low density of intragranular LAGBs. If no static recrystallization had occurred the typical features of a strongly deformed structure would have been observed with a strictly lamellar or fibre-like arrangement of HAGBs. The presence of a cellular structure of

Figure 4: Grain fragmentation of TiIf steel after the 10th cycle at 680 °C

LAGBs also indicates that no static recrystallization occurred during the additional recovery annealing treatment that was carried out before collecting the OIM scan.

Although this sample does obviously not exhibit a nanocrystalline structure such as obtained in other ARB experiments carried out at lower temperatures, the structure is nevertheless much finer as the one compared to conventionally processed IF steels. The average grain size of this structure, as calculated by the OIM post-processing statistical software, is 1.7 µm, with a grain tolerance misorientation of 15°. Apparently the fraction of small recrystallization nuclei has a strong decreasing effect on the average grain size. This indicates that severe rolling strains in the ferritic region can truly operate as a grain refining technique. According to the hardness data of Fig. 3 this material displays a hardness of 190 to 220 HV, which empirically corresponds to a tensile stress of 570 to 660 MPa. Although these values are significantly lower than those reported by Tsuji et al.[8], the present material presumably achieves superior mechanical properties in terms of an appropriate balance between strength and ductility. Moreover, the texture of this material displays a strong <111> // ND fibre, cf. Fig. 5. It is generally known that this type of fibre texture produces excellent deep drawing properties. The combination of a tensile strength of more than 500 MPa, a reasonable degree of ductility and outstanding deep-drawability represents a very attractive mechanical properties profile.

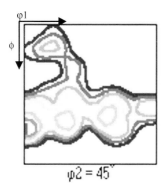

Figure 5: Texture of TiIf steel after the 10th cycle in an ARB process at 680 °C. Levels: 07-1.3-1.7-2.3-3-4

In the present experiment the highest strength levels were obtained for the samples that were rolled at 500 °C. After 10 rolling passes, the Vickers hardness increases to 250 HV, corresponding to an estimated tensile strength of 750 MPa, which is comparable to the data reported by Tsuji et al.[8]. In this case a much more fragmented microstructure is developed, as illustrated by the orientation scan of Figs. 6a and b. These orientation maps, which were measured after the 4th rolling pass (corresponding to a rolling reduction of 93.8 %) also show a much more fibre-like or lamellar structure in the middle of the sheet as compared to a more equi-axed structure near the sheet surface. This difference in structure can be readily interpreted by the role of surface shear introduced during rolling, which was carried out without lubrication.

The appearance of surface shear is also confirmed by the textures shown in Figs. 7a and b, which represent the midlayer and the surface texture, respectively. Near the middle of the sheet a conventional α/γ fibre texture is observed (with <110> // RD and <111> // ND, respectively) whereas near the surface a typical shear texture is observed characterized by the Goss compo-

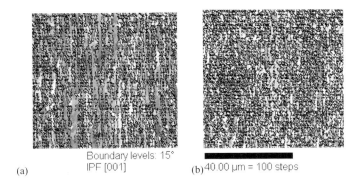

Figure 6: OIM scans of ARB processed sample after 4 rolling passes at 500 °C (93.8 % reduction) measured in the (a) middle and (b) near the surface of the sheet.

nent ({110}<001>). Although the occurrence of plane strain deformation in the midlayer and simple shear near the surface is a normal phenomenon in a conventional rolling experiment, it is not as evident during ARB rolling. Because of the composite nature of the sample consisting of various sheets stacked on top of each other, a more complex distribution of strain modes could be expected. The present texture observations indicate, however, that the conventional strain distribution is largely preserved, which suggests that the shear texture which is produced in one rolling pass is relatively unstable, and is rapidly transformed to the conventional rolling fibres in the next rolling pass. The extent to which surface shear contributes to grain fragmentation and ultimately to grain refinement is not clear as yet, but it is probably an important contributing factor in the shaping of the final structure. Shear deformation might be a technologically attractive alternative to the complexity of severe rolling reductions applied in the ARB process.

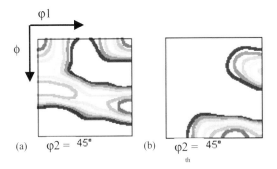

Figure 7: Texture of TiIf steel after the 4th cycle in an ARB process at 500 °C. (a) midlayer texture and (b) surface texture. Levels: 07-1.3-1.7-2.3-3-4

4   Conclusions

In the present paper the results were presented of an accumulative roll bonding experiment carried out in 10 consecutive rolling passes to a total reduction of 99.9 % at various temperatures between 500 and 680 °C. At the latter reheating temperature a steady state hardness was observed throughout the ARB process. This has produced a material with an average grain size of 1.7 µm and a microhardness of 220 H_V in combination with an excellent deep drawing texture. The higher strength level (250 H_V) was obtained with the lower reheating temperature of 500 °C. In the latter case, however, severe microstructural and textural gradients were observed across the thickness of the sheet.

5 Acknowledgement

The authors gratefully acknowledge the financial support for part of this work, granted by the Institute for the Promotion of Innovation by Science and Technology in Flanders (IWT).

6 References

[1] Y. Saito, H. Utsunomiya, N. Tsuji and T. Sakai, Acta mater. 1999, 47, 579–583
[2] S. Takaki and Y. Kimura, Journal of the Japan Society of Powder and Powder Metallurgy, 1999, 46, 1235–1240
[3] T.C.Lowe, R.Z.Valiev., JOM, 2000, 27–28
[4] G.L. Kelly, H. Beladi and P.D. Hodgson, Proceedings: First International Conference on Advanced Structural Steels,.2002, 13–14
[5] M.J. Zehetbauer, Proceedings TMS Conference- Ultrafine Grained Materials II, 2002, 669–678
[6] A.C.C. Reis, L. Kestens and Y. Houbaert, Proceedings of the 22nd Risoe International Symposium on Materials Science: Science of Metastable and Nanocrystalline Alloys. Structure, Properties and Modelling, 2001, 383–388
[7] N. Tsuji, Y. Saito, Y. Ito, H. Utsunomiya and T. Sakai. Ultra Fine Grained Materials, 2000, 207–218
[8] N.Tsuji, Y.Ito, Y.Koizumi, Y.Minamino and Y.Saito. Ultrafine Grained Materials II, 2002, 389–397

X SPD with Ball Milling and Powder Consolidation

Mechanically Activated Powder Metallurgy : A Suitable Way To Dense Nanostructured Materials

E. Gaffet[1,3], S. Paris[1,2,3], F. Bernard[2,3]
[1] UMR 5060 CNRS / UTBM, Sévenans, Belfort, France
[2] LRRS UMR 5613 CNRS / Univ. Bourgogne, Dijon, France
[3] GFA, GdR 2391 CNRS, Dijon, France

1 Introduction

The mechanical activation of powders has been found to lead to the synthesis of nanostructured micrometer sized powders by the so called ERAM / M2AP process (Elaboration par Recuit Activé Mécaniquement / Mechanically Activated Annealing Processing). Homogeneous but porous nanomaterials can be obtained by the *MASHS* process (Mechanically Activated Self Heat Sustaining Reaction). Recently, such an activation step to self heat sustaining reaction has been found to lead to the synthesis of fully dense bulk nanostructured materials (so called *MAFAPAS*, patented process: Mechanically-Activated Field – Activated Pressure – Assisted Synthesis).

The recent development and understanding of the mechanical activated powder metallurgy fields applied to the bulk nanostructured materials synthesis will be discussed.

2 Experimental Mechanical Activation Procedure

The mechanical activation processing of powder mixtures is performed in a high energy planetary milling machine, the so called G5 machine (see Fig. 1.; the G5 is now worldwide distributed by Fritsch as the P4 vario- planetary mill.)

Figure 1 : Typical planetary ball milling G5 machine

The milled particles are trapped between colliding balls or between balls and vial wall. The particles are subjected to severe plastic deformation, which exceed their mechanical strength, accompanied by a temperature increase. During the collisions, powders are subjected to high stresses (from 200 MPa up to 2 GPa) during a very short time (microseconds).

The milling duration (all other parameters being fixed) has to be short enough to avoid any formation of mechanically-induced endproduct phases, but still sufficiently long to form a mechanically activated mixture. Such a mechanically activated mixture is a state for which the micrometer sized powder particles do contain a 3-D nanodistribution of the elemental and initial components

3 Mechanically Activated Powder Metallurgy Processing

The mechanical activation of powders leads to the synthesis of nanostructured micrometer powders (*ERAM / M2AP* process). Homogeneous but porous nanomaterials can be obtained by the *MASHS* process. Recently, such an activation step to self heat sustaining reaction has been found to lead to the synthesis of fully dense bulk nanostructured materials (so called *MAFAPAS*, patented process).

3.1 ERAM / M2AP Processing

In the early 90's, N. Malhouroux- Gaffet and E. Gaffet, proposed the so called Mechanically Activated Annealing Processing (M2AP, in french called "Elaboration par Recuit Activé Mécaniquement / ERAM).

This solid state method which combines short duration mechanical alloying and low temperature annealing has successfully been applied to the synthesis of $FeSi_2$ [1, 2], $MoSi_2$ [3] and WSi_2 [4]. Starting from a mixture of elemental powders, the first M2AP step corresponding to a short duration mechanical alloying leads to the formation of micrometer sized powders in which a 3-D nanodistribution of the elemental components are formed. The solid state reaction leading to the end product phases occurs during the annealing. Both the grain size, which is determined by the milling condition, and the residual stresses modify the phase transformation kinetics during the low temperature isothermal annealing starting the short duration mechanically alloyed elementary components. Therefore, the M2AP is a very suitable powder metallurgy process allowing the direct driving of the solid state reaction occurring for example during a reactive sintering process. Moreover, the M2AP is a very suitable method to produce nanocrystalline $MoSi_2$ phase [3]. Indeed, the first mechanical M2AP step leads to an activation of the 3-D elementary distribution which reacts 400 °C below the temperature of the classical processes, i.e. 800 °C instead of 1200 °C.

3.2 Mechanically Activated Self Heat Sustaining Reaction (Nanostructured Porous Homogeneous Materials)

If a very exothermic reaction between solid-solid or solid - liquid reactants is locally initiated, it may generate enough heat to ensure the propagation of a transformation front, which leads to an end product involving the total initial amount of the elemental components.

Such a process is characterized by a fast moving combustion front (1–100 mm/s) and a self-generating heat which leads to a sharp increase of temperature, sometimes up to several thousands of K/s. The temperature reached inside the reaction front has been found to be effective to volatilize low boiling point impurities, helping to produce purer end product than those obtained by some more conventional techniques.

Combustion synthesis or self-heat-sustaining reactions (SHS) provides a suitable method for producing advanced materials, such as ceramics, composites and intermetallics. The SHS process offers advantages with respect to process economy and simplicity. The basis of such a synthesis method relies on the ability of highly-exothermic reactions to be self sustaining, and therefore, energetically efficient.

The new version of the SHS process introducing a first mechanically activation step (before the ignition of the SHS reaction) – i.e. the so-called MASHS (Mechanically Activated Self Heat Sustaining reaction)process – was firstly proposed by E. Gaffet in 1995 [5] for the synthesis of the nanocrystalline FeAl compound. This MASHS process has been successfully used to obtain various nanocrystalline compounds such as : FeAl [6–10], $MoSi_2$ [11, 12], $FeSi_2$ [13–15], Cu_3Si [16–18], $NbAl_3$ [19–22]. Although the MASHS process has a great potential to produce nanostructured materials, it induces the formation of porous end products containing 40 up to 50 % of porosity (Fig. 2)

To reduce such a residual porosity, a new process has been developed, leading to the formation of fully dense nanocrystalline materials.

SEM - secondary electrons

Figure 2 : Porous nanostructured $MoSi_2$ end-product produced by MASHS process

3.3 Mechanically Activated Field Activated Pressure Assisted Synthesis (Nanostructured Fully Dense Homogeneous Materials)

Some years ago, the simultaneous effect of an electrical field combined with an applied pressure during the combustion was found by Z.A. Munir as a suitable method to produce dense intermetallics compounds in a one step process. This process has been called FAPAS (Field-Activated Pressure Assisted Synthesis) (Figure 3).

The application of the FAPAS process to mechanically activated powders is a new way to produce dense nanostructured materials (so called the MAFAPAS process [23]). The feasibility of this new approach for the simultaneous synthesis and densification of nanomaterials has been successfully demonstrated for the synthesis of FeAl [24], $MoSi_2$ [25] and $NbAl_3$ [26] compounds. This new process combining electric field activation and applied pressure on mechani-

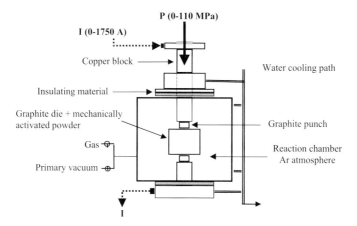

Figure 3 : The FAPAS apparatus

cally activated powder mixtures is demonstrated as a means to simultaneously synthesize and densify nanostructured materials in one step.

The mechanically activated powder mixtures were first cold compacted into well-adjusted cylindrical graphite dies using a uniaxial pressure (from 0 to 200 MPa). The relative density of the green samples ranged from 50 to 65 %. The graphite die containing the cold compacted mixture was placed inside the FAPASA reaction chamber evacuated and back-filled with argon to minimize oxidation phenomena. Then, the mechanically activated reactant powders were subjected to a uniaxial pressure (0–110 MPa) while a high level electric current (0–1750 A) is passed through the die. Under these conditions, a reaction is initiated and completed within a short period of time (2–6 min) (see Table 1 for processing parameters set used in previous works). Temperature was measured on the external surface of the die by means of a K-type thermocouple.

The densified MAFAPAS end-products specimens were disks of 20 mm in diameter and about 2 to 5 mm height.

Table 1 : Typical FAPAS processing parameters set for different systems

System	Intensity	Pressure	Time
Fe/Al	1250–1500 A	70–106 MPa	2.5–3.5 min
Mo/2Si	1600A	106 MPa	3–6 min
Nb/3Al	1500–1650 A	56–84 MPa	3–6 min

The relative density evaluated by weight and geometric measurements, the crystallite size determined from X-ray Line profile analysis and chemical composition are presented on Table 2. The various tested end products were found to be a nanocrystalline compound corresponding to the expected phases with little or no secondary phases.

Table 2 : Typical relative density, crystallite size and phases obtained in each system

System	Density	Size	Phases
Fe/Al	98 to 99,5 %	32–90 nm	$Fe_{0.515}Al_{0.485}$
Mo/2Si	82 to 93 %	58–75 nm	$MoSi_2$
Nb/3Al	85 to 96 %	57–150 nm	$NbAl_3$, Al_2O_3

4 Conclusions and perspectives

The paper has focused attention to a specific way to produce dense nanostructured materials ; namely mechanically activated powder metallurgy. This research field is very active since the highlighting of some particular mechanical behaviour of materials such as nano/micro composites and/or materials exhibiting bimodal grain size distributions [27–34].

5 Acknowledgements

Two of the authors (EG, FB) would like to thank their former students: N. Malhouroux, F. Charlot, H. Souha, V. Gauthier, Ch. Gras. The authors would like to address a special thank to Professor Z.A.Munir for very fruitful contributions.

6 References

[1] E. Gaffet, N. Malhouroux, M. Abdellaoui, J. All. Comp. 1993, 194, 339–360
[2] N. Malhouroux-Gaffet, E. Gaffet, J. All. Comp. 1993, 198, 143–154
[3] E. Gaffet, N. Malhouroux-Gaffet, J. All. Comp. 1994, 205, 27–34
[4] E. Gaffet, N. Malhouroux-Gaffet, M. Abdellaoui, A. Malchère, Rev. Métal. 1994, 757–769
[5] M. Zeghmati, E. Duverger, E. Gaffet, Proc. CANCAM 95, Ed. B. Tabarrock, S. Dost, 15ème Cong. Canad. Mécan. Appl. 1995, 2, p 952–956
[6] F. Charlot, E. Gaffet, B. Zeghmati, F. Bernard, J.-C. Niepce, Mater. Sci Eng. 1999, A262, 279–288
[7] F. Bernard, F. Charlot, E. Gaffet, J.C. Niepce, Int. J. Self Prop. High Temp. Synth. 1998, 7(2), 233–247
[8] F. Charlot, E. Gaffet, F. Bernard, Ch Gras, J.C. Niepce-Mater. Sci. For., 1999, 312–314, 287–292
[9] F. Charlot, F. Bernard, D. Klein, E. Gaffet, J.-C. Niepce, Acta Mater. 1999, 47(2), 619–629
[10] F. Charlot, C. Gras, M. Grammond, F. Bernard, E. Gaffet, J.C. Niepce, J. Phys., IV, Suppl. Coll., 1998, IV(8), 497–504
[11] Ch Gras, D. Vrel, E. Gaffet, F. Bernard, J. All. Comp. 2001, 314 (1–2), 240–250
[12] Ch. Gras, F. Charlot, F. Bernard, E. Gaffet, J.C. Niepce, Acta Mater. 1999, 47(7), 2113–2123
[13] Ch. Gras, N. Bernsten, F. Bernard, E. Gaffet, Intermetallics 2002, 10(3), 271–282

[14] E. Gaffet, F. Bernard-Annales de Chimie / Science des Matériaux, 2002, 27(6), 47–59
[15] Ch. Gras, E. Gaffet, F. Bernard, J.C. Niepce, Mater. Sci. Eng. 1999, A264, 94–107
[16] H. Shouha, E. Gaffet, F. Bernard, JC Niepce, J. Mater. Sci. 2000, 35, 3221–3226
[17] F. Bernard, H. Souha, E. Gaffet, Mater. Sci. Eng. 2000, A284, 301–306
[18] H. Shouha, F. Bernard, E. Gaffet, B. Gillot, Thermochemica Acta 2000, 351, 71–77
[19] V. Gauthier, F. Bernard, E. Gaffet, D. Vrel, J.P. Larpin, Intermetallics 2002, 10(4), 377–389
[20] V. Gauthier, F. Bernard, E. Gaffet, C. Josse, J.P. Larpin, Mater. Sci. Eng. 1999, A272(2), 334–341
[21] V. Gauthier, C. Josse, F. Bernard, E. Gaffet, J.-P. Larpin, Mater. Sci. Eng. 1999, A265, 117–128
[22] V. Gauthier, JP Larpin, M. Vilasi, F. Bernard, E. Gaffet, Mater. Sci. For., 2001, 369–372, 793–800
[23] Z.A. Munir, F. Charlot, F. Bernard, E. Gaffet, One step synthesis and consolidation of nanophase materials, US Pat. 6,200,515 published on the 13/03/2001
[24] F. Charlot, E. Gaffet, F. Bernard, Z.A. Munir, J. Amer. Cer. Soc. , 2001, 84(5), 910–914
[25] Ch. Gras, F. Bernard, F. Charlot, E. Gaffet, Z. A. Munir, J. Mater. Res. 2002, 13(3), 542–549
[26] V. Gauthier, F. Bernard, E. Gaffet, Z. Munir, J.-P. Larpin, Intermetallics. 2001, 9(7), 571–580
[27] S. Van Petegem, F. Dalla Torre, D. Segers, H. Van Swygenhoven, Scripta Mater., 2003, 48, 17–22
[28] L. Lu, M. Sui, K. Lu, Science, 2000, 287(5457), 1463–1466
[29] F. Wakai, Ceram. Inter., 1991, 17, 153–163
[30] R.Z. Valiev, R.K. Islamgaliev, I.V. Alexandrov, Prog. Mater. Sci., 2000, 45, 103–109
[31] Y. Champion, C. Langlois, S. Guerin-Mailly, P. Langlois, J.-L. Bonnetien, M.-J. Hytch, Science, 2003, 300, 310
[32] S.X. McFadden, R.S. Mishra, R.Z. Valiev, A.P. Zhilyaev, A.K. Mukherjee, Nature, 1999, 398(6729), 684–686
[33] Y. Wang, M. Chen, F. Zhou, E. Ma, Nature, 2002, 419, 912–915
[34] R.Z. Valiev, I.V. Alexandrov, Y.T. Zhu, T.C. Lowe, J. Mater. Res., 2002, 17(1), 5–8

Characterization and Mechanical Properties of Nanostructured Copper Obtained by Powder Metallurgy

C. Langlois[1], M.J. Hÿtch[1], P. Langlois[2], S. Lartigue[1] and Y. Champion[1]
[1]Centre d'Etudes de Chimie Métallurgique CECM-CNRS, Vitry-sur-Seine (France)
[2]Laboratoire d'Ingénierie des Matériaux et des Hautes Pressions LIMHP-CNRS, Villetaneuse (France)

1 Introduction

Nanostructured materials are of considerable interest from a fundamental viewpoint but also due to their high potential for applications in new technologies. For metals, progress in both strength and ductility are opening new perspectives for specific applications. However, the properties of these materials are strongly related to their microstructure and hence on the synthesis route chosen for fabrication. The wide variety of results obtained for the mechanical properties of nanomaterials is typical of this problem [1]. Hence, a careful characterization of the material is necessary to understand the behaviour and to identify the structural parameters controlling the properties. This paper presents the microstructure obtained for each processing step of bulk nanocrystalline copper after cold isostatic compaction, sintering and differential room temperature extrusion of nanocrystalline powders.

2 Specimen Preparation

Ultrafine copper powders are synthesized by evaporation of a molten copper droplet and condensation in a cryogenic liquid (for full details see [2]). The powders are hydrostatically compacted at room temperature (400 MPa) to a density of 70 % and then sintered at constant heating rate of 0.5 K/min until 230 °C under hydrogen flow, followed by a slow cooling inside the oven. The density after sintering is 90 %. Specimens are machined into a cylindrical shape and introduced into a copper billet. Hydrostatic differential extrusion is carried out through a tungsten carbide die. Pressurizing the container after the die (400 MPa) prevents the rupture of the nCu specimen inside the billet by containing the stress release in the sample just after the die. A pressure differential of 350 MPa is necessary to extrude the billet, thus the pressure in the container before the die is experimentally around 800 MPa during the extrusion, which allows a better codeformation between the billet and the specimen and thus a complete densification of the sample (> 99 %) without breaking. Cylindrical samples had a diameter of 5mm and a height of 60 mm after extrusion.

3 Characterization for Each Processing Steps

3.1 Relevant Characterization Techniques

In order to determine accurately the grain sizes and investigate structural features like the residual porosity and grain boundary character, X-ray diffraction (XRD) experiments, SEM and TEM observations were carried out. For XRD analysis, the Warren theory on planar fault line broadening [3], extended by Ungár et al. for dislocation contribution [4], was used to isolate the coherent diffracting domain size with good accuracy. Results from the XRD analysis are presented with the as-modified Williamson-Hall plot for each processing step. The value of the grain size is extracted from the intercept on the modified Williamson-Hall plots.

SEM images were performed with a FEG-SEM LEO 1530 with a spatial resolution of 15 nm. TEM observations were carried out on a JEOL 2000EX operating at 200 kV. XRD measurements were obtained with a Philips 1049 diffractometer with a Co anode. EVA software (BRUKER) was used for the line broadening analysis and image processing carried out using the VISILOG software package.

3.2 nc-Cu Powders

The nc-Cu powders used as material for the fabrication of the bulk material have been thoroughly analyzed in previous work [5]. They present the following main features: a narrow distribution with a mean grain size of 38 nm (value from XRD analysis and TEM observations) and a 2 nm oxide layer covering each particle. Particles also reveal a certain amount of twinning.

3.3 nc-Cu Compacted Powders

The structure of the compact sample consists of agglomerates of grains, loosely compacted (see Figure 1).

Figure 1: SEM images of the compacted sample

The density is limited to around 70 % in order to keep an open porosity, necessary for the reduction of the oxide layer by the hydrogen flow during sintering. Figure 2 presents the Williamson-Hall plots of the compacted powder (a) without any correction, (b) corrected from planar faults according to the Warren theory and (c) with full correction of the planar faults and dislocation contribution. The integral breadths of the first five reflections of copper have been taken into account for the XRD analysis.

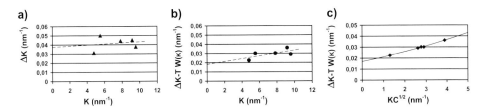

Figure 2: Williamson-Hall plots for the compacted powder (a) without correction, (b) corrected from planar faults contribution and (c) corrected from planar faults and dislocation contribution

The perfect fit by a quadratic polynomial regression on Figure 2(c) shows that both planar faults and strain fields from dislocations are present in the material. This means that, during the compaction, the nanoparticles undergo a plastic deformation which may be accommodated by dislocation activity. The particle size determined by the XRD analysis is 59 nm, close to the value of 50 nm determined by cross-counting on TEM images. The difference between the values may arise from overlapping of grains on the micrographs.

3.4 Sintered Material

SEM observations (Figure 3) show that the material after sintering turns to a polycrystalline metal with well defined grain boundaries and a relative density of 90 %. The residual porosity consists of spherical cavities with approximately the same size as the grain, homogeneously distributed in the structure.

Figure 3: SEM fracture image showing the cavities present in the sintered material structure

The grain size cannot be determined from the SEM observations as not all boundaries are visible in the images. XRD analysis gives a value of 83 nm. The true value is slightly higher, as revealed by TEM micrographs (Figure 4). This is due to the fact that the XRD measurements do not account for coherent diffracting domains larger than 1000 λ (~180 nm). Some grain growth has therefore occurred during the sintering, which in the future can be limited by a better control of the temperature.

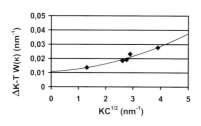 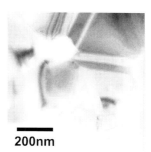

Figure 4: (a) XRD analysis of the sintered material (grain size 83 nm) and (b) TEM micrograph supporting a larger value for the grain size

3.5 Extruded Material

The most important aspects to asses at the end of the whole process is the residual porosity and grain size. Extrusion is intended to close the porosity observed in the sintered material and to increase the density towards 100 %. The residual porosity is characterized here in terms of volume, shape and number of cavities by SEM observations on axial surfaces, parallel to the extrusion axis (Figure 5).

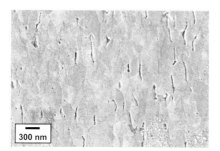

Figure 5: SEM image showing the elongation and size of the pores after extrusion. The extrusion axis is vertical.

Pores are elongated along the extrusion axis, the mean length being around 150 nm. Appropriate numerical image processing gives a density of around 98.5 %. As the electrochemical polishing slightly enlarges the pores visible on the SEM images, the true density of the samples will be higher.

TEM observations reveal that significant changes in the microstructure are induced by extrusion. Figure 6 shows a grain representative of the transformation occurring during extrusion.

Dislocation nucleation in the largest grains results in the formation of sub-grains with low misorientation. The two straight boundaries visible on the lower part of the micrograph are dis-

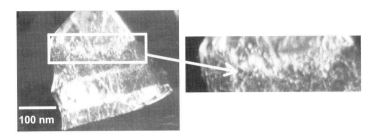

Figure 6: TEM dark field micrograph showing the substructure inside the largest grains

location walls or twin boundaries perturbed by intergranular dislocations. These configurations are characteristic of the material. Matrix dislocations are also visible in grains of all sizes and are always witnessed in interaction with the twins and sub-boundaries.

Due to the substructure observed in the grains, the grain size can be defined as the size of the regions surrounded by large angle grain boundaries (140 nm from TEM observations) or as the size of the regions surrounded by all types of boundaries, including twins and low angle boundaries (90nm from XRD analysis and TEM observations, both reported on Figure 7). The most important for dislocation activity and mobility is the latter.

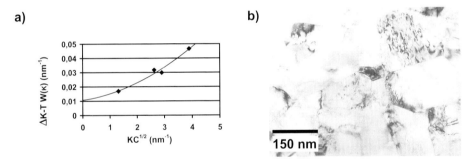

Figure 7: (a) XRD analysis of the extruded sample and (b) TEM bright field of the extruded microstructure

4 Discussion and Conclusions

Dislocation activity is witnessed in all the structures obtained during processing, even in the compacted material with an individual particle size in the range 50–60 nm. Copper nanograins obtained by mechanical alloying also show dislocation activity [6]. Moreover, the deformation induced by the extrusion step is accommodated by dislocations (TEM observations and XRD analysis). Dislocation sources in the grain boundaries [7] could be activated during the extrusion according to the configuration of matrix dislocations observed in the extruded material. Due to this dislocation activity, the largest grains of the distribution are divided into smaller domains by dislocation walls but in the whole microstructure, the grain boundaries are in most cases large angles grain boundaries. The distribution is not bimodal, in the sense of the material reported by Wang *et al.* [8]. If we assume that the mean free path for dislocations is determined by grain size, the effective one amounts to about 90 nm. These aspects and the different types of grain boundary, have to be considered to understand the particular mechanical properties recently reported for this material [9].

Dislocation activity may not be the only active deformation mechanism in the range of grain size reported here. In copper obtained by severe plastic deformation (SPD), grain boundary sliding accommodated by dislocations in the grain boundaries, as well as grain rotation, may have an important role during deformation, coexisting with dislocation slip [10]. The grain boundary character is therefore a key parameter. Grain boundary characteristics are the main differences between SPD nc-Cu and nc-Cu obtained from powder metallurgy: firstly, the intergranular dislocation density is much higher in the SPD Cu due to the large strains undergone by the material, and secondly, the frequency and repartition of large angle and low angle grain boundaries is completely different. Finally, the effective grain size relevant for dislocation activity is 90 nm for our nc-Cu which is smaller than that of SPD material. The differences in the tensile test behavior (strain hardening, necking, ductility, strain rate sensitivity…) between the two materials should originate from these aspects.

The process employed in this study was designed to obtain samples with large dimensions and small grain size. Nanostructured copper obtained by Equal Channel Angular Extrusion (ECAE) typically shows a grain size not below 180 nm [11], except when high pressure is employed (2 GPa [10]). High pressure torsion is more efficient in refining the grain size [10] but the sample dimensions are small (<1 mm thickness). The main drawback of powder metallurgy is that the processing is more difficult to implement and that porosity remains. However, the possibility to carry out tensile tests with large specimens, despite of these flaws in the microstructure, is of particular interest as an alternative approach. The comparison between the tensile properties of two different microstructures (SPD and powder consolidation) in this range of grain size will shed a new light on the mechanisms involved in the deformation of the nanomaterials.

5   References

[1]   K. A. Padmanabhan, Mat. Sci. Eng. 2001, A304–306, 200
[2]   Y. Champion, J. Bigot, Nanostruct. Mater. 1998, 10, 1097
[3]   B. E. Warren, Progr. Metal Phys. 1959, 8, 147
[4]   T. Ungár, S. Ott, P. J. Sanders, A. Borbély, J. R. Weertman, Acta Mater. 1998, 46, 3693
[5]   Y. Champion, J. Bigot, Scripta Mater. 1996, 35, 517
[6]   J. Y. Huang, Y. K. Wu, H. Q. Ye, Acta Mater. 1996, 44, 1211
[7]   S. Cheng, J. A. Spencer, W. W. Milligan, Acta Mater. 2003, 51, 4505
[8]   Y. Wang, M. Chen, F. Zhou, E. Ma, Nature 2002, 419, 912
[9]   Y. Champion, C. Langlois, S. Guérin-Mailly, P. Langlois, J-L Bonnentien, M. J. Hÿtch, Science 2003, 300, 310
[10] R. Z. Valiev, I. V. Alexandrov, Y. T. Zhu, T. C. Lowe, J. Mater. Res. 2002, 17, 5
[11] S. Komura, Z. Horita, M. Nemoto, T. G. Langdon, J. Mater. Res. 1999, 14, 4044

Densification of Magnesium Particles by ECAP with a Back-Pressure

R.Ye. Lapovok, P.F. Thomson
CAST CRC, School of Physics and Materials Engineering, Monash University, Melbourne, Australia

1 Introduction

The increasing application of magnesium alloys in automotive, aerospace and other industries leads to the loss of a large amount of magnesium during production processes. It has become important in the magnesium industry to minimise the environmental impact of waste products and to utilise an economical source of secondary magnesium. It is shown that in typical magnesium die-casting operations only about 50 % of the metal input ends up as a finished product. The remaining 50 % goes to scrap. Finely divided waste in the form of machining swarf is seen as an acute problem in case of fire and it is urgent to convert this to a less potentially incendiary form.

Government regulations and legislation on waste disposal in Europe have highlighted the importance of finding new, low-cost methods of recycling waste materials [1]. The recycling of magnesium not only produces feedstock for primary metal production, but decreases energy consumption. A kilogram of magnesium produced from raw materials requires the equivalent of 35 kWh of energy whereas a kilogram of refined-recycled metal would consume less than 3 kWh [2]. Ongoing studies of the handling of magnesium swarf must be undertaken to yield the solution of these problems.

The conventional technology of briquetting of swarf [3], which allows production of batches of density 1.4–1.5 g/cm^3 still has the disadvantage of high surface-to-volume ratio leading to melt losses, high oxide content and hazards during transportation. The work reported here was aimed on investigation of swarf compaction technique leading to increase recovery rate of Mg in re-melting and decrease hazards of transportation of swarf.

The production of solid rod from magnesium alloy swarf by Equal Channel Angular Extrusion (ECAE) with back-pressure was investigated. It was found that ECAE promotes compaction very effectively. The optimal parameters of the process have been defined.

2 Material and Procedure

Magnesium swarf supplied by Queensland Manufacturing Institute was produced principally by milling and drilling magnesium components on a Computer Numerical Controlled machine centre. The cutting fluid used in the machining process was Clearedge EP690, a soluble synthetic metal cutting lubricant produced by Castrol. The amount of residual lubricant estimated by spectrographic and chemical analysis was around 0.55 % of the weight.

The swarf was principally AZ91D produced at the Australian Magnesium Corporation plant though some of the swarf came from the machining of AZ91D produced by Timminco in Canada. There may also have been a small amount of pure magnesium swarf. The swarf was collected as it came off the machine centre and was not segregated as various machining jobs were undertaken.

The machine for Equal Channel Extrusion with controlled back-pressure, designed and manufactured in Monash University was used in experiments (Fig 1a). Optimal die angle, back pressure and velocity of extrusion were investigated. The entry channel of the ECAE machine was crammed with swarf while a back pressure punch blocked the exit channel. During operation, the trust of the forward punch created a large hydrostatic pressure. When the pressure exceeded the preset backward pressure, plastic deformation commenced and the compact moved into the second channel against the back pressure punch (Fig. 1b).

Swarf has been compacted using two die angles 90 and 120° and two velocities of the forward punch 1 and 6 mm/sec. The quality of the compact has been evaluated by optical microscopy, and measurements of hardness and density. To characterise the hardness of each compact, the Vickers Hardness under a 1kg load was calculated from five points measurements in the deformed centre of the specimen and in the back and front corners not been subjected to shear. The influence of particle size was studied by separating the swarf into three sieve sizes before compaction 0.5–1.0, 1.0–2.4, 2.4 – 4.0 mm.

Density of the samples was measured by weighting in the water and air using formula:

$$\rho = \rho_{wat} \frac{W_{air}}{W_{air} - W_{wat}} \qquad (1)$$

where W is weight and ρ is density.

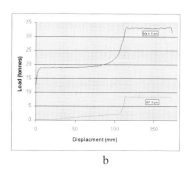

a
b

Figure 1: The ECAE machine used for experiments and loading graphs. a – machine; b – forward and backward forces versus displacement

The role of grease and other contaminants, was considered by performing ECAE with- and without a lubricant. The role of machining lubricants, oxide films and the need to remove them, were also investigated. Swarf was degreased by soaking in ethanol bath for 10 and 5 min., sequentially and dried. The surface of the swarf had a shiny appearance after cleaning whereas it was lustre in the as-received conditions. Samples were compacted from cleaned swarf without using a lubricant. The improvement in the quality of compact was estimated by remelting of these samples and determining the recovery rate of magnesium. To investigate the effect of oxidation of the surface, the same cleaning technique was followed by immersion in a solution of 0.5 % acetic acid, before compaction and re-melting.

3 Results and Discussion

3.1 The Influence of a Die Angle

Despite the high level of back pressure the amount of plastic deformation introduced into the material during extrusion through the 90° die exceeded the critical strain at fracture and a visible network of cracks developed on the surface and within the sample (Fig. 2a). However, the critical level of strain was not reached at any values of the back pressure used when extrusion was performed through an angle of 120° (Fig. 2b). In preliminary experiments through a die with an angle of 150° did not introduce sufficient plastic deformation to compact the swarf and an additional pass was required.

a b

Figure 2: The compacts produced by ECAE of swarf in as-received conditions. a – extrusion through 90° die; b – extrusion through 120° die

3.2 The Influence of the Back-Pressure and the Velocity of Extrusion

The microstructure of samples processed by ECAE through the 120° die with two different levels of back-pressure at a ram speed of 1 mm/sec is shown in Fig.3. As a result of shear deformation, particles of magnesium swarf took a preferred direction with an angle about 21° to the line of intersection of the channels. The spacing between particles oriented in the same manner diminished with increasing back-pressure.

a b

Figure 3: Microstructure of samples after ECAE through 120° die with the back-pressure of a – 20 bar; b – 60 bar

Change the punch speed in ECAE from 1 to 6 mm/sec increased the average strain rate at the shear plane from 0.06 s^{-1} to 0.4 s^{-1} with an increase in temperature from 12 °C to 32 °C, but this was not sufficient to promote diffusion bonding of particles.

The effects of die angle, velocity of extrusion and back-pressure are summarised in Fig. 4 showing the Vickers hardness of the compacted samples. Within the range of the variables examined, the best compaction has been obtained when ECAE has been performed through the 120° die at a ram speed of 6mm/sec against a back-pressure of 60 bar.

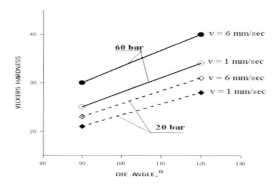

Figure 4: The effect of die angle, velocity of extrusion and back-pressure or Vickers hardness of samples compacted by ECAE

3.3 The Influence of Particles Size

Particles sieved to three size fractions as described previously were extruded through the 120° die at a ram speed of 1 mm/sec against a back pressure of 60 bar. The Vickers hardness versus diameter of particles and the typical microstructure of compacts formed from fraction of particles are shown in Fig. 5. The hardness and density of samples increased as particle size decreased because the size of cavities between particles also decreased. There was a greater contact surface formed between smaller particles, favouring diffusion bonding.

3.4 The Influence of the Lubricant

Samples compacted with a use of the molybdenum di-sulphide grease lubricant had a density equal to 1.67 g/cm^3 compared with that of pure magnesium, 1.741 g/cm^3. Vickers Hardness (1kg load) measured at five points was 32.8 HV in the deformed middle part of the specimen and 21.2 HV and 23.4 HV in the back and front non-deformed corners subsequently. The use of lubricant leads to preventing bonding between particles and creates cavities with trapped grease.

Samples compacted without a lubricant (Fig. 6) had a density equal to 1.74 g/cm^3 compared with the density of pure magnesium, 1.741 g/cm^3. Vickers Hardness (1kg load) measured at five points was 39.4 HV in the deformed middle part of the specimen and 25.6 HV in the non-deformed corners. The of Vickers Hardness obtained on samples compacted with and without lubricant is shown in Fig. 7. It was shown by microstuctural observation and measurements of density and hardness that ECAE without a lubricant produced more dense and cleaner compact.

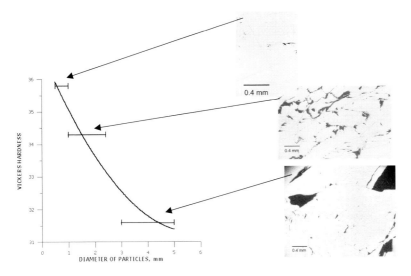

Figure 5: Hardness and microstructure of compacts produced from particles of different diameters

Remelting of samples showed that the recovery rate of magnesium also improved with increasing density of the compact from 46 % for briquettes of uncleaned swarf and 71 % and 77 % respectively for samples compacted by ECAE with and without a lubricant. Cleaning the swarf as described above, followed by compaction without a lubricant, resulted in a further increase in recovery rate of magnesium after re-melting up to 92–96 % .

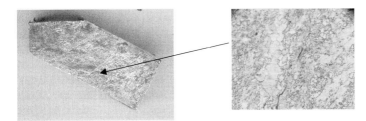

Figure 6: Samples compacted by ECAE without a lubricant

3.5 The Role of Shear Deformation in Improvement of Compaction Process

The improvement of compaction due to shear deformation was estimated by interrupting compaction by ECAE and measuring the Brinnel hardness in the two parts before and after the shear plane (Table 1) using a spherical indentor of 5 mm diameter. It was necessary to use Brinnel hardness test because of the scale of irregularity caused by the partial compaction of the swarf. The swarf in the first channel is compacted only by hydrostatic pressure similar to that in briquetting technique, while in the second channel this compact had been subjected to superimposed plastic deformation by shear. Multiple cracks surrounded the indented area in the region before shear plane, while only a few appeared around the indented area in the part following shear plane. The improvement in hardness due to simple shear deformation with and without lubrication is clearly seen in Fig. 7.

Table 1: Brinell hardness of the compact before and after the shear plane

	Before shear plane		After shear plane	
P (gf)	261.5	116.2	261.9	109.5
D (mm)	3.4	2.4	2.9	2.0
BHN	24.96	24.11	35.98	33.40

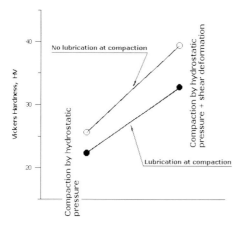

Figure 7: Comparison of Vickers hardness for samples compacted by hydrostatic pressure and by ECAE with and without lubricant

Those results were confirmed by Finite Element Simulation of ECAE of porous material. The relative density approaches a value of 1 after material undergoes shear deformation while it is between 0.8 and 0.9 before the shear plane which corresponds to a density 1.4 g/cm^3–1.56 g/cm^3

4 Conclusions

The capacity of Equal Channel Angular Extrusion to enhance compaction of extruded magnesium is investigated. The production of solid rod from magnesium alloy particles by Equal Channel Angular Extrusion (ECAE) with back-pressure is reported here. It was found that ECAE promotes compaction very effectively. Parameters affecting the compaction process die angle, back-pressure, velocity of extrusion and particle size were investigated. The role of grease and other contaminants, lubricants, oxide films and the need to remove them, were also considered. The parameters of the process to produce a fully-dense magnesium rod are found and characteristics of the product, namely density, hardness and microstructure were investigated.

5   References

[1]   H. Antrekowitsch, G. Hanko, P. Ebner, in Magnesium Technology 2002 (Ed. H.I. Kaplan), The Minerals, Metals & Materials Society, USA, 2002, 43–48
[2]   L. Riopelle, Journal of Metals, 1996, October, 44–46
[3]   D.J. Roth, G. Von-Aschwege, Magnesium Technology 2002 (Ed. H.I. Kaplan), The Minerals, Metals & Materials Society, USA, 2002, 317

Annealed Microstructures in Mechanically Milled Fe-0.6%O Powders

A. Belyakov, Y. Sakai, T. Hara, Y. Kimura and K. Tsuzaki
Steel Research Center, National Institute for Materials Science, Tsukuba, Japan

1 Abstract

The structural changes taking place in mechanically milled Fe-0.6%O powders during annealing at 700 °C for an hour were studied. The iron-iron oxide powders were mechanically milled in an argon atmosphere for 20 to 300 hours. An increase in the milling time promoted the development of ultra fine-grained microstructures after the heating. The average annealed grain size decreased from about 0.8 to 0.13 µm when the milling time was increased from 20 to 300 hours. Such difference in the grain sizes was caused by the variation of oxide distributions. The volume fraction of dispersed oxides (the size of about 10 nm) increased from 0.3 % to 2.5 % with increase in the milling time from 20 to 300 hours.

2 Introduction

One of the promising methods for the production of advanced structural materials is mechanical milling followed by consolidating working [1, 2]. This processing method combines the advantages of two techniques: severe plastic deformation and powder metallurgy. The method allows us to make sizeable products and it is applicable to a wide variety of materials, which may be difficult to produce by any other processing. Recently, mechanical milling followed by warm consolidating rolling was applied to the production of oxide-bearing steels with submicrocrystalline structures [3–5]. The dispersed oxides were introduced into ferrite matrix to increase the strength of steels as well as to promote the grain refinement in final products.

This processing method is rather complex one that consists of two sequential stages, i.e. mechanical milling and consolidating rolling. The latter is carried out at elevated temperatures and, therefore, it can significantly affect the deformation microstructures evolved in the milled powders. It was shown that increasing the milling time suppressed any discontinuous recrystallization processes during subsequent annealing and provided the development of ultra fine-grained microstructure [6, 7]. The main structural evolution mechanism operating in severe milled powders upon the heating was discussed as a normal grain growth. However, the annealed structures in mechanically milled powders have not been clarified in detail; effect of dispersed oxides on the annealing behavior of milled powders is still unclear. The aim of the present work is to define the structural changes in mechanically milled Fe-0.6%O powders during subsequent heating, i.e. to study the changes in dislocation substructures, (sub)grain sizes, (sub)grain boundary characters, distribution of dispersed oxides.

3 Experimental Procedure

An Fe-0.6%O powder (namely 0.58%O, 0.005N, 0.01%C, 0.009%Si, 0.11%Mn, 0.009%P, 0.007%S, all in mass%, and the balance Fe) was used as the starting material. The oxygen was in form of Fe_3O_4 surface oxide comprising of about 3 vol.%. (The volume fraction of oxides was derived from the total oxygen content in the steel, assuming that all the oxygen is Fe_3O_4 oxide [4].) The powders were planetary ball milled using a stainless steel pot and balls in an argon atmosphere for 20 ~ 300 hours. After mechanical milling, the worked powders were compacted in a steel cylindrical vessel by cold-pressing method [8, 9] and then annealed at 700 °C for one hour using an argon muffle furnace. The structural observations were carried out using a JEM-2010F transmission electron microscope (TEM) operating at 200 kV. The (sub)grain size was measured by the linear intercept method. The misorientations across the (sub)grain boundaries were studied by the Kikuchi-line technique. About 100 boundaries were studied for each sample.

4 Results and Discussion

Fig. 1 shows typical TEM micrographs for the microstructures evolved in the Fe-0.6%O after mechanical milling for 20 and 300 hours. These micrographs are similar to each other and look like those reported after severe plastic working [6, 10]. The milled microstructures are characterized by high dislocation densities and rather complex internal distortions as suggested by the rich contours of local lattice bending. The many scattered diffraction spots also suggest the large crystalline misorientations that develop in the milled powders.

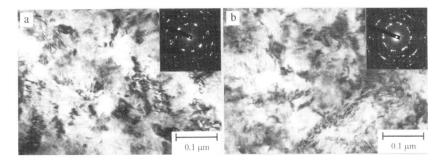

Figure 1: Typical structures evolved in the Fe-0.6%O powders after mechanical milling for (a) 20 hours and (b) 300 hours. Electron diffraction patterns were obtained from selected areas of 0.25 μm.

The structural changes in the mechanically milled powders caused by the heating depend significantly on the milling time (Fig. 2). Annealing of the powders milled for 20 hours results in the development of relatively coarse-grained microstructure with the grain size of 0.8 μm. On the other hand, increase in the milling time promotes the development of much finer grains upon following annealing. The annealed grain sizes are about 0.2 and 0.1 μm in the powders milled for 100 and 300 hours, respectively. It should be noted that the annealed powders are quite different in the substructures. Annealed microstructure in the 20 hours milled powder is almost free of interior dislocation substructure, while those in the powders milled for 100 and 300

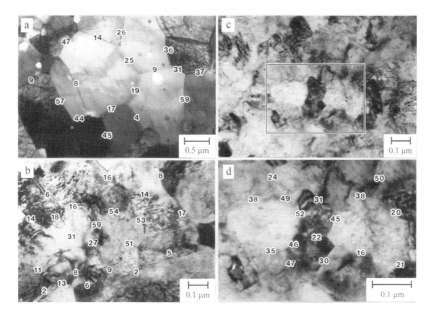

Figure 2: Typical microstructures developed in the Fe-0.6%O powders after annealing for an hour at 700 °C following the mechanical milling for (a) 20 hours, (b) 100 hours, (c) and (d) 300 hours. The numbers indicate the (sub)boundary misorientations in degrees.

hours contain high dislocation densities in the grain interiors. It is also interesting that the 100 hours milled powder includes many low-angle dislocation subboundaries after annealing.

After annealing, the distribution of the oxide particles in the milled powders is also strongly affected by the milling time (Fig. 3). The size of the dispersed oxides of about 10 nm does not depend on the milling time, while their volume fraction increases form about 0.3 to 2.5 % when the milling time increases from 20 to 300 hours. (Note here that only fine oxide particles with a size of below 100 nm were counted as the dispersed oxides). That is to say, in case of a relatively short milling time, the oxides are not dispersed homogeneously throughout the ferrite matrix, and the certain amount of oxygen is still in the form of coarse initial oxides.

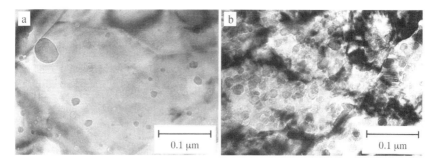

Figure 3: Dispersed oxides in the mechanical milled Fe-0.6%O powders after annealing for an hour at 700 °C. Milling time: (a) 20 hours, (b) 300 hours.

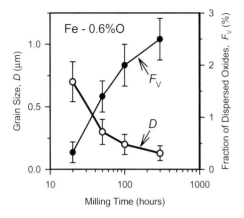

Figure 4: Effect of the mechanical milling time on the grain size and volume fraction of dispersed oxides (average oxide size is about 10 nm) that evolve in the Fe-0.6%O powders after one hour annealing at 700 °C

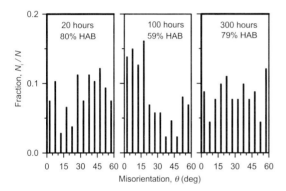

Figure 5: Misorientation distributions for the (sub)grain boundaries in the Fe-0.6%O powders, which were annealed for one hour at 700 °C following mechanical milling for various times

Effect of the milling time on the microstructures developed in the Fe-0.6% powders after annealing at 700 °C is represented in Figs. 4 and 5. The low fraction of dispersed oxides in the powder milled for 20 hours is not enough to suppress a static recrystallization during annealing. The resulted microstructure is characterized by a relatively large grain size and a large fraction of high-angle grain boundaries. On the other hand, increasing the fraction of dispersed oxides in powders milled for longer times diminishes the annealing effect on the final grain size. Annealing results in formation of sharp (sub)grain boundaries but does not lead to complete recrystallization. Such mechanism of microstructure evolution can be discussed as a polyogonization [11]. Therefore, the misorientations between (sub)grains correspond directly to those evolved by previous severe deformation. It is clearly seen in Fig. 5 that the fraction of high-angle grain boundaries increases with increasing the milling time from 100 to 300 hours. This is a general tendency for the strain-induced grain boundaries [10, 12–15].

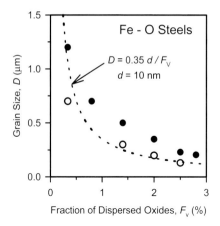

Figure 6: Relationship between the grain size and the volume fraction of dispersed oxides in the mechanically milled Fe-0.6%O powders after one hour annealing at 700 °C (open symbols) and consolidating rolling at 700 °C [5, 7] (closed symbols)

Finally, let us consider the pinning effect of dispersed oxides on the annealed grain size (Fig. 6). Fig. 6 represents also the data for powder samples consolidated by warm rolling at 700 °C [5, 7]. The maximum retarding force results in a cessation of grain growth when $D = K d / F_V$, where D is the grain size, K is a structural factor, d and F_V are the size and the volume fraction of dispersed particles [16]. According to numerous studies, the K may vary from 0.15 to 0.35. The present results are consistent with the pinning relationship with $K = 0.35$. Therefore, for the studied material, the pinning has a general effect on annealed microstructures either or both recrystallization and polygonization operates upon annealing. Little bit coarser grains developed after consolidating rolling may result from a strain hardening, i. e. an additional driving force for grain growth.

5   Conclusions

The microstructures developed in the mechanically milled Fe-0.6%O powders after annealing at 700 °C for an hour were studied.

1. Increase in the milling time promotes the development of submicrocrystalline structures after annealing. The annealed grain size decreases from 0.8 to 0.13 µm as the milling time increases from 20 to 300 hours.
2. The annealed microstructures are greatly affected by the pinning of dispersed oxides. The size of dispersed oxides is about 10 nm, while their volume fraction gradually increases from about 0.3 % to 2.5 % with increasing the milling time from 20 to 300 hours.
3. The annealed grains in the powders milled for 20 hours result from a static recrystallization and can be characterized by near random orientations. On the other hand, in the powders milled for longer times over about 100 hours, the annealed microstructures result mainly

from a polygonization; the (sub)boundary characteristics depend on the intensity of previous deformation.

6 References

[1] C. Suryanarayana, Int. Mater. Rev. 1995, 40, 41–64
[2] C.C. Koch, NanoStructured Mater. 1997, 9, 13–22
[3] M. Ohtaguchi, K. Tsuzaki, K. Nagai, in Recrystallization and Related Phenomena (Eds.: T. Sakai, H.G. Suzuki), Japan Institute of Metals, Sendai, Japan, 1999, p. 495
[4] Y. Sakai, M. Ohtaguchi, Y. Kimura, K. Tsuzaki, in Ultrafine Grained Materials (Eds.: R.S. Mishra et al.), TMS, Warrendale, PA, 2000, p. 361
[5] Belyakov, Y. Sakai, T. Hara, Y. Kimura, K. Tsuzaki, Metall. Mater. Trans. A 2001, 32A, 1769–1776
[6] Y. Kimura, S. Takaki, Mater. Trans. JIM 1995, 36, 289–296
[7] Belyakov, Y. Sakai, T. Hara, Y. Kimura, K. Tsuzaki, Metall. Mater. Trans. A 2002, 33A, 3241–3248
[8] K. Ameyama, O. Okada, K. Hirai, N. Nakabo, Mater. Trans. JIM 1995, 36, 269–275
[9] Y. Kimura, H. Hidaka, S. Takaki, Mater. Trans. JIM 1999, 40, 1149–1157
[10] R.Z. Valiev, R.K. Islamgaliev, I.V. Alexandrov, Progr. Mat. Sci. 2000, 45, 103–189
[11] F. Haessner, in Recrystallization of Metallic Materials (Ed.: F. Haessner), Verlag, Stuttgart, Germany, 1978, p. 1
[12] K. Tsuzaki, X. Huang, T. Maki, Acta Mater. 1996, 44, 4491–4499
[13] D.A. Hughes, Q. Liu, D.C. Chrzan, N. Hancen, Acta Mater. 1997, 45, 105–112
[14] Belyakov, T. Sakai, H. Miura, Mater. Trans. JIM 2000, 41, 476–484
[15] Belyakov, T. Sakai, H. Miura, K. Tsuzaki, Phil. Mag. A 2001, 81, 2629–2643
[16] F.J. Humphreys, M. Hatherly, Recrystallization and Related Annealing Phenomena, Pergamon Press, Oxford, UK, 1996, p. 306

Processing and Characterization of Nanocrystalline Aluminum obtained by Hot Isostatic Pressing (HIP)

S. Billard, G. Dirras, J.P. Fondere, B. Bacroix
Laboratoire des Propriétés Mécaniques et Thermodynamiques des Matériaux (LPMTM) CNRS, Institut Galilée
Université Paris 13, Villetaneuse FRANCE

1 Introduction

Nanocrystalline materials are the object of an increasing attention since the first works of Gleiter [1]. This class of materials offers a range of very promising mechanical properties, mostly because of the important reduction in the grain size. Due to the difficulties in obtaining reasonable amounts of material, studies devoted to measuring the mechanical properties of nanocrystalline materials other than by microhardness tests are scarce [2–4]. It appears then that the optimization of the methods for processing fully dense nanocrystalline materials is a crucial step for understanding their latent properties. Nevertheless, in spite of the variety of techniques to produce these materials in sufficient quantities (see for example [4–7] for nanocrystalline aluminum), the HIP process [8] is an interesting method to obtain large bulk and fully dense material from nanometric metallic powders.

Actually, HIP is a process that permits the consolidation of porous materials such as metal powders. Bulk materials are then obtained with more or less complex shapes and sizes, and exhibit in general a uniform microstructure. Practically, HIP consists in subjecting an envelope or capsule, previously filled with powder, to a thermal treatment under isostatic pressure; the pressure can reach 400MPa and the temperature 2000 °C. This pressure is applied through an inert gas (argon or nitrogen). Under the simultaneous action of the pressure and the temperature, the powder density increases to reach the bulk solid theoretical value. Contrary to sintering, HIP allows the densification to occur at lower temperatures limiting the grain coarsening, which is an essential condition in processing nano-sized grains.

The present work was conducted to optimize the HIP process for producing large and bulk nanocrystalline compounds from nanocrystalline aluminum powder. The microstructural evolution after different HIP conditions and the mechanical properties were studied by means of transmission electron microscopy (TEM) and compression tests completed by Vickers microhardness measurements.

2 Experimental Procedures

2.1 The as-received Material

The Al nano-powder (n-Al powder) was supplied by Argonide Corporation (USA), obtained by electro-explosion of wire ('ELEX' process) [9]. This process involves applying a very high current over a very short time through thin metallic wires, to convert them into a plasma contained by the very high electro-magnetic field. When the field disappears the plasma expands with supersonic velocity. The extremely fast cooling rate (10^{-6} to 10^{-8} °C/s) provides conditions for the creation of nanometric powder. The main characteristics of powder particles are a spherical shape and an average diameter of about 80 nm.

2.2 The Process of Hot Isostatic Pressing (HIP)

2.2.1 The Capsule Preparation

To get a homogeneous distortion and therefore a uniform densification during hipping, the capsule is designed to have:

- a uniform temperature in the powder.
- a sufficiently weak capsule thickness to allow isotropic contraction and the transmission of the pressure to the powder.

304L and 316L stainless steels were used as containers of the powder during hipping (figure 1). The container (1) is 20 mm in diameter, 30 mm in height and 0.5 mm in thickness. It has a top plug (A) with a small ventilation tube (3) for degassing and pumping, and a bottom part (B). The powder height in the capsule is 10mm.

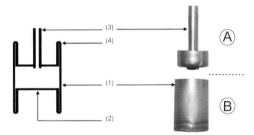

Figure 1: HIP capsule

After having filled the bottom part with powder (about 4.5 g) using an hydraulic press – in order to force the out-flow of powder grains to reach minimum initial relative density of 50 % – the container was covered with the top plug and welded.

Then, the capsule was pumped to extract gases introduced during filling, because the presence of such gases within the powder during hipping modifies the densification kinetics with the creation of porosity. It should be noted that with nanometric powders, paths for gas molecules to escape from the capsule are long and complex. To reduce the pumping time and to increase degassing, the container is heated to 100 °C during pumping. It takes about 120 hours to reach a vacuum lower than 10^{-5} mbar.

This geometry of capsule is used for different reasons:

- The small quantity of matter used for a test.
- The possibility to follow the distortion of the capsule during the HIP process.

2.2.2 The Hot Isostatic Press

The main element of the device of HIP is the high pressure head that can support a pressure of 300 MPa (figure 2). The pressure is applied by the inert gas argon. The internal equipment of the press is composed of three elements: a furnace (A), a basis of the furnace support (B) and a candle containing the expansion device (C). The particularity of the compaction device is the presence of an expansion cell [10] within the hot press. It permits the measurement of the capsule height or width during compaction and information on the sample density. It represents a very large time saving for the determination of development conditions and for monitoring HIP cycles.

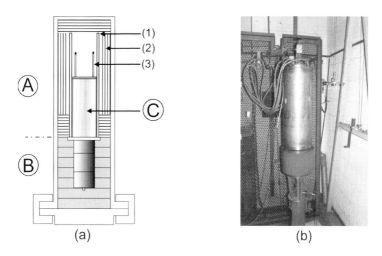

Figure 2: Diagram of the internal equipment of the HIP device (a) and general view of closed hot press (b)

The furnace (A) is constituted of a kanthal coil (a commercial heating fibre) (1) allowing for a temperature of 125 °C with a maximum heating rate of 600 °C/hour. The thermal insulation is assured by several metallic parts (2). The temperature is measured by two thermocouples (3). The diameter and the height of the useful zone is respectively 40 and 100mm. The basis for the support of the furnace (B) is closed by the metallic elements; it constitutes a cold zone of homogeneous temperature in which is arranged the set of electric connections for measurements. The candle (C) is constituted of two thermal insulator materials. In the hot zone, it acts as a support to the expansion cell and to the capsule. The displacement sensor of is arranged in its bottom part.

2.2.3 Experimental Characterization Techniques

The microstructure and the mechanical properties of the as-processed samples were investigated. TEM characterization was conducted via a Philips EM300 apparatus operating at 100 kV and on a JEOL 200CX operating at 200 kV. The mechanical properties were conducted mainly by compressive tests at a strain rate of $10^{-4}\,s^{-1}$ and completed by Vickers microhardness measurements.

3 Results and Discussion

3.1 Powder Characterization

The chemical composition of n-Al powder is Al-0.16Si-0.073Fe-0.028Ti-0.027Cu-0.01Mn-0.013Mg. The size (diameter) distribution was measured by manual counting from a population of about 500 particles. So, the average particle size measured is 80 nm but coarse particles of few µm in diameter were occasionally observed. Figure 3a is a TEM micrograph of the powder, and figure 3b represents the size distribution. The spherical shape of the particles appears on figure 3a, and TEM selected area diffraction patterns indicate that each individual particle is a single crystal.

(a)

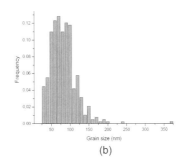
(b)

Figure 3: TEM micrograph of Al particles (a) and experimental particles size distribution (b)

Moreover, the presence of a 2 to 5 nm thick amorphous oxide layer on the particles surface has been reported [4,9,11,12]. This oxide layer is amorphous alumina (Al_2O_3), formed by contact with water or humid atmosphere. A recent work on molecular-dynamics simulation on the dynamics of oxidation of aluminum nanoclusters [13] shows that the thickness saturates at about 4 nm. Whereas the density is close to 3.18 g cm^{-3} for amorphous alumina, 3.2 g cm^{-3} for γ-Al_2O_3 and 3.98 g cm^{-3} for α-Al_2O_3, this study gave a value of 2.9 g cm^{-3} for the oxide layer over Al nanoclusters.

A calculation based on the particles size distribution, with an oxide layer thickness of 4nm, gives a volume percentage close to 30 % of oxide. Taking 2.9 g cm^{-3} as density of the oxyde, the average mass density for the aluminum powder is found to be about 2.76 g cm^{-3}.

3.2 Compacted Samples

For the capsule geometry used in this study, sample's shape after HIP are discs of 5mm height and 16mm in diameter. But HIP process permits to have larger samples.

During HIP there occur 4 different stages which are:

- The cold compression stage.
- The temperature increase ramp (inducing an increase of pressure).
- Temperature and pressure plateau stages.
- The temperature and pressure decrease stages.

The pressure of cold compaction is not an independent variable of the HIP cycle. Its determination takes into account the plateau's pressure and temperature. As the gas volume for HIP is constant, it is its dilation that increase the pressure from cold pressure to plateau's pressure. So, there are 4 process variables to monitor HIP; these are the temperature ramp rate, the plateau's temperature (PT °C), the plateau's pressure (PP) and plateau duration (Pt). Table 1 shows the 2 types of HIP cycles used here.

Table 1: HIP samples elaboration conditions

Sample	Ramp rate (°C/hour)	PT °C (°C)	PP (MPa)	Pt (min)
Type A	375	250	220	100
Type B	375	550	100	210

Type A is a low temperature and high pressure HIP cycle and type B is a high temperature and low pressure cycle. Plateau's duration is fixed when the capsule distortion doesn't change which means that the maximum densification is reached.

To measure the density of HIP samples an Archimedes method was used. The determined densities were close to 97 % (relative density) for type A and higher than 99 % for type B samples.

3.3 TEM Observations

The microstructure of the processed bulk materials are displayed in Figure 4 where bright field images are presented along with dark field ones for more insight.

The striking feature in samples from type A is the presence of pores and the still spherical shape of the grains. This suggests the samples to be not fully compacted under the operative conditions. Some large grains could also be seen within which dislocation entanglement oc-

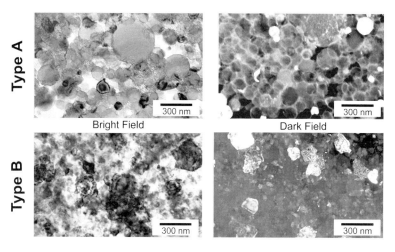

Figure 4: TEM micrographs of the 2 type samples obtained by HIP

cured. Samples of type B, on the other hand, showed a microstructure departing from type A. It consists of polygonal grains of about 150 nm which are randomly oriented as shown by the image contrast. No porosity was observed, in accordance with the calculated theoretical density of 100 %. Some large grains were also observed but in very small proportions. Moreover, a close inspection of the microstructure of samples B showed that a phase transformation from amorphous alumina to cristalline γ-Al_2O_3 has occurred during the processing, in accordance with earlier work [9]. This transformation which occurs at about 500 °C permits the densification of the Al powder through the breaking of the amorphous film that surrounds Al powder but is also accompanied by a slight grain growth.

3.4 Mechanical Characteristics

Compressive and Vickers microhardness tests were performed to characterize the mechanical properties of the HIPed samples. The results are shown in table 2 below.

Table 2: mechanical characteristics of HIP samples

Sample	Relative Density	Grain size(nm)	$\sigma_{0.2}$ (MPa)	Maximum stress (MPa)	Strain to failure	Vickers microhardness (HV)
Type A	97 %	70*	480	480	0.2 %	110
Type B	> 99 %	~150**	270	400	> 10 %	120
C-Al	100 %	50	50	20		

* X-Ray Diffraction with Warren-Averbach method, ** TEM, grain between 50–300nm

The compressive and the Vickers micro hardness tests confirm that the decrease of the grain size induced an increase of the yield strength. An increase of strain to failure is observed, between type A and type B. The behavior of type A sample is brittle, but for type B a plastic deformation is observed. This may be due to the slight increase in the grain sizes in sample B along with the crystallization of the amorphous alumina film.

4 Conclusions

The ALEX nano-powder supplied by Argonide has an average grain size of 80 nm. The oxide layer on the particles surface prevents the densification of the powder if optimum conditions are not reach. Nevertheless we have successfully processed bulk nanocrystalline aluminium-alumina composites the mechanical properties of which are well above that of conventional grain-sized aluminium sample showing that the HIP method is a mature alternative method of producing such materials.

5 References

[1] H. Gleiter., Progress in Mater Science 1989, 33, p. 223
[2] M. Legros, B.R. Elliott, M.N. Rittner, J.R. Weertman, K.J. Hemker, Phil. Mag. 2000, 1017–1026
[3] D. Jia, K.T. Ramesh, E. Ma, Scripta Mat. 2000, 73–78
[4] X.K. Sun, H.T. Cong, M. Sun, M.C. Yang, NanoStructured Materials 1999, 11, 917–923
[5] V.L. Tellkamp, E.J. Lavernia, NanoStructured Materials 1999, 12, 24–252
[6] M.N. Rittner, J.A. Eastman, J.R. Weertman, Scripta Metall. et Mater. 1994, 31, 841–846
[7] E. Bonetti, L. Pasquini, E. Sampaolesi, NanoStructured Materials 1997, 9, 611–614
[8] H.V. Atkinson, S. Davies, Metall. Mater. Trans. A 2000, 31A, 2981–3000
[9] F. Tepper, Metal Powder Report 1998, 31–33
[10] C. Rizkallah, J.P. Fondère, H.F. Raynaud, A. Vignes, La revue de métallurgie 2001, 1109–1128
[11] Y. Champion, J. Bigot, NanoStructured Materials 1998, 10, 1097–1110
[12] T.G. Nieh, P. Luo, W. Nellis, D. Lesuer, D. Benson, Acta Metall. 1996, 44, 3781–3788
[13] T. Campbell, R.K. Kalia, A. Nakano, P. Vashishta, S. Ogata, S. Rodgers, Physical Review Letters 1999, 82, 4866–4869

Characteristics of Nano Grain Structure in SPD-PM Processed AISI304L Stainless Steel Powder

H. Inomoto[1], H. Fujiwara[2] and K. Ameyama[3]
[1] Ritsumeikan University, Kusatsu, Shiga, Japan
[2] Department of Environmental Systems Engineering, Kochi University of Technology, Tosayamada, Kochi, Japan
[3] Deptartment of Mechanical Engineering, Ritsumeikan University, Kusatsu, Shiga, Japan

1 Introduction

Grain refinement is very effective for improving mechanical properties as well as the workability of materials. There are some severe plastic deformation processes used to produce nanocrystalline materials, such as ARB (Accumulative Roll Bonding) [1], ECAP (Equal Channel Angular Process) [2], HPT (High Pressure Torsion) [3] and SPD-PM (Severe Plastic Deformation – Powder Metallurgy) Process [4]. The SPD-PM process is a novel powder metallurgy process combining Mechanical Milling (MM), Mechanical Alloying (MA), heat treatment and sintering processes. It enables one to produce a nano grain structure quite easily. However, the formation of nanocrystalline grains in the SPD-PM process is not very well understood. The main reason is that MM powder can coalescence easily and deform in a complicated manner during MM. The objective of the present paper is to clarify the mechanisms of nano grain formation and the phenomena which accompany the grain refinement in AISI304L austenitic stainless steel. To reveal the microstructural changes during MM process, a comparatively large powder of about 0.9mm diameter is used, and reduced MM energy was provided.

2 Experimental Procedure

AISI304L (C: 0.019, Si: 0.22, Mn: 1.89, P: 0.035, S: 0.014, Ni: 9.61, Cr: 19.80, Fe: Bal. (mass%)) powder with the particle size of approximately 0.9 mm were produced by PREP (Plasma Rotating Electrode Process). The PREP has the advantage that the product powder is hardly contaminated by impurities such as oxygen or nitrogen gases during the process [5]. A Fritsch P-5 planetary ball mill with AISI304 stainless steel vials and balls were used for mechanical milling under an Ar atmosphere. The PREP powder was loaded into the vial, inside a glove box kept under an atmosphere of purified Ar gas. Owing to cooling fins attached to the outside wall of the vials and a strong ventilating system, the temperature of the vials as well as the milling powder was kept below 323 K [4]. A ball-to-powder weight ratio of 1.8:1 was chosen and the milling intensity was adjusted to a rotation speed of 4.2 rps (250 rpm). MM powders after milling times of 360 ks and 720 ks were characterized by means of X-ray diffraction (XRD), Scanning Electron Microscopy (SEM), Electron Probe Micro Analyzer (EPMA) and Transmission Electron Microscopy (TEM).

3 Results and Discussion

3.1 Microstructural Changes during MM

Figure 1 shows SEM micrographs of PREP powder before MM process {(a), (b)} and after milling for 720ks {(c), (d)}. Figs. 1(b) and (d) are magnified sections near the powder surface indicated in Figs. 1 (a) and (c) respectively. As can be seen in Fig. 1 (d), a surface layer with the thickness of approximately 100μm has formed after the MM process. Figure 2 indicates changes of the width of the surface layer and the particle size of the powder with MM time. Although the particle size is kept almost constant, the width of the surface layer increases with increasing MM time. Usually, MA or MM process produces fine particles by crushing the powder when the powder is as small as tens of microns, and those crushed fine particles tend to stick together. However, in the present study we used a large size of PREP powder and the milling energy is relatively small, and we could not find any fine particles in the vial after MM, so that it can be concluded that the growth of the surface layer took place not by adhesion of the peeled off MM fine particles but by severe plastic deformation of the powder surface. These imply that no coalescence of the powders occurred under the given MM conditions and deformation progresses from the surface to the inside of the powder. Therefore, a careful examination of the microstructure from the surface to the inside region of powder enables to make the microstructural change more clear. Figure 3 demonstrates the microstructure and Vickers hardness profile of the AISI304L PREP powder milled for 720 ks. The hardness of the surface layer was twice as large as that in the inner region of the powder. These results indicate that the surface layer structure formed not by contaminations from outside of the powder but by the severe plastic defor1mation of the powder itself.

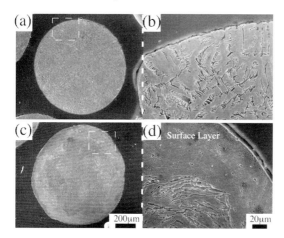

Figure 1: AISI304L PREP Powder MM for 0 ks (A, b) and 720 ks (c, d)

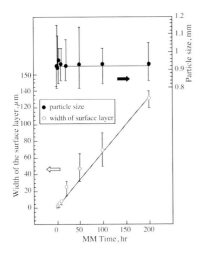

Figure 2: changes in the width of the surface layer and particle size with MM time

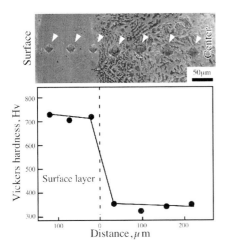

Figure 3: Microstructure and Vickers hardness profile in AISI304L PREP Powder MM for 720 ks

Figure 4 shows a TEM micrograph of the surface layer of the powder milled for 720 ks. A BCC nano equiaxed grain structure with grain sizes less than 15 nm are observed. Figure 5 shows a TEM micrograph of the boundary region of the surface layer (left side of the photo) and the inner area (right side) of the powder. In the surface layer side, elongated grains ranging from 20 nm to 100 nm and aspect ratios from 3 to 5 are observed. These grains have large misorientation angles. Inside of the elongated nano layered grains, high angle boundaries formed by the fragmentation of the grains. Additionally, subgrain boundaries were often observed. Such a nano layered grain structure was observed in the vicinity of the grains with Dislocation Tangled

High-angle Boundaries (DTHB) in all specimens. In the interior part of the powder, a microstructure with smaller dislocation density and coarse grains appears. In other words, the closer the particle center is approached, the more the earlier stages of the MM process are observed.

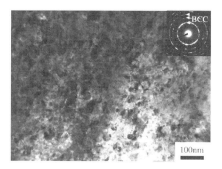

Figure 4: TEM Micrograph of AISI304L PREP Powder MM for 720 ks (surface layer)

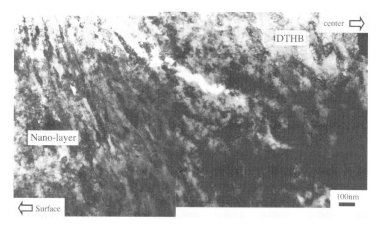

Figure 5: TEM Micrograph of AISI304L PREP Powder MM for 720 ks (inner layer)

Figure 6 shows TEM micrograph and the selected area diffraction pattern (SADP) of the inner area of the powder (several μm from the boundary) showing that a lath martensite structure with low angle boundaries has formed.

Figure 7 summarizes the formation mechanism of the nano-grain structure. At the initial stage of MM {Fig. 7 (a)}, martensite transformation takes place producing a lath grain structure with low angle boundaries. As the MM proceeds {Fig. 7 (b)}, the deformed lath grains have high angle boundaries due to the fragmentation or sub-division of the lath grains. High angle grain boundary formation by deformation is also observed in the heavily cold rolled AISI304 bulk material. In the middle stage of MM {Fig. 7 (c)}, those grains become a cell-like structure with a high angle boundary, that is, a Dislocation Tangled High Angle Boundary (DTHB) structure. The DTHB grain diameter is approximately 100 nm, and it changes to a pan-cake like structure, i.e., nano layered structure, in the latter stage {Fig. 7 (d)}. Finally, the nano equiaxed grain structure forms by fragmentation of the nano layered grains {Fig. 7 (e)}. Fragmentation or

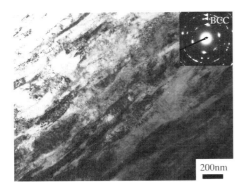

Figure 6: Micrograph of SUS304L PREP Powder MM for 720 ks (boundary region of the surface layer)

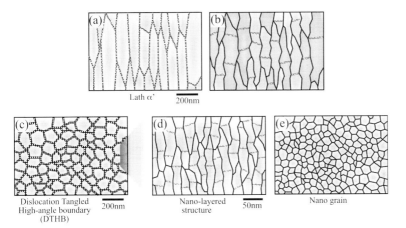

Figure 7: Scheme of grain refinement in SUS304L by MM

sub-division of the spherical DTHB grains with diameter less than several tens nm seems to be difficult because dislocations are unstable to stay inside of DTHB grain and eliminate to the DTHB boundary due to the elastic energy when the DTHB grain becomes finer. On the other hand, formation of a pan-cake shaped grain such as a nano layered grain from the DTHB grain by compressive stress is much easier. Therefore, nano layered grain formation and its fragmentation is necessary for the nano equiaxed grain structure formation. It can be emphasized that the nano layered structure is essential for the nano equiaxed grain structure formation. The observation that the nano equiaxed grain structure is always in the neighborhood of the nano layered structure strongly supports this hypothesis.

3.2 Other Phenomena Enhanced by MM

Figure 8 shows the relationship between milling time and the Vickers hardness of the powder. Circled and squared symbols correspond to hardness of the inner side and surface side of the

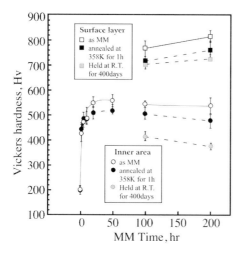

Figure 8: Change in Vickers hardness with milling time

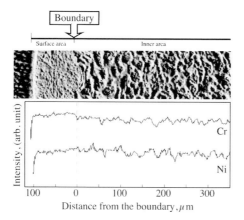

Figure 9: Change in Cr, Ni contents in AISI302L Prep powder MM for 720 ks (EPMA)

boundary, respectively. As mentioned above, hardness in the surface layer was about twice as large as that in the inner region of the powder. The hardness of the inner area of milled powders reaches approximately the same value of the cold rolled specimen after milling for 10 hours and finally becomes constant after 20 hours of milling. It is noted, as indicated in the figure, that the hardness of the inner area of powders milled for longer than 20 hours decreases with MM time after annealing at 358 K (85 °C) for 60 minutes. Moreover, a significant decrease of the hardness was observed after storage at room temperature for 400 days. These results strongly indicate that recovery of dislocations took place even close to room temperature in the steels, and extremely high density of vacancies have been introduced by the MM process.

Figure 9 shows Cr and Ni contents obtained by EPMA. The inside area of the powder indicates inhomogeneous concentration profiles of Cr and Ni caused by solidification of a liquid

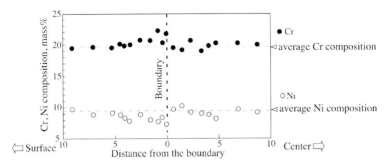

Figure 10: EDS results of AISI304L PREP powder MM for 720 ks (Each analyzed area ~ 300 nm ϕ)

phase at during powder production, while homogeneous Cr and Ni profiles are observed in the surface layer. The homogeneity of the alloying elements in the surface layer can be attributed to the increase of diffusion path, such as grain boundaries and point defects. Nano grain formation enhances a homogeneous microstructure formation.

Figure 10 shows Cr and Ni profiles in the vicinity of the boundary region obtained by TEM/EDS analysis. A gradual change of the Cr and Ni contents was observed through the boundary area. Increase of Cr and decrease of Ni in the nano layered structure were observed, and it reversed in the DTHB region. This composition change presumably occurred by the heat of friction at the boundary since the hardness is quite different between the nano layered structure and the DTHB structure. In addition, it is important to point that extremely high density of defects possibly accelerates diffusion of those elements. Moreover, the nano grain structure also increases the free energy of the matrix phase, and thus it produces enough driving force for the transformation. Recently, the authors indicated that nano grain formation induced ferrite transformation in the austenitic stainless steel. When the grain size is reduced to 15 nm, the driving force more than 1000 kJ/mol arises only by the increase of grain boundary area [6]. The amount is almost comparative to the martensitic transformation in steels.

4 Summary

The SPD-PM process such as a Mechanical Milling (MM) process is very effective on nano grain formation. The MM process was applied to an AISI 304L PREP powder with the particle size of approximately 0.9mm. The microstructure change and other phenomena enhanced by the MM were examined by means of TEM/EDS, SEM, EPMA and XRD. The obtained results are as follows;

1. Nano equiaxed grains, with grain sizes approximately 15 nm, are formed in the surface region of the MM powder. The surface layer grows with MM time not by the contamination, but by the severe plastic deformation of the powder itself.
2. Nano exquiaxed grain structure essentially requires the nano layered grain formation and its fragmentation.

3. Decrease of the hardness of the MM powder annealed at 358K for 60min suggests that the low temperature (~0.2 Tm) recovery easily occurs by introduction of extremely high density of point defects.
4. Local partitioning of Cr and Ni elements arose at the boundary region implies that the extremely high density of defects such as nano grain boundaries as well as point defects possibly accelerate the diffusion of these elements.

5   References

[1]   Y. Saito, N. Tsuji, H. Utsunomiya, T. Sakai and R. G. Hong: Scripta Mater., 39, 1998, 1221
[2]   Z. Horita, M. Furukawa, T. G. Langdon, and M. Nemoto: Materia Jpn., 47, 1998, 767
[3]   Z. Horita, D. J. Smith, M. Furukawa, M. Nemoto, R. Z. Variev and T. G. Langdon: J. Mater. Res., 11, 1996, 1880
[4]   K. Ameyama, N. Imai and M. Hiromitsu: Tetsu-to-Hagane, 84, 1998, 357
[5]   K. Isonishi, M. Tokizane: Tetsu-To-Hagane, 76, 1990, 50
[6]   H. Fujiwara, H. Inomoto, R. Sanada and K. Ameyama; Scripta Materialia, 44, 2001, 2039

Formation of Powder and Bulk Al-Cu-Fe Quasicrystals, and of Related Phases During Mechanical Alloying and Sintering

S.D. Kaloshkin[1], V.V. Tcherdyntsev[1], G. Principi[2], A.I. Laptev[1], E.V. Shelekhov[1], T.Spataru[2]
[1]Moscow State Institute of Steel and Alloys, Moscow, Russia
[2]Settore Materiali and INFM, DIM, Padova, Italy

1 Introduction

In recent years, quasicrystalline (QC) phases, due to complex of unusual properties, have attracted much attention as promising base for new materials. At present various kinds of techniques are considered for QC production [1–4]. A number of papers report that QC phases can be produced by mechanical alloying (MA) of elemental components or by subsequent annealing of MA powders [5, 6].

Due to particularities of crystallographic and electronic structure properties of QC phases are close to those of ceramics, they are brittle, hard, have low friction coefficient [7]. So, QC materials could find application fields at least similar to that of ceramics. Application of powdered QC may be realised in different types of composite materials with metallic [8, 9] or some organic matrixes [10]. Nowadays the MA process is applied for creation of new composite materials with high performance properties, it allows significant structure refinement opening the way to produce composite materials with very fine structure. Such in general faulty properties of MA QC powders as defective structure and irregular shape of particles could be insignificant or even desirable in the case of using the MA technique for composites preparation. Due to these properties the QC powders can be easily refined with establishment of good bonding with the composite matrix.

For Al-Cu-Fe system the complicated chain of solid state transformations develops at heating of as-milled compositions, finally resulting compete transformation to QC phase or, because of little deviations of alloy composition from the nominal one, to its mixture with other approximant phases of this system. It was found that the increase of time of prior MA treatment led to formation of QC phase at lower temperature during postheating [11]. However, according to Mössbauer study structure of the MA QC samples is very defective and becomes more regular only after rather high temperature annealing [11, 12].

This paper, in continuation of the previous research on Al-Cu-Fe QC preparation by MA technique [11–14] and investigation of possibility to obtain bulk materials by precursor powder compaction [14], presents the first results on obtaining the Al-based composite materials reinforced by the Al-Cu-Fe QC phase. As the final product of the MA technique is powder, the bulk material can be obtained by low temperature consolidation with subsequent annealing or by hot consolidation. The higher the temperature of thermal treatment of bulk specimens, the better sintering of powder particles and higher the mechanical properties. However, the Al-Cu-Fe QC phase can dissolve in aluminium matrix at annealing with failure of useful properties. This is the point of special interest of present investigation.

2 Experimental Procedure

Powders of carbonyl iron annealed in hydrogen (99.95 %), copper (99.9 %) and aluminium (99.0 %) were used for the experiments. A MAPF-1 planetary ball mill with two vials (volume 1000 cm^3 each) of hardened carbon-chromium steel, running at a vial rotation velocity of 600 rpm was employed for composites preparation. Balls of 9 mm in diameter were used, with a ball-to-powder mass ratio of 10:1. MA was proceeded in Ar atmosphere.

A DRON-3 diffractometer with Co–K$_\alpha$ radiation was used for the X-ray diffraction (XRD) experiments. The phase composition was determined by the reduced Rietveld method, i.e. by fitting of the full experimental X-ray powder diffraction patterns.

A CamScan scanning electron microscope (SEM) was used to investigate the microstructure of the samples. The microhardness of the consolidated samples was measured by Vickers indentation using a PMT-3 machine with a tetrahedral diamond pyramid, applying the load for a duration of 5 s.

Room temperature Mössbauer measurements were carried out by means of a constant acceleration spectrometer with a ^{57}Co:Rh source. The isomer shifts were calculated with respect to the centroid of the α-Fe spectrum. Current minimization routines were used to obtain the best fit of the spectra.

Hot consolidation of samples of 4 mm in diameter was performed under the pressure of 4.5 GPa with heating up to the temperatures of 800, 600 or 300 °C, and holding at these elevated temperatures for 10 s. Microhardness of the consolidated samples was measured by Vickers indentation using a PMT-3 machine with a tetrahedral diamond pyramid, applying the load during 5 s. Compression tests were performed using Zwick 1474 machine. Tribological characterization of samples was performed using the IMASh setup under loading of 45.4 N, the wear intensity was calculated as the ratio between the linear wear and the total wear path of rotating counterpart.

3 Results and Discussion

3.1 Consolidation of Al$_{65}$Cu$_{23}$Fe$_{12}$ Milled Powders

The Al$_{65}$Cu$_{23}$Fe$_{12}$ mixtures were prepared by milling of elemental powders during 1 h. The as-milled powders represents a mixture of pure elements and some amounts of intermetallic phases, namely Al$_2$Cu and metastable Al(Cu,Fe) cubic phase. As we showed previously [11–14], annealing of such as-milled samples in furnace at 600–800 °C results in the transformation of this mixture of phases into single-phase QC powder. The aim of the present study was to investigate the possibility of bulk QC samples formation by hot consolidation of the as-milled powders. Before the consolidation, the powder was classified into two fractions: the coarse one, with the powder size of 60–100 m and the fine fraction, with the powder size of 20–60 m. Figure 1a and b show the X-ray diffraction patterns for the samples consolidated at 800 °C from these fraction, correspondingly. Formation of the icosahedral quasicrystals was observed for both powder fractions during the hot consolidation process. Quantitative phase analysis gives the appearance of 88.3 and 91.1 vol.% of quasicrystalline phase for coarse and fine fractions, correspondingly.

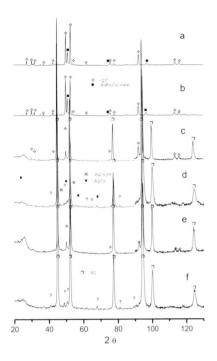

Figure 1: X-ray diffraction patterns for the consolidated samples. (a) – QC, coarse fraction; (b) – QC, fine fraction; (c) Al + 20 wt.% QC, 300 °C; (d) Al + 20 wt.% QC, 600 °C; (e) Al + 10 wt.% QC, 300 °C; Al + 10 wt.% QC, 600 °

Mössbauer spectra of these samples are presented in Fig. 2 a and b, correspondingly. They represents a symmetrical doublet, which is typical for the Al-Cu-Fe quasicrystals [15]. The quadrupole splitting distribution has the shape close to the obtained for spectra of the mechanically alloyed Al-Cu-Fe powder, annealed at 700 °C [12].

The results of microhardness measurement are presented in Table 1. The Vickers microhardness values for both coarse and fine fraction samples were obtained to be of about 5 GPa. This value is significantly higher than obtained previously magnitude 2.3 GPa [40] for the similar samples prepared by cold and hot consolidation at relatively low pressure. On the other hand, the values obtained in the present study are smaller than the magnitude of 7.85 GPa, which was reported for cast bulk samples [16]. From this comparison, we can assume that even applying of such high pressure do not allow to avoid the porosity of samples. Figure 3 a shows the SEM micrograph, which confirms the porous structure of the bulk QC sample. In contrast to the previously observed results for samples consolidated al low pressure [14], where the pore size reached 20 μm, now the pore size is less than 1 μm. This may be a reason that we obtained the higher value of microhardness in the present study.

Light fields in Fig. 3 belong to the regions enriched in Cu and Fe, and dark fields – to those enriched in Al. So, we can conclude that the dark fields corresponds to the QC phase and light field – to the metastable cubic phase Al(Fe, Cu) which contains about 50 at.% of Al. The location of light fields indicates that the cubic phase forms predominantly on the interparticle boundaries or close to pores.

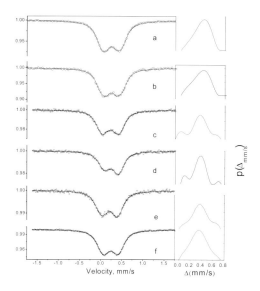

Figure 2: Mössbauer spectra of the consolidated samples. (a) – QC, coarse fraction; (b) – QC, fine fraction; (c) Al + 20 wt.% QC, 300 °C; (d) Al + 20 wt.% QC, 600 ° (e) Al + 10 wt.% QC, 300 °C; Al + 10 wt.% QC, 600 °C

3.2 Formation and Structure of Al – QC Compositions

The Al powder was milled together with the preliminary obtained quasicrystalline powder of fine fraction in the ratios of Al-20 wt.% QC and Al-10 wt.% QC. The duration of milling was 15 min. The consolidation of samples was performed at temperatures 300 and 600 °C. Figure 1 c–g shows the X-ray diffraction patterns of these samples. For both tested compositions, after consolidation at 300 °C the quasicrystalline phase remained in the samples structure, whereas the consolidation at 600 °C results in the disappearing of the QC diffraction peaks and formation of crystalline phases, such as Al_2Cu and Al_7Cu_2Fe. Also some diffraction peaks, which cannot be related to any known phase of this system, appeared after consolidation at 600 °C.

Table 1: Mechanical and tribological properties of the consolidated samples (for definition of quantities, see text)

Sample	H_v, GPa	σ, MPa	f	$I, \times 10^7$
QC, fine fraction	4.66 ± 0.47	–	–	–
QC, coarse fraction	5.17 ± 0.52	–	–	–
Al + 10 wt.% QC, 300 °C	0.63 ± 0.06	178 ± 23	0.24 ± 0.02	4.79 ± 0.12
Al + 10 wt.% QC, 600 °C	0.98 ± 0.10	228 ± 7	0.21 ± 0.02	3.67 ± 0.11
Al + 20 wt.% QC, 300 °C	0.51 ± 0.05	–	–	–
Al + 20 wt.% QC, 300 °C	0.87 ± 0.09	–	–	–
Pure Al	0.28 ± 0.06	76 ± 5	–	–

Mössbauer spectra of the composite samples are presented in Fig. 2 c–f. The spectra for samples consolidated at 300 °C remains symmetrical. For the sample Al-10 wt.% QC the consolidation at 600 °C results in the appearance of significant asymmetry of the spectra, which is typical for the cubic phase of this system [17]. So, the disappearance of QC phase by consolidation at 600 °C may be also confirmed by Mössbauer spectroscopy.

The microstructure of the composite samples is shown in Fig.3, b–d. One can see that besides rather large particles of reinforcing phase of about 20 µm there are also very small particles of less than 1 µm. These particles are distributed in the structure not uniformly – the micrograph on Fig.2,c shows the fields with relatively low and relatively high density of particles. The difference in compositions of samples obtained at 300 °C and 600 °C can be hardly found from the samples microstructure (compare Fig. 3, b and d). However, the sample annealed at 600 °C has no quasicrystalline phase, it transformed into the crystalline one because of interaction with aluminium matrix. Both alloys prepared by consolidation at 300 °C contain almost untransformed quasicrystalline phase.

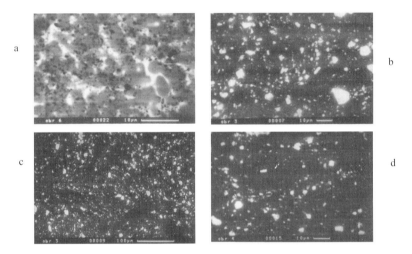

Figure 3: SEM micrographs (back-scattering mode) of the consolidated samples. (a) – QC, coarse fraction; (b and c) Al + 20 wt.% QC, 300 °C; (d) Al + 10 wt.% QC, 600 °C

The values of Vickers microhardness for the composite samples are presented in Table 1 (see column "H_v"). The values for samples consolidated at 600 °C are higher than for that consolidated at 300 °C, which may be associated with the dissolution of Fe and Cu in Al matrix at high temperature. Compression experiments show that crushing of samples proceeds by formation of the extensive cracks in the direction of applied loading. Increase in the consolidation temperature not only results in an increase of compression strength σ, but also in an improvement of triboligical properties, namely, decrease in the friction coefficient f and wear intensity I, as it is seen from Table 1. These temperature-induced changes of properties seem to be associated with the change in phase composition of the samples, too.

4   Summary

Materials with bulk quasicrystalline Al-Cu-Fe phase as well as Al-based composite materials reinforced by quasicrystals were prepared. The MA technique was used for preparation of precursor powder for QC phase from elemental powders and for aluminum based composites preparation. Hot high-pressure consolidation of powders into bulk samples was performed. It was found that the bulk quasicrystalline samples had porous structure with Vickers microhardness of 5 GPa. The composite bulk samples had dense structure, however the distribution of reinforcing phase was not uniform for used regimes of MA processing. Increase of the temperature of hot consolidation resulted in interaction of quasicrystalline phase with aluminium matrix and its transformation into the crystalline one.

5   Acknowledgements

The authors wish to thank the ISTC organization for financial support wihtin project 1968.

6   References

[1]   A. P. Tsai, A. Inoue, T. Matsumoto, Mater. Trans. JIM. 1989, 40, 666–676
[2]   T. J. Sato, H. Takakura, A.P. Tsai, Jpn. J. Appl. Phys. 1998, 37, L663–L665
[3]   T. Klein, O.G. Simko, Appl. Phys. Lett. 1994, 64, 431–433
[4]   A. P. Tsai, MRS Bulletin 1997, 22, 44–47
[5]   E. Yu. Ivanov, I.G. Konstanchuk, B.D. Bokhonov, V.V. Boldyrev, Reactivity of Solids. 1989, 7, 167–172
[6]   N. Asahi, T. Maki, S. Matsumoto, T. Sawai, Mater. Sci. Eng. A. 1994, 181, 841–844
[7]   K. Urban, M. Feuerbacher, M. Wollgarten, MRS Bulletin. 1997, 22, P. 65-68
[8]   A. Inoue in New Horizons in Qusicrystalline Research and Application (ed. A.L. Goldman, D. J. Sordelet, P.A. Thiel, J.M. Dibous), World Scientific, Singapore, 1997, p. 256
[9]   A .P. Tsai, K. Aoki, A. Inoue, T. Masumoto, J. Mater. Res. 1993, 8, 5–7
[10] P. D. Bloom, K.G. Baikerikar, J.U. Otaigbe, V.V. Sheares, Mater. Sci. Eng. A. 2000. 294-296. 156-159
[11] V. Tcherdyntsev, S.D. Kaloshkin, A.I. Salimon, I.A. Tomilin, A.M. Korsunsky, J. Non-Cryst. Solids, 2002, 312-314, 522-526
[12] S. D. Kaloshkin, V.V. Tcherdyntsev, A.I. Salimon, I.A. Tomilin, T. Spataru, G. Principi, Hyp. Inter. 2002, 139/140, 399-405
[13] I. Salimon, A.M. Korsunsky, S.D. Kaloshkin, V.V.Tcherdyntsev, E.V.Shelekhov, T.A.Sviridova, Mater. Sci. Forum. 2001, 360-362, 373–378
[14] V. V. Tcherdyntsev, S.D. Kaloshkin, A.I. Salimon, E. A. Leonova, I. A. Tomilin, J. Eckert, F. Schurack, V.D. Rogozin, S.P. Pisarev, Yu. P. Trykov, Mater. Manufact. Proc. 2002, 17, 825–841
[15] R. A. Dunlap, D.W. Lawther, Mater. Sci. Eng. R. 1993, 10, 141–185
[16] E. Giaccometti, N. Baluc, J. Bonneville, G. Rabier, Scr. Mater. 1999, 41, 989–994
[17] C. L. Chien, M. Lu, Phys. Rev. B. 1992, 45, 12793–12796

Production and Consolidation of Nanocrystalline Fe Based Alloy Powders

M. Rombouts[1], L. Froyen[1], A.C.C. Reis[2] and L. Kestens[2]
[1] Department of Metallurgy and Materials Engineering (MTM), Katholieke Universiteit Leuven, Belgium
[2] Department of Metallurgy, Ghent University, Belgium

1 Abstract

The consolidation of nanostructured Fe-10 at% Cu and Fe-10 at% Mo powder produced by mechanical milling is investigated. In previous work, it was found that the thermal stability of the nanostructured Fe-10 at% Cu powder is rather low, making hot compaction impossible. Therefore, the powder is first annealed at relatively low temperature (400-500 °C), then cold pressed and finally warm rolled at 420 °C. The thermal stabilty of nanostructured Fe-10 at% Mo was found to be better. However, full densification by Hot Uniaxial Pressing (HUP) leads to the loss of the nanostructure. Therefore, partial densification by HUP followed by hot rolling or by Equal Channel Angular Extrusion, is used as an alternative processing route, in order to keep the grain growth under control.

2 Introduction

Nanocrystalline materials are widely used due to their enhanced properties when compared with coarse-grained polycrystalline materials [1–2]. Mechanical milling offers an opportunity to produce relatively large powders (> 10 µm) with a homogeneous nanoscale grain size for almost any material system.

Consolidating the powders into fully dense components while retaining their nanostructure remains a challenging subject. This is due to their high hardness making cold consolidation difficult [3]. It can also lead to grain growth and recrystallisation during hot compaction. Several consolidation methods have already been investigated like cold isostatic pressing (CIP) followed by hot isostatic pressing (HIP) [3], hot uniaxial pressing (HUP) [4], rapid forging [5], hot extrusion [6] and hot rolling (HR) [7]. In the present study the consolidation of mechanical alloyed Fe-10 at% Cu and Fe-10 at% Mo nanostructured powder is investigated. The addition of these alloying elements leads to a two-phase structure that enhances the thermal stability and thus the ability for compaction.

3 Experimental Procedures

The consolidation is focussed on nanostructured Fe-10 at% Cu and Fe-10 at% Mo powder, produced by mechanical alloying [8]. The applied processing routes are different due to their difference in thermal stability (Figure 1–2).

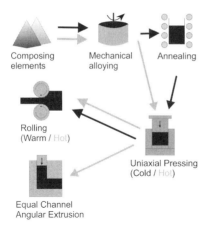

Figure 1: Investigated processing routes for the Fe-10 at% Cu alloy (black arrows) and the Fe-10 at% Mo alloy (gray arrows)

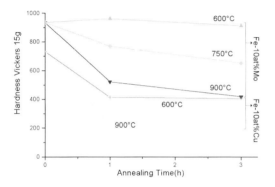

Figure 2: Thermal stability of Fe-10 at% Cu and Fe-10 at% Mo obtained by measuring the hardness after different annealing times [8]

3.1   Nanostructured Fe-10 at% Cu Powder

The Fe-10 at% Cu powder is produced by alloying in the horizontal high speed attritor (type ZOZ Simoloyer CM01-2l) under an Ar atmosphere using ethanol as PCA. The following cycle is repeated 18 times: 8 min at 1600 rpm + 2 min at 1400 rpm, leading to a total milling time of 3 h. The milling media are stainless steel balls (ϕ = 5 mm) with a powder to ball ratio of 1/10. The alloy has a high hardness, but a poor thermal stability (figure 2). Cold compaction is not possible due to the high hardness. The powder is therefore annealed at 450 °C for 30 min in an Ar atmosphere and then Cold Uniaxial Pressed (CUP) at room temperature at a pressure of 800 MPa. This compact is encapsulated in a plate and further processed by warm rolling at 420 °C with a reduction of resp. 25 % and 40 % in a single step.

3.2 Nanostructured Fe-10 at% Mo Powder

The Fe-10 at% Mo powder is produced using a high-speed horizontal attritor ball mill (Zoz Simoloyer CM01-2l) in an Ar atmosphere with ethanol as PCA. The following cycle is repeated 18 times: 6 min at 1600 rpm + 4 min at 1200 rpm, leading to a total time of 3 h. The milling media are stainless steel balls (ϕ = 5 mm) added with a powder to ball ratio of 1/10. These powders are very hard which makes cold pressing extremely difficult. However its structure is very well maintained at higher temperatures (Figure 2) making hot compaction attractive. HUP using a pressure of 40 MPa, a heating rate of 50 °C/min until maximum temperatures ranging from 800 °C to 950 °C and a holding time at this maximum temperature ranging from 3 to 15 min is performed.

After HUP the material is encapsulated in a plate, heated to 900 °C, held at that temperature for 5 to 10 min, hot rolled (25–75 % reduction in one step) and finally water quenched. Cross rolling, that means rotating the part 90° between 2 passes is also executed. Another way of post consolidation by Equal-Channel Angular Extrusion (ECAE) is also performed [9]. The HUP sample is encapsulated in a cylinder of IF-steel in order to fill the circular channels of the die which intersect at an angle of 90°. Up to 4 passes through the die are performed following route A and C: in the former the material is not rotated while in the latter the material is rotated over an angle of 180 °C between consecutive passes [10]. The processing temperature is 400 °C.

Investigation of the microstructure is performed by a Philips SEM XL 30 FEG scanning electron microscope. X-ray diffraction measurements are carried out with a Seifert diffractometer using Cu K$_\alpha$ radiation to determine the grain size out of the peak broadening. The measured data are fit to a Pseudo-Voigt function and the FWHM (full width at half maximum intensity of the peaks) are derived from the fit parameters of this function. The average crystallite size d is calculated using the Williamson-Hall method for size/strain separation.

$$\Delta B \cdot \cos\theta = \left(\frac{k\lambda}{d}\right) + \eta \sin\theta \text{ and } \Delta B = B_{exp} - \left(\frac{B_{instr}^2}{B_{exp}}\right) \quad (1)$$

with B_{exp} and B_{instr} the FWHM of respectively the measured and the reference material; λ is the wavelength of the XRD source; θ is the diffraction angle; k = 1; η = internal strain. As input the peaks resulting from the diffraction of the {110} and the {220} planes are used.

4 Results and Discussion

4.1 Consolidation of Nanostructured Fe-Cu Powder

Earlier investigations on hot isostatic pressing of Fe-10wt%Cu revealed that full density is achievable, but the median grain size increased from 18 nm to 130 nm [11]. After rapid forging a grain size of 50 nm was obtained [12].

The as-produced Fe-10at%Cu powder can not be cold compacted even not at a pressure of 800 MPa due to the high hardness which prevents the particles to be deformed plastically. The powder has to be annealed (at 450 °C for 30 min) in order to be able to perform CUP. The grain size increases from 21 nm to 47 nm [XRD] and the Vickers hardness decreases from 720 to 350 during annealing. The density after CUP is very low. Post compaction by warm rolling at 420 °C leads to insufficient densification due to the low temperature used. The Vickers hardness decreases further from 350 to 200 after a reduction of 25 % and to 300 after a reduction of 40 %.

4.2 Consolidation of Nanostructured Fe-Mo Powder

The as-produced nanocrystalline Fe-Mo powder is similar to the Fe-Cu powder too hard to be cold compacted. The investigated consolidation process is represented in figure 1.

4.2.1 Hot Uniaxial Pressing

The as-produced Fe-10at%Mo powder has an average grain size of 47 nm [XRD]. It is not possible to keep the nanostructure and attain a fully dense material by HUP alone (table 1).

Table 1: Density and hardness after HUP of Fe-10 at% Mo (original hardness of the powder is $H_v = 935$) at different temperatures and times and a pressure of 40 MPa.

Sample	temperature (°C)	time (min)	Relative density (%)	Vickers hardness
HUP1	950	15	92	400
HUP2	850	10	75	415
HUP3	850	3	53	530
HUP4	900	3	73	550

4.2.2 Hot Uniaxial Pressing Followed by Hot Rolling

As alternate compaction route, the material is first only partially densified by HUP at 850 °C during 3 min (HUP2, table 1, $d = 130$ nm [XRD]). Next, further consolidation by hot rolling at 900°C is performed (figure 3, table 2). The hardness is generally higher, the grain size lower (table 2) and the structure denser (figure 3) after hot rolling in comparison with the hot uniaxial pressed samples. A reduction of 75 % is too high to be applied in one pass resulting in a lower hardness and density in comparison with a reduction of 50 %. There are however still cracks present at the surface and inside the material, especially when the material is water quenched after hot rolling.

Table 2: The grain size [XRD] and the hardness after hot rolling of HUP Fe-10 at% Mo ($H_v = 415$, $d = 56$ nm [XRD]) as a function of the hot rolling parameters.

Sample	Temperature (°C)	time (min)	force (kN)	Reduction (%)	grain size (nm)	Vickers hardness
HUP2HR1	900	5	95–100	25	–	340
HUP2HR2	900	10	95–100	50	67	460
HUP2HR3	900	10	95–100	50 [CR2 [a]]	86	450
HUP2HR4	900	10	200	75	120	414

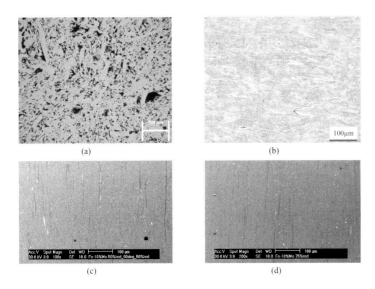

Figure 3: Fe-10 at% Mo powder after (a) HUP at 850 °C during 3 min and subsequent hot rolling at 900 °C, (b) 1 pass with a reduction of 50 %, (c) 2 passes each with a reduction of 50 % by cross rolling and (d) 1 pass with a reduction of 75 %

4.2.3 Hot Uniaxial Pressing Followed by Equal-channel Angular Extrusion (ECAE)

ECAE is also applied as a post consolidation step. The starting material for ECAE is the partially densified HUP2 sample of table 1. The main part of the material, especially near the inner corner of the die, shows already after 1 pass many cracks at prior particle boundaries (figure 4a). After 4 passes via route C, the material completely broke up. After 4 passes via rou-

Figure 4: Microstructure of Fe-10 at% Mo after HUP at 850 °C during 3 minutes and subsequent 1 pass through the ECAE die, (b) 4 passes through the ECAE die via route A

te A, local densification and deformation according to the extrusion direction is observed (figure 4b), but insufficient at the global level.

5 Conclusions

1. The poor thermal stability of the Fe-10 at% Cu alloy limits the temperature during consolidation. The thermal stability of the Fe-10 at% Mo alloy is good up to temperatures in the range of 750–900 °C.
2. In order to maintain a nanostructural aspect, the processing temperature of the Fe-10 at% Cu alloy has to remain rather low (~450 °C) leading to insufficient densification after cold pressing and warm rolling.
3. Hot uniaxial pressing at 850 °C during 3 min does not lead to a fully dense, nanostructured Fe-10 at% Mo alloy. However, post consolidation by hot rolling densifies the material while retaining the nanostructure. Alternatively, Equal Channel Angular Extrusion of partially densified Fe-10 at% Mo leads to many prior particle boundary cracks due to the high hardness of the particles and the poor bonding between the particles.

6 References

[1] Gil Sevillano J., Van Houtte P., Aernoudt E., Progress in Materials Science 1980, 25, p.69–412
[2] Valiev R.Z., Materials Science and Engineering 1997, A234–236, p.59–66
[3] Munitz A., Fields R.J., Powder Metallurgy 2001, 44, No.2, p.139–147
[4] Cao H.S., Hunsinger J.J., Elkedim O., Scripta Materialia 2002, 46, p.55–60
[5] Shaik G.R., Milligan W.W., Metallurgical and Materials Transactions 1997, 28A, p.895 904
[6] Morris M.A., Morris D.G., Material Science and Engineering 1989, A111, p.115–127
[7] Takaki S., Kimura Y., Journal of the Japan Society and Powder Metallurgy 1999, 46, No.12, p.1235–1240
[8] Moons C., Froyen L., in Proceedings of EuroPM 2001, 22-24 October 2001, 2001, Volume 2, p.376–381
[9] Segal V.M., Materials Science and Engineering 1999, A271, p.322–333
[10] Iwahashi Y., Horita Z., Nemoto M., Langdon T.G., Acta Materialia 1998, 46, No.9, p.3317–3331
[11] Carsley J.E., Milligan W.W., Hackney S.A., Aifantis E.C., Metallurgical and Materials Transactions 1995, 26A, p.2479–2481
[12] Shaik G.R., Milligan W.W., Metallurgical and Materials Transactions 1997, 28A, p.895–904

Strain Measurement in the ECAP Process

J. C. Werenskiold, H. J. Roven
Norwegian University of Science and Technology, Trondheim, Norway

1 Introduction

Material processing by severe plastic deformation, SPD is currently receiving much attention due to the promising results in regard to creating sub-micron grained structures in commercial alloys. So far, one of the most promising SPD techniques has been equal channel angular pressing, ECAP. In this process, high shear strains can be obtained by multiple passes through a die without any reduction in the cross sectional area. Nano-scaled microstructures and high strain rate superplasticity are the main achievements from this process, and has been reported frequently the last few years [1–6].

Common in the ECAP process is multiple passes, from 4 to 12 times through the die. With the most common die designs, there is no way of removing the sample from the die other than pressing it out by the next sample. The deformation at both ends of the sample can prove to have a large effect on the overall deformation of the sample.

FE-simulations of the strain distribution in ECAP processed samples have been reported by Srinivasan [7], Kim [8], Pragnell et al. [9] and Shu et al. [10].

Direct measurements of the strain in ECAP processed samples have not yet been reported for multiple passes through the die. In this work the strains in ECAP deformed samples have been measured by ASAME (Automated Strain Analysis and Measurement Environment) equipment for samples pressed up to 4 times by route A and two passes by route C. The results are compared to the simulations carried out by Srinivasan [7].

The shear strain introduced by the ECAP process is given by Iwahashi et al. [11]:

$$\gamma = \psi \operatorname{cosec}\left(\frac{\phi}{2} + \frac{\psi}{2}\right) + 2\cot\left(\frac{\phi}{2} + \frac{\psi}{2}\right) \qquad (1)$$

where γ is the shear strain, ϕ is the die angle and ψ is the outer angle of curvature. The die characteristics used in the strain measurement experiments has $\phi = 90°$ and $\psi = 20°$. This gives $\gamma = 1.83$ or an effective strain, $\varepsilon_{\text{eff}} = 1.05$. This counts for ideal deformation, ignoring friction and end effects.

In the present experiments the end effects on the overall strain has been investigated. It has been shown that the deformation is not homogeneous and the strains per pass cannot simply be summed together to get the total accumulated strain.

2 Experimental Procedure

2.1 Die Design

The ECAP die has a square cross section with dimensions 20×20 mm², a die angle of 90° and an outer curvature of 20°. The exit cross section is 19.5×19.5 mm². This design should give the samples a shear strain $\gamma = 1.83$ or an effective strain $\varepsilon_{eff} = 1.05$.

The die is fitted on a 60 tons hydraulic press.

2.2 Strain Measurements

The material used is AA6082.50, Al-1.01Si-0.65Mg-0.55Mn-0.20Fe. The samples were cut from homogenized billet. The sample dimensions are $19.5 \times 19.5 \times 85$ mm³.

The samples have been deformed at room temperature and with graphite based lubricant by route A: no rotation between each pass. In order to measure the strain, one uses split samples with an overlaid grid. The grid size is 2×2 mm² and is electro-etched on the surface.

After ECAP the sample is analyzed by ASAME equipment, which calculates the strains from the deformed grid. The maximum shear strain is calculated as half the difference between the first and second principal strains. The effective strain is also calculated as:

$$\varepsilon_{eff} = \frac{2\sqrt{\varepsilon_1^2 + \varepsilon_2^2 + \varepsilon_1 \varepsilon_2}}{\sqrt{3}} \qquad (2)$$

where ε_1 and ε_2 are the first and second principal strains.

The samples are analyzed between each pass through the ECAP die. Since the deformation is very large, a new grid has to be etched on the sample after each pass in order to carry out the analysis, i.e. the strain per pass is measured, not the total strain, unless an other procedure has been specified.

The experiments has been performed with a filled die unless other is specified, i.e. the last sample is still present in the die, providing a certain amount of back pressure.

3 Results and Discussion

Figures 2 to 9 show strain contours for samples pressed 1 to 4 times by route A. Pressing a sample with rectangular cross section the first time vs. pressing a deformed sample with non-planar ends, shows large differences in strain distribution and material flow.

When comparing the measured strain distribution to the simulation from Srinivasan, figure 1 and 2, there are some differences and some similarities. For one, the sample has been pressed with a filled die, while the simulation is for an empty die. This gives a different shape at the front of the sample. The homogeneous area has more or less the same shape and dimension for the measured and the simulated sample. The simulation gives a slightly higher value for the strain in the homogeneous area than equation 1, while for the measured sample this value is de-

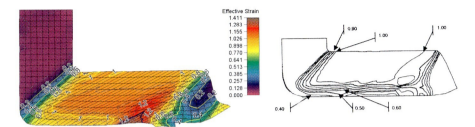

Figure 1: $N = 1$ Sample removed from die, halfway through the press. The width of the shear zone is clearly shown

Figure 2: FE-simulation from Srinivasan [7]

viating only by ± 2 % as compared to that predicted by equation 1. The difference in the width of the lower area with lower strains may be due to the different curvature in the two cases. The shear zone is shown with good detail in the measured sample and it is worth noting the width of the zone. This can give an idea of the strain rates involved. Figure 3 shows the ideal case, a sample after a single pass through an empty die. There is a fairly large area with homogeneous deformation at $\gamma = 1.8$, which is in accordance with the theory.

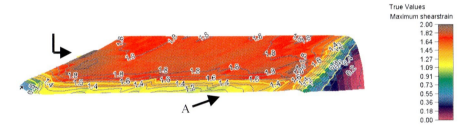

Figure 3: Sample pressed once through an empty die. Showing the measured shear strain distribution, varying from $\gamma = 0$ to $\gamma = 2$

It is clear that the lower area, marked A, has undergone less shear than the rest of the sample. This area stretches almost one third into the sample. This is not the case when the sample is pressed with a filled die (last sample still present), as shown in figure 4.

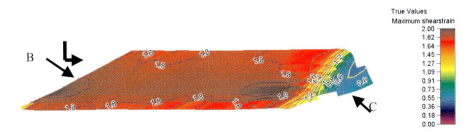

Figure 4: N=1 with a filled die. Shear strain varying from 0.6 to 2

In this case, the homogeneous area is larger, and the maximum shear strain is still around 1.8. The side marked B shows homogeneous deformation after the first pass, while the area marked C still has a lower strain than the homogeneous area, $\gamma \approx 0.7$.

Figure 5 shows the sample after two passes (shown here overlaying the sample). We can clearly see non-homogeneous deformation at both ends of the sample. This is referred to as end effects: the sample will be deformed in the die to a near rectangular shape before the ECAP shearing process starts. This causes the typical wave form on the horizontal lines near the ends of the sample. This normally gives a larger strain in the B area, typically $\gamma = 2.5$ to 3, and is approximately 0.6 times wide as the samples cross section. Again, the A-region has a lower strain at approximately 1.4, and its width is 30% of the total width of the sample. The remaining area, approximately 45 % ± 5 % of the sample surface, has a relatively homogeneous deformation, with γ in the range of 1.7 to 1.9. This counts for $N > 1$. For $N = 1$ with a full die, the homogeneous fraction with γ in the range of 1.7 to 1.9 is 77 % ± 5 %. From these observations one can clearly see that there is a large difference between the first pass and the following ones due to the end effects.

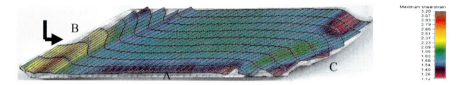

Figure 5: $N = 2$. The bending of the horizontal lines outlines the heterogeneity of the deformation

The deformation between each pass, from pass number 2, carries about the same characteristics. Measurements from pass number 2, 3 and 4 are show in figures 6 to 8. The homogeneously deformed area has approximately the same size and location for the three cases. The C side shows some difference in shape, but the strains are generally lower than in the homogeneous area. The B side generally has the same dimensions and strain.

Figure 6: $N = 2$. The end effects are clearly visible. The shear strain has a maximum at 1.8 in the homogeneous area. Same sample as in figure 5

We observe a formation of cracks in the lower surface as shown by the arrows in figure 8 and 9. These cracks are formed by folding of the material during the compression of the samples inside the upper die channel. They are observed in samples pressed 3 times or more. These cracks may not be present in other die designs or with other materials and may be due to low formability of the 6082 alloy at room temperature

Figure 7: $N = 3$. The results shows similar strain distribution as for $N = 2$.

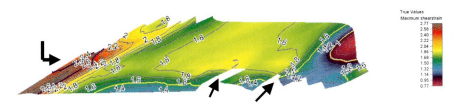

Figure 8: $N = 4$. The shear strain distribution is similar to $N = 2$ and $N = 3$. Cracks are marked by the arrows. The right end shows a different deformation characteristic than the other samples.

What characterizes the deformation is a rotational flow in the material due to the end effects. This is clearly shown in figure 9. The grid shows the resulting deformation from 3rd and 4th pass. This can give an idea of the complexity involved when attempting FE-simulations of the ECAP process. A simple shearing model is clearly not good enough for modeling more than a single pass through the die, and might at best describe a very small area in the center of the sample for higher number of passes.

Figure 9: $N = 4$. Grid deformed in 3rd and 4th pass. One can clearly see the rotational flow. A crack is also seen at the same place as in the other samples. The grid is too heavily deformed to be analyzed by quantitative methods.

A last experiment was performed by pressing a sample twice by route C, 180° rotation of the sample between each pass. The grid was deformed in 1st and 2nd pass and shows the resulting deformation. In an ideal case, the grid would be sheared in the first pass and then sheared back to the original shape in the second pass. Figure 10 shows the resulting grid. The extension of the end effects are clearly shown at both ends. A small area in the middle section shows zero strain, i.e. the material has been sheared back to its original texture.

Figure 10: $N = 2$ Route C, Same grid for both presses. The Process is fairly reversible in the mid-section of the sample

4 Conclusion

Strain distribution measurements by ASAME have proven to be useful in the comparison between experimental results and FE-simulations. The experimental results give a clearer view of the end effects and the heterogeneity of strain distribution after multiple passes through the die. The complexity involved in FE-simulations of multiple passes may prove to be very difficult.

5 References

[1] Iwahashi Y.; Horita Z.; Nemoto M.; Langdon T. G., Acta Mater 1997, 45, 4733–4741
[2] Valiev R. Z.; Salimonenko D. A.; Tsenev N. K.; Berbon P. B.; Langdon T. G., Scripta Mater 1997, 37, 1945–1950
[3] Komura S.; Berbon P. B.; Furukawa M.; Horita Z.; Nemoto M.; Langdon T. G., Scripta Mater 1998, 38, 1851–1856
[4] Furukawa M.; Iwahashi Y.; Horita Z.; Nemoto M.; Tsenev N. K.; Valiev R.Z.; Langdon T. G., Acta Mater 1997, 45, 4751–4757
[5] Iwahashi Y.; Horita Z.; Nemoto M.; Langdon T. G., Acta Mater 1998, 46, 3317–3331
[6] Lee S.; Utsunomiya A.; Akamatsu H.; Neishi K.; Furukawa M.; Horita Z.; Langdon T. G., Acta Mater 2002, 50, 553–564
[7] Srinivasan R., Scripta Mater 2001, 44, 91–96
[8] Kim H. S., Materials Science and Engineering 2001, A315, 122–128
[9] Prangnell P. B.; Harris C.;Roberts S. M., Scripta Mater 1997, 37, 983–989
[10] Shu J. Y.; Kim H. S.; Park J. W.; Chang J. Y., Scripta Mater 2001, 44, 677–681
[11] Iwahashi Y.; Wang J.; Horita Z.;Nemoto M.; Langdon T. G., Scripta Mater 1996, 35, 143

XI Mechanical Properties and Thermostability of Nanocrystalline Structures

Atomistic Modeling of Strength of Nanocrystalline Metals

H. Van Swygenhoven, P. M. Derlet and A. Hasnaoui
Paul Scherrer Institute, Villigen-PSI, Switzerland

1 Abstract

Large scale atomistic simulations of model nanocrystalline materials are used to investigate the plastic deformation mechnisms active in interface dominated materials, with the view to understanding the origin of the related high strength seen in experiment. Results are presented detailing both inter- and intra-granular deformation processes under uniaxial tensile and nano-indentation loading conditions.

2 Introduction

With a reduction of the grain size to the nanometer scale and a corresponding increase in the percentage of grain boundary atoms, the traditional view of dislocation driven plasticity in polycrystalline metals needs to be reconsidered. [1] In coarse-grained metals dislocation sources are active within the grains, and often grain boundaries (GBs) hinder dislocation transmission, creating a dislocation pile-up at the boundary, thus making the material harder to deform. In a 20 nm-grain sized sample however, ~5 % of atoms are sitting in or affected by GBs and therefore it is believed that dislocation sources and pile-up can hardly exist and that therefore from a certain grain size on down, a large part, if not the majority of deformation is carried by the grain boundaries. There is a lack in understanding at the atomic level of the mechanisms by which GBs could accommodate deformation, and unfortunately no direct visualization technique is available that allows an investigation of grain boundary structures during deformation without interaction. Indeed, transmission electron microscopy requires samples to be thinned down to a thickness comparable to the grain size, which undoubtedly induces structural relaxations, thus changing the GB structure. [2]

Tensile deformation studies show an increase in strength of up to 6 times the strength for the coarse grained counterpart. The observed increase is dependent on the synthesis method as a result of the different obtained microstructures. [1,3,4] The increase in strength is however accompanied by a dramatic loss in ductility, and no big fluctuations are observed among the samples synthesized by different techniques, except for a few very special microstructures such as a bimodal Cu sample. [5] In this case, however, the grain sizes limited by high angle GBs are of the order of a few 100 nm. In our recent study on nanocrystalline (nc) Ni synthesized by electrodeposition and High Pressure Torsion, [3,4] other typical features characterizing the deformation mechanism of nc-metals were found such as 1) the increased strain rate sensitivity, up to 10 times higher than the value of the coarse grained material but still low compared to what is observed during superplastic deformation 2) a relatively low activation volume measured by strain rate jump tests 3) a fast decrease in the work hardening, leading to limited uniform deformation and the onset of instabilities, resulting in shear bands at higher strain rates.

The above mentioned experimental results show that we are far from understanding the deformation mechanism at the atomic level. Massively parallel computers offer the possibility to perform atomistic simulations involving millions of atoms, which in terms of grain sizes means that the computer samples contain a fully three dimensional network with up to 15 grains of 20 nm or 150 grains of 6 nm. The use of large scale Molecular Dynamics (MD) in the study of the mechanical properties provides insight into the atomic scale processes that occur during plastic deformation. In the present paper, an overview of recent results obtained from simulations of tensile deformation performed on a full three-dimensional (3D) GB network is given. It will be shown that the results provide some basic insight and at the same time suggest further experimental work.

3   Simulation Technique

All nanocrystalline samples are created using the Voronoi construction with random nucleated seeds and random crystallographic orientations. It has been previously shown that the samples resulting from this type of synthesis, have grain boundaries that are fundamentally not different from their coarse grained counterparts. [6] For all types of misfit, a large degree of structural coherence is observed and misfit accommodation occurs in quite regular patterns.

Recently, is has also been shown that only simulations of a fully 3D grain boundary network, in contrast to two-dimensional (2D) columnar networks, [7] guarantee the possibility of grain boundary relaxations during deformation and the non-geometrical restriction of available slip systems. [8]

Two types of samples are used in this study: 1) a series of self-similar structures having 15 grains with mean grain sizes of 5, 12, and 20 nm (Fig. 1a showing the 12 nm sample) and 2) a sample with 125 grains with a mean grain size of 6 nm (Fig. 1b). All samples have been equilibriated at 300 K. The series of self-similar structures are deformed at 300 K, whereas the sample with 125 grains is deformed at 800 K.

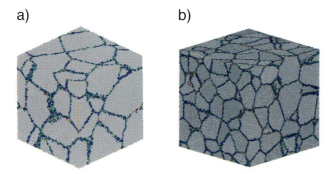

Figure 1: Fully 3D nc samples used in the present study a) A sample containing 15 grains with mean diameter of 12 nm b) A sample containing 125 grains with mean diameter of 6 nm.

All molecular dynamics are performed within the Parrinello–Rahman approach with periodic boundary conditions and fixed orthorhombic angles. Samples are deformed under uniaxial tensile loading. We used the second moment (tight binding) potential of Cleri and Rosato for "model" face-centered cubic (fcc) Ni. [9] Atomic visualization is aided by determining the local crystalline order according to the Honneycutt analysis. [10] For more details about this procedure and the simulations we refer to. [6,11–16] Using this classification scheme, we define four classes of atoms via a color-code system: gray = fcc, red = hexagonal close-packed (hcp), green = other twelve, and blue non-twelve coordinated atoms. Figure 1 uses this scheme demonstrating that the grain and grain boundary regions can clearly be identified.

4 The Deformation Mechanism

In polycrystalline metals we can expect several types of deformation processes according to the temperature, grain size and other microstructural parameters. Based on this knowledge, we consider three inter-related classes of deformation processes 1) inter-grain deformation such as GB sliding (GBs) 2) intra-grain deformation processes such as dislocation activity and finally 3) collective processes as a result of the formation of mesoscopic shear planes.

4.1. Inter-grain Deformation Mechanism

In all samples with mean grain sizes up to 20 nm, grain boundary sliding is observed as being the main contribution to the observed plasticity. Careful analysis of the GB structure during sliding under constant tensile load shows that sliding includes a significant amount of discrete atomic activity, either through uncorrelated shuffling of individual atoms or, in some cases, through shuffling involving several atoms acting with a degree of correlation. [13] In all cases, the excess free volume present in the disordered regions plays an important role. In addition to the shuffling, we have observed hopping sequences involving several GB atoms. This type of atomic activity may be regarded as stress assisted free volume migration. Together with the uncorrelated atomic shuffling they constitute the rate controlling process responsible for the GBS. A detailed description of these processes is given van Swygenhoven and Derlet. [13]

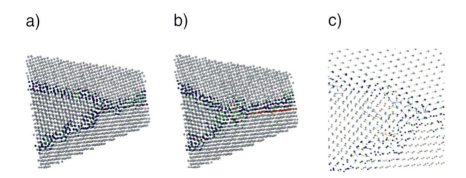

Figure 2: TJ between three grains of a 12 nm sample a) at 1.7 % plastic deformation b) at 3 % plastic deformation c) more detailed view in slightly different orientation, showing the atomic displacement vectors with the color set to their final crystallinity at 3 % plastic strain

Sliding is accompanied by stress build up across neighboring grains. A local stress increase across a grain will lead indirectly to accommodation by GB and triple junction migration, as is shown in detail in Figure 2. Figure 2 shows a triple junction between 3 grains, which slide significantly relative to each other during the deformation. The left picture is at 1.7 % plastic strain and the middle is at 3 % plastic strain. Here, we see that there is a clear migration of the GB structure, corresponding to a displacement of the triple junction region by several atomic distances. The right-most picture in Figure 2 displays the atomic displacement vectors between the two levels of strain. The color of the vectors is determined by the magnitude of the displacement as indicated by the corresponding colour bar. The atom positions are at the initial positions at 1.7 % strain.

Analysis of individual atomic activity using a number of intermediate strains (times) reveals that atomic shuffling facilitates migration. The nature of the atomic shuffling could be classed into two groups. There was short range shuffling involving distances of less than 1.5 Å in which primary grain boundary migration could be resolved. In addition atomic shuffling involving atomic distances of the order of the fcc nearest neighbor can be clearly seen. The displacement vectors for such shuffling was largely random with respect to the grain boundary migration and also occurred inhomogeneously both in time and space. In Figure 2 at 3 % strain, twin hcp planes can be seen due to the emission of a partial dislocation within the lower grain. Its point of emission is from a region not directly related to the triple junction of present interest.

4.2 Intra-grain Deformation Mechanism

At larger grain sizes, dislocation activity is observed. In fully 3D GB networks, which now have been modeled up to 20 nm grain sizes, only partial dislocations have been observed.

MD simulations have shown that a GB dislocation emits a partial lattice dislocation meanwhile changing the grain boundary structure and its dislocation distribution. [13] This mechanism is the reverse of what is often observed during absorption of a lattice dislocation, where the impinging dislocation is fully or partially absorbed in the GB, creating local changes in the structure and GB dislocation network.

Figure 3 represents the atomic configuration of a section of grains 12 and 13 in the 12 nm sample, before and after the emission of a partial dislocation The GB and the two neighboring triple junctions can be identified by the blue and green colored atoms. Figure 3a is at 318.7 ps of deformation and Figure 3b is at 334.7 ps, where the dislocation nucleated at approximately 326.7 ps. The view is along a [11°0] direction of grain 13, where for this grain the unit cell has been highlighted by a filled yellow rectangle. The grain boundary plane is close to a (1,1°,13) plane of grain 13 and the tilt angle between the observed (111) planes in grain 13 and 12 is approximately 24°. The twist angle was found to be approximately 18°. The GB structure accommodates the above mentioned misfit through a GB dislocation (GBD) network. This is evidenced in Figure 3 where a clear coherence across the GB "between a set of (111) planes in grain 12 and in grain 13 is seen. For this orientation, the extra (111) planes identifying the GBDs are in grain 13. In Figure 3a before dislocation "nucleation, the GBDs can be identified by the six indicated extra (111) planes in grain 13, relative to the (111) planes of grain 12. To locate their position, follow the indicated (111) planes of grain 13 to the GB region.

Upon the emission of the partial (where the two (111) hcp planes of the resulting stacking fault have been indicated in Fig. 3b by arrows) there is a clear change in the GBD distribution.

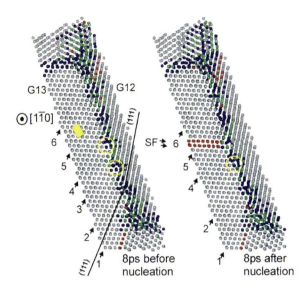

Figure 3: View of the GB between grains 12 and 13 before and after partial dislocation "nucleation. The yellow circles indicate two GBDs and the two red (111) planes after nucleation indicate the stacking fault left behind after the partial was emitted from the GB.

Specifically the partial dislocation nucleates in the region of GBD 3 (in Fig. 3a) resulting in the removal of this region of misfit. Moreover the location of GBD 1 and 2 moves up the page one (111) lattice plane of grain 13. There is also a change in the locations of GBDs 4 and 5. The arrows for these GBDs indicate the extra 111 planes above the stacking fault defect, and not those below. In part this change arises from the reorientation of grain 13 resulting from the slip due to the partial dislocation. Inspection of the atomic displacements of the atoms above and below the stacking fault, indicate that relative to the surrounding grain structure, they move equally in opposite directions, the difference of which is a partial Burgers vector. The corresponding average Burgers vector magnitude over the entire stacking fault region was found to be 1.47 Å, which is close to the ideal value of the Burgers vector magnitude of a Shockley Partial, $a_o/\sqrt{6}$ = 1.44 Å, where a_o is the lattice constant of Ni. The discrepancy in the last digit of the measured Burgers vector magnitude is generally due to thermal noise.

From a detailed analysis of the atomic motion between 318.7 ps and 334.7 ps in the vicinity of the partial dislocation nucleation region, a clear sequence of atomic activity at a two picosecond time resolution has been identified, as is shown in Figure 4. In the period of time approximately eight picoseconds before the nucleation event at 326.7 ps, there is free volume migration from the region of GBD 3 (the nucleation region) to GBD 2, with additional atomic activity at GBD 3. The configuration where we have the earliest indication of a nucleation event shows a partial dislocation core (and its associated stacking fault defect) extending already a nanometer in length. The atomic activity associated with the above mentioned free volume migration away from GBD 3, starts in a "region close to one of the dislocation core ends. In the ten picoseconds after nucleation, there is further free volume "migration from GBD 3 to GBD 2. In this case, however, the free volume originates from the other partial dislocation core intersecting the GB. The orientation of Figure 4 differs from that of Figure 3, in that the viewing direction is now

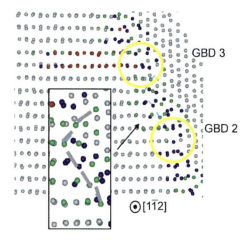

Figure 4: Detailed view of GB12-13, with in the inset the path followed by the free volume when diffusing from GBD2 towards GBD3

along a [11°2] direction of grain 13. The fcc atoms of G12 are not indicated. The atomic positions are those of just after the emission of the partial dislocation, indicated in grain 13 by the red 111 planes constituting the stacking fault in the fcc region. In this orientation, the GBD 2 and GBD 3 are not clearly evident and have been indicated by the large open circles. What are also shown are the atomic displacement vectors greater than 1 Å between the configurations two picoseconds after nucleation and 8 ps later. What is seen is a co-linear sequence of atomic hopping corresponding to free volume migration from GBD3 to GBD2 occurring after the nucleation of the partial dislocation, resulting in part also in the movement of GBD 2 down the figure one (111) plane of grain 13.

It is shown by Derlet and van Swygenhoven [13] that for "increased grain diameters, an increase in partial dislocation activity is seen, but no full dislocations are observed probably due to subsequent structural relaxation after the emission of the partial. Only in quasi 2D-samples with columnar diameters greater than 20 nm full dislocations have been observed. [7] The amount and the character of the dislocation "activity is dependent on how the resulting shear can be "accommodated by the surrounding grains and GB network. In the 2D columnar network where GBs are perpendicular to the tensile direction due to the sample design, shear is predominantly accommodated by dislocation activity in the surrounding grains and not by slower mechanisms such as atomic scale GB activity. This probably is the reason why a much higher dislocation activity is observed in the 2D geometry. In a full 3D-GB network the situation is, however, different: extensive atomic scale activity is observed in the timescale a dislocation needs to travel across the grain, leading to GB migration as a result of atomic shuffling and stress assisted free volume migration. The latter allow local structural relaxation after emission of the partial dislocation in that particular GB, but also in the surrounding GBs. This process along with stress relief within the grain compensates to a large extend the energy increase resulting from the creation of the extended stacking fault. A detailed study of the difference between a 2D columnar and a fully 3D network has been given by Derlet and van Swygenhoven [8] and the energetics of this process will be published elsewhere.

That the GB activity plays an important role in the emission process of the partial dislocation is shown during the simulation of dislocations emitted under an indenter. Figure 5 shows a snapshot of the dislocation activity when a hard indenter is pushed into a nc-substrate with a 12 nm mean grain size. The GBs below the indenter attract the partial dislocations emitted under the indenter; this is demonstrated by the red hcp planes extending outwards from the indenter towards the GBs. The grain below the indenter contains also an example of a full dislocation, where a trailing partial dislocation annihilates the large stacking fault extending throughout the principal grain. In the present case, the partials are emitted below a lowering indenter where local stress is immediately renewed and local structural relaxation between surface and indenter is not possible; therefore, emission of the trailing partial is necessary to accommodate the energy increase due to the extended stacking fault. For more details concerning the simulation conditions of the nanoindentation, we refer to. [17]

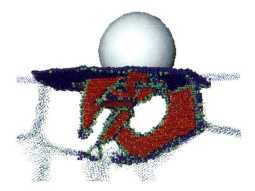

Figure 5: Zone beneath the indenter for a 12 nm grain sample at an indenter depth of 11.9 Å. For clarity of the image, the fcc atoms are left out.

4.3 Collective Mechanism

In order to investigate collective processes such as cooperative grain boundary sliding via the formation of shear planes spanning several grains, a sample with 125 grains and a mean grain size of 6 nm was deformed. A large number of grains are necessary in order to minimize the effects imposed by the periodicity used to simulate bulk conditions. The small grain size is chosen to reduce the total number of atoms in the sample (to 1.2 million) so that longer deformation times are possible at acceptable strain rates. In order to increase grain boundary activity the deformation was done at 800 K.

A typical example of formation of common shear planes during deformation is presented in Figure 6. Figure 6a displays the atomic positions prior to deformation and Figure 6b shows the atomic positions after 3.6 % plastic deformation. Comparison of the two plots demonstrates that after plastic deformation this section of the sample underwent a reorganization of the GB regions (in particular GB (85,108), GB (10,17), and GB (8,117)) resulting in an alignment of multiple GBs to form two common shear planes indicated by the large black arrows in Figure 6b. The nature of the atomic activity that facilitates such migration is predominantly of the form of atomic shuffling with some free volume migration resulting in an increase in GB structural order similar to what we have seen in past work. [13,16] The orientation of shear plane 1 is inclined at about 53° to the tensile axis, whereas shear plane 2 is orientated at 46° to the tensile axis, which is close to the maximum resolved applied shear at 45°.

It is also observed that some grains move collectively relatively to some others. Arrows in Figure 6b represent the relative sliding of some grains involved in this collective motion. The directions of the three nearly vertical arrows in Figure 6b for shear plane 1, are derived by the displacement of atoms in grains 108, 17, and 117 relative to the center of mass of the respective grains 85, 10, and 8. The length of the arrows does not represent the absolute degree of sliding, which for all grains was typically between 1.5 and 2 Å.

The underlying mechanisms that have been observed for the formation of shear planes are 1) pure GBS induced migration of parallel and perpendicular GBs to form a single shear interface 2) Coalescence of neighboring grains that have low angle GB facilitated by the propagation of Shockley partials. 3) Continuity of the shear plane by intragranular slip. A detailed description of the processes is given in. [15]

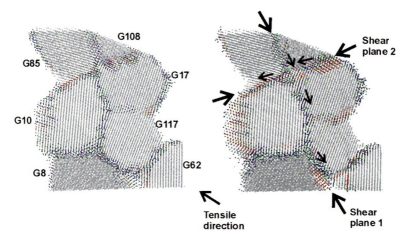

Figure 6: Region of nc-Ni sample (6 nm mean grain size) in which two shear planes (1'and 2) have been identified. The small arrows display the direction of the relative sliding between grains. a) is the atomic configuration before loading, b) after 3.6 % plastic deformation.

5 Discussion and Conclusion

The atomistic simulations show that in grains below 20 nm, both GB sliding and dislocation activity contribute to the deformation mechanism. The sliding is triggered by atomic shuffling and to some extent free volume migration. The dislocation activity is different from what we know from coarse grained metals. Here, it is the GB that acts as source and as sink for the dislocation activity, which suggests the importance of the GB structure. The dislocations observed so far, are emitted from GB dislocations and result in a complete reorganization of the remaining GB dislocations. The emission process is continuously accompanied by atomic shuffling and stress assisted free volume migration in other words the two deformation mechanism are not independent. There are two important issues concerning the dislocation activity: first, it seems to be that they are related to GB dislocations and that the GBs relax (decrease their excess energy) by this process. This means that the effectiveness of this deformation mechanisms will be very dependent on the GB structure. For instance, it is to be expected that this process will be more effective in the presence of unrelaxed high angle GBs containing many GB dislocations. Second, in 20 nm grained samples, only partial dislocations are emitted from GBs. Full dislocations are only observed when a 2D GB network is simulated in a columnar structure [7] but this can not be compared with full 3D networks since on one hand GB accommodation processes are very limited, and on the other hand these configurations have a limited number of slip systems. Following the evolution of the energy of a grain plus its GBs before, during and after emission of the partial dislocation, shows that the increase expected due to the extended stacking fault is partially recovered by the GB relaxations (to be published). Since the local GB structure changes after emission by a reorganization of the remaining GB dislocations, there might be neither energetically, nor structurally no need or possibility to emit the trailing partial. In-situ deformation studies in the electron microscope (TEM) show limited or no dislocation activity in grains below 50 nm. Often, however, "contrast changes" while increasing the load, which at stationary load or unloading do not show the presence of dislocations. This might be a sign that dislocations are absorbed in the opposite GBs, leaving no traceable track. Moreover, usually the internal stress in the samples is very high, obscuring the quality of the TEM image so that even single stacking faults left behind after the passage of a partial dislocation in a grain with a size below 20 nm, would be difficult to observe during the in-situ measurement. Until now there is no experimental evidence of the dislocation activity from GBs and this should be a matter of research in future experiments.

6 Acknowledgement

Work is supported by the Swiss NSF grant (2000-056835.99) and TOPNANO21 (CTI 5443.2).

7 References

[1] J. R. Weertman, Mechanical Behaviour of Nanocrystalline Metals, Nanostructured Materials: Processing, Properties, and Potential Applications, William Andrew Publishing, Norwich 2002
[2] P. M. Derlet, H. Van Swygenhoven, Phil. Mag. A. 2001, 82, 1
[3] F. Dalla Torre, H. Van Swygenhoven, M. Victoria, Acta Mat. 2002, 50, 3957
[4] F. Dalla Torre, Ph.D. Thesis, EPFL, Switzerland, 2003
[5] Y. Wang, M. Chen, F. Zhou, E. Ma, Nature 2002, 419, 912
[6] H. Van Swygenhoven, D. Farkas, A. Caro, Phys. Rev. B 2000, 62, 831
[7] V. Yamakov, D. Wolf, M. Salazar, S. R. Phillpot, H. Gleiter, Acta Mater. 2001, 49, 2713
[8] P. M. Derlet, H. Van Swygenhoven, Scripta Mater. 2002, 47, 719
[9] F. Cleri, V. Rosato, Phys. Rev. B 1993, 48, 22
[10] D. J. Honeycutt, H. C. Andersen, J. Phys. Chem. 1987, 91, 4950
[11] H. Van Swygenhoven, Science 2002, 296, 66
[12] P. M. Derlet, H. Van Swygenhoven, Phys. Rev. B. 2002, 67, 014202-8

[13] H. Van Swygenhoven, P. M. Derlet, Phys. Rev. B 2001, 64, 224105-9
[14] H. Van Swygenhoven, P. M. Derlet, A. Hasnaoui, Phys. Rev. B 2002, 66, 024101-8
[15] A. Hasnaoui, H. Van Swygenhoven, P. M. Derlet, Phys. Rev. B 2002, 66 184112-8
[16] A. Hasnaoui, H. Van Swygenhoven, P. M. Derlet, Acta Mater. 2002, 50, 3927
[17] D. Feichtinger, P. M. Derlet, H. Van Swygenhoven, Phys. Rev. B. 1991, 67, 024113-4.

Multiscale Studies and Modeling of SPD Materials

I.V. Alexandrov

Ufa State Aviation Technical University, Ufa, Russia

1 Introduction

Recent years are characterized by a great interest of scientists to a problem of nanomaterials and nanotechnologies [1–3]. At the same time numerical investigations, conducted in this field, revealed a significant potential for applying the severe plastic deformation (SPD) technique to form bulk ultrafine-grained and even nanostructured states in different metals and alloys.

Nanostructured materials, processed by SPD, are very challenging for research due to a fact, that their grain size constitutes only hundreds or thousands interatomic spacings and is comparable, for instance, with a length of the free dislocation path. High-angle grain-boundaries in the given materials contain a lot of extrinsic grain boundary dislocations (EGBDs). That is why they represent very nonequilibrium boundaries. Fields of elastic long-range stresses, being created around the latter ones, lead to a significant atomic shifting from the equilibrium position in a crystal lattice. As a result, the given materials are characterized by a considerable decrease in the Debye-temperature, a multiple increase in the coefficient of grain-boundary diffusion, manifestation of low-temperature and high-rate superplasticity [2].

Bulk nanostructured materials, processed by SPD, are porousless. They demonstrate enhanced physical and mechanical properties. They are as a result very attractive objects in the view of applying them as industrial and functional materials, nanotechnologies included.

However, it should be noted, that the mentioned problem of obtaining, researching and applying of bulk nanostructured materials has been developing actively only in the recent time. In this connection a lot of problems, concerning optimization of technological processes of obtaining, adequate analyzing of defect microstructure peculiarities and investigation of properties' evolution in SPD materials, as well as development of industrial technologies for processing of bulk nanostructured billets and applying of billets and devices made out of this material still remain unsolved. Aspects, connected with understanding of physical mechanisms of grain refinement, non-equilibrium high-angle grain boundaries formation, reasons for decreasing the Debye-temperature, mechanisms of extraordinary deformation behavior and realization of low-temperature and high-rate superplasticity, etc. are also very important problems.

Answers to these questions, to our mind, can be received (as in case of studying a conventional plastic deformation) by conducting careful systematic experimental investigations and computer simulation of processes, which occur at SPD, on different structure levels. They are: macrolevel (distribution of the accumulated strain fields, cooperative deformation mechanisms, microstructure homogeneity, preferred grain orientation, etc. in the bulk of the whole billet or its definite parts), mesolevel (long-range fields of internal elastic stresses in the bulk of grains, local texture, misorientation angles of grain boundaries), microlevel (dislocation accumulation in the bulk of grains and subgrains, as well as in their boundaries; disclinations; interaction between the elements of defect structure, etc.).

In the present work one can observe some examples' results of such recently conducted experimental investigations and computer simulation, dedicated to the problem of nanostructured materials, processed by SPD.

2   Macrolevel

High-pressure torsion (HPT) and equal-channel angular (ECA) pressing are the principal and most studied means for realizing the SPD technique [2]. However, special attention at investigating SPD processes is paid to the ECA pressing technique due to its capability to process bulk billets of large size and to permit conducting of nature experiments and standard mechanical testing.

Experimental investigations of material flow on macrolevel was conducted with a use of such techniques as insertion inside of bulk billet extrinsic posts and plotting of a scale on the interior surface of a billet, which has been cut along its longitudinal axes. The obtained results point out to a heterogeneous deformation character in different parts of a billet at ECA pressing [4-6]. At the same time homogeneous deformation by means of simple shearing in the main bulk of a billet leads to a regular inclination of the above-mentioned markers in respect to the billet's axis.

Analysis of the experimental simulation results allowed realizing ECAP of bulk billets out of ductile [1] and even hard-to-deform low-ductility materials [10]. However, conduction of experimental investigations, aimed at optimizing of multifactor experiment (such as ECA pressing), is a very difficult and challenging task, which requires considerable expenses. In this connection computer simulation plays here a special role.

Computer simulation, conducted by means of the finite element method showed that peripheral areas of a billet, which contact the die-set, experience heterogeneous deformation. It is explained by friction forces, a change in the flow character of the material near external and internal angles of channels intersection, influence of the die-set's geometry, turn of the billet between the passes, etc [11, 12]. It has been revealed that a formation of stagnation zone near the external angle of the die-set's intersection, as well as lack of contact between the die-set and the upper part of a billet at first pass of ECA pressing, take place. At last, special conditions, the ends of the deformed billet are put in, lead to an extremely different character of material flow in them in comparison to the major bulk of the billet. Thus, preserving the homogeneity of the material flow at ECA pressing is a rather important problem, requiring investigation of the influence of the die-set's geometry, friction coefficient, rate, number of passes, routes and deformation temperature, contact pressures, presence of backpressure. For instance, analysis of the kinetics of plastic flow at ECAP simulation of commercially pure W billet at the temperature of 1100 °C revealed the following: 23 % higher value of the accumulated strain, in case of the route A (lack of a turn of the billet between the subsequent passes) than for the route C (180° turn of the billet round the longitudinal axes) [12]. Meanwhile, the route leads to a significantly more homogeneous distribution of the accumulated strain than the route . On the other hand, the analysis of stresses occurring in the die-set, conducted considering the computer simulation, allowed to reveal the areas of their concentration and to consider the influence of the friction coefficient on the character of their distribution (Figure 1).

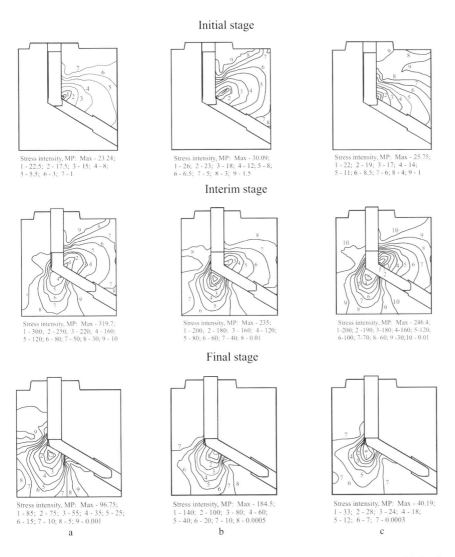

Figure 1: Change of stress intensity in the tool at different stages of ECAP depending on the friction factor m: 0.05 (a); 0.1 (b); 0.2 (c). ECAP W. The ingot temperature = 1100 °C. The die-set temperature = 900 °C.

The received simulation results can be reliable due to the fact that numerical forecasts of the character of the material flow conform to reality when using more simple deformation schemes and when correlating with the results of experimental investigations of the ECA pressing process itself. Peculiarities of structure and deformation behavior undoubtedly complicate the process of computer simulation and require a careful selection of an adequate model of SPD nanomaterials.

When analyzing SPD processes on macrolevel, a conduction of experimental investigations of microstructure homogeneity, crystallographic texture, mechanical properties in different areas, as concerning the length, as well as the cross section of the processed bulk nanostructured billets still remains topical. A discovered positive effect of backpressure on the integrity and mechanical properties of bulk nanostructured billets requires further systematic investigations [13]. The computer simulation method should be widely applied when developing different, especially new SPD schemes, such as accumulative roll bonding [14], repetitive corrugation and straightening [15], multi-axis deformation [16], twist extrusion [17] and others. At the same time special attention should be paid to an experimental development of new SPD schemes, based on the simulation results. These new schemes should meet the requirements made for industrial technological processes.

3 Mesolevel

It has been established that bulk nanostructured states in different metals and alloys can be formed as a result of SPD [1-3]. At the same time an average grain size in pure metals reaches from several dozens to several hundreds of nanometers. In alloys it is smaller. The HPT method provides a smaller grain size than the ECA pressing does. The latter fact, obviously, is determined by achieving higher accumulated strain degrees, realized at higher implied pressures, which protect the destruction of billets.

The conducted systematic experimental investigations point out to decreasing of an average size of fragments, growing of an average misorientation angle between them, developing of strong axial crystallographic texture <111> [18,19]. For pure SPD copper and nickel it is typical to possess also enhanced values of linear extension coefficient, stronger atomic shifts from equilibrium state in an ideal crystal lattice, lowered Debye-temperature values, especially in boundary areas [20,21].

Investigations of pure copper, subjected to ECA pressing, also testify to fundamental microstructure reorganization as a result of SPD [22–24]. At the same time, in particular, besides structure refining, increasing of the share of high-angle grain boundaries, enhancing of static and dynamic atomic shifts there have been revealed facts of increasing the share of edge dislocations in respect to a share of screw dislocations alongside with increasing the accumulated strain degree, as well as falling of a general dislocation density when the grain size in pure copper becomes lower than 100 nm [24].

In spite of a large number of experimental results obtained, it is still topical to investigate the effect of applied pressure to plot histograms of grain distribution, taking into account the sizes, and grain boundaries, taking account misorientations, analysis of local textures and peculiarities of defect structure in grains, which refer to different texture components, etc. The method of electron backscattering should play an important role here. However, it should be mentioned, that a resolution, achieved with its help nowadays is insufficient in a number of cases.

The results of conducted texture investigations testify to a formation of complex shear textures as a result of ECA pressing [19,25–27]. At the same time one can observe some shift of texture maxima in respect to the symmetry axes of pole figures. There are various ways of treating the received results. In particular, in work [26] on the basis of computer simulation with the use of the finite-element method the discovered effect was explained by the appearance of additional compressive pressures in a billet, subjected to ECA pressing. The major simulation's result of microstructure evolution processes on the mesolevel should be an explanation of reasons of microstructure refinement and increase in misorientation angles of neighboring grains during

SPD. By the present time there has been made just first efforts to answer this question [27]. At that, simulation of texture formation on the basis of visco-plastic model has been taken into account. There has been also considered purely geometric criterion of refinement, connected with acquiring by grains of a certain form during ECA pressing. A successful solution of the problem is impossible without analyzing the processes which take place during SPD on microlevel.

4 Microlevel

At the present time many peculiarities of structure and behavior of nanomaterials can be explained by a formation, as a result of SPD, of non-equilibrium high-angle grain boundaries, possessing an enhanced density of EGBDs [28]. However, high resolutionability of TEM makes it possible to evaluate mainly qualitatively the EGBDs density, limited above by the value $\rho = 1 \cdot 10^9 \, m^{-1}$. The proposed approach to computer simulation of X-ray scattering in nanomaterials allowed to estimate quantitevely the EGBDs density in nanostructured Cu produced by the SPD methods [29]. On the basis of comparison of simulation results and experimental data it has been shown that the observed changes of SPD Cu X-ray profiles (as compared to those of coarse-grained Cu) are caused by EGBDs of high density (about $\rho = 0.5 \cdot 10^8 \, m^{-1}$). Such an approach could be very useful when investigating the defect structure of nanomaterilas with highly distorted crystal lattice. It is impossible to explain extraordinary mechanical properties and deformation behavior of nanostructured materials without analyzing the processes running on the microlevel. In particular, a comparative analysis [30, 31] of the peculiarities of deformation behavior of coarse-grained and nanostructured copper, taking into account the development and application of dislocation model of Estrin-Toth, points out to activation of sources and sinks of dislocations in non-equilibrium grain boundaries of bulk nanostructured copper, processed by ECA pressing, which possesses high density of EGBDs. Presently, the mentioned model has been widely developing and, apparently, in the nearest time we can expect further achievements at developing the structure model of nanomaterials, processes by means of SPD and at explaining their extraordinary properties.

5 Conclusion

The obtained results testify that the complex approach to investigation of nanostructured materials and their properties by means of mutually complementary experimental investigations and computer simulation on different structure levels appears to be very productive.

6 Acknowledgement

The work was conducted with the support of the international project INTAS 01-0320.

7 References

[1] Investigations and Applications of Severe Plastic Deformation, NATO Sci. Series, 80, Kluwer Publ., Dordrecht/Boston/London, 2000, p. 394
[2] Valiev, R.Z.; Islamgaliev, R.K.; Alexandrov, I.V., Progr. Mater. Sci. 2000, 45, p. 103–189
[3] Ultrafine Grained Materials II, TMS, Warrendale, PA, 2002, p. 685
[4] Zhernakov, V.S.; Budilov, I.N.; Raab, G.I.; Alexandrov, I.V.; Valiev, R.Z., Scripta mater. 2001, 44, p. 1765–1769
[5] Segal, V.M., Mater. Sci. Engng. 1999, A271, p. 322–333
[6] Gholinia, A.; Bowen, J.R.; Prangnell, P.B.; Humphreys, F.J., Proceedings 6th Int. Conf. On Aluminium Alloys (ICAA6), 1998, 1, Tokyo, Japan, p. 577–582
[7] Kim, H.S.; Seo, M.H.; Hong, S.I., J. Mater. Proc. Techn. 2001, 113, p. 622–626
[8] Goforth, R.E.; Hartwig, K.T.; Cornwell, L.R., Investigations and Applications of Severe Plastic Deformation, NATO Sci. Series, 80, Kluwer Publ., Dordrecht/Boston/London, 2000, p. 3–12
[9] Kim, H.S., Mater. Sci. Engng. 2002, A328, p. 317–323
[10] Alexandrov, I.V.; Raab, G.I.; Shestakova, L.O.; Kilmametov, A.R.; Valiev, R.Z.; Phys. Met. Metallogr. 2002, 93, p. 493–500
[11] Budilov, I.; Alexandrov, I.V., in Proceedings 20th CAD-FEM users' meeting 2002, Intern. Congress on FEM Technology, 9-11 October 2002, Lake Constance, Germany, 2002, paper 2.10.6, p. 1–9
[12] Krallics, G.; Budilov, I.N.; Alexandrov, I.V.; Raab, G.I.; Zhernakov, V.S., this volume
[13] Raab, G.I.; Krasilnikov, N.A.; Alexandrov, I.V.; Valiev, R.Z., Memory shape alloys and another advanced materials, Saint Petersburg, Russia, 2001, p. 409–413, in Russian
[14] Saito, Y.; Utsunomiya, H.; Tsuji, N.; Sakai, T., Acta Mater. 1999, 47, p. 579–583
[15] Zhu, Y.T.; Huang J., Ultrafine Grained Materials II, TMS, Warrendale, PA, 2002, p. 331–340
[16] Salishchev, G.A.; Zherebtsov, S.V.; Galeyev, R.M., Ultrafine Grained Materials II, TMS, Warrendale, PA, 2002, p. 123–131
[17] Beygelzimer, Y.; Varyukhin, V.; Orlov, D.; Efros, B.; Stolyarov, V.; Salimgareyev, H., Ultrafine Grained Materials II, TMS, Warrendale, PA, 2002, p. 297–304
[18] Alexandrov, I.V.; Dubravina, A.A.; Kim, H.S., Defect and Diffusion Forum 2002, 208–209, p. 229–232
[19] Alexandrov, I.V.; Dubravina, A.A.; Kilmametov, A.R.; Kazykanov, V.U.; Valiev, R.Z., KIMM, submitted
[20] Alexandrov, I.V., Mazitov, R.M.; Kilmametov, A.R.; Zhang, K.; Lu, K.; Valiev, R.Z., Phys. Met. Metal. 2000, 90, p. 77–82
[21] Zhang, K.; Alexandrov, I.V.; Valiev, R.Z.; Lu, K., J. Appl. Phys. 1998, 84, p. 1924–1927
[22] Ungár, T.; Alexandrov, I.; Zehetbauer, M., JOM 2000, April, p. 34–36
[23] Alexandrov, I.V.; Valiev, R.Z., Mater. Sci. Forum 2000, 321-324, p. 577–582
[24] Ungár, T.; Alexandrov, I.V.; Hanák, P., Investigations and Applications of Severe Plastic Deformation, NATO Sci. Series, 80, Kluwer Publ., Dordrecht/Boston/London, 2000, p. 133–138
[25] Zhu, Y.T.; Lowe, T.C., Mater. Sci. Engng. 2000, A291, p. 46–53
[26] Gholinia, A.; Bate, P.; Prangnell, P.B., Acta Mater. 2002, 50, 2121–2136

[27] Beyerlein, I.J.; Lebensohn, R.A.; Tomé; C.N., Ultrafine Grained Materials II, TMS, Warrendale, PA, 2002, p. 585–594
[28] Horita, Z.; Smith, D.J.; Furukawa, M.; Nemoto, M.; Valiev, R.Z.; Langdon, T.G., J. Mater. Res. 1996, 11, 1880–1890
[29] Enikeev, N.A.; Aleksandrov, I.V., Valiev, R.Z., Phys. Met. Metallogr. 2002, 93, 515–524
[30] Alexandrov, I.V.; Zehetbauer, M.; Tóth, L.S., in 4th Euromech. Book of Abstracts II, Solid Mechanics Conference, 26-30 June, 2000, Metz, France, p. 474
[31] Enikeev, N.A.; Kim, H.S.; Alexandrov, I.V., this volume

Microstructural Stability and Tensile Properties of Nanostructured Low Carbon Steels Processed by ECAP

Dong Hyuk Shin[1] and Kyung-Tae Park[2]
[1]Department of Metallurgy and Materials Science, Hanyang University, Ansan, Korea
[2]Division of Advanced Materials Science & Engineering, Hanbat National University, Taejon, Korea

1 Introduction

At present, in view of metallurgy, it is the grain refinement strengthening that is the most viable methodology to achieve an excellent combination of the ultrahigh strength and enhanced toughness in most structural metallic materials. Particularly, for the low carbon steel which is the most widely used structural material, the strength can be doubled without compositional modification when the ferrite grain size of 10 μm is refined to 1 μm, if the well-known Pickering's equation[1] is applied. In addition, the Pickering's equation predicts that this ferrite grain refinement results in a simultaneous decrement of the ductile-brittle transition temperature as much as 250 K. This fact initiated the recent researches on manufacturing ultrahigh strength low carbon steels with moderate ductility, toughness, weldability and extended lifetime for the use of gigantic infrastructure construction, that are being conducted vigorously as the national research projects in the world leading laboratories and steel companies.

Of several processing routes for fabricating ultrafine grained (UFG) materials, severe plastic deformation (SPD), at present, is the most developed and viable method in terms of an ability producing fully-dense bulk UFG materials[2]. Accordingly, several investigations have been performed to fabricate UFG low carbon steels with ultrahigh strength by SPD[3,4]. However, the microstructures of UFG materials produced by SPD, including low carbon steels, are known to be thermally unstable due to their non-equilibrium nature[5,6]. This implies that grain growth easily occurs in these materials, causing the loss of mechanical superiority associated with a UFG structure. Therefore, for full utilization of UFG low carbon steels produced by SPD, it is essential to ensure a thermal stability. Along with the motivations addressed above, the present investigation was performed to examine the microstructural evolution of low carbon steel during SPD and subsequent heat treatment and the corresponding tensile properties, and to find feasibilities to obtain a thermally stable UFG low carbon steels.

2 Experimental

In the course of investigation, the two grades of low carbon steel were used: Fe-0.15C-0.25 Si-1.1Mn (in wt.%) (hereafter, CS steel) and Fe-0.15C-0.25Si-1.1Mn-0.06V-0.008N (in wt.%) (hereafter, CSV steel), were used. The CS steel was austenitized at 1473 K for 1 hr and then air-cooled. The CSV steel was oil-quenched to room temperature after the same austenitization treatment as the CS steel and then normalized. Normalization for the CSV steel consisted of soaking at 1223 K for 1 hr and the subsequent air-cooling. After heat treatment, a typical ferrite-pearlite structure was obtained in both steels. The ferrite grain size was ~ 30 μm and ~10 μm for the CS and CSV steels, respectively.

Equal channel angular pressing (ECAP) was used as a SPD process. The present ECAP die was designed to yield an effective strain of ~ 1 per pass[5]: the inner contact angle and the arc of curvature at the outer point of contact between channels of the die were 90° and 20°, respectively. During ECAP, the sample was rotated by 180° around its longitudinal axis between the passages. After machining the cylindrical samples of 18 mm ϕ × 130 mm from the heat-treated plates of both steels, ECAP was carried out on the samples at 623 K up to 8 passes.

In order to examine thermal stability of the ECAP deformed steels, 1 hr static annealing was conducted at temperatures of 373~873 K using silicone oil bath or molten salt bath and the resultant microstructural changes were examined by field emission scanning electron microscope (FE-SEM, JEOL6330F) and transmission electron microscope (TEM, JEOL2010). Room temperature tensile tests were performed on the samples with the gage length of 25.4 mm using an Instron machine (Model 1125) at an initial strain rate of $1.33 \cdot 10^{-3}$ s^{-1}.

3 Results and Discussion

3.1 Microstructural Evolution

Figure 1 shows TEM micrographs of the CS steel by the ECAP passage. After 4-passes ECAP, the ferrite grains with the initial size of ~ 30 μm (Fig. 1a) were refined to 0.2~0.3 μm (Fig. 1b). Further repeating ECAP up to 8 passes did not result in significant grain refinement (Fig. 1c), but an inspection of the corresponding selected area diffraction pattern revealed that it contributed to an increase in the misorientation angle across the grain boundaries.

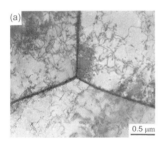  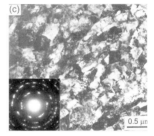

Figure 1: TEM micrographs showing the microstructural evolution in the CS steel after ECAP. (a) ferrite before ECAP, (b) ferrite after 4 passes, (c) ferrite after 8 passes

Many previous investigations [2,7] revealed enhanced grain growth behavior of ECAP deformed metals and alloys compared to those produced by conventional working processes due to relatively large accumulated strain. For full utilization of advantages arising from an ultrafine grain structure, it is essential to ensure thermal stability of these materials. For this purpose, the two approaches were attempted to improve thermal stability of the ECAP deformed steels: the increment of ECAP strain and microalloying. The first concept is related to the fact that carbon dissolves from pearlitic cementite by severe working such as cold drawing and excessive carbon higher than the equilibrium content exists in ferrite [8]. It would cause retardation of recovery, recrystallization and grain growth. In addition, carbon dissolution becomes more considerable with increasing strain. Accordingly, it may be expected that higher ECAP strain is effective in enhancing thermal stability of ECAP deformed steel. Second, dilute addition of carbide and/or nitride forming elements to steel is also very effective on increasing recrystallization temperature and on suppressing grain growth [9].

Figure 2 shows the microstructural evolution of the CS steel annealed at 813 K and 873 K for 1 hr after either 4 passes or 8 passes ECAP, i.e. effective accumulated strain of 4 or 8. At 813 K, the microstructure of the 4 passes ECAP deformed CS steel (Fig, 2a) consisted of coarse recrystallized grains and fine unrecrystallized grains. By contrast, the grains of the 8 passes CS steel (Fig. 2b) were uniform in the size and the dislocation density remained high. At 873 K, the ferrite grains were completely recrystallized and pearlite colonies were well-defined with a dark contrast in the 4 passes ECPA CS steel (Fig. 2c). However, for the 8 passes ECAP deformed CS steel, pearlite colonies were ill-defined (Fig. 2d). In addition, as shown in Fig. 3a, the ferrite grains remained ultrafine and nano-sized particles, identified as cementite by energy dispersive spectrometer analysis (Fig. 3b), existed mainly at ferrite grain boundaries in the vicinity of pearlite colonies. These particles resulted from precipitation of carbon atoms dissolved from pearlitic cementite during ECAP as cementite by subsequent annealing and prevented grain boundaries from migrating by pinning effect. From the above findings, it is obvious that higher ECAP strain may enhance thermal stability in the present CS steel.

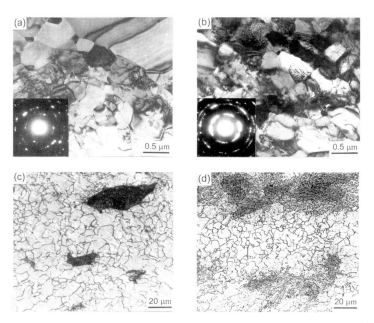

Figure 2: Examples of microstructures of the ECAP deformed CS steel after 1 hr static annealing. (a) 4 passes, 813 K, (b) 8 passes, 813 K, (c) 4 passes, 873 K, (d) 8 passes, 873 K

Figure 4 presents the micrographs of the CSV steel, containing 0.06 wt.% V and 0.008 wt.% N, annealed at 873 K for 1 hr after 4 passes ECAP. Unlike the CS steel processed by the same ECAP conditions (Fig. 3c), pearlite colonies were hardly defined (Fig. 4a) and nano-sized ce-

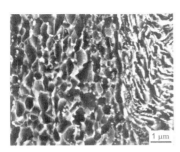 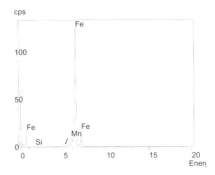

Figure 3: (a) SEM micrograph showing the microstructure in the vicinity of pearlite in the 8 passes ECAP deformed CS steel after annealing of 873 K ×1 hr. The nano-sized cementite particles were precipitated mainly at the ferrite grain boundaries. (b) Energy dispersive spectrometer profile of the nano-sized particle, identified as cementite

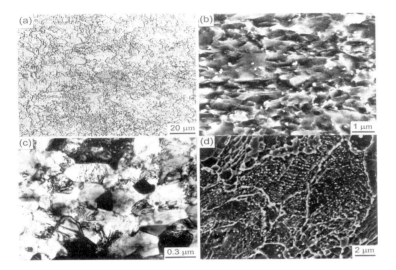

Figure 4: Microstructure of the 4 pass ECAP deformed CSV steel after annealing of 873 K ×1 hr. optical microstructure, (b) SEM microstructure of ferrite, (c) TEM microstructure of ferrite, (d) SEM microstructure of pearlite

mentite particles were uniformly distributed throughout UFG ferrite matix (Fig. 4b). It is also noticeable that dislocation density remained high in the ferrite grains (Fig. 4c) and pearlitic cementite which was initially in the form of the lamellar plates became particle-like (Fig. 4d). This thermally stable microstructure of the CSV steel resulted from the following effects. The presence of a small amount of vanadium increases recrystallization temperature. So, the dislocation density remained high and grain boundaries were still in non equilibrium state. Dislocations and non-equilibrium boundaries provided effective diffusion path so that carbon atoms dissolved from pearlitic cementite could diffuse throughout the ferrite matrix. In the CS steel, dissolved carbon atoms could not diffuse away from pearlite colonies due to the lack of diffusion path and so the distribution of nano-sized cementite particles is restricted in the vicinity of pearlite colonies. Then, nano-sized cementite particles precipitated at the grain boundaries throughout ferrite matrix during annealing suppressed grain growth.

3.2 Tensile Properties

The variation of the room-temperature tensile properties of the CS and CSV steels after 4 passes with annealing temperature is plotted in Fig. 5. For the CS steel, YS and UTS of the sample annealed at 693 K are slightly less than those of the as-ECAP deformed sample, and decrease rapidly with further increment of annealing temperature. The loss of strength with increasing annealing temperature is attributed to two factors: the appearance of coarse recrystallized ferrite grains (see Fig. 2a), and softening of the pearlite due to the spheroidization of pearlitic cementite. On the contrary, the strength of the CSV steel annealed at 693 K is higher than those of the as-ECAP deformed sample, probably due to the precipitation of very fine Fe_3C particles. In the temperature range of 693 ~ 813 K, the strength of the CSV steel decreased gradually with increasing annealing temperature. It is noted that the strength of the CSV steel annealed at 813 K is comparable to that of the CS steel annealed at 693 K. The extension of the mechanical stability in the CSV steel to higher temperatures resulted mainly from the preservation of a UFG ferrite grains associated with homogeneously distributed nano-sized cementite particles. The elongation to failure, e_f, of the CS steel remained unchanged up to 753 K. Then, it increased rapidly with increasing annealing temperature and finally recovered to a value close to the as-received state at 873 K. For the CSV steel, e_f gradually increased with increasing annealing temperature. Smaller e_f of the CSV steel than that of the CS steel above 813 K would result from its higher strength.

In Fig. 6, YS of the CS and CSV steels was plotted against $d^{-1/2}$ (d: ferrite grain size). In order to show the validity of the present tensile data, the YS data of UFG CS steel fabricated by warm groove rolling[10] was included in the same plot. All data coalesce into a single straight line, proving the validity of Hall-Petch relation in the present ferrite grain size range of 0.2 ~ 30 μm. For the prediction of YS of low carbon - manganese steel with the ferrite grain size larger than 10 μm, the following equation[1] is often quoted.

$$YS \text{ (MPa)} = 15.4 \, (\, 3.5 + 2.1 \, [Mn] + 5.4 \, [Si] + 23 \, [N_f] + 1.13 \, d^{-1/2} \,) \quad (1)$$

where [YS] is in wt.%, d is in mm and N_f is the free nitrogen content. The Hall-Petch constant estimated from the slope of Fig. 8, 10.4 MPa $mm^{1/2}$ (= 328 MPa $\mu m^{1/2}$), is lower than that in Eq. 1, ~ 17.4 MPa $mm^{1/2}$. Although, at present, the origin of this difference is not clear and to be further investigated, the following postulation is worth considering. The Hall-Petch constant is regarded as a measure of resistance against yield propagation from the yielded grain to the adjacent unyielded grain. Li [11] suggested that yield propagation occurs by emitting dislocations at ledges of grain boundary between the yielded grain and the adjacent unyielded grain. The large portion of grain boundaries of UFG materials fabricated by severe plastic straining remains non-equilibrium even after prolonged annealing. The Hall-Petch equation or Eq. 1 assumes that grain boundaries are equilibrium and high-angled. Accordingly, dislocation emission from non-equilibrium low-angled boundaries would be relatively easy compared to that from equilibrium high-angled grain boundaries, causing the low value of the Hall-Petch constant.

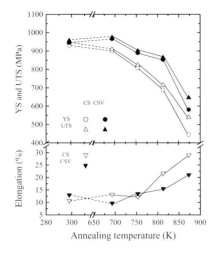
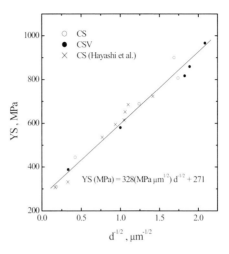

Figure 5: The variation of the room-temperature tensile properties of the CS and CSV steels with annealing temperature

Figure 6: The Hall-Petch plot for the yield strength of the UFG CS and CSV steels

4 Conclusions

1. An ultrafine ferrite grain structure of 0.2 ~ 0.3 µm was obtained by conducting equal channel angular pressing at 623 K to an effective strain of ~ 4 or ~ 8 in the two grades of low carbon steel, one without vanadium and the other containing vanadium. The as-ECAP deformed steels exhibited very high strength, more than twice of the strength of the steels before ECAP.
2. Ultrafine grained ferrite in the more severely pressed steel exhibited more sluggish recovery and recrstallization kinetics due to the presence of excessive carbon content in the ferrite matrix by carbon dissolution from pearlitic cementite during pressing.
3. Under the identical annealing conditions, ultrafine ferrite grain size and the high room-temperature strength were preserved at higher annealing temperatures in the steel containing vanadium compared to the steel without vanadium.
4. The yield strength of CS and CSV steels were shown to be well fitted to the standard Hall-Petch type relationship in the ferrite grain size range of 0.2 ~ 30 µm.

5 References

[1] F. B. Pickering, Physical Metallurgy and Design of Steels, Appl. Sci. Pub. Ltd., London, 1978, p. 63
[2] R. Z. Valiev, R. K. Islamgaliev, I. V. Alexandrov, Prog. Mat. Sci., 45, 2000, 103
[3] D. H. Shin, B. C. Kim, Y. S. Kim, K. -T. Park, Acta Mat., 48, 2000, 2247
[4] Y. Fukuda, K. Oh-ishi, Z. Horita, T. G. Langdon, Acta Mat., 50, 2002, 1359

[5] D. H. Shin, B. C. Kim, K. -T. Park, W. Y. Choo, Acta Mat., 48, 2000, 3245
[6] K-. T. Park, Y. S. Kim, J. G. Lee, D. H. Shin, Mater. Sci. Eng., A293, 2000, 165
[7] Z. Horita, D. J. Smith, M. Furukawa, M. Nemoto, R. Z. Valiev, T. G. Langdon, J. Mat. Res., 11, 1996, 1880
[8] M. H. Hong, W. T. Reynolds Jr., T. Tauri, K. Hono: Metall. Mat. Trans., 30A, 1999, 717
[9] T. Gladman, The Physical Metallurgy of Microalloyed Steels, The Ins. of Mater., London, 1997, p. 213
[10] T. Hayashi. O. Umezawa, S. Torizuka, T. Mitsui, K. Tsuzaki, K. Nagai, CAMP-ISIJ, 12, 1999, 385
[11] J. C. M. Li, Trans. AIME., 227, 1963, 239

Microstructure and Mechanical Properties of Severely Deformed Al-3Mg and its Evolution during subsequent Annealing Treatment

M.A. Morris-Muñoz, C. Garcia Oca, G. Gonzalez Doncel and D.G. Morris
CENIM, CSIC, Madrid, Spain

1 Abstract

During the intense deformation of Al-3Mg alloy by ECAP, an elongated grain/subgrain/ cellular structure is set up of width about 100 nm and length close to the micron. Analysis of grain boundary misorientations by converging beam electron diffraction shows that many of the boundaries are of very low angle. On annealing these materials, processes of dislocation recovery, of grain growth, and also of recrystallization occur. During the early stages of annealing, dislocation recovery into transverse grain boundaries can lead to some refinement of the microstructure. Recrystallization at higher temperatures can lead to a duplex microstructure, of recrystallized and unrecrystallized regions, depending on the initial level of strain and the annealing treatments given. Other studies are examining changes of texture during processing and annealing, and also the role of the various structural features on mechanical behaviour. Mechanical properties have been measured. It has been found difficult to interpret the strengthening occurring because of the variety of boundaries present (low-angle to high-angle) and their different contributions to material strength.

2 Introduction

Severe plastic deformation by processes such as ECAP has been claimed to produce highly refined microstructures, below the micron level and into the nano-scale, depending on the amount of strain and the material examined. Only rarely have the misorientations across the boundaries produced been examined, and the proportion of low-angle, medium-angle and high-angle boundaries determined. This has been one of the main objectives of the present study. Interpretation of material strengthening in terms of strengthening due to such boundaries, or other factors such as retained dislocation density, clearly also requires a precise description of the microstructures present, which has been the second objective of this study.

3 Experimental Details

Cast and homogenised ingots of Al-3 wt%Mg have been processed to a total strain of 5.6 by passing 8 times through a round cross-section ECAP die with a die-rotation angle of about 120° giving a true strain of 0.7 per pass. No ingot rotation about its cylindrical axis was given between passes, corresponding to route A in published convention [1]. Full details have been given elsewhere [2]. Some materials were also annealed after deformation. Microstructures were determined by quantitative image analysis of micrographs obtained by SEM and TEM. Grain orientations and boundary misorientations were determined from converging beam electron diffraction patterns, using the Kikuchi patterns observed. Mechanical properties were determined by compression testing on cylinders cut from deformed materials perpendicular to both inlet and outlet channel axes [3].

4 Results

4.1 Microstructure Evolution during Deformation and Subsequent Annealing

On deforming Al-3Mg materials, the grain size gradually reduces to dimensions of about 1 µm (grain length) and 0.1µm (grain width), reaching these dimensions after strains of about >5 and stabilising slowly at these values [4]. At this point there is a high density of dislocations both within grains and at grain boundaries, Fig. 1a, and the grains are elongated in a direction close to that of the intense ECAP shear direction. Weak beam imaging of these dislocations shows that many are arranged in arrays as sub-boundaries. On annealing highly-deformed materials at moderate temperatures, 150–200 °C for 1h, or at 250 °C for 15min, there is considerable loss of dislocations, Fig. 1b, some increase in grain width and reduction of grain length. The increase of grain width is believed to occur by the loss of some of the longitudinal boundaries, and the reduction of grain length occurs as some of the dislocations rearrange into transverse boundaries across the longer grains. Eventually, at sufficiently high annealing temperatures for sufficiently long times (e.g. 1h at 250 °C) recrystallization occurs at some locations throughout the material, leading to a duplex structure, Fig. 1c, before full recrystallization eventually occurs after more severe heat treatments.

4.2 Boundary Misorientations

The misorientation across boundaries has been determined on material deformed to true strains of 4 to 5.6, and after subsequently annealing at temperatures of 150–200 °C the materials initially strained to 5.6. The important results are shown in the histogram of Fig. 2, which shows that the vast majority of boundaries present after deforming to such strains are very low angle boundaries (misorientations mostly below 10°). Similar results have been reported elsewhere [e.g. 4], where it has been shown that grain boundary misorientations only become as high as for grains in a randomly oriented polycrystalline material after much higher strains, at strain levels up to about 10. It is also seen in Fig. 2 that mild anneals, that lead only to dislocation recovery and some slight changes in grain size, have no significant effect on the boundary misorientations.

It is commonly considered that (high-angle) grain boundaries are those with misorientations greater than about 15°. As such, it follows from the histogram above that the majority of the boundaries (about two thirds) present in the highly deformed Al-Mg alloy are not true grain boundaries but are really low-angle boundaries, or sub-boundaries. Equally the grain size is not as small as initially reported for such an Al alloys after such a level of straining (about 0.1–0.2 µm), but is a factor of about three higher. Fig. 3 shows a line drawing (left) of all the

Figure 1:. (a) Transmission electron micrograph showing material after ECAP deformation to a strain of 5.6; (b) Transmission electron micrograph of same material after annealing for 10 min at 250 °C; and (c) Back-scattered electron, scanning electron micrograph showing same material after annealing for 1 h at 250 °C.

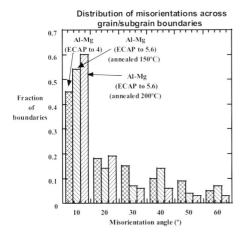

Figure 2: Histogram showing distribution of grain boundary misorientations in Al-3Mg deformed by ECAP to strains of 4 to 5.6, and also after annealing for recovery at 150–200 °C

boundaries detected in highly deformed Al-3Mg, and also (right) of only those boundaries with a misorientation greater than 10°. The density of grain boundaries is approximately one third of the total density of boundaries of all types. In addition, as reported elsewhere [5], there exists a hierarchy of boundary types inside the grain boundaries, composed of features like Incidental Dislocation Boundaries, Geometrically Necessary Boundaries, and so on, and these boundaries are distributed in well-defined patterns inside a deformed material. Fig. 3 gives the impression that the high-angle boundaries formed during deformation tend to be concentrated in bands inside the original grains (The original grain size of the Al-3Mg samples was about 200-300µm in the present study.).

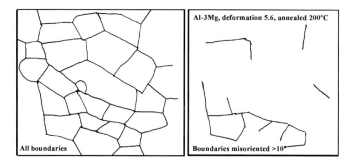

Figure 3: Line drawing showing outline of boundaries in Al-3Mg deformed to a strain of 5.6 and annealed for 1h at 200 °C: left shows all boundaries detected by TEM, and right shows only those boundaries with misorientations greater than 10°

4.3 Mechanical Behaviour

The general mechanical behaviour of deformed Al-3Mg and of subsequently annealed material has been reported elsewhere [3]. As-deformed materials are strong, and after yielding show a rather low work hardening rate. Annealing leads to a steady fall in yield stress and increase of work hardening rate: such changes occur gradually as dislocation recovery begins during low temperature anneals, and take place rapidly at the higher annealing temperatures as extensive recovery occurs, and later as recrystallization begins. It is clear that mechanical behaviour depends both on the extent of recovery - since reduction of yield stress and increase of work hardening rate occur for annealing temperatures where dislocation density is reduced but grain/subgrain size hardly changes - and also on the extent of grain growth – since further reduction of yield stress and increase of work hardening rate occur after dislocations have been almost completely lost and major changes of grain size take place.

Fig. 4 shows an attempt to relate the yield stress to the grain size, taking account of all the boundaries detected (at left) or taking account of only those boundaries with high misorientations (>15°) (at right). The diagram at left shows also the Hall-Petch plots corresponding to conventional polycrystalline Al-3Mg materials, as well as data from previous studies of ECAP-processed Al-3Mg materials [6–8]. The present yield stress data fall only loosely on a Hall-Petch plot at left, considering all boundaries present as "grain" boundaries, with a Hall-Petch slope slightly less than that found previously by ECAP processing [7,8], and also with a slope somewhat less than that reported for conventional Al-3Mg polycrystalline material of various grain sizes [6]. Note that the dimension chosen as "grain size" is the width of the elongated regions defined by considering all the boundaries present – high and low angle varieties. The diagram at right instead correlates the yield stress with various measures of grain size: the same "grain width" as at left, taking into account all the boundaries present; the average "grain size" of these elongated regions, again taking account of all the boundaries present; and the "real grain width", taking into account only those boundaries with high angles of misorientation (>15°). For this final parameter – transverse separation of the high-angle grain boundaries – there is the closest agreement with previous studies of grain size dependence of strengthening in Al-3Mg. (Note that the previous studies of ECAP-processed Al-3Mg [7,8] had measured only the high-angle boundaries.) This close agreement with the standard Hall-Petch strengthening

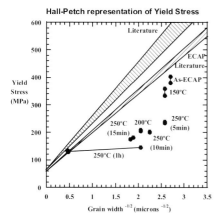 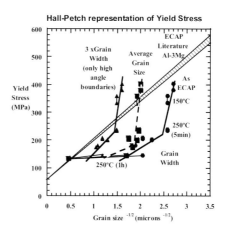

Figure 4: Correlation of yield stress measured in compression (for Al-3Mg deformed to a strain of 5.6 by ECAP, and subsequently annealed at temperatures in the range 150–250 °C for times of 5 min to 1 h) with the square root of boundary separation (i.e. as in a Hall-Petch plot). Left shows a correlation of yield stress with the transverse separation of all boundaries (i.e. the width of the somewhat elongated grains), and literature data of Hall-Petch dependencies of strength on grain size for Al-3Mg alloy. Right correlates yield stress with this same grain width (considering all boundaries), with the average grain size (about the average of grain length and grain width, considering again all boundaries present), and the grain width considering only high angle boundaries (those with misorientations greater than 15°). Note that material annealed for 1h at 250 °C has a duplex structure of unrecrystallized and recrystallized regions, and hence its strength is shown related to two grain size values.

could explain strengthening as due to grain-size hardening for well-annealed materials, and as combined grain-size hardening and dislocation hardening for as-deformed and for slightly-annealed materials, where a significant dislocation density still remains.

Given that the majority of the boundaries are low-angle boundaries for most of the materials studied (all with the exception of the recrystallized parts of the material annealed for 1h at 250 °C) an alternative description of strengthening may be attempted using theories developed for low-angle boundaries or, specifically, for dislocation cell boundaries [9]. For these, the material strengthening due to randomly arranged dislocations can be written as:

$$\sigma = 3 \cdot \mu\, b\, \rho^{0.5}/(2\pi), \tag{1}$$

where μ is the shear modulus, b the Burgers vector, and ρ the dislocation density. Noting that the average separation of such dislocations is $\lambda = 1/\rho^{0.5}$, and the dislocation cell wall spacing is approximately $D_C = 10 \cdot \lambda$, equation (1) can be rewritten as:

$$\sigma = 5\, \mu\, b/(D_C). \tag{2}$$

Thus, we expect a linear relationship between the strength increment and the reciprocal of the dislocation cell (sub-grain) spacing. The slope of the line expected has a value of $5\,\mu b$, i.e. about 28–35 N/m for Al. Fig. 5 shows the experimental yield stress of deformed and annealed materials related to the reciprocal of the cell (all boundaries previously included plus additional dislocation cell walls not included before) width. A reasonable straight line is seen, apart perhaps from the material annealed for 1h at 150 °C after deformation, which shows large strengthening for the measured cell/low-angle boundary spacing. The slope of the line shown in Fig. 5 has a value of about 24 N/m, which is somewhat lower than the expected value, but not excessively low to be inconsistent with this explanation of strengthening.

5 Discussion and Conclusions

Severe plastic deformation by ECAP of an Al-3Mg alloy to a total strain of 5.6 is seen to produce material with a very high dislocation density and a high density of boundaries. Most of these boundaries are low-angle boundaries with typical misorientations of about 5–15°. The fraction of high-angle boundaries, i.e. true grain boundaries, is still low, at about 1/3 of the total. On annealing at low temperatures, recovery occurs with dislocation loss, the loss of some longitudinal boundaries inside the elongated microstructure, and the formation of new transverse boundaries: the microstructure thus appears slightly wider and shorter, and the proportion of low-angle boundaries remains essentially unchanged. After more severe annealing, recrystallization occurs at a few locations through the material, leading to a duplex structure of unrecrystallized and recrystallized regions. Strengthening by the fine microstructure is difficult to explain satisfactorily: combined contributions of grain-size hardening and dislocation hardening provide an incomplete description of the hardening observed; hardening by dislocation cell/sub-grain boundaries may provide a better description.

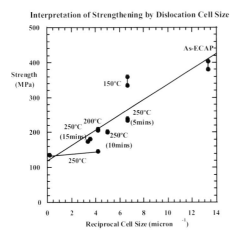

Figure 5: Relationship of yield stress to reciprocal of cell size for as-deformed and for subsequently annealed Al-3Mg samples

6 References

[1] Iwahashi, Y.; Horita, Z.; Nemoto, M.; Langdon, T.G., Acta Mater. 1997, 45, 4733
[2] Morris, D.G.; Muñoz-Morris, M.A., Acta Mater. 2002, 50, 4047
[3] Muñoz-Morris, M.A.; Garcia Oca, C.; Morris, D.G., Scripta Mater. 2002, in press

[4] Prangnell, P.B.; Bowen, J.R.; Gholinia, A., in Proceedings 22nd Riso International Symposium, Riso National Laboratory, Roskilde, Denmark 2001, p. 105
[5] Hughes, D.A., Mater. Sci. Eng. 2001, A319-321, 46
[6] Armstrong, R.W., Douthwaite, R.M., in Proceedings Materials Research Society Symposium, Vol. 362, Mater. Res. Soc., Pittsburgh, Pennsylvania, USA 1995, p. 41
[7] Furukawa, M., Horita, Z., Nemoto, M., Valiev, R.Z., Langdon, T.G., Philos. Mag. 1998, A78, 203
[8] Hayes, J.S., Keyte, R., Prangnell, P.B., Mater. Sci. Tech. 2000, 16, 1259
[9] Weertman, J., Weertman, J.R., in: Physical Metallurgy, Elsevier Science Publ., Amsterdam, Holland, 3rd. Edition, 1983, p. 1260

Dependence of Thermal Stability of Ultra Fine Grained Metals on Grain Size

J. Čížek, I. Procházka, R. Kužel, M. Cieslar, I. Stulíková
Faculty of Mathematics and Physics, Charles University, Prague, Czech Republic
R.K. Islamgaliev
Institut of Physics of Advanced Materials, Ufa State Aviation Technical University, Ufa, Russia

1 Introduction

High pressure torsion (HPT) is a technique based on severe plastic deformation (see [1] for review). The initial coarse-grained sample is deformed by torsion and simultaneously by high pressure of several GPa is applied. Ultra fine grained (UFG) metals with grain size around 100 nm can be produced by HPT. The mean grain size of HPT prepared sample depends on material and parameters of the deformation. Different grain size can be obtained, e.g. by varying the applied hydrostatic pressure.

UFG structure of as-deformed samples represents highly non-equilibrium kind of material structure from thermodynamic point of view, therefore, processes of its recovery towards more equilibrium one can be observed with increasing temperature. In the present work we studied how grain size of as-deformed UFG structure influences its thermal recovery. UFG microstructure and its evolution with increasing temperature were studied by positron lifetime (PL) spectroscopy, which represents non-destructive technique with very high sensitivity to open-volume defects such as vacancies, vacancy clusters, dislocations etc [2]. PL spectroscopy was combined with X-ray diffraction (XRD) and transmission electron microscopy (TEM).

2 Experimental Details

Two samples of UFG Cu (purity 99.99 %) denoted as A and B were prepared by HPT. In order to obtain different grain sizes the sample A was made using hydrostatic pressure $p = 3$ GPa, while the sample B using $p = 6$ GPa. Microstructure of the as-deformed samples was characterized. Subsequently, the samples were isochronally annealed with effective heating rate 1 K/min. The heating was carried out in silicon oil bath up to 250 °C and in vertical furnace with protective argon atmosphere above this temperature. Each annealing step was finished by rapid quenching into water of room temperature. A PL spectrometer with timing resolution of 150 ps (FWHM ^{22}Na) at coincidence count rate of 80 counts/s was used. See [3,4] for its detailed description. XRD studies were carried out with the aid of XRD7 and HZG4 (Seifert-FPM) powder diffractometers using Cu K_α radiation. TEM observations were performed on the JEOL 2000 FX electron microscope operating at 200 kV.

3 Results and Discussion

3.1 As-Deformed Samples

The mean grain sizes d determined by TEM for samples A and B are shown in Table 1. Clearly, the sample A prepared using lower pressure exhibits larger grain size.

Table 1: Structure properties of the as-deformed UFG Cu samples. Hydrostatic pressure p used in HPT is given in the second column. The mean grain size determined by TEM and domain size obtained from XRD and PL spectroscopy, respectively, are given in the next columns. The values in parenthesis represent one standard deviation.

Sample	p [GPa]	grain size [nm]	Domain size [nm]	
		TEM	XRD	PL
A	3	150(10)	120(20)	90(5)
B	6	105(10)	80(20)	70(10)

Bright field TEM image of sample A is shown in Fig. 1. One can see that it contains relatively high number of dislocations. Spatial distribution of the dislocations is strongly non-homogeneous. Grain interiors (non-distorted regions) almost free of dislocations are separated by distorted layers with high dislocation density. The distorted layers with thickness around 10 nm are situated along grain boundaries (GBs). Rather "diffusive" contrast testifies non-equilibrium state of majority of GBs. Microstructure of the sample B exhibits the same features as that of sample A. The only difference is smaller grain size.

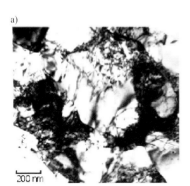
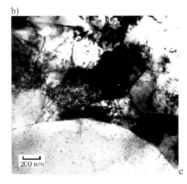

Figure 1: Bright field TEM image of sample A a) as-deformed state b) after annealing up to 250 °C

Broadening of X-ray diffraction profiles with characteristic anisotropy of the type $\beta_{h00} < \beta_{hhh}$ was observed in both samples. It can be explained by dislocation-induced line broadening. We used a simple procedure described in [5] to obtain coherent domain size from analysis of x-ray diffraction profiles. This procedure is similar to more complex procedure introduced by Ungár et al. [6,7]. The mean coherent domain sizes for both samples are given in Table 1.

PL spectra of both samples can be well fitted by two components. Lifetimes and relative intensities of these components are listed in Table 2. Detailed discussion of PL results for sample A can be found in [8]. Therefore, only brief description is given here and we will focus on differences in thermal recovery of both samples. The first component with lifetime $\tau_1 \approx 163$ ps comes from positrons trapped at dislocations inside the distorted regions. The longer component can be attributed to positrons trapped in microvoids, i.e. point defects with size of 4–5 vacancies [8]. No free positron component was found. It means that all positrons are trapped at defects in both samples. Sample B exhibits higher concentration of microvoids compared to sample A.

Table 2: Lifetimes and relative intensities of components resolved in PL spectra of as-deformed samples A and B. The values in parenthesis represent one standard deviation.

Sample	τ_1 [ps]	I_1 [%]	τ_2 [ps]	I_2 [%]
A	164(1)	83(4)	255(4)	17(3)
B	161(3)	64(4)	249(2)	36(5)

We developed a diffusion trapping model, which allows determination of size of the non-distorted regions (grain interiors) and volume fraction of the distorted regions from experimental PL spectra. The model is based on positron diffusion model introduced by Dupasquier et al. [9] modified for the structure of UFG metals. Detailed description of the model is out of scope of this paper and can be found in [8]. Size of the non-distorted regions obtained using the model is given in Table 1. One can see that the size of the non-distorted regions agrees well with the domain size determined by XRD, as both quantities represents size of dislocation-free grain interiors. Both these values are slightly lower than grain size obtained from TEM. It is because the former quantities are closely related to dislocation density, while grain size determined by TEM is connected with change of contrast between grain and GB. For detailed discussion see [8].

3.2 Thermal Recovery

Recovery of UFG structure is realized by similar processes in both samples. Therefore, we will illustrate them on an example of sample A. No change of microstructure of sample A was detected by TEM up to 190 °C. In temperature range from 190 °C to 250 °C isolated recrystallized grains with size ≈ 3 μm appeared in virtually unchanged deformed matrix. This so called abnormal grain growth was observed on UFG Cu also by Islamgaliev et al. [10]. Interface between the recrystallized grain and the deformed matrix is shown in Fig. 1b), which represents TEM image of sample A annealed up to 250 °C. Finally, from 280 °C to 400 °C, recrystallization in whole volume of sample takes place. Bright field TEM image of sample A annealed up to 400 °C is shown in Fig. 2a). Clearly, the material is fully recrystallized and exhibits mean grain size around 3 μm.

Temperature dependence of the mean positron lifetime is plotted in Fig. 2b). Two stages of recovery can be distinguished: (i) a drop in temperature interval 190–250 °C, which corresponds to the abnormal grain growth and (ii) radical decrease of the mean lifetime in temperature range 280-400 °C, which reflects the recrystallization.

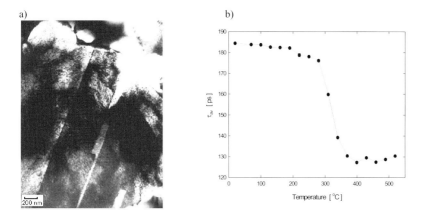

Figure 2: a) Bright field TEM image of sample A after annealing up to 400 °C. b) Mean positron lifetime for sample A as a function of annealing temperature

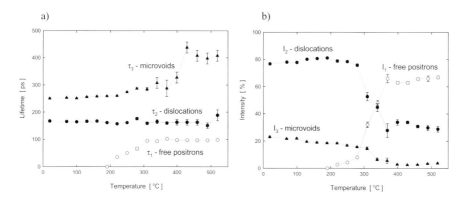

Figure 3: Temperature dependence of a) positron lifetimes and b) relative intensities of individual components for sample A

Temperature dependences of lifetimes and corresponding relative intensities for the sample A are shown in Fig 3 a) and b), respectively. Free positron component appeared in PL spectra from 190 °C when the abnormal grain growth occurs. It represents contribution of free positrons, which annihilate inside the recrystallized grains. Intensity of the free positron component radically increases during recrystalization (temperature interval 280–400 °C). It is accompanied by dramatic decrease of intensity I_2 of positrons trapped at dislocations inside the distorted regions. It reflects the situation when the distorted regions with high dislocation density are consumed by dislocation-free recrystallized grains. Lifetime τ_3 of microvoids increases during recrystallization, while intensity I_3 of this component decreases. It indicates that small microvoids become mobile and are annealed out or cluster to larger ones.

The volume fraction η of the distorted regions obtained by application of the diffusion model on PL spectra of isochronally annealed samples A and B is plotted in Fig 4a) as a function of annealing temperature. The volume fraction η exhibits abrupt decrease during the recrystallization because the distorted regions are replaced by the recrystallized grains. It is clear from Fig. 4a) that in the case of sample A the abrupt decrease of η occurs at temperature interval 300–400 °C. It means in the temperature range when the recrystallization was observed in this sample by TEM, as was shown above. On the other hand, the abrupt decrease of η takes place from 190 °C to 250 °C in sample B. It means that the recrystallization is shifted to significantly lower temperatures in sample B, i.e. the sample with smaller grain size.

The mean size of the non-distorted regions (which is equivalent to domain size) obtained from PL results using the diffusion model is shown in Fig. 4b). As one can see in the figure the domain size substantially increases during the recrystallization. The increase of domain size starts around 300 °C in sample A in concordance with start of the recrystallization. On the other hand, in sample B the domain size begins to increase already from 190 °C. It clearly indicates that the recrystallization takes place at significantly lower temperatures in sample B due to smaller initial grain size. The domain size determined from XRD for sample B is also shown in Fig. 4b). It agrees well with the size of the non-distorted regions obtained from PL spectra. It was not possible to determine domain size at 200 °C and higher temperatures by XRD because the peak broadening was too small. Thus, domain size at 200 °C is higher than $\approx$ 300 nm, i.e. maximum domain size, which can be determined with resolution of the diffractometer used. It agrees well with the onset of the recrystallization at 190 °C determined by PL spectroscopy. Temperature dependence of the mean grain size obtained by TEM for sample B is shown in Fig. 4b) as well. TEM results are in reasonable agreement with PL spectroscopy and XRD. The mean grain size starts to significantly increase from 190 °C when the recrystallization occurs in sample B. Thus, grain growth takes place at about of 100 °C lower temperatures in sample B with smaller grain size of 105 nm compared to the sample A with grain size of 150 nm. We can, therefore, conclude that smaller initial grain size leads to lower thermal stability of corresponding UFG microstructure.

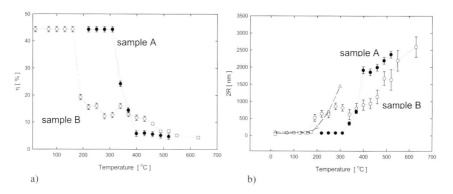

Figure 4: a) Temperature dependence of volume fraction η of the distorted regions obtained from PL spectra using the positron diffusion model. Full circles-sample A, open circles-sample B. b) Mean grain/domain size as a function of annealing temperature. Full circles – size of dislocation free grain interiors obtained from PL spectra of sample A, Open circles – size of the grain interiors obtained from PL spectra of sample B, Open triangles – grain size determined by TEM for sample B, Open squares – domain size determined from XRD for sample B.

4 Conclusions

Thermal stability of two samples of UFG Cu prepared by HPT with initial grain size of 150 nm and of 105 nm was compared in the present work. Recovery of the UFG structure is realized by the same processes in both samples. The abnormal grain growth when isolated recrystallized grains appeared in virtually unchanged deformed matrix is followed at higher temperatures by recrystallization in whole volume of sample. Temperature range when the recrystallization takes place is shifted to about of 100 °C lower temperatures in the sample with smaller grain size of 105 nm compared to that with grain size 150 nm. Hence, we can conclude that smaller initial grain size leads to lower thermal stability of UFG structure.

5 Acknowledgement

This work was partially supported by The Grant Agency of Czech Republics (contract 106/01/D049), The Ministry of Education, Youth and Sports of Czech Republics (project COST OC523.50) and The Grant Agency of Charles University (project 187/2001).

6 References

[1] Valiev, R.Z., Islamgaliev, R.K., Alexandrov, I.V., Prog. Mat. Sci., 2000, 45, 103
[2] Hautojärvi, P., Corbel, C. in Proceedings of the International School of Physics "Enrico Fermi", Course CXXV, ed. Dupasquier, A., Mills, A.P., IOS Press, Varena, 1995, p. 491
[3] Becvár, F., Cižek, J., Lešták, L., Novotný, I., Procházka, I., Šebesta, F., Nucl. Instr. Meth. A, 2000, 443, 557
[4] Becvár, F., Cižek, J., Procházka, I., Acta Phys. Pol. A, 1999, 95, 448
[5] Kužel, R., Cižek, J., Procházka, I., Chmelík, F., Islamgaliev, R.K., Materials Science Forum, 2001, 378-381, 463
[6] Ungár, T., Tichy, G., phys stat. sol. (a), 1999, 171, 425
[7] Ungár, T. in Investigations and Applications of Severe Plastic Deformation, NATO Science Series 3. High Technology -Vol. 80, ed. Lowe, T.C., Valiev, R.Z., Kluwer Academic Publishers, Dordrecht 2000, p. 93
[8] Cižek, J., Procházka, I., Cieslar, M., Ku_el, R., Kuriplach, J., Chmelík, F., Stulíková, I., Becvár, F., Islamgaliev, R.K., Phys. Rev. B, 2002, 65, 094106
[9] Dupasquier, A., Romero, R., Somoza, S., Phys. Rev. B, 1993, 48, 9235
[10] Islamgaliev, R.K., Amirkhanov, N.M., Kurzydlowski, K.J., Bucki, J.J., in Investigations and Applications of Severe Plastic Deformation, NATO Science Series 3. High Technology -Vol. 80, ed. Lowe, T.C., Valiev, R.Z. Kluwer Academic Publishers, Dordrecht 2000, p. 297

Thermo-mechanical Properties of Electrodeposited Ultrafine-grained Cu-Foils for Printed Wiring Boards

A. Betzwar-Kotas[1], V. Gröger[1], G. Khatibi[1,2], B. Weiss[1], I. Wottle[1], P. Zimprich[1]
[1]) Institute of Material Physics, University of Vienna, Austria
[2]) Institute of Physical Chemistry, Material Science, University of Vienna, Austria

1 Introduction

Motivated by the remarkable mechanical properties of metals with grains in the submicrometer range there is a strong interest in methods for industrial production of nanocrystalline materials with favourable defect structures [1]. For severe plastic deformation under high hydrostatic pressure (up to now only laboratory production) this condition is certainly met. On the other hand electro-deposition methods are industrially widely in use and are able to produce nanocrystalline metals if the deposition parameters are chosen suitably which also causes a very particular defect structure. This investigation aims to contribute to find out if nanocrystalline electrodeposited foils might have mechanical advantages over similar materials with grains in the micrometer range. This will be done for the example of Cu foils of 35 µm thickness which are widely in use for printed wiring boards. Detailed information about their microstructure is available and will be shortly outlined. For estimating the reliability Young's modulus and the coefficient of linear thermal expansion are of particular interest but as they are not easily determined for thin foils there is a severe lack of data. For the measurement of strain for determinations of Young's modulus (YM) and coefficient of thermal expansion (CTE) we use a contact free laser speckle correlation method.

2 Experiment and Material Characterization

2.1 Sample Material

Electrodeposited foils of 35 µm thickness have been produced commercially on Ni foils at very different deposition conditions by Gould Electronics Inc.,[2], have been stripped off the substrate and have been subjected to a subsequent stabilization annealing for 30 min at 180 °C (conditions similar as in the Cu/FR-4 laminating process). The commercial electro-deposition process is known to produce a characteristic texture, to introduce hydrogen evolved at the cathode, to insert special additives into the material and to cause extraordinary high vacancy and void concentrations at the grain boundaries. A wealth of systematic investigations of the structure of these deposits [2, 6–9] is used as an additional source. The defect concentration, the crystallographic texture and the grain size are governed mainly by the overpotential (excess of deposition potential over equilibrium value) and the surface active electrolytic additives. The void concentration will thus be inevitably connected with the grain size. Characteristic structural features of electrodeposits are given in Fig.1 schematically.

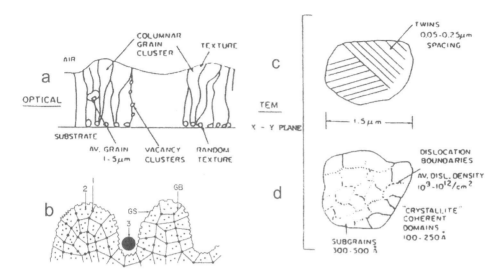

Figure 1: Schematic Structure of Electrodeposit: a) Cross section perpendicular to plane normal b) Voids produced at: 1– grain boundaries, 2 .. growth steps, 3 .. morphological valleys c) Twins, d) Subgrains and coherent

The texture is a fiber texture perpendicular to the surface but random in the surface plane and remains so after thermal treatment. Grain size distributions have been obtained by TEM (2). Contrary to rolling, vacancies and voids produced at grain boundaries, growth steps and morphological valleys (Fig.1b) seem to be stabilized by hydrogen and tend to agglomerate instead of annealing out resulting in local vacancy and void densities up to 1 % [7,8]. The defect densities inside the grains are strongly reduced, sub-grains and special grain boundary structures develop as result of the deposition process (Fig.1d). For comparison also a rolled Cu foil of the same thickness with an in-plane texture of type <100> and <111> in the rolling direction has been included in this investigation. The characterization of the sample materials used in this investigation is given in Table 1. With increasing grain size an increased twin spacing and a reduced vacancy concentration occurs. The structures are far from equilibrium with the strongest instability in the most fine grained sample. The anneal at 180 °C sharpens the sub-grain boundaries, coarsens the vacancy agglomerates and increases grain size, while twin structures remain essentially unchanged. Due to the high concentration of vacancy type defects recrystallization occurs at low activation energies [7,8] up to 250 °C. Thus the samples recrystallize during the measurement when testing temperatures up to 300 °C are applied. The recrystallization rate for samples with smaller grains is increased due to their higher vacancy and void content. This can be seen from the grain sizes after testing given also in Table 1.

Table 1: Characterization of the copper foils (thickness 35 µm, thermal anneal 180 °C for 30 min)

Foils	Chemical Comp.	Microstructure	Grain size (µm) before / after testing at 300 °C		Texture	Remarks
AM	99,99	Equiaxed, no growth twins high disl. density	0,3	2	Random in (z) and in x–y planes	High point defect conc. near GB and in GB 0,1%,
GR 3 (high profile)	99,99	Columnar morphology // z, numerous growth twins x–y plane equiaxed	2	2–3	<220> fibre in z x–y plane: random	vacancy clusters, microvoids, gas bubbles
GR 3 (low profile)	99,99	Few growth twins x–y plane equiaxed	0,5	2	<220> fibre in z x–y plane: random	
GR 8 (rolled)	99,9 (> 100 ppm Oxygen) (ETP grade)	Pancaked in the x–y plane, disl. cells, subgrains	4,5	30–40	Mixed <100> plus <111> in x-y plane	Large flattened as-cast grains, highly anisotropic microstructure

2.2 Mechanical and Thermal Measurement

For the determination of Young's modulus displacement controlled tests in a microtensile machine (load capacity 10–50 N) have been carried out at a strain rate of $5 \cdot 10^{-5}\,\text{s}^{-1}$ making use of a laser speckle correlation strain sensor with CCD plane cameras [3]. Different regions of the sample surface being a known distance (20 mm) apart in measurement direction are illuminated coherently by collimated laser beams of a few mW power. The images of the surface regions obtained by lenses are recorded by CCD cameras. Due to the natural surface roughness of the sample and the scattering at the lens aperture an interference (speckle) pattern in the image occurs which is characteristic for a particular surface region. From the shifts of these patterns in subsequent measurements and the magnification of the system the surface shifts are determined. From the simultaneous shifts of two areas Δ_1 and Δ_2 separated the base-length l, the strain is obtained as $\varepsilon = (\Delta_1 - \Delta_2)/l$. Foils are glued to specially designed grips of the microtensile machine and aligned by an x–y die. For determination of Young's modulus a loading-unloading technique during the tensile test had to be used as the initial slope portion was usually distorted due to unavoidable mounting imperfections. The average slopes of the loading parts of the curve consecutive to unloading were used for the determination [4]. The errors rise from 5 % at room temperature up to 10 % at 300 °C due to the reduced stress level and some creep influence. The setup of the laser speckle based dilatometer (LSBD) used for a measurement of the thermal expansion is shown in Fig.2. It consists of a specially designed heating chamber filled with Ar atmosphere avoiding oxidation above 230 °C, which was observerd to be typical for the foils under investigation [5]. The heating surface supports the sample which is contacted by use of a thin film of silicon thermal paste in the case of thin foils. As the paste remains viscous over the whole temperature range it did not influence the measurement mechanically but reduces also wrinkling of the foil. With an equilibrium time of 10 min, a reproducibility of 1,5 % could be achieved. A temperature stability of ± 0.5 °C was achieved by a PID-controlled furnace. From three successive measuring runs for each foil a fourth order polynomial was fitted to the data whose derivation gives the coefficient of thermal expansion with an error of about 1 ppm per °C.

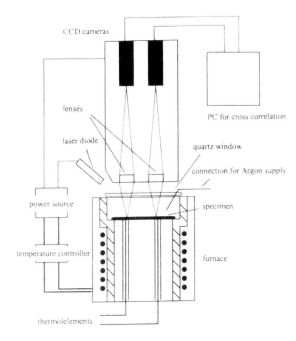

Figure 2: Laser-Speckle-Based Dilatometer (LSBD) formed by the combination of a Laser-Speckle-Correlation-System and a special designed heating chamber for thermal strain measurements of thin foils [5]

3 Results and Discussion

The dependence of Young's modulus (YM) on measuring temperature for electrodeposited materials with different grain sizes, for the rolled foil and for bulk copper with random texture are shown in Fig.3. The temperature dependence of the coefficient of linear thermal expansion (CTE) for electrodeposited materials and a standard copper material (National Institute of Standards and Technology, SRM 736) is given in Fig.4. A considerable change in grain size (indicated by the inserted grain size before and after the measurement) occured during the measurements, most pronounced for the materials with the finest grains in a way that all electrodeposited samples ended up at comparable grain sizes. This effect was similar in both measurements due to the comparable measuring times.

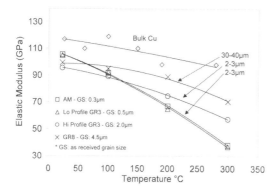

Figure 3: Influence of test temperature on Young's modulus for copper foils of 35 μm thickness (AM and GR3 electrodeposits, GR8 rolled foil). The grain sizes after the 300 °C test (right insert) differ from the initial ones (left insert) due to microstructural instabilities [4]

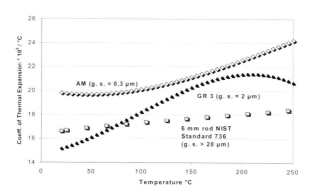

Figure 4: Coefficient of linear thermal expansion for electrodeposited Cu foils and a standard bulk sample [5]

Thus the dependences of YM and CTE on temperature are strongly influenced by the microstructural changes in addition to a direct temperature dependence. Assuming that the direct temperature dependence is not too different from the bulk values one can attribute the change of the relative values with respect to the bulk mainly to microstructural changes being strongest in the most fine-grained samples. The influences on YM and CTE stem from the particular structure of the grain boundaries, the agglomerated vacancies, voids and impurities [10]. YM could be additionally reduced by dislocation bow-out inside the grains. It seems that the extraordinary reduction of YM for the fine-grained electrodeposited samples requires a crack mechanism. Cracks could be nucleated at voids the crack closing could be inhibited by hydrogen. It can be seen that the CTE-data do not obey the Grüneisen relation, which is characteristic for inhomogeneus material with e.g. a high concentration of voids and cracks [11]. Also some phase transition-like mechanism can not be excluded, because of the special non-equilibirium microstructure of the foils.

4 Acknowledgement

The authors would like to thank the Fonds zur Förderung der wissenschaftlichen Forschung, Wien (P12311 TEC, P14732 TEC) for financial support. We thank H. Merchant for providing the copper foils and for critical and helpful discussions.

5 References

[1] Valiev, R.; Islamgaliev, R.K.; Alexandrov, I.V., Progress in Material Science, Vol. 45, 2000, p. 103–189

[2] Merchant H. D.; Defect Structure of Electrodeposits, Defect Structure, Morphology and Properties of Deposits, Ed. H. Merchant, The Minerals, Metals & Materials Society 1995, p. 1–59

[3] Anwander M., Zagar B., Weiss B. and Weiss H.; Non-contacting strain measurements at high temperatures by digital laserspeckle technique, Experimental Mechanics, Vol. 40, No. 1, March 2000, p. 697–702

[4] Khatibi G., Gröger V., Weiss B., Merchant H. and Wiechmann R.; Thermo-Mechanical Response of Cu Foils for Microsystems, "MATERIALS WEEK 2001 – Proccedings", Ed. Werkstoffwoche-Partnerschaft GbR, Publisher: Werkstoff-Informationsgesellschaft mbH, Frankfurt 2002

[5] Zimprich P., Wottle I., Merchant H., Wiechmann R., Zagar B. and Weiss B.; Coefficient of Thermal Expansion (CTE) of Thin Copper Foils for Electronic Laminate Structures, "MATERIALS WEEK 2001 – Proccedings", Ed. Werkstoffwoche-Partnerschaft GbR, Publisher: Werkstoff-Informationsgesellschaft mbH, Frankfurt 2002

[6] Merchant H. D.; Annealing Kinetics and Embrittlement of Electrodeposited Copper, Journal of Electronic Materials, Vol. 22, No. 6, 1993, p. 631–637

[7] Merchant H. D. and Girin O. B.; Defect Structure and Crystallographic Texture of Polycrystalline Electrodeposits, „Electrochemical Synthesis and Modification of Materials", MRS Proccedings, Vol. 431, 1997, p. 433–444

[8] Merchant H. D.; Thermal Response of Electrodeposited Copper, Journal of Electronic Materials, Vol. 24, No. 8, 1995, p. 919–925

[9] De Angelis R. J., Knorr D. B. and Merchant H. D.; Through-Thickness Characterization of Copper Electrodeposits, Journal of Electronic Materials, Vol. 24, No. 8, 1995, p. 927–933

[10] Kristc V., Erb U. and Palumbo G.; Effect of Porosity on Young`s modulus of nanocrystalline materials, Scripta Metallurgica et Materialia, Vol. 29, 1993, p. 1501–1504

[11] Pampuch R.; Constitution and Properties of Ceramic Materials, Materials Science Monographs 58, Elsevier, Amsterdam, 1991, p. 252–257

Effect of Grain Boundary Phase Transitions on the Superplasticity in the Al–Zn system

G.A. López, B.B. Straumal[1], W. Gust, E.J. Mittemeijer
Max Planck Institute for Metals Research and Institute of Physical Metallurgy, University of Stuttgart, Stuttgart, Germany
[1]Institute of Solid State Physics RAS, Chernogolovka, Russia

1 Introduction

Most important properties of modern materials in high-technology applications are strongly influenced by the occurrence of interfaces such as grain boundaries (GBs) [1]. In the last decade the study of nanocrystalline solids has increased considerably. The inherently high concentration of GBs in these materials makes the understanding of interface properties of great importance. Processes that can modify the properties of the GBs affect significantly the bulk behavior of polycrystalline materials. Among these processes the GB phase transitions can be mentioned as important examples [2–7]. In recent times, GB wetting phase transitions have been included in the traditional equilibrium diagrams of several systems [7–13].

The GB energy, σ_{GB}, plays a critical role for the occurrence of wetting. Figure 1 shows the contact angle Θ formed between a bicrystal and a liquid phase. When σ_{GB} is lower than $2\sigma_{SL}$, where σ_{SL} is the energy of the solid/liquid interface, the GB is nonwetted and $\Theta > 0°$ (Fig. 1a). However, the GB is wetted and the contact angle $\Theta = 0°$ if $\sigma_{GB} \geq 2\sigma_{SL}$ (Fig. 1b). Taking into account the temperature dependences of σ_{GB} and $2\sigma_{SL}$, where the two curves intersect the GB wetting phase transition will take place upon heating (Fig. 1c). The temperature of the intersection is identified as the wetting temperature, T_W, and for every temperature higher than T_W the contact angle is $\Theta = 0°$.

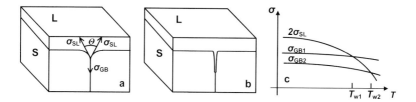

Figure 1: A nonwetted GB in contact with a liquid phase at $T < T_W$, $\Theta > 0°$ (a). A completely wetted GB, $\Theta = 0°$, $T \geq T_W$ (b). Schematic dependences of $\sigma_{GB}(T)$ and $2\sigma_{SL}(T)$ for two different GBs. They intersect at the T_{W1} and T_{W2} of the GB wetting phase transition (c). L = liquid; S = solid

The superplasticity has drawn much interest in recent years. Usually this property was observed at relatively low strain rates, typically about 10^{-4} to 10^{-3} s^{-1}. Sometimes it occurs at extremely high strain rates (up to 10^2 s^{-1}) and in this case it is referred as high-strain rate superplasticity (HSRS) [14–19]. There is agreement that a small grain size is important for the occurrence of HSRS. Additionally, this phenomenon has often been observed at temperatures close to the matrix solidus temperature [20,21]. The phenomenon of HSRS is most pronounced in the Al–Mg–Zn ternary alloys. The reason for HSRS could be the GB phase transitions. Therefore, we decided to study the GB phase transitions in the Al–Mg and Al–Zn binary systems. Recently, the grain boundary wetting phase transition in the two-phase (S+L) region of the Al–Mg system has been investigated [22]. In this work, the occurrence of *prewetting* or *premelting* was given as the reason for the HSRS. In the present contribution, an analogous study has been performed for the Al–Zn system.

2 Experimental

Cylindrical samples (diameter 7 mm) of seven Al–Zn alloys with Zn contents of 10, 20, 30, 40, 60, 75 and 85 wt.% were produced from Al (99.999 wt.%) and Zn (99.995 wt.%). Slices (2 mm thick) of the different alloys were cut and sealed into evacuated silica ampoules with a residual pressure of approximately $4 \cdot 10^{-4}$ Pa at room temperature. Then, several samples were annealed in furnaces for three days at temperatures between 390 and 630 °C, in steps of 20 °C, and subsequently quenched in water. The accuracy in the annealing temperature was ±1 °C. After quenching, the specimens were embedded in resin and then mechanically ground and polished, using 1 μm diamond paste in the last polishing step, for the metallographic study. The samples were then etched and investigated by means of light microscopy.

A quantitative analysis of the wetting transition was performed adopting the following criterion: every GB was considered to be wetted only when a liquid layer had covered the whole GB; if such a layer appeared to be interrupted, the GB was regarded as a nonwetted GB. Accordingly, the percentage of wetted GBs was determined on the basis of light microscopy analysis. At least 100 GBs were analysed at each temperature.

3 Results and Discussion

Figure 2 shows optical micrographs of samples annealed at four different temperatures. As can be seen almost all the GBs have been covered by a liquid layer after annealing at 620 °C (dark layers at the original GBs in Fig. 2a). Upon annealing at 560 °C the number of wetted GBs is considerably lower (Fig. 2b), reaching 66.7 % of the total, whereas in the sample annealed at 480°C this fraction was just 35 % (Fig. 2c). In the samples treated at 440 °C and lower temperatures no GBs were wetted (Fig. 2d). Light gray in the interior of the grain indicates that partial melting has occurred (Fig. 2b and c); however, such a liquid in the solid remains isolated. Pores (see black spots), probably produced by the rapid quenching, can also be observed. The fraction of wetted GBs is shown as a function of temperature in Fig. 3a. A gradual increase of this fraction can be observed between 440 and 620 °C from 0 to 100 %. Therefore, the GB wetting phase transition proceeds in the Al–Zn system between 440 and 620 °C. All GB wetting phase transition tie-lines corresponding to each individual GB lie between these two temperatures. In other words, the minimal (T_{wmin}) and maximal (T_{wmax}) temperatures of GB wetting phase transition in the Al–Zn system are 440 and 620 °C, respectively. Assuming that the bulk solidus and bulk liquidus do also hold for the GB phases at their interface, the respective tie-lines at T_{wmin} and T_{wmax} have been drawn in the two-phase (S+L) region area of the Al–Zn bulk phase diagram (Fig. 3b).

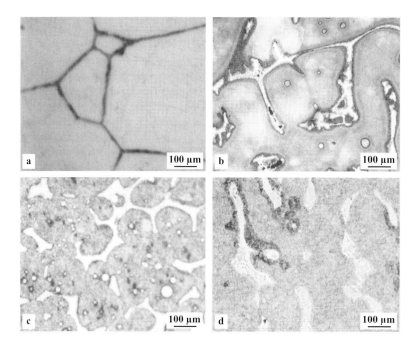

Figure 2: Optical micrographs of samples annealed at 620 °C (a), 560 °C (b), 480 °C (c) and 440 °C (d)

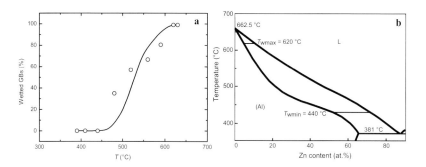

Figure 3: Temperature dependence of the fraction of wetted GBs in the Al–Zn system (a). Al–Zn equilibrium phase diagram with the lines of bulk phase transitions [23] and the tie-lines of GB wetting phase transitions (thin lines) (b)

Taking into account that wetting of GBs has been observed in the bulk (S+L) two-phase region of the Al–Zn system, it can be suggested that a *premelting* or *prewetting* of the GBs may also take place in the S single-phase region near the bulk solidus line. This kind of GB phase transitions has extensively been observed for several systems [2–6, 11, 24, 25] and is interpreted according to the theory developed by Cahn [26, 27]. Due to this phenomenon a thin liquid-like layer can be formed at the GBs.

The presence of a liquid layer at the original positions of the GBs will critically affect the mechanical properties of an alloy, as was observed by Iwasaki et al. in the study of pure shear of a commercial Al–Mg alloy [28].

In several nanostructured Al ternary alloys and nanostructured Al metal-matrix composites, containing Zn and Mg, high-strain rate superplasticity has been observed [16–21, 28–32]. The maximal elongation-to-failure increases drastically from 200–300 % up to 2000–2500 % in a very narrow temperature interval of about 10 °C just below the respective solidus temperature. Until now no satisfactory explanation has been offered for this phenomenon.

Due to the fact that the Al–Zn and Al–Mg systems are the basis of multicomponent alloys which present HSRS, and having observed wetting of GBs for these systems, it is suggested that GB premelting or prewetting is responsible for the HSRS. In that case, a liquid-like thin layer would cover the GBs, leading to an enhanced plasticity of the materials. Considering this hypothesis and using results published on HSRS for Al–Zn–Mg alloys, it could be observed that GB wetting proceeds in multicomponent alloys as well as in binary systems [20,21, 29–32]. From the micrographs published in [20, 21, 29, 30] it was estimated that T_{wmax} = 535 °C for the 7xxx Al–Mg–Zn alloys (Fig. 4). At 475 °C about 50 % of GBs are still wetted (see $T_{w50\%}$ in Fig. 4). Unfortunately, the micrographs of the Al–Mg–Zn alloys published in [20, 21, 29, 30] do not permit us to estimate $T_{wmin} < T_{w50\%}$. Comparing the T_w values of multicomponent alloys with those of binary alloys, it seems to be that the presence of third or fourth elements lowers the wetting temperature. But further investigations must be performed to clarify this tendency.

Finally, most of the HSRS tests were performed at temperatures slightly below the bulk solidus and the temperature of maximal elongation-to-fracture is also close to the bulk solidus temperature. Therefore, we conclude that *premelting* or *prewetting* can be the reason for the HSRS in these alloys.

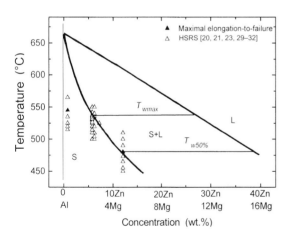

Figure 4: Pseudobinary phase diagram for 7xxx Al–Zn–Mg alloys (thick lines), containing GB wetting phase transition lines (thin lines)

4 Conclusions

Grain boundary wetting has been observed in Al–Zn alloys. On the basis of grain boundary wetting data the maximal and minimal temperatures, T_{wmax} and T_{wmin}, of grain boundary wetting phase transitions have been indicated and the corresponding tie-lines in the two-phase (S + L) region of the respective bulk phase diagram have been drawn. The occurrence of grain boundary prewetting or premelting is considered as the origin of high-strain rate superplasticity.

5 Acknowledgements

The financial support from the Russian Foundation for Basic Research (contract 01-02-16473) and the German Federal Ministry for Education and Research (contract WTZ RUS 00/209) is acknowledged.

6 References

[1] Langdon, T.G.; Watanabe, T.; Wadsworth, J.; Mayo, M.J.; Nutt, S.R.; Kassner, M.E., Mater. Sci. Eng. A 1993, 166, 237–241
[2] Rabkin, E.I.; Semenov, V.N.; Shvindlerman, L.S.; Straumal, B.B., Acta metall. mater. 1991, 39, 627–639
[3] Noskovich, O.I.; Rabkin, E.I.; Semenov, V.N.; Straumal, B.B.; Shvindlerman, L.S., Acta metall. mater. 1991, 39, 3091
[4] Straumal, B.B.; Noskovich, O.I.; Semenov, V.N.; Shvindlerman, L.S.; Gust, W.; Predel, B., Acta metall. mater. 1992, 40, 795–801
[5] Straumal, B.; Rabkin, E.; Lojkowski, W.; Gust, W.; Shvindlerman, L.S., Acta mater. 1997, 45, 1931–1940
[6] Chang, L.-S.; Rabkin, E.; Straumal, B.B.; Baretzky, B.; Gust, W., Acta mater. 1999, 47, 4041–4046
[7] Straumal B.B.; Gust, W., Mater. Sci. Forum 1996, 207-209, 59–68
[8] Straumal, B.; Muschik, T.; Gust, W.; Predel, B., Acta metall. mater. 1992, 40, 939–945
[9] Straumal, B.; Molodov, D.; Gust, W., J. Phase Equilibria 1994, 45, 386–391
[10] Straumal, B.; Semenov, V.; Glebovsky, V.; Gust, W., Defect Diff. Forum 1997, 143–147, 1517–1522
[11] Chang, L.-S.; Rabkin, E.; Straumal, B.B.; Hofmann, S.; Baretzky, B.; Gust, W., Defect Diff. Forum 1998, 156, 135–146
[12] Straumal, B.; Gust, W.; Watanabe, T., Mater. Sci. Forum 1999, 294–296, 411–414
[13] A.P. Sutton, R.W. Balluffi, Interfaces in Crystalline Materials, Oxford University Press, New York, 1995, p. 400–413
[14] Higashi, K.; Tanimura, S.; Ito, T., MRS Proc. 1990, 196, 385–390
[15] Mabuchi, M.; Imai, T., J. Mater. Sci. Lett. 1990, 9, 762–763
[16] Nieh, T.G.; Henshall, C.A.; Wadsworth, J., Scripta metall. 1984, 18, 1405–1408
[17] Mabuchi, M.; Higashi, K.; Okada, Y.; Tanimura, S.; Imai, T.; Kubo, K., Scripta metall. 1991, 25, 2003–2006

[18] Nieh, T.G.; Gilman, P.S.; Wadsworth, J., Scripta metall. 1985, 19, 1375–1378
[19] Higashi, K.; Okada, Y.; Mukai, T.; Tanimura, S., Scripta metall. 1991, 25, 2053–2057
[20] Higashi, K.; Nieh, T.G.; Mabuchi, M.; Wadsworth. J., Scripta metall. mater. 1995, 32, 1079–1084
[21] Takayama, Y.; Tozawa, T.; Kato, H., Acta mater. 1999, 47, 1263–1270
[22] Straumal, B.B.; López, G.A.; Mittemeijer, E.J.; Gust, W.; Zhilyaev, A.P., Defect Diff. Forum 2003, 216–217, 307–312
[23] Massalski, T.B. et al. (Eds.), Binary Alloy Phase Diagrams, ASM International, Materials Park, OH, 1993, p. 239–240
[24] Schölhammer, J.; Baretzky, B.; Gust, W.; Mittemeijer, E.; Straumal, B., Interf. Sci. 2001, 9, 43–53
[25] Molodov, D.A.; Czubayko, U.; Gottstein, G.; Shvindlerman, L.S.; Straumal, B.B.; Gust, W., Phil. Mag. Lett. 1995, 361–368
[26] Cahn, J.W., J. Chem. Phys. 1977, 66, 3667–3672
[27] Cahn, J.W., J. Phys. Colloq. 1982, 43-C6, 199–213
[28] Iwasaki, H.; Mori, T.; Mabuchi, M.; Higashi, K., Acta mater. 1998, 46, 6351–6360
[29] Baudelet, B.; Dang, M.C.; Bordeux, F., Scripta metall. mater. 1992, 26, 573–578
[30] Imai, T.; Mabuchi, M.; Tozawa, Y.; Murase, Y.; Kusul, J., in: Metal & Ceramic Matrix Composites: Processing, Modelling & Mechanical Behaviour, Bhagat R.B. et al. (Eds.), TSM-ASME, Warrendale, PA, 1990, p. 235–242
[31] Mabuchi, M.; Higashi, K.; Imai, T.; Kubo, K., Scripta metall. 1991, 25, 1675–1680
[32] Furushiro, N.; Hori, S.; Miyake, Y., in: Proc. Int. Conf. Superplast. Adv. Mater., ICSAM-91, Hori, S. et al. (Eds.), Japan Soc. Res. Superplast., Sendai, 1991, p. 557–563

Microstructure and Thermal Stability of Tungsten based Materials after Severe Plastic Deformation

A. Vorhauer[1,2], W. Knabl[3], R. Pippan[1,2]
[1]Erich Schmid-Institute of Materials Science of the Austrian Academy of Sciences, Leoben, Austria;
[2]Christian Doppler Laboratory of Local Analysis of Deformation and Fracture, Leoben, Austria;
[3]PLANSEE Aktiengesellschaft, Reutte, Austria.

1 Introduction

Tungsten and its alloys have a wide range of application for plasma-facing components. In order to refine the initial microstructure of these materials, Severe Plastic Deformation (SPD) has been applied to three different types of tungsten-based materials: pure tungsten (W), a tungsten-rhenium alloy (W26 wt-%Re) and a lanthanum-oxide (La_2O_3) dispersion strengthened tungsten (WL10; tungsten with 1 wt-% La_2O_3).

An effective refinement of the initial microstructure (forged condition) can only be reached by SPD at low homologous temperatures [1] in order to prevent (dynamic) recovery or (dynamic) recrystallization. For common degrees of deformations (max. effective stain $\varepsilon_{vM} \sim 1$) the temperature limits are given as $\sim 0.3\ T_m$ for recovery and $\sim 0.5\ T_m$ for recrystallization. The formation of very fine microstructures with homogenously distributed size of the structural elements necessitates effective von Mises strains ε_{vM} of more than 6–8 [2,3].

The present paper describes the microstructural changes in the three different types of tungsten based materials after Severe Plastic Deformation (SPD) up to effective strains of $\varepsilon_{vM} = 12$. Because of the large density of lattice defects in such materials, the microstructures are very metastable and thus the thermal stability of the SPD microstructures is investigated. Measurements of the micro hardness at different states of materials processing provide relation between microstructural features (size of structural elements) and strength of the material.

2 Materials and Processing

Many different types of SPD are commonly known. The actually most popular variants are Equal Channel Angular Extrusion and High Pressure Torsion (HPT) [3]. In order to do SPD at temperatures as low as possible, we have chosen HPT for this purpose, because this method provides a material processing at high hydrostatic pressures and therefore a processing of brittle (tungsten based) materials at low homologous temperatures without fracture.

A sketch of a HPT tool is shown in Fig. 1. In principal a disc shaped specimen is deformed under a hydrostatic pressure by rotating the two plungers with respect to each other. In order to prevent a sliding of the specimen all contact areas between specimen and plunger were sandblasted before material processing.

Discs, 6mm in diameter were subjected to HPT at a hydrostatic pressure of about 7 GPa. A special heating unit allowed material processing at elevated temperatures (max. 370 °C). Table 1 gives an overview about the investigated specimens and the applied temperature intervals.

Table 1: Summary of the processed specimens

Material	Processing temperature °C	T/T_m
W	330–290	0.164–0.153
W26Re	370–290	0.175–0.153
WL10	210–185	0.131–0.124

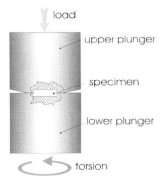

Figure 1: Principal of Severe Plastic Deformation by High Pressure Torsion

3 Microstructure after Severe Plastic Deformation

The microstructural investigation was performed in a Scanning Electron Microscope (SEM), by capturing micrographs with a detector for Back Scattered Electrons (BSE). In a single-phase material, the energy of the detected backscattered electrons depends on both the orientation of the crystallites with respect to the direction of the incident electron beam and the density of dislocations. Thus, differently oriented structural elements appear in these micrographs as regions of different grey-values, which allows an estimation of the typical microstructural sizes.

A comparison of BSE micrographs of the investigated microstructures at $e_{vM} \sim 2$ and 12 is given in Figure 2. These micrographs were all captured at the same SEM magnification of 60.000.

3.1 BSE Micrographs of SPD W

The comparison of the two BSE micrographs of W at effective strains of 2 and 12 shows the inherent subdivision of the initial grains into very small crystallites. At $\varepsilon_{vM} \sim 2$, the size of the structural elements is wide spread within the range of 50–700 nm; the mean size of these elements is about 400 nm. At $\varepsilon_{vM} \sim 12$ the mean size of the structural elements is somewhat smaller than 100 nm and the distribution of microstructural sizes is much narrower (40–200 nm) than at lower degrees of deformation.

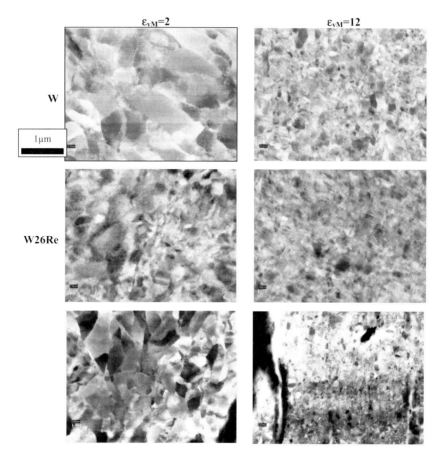

Figure 2: BSE micrographs of W, W26Re and WL10 at effective strains of 2 and 12. All micrographs are of the same magnification

3.2 BSE Micrographs of SPD W26Re

The microstructure of W26Re after HPT shows the same tendencies as W. Again the uniformity in structural size increases with increasing strain and also the mean size of these elements decreases with increasing effective strain. But now the size of the structural elements is smaller than in W. The spread in structural size decreases from values of 50–500 nm at $\varepsilon_{vM} \sim 2$ to values of somewhat smaller than 30–200 nm at $\varepsilon_{vM} \sim 12$. In the case of W, the dislocations tend to form Low Energy Dislocation (LED) structures in order to minimize the system-energy by decreasing long-range stress fields. The large concentration of Re atoms in W26Re reduces the mobility of dislocations. Thus LEDs cannot be formed as efficiently as in W which leads to larger internal stresses and therefore to a lower sharpness of the BSE micrographs of W26Re.

3.3 BSE Micrographs of SPD WL10

Detailed information about the substructure of the tungsten-matrix is given in Fig. 3. At a strain $\varepsilon_{vM} \sim 2$, the substructure of the matrix in WL10 is more inhomogeneous as regards size of the structural elements than that of W at the same effective strain. This larger inhomogeniety possibly is induced by the interaction of slip bands with the lanthanum-oxides (La_2O_3) during deformation, resulting in the formation of dislocation pile-ups at the matrix-oxide interfaces. Furthermore it should be noted that the slightly lower homologous processing temperature of the WL10 also could lead to a smaller mean size of the structural elements.

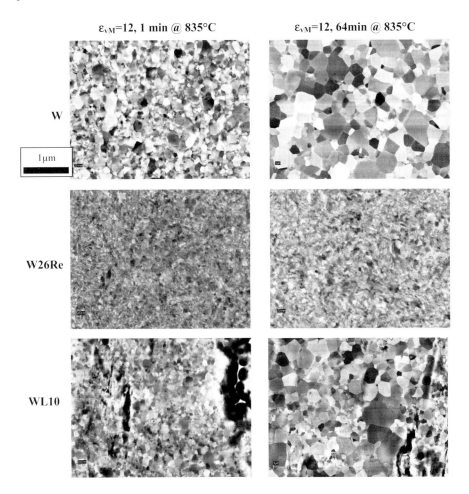

Figure 3: Comparison of BSE micrographs of W, W26Re and WL10 at a strain $\varepsilon_{vM} \sim 12$ after different times of thermal treatment at 835 °C. All micrographs are of the same magnification.

4 Thermal stability

Thermal treatments for 1, 4, 16, 32 and 64 minutes at 835 °C (~ 0.3 T_m) of the produced microstructures were carried out in vacuum at 9.333 mbar in order to prevent oxidation.
The following features are clearly visible:

W: The size distribution of the structural elements at $\varepsilon_{vM} \sim 12$ is about 40–350 nm after 1 minute and about 60–1200 nm after 64 minutes at 835 °C. In comparison to the BSE micrographs of the SPD state, now the structural elements are separated by sharp boundaries. This is an indication that recovery processes have taken place that decreased internal stresses.

WL10: The structural elements in WL10 at $\varepsilon_{vM} \sim 12$ after 1 and 64 minutes of thermal treatment are somewhat smaller than in W. The smaller HPT temperature again probably induces this. The distribution of the size of structural elements is more wide spread than in W.

W26Re: The comparison of the BSE micrographs after 1 and 64 minutes shows clearly that the microstructure of W26Re remains almost equal during this treatment. The annealing procedure at 835 °C in W26Re does not lead to microstructures consisting of structural elements with clear boundaries. With increasing time of heat treatment more and more structural elements are clearly visible, but they are not comparable to the microstructures of W or the W matrix of WL10. The large amount of Re in W26Re defers the recovery processes efficiently to longer times and/or to higher temperatures.

A first impression of the effect of microstructure on the mechanical properties is given by an analysis of the change in microhardness after different states of materials processing. The measurement of the Vickers microhardness was done at loads of 0.637, 1.089 and 1.500 N as a function of the distance from the center of the HPT specimen (effective strain!) at all states of materials treatment. For a better comparability, microhardness values are given for indent sizes of $\delta = 20$ µm at distances of the center according to effective strains ε_{vM} of 6 and 12.

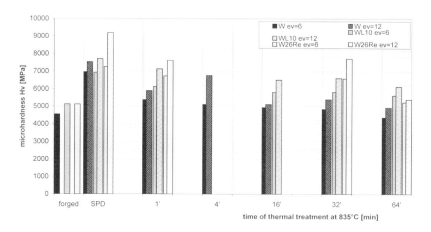

Figure 4: Microhardness at different states of materials processing

In the forged state, the micro hardness of the three materials is about equal within a range of 10 percent. The same situation is present in the SPD state but now the microhardness values are about a factor 1.35–2 larger than in the forged state. The micro hardness at $\varepsilon_{vM} \sim 12$ is about 10 percent large than at $\varepsilon_{vM} \sim 6$. With increasing time of thermal treatment the microhardness decreases. An inherent drop is present after 1 minute at 835 °C for all materials and degrees of deformation. Increasing time of treatment decreases the microhardness continuously. The smallest differences between 1 and 64 minutes are given for WL10. After 64 minutes the microhardness of W and W26Re is about the same as in the forged state. WL10 shows at this time microhardness values that are about 10–20 percent larger than in the forged state.

5 Summary

1. SPD at low homologous temperatures leads in the three materials – W, WL10 and W26Re – to very small structural elements. BSE investigations have shown that the internal stresses within W26Re after SPD are larger than in W and WL10. The distribution in structural size is nearly the same in W and WL10. The WL10 structures are somewhat smaller than in W. Elongated lanthanum-oxide dispersoids do not take part in the SPD deformation but their alignment with the shear direction increases with increasing strain. In W26Re it was not possible to estimate the structural size from BSE micrographs.
2. Thermal treatment leads in all three materials to recovery processes, which decrease internal stresses (sharpness of boundaries between structural elements increases). This effect is not so distinct in W26Re as in W and the W matrix of WL10; only a few structural elements are clearly visible in W26Re.
3. The microhardness of the investigated materials in the SPD state is about 1.35–2 times larger than in the forged state. An inherent drop in microhardness occurs after 1 minute at T_{wmax} 835 °C for all materials and degrees of deformation. Further thermal treatment at the same temperature leads to a further coarsening of the microstructure and a somewhat smaller reduction in microhardness.

6 Acknowledgement

This work has been carried out within Association EURATOM-ÖAW and is supported by the Bundesministerium für Bildung, Wissenschaft und Kultur. The content of the publication is the sole responsibility of its authors and does not necessarily represent the views of the European Commission or its services.

7 References

[1] Valiev, R.Z., Islamgaliev, R.K., Alexandrov, I.V., Progress in Materials Science, Vol.45, 2000, 103–189
[2] Chang, C.P., Sun, P.L., Kao, P.W., Acta Mater., Vol 48, 2000, 3377–3385
[3] Vorhauer, A., diploma thesis, university of Leoben, 2000
[4] Jiang, H., Zhu, T. Butt, D.O., Alexandrov, V., Lowe, T.C., Materials Sci. Eng., A290, 2000, 128–138

Development of Microstructure and Thermal Stability of Nanostructured Chromium Processed by Severe Plastic Deformation

R. Wadsack[1], R. Pippan[1,2], B. Schedler[3]
[1]Erich Schmid Institute for Materials Science of the Austrian Academy of Sciences, Leoben (A)
[2]Christian Doppler Laboratory for Local Analysis of Deformation and Fracture, Leoben (A)
[3]Plansee AG, Reutte (A)

1 Introduction

Since chromium is superior to most materials with regard to the low neutron-induced radioactivity, it is considered as material in fusion technology. Limitations for the structural application in industry are the low ductility at room temperature and a Ductile to Brittle Transition Temperature (DBTT) which lies significantly above room temperature. In the last years intense research has started to produce nanostructured materials by Severe Plastic Deformation (SPD) [1]. Compared with the undeformed materials these materials with grain sizes clearly smaller than 1 µm are distinguished by an increase in strengths without loosing ductility.

In the present study unirradiated chromium with a purity of 99,97 % (DUCROPUR) has been deformed by High Pressure Torsion (HPT) and Cyclic Channel Die Compression (CCDC) [2–4]. The chemical composition and the mechanical characteristics of the undeformed chromium are given in [4, 5]. This paper is mainly focused on the structural refinement and the thermal stability of the achieved SPD microstructures.

2 SPD of pure chromium

2.1 High Pressure Torsion (HPT)

HPT has been used for the fabrication of disk samples ($\varnothing$ = 8 mm, h = 0.6 mm) with a nanostructured microstructure. The undeformed sample is held between two anvils which are pressed together. The lower anvil rotates and due to the friction between the anvils and the sample it is strained in torsion. For the deformation process it is very important to avoid slipping between the sample and the anvils. This has been ensured by sandblasting both the sample and the anvils before rotating.

In this study the deformation by HPT has been performed above the DBTT ($\approx$350 °C) of DUCROPUR in the undeformed condition. To investigate the influence of temperature on the development of microstructure the same degrees of deformation have been applied at room temperature. Specimens with a half and two revolutions have been produced under an applied stress of 7.8 GPa.

2.2 Cyclic Channel Die Compression (CCDC)

To investigate the development of microstructure at lower strains (<6), specimens have been deformed also by CCDC. The deformation during one cycle is depicted in Fig. 1. The sample has a prismatic shape (15 × 15 × 30 mm) with a height equal to the length of the die. Since the dimensions of the sample don't change before and after deformation the sample has been rotated by turning it 90° around the axis perpendicular to the depicted plane and by 90° around the length axis, then the deformation is repeated.

The single deformation has been done at room temperature, the multiple deformations again above the DBTT (≈350 °C) of DUCROPUR in the undeformed condition.

Figure 1: Illustration of the deformation during one CCDC cycle; the front wall of the channel is removed here

3 Developed Microstructures

The developed microstructures have been investigated in the Scanning Electron Microscope (SEM) by means of Back Scattered Electrons (BSE) and Electron Back Scatter Diffraction (EBSD). A BSE image gives an impression of the typical sizes of the developed microstructure. In a single phase material the intensity of the back scattered electrons depends on the crystallographic orientation of the scanned area and the dislocation density. It is not possible to identify the degree of misorientation between the grains from such micrographs. To distinguish if the boundaries are large or low angle boundaries the degree of misorientation has been determined with the EBSD method. The results are presented in colored Orientation Maps (OM). The different colors show the different crystallographic orientations with respect to a certain specimen coordinate as a function of location [6, 7]. Furthermore, in the colored OM, boundaries are sketched in, which are divided in misorientation groups, 3°–10° (yellow), 10°–20° (red), and >20° (black).

3.1 Developed Microstructures at Small Strains (<6)

The original grain size of DUCROPUR has been about 80 μm. In Fig. 2 BSE micrographs and OM illustrate the formation of a new fine microstructure during CCDC. After a strain of $\Phi \approx 0.72$ the original grains are still visible. Within these grains a continuous change superimposed with a very small fluctuation of the original orientation prevails. With increasing strain the refinement increases, too. The BSE micrograph after a strain of $\Phi \approx 2.77$ shows a band structure. The result of an EBSD scan indicates that beside boundaries with small misorientations (yellow lines) also boundaries with larger misorientations (red and black lines) have generated. At a strain of $\Phi \approx 5.55$ the elongated grains or cells have disappeared and an equiaxed microstructure is observed. The sizes of the structural elements are about 1 µm but many boundaries still have a misorientation smaller than 10°.

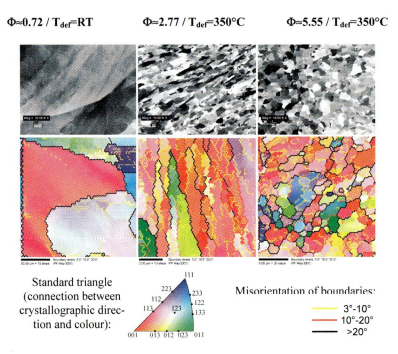

Figure 2: BSE micrographs and OM of DUCROPUR at a strain of $\Phi \approx 0.72$ (T_{def} = RT), $\Phi \approx 2.77$ (T_{def} = 350 °C), and $\Phi \approx 5.55$ (T_{def} = 350 °C)

3.2 Effect of Strain and Temperature

Specimens processed by HPT have been used to investigate the impact of larger strains and deformation temperature on the developed microstructure. The results are summarized in the schematic illustration (Fig. 3) which shows the dependence of structural size (d) and mean misorientation between structural elements (MO) on the degree of deformation (Φ).

With increasing strain the structural size decreases. After a certain deformation a saturation size of the structural elements is reached and only a small further refinement of the microstructure with increasing strain is observed. The saturation size depends on the deformation temperature. At room temperature a structural size of about 200 nm is found (see Fig. 4), whereas at 350 °C a characteristic size of the microstructure of about 500 nm has been developed. Although the structural sizes between a strain of $\Phi = 5.5$ and $\Phi = 25.8$ decrease only slightly, a significant change between the microstructures has been detected. Within this strain range the mean misorientation (MO) between structural elements increases with increasing degree of deformation until nearly a random distribution of misorientations is reached.

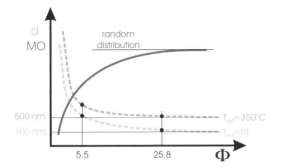

Figure 3: Schematic illustration of the structural size (d) and mean misorientation between structural elements (MO) as a function of strain (Φ)

4   Thermal Stability of the Microstructures

The HPT processed specimens have been annealed for 10 hours at 500 °C, 600 °C, and 700 °C, respectively. Between the heat treatments the microstructures have been examined in the SEM and the micro hardness (load = 1.38 N) have been measured.

After the heat treatments at 500 °C only small changes in microstructure are found. Very fine structured regions have "recrystallized" whereas the sizes of the larger elements remained about constant. During the following heat treatments at 600 °C and 700 °C different structural coarsening behaviours are observed (Fig. 4). The "recrystallization" may start in triple points of boundaries or very small structural elements with large misorientations or by bulging of boundaries in the direction of sub-grains (grain growth). Samples with a lower percentage of boundaries with large misorientations (samples deformed to a strain of $\Phi = 5.55$), have a lower number of nucleuses and hence tend to a non uniform "recrystallization". In these samples some large grains have been developed, surrounded by small grains. In samples where boundaries with large misorientations prevail (samples deformed to a strain of $\Phi = 25.8$), a very uniform "recrystallization" is observed. These samples show finer microstructures after the heat treatments.

The measured micro hardness (Fig. 5) are in good agreement with the observed microstructures. The finer the microstructure, the higher the micro hardness. After the heat treatments at 500 °C the samples, where many fine grained regions have "recrystallized", show the steepest decreases of micro hardness. Also the internal stress fields which have been generated during HPT have been reduced. Specimens with an uniform coarsening (finer microstructure) have higher hardness than those with a non uniform coarsening.

Compared with the original grain size of 80 µm, the observed microstructures after all heat treatments are still fine . Also the micro hardness is still about two times larger than that of undeformed DUCROPUR.

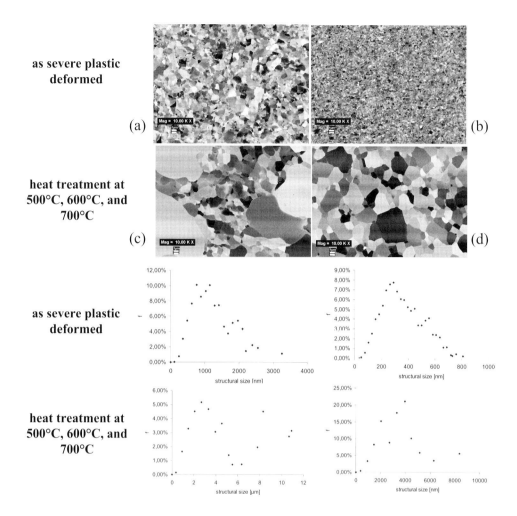

Figure 4: BSE micrographs of HPT samples deformed at room temperature before ($\Phi = 5.5$ (a), $\Phi=25$ (b)) and after heat treatments ((c),(d)) and the corresponding size distribution (area fraction) of the structural elements

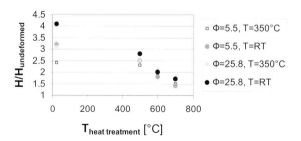

Figure 5: Change of the relative hardness (hardness related to the hardness of undeformed DUCROPUR) after heat treatments

5 Summary

In pure chromium the development of a fine equiaxed microstructure with increasing strain starts with the formation of cell structures within the original grains, followed by the generation of a band structure and reaching finally a nearly equiaxed grain like structure.

Microstructures smaller than 1 µm have been produced by HPT. The larger the strain and the lower the deformation temperature, the finer the developed microstructure. After heat treatments at 500 °C only the very fine structured regions have coarsen. The coarsening behaviour at 600 °C and 700 °C depends on the number of boundaries with large misorientations and hence on the applied degree of deformation.

After the heat treatments at 700 °C the micro hardness are still about two times larger than the hardness of the undeformed samples and compared with the original grain size (80 µm) still a fine grained microstructure with sizes below 5 µm is found.

6 Acknowledgement

This work has been carried out within Association EURATOM-ÖAW and is supported by the Bundesministerium für Bildung, Wissenschaft und Kunst. The content of the publication is the sole responsibility of its authors and does not necessarily represent the views of the European Commission or its services.

7 References

[1] R. Z. Valiev, R. K. Islamgaliev, I. V. Alexandrov, Progress in Materials Science 45 2000, p. 103–189
[2] T. Hebesberger, Entwicklung der Mikrostruktur bei Hochverformung kubisch flächenzentrierter Metalle, Doctor thesis, Institute of Metal Physics – University of Leoben 2001
[3] Patentanmeldung A1516/2001
[4] R. Wadsack, Fracture behaviour and Severe Plastic Deformation of Chromium, Doctor thesis, Institute of Metal Physics – University of Leoben 2002
[5] R. Wadsack, R. Pippan, B. Schedler, Chromium – a material for fusion technology, Fusion Engineering and Design 58-59 2001, p. 743–748
[6] C. Semprimoschnig, Die kristallographische Fraktometrie – Entwicklung einer Methode zur quantitativen Analyse von Spaltbruchflächen, Doctor thesis, Institute of Metal Physics – University of Leoben 1996
[7] A. Tatschl, Neue experimentelle Methoden zur Charakterisierung von Verformungsvorgängen, Doctor thesis, Institute of Metal Physics – University of Leoben 2000

XII Influence of Deformation Path to Properties of SPD Materials

Fatigue of Severely Deformed Metals

A. Vinogradov and S. Hashimoto
Department of Intelligent Materials Engineering, Osaka City University, Osaka, Japan

1   Abstract

In this brief communication, we would like to review present data on fatigue performance of ultra-fine grain materials fabricated by severe plastic deformation (SPD) and to discuss the possible mechanisms of their plastic deformation and degradation in light of currently available experimental data. The most prominent effect of SPD is often associated with significant grain refinement down to the nanoscopic scale. The other evident effect, which accompanies intensive plastic straining, is the dislocation accumulation up to limiting densities of 10^{16} m^{-2}. Since namely these two factors, the grain size and the dislocation density, govern the strengthening of polycrystalline materials, we shall primarily confine ourselves to their role in cyclic deformation of severely pre-deformed metals.

2   Effect of Grain Size on Fatigue

A variety of SPD techniques has been developed in the past decade. The equal channel-angular pressing (ECAP) technology was introduced by Segal [1] as a cold- (or warm-) working technique that allowed achieving extremely large imposed strains in bulk samples without fracture. This technique is of particular interest because of its flexibility in fabrication of massive samples with a broad variety of fine-grain structures. For this reason, the majority of relevant experimental data have been obtained on ECAP materials so far. The most promising feature of SPD materials for enhanced "fatigue properties is associated with the often claimed combination of the high tensile strength with good ductility. [2] Low-cycle fatigue (LCF) and high-cycle fatigue (HCF) regimes are conventionally distinguished in accord with applied strain amplitude. At high strains corresponding to short lives, the plastic strain component is dominant in the total applied strain and the fatigue life is determined by ductility. At long fatigue lives, the elastic strain amplitude is more significant than plastic and the fatigue life is dictated by the fracture strength so that the fatigue limit increases with strength. [3] Unfortunately, the highest strength is often archived when the ductility is sacrificed: high strength materials are usually brittle. Therefore, the simultaneous enhancement of both HCF and LCF properties is tricky and the improvement of one property may result in degradation of the other. As underlined by Suresh, [3] optimizing the overall fatigue properties requires inevitably a balance between strength and ductility.

The influence of the grain size on fatigue of conventional polycrystalline materials has been reviewed by many investigators (see, for example Suresh and Pelloux). [3,4] Most observations can be summarized in two sentences. 1) The fatigue strength of pure face-centered cubic (fcc) metals is not affected by the grain size. 2) The fatigue strength of materials having planar slip increases with decreasing grain size and follows the Hall-Petch relationship. It has been concluded that one of the most important structure parameters in fatigue is the slip character. In the low-cyclic fatigue, when the specimen is loaded under relatively high strain amplitude, the wavy slip materials form a well-defined cell structure with the cell size being dependent upon the saturation stress and independent of the preliminary strain history. [3–5] Materials with a planar slip do not form a cell structure and the dislocations are arranged in planar arrays extending across a grain. Hence, it is apparent that the effect of the grain size is more pronounced for material with the slip mode other than wavy. This standpoint is well justified for conventional materials and Mughrabi has successfully extended it to ultra-fine grain (UFG) metals. [6] Indeed, the fatigue limit of Al-alloys (5056 [7] and 6061 [8]) did not improve after ECAP despite notable enhancement of the monotonic strength. However, one should bear in mind that the above-mentioned cyclic strength independence of the grain size in fcc metals is caused by a specific cell-like structure formed during cycling. The grain size of SPD metals is considerably smaller than the typical cell size of 0.5 µm and, hence, no dislocation patterning could be expected in ultra-fine grains. The fatigue limit of UFG Cu does show the noteworthy improvement depending on processing. The most impressive enhancement of the high-cyclic fatigue life has been observed in ECAP Ti [9] and in the peak-aged CuCrZr alloy [10] when compared with their conventionally fabricated analogues.

A small grain size prompts a more homogeneous deformation that helps to retard the crack nucleation by reducing stress concentrators and to raise the fatigue limit of the material. On the other hand, the truly uniform deformation in SPD materials is hardly realized at low temperatures because of stress concentrators inherited from inhomogeneity of plastic deformation during processing. To get rid of possible gradients of plastic flow it is important to provide the uniformity of simple shear in the course of ECAP, i.e., the required boundary conditions [2] have to be fulfilled by minimizing the contact friction and ensuring a hydrostatic pressure in the deforming region. It has been shown by Vinogradov et al. [11] that with decreasing grain size the fatigue limit obeys the Hall-Petch relationship in the same way as the ultimate tensile strength ρ_{UTS} until a certain critical grain size is attained below which the slop of the $\rho \angle d^{-1/2}$ curve alters. Hence, there appears to be an optimum grain size in the nanoscale region, when the maximum monotonic and cyclic strength is reached. At larger grain sizes, the dislocation-based plasticity occurs during straining. The grain boundary sources operate and the dislocations move through the grain interior. At the critical grain size the mechanism of deformation is supposed to turn into the other type. Since for very small grain sizes, the volume fraction of grain boundaries is comparable with the volume of grains, the deformation should localize in the grain boundary region. Although it is apparent that the grain boundaries play increasing role in deformation of nanomaterials, Youngdahl et al. [12] have concluded that deformation of nanocrystalline copper with a mean grain size ranged from 30 to 100 nm produced by gas condensation and subsequent compaction, occurs due to dislocation activity. No evidence for grain boundary sliding or rotation was found in the in situ experiments. [12]

Undoubtedly the grain boundaries play significant role in deformation and cyclic degradation. During fabrication of UFG materials by SPD, the grain boundaries experience significant distortion due to their interaction with a large number of lattice dislocations. Transformation of lattice dislocations to grain boundary dislocations results in increasing free volume in the grain boundary. A high volume fraction of greatly distorted grain boundaries makes it possible to consider nanomaterials as composites or two-phase solids consisting of a crystalline grain interior and amorphous-like grain boundary region. [13]

Before making an attempt to rationalize the cyclic behavior of SPD materials we shall address the following issues: microstructure prior to and after fatigue, fatigue life, cyclic hardening/softening and its dependence on the initial structure and loading conditions.

The experimental results concerning the cyclic response are currently available for various UFG SPD metals such as i) pure Cu [14–21] which is representative of fcc wavy-slip materials; ii) single phase solid solutions, AA5056 Al-Mg alloy [7,22] and Fe-36Ni (Invar); [23] iii) Ti [9,24] which is a typical example of hexagonal close-packed (hcp) metals with planar-slip; iv) precipitation hardenable CuCrZr alloy, [10] and 6061 Al-Mg alloy. [8] A brief review of findings regarding the fatigue life of ECAP metals has been reported by Vinogradov and Hashimoto. [25] The details of sample preparation have been reported in the above-cited papers.

3   Survey of Experimental Results on Fatigue of SPD Metals

3.1   Structure of ECAP Metals

A typical UFG structure after ECAP is shown in Figure 1a. The structures achieved during ECAP differ in many aspects such as distribution of grain sizes and shapes, texture, misorientations of adjacent grains, dislocation density, and arrangement of dislocations, structure of grain boundaries, etc. (see Valiev et al. [1] for a comprehensive review). Obviously, all these aspects affect the resultant properties in general and fatigue in particular. Let us briefly summarize those structural features of SPD metals, which are most relevant to fatigue.

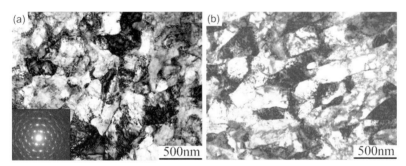

Figure 1: Fine structure of ECAP Invar (8 ECA-passes, Bc) before a) and after b) fatigue at $\Delta\varepsilon_{pl}/2 = 10^{-3}$

- The average grain size is small ranging between 150–350 nm in different metals. The grain size distribution is fairly broad and the largest grains can have 1–5 µm dimensions.
- The grains are separated by grain boundaries with a high angle of misorientation primarily as is evidenced by the SAEDP or EBSD analysis. [1,26] However, the grain boundaries are often not well defined on TEM images and therefore, the term "grain" should be used with caution to avoid possible confusion with cells, blocks or fragments.
- Two characteristic structures can be distinguished by the grain shape: a rather uniform structure with nearly equiaxed grains as exemplified in Figure 1a and a kind of fragmented structure with significantly elongated grains. [1,25]

- High internal stresses are revealed by i) the azimuth spreading of the spots in the SAEDP, ii) the bend extinction contours inside the grains and iii) X-ray analysis. [1,27]
- The relatively high average dislocation density and non-uniform dislocation arrangement is commonly revealed by TEM. The average dislocation density is close to 10^{14}–10^{15} m^{-2}. [1,28,29] When the dislocation density is high, the chaotic dislocation distribution is energetically unfavorable. The smaller grains in their central parts are rather dislocation free and most dislocations are attracted to the grain boundaries.

After cyclic deformation, the following structural features are common for ECAP metals.

- The most striking attribute of the post-fatigue structure of pure SPD metals wit wavy slip is the abnormal grain growth triggered by cyclic deformation. This phenomenon has been reported in many publications. [17–20,25,30] The most detailed investigations have been performed by Höppel et al. [20] on UFG Cu and Thiele et al. on UFG Ni. [27] Recovery, recrystallization and grain growth can be largely suppressed by limiting dislocation and grain boundary mobility. The dislocation mobility is controlled by the slip mode and hcp or low-stacking fault fcc materials are supposed to be more stable than wavy-slip metals. The grain boundary can be pinned by impurities and "precipitates so that alloying and precipitation hardening can be effective for SPD structure "stabilization.
- The internal stresses are reduced during fatigue as has been carefully shown by Thiele et al. on UFG Ni by using the X-ray technique. [27]
- Solid solutions and hcp metals are more stable and no gross grain growth is observed in the "Invar alloy [24] and Ti after fatigue. [23] The grain growth in the AA5056 Al-Mg alloy depends on the processing and can be rather small. [7]
- No gross change in the dislocation arrangement is noticed after fatigue. While the average dislocation density in the central part of a grain remains unaltered, being of the order of 10^{13}–10^{15} m^{-2}, the grain boundaries appear more clearly visible in all materials examined (e.g., Fig. 1b). This seems to be the most noticeable structural change in fatigued Ti, Al-Mg, CuCrZr, and Invar "alloys and it can be attributed to the dislocation density reduction in the grain boundary vicinity

3.2 Fatigue Limit

Tensile and HCF properties of some ECAP materials are summarized in Table 1.

The significant enhancement of the HCF life in terms of fatigue limit is achieved for all materials but aluminum alloys. Markushev and Murashkin [31] reviewed the effect of SPD on submicrocrystaline structure formation and mechanical properties of Al-alloys. They concluded that SPD is ineffective for the strength and fatigue improvement of Al-alloys. The presently available experimental data reveal that the ultimate tensile strength and the fatigue limit follow the Hall-Petch relationship, Fig. 2 (see also Vinogradov et al. [23] for Ti). The LCF life can be represented by the Coffin-Manson plot and quantified by the fatigue ductility and fatigue exponent. Most SPD metals demonstrate the shorter LCF life in comparison with their coarse grain (CG) counterparts because some ductility is lost after SPD. [10,15,17,18]

Table 1: Grain size and mechanical properties of SPD alloys.

Material	Processing	d [μm]	$\sigma_{0.2}$ [MPa]	σ_{UTS} [MPa]	δ [%]	σ_{f0} [MPa]
Cu 99.96%	CR75%, HT 550°C 2h	35	140	240	46	65
Cu 99.96% [18]	ECAP, B 8	0.2eq	390	440	22	80
Cu-0.8Cr-0.07Zr	Q, Drawing, A 500°C,1h	N/A	100			100
Cu-0.44Cr-0.2Zr [10]	ECAP Bc, 8, A 500°C, 1h	0.16eq	650	720	12	285
Ti VT1-0	CR	15	380	460	26	240
Ti VT1-0 [23]	ECAP Bc, 8 400°C	0.3eq	640	810	15	380
Ti VT1-0 [9]	ECAP Bc, 8 400°C, CR75%	0.15el	970	1050	8	420
Fe-36Ni Invar [24]	CR 75%	N/A	275	490	40	137
Fe-36Ni Invar [24]	ECAP Bc 2,	0.30eq	570	732	47	280
Fe-36Ni Invar [24]	ECAP Bc 8,	0.26eq	690	790	35	290
Fe-36Ni Invar [24]	ECAP Bc 12,	0.18el	835	912	52	330
5056 Al alloy [7]	O-temper	25	122	290	43	116
5056 Al alloy	H18		407	434	10	152
5056 Al alloy [22]	ECAP C, 4, 150°C	0.35el	280	340	25	116
5056 Al alloy [7]	ECAP Bc, 8, 110°C	0.22el	392	442	7	116

d: initial grain size; ρ_{02}: conventional yield stress; ρ_{UTS}: ultimate tensile strength; ρ: elongation to failure in tension; ρ_{f0}: endurance limit based on 10^7 cycles; CR: cold-rolling; Q: quenching; A: aging; HT: heat treatment; eq and el: equiaxed and elongated grain structure respectively.

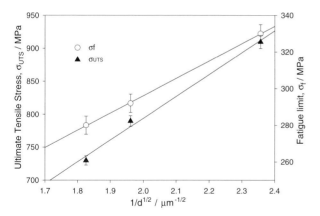

Figure 2: The Hall-Petch behavior of monotonic strength ρUTS and the fatigue limit ρ_f of ECAP (Bc) Invar alloy

3.3 Cyclic Softening/Hardening

Very first experiments revealed that the cyclic response of ECAP materials is strongly dependent on materials purity, processing and the initial UFG structure. Vinogradov et al. [14] found that the cyclic hardening curve of UFG Cu was nearly flat during most fatigue life. No softening was observed under plastic strain amplitudes $\Delta\varepsilon pl/2 = 5 \cdot 10^{-4}$ and 10^{-3}. Furthermore, some light hardening was noticed on the early stage of straining (Fig. 3). However, Agnew and Weertman [18] have demonstrated pronounced cyclic softening in UFG Cu produced by ECAP. The mechanisms of cyclic softening of UFG fcc metals have been largely understood and associated with a complex effect of dislocation recovery, dynamic "recrystallization" and grain coarsening. For further details the readers are addressed to relevant publications. [17–21,25,27] A feature of UFG SPD metals is that during cyclic softening the abnormal grain growth occurs facilitating formation of dislocation structures typical for ordinary metals, i.e., cellular and ladder-like dislocation arrangements are observed in coarsened grains. Thiele et al. [27] have performed the detailed structural investigations of the fatigue-induced structures in UFG Ni in the dependence on the grain size. Using the X-ray technique they found some reduction of internal stresses in the course of cycling. It has been demonstrated that there is a lower threshold grain size dth of 1 μm above which the dislocation patterning takes place with a length scale of 500 nm "independently of the grain size. For the materials with $d < d_{th}$ the cyclic stress-strain curve obeys the Hall-Petch relation.

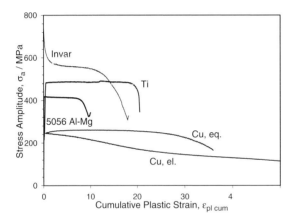

Figure 3: Cyclic hardening/softening curves showing variability of cyclic behavior of ECAP metals

The SPD structure can be stabilized and the rate of cyclic softening can be reduced by i) annealing at an intermediate temperature prior to cyclic loading, i.e., be reducing the internal stresses, [7,19,21] ii) using the solid solution alloys instead of pure metals, [7,24] i.e., by reducing the dislocation mobility and iii) stabilizing the grain and grain boundary structure by precipitations. [10] The behavior of UFG Invar during stain-control fatigue differs from Cu and Ni in that we do not observe the significant grain growth. Although some grain coarsening was noticed by TEM (Fig. 1b), this effect was observed only on the late stage of fatigue and could account for rapid softening commenced shortly after the beginning of deformation. Cyclic softening is observed in the precipitation hardened ECAP metals (CuCrZr). [10] As has been shown in [10] the main reason for rapid cyclic softening of the UFG peak-aged CuCrZr alloy is related to dislocation cutting through the fine precipitates.

3.4 Hysteresis Loops, Bauschinger Effect, and Back Stresses

The hysteresis loops in UFG SPD metals are rather pointed and indicative of a large Bauschinger effect (Fig. 4). [14] Abel has suggested that there is a correlation between the fatigue life and the Bauschinger effect. [32] Two parameters have been used to characterize the hysteresis loop shape: [4] the stress amplitude ρ_a and the stress ρ_y corresponding to the onset of reverse plastic flow. One can introduce a friction stress ρ_f and back stress ρ_b as

$$\sigma_f = (\sigma_a - \sigma_y)/2$$
$$\sigma_f = (\sigma_a + \sigma_y)/2 \qquad (1)$$

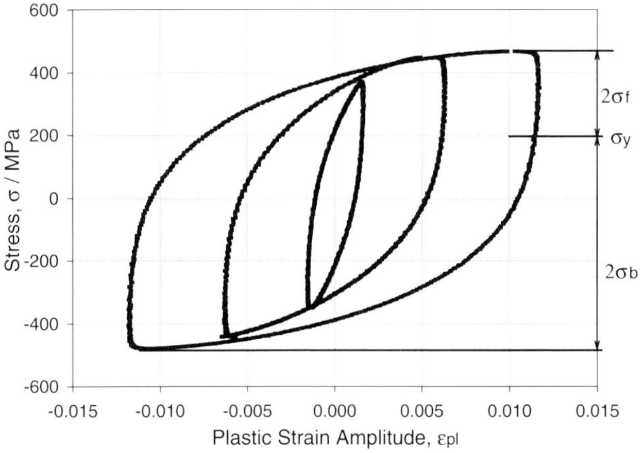

Figure 4: Hysteresis loops of UFG Cu tested with different $\Delta\varepsilon_{pl}/2$, illustrating the Cottrell scheme

In this description, the onset of plastic flow would be assisted by conventional long-range back stresses associated with crystal lattice strains detectable by diffraction. The dislocations dynamically stored during loading also create a back stress which facilitates their backward motion upon reversal loading. Carstensen and Pedersen [33] have distinguished between these two components of the phenomenological Cottrell back stresses: a static back stress associated with lattice strains detectable by diffraction methods and a dynamic back stress merely representing the directional dislocation dynamics in association with the Bauschinger effect. Abel and Muir [34] have proposed that the hysteresis loop shape can be evaluated by the so-called Bauschinger energy parameter

$$\beta_E = \frac{4\sigma_a \varepsilon_{pl} - \int_{loop} \sigma d\varepsilon}{\int_{loop} \sigma d\varepsilon} \qquad (2)$$

where ρ_a is the peak stress amplitude. A larger value of β_E corresponds to a "pointer" hysteresis loop and to the larger Bauschinger effect. Clearly, β_E serves as a measure of the overall reversibility of the energy storage mechanisms during fatigue. [32] According to Abel, [32] the higher the reversibility of to-and-fro dislocation motion under otherwise the same conditions – the smaller the fatigue damage exerted per cycle and the longer the fatigue life. The β_E magnitude can be related to the back stress which cause this reversibility as [32]

$$\beta_E = \bar{\sigma}_b / \bar{\sigma}_a \tag{3}$$

In this way, the β_E parameter is a direct measure of the average elastic stress $\bar{\sigma}_b$ built up during the previous directionally opposite deformation cycle. Hence, the β_E parameter is particularly convenient for estimation of the back stresses when it is difficult to quantify the onset of plastic flow in the Cottrell scheme because of the roundness of a hysteresis loop. We have utilized both methods of the hysteresis loop analysis (the Cottrell scheme and the Bauschinger energy parameter) and found a satisfactory agreement between them.

The example of the friction and back stress behavior in UFG copper with equiaxed grain structure is shown in Figure 5a. Figure 5b demonstrates the change in the ρ_b and ρ_f magnitude after short time annealing (200 °C, 3 min) of the same Cu.

One can notice that while ρ_b behaves in the same manner as the ρ_a, the ρ_f magnitude experiences only little or no change during cycling. In other words, the development of back stresses

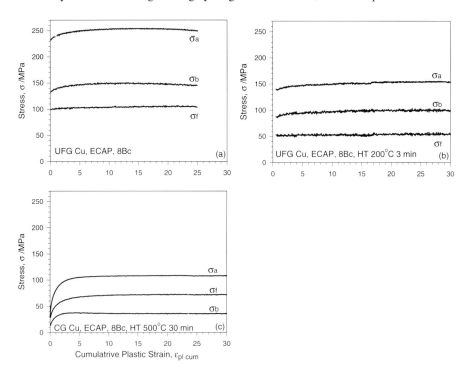

Figure 5: Cyclic stress amplitude, ρ_a and its components - friction stress ρ_f and back stress ρ_b in ECAP Cu tested at $\Delta\varepsilon_{pl}/2 = 10^{-3}$

controls the cyclic response of the UFG materials primarily. Although we exemplify our results solely for ECAP Cu, the similar conclusions can be drawn for other materials: β_b does not change if the material exhibits neither hardening nor softening, β_b decreases together with β_a during cyclic softening while the β_f value remains almost unaltered. Figure 5b demonstrates that the grain size is not the main factor that affects the back stress because the grain size of the as-fabricated ECAP specimens (Fig. 5a) and the specimen after short term annealing is almost the same. Since the recovery starts in the most heavily distorted regions, the reduction of ρ_b and ρ_f magnitude after short term annealing should be "associated either with non-equilibrium grain boundaries or with recovery of the dense dislocation arrays in the grain boundary vicinity. In the coarse grain (CG) copper (Fig. 5c) the friction stress increases during initial hardening due to accumulation of dislocations. Among other experimental findings related to the back stress behavior of UFG metals we can mention that i) the ρ_b value increases the number of ECA-passes, i.e., with structure refinement, ii) ρ_a increases whereas ρ_b decreases with the plastic strain amplitude $\Delta\varepsilon_{pl}/2$. For all ECAP materials the stress amplitude ρ_a reaches its maximum or saturates quickly (Figs. 3,5) showing that multiplication of dislocations is not remarkable, and prior to testing the material had been hardened to the maximum (or almost maximum) possible extent. The cyclic stress amplitude saturation indicates that there is equilibrium in the dislocation ensemble and dislocation generation is compensated by annihilation (dynamic recovery). We shall show that these findings shed some light on the microscopic fatigue mechanisms and any model which attempts to account for the cyclic response of SPD metals should be able to describe the hysteresis loop shape as well as the Bauschinger effect and back stress nature.

3.5 Fatigue Damage

Strain localization in ECA-processed materials is often observed during both monotonic and cyclic deformation. [10–15,40] Figure 6 shows the shear bands oriented at 45° to the loading axis. They often appear shortly after yielding in tensile deformation or at the end of saturation in cyclic testing. The cracks initiate and propagate along the shear bands. [25] Apparently, the shear banding is the major form of fatigue damage of wavy-slip UFG materials. The fracture surface analysis and analysis of the morphology of the surface cracks shows that failure in the SPD metals occurs intergranularly.

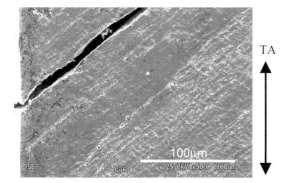

Figure 6: Shear bands on the surface of ECAP Cu deformed cyclically at $\Delta\varepsilon_{pl}/2 = 10^{-3}$

The strain localization occurs primarily along the grain boundaries. Figure 7 shows the AFM image of the shear band fragment on the surface ECAP Cu deformed in tension. The possibility of grain boundary sliding at ambient temperature in UFG SPD materials has been discussed in [1] in terms of enhanced grain boundary mobility due to intensive diffusion processes within the non-equilibrium heavily distorted grain boundaries. For the purpose of the present survey, it is of interest to notice the traces of dislocation slip of 1–10 nm height (AFM evaluated) in the grain interior (Fig. 7). [16] The slip lines are confined to a single grain and the slip does not transfer through the boundary. The AFM observations justify that that the conventional dislocation activity cannot be neglected in plastic deformation of UFG metals.

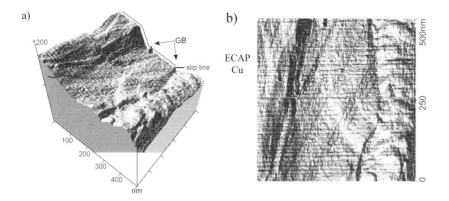

Figure 7: AFM view of the surface after deformation

3.6 Possibility to Enhance the Fatigue Performance of SPD Materials

The fatigue properties of UFG metals can be improved by gaining ductility and reducing constrains for dislocation motion, i.e., by decreasing the trend to shear banding and strain localization along the grain boundaries in severely hardened metals. Thus, it can be rewarding for fatigue properties to employ materials with partly recovered structure. It has been shown that that the susceptibility to shear banding decreases dramatically after heat treatment. [15,25] The LCF life can be improved by a factor of 5–10 after annealing at relatively low temperatures. [19,21,23] Hence, one can conclude that there is still some potential for further enhancement of fatigue performance, particularly in the low-cyclic regime, of high strength ECAP materials through partial relaxation of the severely pre-deformed state.

Additional improvement of tensile and high cyclic strength is observed if the UFG material obtained by SPD is subjected to further conventional cold rolling with or without intermediate annealing at moderate temperature. This has been shown for several commercial Al-Mg alloys [31] and pure Ti. [23,25]

As a guideline for further research, we should notice that only scarce results are available on the effect of manufacturing routine on fatigue performance. No systematic data have been reported on the role of strain path, temperature, velocity, and other parameters. Apparently, the efforts have to be invested in development of optimum processing scheme for desired fatigue properties of SPD materials.

4 On Fatigue Mechanisms in UFG Materials

The first attempt to model the fatigue life of SPD metals has been performed by Ding et al. [35] The microstructure of the UFG metal was treated as a "composite" consisting of the soft matrix associated with dislocation free grains and hard reinforcement associated with grain boundaries. Both LCF and HCF properties have been predicted successfully using mechanistic approach. The details of microscopic mechanism of fatigue damage remain uncovered. Plenty of discussions have been put forward in the literature about the particular role of grain boundaries in the properties of UFG materials. [1] The specific non-equilibrium state of grain-boundaries cannot be disregarded for many phenomena (particularly in nanocrystalline materials). In fatigue of SPD metals, for instance, the grain boundaries play a doubtful role. On one hand, the fine-grained structure usually shows the higher fatigue life (at least under stress controlled cycling) than the coarse grain structure. On the other hand, the grain boundaries are most heavily stressed elements that determine a relatively low stability of the UFG structure and the tendency towards recovery and grain coarsening during cycling and/or temperature. Furthermore, namely grain boundaries are responsible for frequent shear banding via grain boundary sliding, crack initiation and propagation, representing, therefore, the most "dangerous" element of structure.

In our opinion, despite numerous factors complicating the UFG structure of SPD metals, the fatigue behavior can be described, in principle in a more simple way than that in ordinary poly- and single-crystals. The main reason for possible simplification is the lack of dislocation patterning in UFG structures. It has been suggested that, within the framework of a one-parameter model, the shape of a stable hysteresis loop can be predicted if only the kinetics of the average dislocation density is taken into account. [14,23] It was assumed that the mobile dislocations pass through the grain and disappear in the grain boundary region (the grain boundaries act in this sense as an effective sink for dislocations). The kinetic equation for dislocation density, ρ, can be written in its simplest form as (after Essmann and Mughrabi, [36] and Estrin and Kubin [37])

$$\frac{d\rho}{d\gamma} = \frac{2}{bL} - \frac{2}{b}y\rho \qquad (4)$$

where L is the slip path of dislocations (L is approximately equal to the grain size) with the Burgers vector b and y is the so-called annihilation length. The first term on the right hand side of Equation 4 describes the rate of dislocation multiplication with the strain increase while the second term accounts for the strain induced decrease of dislocation density. One can show that using a similar approach for two kinds of dislocations – mobile and immobile (the former are responsible the intergranular slip and the latter are attracted to the grain boundaries) – it is possible to account for the cyclic hardening/softening behavior, Bauschinger effect and cyclic stress-strain curves. The relevant detailed modeling and discussion will be given in the forthcoming publication because of the limited length of the present communication.

Before concluding this brief survey, we would like to highlight possible alternative approaches, which are worth exploring in attempt to explain and to quantify the fatigue life and cyclic behavior of UFG materials. Since as has been mentioned above, the grain size distribution in UFG SPD metals is often quite broad (Fig. 8) a kind of "weakest link" concept can be plausibly applied. Apparently, different grains are not equal with regard to their response to the applied load: because of the grain size distribution there is a resultant distribution of yield stresses and the larger grains will tend to flow first while the finer ones are hard to deform. Early deformation and possible dislocation patterning in the grains with $d > d\text{th}$ gives rise to the gradients of plastic deformation which may lead to premature failure. In this sense, the key elements responsible for early fracture are related to the largest grains existing in the structure whereas most attractive structural features and advantages of fine grains can be almost irrelevant for fatigue life. Thus, it seems that to take most advantage from grain refinement for fatigue performance it is beneficial to manufacture a structure with a small variance of grain size distribution. The presence of large grains having $d > d_{th}$ corresponding to the tail of the grain size distribution promotes early strain localization via dislocation mechanisms typical for fatigue of conventional metals and, hence, should be avoided. However, further systematic investigations are still required employing the specimens which have different structures after the same amount of prestrain, i.e., manufactured by different routes and/or at different temperatures.

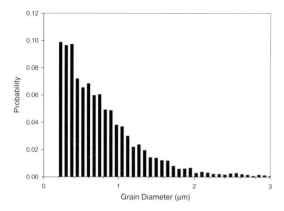

Figure 8: Grain size distribution in the ECAP (8Bc) CuCrZr alloy (from EBSD analysis)

5   Summary and Conclusions

- The significant enhancement of high cyclic fatigue life of SPD metals can be achieved during testing under constant stress amplitude.
- Most SPD metals demonstrate the shorter LCF life in comparison with their coarse grain counterparts because some ductility is lost during manufacturing. The expecting improvement of ductility after SPD, which is being broadly discussed in the literature, provides some reason for more optimism in obtaining superior fatigue performance in both high-cyclic and low-cyclic regions.
- The fatigue damage occurs on different scale levels. The susceptibility of the ECAP materials to early strain localization and microcracking can be a main factor which limits their tensile and fatigue ductility and to a large extent determines their fatigue performance.
- The intensity of shear banding and the crack nucleation depends on the material and its SPD processing. The higher the thermal stability of the UFG SPD structure, the lower the susceptibility to shear banding.

- Both grain-refinement and work hardening play important role in resultant properties of SPD materials. Most "cyclic properties of ECA-pressed materials can be rationalized, at least qualitatively, in terms of Hall-Petch and dislocation hardening within the framework of a simple approach involving one- or two-parametric dislocation generation–annihilation kinetics.

6 Acknowledgement

The ECAP 5056 Al-Mg alloy was kindly provided by YKK Corporation, Japan. Copper and titanium were processed by ECAP by R. Z. Valiev and co-workers at Ufa State Aviation technical University (Russia). The authors are indebted to Dr. V. I. Kopylov (Minsk Physico-Technical Institute of the National Academy of Science, Belarus) for supplying ECAP Cu, CuCrZr and Invar alloy, and to Dr. Y. Kaneko (Osaka City University) for his co-operation in experiments and "discussions. Sincere thanks are due to Prof. K. Kitagawa (Kanazawa University) for encouragement and stimulating interest. The authors are grateful to Prof. S. Dobatkin (Institute of Metal Science RAS, Moscow, Russia) for his constant stimulating interest and encouragement.

7 References

[1] V. M. Segal, Mat. Sci. Eng. 1995, A197, 157
[2] R. Z. Valiev, R. K. Islamgaliev, I. V. Alexandrov, Prog. Mater. Sci. 2000, 45, 103
[3] S. Suresh, Fatigue of Materials, Cambridge University Press, New York 1991, p. 617
[4] R. M. Pelloux, in Ultrafine-Grain Metals (Eds: J. J. Burke, V. Weiss), Syracuse Univ. Press, New York 1970, p. 231
[5] a) C. E. Feltner, C. Laird, Acta Metall. 1967, 15, 1621. b) C. E. Feltner, C. Laird, Acta Metall. 1967, 15, 1633
[6] H. Mughrabi, in Investigations and Applications of Severe Plastic Deformation (Eds: T. C. Lowe, R. Z. Valiev), NATO Science Series, v.3/80, Kluwer Publishers, Norwell 2000, p. 241
[7] V. Patlan, A. Vinogradov, K. Higashi, K. Kitagawa, Mater. Sci Eng, A 2001, 300, 171
[8] C. S. Chung, J. K. Kim, H. K. Kim, W. J. Kim, Mat. Sci. Eng, A 2002, 337, 39
[9] V. V. Stolyarov, I. V. Alexandrov, Yu. R. Kolobov, M. Zhu, Y. Zhu, T. Lowe, in Proc. of the 7th Int. Fatigue Congress, Beijing, P.R. China (Eds: X. R. Wu, Z. G. Wang), vol. 3, Higher Education Press, Beijing, China 1999, p. 1345
[10] A. Vinogradov, Y. Suzuki, V. I. Kopylov, V. Patlan, K. Kitagawa, Acta Metal. 2002, 50, 1636
[11] S. Li, K. Lu, F. Guo, R. Chu, Z. Wang, Mater. Lett. 1997, 30, 305
[12] C. J. Youngdahl, J. R. Weertman, R. C. Hugo, H. H. Kung, Scr. Mater. 2001, 44, 1475
[13] H. S. Kim, Y. Estrin, M. B. Bush, Acta Mater. 2000, 48, 493
[14] A. Vinogradov, Y. Kaneko, K. Kitagawa, S. Hashimoto, V. Stolyarov, R. Valiev, Scr. Mater. 1997, 36, 1345
[15] A. Vinogradov, Scr. Mater. 1998, 38, 797
[16] A. Vinogradov, V. Patlan, K. Kitagawa, Mater. Sci. Forum 1998, 312–314, 607
[17] S. R. Agnew, J. R. Weertman, Mat. Sci. Eng, A 1998, 244, 145

[18] S. R. Agnew, A. Yu. Vinogradov, S. Hashimoto, J. R. Weertman, J. Electron. Mater. 1999, 28, 1038
[19] H. Mughrabi, H. W. Höppel, MRS Proc. 2001, 634, B2.1
[20] H. W. Höppel, Z. M. Zhu, H. Mughrabi, R. Z. Valiev, Phil. Mag. A 2002, 82, 1781
[21] H. W. Höppel, R. Z. Valiev, Z. Metallkd. 2002, 93, 641
[22] A. Vinogradov, V. Patlan. K. Kitagawa, M. Kawazoe, Nanostruct. Mater. 1999, 11, 925
[23] A. Vinogradov, V. V. Stolyarov, S. Hashimoto, R. Z. Valiev, Mat. Sci. Eng. A 2001, 318, 163
[24] A. Vinogradov, V. Kopylov, S. Hashimoto, Mat. Sci. Eng., A, in press
[25] A. Vinogradov, S. Hashimoto, Mater. Trans., JIM 2001, 42, 74
[26] A. Gholina, P. B. Prangnell, M. V. Markushev, Acta Mater. 2000, 48, 1115
[27] E. Thiele, C. Holste, R. Klemm, Z. Metallkd. 2002, 93, 730
[28] V. Y. Gertsman, B. Birringer, R. Z. Valiev, H. Gleiter, Scr. Metall. 1994, 30, 229
[29] R. Z. Valiev, E. V. Kozlov, Y. F. Ivanov, J. Lian, A. A. Nazarov, B. Baudelet, Acta Metall. 1994, 42, 2467
[30] T. Yamasaki, H. Miyamoto, T. Mimaki, A. Vinogradov, S. Hashimoto, in Ultrafine Grain Metals II (Eds: Y. T. Zhu, T. G. Langdon, R. S. Mishra, S. L. Semiatin, M. J. Saran, T. C. Lowe), TMS, USA 2002
[31] M. V. Markushev, M. Yu. Murashkin, Phys. Met. Metallogr. 2000, 90, 506
[32] A. Abel., Mat. Sci. Eng. 1979, 37, 187
[33] O. B. Pedersen, J. V. Carstensen, Mat. Sci. Eng., A 2000, 285, 253
[34] A. Abel, H. Muir, Phil. Mag. 1972, 26, 489
[35] H. Z. Ding, H. Mughrabi, H. W. Höppel, Fatigue Fract. Eng. Mater. Struct. 2002, 25, 975
[36] U. Essmann, H. Mughrabi, Phil. Mag. A 1979, 40, 731
[37] Y. Estrin, L. Kubin, Acta Metall. 1986, 34, 2455.

Cyclic Deformation Behaviour and Possibilities for Enhancing the Fatigue Properties of Ultrafine-Grained Metals

H. W. Höppel[1], C. Xu[2], M. Kautz[1], N. Barta-Schreiber[1], T.G. Langdon[2] and H. Mughrabi[1]
[1]Friedrich-Alexander-Universität Erlangen-Nürnberg, Erlangen, Germany.
[2]University of Southern California, Los Angeles, U.S.A.

1   Introduction

Refining the grain size of metallic materials is of high technological relevance. Not only is a strongly increased monotonic strength, an enhanced superplastic forming capability and a partially increased fatigue resistance achievable but, in addition, ultrafine-grained (UFG) materials are also necessary for the realisation of new, miniaturised (electromechanical) (MEMS) devices with predictable as well as reliable mechanical properties. For these reasons, severe plastic deformation methods, for example Lowe and Valiev [1], and in particular the Equal-Channel Angular Pressing (ECAP) process, for example Segal [2, 3], have attracted considerable attention in recent years [4–8].

Although there have been some limited investigations of the cyclic deformation behaviour of several UFG-materials [9–13], the deformation mechanisms and the dominating damage mechanisms during fatigue are not yet clear. Since the ECAP process introduces a very high dislocation density and large internal stresses into the material and, due to the shear process, highly distorted grain boundaries, the UFG microstructure is prone to be very metastable. In particular, the microstructural stability of the UFG material plays an important role with respect to the cyclic deformation behaviour and the fatigue life. In this context, and bearing in mind the objective to enhance the fatigue behaviour of UFG materials, the following aspects will be addressed:

- Since the relevant information from monotonic deformation experiments with respect to cyclic deformation is available, how does the micro-yielding behaviour of UFG materials differ from that of materials with conventional grain sizes?
- What are the responsible microstructural mechanisms for the cyclic softening which has been reported for almost all UFG materials investigated to date [8–14]? Related to this question, it is necessary to discuss also the effect of dislocation interactions with the highly deformed microstructure and the highly distorted grain boundaries. In this context, the question also arises whether grain boundary sliding must be considered as a relevant deformation mechanism in UFG materials?
- Together with the problem of microstructural stability during cyclic loading, the mechanisms for the formation of slip bands must be identified.
- What possibilities exist to stabilise the metastable microstructural arrangements prevailing after ECAP processing by a recovery heat treatment without any significant increase of grain size and loss of strength?

Through the work on the fatigue behaviour of UFG copper by Vinogradov et al. [9], Vinogradov and Hashimoto [12], Agnew and Weertman [11], Agnew et al. [10] and also the work of the group of authors in Erlangen [13–17], a detailed understanding was achieved during recent years of the basic cyclic deformation processes in UFG copper as a model material. Although the microstructures related to different ECAP processing routes as well as the purity of the materials differed in these various studies, a common finding was that the fatigue lives were increased in the regime of intermediate to low plastic strain amplitudes, whereas in the regime of intermediate to high plastic strain amplitudes the fatigue lives were reduced by comparison with copper of conventional grain (CG) size. This apparent contradiction can be resolved by examining the total strain fatigue life diagram. As summarised earlier [13], the total strain amplitude can be resolved into the elastic strain amplitude $\Delta\varepsilon_{el}/2$ which governs primarily the High-Cycle Fatigue (HCF) regime (Basquin law) and the plastic strain amplitude $\Delta\varepsilon_{pl}/2$ which governs the Low-Cycle Fatigue (LCF) regime (Coffin-Manson law). The dependence of fatigue life N_f on the total strain amplitude $\Delta\varepsilon_t/2$ is represented by the summation of the elastic and plastic strain resistance. For UFG copper, a strongly decreased fatigue ductility coefficient ε'_f is held responsible for the lower LCF resistance and an increased fatigue strength coefficient σ'_f for the enhanced HCF resistance. At the same time, it is noted that fatigue ductility as well as the fatigue strength exponent do not change significantly [16]. With the exception of UFG titanium, this tendency is also true for the fatigue behaviour of other ECAP processed materials as in, for example, the data summarised by Vinogradov and Hashimoto [12].

The formation of macroscopic shear bands and the onset of intense cyclic softening have been reported not only for UFG copper but also for other UFG materials [10–15]. As reported earlier for UFG copper fatigued at room-temperature [15], the intense cyclic softening can be related to a partial dynamic grain growth (dynamic recrystallization) process, which itself was shown to be dependent on time, temperature and, less specifically, on the amplitude. These findings suggest that classical "high-temperature" mechanisms are probably active at reduced homologous temperatures T/T_m (T_m: absolute melting temperature). Although macroscopic shear banding is found in several UFG materials, the mechanisms of formation are not yet clear. As proposed earlier [13], two possible mechanisms should be taken into account: Firstly, the shear band starts to form at a coarsened grain/patch in the interior of the material and then spreads out, triggered by the cyclic deformation process, and accompanied by local grain coarsening. Secondly, catastrophic shear localization could occur due to the strain path change from ECAP to fatigue. Recently, Wu et al. [18] suggested that large-scale heterogeneities, such as in particular shear bands inherently formed during the ECAP process, act as nuclei which promote the localization of deformation and the dynamic recrystallization process. In our opinion, this mechanism is confined to a special kind of ECAP-induced microstructure observed only after small (and odd) numbers of ECAP passes. Very large shear bands, several 100 μm long and more than 50 μm broad, were never found in the UFG copper investigated by the authors.

2 Experimental

Rods of commercial purity (99.5 %) Al were processed by ECAP for 12 passes using route B_c [19] with a die having an angle of $\Phi = 90°$ between the two channels. High-purity UFG copper (99.99 %) was processed using route C [19], also with a die having $\Phi = 90°$, by Prof. Valiev and co-workers at the Institute of Physics of Advanced Materials, Ufa State Aviation Technical University, Russia: further details of this material were given earlier [14, 16]. All of the fatigue testing and microstructural analysis was conducted in Erlangen.

The fatigue and the micro-yielding experiments were performed on a servohydraulic testing system (MTS 810) at room temperature (20 °C) to investigate the fatigue lives and cyclic deformation behaviour. The fatigue specimens, which had a cylindrical gauge length with a diameter of 5 mm, were machined from the central part of the ECAP rods. To avoid any influence of mechanical pre-treatment and of surface deformation, the specimens were chemically etched and electrolytically polished. More details on the specimen preparation and the cyclic deformation conditions are available elsewhere [13, 14, 17]. The fatigue tests were carried out in dried air under strain control (with constant plastic strain amplitude) with a constant mean plastic strain rate of $1 \cdot 10^{-3}$ s^{-1} at a plastic strain ratio of $R(\varepsilon_{pl}) = -1$. The fatigue failure criterion was defined as a 20 % drop in maximum tensile load or specimen fracture. For the electron-channelling-contrast (ECC) images, a scanning electron microscope (SEM, Jeol JSM 6400) was used. The micro-hardness measurements were carried out on a Leica Microhardness system.

3 Results and Discussion

In general, recovered materials with a conventional grain size, exhibit a "normal" micro-yielding and unloading behaviour, as displayed in fig. 1c. The unloading and the reloading branch are almost coincident. Compared to this normal behaviour, the ECAP-processed copper exhibits an unusually large inelastic back strain during unloading (fig. 1a) which is considered to reflect the large internal strains in the highly pre-deformed microstructure. Similar tests on heat-treated UFG copper (170 °C/ 2 h), which has a bimodal microstructure consisting of larger grains (some µm) embedded in an ultrafine-grained matrix [14, 16], revealed that, while the stress level was reduced considerably due to recovery, the inelastic back strain during unloading remained practically undiminished (fig. 1b). The observed micro-yielding and unloading behaviour after recovery probably reflect the interactions of the dislocations with the grain boundaries. Comparing the three micro-yielding and unloading curves and, in particular, figs. 1b and 1c, a strong grain size effect is noted. Based on the current investigations and referring to the differences in the micro-yielding behaviour, it is evident that different cyclic deformation behaviour must be expected, depending on the grain size and on the heat treatment. More specifically, the different types of cyclic deformation behaviour shown schematically in fig. 2 for the different microstructures at intermediate to high plastic strain amplitudes are proposed and these curves are explained as follows. UFG materials produced by the ECAP process have small grain sizes of the order of some 100 nm and they start to soften dramatically after a very short and not pronounced period of cyclic hardening. On the other hand, annealed CG materials, with grain sizes of some 10 µm, exhibit an extended period of cyclic hardening followed by pronounced cyclic saturation. In the case of fine-grained (FG) materials, also produced by the ECAP process but with a grain size of the order of about 1 µm, as well as for the heat-treated UFG material, cyclic softening occurs after a period of intense cyclic hardening. This cyclic softening is more pronounced for the FG material than for the heat-treated UFG material. Due to the very limited number of investigations on heat-treated UFG materials, the effect of the heat treatment on the relative numbers of cycles to failure N_f for the FG and the CG materials is not yet clear. Nevertheless, as was shown earlier for heat-treated UFG copper [14], and as the current investigations on heat-treated UFG Al reveal (compare also fig. 5), heat-treated UFG materials exhibit a clearly extended fatigue life compared to the non-heat-treated UFG material and can reach the fatigue life of the CG material (but at a higher stress level). Further investigations must clarify whether this behaviour can be generalised.

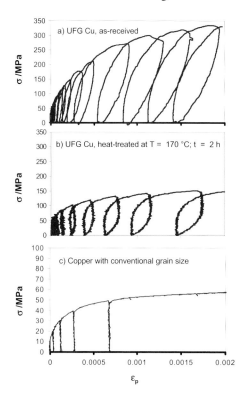

Figure 1: Micro-yielding curves, stress σ vs. plastic strain ε_p, with unloading and reloading for a) UFG copper as-received, b) heat-treated UFG copper, c) copper with conventional grain size (40 µm) [20]

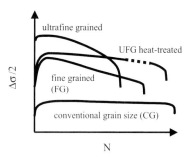

Figure 2: Schematic cyclic deformation curves, stress amplitude $\Delta\sigma/2$ vs. number of cycles to failure N, of UFG material compared to CG, FG and heat-treated UFG materials.

As discussed earlier, the mechanisms of shear band formation are not clear at the present time. Current investigations on UFG Cu show many frequently intersecting shear bands on the surface. The formation of these shear bands has been shown to be related to a local grain coarsening in the interior of the material. The shear bands form by spreading out from the coarsened patches, triggered by the cyclic deformation, as suggested earlier [13]. Fig. 3 shows two intersecting micro-shear bands obtained by using the ECC-technique. The observation of intersection leads to two conclusions. Firstly, the intersection can be interpreted as a "geometrically necessary" shear-banding mechanism keeping the material aligned under axial loading. Secondly, the formation of the shear bands in the UFG Cu investigated here is not induced by inherent pre-formed shear bands as suggested by Wu *et al.* [18]. Moreover, the mechanism proposed by Wu *et al.* [18] does not take into account the formation of intersecting shear bands. It is concluded that their model should be confined to their UFG copper and to a specific microstructural condition.

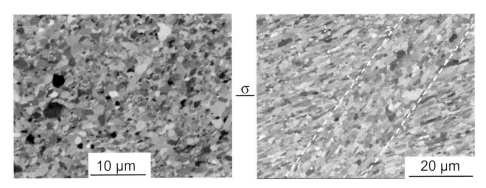

Figure 3: Intersecting shear bands in UFG Cu fatigued at $D\varepsilon_{pl}/2 = 1 \cdot 10^{-3}$. ECC-SEM image

Figure 4: Shear band and grain rotation in UFG Al fatigued at $\Delta\varepsilon_{pl}/2 = 1 \cdot 10^{-3}$. ECC-SEM image

In fatigued UFG Al (grain size: 800–900 nm), macroscopic shear banding was also found, as shown in fig. 4. In contrast to the investigations on UFG copper, only one large-scale shear band was found and this contained only slightly coarsened grains in the interior. But grain rotation has taken place as is also visible in fig. 4. Micro-hardness measurements reveal that the hardness within the shear band is lower by 20–30 % than the hardness in the neighbouring material. This observation is consistent with shear strain localisation in the band. The less marked grain coarsening compared to the high-purity UFG copper is probably a result of the smaller mobility of the grain boundaries in the less pure aluminium.

Based on these findings, it is reasonable to conclude that the mechanism responsible for the formation of macroscopic shear bands is an interaction of cyclically-induced grain coarsening and grain rotation which both lead to a locally reduced strength of the material and, as a consequence, to a concentration of plastic deformation in the coarsened and/or rotated regions. This localisation itself triggers the grain coarsening and/or rotation and the shear band spreads out. Comparing the results on fatigued high-purity UFG Cu and commercial-purity UFG Al, it becomes evident that the purity of the material can have a strong effect on the mechanism of the shear band formation.

The beneficial effect of a recovery heat treatment on the fatigue behaviour was already shown for UFG copper [16]. Hence, a similar study was performed on heat-treated UFG Al. Applying a special heat treatment up to temperatures of 220 °C [16] and also 350 °C, a recovered microstructure with negligibly coarsened grains was obtained. The first preliminary investigations (fig. 5) reveal that, at the given plastic strain amplitude, the recovery heat treatment leads to an enhanced fatigue life compared to the UFG material in the as-ECAP processed condition. The scatter in the fatigue lives of the two UFG Al in the ECAP-processed condition is not yet clear. Otherwise, the cyclic deformation behaviour is similar with respect to the marked initial cyclic hardening and differs very strongly from the behaviour of the heat-treated UFG Al. Since the UFG Al specimens have a rather large grain size (800–900 nm), the material belongs to the FG-type in terms of the cyclic deformation behaviour discussed earlier. The heat-treated material shows no cyclic hardening, but continuous cyclic softening, in agreement with the schematic behaviour proposed in fig. 2.

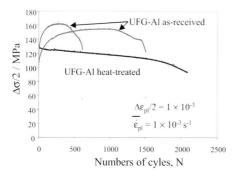

Figure 5: Cyclic deformation curves of UFG Al and UFG Al heat-treated at 220 °C for 1 h

4 Conclusions

Based on micro-yielding loading and unloading curves and on investigations of the cyclic deformation behaviour of materials with different grain sizes, the cyclic deformation behaviour is discussed. The responsible mechanisms for macroscopic shear banding in UFG materials are clarified. It is shown that grain coarsening and/or grain rotation play a dominant role in the formation of the shear bands. As found earlier for UFG Cu, a recovery heat treatment is beneficial for the LCF behaviour of UFG Al.

5 Acknowledgments

The authors are very grateful to Deutsche Forschungsgemeinschaft DFG for their financial support.

6 References

[1] T.C. Lowe, R.Z. Valiev, eds., Investigations and Applications of Severe Plastic Deformation, Kluwer Academic Publishers, Dordrecht, The Netherlands, 2000
[2] V.M. Segal, Mat. Sci. Eng. 1995, A197, 157–164
[3] V.M Segal, Mat. Sci. Eng. 1995, A338, 331–344
[4] Y. Iwahashi, Z. Horita, M. Nemoto, T.G. Langdon, Acta Mater. 1998, 46, 317–3331
[5] R.Z. Valiev, R.K. Islamgaliev, I.V. Alexandrov, Prog. Mater. Sci., 2000, 45, 103–189
[6] T.C. Lowe, R.Z. Valiev, JOM, 2000, 52 (4), 27–40
[7] M. Furukawa, Z. Horita, T.G. Langdon, Adv. Eng. Mat. 2001, 3, 121–125
[8] H. Mughrabi in Investigations and Applications of Severe Plastic Deformation (Eds.: T. C. Lowe and R. Z. Valiev), Kluwer Acad. Publishers, Dordrecht, The Netherlands. 2000, pp. 241–253
[9] A. Vinogradov, Y. Kaneko, K. Kitagawa, S. Hashimoto, R.Z. Valiev, Mat. Sci. Forum 1998, 269-272, 987–992
[10] S.R. Agnew, A. Vinogradov, S. Hashimoto, J.R. Weertman, J. Electron. Mater. 1999, 28, 1038–1044
[11] S.R. Agnew, J.R. Weertman, Mat. Sci. Eng. 1998, A244, 145–153
[12] A. Vinogradov, S. Hashimoto, Mat. Trans. 2001, 42, 74–84
[13] H. Mughrabi, H.W. Hoeppel in Mater. Res. Soc. Symp. Proc., Vol. 634, (Eds. D. Farkas, H. Kung, M. Mayo, H. v. Swygenhoven and J. Weertman) MRS, Warrendale, USA, 2001, p. B2.1.1–B2.1.12
[14] H.W. Hoeppel, M. Brunnbauer, H. Mughrabi, R.Z. Valiev, A.P. Zhilyaev in Werkstoffwoche 2000, http://www.materialsweek.org/proceedings, 2001
[15] H.W. Hoeppel, Z.M. Zhou, H. Mughrabi, R.Z. Valiev, Phil. Mag. A 2002, 82, 1781–1794
[16] H.W. Hoeppel, R.Z. Valiev, Z. Metallkd. 2002, 93, 641–648
[17] H.W. Hoeppel, Z.M. Zhou, M. Kautz, H. Mughrabi, R.Z. Valiev in Proc. of 8th Int. Fatigue Congress (FATIGUE 2002), (Ed. A. Blom), vol. 3, EMAS Ltd., Cradley Heath, UK, 2002, p. 1617–1624
[18] S.D. Wu, Z.G. Wang, C.B. Jiang, G.Y. Li, Phil. Mag. Lett. 2002, 82, 559–565
[19] M. Furukawa, Z. Horita, M. Nemoto, T.G. Langdon, J. Mat. Sci. 2001, 36, 2835–2843
[20] N. Barta-Schreiber, Diplomarbeit, Universität Erlangen-Nürnberg, 2001

The Influence of Type and Path of Deformation on the Microstructural Evolution During Severe Plastic Deformation

A. Vorhauer, R. Pippan

Erich Schmid-Institute of Materials Science of the Austrian Academy of Sciences, Leoben, Austria;
Christian Doppler Laboratory of Local Analysis of Deformation and Fracture, Leoben, Austria.

1   Introduction

It is well known that very large plastic deformation lead to an enormous subdivision of the initial grain structure into crystallites that are rotated with respect to each other. Microstructural features like size of structural elements decreases and angle of disorientation between adjacent structural elements increases with increasing strain [1]. In order to receive nanometer or submicron grained microstructures, effective strains of about 6-8 [2,3] are necessary and the materials processing have to be performed at low homologous temperatures – typically lower than 0.3 T_m (T_m: melting temperature in K) [4].

Many different processing techniques are known which allow imposing the necessary effective strains without failure of the material and are summarized by the term Severe Plastic Deformation (SPD). The most popular variants are Equal Channel Angular Extrusion (ECAE) and High Pressure Torsion (HPT). Reports about processing of different metallic materials at different conditions and features of the resulting microstructures are available. Some of them report about the influence of ECAE processing route on the microstructural evolution [e.g. 5].

In this paper we compare the microstructures formed in pure aluminum (99.99 %) and pure copper (OFE, 99.99 %) after SPD treatment by ECAE, HPT and Cyclic Channel Die Compression (CCDC) – a special variant of Channel Die Compression – at room temperature. The aim of this paper is to decide from a microstructural point of view, if some types of SPD are more efficient in forming small scaled and equiaxed microstructures or if the subdivision of the initial grains is independent in type and path of materials processing.

2   Materials Processing

The materials processing by **ECAE** is well and extensively described in [e.g. 6]. In principal the tool consists of two intersecting channels of same cross section, which meet at an angle 2ϕ (see Fig. 1). The geometry of this tool provides that the material is deformed by simple shear at ideal conditions (without friction). The cross section of the specimen remains about equal before and after a processing step, thus it is possible to subject one specimen several times to ECAE in order to reach high degrees of plastic deformation. A circular or squared cross section of the channel provides the possibility of a materials processing at different routes that are distinguished by their different combinations of sample rotation around the channel axes between consecutive processing steps [7]. In this paper mainly ECAE ($\phi = 120°$) route A and C are used. The size of the investigated specimen was the 15 × 15 × 40 mm³.

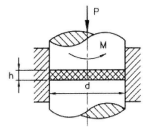
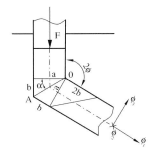

Figure 1: Principle of ECAE

Figure 2: Principal of HPT

HPT provides in comparison to ECAE a continuous shear deformation of a disc shaped specimen [8]. A sketch of this process is shown in Fig. 2. The principal is to deform a disc shaped specimen under a hydrostatic pressure by rotating the two plungers with respect to each other. In order to prevent a sliding of the specimen all contact areas between specimen and plunger were sandblasted before materials processing. Discs, 11 mm in diameter are subjected to HPT at a hydrostatic pressure of about 4 GPa.

The **CCDC** (see Fig. 3) is a special variant of the Channel Die Compression process that is used since several decades to simulate rolling [e.g. 9]. The geometries of specimen and the CCDC channel are in principal syntonized in that way, that the material flows only in direction of the channel at applied compressive load. The geometry of the specimen remains about equal before and after a processing step if initial height of the specimen and length of channel are identical. Thus, it is possible to subject the same specimen several times to the CCDC process by rotating the material by 90° around the transverse direction (TD) of the channel (named 'route I'). If the specimen has additionally a squared cross section a second CCDC processing route is possible that is given by two consecutive rotations of each 90°: first around TD (like route I) followed by a rotation around the normal direction (ND). This combination of materials rotation is named 'route II'. A comparison of these two processing routes is shown in Fig. 4. In this paper specimen of the size $15 \times 15 \times 30$ mm³ (Al) and $6 \times 6 \times 12$ mm³ (Cu) are analyzed.

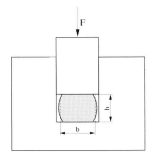
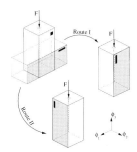

Figure 3: Principle of CCDC

Figure 4: Possible CCDC processing routes.

3 Microstructural Investigations

The microstructural investigation was performed in a Scanning Electron Microscope (SEM) by means of different techniques. Micrographs have been taken with a detector for Back Scattered Electrons (BSE). These micrographs allow an estimation of the typical microstructural sizes. In a single-phase material, the energy of the detected backscattered electrons depends on both, the orientation of the crystallites with respect to the direction of the insistent electron beam, and the density of dislocations. Thus, different orientated structural elements appear in these micrographs as regions of different grey-values. As this technique does not allow the definition of crystallographic orientations, the Electron Back Scatter Diffraction (EBSD) technique was applied to quantify the development of the distribution of crystallographic orientation. The result of such investigations can be plotted as orientation maps. Each data point shows a characteristic color depending on its crystallographic orientation; the color designates the crystallographic orientation with respect to a selected specimen direction.

3.1 BSE Micrographs

BSE micrographs are captured for both investigated materials in correlation to the applied SPD method, strain path and degree of deformation. An overview about the principal of microstructural subdivision during SPD is given in Fig. 5 for Cu after processing by CCDC route II.

Figure 5: BSE micorgraphs of Cu at different degrees of CCDC route II deformation. All micrographs are of the same magnification

The deformation is mainly localized in shear bands. This leads to a very inhomogeneous distribution of the size of structural elements at low plastic deformations. Fig. 5a and 5b show intersecting shear bands at effective von Mises strains of 1.38 and 2.76. The size of the structural elements is smaller in shear bands than between. The microstructure between shear bands is subdivided in smaller structural elements when the effective strain increases. Thus the homogeneity in distribution of the size of structural elements increases with increasing strain. No large differences in mean microstructural size can be found by comparing Fig. 5c ($\varepsilon_{vM} \sim 5.52$) and Fig. 6a ($\varepsilon_{vM} \sim 11.04$) – only the size distribution becomes narrower and the equiaxiality increases with increasing strain.

Fig. 6 shows BSE micrographs at nearly the same degree of plastic deformation ($\varepsilon_{vM} \sim 11$) after materials processing by CCDC RI, RII and ECAE RC. The qualitative comparison of these

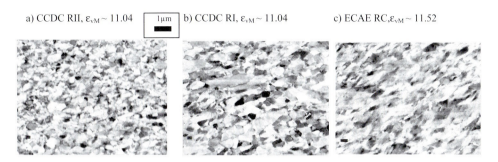

Figure 6: BSE micorgraphs of Cu at same strain $\varepsilon_{vM} \sim 11$ after different types of SPD treatment. All micrographs are of the same magnification

micrographs shows, that the size of the structural elements is about the same after different types of SPD. It seems that structural elements after SPD by CCDC RI and ECAE RC are more elongated than in the case of SPD by CCDC RII.

3.2 Orientation Maps

EBSD measurements are performed in order to quantify the microstructural differences between different types and paths of SPD with respect to the size of structural elements and the distribution of disorientation angle between neighboring structural elements (see Fig. 7). The definition of both microstructural parameters is done on the basis of measured data without any data-cleanup (removing of undefined points) in order to prevent a clean up induced falsifying. Therefore we use a new analysis procedure, which is described in [10].

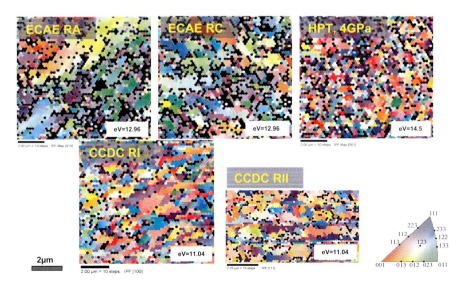

Figure 7: Orientation maps at effective strain of about 11 for different variants of SPD

Fig. 8 shows the distribution of the disorientation angle between structural elements after different SPD treatments at an effective strain of ~11. The black curve in each diagram gives the distribution of the disorientation angle for an arrangement of randomly distributed orientations with cubic crystal system (Mackenzie-distribution) [11]. In principal all measured distributions are similar: mainly neighboring structural elements are separated by high angle boundaries (fraction above 70 %) and the frequency maximum is situated at angles of disorientation of about 45°. The best accordance with the Mackenzie-distribution is given for the HPT deformed samples.

In order to quantify the differences between measured distribution of disorientation angle and the Mackenzie-distribution we use the χ^2-parameter that is usually applied in statistics to decide whether the differences between a measured and a hypothetical distribution of frequency values are significant or of random nature. The calculation of the χ^2-parameter is performed according to Eqn. 1; $p_{r,i}$ and p_i are the calculated and hypothetical (Mackenzie) relative frequency values at a certain class of disorientation i.

$$\chi^2 = \sum_{i=1}^{k} \frac{(p_{r,i} - p_i)^2}{p_i} \tag{1}$$

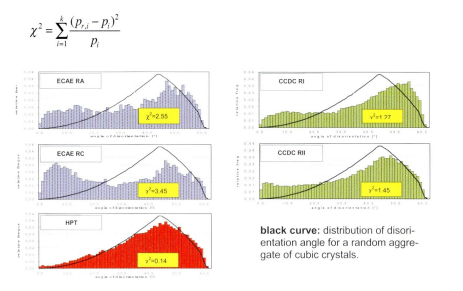

Figure 8: Distribution of disorientation angle for structural elements after SPD by different techniques and routes at the same effective strain of about 11

Fig. 9 shows the evolution of χ^2_{rs} as a function of effective strain. The graphs for all investigated SPD variants show the same tendencies: an inherent drop at small degrees of plastic deformation and a reaching of a saturation value of about 1–2.8 at a strain level of 6. The differences between different SPD variants are large at lower strains and decreases with increasing strain. This indicates that the accordance of the measured distributions of the disorientation angle with the Mackenzie-distribution is nearly the same for all SPD variants at strain levels larger than 6.

Fig. 10 shows the change in microstructural size in Cu after a materials processing variants of SPD as a function of effective strain. It can be seen that the structural sizes of all SPD microstructures decreases rapidly with increasing strain till they reach a kind of a saturation value of

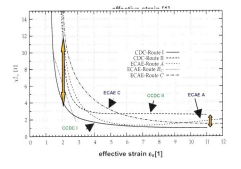 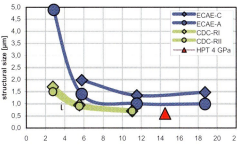

Figure 9: χ^2 as a function of effective strain for the investigated SPD variants

Figure 10: Evolution of structural sizes in Cu as a function of effective strain and SPD variant

about 1 µm at an effective strain of about 6. This drop is most distinct for CCDC microstructures and additionally these microstructures for both routes are smaller in size at all investigated strain levels than after materials processing by ECAE. The smallest structural elements (mean size about 600 nm) are generated by HPT.

4 Summary and Conclusion

Pure Al and Cu are deformed up to effective strains of ~14.5 by ECAE, HPT and CCDC at different processing routes. The comparison of BSE micrographs and the analysis of orientation maps at different effective strains revealed the following key features:

- Large microstructural differences between different types of SPD are present at lower degrees of deformation. With increasing strain this differences decrease.
- The smallest size of the structural elements is found after HPT, the largest structural elements are present in the case of SPD treatment by ECAE (especially route C).
- The accordance of measured distributions of disorientation angle with the Mackenzie-distribution increases with increasing strain for all investigated SPD variants. This indicates that the present texture components are of low intensity at effective strains larger than 6.

5 Acknowledgement

The financial support by the 'Österreichischer Fond zur Förderung der wissenschaftlichen Forschung' (Project: P 12944-PHY) is gratefully acknowledged.

6 References

[1] Liu, Q., Hansen, N. , Scripta Metall. Mater., 1995, 1289–1295
[2] Chang, C.P., Sun, P.L., Kao, P.W., Acta Mater., Vol. 48, 2000, 3377–3385
[3] Vorhauer, A., diploma thesis, University of Leoben, 2000.
[4] Valiev, R.Z., Islamgaliev, R.K., Alexandrov, I.V., Progress in Materials Science, Vol. 45,2000,103–189
[5] Gholinia, A., Prangnell, P.B., Markushev, M.V., Acta metall. Mater., Vol. 48, 2000, 1115–1130
[6] Segal, V.M., Materials Sci. Eng., A271, 1999, 322–333
[7] Furukawa, M., Iwahashi, Y., Horita, Z., Remoto, M., Materials Sci. Eng., A257, 1998, 328–332
[8] Jiang, H., Zhu, T. Butt, D.O., Alexandrov, V., Lowe, T.C., Materials Sci. Eng., A290, 2000, 128–138
[9] Maurice, Cl., Driver, J.H., Acta metal mater, Vol. 41, No. 6, 1993, 1653–1664
[10] Vorhauer, A., Hebesberger, T., Pippan, R., Acta Mater., Vol. 51, 2003, 677–686
[11] Mackenzie, J.K., Acta Metall., Vol. 12, 1963, 223–225

Formation of a Submicrocrystalline Structure in Titanium during Successive Uniaxial Compression in Three Orthogonal Directions

[1]G.A. Salishchev, [1]S.V. Zherebtsov, [1]S.Yu. Mironov, [2]M.M. Myshlayev and [3]R. Pippan
[1]Institute for Metals Superplasticity Problems, Ufa, Russia
[2]Baikov Institute of Metallurgy and Materials Science, Moscow, Russia
[3]Institute of Material Science, Leoben, Austria

1 Introduction

Severe plastic deformation (SPD) of metals and alloys attracts increasing interest as a method for producing submicrocrystalline structure (SMC) in large-scale billets. Two main techniques such as equal channel angular pressing (ECAP) [1] and multiple forging [2] performed at cold or warm deformation can be used. The feature of the both techniques is that, for attaining large strains, the initial shape of billets should remain almost unchanged. To increase the uniformity of plastic flow one of the most useful operations to be used is rotation of a billet that changes the direction of deformation. In this case slip and twining systems are involved in operation, which did not take part in plastic flow during the previous deformation. However, rotation can cause some other effects. In particular, the high stresses and the density of defects resulted from large strains change the interaction between defects. The mobility of dislocations is increased essentially and deformation induced dislocation boundaries (DIDB) are formed [3]. High gradients of external stresses are responsible for macroscopic direction of these boundaries that leads to formation of a lamellar structure being finer than the initial microstructure [3]. Evidently the change in the strain path may assume both the interaction between "new" and "old" DIDB at their intersection and the considerable changes in the microstructure formed at previous stage of deformation (because of Baushinger effect for example) [3]. The investigations of evolution microstructure at ECAP using different loading schemes and various deformation direction showed that microstructure changed rather essentially [5].

The aim of the present paper is to study the microstructure evolution and mechanical behavior of the commercial pure titanium during successive compression of prismatic samples in three orthogonal directions.

2 Experimental

The commercial pure titanium (impurities content, wt%: 0.25Al; 0.15Fe; 0.07Si; 0.05C; 0.005H; 0.02N; 0.12O) with a mean grain size of 35 µm was used. Mechanical behavior of titanium was studied during compressions at 400 °C of prismatic samples ($14 \times 16 \times 20$ mm^3) along three orthogonal directions. Prior to each rotation a prismatic shape was imparted to the sample by cutting the curved faces. The initial strain rate and true strain per each compression step were 10^{-3} s^{-1} and about 0.5, respectively. The amount of deformation steps was 12.

Compression tests were performed in air using a SCHENK machine. The deformation relief occurring on the surface of samples at different stages of loading was studied after additional compression of the samples by $\varepsilon = 15\%$. The microstructure was studied using a transmission electron microscope (TEM) JEOL JEM-2000EX and scanning electron microscope (SEM) JEOL JSM-840. Crystallographic analysis of structure parameters was made by EBSD-technique on a SEM LEO-440 Stereoscan at the Institute of Material Science in Leoben. The textures of specimens were determined by measuring incomplete pole figures (002), (100), (101) by means of the X-ray diffraction goniometer "DRON-3M". Experimental data were processed via the POPLA program using harmonic analysis.

3 Results

Figure 1 a shows the set of true stress-true strain curves for all deformation steps. One can see that the change in the deformation axis leads to a change in flow stresses from one step to another. The first steps of deformation are characterized by the growth of yield stress. A sharp decrease of yield stress took place at the 6th compression step. Then the yield stress increase step by step again. The considerable drop in the yield stress at each succeeding step as compared to flow stress at the end of the previous one and the strain hardening were revealed on these curves.

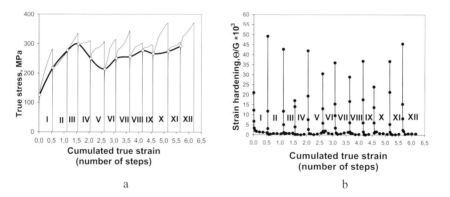

Figure 1:. Mechanical behavior of titanium during successive compressive strain (by a true strain of 0.5 at each step) in three orthogonal directions with total step number of 12 at 400 °C and $\dot{\varepsilon} = 10^{-3}$ s^{-1} : a – true stress vs true strain; b –strain hardening vs true strain. (the 1st, 2nd, 3rd point of each step is obtained at a strain of 0.2 %, 0.5 % and 1 % respectively).

As seen from Figure 1b, strain hardening at each step is not equal to zero and characterized by the presence of two stages. At strains up to 5% it decreases sharply. After that the dependence of strain hardening on strain becomes less pronounced and then remains almost unchanged.

Investigation of texture evolution under loading has shown that both the initial and the deformed structures are characterized by the axial texture; namely, planes (002) were azimuth deflected through 30°–60° from the axis of the rod (Fig. 2a–d). It is seen that deformation accompanied by successive changes in the strain path only slightly changes the intensity and scattering of texture and does not change it essentially. These results show that the change in the direction of deformation causes such structure transformation, which adapt the microstructure to new deformation conditions.

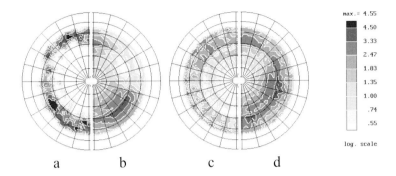

Figure 2: (002) pole figures of titanium in initial condition (a) and after different steps of deformation: b – 3, c – 5 and d – 12 steps.

Investigation of deformation relief in all stages of plastic flow revealed the presence of parallel almost bands (shear bands). However, if, at small strains, these bands were arranged within the grains and in different grains their direction is different, then, with increasing the number of steps above 3 they became rectilinear (in spite of local bends) and their length was about few hundred micrometers. The initial polyhedral structure, which was quite dominant at small strain, was not revealed after 3 compression steps. At strain above 5 steps two systems of parallel bands intersected at an angle of ~ 90° are revealed in the structure (Fig. 3a). The angle between deformation axis and the bands was approximately 45°. In sites of mutual intersection the local bends, shears and even gaps at boundaries were observed. The final structure after 12 steps is characterized by a significant smoothing of the relief, namely, "smoother" and shorter bands were observed (Fig. 3b).

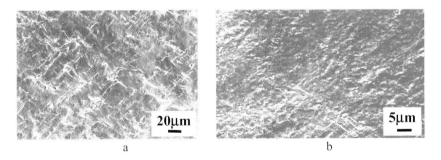

Figure 3: SEM images of deformation relief on titanium sample surface during different steps of compression strain at 400 °C and strain rate of 10^{-3} s^{-1}: - 5 steps + ε = 15%; b –12 steps + ε = 15%.. The deformation axes were vertical in the both cases

Results of TEM and EBSD analysis show that at all stages the deformation occurs by slip and twining. The main feature of the microstructure at small strains (Fig.4a) is a grain subdivision by formation of DIDB. These boundaries can be identified as dense dislocation walls and double wall microbands [3, 4]. At the further deformation the extended dislocation boundaries look like as lamellar type boundaries [3, 4]. Their length achieves tens of micrometers and they are rectilinear (Fig.4b). Lamellar boundaries structure replaces the initial equiaxed structure and becomes dominating after these deformation steps. The feature of this structure is the absence of a distinct system of boundaries, the presence of numerical bending extinction contours inside of bands, essential heterogeneity of their transverse sizes (Fig. 4c). At the intersection of these boundaries local bends and shears take place along it. The further increase in the number of deformation steps up to 12 leads to the formation of SMC structure (mean grain size of about 0.3 µm) (Fig.4 d).

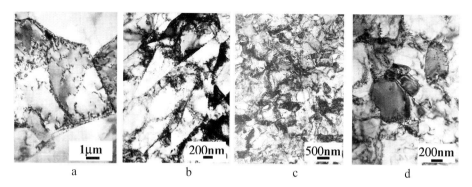

Figure 4: Microstructure evolution of titanium during different steps of compression strain at 400 °C and strain rate of 10^{-3} s^{-1}: – ε = 15%; b – 3 steps; c – 5 steps; d – 12 steps. TEM images

The TEM analysis indicates that the volume fraction of low angle grain boundaries decreases with strain. The study of conjugated crystallites misorientation via EBSD-scanning (Table 1) confirm the decrease in the volume fraction of low angle grain boundaries with strain and this decrease is extremely pronounced at the 5th step.

Table 1: Effect of the cumulative strain on the fraction of different types of boundaries

Type of boundaries	Number of compression steps (ε)				
	1 (30%)	1 (60%)	3 (by 40%)	5 (by 40%)	12 (by 40%)
Low-angle*(from 2° to 15°)	90%	90.3%	81.1%	54.2%	31.1%
High-angle*(from 15° to 93.8°)	10%	9.7%	18.9%	45.8%	68.9%
Twins**(Σ7b, Σ11a, Σ11b, Σ13b, Σ13c, Σ19c)	1.3%	4%	2.1%	2.2%	2.7%

*Step of scanning – 0.1 – 0.2 µm, **Step of scanning – 0.1 – 5 µm

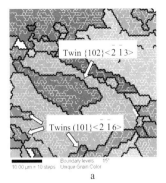
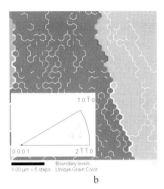

Figure 5: Illustration of the changes of crystallographic parameters of boundaries: a – twined ($\varepsilon = 60\%$; special boundaries are indicated by color); b – random high angle ($\varepsilon = 40\%$; low angle boundaries with the angle of misorientation up to 15° are indicated by white color)

EBSD-analysis has shown that at the initial stage of deformation the twinning at the $(1 0 1)\langle \bar{2}\bar{1} 6\rangle$ system is prevailing. As strain increases the twinning at the $\{1 0 2\}\langle 1 1 2\rangle$ and $1 0 2$ systems become more active. The twinning on the last two systems takes place over the whole investigated range of strain up to the 12th step. As seen from Table 1, the volume fraction of twins depends rather weakly on the amount of compression steps. However, one should take into account the following: Disorientations axes and angles of boundaries, which surround twins, do not correspond presciently to twin-matrix conjugation (Fig. 4a). Its values can be deviated in very broad range and even leave Brandon's interval (interval of angles and axes of disorientations, inside that boundary is close to coincidence site lattice boundary).

It should be noted that the inconstancy of crystallographic parameters along of the boundary plane is apt to any boundaries and not only to twin ones. In Fig. 5b it can be seen the angles and the axis of misorientation are not discrete and constant along the boundary but they vary within rather wide range.

4 Discussion and Summary

A cyclic character of loading with successive rotations of a sample of commercial pure titanium carried out at warm deformation ($T = 400$ °C) results in some unusual mechanical behavior: 1) The yield stress dependence is non-monotonic: For the first three steps an increase is observed, then after the 4th step a sharp decrease and after that a further increase is observed; 2) The yield stress of each following step is always less than the flow stress at the end of the previous one; 3) The strain hardening as a function of strain in each deformation step is similar: The third stage completed very early and the forth stage of strain hardening prevails. Though the first fact indicates the evident change in the microstructure with increasing total strain, both the second and the third facts shows some definite repetition in microstructure evolution on each step of deformation.

Structure formation of titanium occurs by slip and twining. The limited number of slip systems in the h.c.p. lattice is insufficient for fully consistency of deformation of adjacent crystallites. This is a reason for occurrence of significant internal stresses in the grain boundaries area.

It leads to local acceleration of deformation processes and appearance of DIDB. A competitive mechanism to stress relaxation is twining.

The low mobility of dislocations evidently results in the propagation of DIDB in the direction of maximum gradient of stresses (and strains). The given macroscopic directions of these boundaries leads to their «germination» into neighboring grains and changes in crystallographic parameters of initial boundaries [3]. In accordance with the principle of "orientation instability" of structure [3, 6] the changes in misorientations of conjugating crystallites correspond to the best adaptation to deformation conditions: attaining of maximum Schmid factor for operating slip systems and the best consistency of deformation**.** Due to the change of strain path the structure formed in the previous strain step becomes "non-stable" in respect to the new conditions of deformation. As it was found in [7], in the case of changing the direction of deformation "scattering" of low-angle dislocation boundaries on dislocations is possible. "Old" dislocation structure may "scatter" during the change of strain path and due to that the yield stress decreases as compared to the flow stress at the previous end step (Fig.1a). However, the observed effect is rather high (in some cases it is up to 100 MPa) that supports the assumption that there exist more powerful "sources" of stock of dislocations. It is clear that the texture formed in the course of previous compression step exerts a significant influence on flow stress of next step. But if we take into account only this reason the decrease in the flow stress should be in equal proportion to their increase, but this does not fit experimental data.

During deformation certain part of dislocations can be transferred into new quality – dislocation boundary. This change and existence of these boundaries was conditioned by local stress-strain state of structure. It is very possible that change of deformation axis will rise the change of local stress-strain state and subboundaries, formed during previous deformation will be not stable in new deformation condition. Probably, that part of its will be "scattering" on individual dislocations (Baushinger effect). Formed "avalanche" of dislocations rise stress decreasing. This is confirmed indirectly by the recent results of EBSD analysis of the effect of strain path on kinetics of microstructure refinement [5].

The occurrence of a high dislocation density from scattering of low angle boundaries may be responsible for decreasing yield stress within the interval between step 4 and step 6. This leads to a significant growth in the volume fraction of high angle boundaries as deformation increases further. Deformation processes accelerate essentially and, as a result, the intense growth of a misorientation angle at DIDB and the appearance of a "new" system of bands crossing the "old" one (Fig. 3a). Interaction of dislocations with twin boundaries transforms them to random high angle boundaries. The further increase in the volume fraction of high angle grain boundaries (Table 1) evidently influences the increase in the contribution of grain boundary strengthening and causes the increase the yield stress.

The process of plastic flow very often completes with fracture (High Pressure Torsion excluding). That is why, local exhaustion of accommodation mechanisms will lead to formation of microcracks and, finally, to fracture of a sample. In this context, the relaxation of local overstresses by changing macroscopic deformation conditions (a change of strain path) is an efficient method for attaining severe plastic deformations in a material, and, as a result, a decrease in the grain size to some critical value.

5 References

[1] Valiev R.Z., Islamgaliev R.K., Alexandrov I.V., Progress in Materials Science 2000, 45, p. 102–189
[2] Patent /US97/18642, WO 9817836, 1998
[3] Rybin V.V., Large Plastic Strains and Fracture of Metals, Moscow, Metallurgy, 1986, p. 224
[4] Hughes D.A., Hansen N., Acta Mater. 1997, 45, p. 3871-3886
[5] Gholinia A., Prangnell P.B., Markushev M.V., Acta Mater. 2000, 48, p. 1115–1130
[6] Rybin V.V., Likhachev V.A., Vergazov A.N., Fizika Metallov i Metallovedenie 1974, 37, p. 620–624 (in Russian)
[7] Mori H., Fujita H., J. Phys. Soc. Japan. 1975, 38, p. 1342–1348

XIII Features and Mechanisms of Superplasticity in SPD Materials

Achieving a Superplastic Forming Capability through Severe Plastic Deformation

By Cheng Xu[1], Minoru Furukawa[2], Zenji Horita[3], and Terence G. Langdon[1]
[1] University of Southern California, Los Angeles, CA, USA
[2] Fukuoka University of Education, Munakata, Japan
[3] Kyushu University, Fukuoka, Japan

1 Abstract

Processing by severe plastic deformation (SPD) leads to very significant grain refinement in metallic alloys. Furthermore, if these ultrafine grains are reasonably stable at elevated temperatures, there is a potential for achieving high tensile ductilities, and superplastic elongations, in alloys that are generally not superplastic. In addition, the production of ultrafine grains leads to the occurrence of superplastic flow at strain rates that are significantly faster than in conventional alloys so that processing by SPD introduces the possibility of using these alloys for the rapid fabrication of complex parts through superplastic forming operations. This paper examines the development of superplasticity in various aluminum alloys processed by equal-channel angular pressing (ECAP).

2 Introduction

Superplasticity refers to the ability of a polycrystalline material to pull out in tension to a very high elongation prior to failure. It is now well established that two basic requirements must be fulfilled in order to establish superplastic flow. [1] First, the grain size of the material must be very small and typically less than ~10 μm. Second, the tensile testing must be conducted at a relatively high temperature so that diffusion-controlled processes, such as superplasticity, can occur reasonably rapidly. The latter requirement means in practice that the temperature should be no less than ~0.5 T_m, where T_m is the melting temperature of the alloy in degrees Kelvin.

The processing of materials by severe plastic deformation (SPD) provides an opportunity for producing materials having grain sizes in the submicrometer or nanometer range [2] where these grain sizes are significantly smaller than those generally produced using conventional thermo-mechanical processing. Accordingly, the application of SPD to bulk materials provides an excellent opportunity for achieving superplastic elongations in materials that are normally not superplastic. Processing by equal-channel angular pressing (ECAP), in which a sample is pressed through a die within a channel bent through an abrupt angle, [3,4] is an especially attractive processing route because it yields reasonably large bulk samples that are readily amenable for tensile testing.

The objective of this paper is to examine the potential for achieving superplasticity with reference to three different aluminum alloys: an Al-3 % Mg-0.2 % Sc alloy fabricated in the laboratory, a commercial wrought Al-2024 alloy and an Al-7034 alloy produced by spray-casting. As will be demonstrated, good tensile ductilities can be achieved in each of these alloys after processing by SPD and, in addition, these high elongations are achieved at strain rates that are significantly faster than those generally associated with superplastic flow.

3 Principles of Superplasticity

In superplastic flow, the polycrystalline grains move relative to each other so that they retain an essentially equiaxed configuration even at very high elongations. The flow process in superplasticity is grain boundary sliding and it can be shown that sliding accounts for essentially all of the strain under optimum superplastic conditions. [5] In practice, sliding cannot occur without some accommodation process within the grains and the theoretical models for superplasticity are therefore based on a process in which sliding is accommodated by the movement of intragranular dislocations across the adjacent grains and the rate of sliding is controlled by the rate of climb of these dislocations into the opposing grain boundaries. [6,7] There is direct experimental evidence demonstrating the movement of these intragranular dislocations during superplasticity. [8,9] Under superplastic conditions, the strain rate varies both with the applied stress raised to a power of 2 and inversely with the grain size also raised to a power of 2.

The implication of this behavior is illustrated schematically in Figure 1 where the strain rate, $\dot{\varepsilon}$, is plotted logarithmically against the applied stress, r. The superplastic regime thus occurs at intermediate strain rates and there are transitions to dislocation climb with a stress exponent of ~5 at higher stresses and to an impurity-controlled regime with a stress exponent close to ~5 at lower stresses: this latter region is due to the inhibition of sliding when impurities are present along the grain boundaries and experiments show the impurity region is no longer present when using materials of very high purity. [10,11]

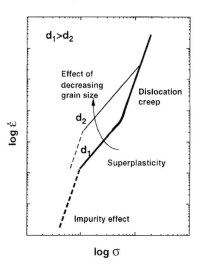

Figure 1: Schematic illustration of strain rate versus stress showing the occurrence of superplasticity at intermediate strain rates and the displacement to faster rates when the grain size is reduced

The strain rate varies with $1/d^2$ in the superplastic region, where d is the grain size, but there is no dependence on grain size in the region of dislocation creep because the deformation process then occurs intragranularly. Thus, a reduction in grain size from d_1 to d_2 effectively displaces the superplastic region to faster strain rates, as illustrated in Figure 1 where it is apparent that the impurity region is also displaced to faster rates when the grain size is reduced. The production of materials with smaller grain sizes therefore provides an opportunity for achieving superplasticity at faster rates.

Since superplasticity leads to high elongations prior to failure, the effect of a decrease in grain size may be illustrated schematically as in Figure 2. Thus, the decrease in grain size displaces the peak elongation to a faster rate and, in addition, there is a tendency to achieve even higher elongations at these faster rates because less time is then available for the development and growth of internal cavities. Experimental examples are available to confirm the validity of the trend illustrated in Figure 2. [12]

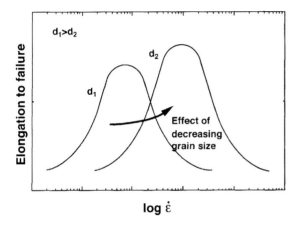

Figure 2: Schematic illustration of the variation of elongation to failure with strain rate showing the effect of a reduction in grain size

The conclusion from this analysis is, therefore, that the use of SPD processing, as in ECAP, should lead both to ultrafine grain sizes and to materials that are capable of exhibiting exceptionally high tensile ductilities at rates that are faster than those generally associated with the superplastic regime. [13] Typically, commercial superplastic alloys have grain sizes of the order of ~5 µm and the superplastic regime is optimized at strain rates in the range of $\sim 10^{-3} - 10^{-2}$ s^{-1} so that superplastic forming times are generally of the order of ~20–30 minutes for each separate component. Since the superplastic strain rate is proportional to d^{-2}, it follows that a reduction in grain size by one order of magnitude, to ~500 nm, will increase the optimum superplastic strain rate to $\sim 10^{-1} - 1$ s^{-1} and this has the potential for reducing the forming time to < 60 s for each component. The use of SPD processing therefore appears to provide an opportunity for expanding the use of superplastic forming into the production of high-volume components as in the automotive or consumer product industries.

4 Production of Ultrafine Grain Sizes Using ECAP

In order to investigate the feasibility of achieving superplastic characteristics through SPD processing, experiments were conducted on three different aluminum-based alloys: an Al-3 % Mg-0.2 % Sc alloy, a commercial Al-2024 alloy, and a spray-cast Al-7034 alloy. All of these materials were subjected to ECAP using dies having an angle, Φ, of 90° between the two parts of the channel and angles, Ψ, of either 45° for the Al–Mg–Sc and the Al-2024 alloys or 20° for the Al-7034 alloy at the outer arc of curvature where the two channels intersect. When $\Phi = 90°$, it can be shown that the strain imposed on each passage through the die is ~1 and this strain varies to only a minor extent with the precise value of Ψ. [14]

All samples were pressed using route B_C where the sample is rotated by 90° in the same direction between each pass through the die. [15,16] This processing route was selected because it leads most expeditiously to an array of reasonably equiaxed ultrafine grains separated by grain boundaries having high angles of misorientation [17] and, therefore, it is the optimum processing route for producing materials that are capable of exhibiting high superplastic elongations. [18] Figure 3 shows the results obtained by subjecting these three "alloys to ECAP and then annealing samples for one hour at selected elevated temperatures: for comparison purposes, data are included for pure Al [19] and an Al-3 % Mg solid solution alloy [19] in addition to the Al-3% Mg-0.2% Sc alloy, [20] and the Al-2024, [21] and Al-7034 alloys. All of these materials were processed using solid dies with the pure Al and the Al-3% Mg alloy pressed at room temperature (RT) for 4 and 8 passes, respectively, the Al-3% Mg-0.2% Sc alloy and the Al-"2024 alloy pressed for eight passes at RT and the Al-7034 alloy pressed for six passes at 473 K.

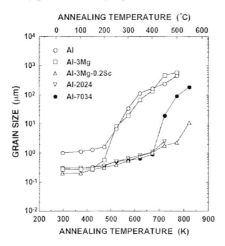

Figure 3: Grain size versus annealing temperature for samples annealed for one hour after ECAP: pure Al [19], Al-3 % Mg [19], Al-3 % Mg-0.2 % Sc [20], Al-2024 alloy [21] and Al-7034 alloy

Inspection of Figure 3 shows that, with the exception only of pure Al where the as-pressed grain size is slightly larger than 1 μm, all of the as-pressed grain sizes in the four alloys are within the range of ~200–300 nm. However, in the absence of any precipitates within the matrix, these ultrafine grains are not stable at elevated temperatures in either pure Al or the Al-3% Mg solid solution alloy and in both of these materials there is rapid grain growth at temperatures above ~500 K. By contrast, the submicrometer grains are remarkably stable in the three alloys where precipitates are present to inhibit grain growth. Thus, grain growth is restricted due to the presence of Al_3Sc precipitates in the Al-3% Mg-0.2% Sc alloy, $CuMgAl_2$ precipitates in the Al-2024 alloy and Al_3Zr, and $MgZn_2$ precipitates in the Al-7034 alloy and in all three alloys the grains remain within the submicrometer range up to temperatures in the vicinity of ~700 K. These three alloys are therefore suitable candidate materials for achieving high temperature superplasticity.

5 Experimental Results with Aluminum-Based Alloys

5.1 Cast Al-3 wt.-% Mg-0.2 wt.-% Sc Alloy

Although the as-pressed grain size in pure aluminum is ~1.3 μm when the ECAP is conducted at RT, [22] it has been shown that the grain size may be substantially refined through the addition of magnesium in solid solution. [23] Thus, the grain size was reduced to ~0.45 μm with the addition of only 1 % Mg and there was a further reduction to ~0.27 μm with the addition of 3 % Mg. [23] Furthermore, through the addition of 0.2 % Sc to the Al-3 % Mg alloy, early experiments showed the as-pressed grain size was further reduced to ~200 nm and there was a potential for achieving superplasticity. [24] Accordingly, detailed experiments were conducted to investigate the high temperature mechanical properties and the flow behavior of the Al-3 % Mg-0.2 % Sc alloy.

The alloy was produced in the laboratory by casting followed by a solution treatment for one hour at 883 K where this temperature was chosen because differential scanning calorimetry (DSC) revealed the onset of partial melting at a temperature slightly above 883 K. [25] Accordingly, this treatment was adopted in order to attain an extensive precipitation of fine secondary Al_3Sc particles to provide the maximum inhibition to grain growth at the higher testing temperatures. Figure 4 shows an example of the experimental results achieved after ECAP and subse-

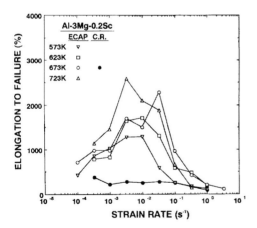

Figure 4: Elongation to failure versus strain rate for an Al-3 % Mg-0.2 % Sc alloy after ECAP and after cold rolling [26]

quent testing in tension at temperatures from 573 to 723 K. [26] Also shown in Figure 4 are the results obtained from samples of the same alloy subjected to cold rolling (CR) at RT to an equivalent strain of ~2.4 without ECAP. Inspection shows that processing by ECAP is exceptionally successful in producing a superplastic capability in this alloy with tensile elongations up to and exceeding 2000 % under the optimum conditions. Furthermore, very high elongations are achieved even at strain rates as high as 10^{-1} s^{-1}, thereby confirming the advent of high strain rate superplasticity. By contrast, the elongations after CR are consistently < 400 % because CR produces a subgrain structure of low-angle boundaries rather than an array of boundaries having high angles of misorientation.

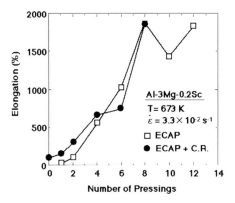

Figure 5: Elongation to failure versus number of passes in ECAP for an Al-3 % Mg-0.2 % Sc alloy after ECAP (open points) and after ECAP + CR (solid points) [27]

The results documented in Figure 4 demonstrate an important difference between the properties attained after ECAP and those attained after CR. The subgrain structure produced in CR is not conducive to superplasticity because high-angle grain boundaries are needed to achieve grain boundary sliding and consequently these boundaries are necessary for the very high tensile elongations associated with superplastic flow. Nevertheless, industrial superplastic forming operations generally utilize materials in the form of thin sheets and it is important, therefore, to determine whether the remarkable superplastic properties achieved through ECAP are also retained when the as-pressed material is rolled into sheets.

To check this requirement, as-pressed samples of the Al-3 % Mg-0.2 % Sc alloy were rolled at RT to a final thickness of 2.2 mm, representing an equivalent strain of ~1.3, and this material was then tested in tension for a direct comparison with the same alloy in the as-pressed condition. [27] The results are shown in Figure 5 where it is apparent that there is no loss in the superplastic properties through a subsequent rolling treatment despite the development of slightly elongated grains and a banded microstructure.

It is reasonable to conclude from these experiments that ECAP and subsequent cold rolling can be used effectively to produce highly superplastic sheet metals. Furthermore, a direct demonstration of the superplastic forming capability of this alloy was provided by cutting small disks from the as-pressed billets and then inserting them into a biaxial gas-pressure forming facility. Using this procedure, it was demonstrated that the disks may be blown into domes under the relatively low pressure of ~1 MPa in very short times from 30 to 60 s. [28] In addition, sectioning of the domes revealed reasonably uniform thinning which is consistent with a genuine superplastic flow behavior.

5.2 Commercial Al-2024 Alloy

The Al-2024 alloy used in this investigation was obtained commercially and contained, in wt.-%, 4.4 % Cu, 1.5 % Mg, and 0.6 % Mn. In the as-received and fully-annealed extruded condition, the alloy contained a dispersion of small (~50–100 nm) $CuMgAl_2$ particles and the grains had a plate-like appearance with approximate dimensions of ~500 × 300 × 10 μm³. The ECAP was conducted for eight passes either at RT (298 K) or at 373 K.

An important characteristic of the Al-2024 alloy is that it contains no additions of either scandium or zirconium. This contrasts with all other aluminum-based alloys where superplastic properties have been reported since, without exception, all of these alloys contain minor additions of Sc and/or Zr to assist in inhibiting grain growth at the elevated temperatures where superplastic flow becomes feasible. Despite the absence of both Sc and Zr in the Al-2024 alloy, the material exhibits good superplastic properties after processing by ECAP at either 298 or 373 K. [21] The results are shown in Figure 6 where the lower points were obtained on unpressed samples and the upper points were achieved after pressing for eight passes at the two different pressing temperatures. Thus, whereas the unpressed material exhibits elongations to failure of < 200 % at all testing strain rates, the samples processed by ECAP give maximum elongations of ~460 % and ~500 % at strain rates of 1.0×10^{-3} s^{-1} and 1.0×10^{-2} s^{-1} after pressing at RT and 373 K, respectively. The low elongations attained in the unpressed material are characteristic of the enhanced but non-superplastic ductilities observed in many Al–Mg alloys [29] and the high elongations achieved after ECAP demonstrate the potential for developing superplastic characteristics in aluminum-based alloys even in the absence of Sc and Zr additions.

An important feature of Figure 6 is that the peak elongation is displaced to a faster strain rate when the samples are processed by ECAP at the higher temperature of 373 K. This displace-

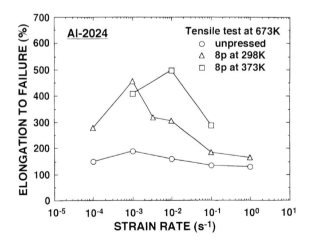

Figure 6: Elongation to failure versus strain rate for an Al-2024 alloy in the unpressed condition and after ECAP for eight passes at either 298 or 373 K [21]

ment occurs despite the fact that the measured grain size after ECAP at 373 K is ~0.5 µm whereas the grain size after ECAP at RT is only ~0.3 µm. This displacement to faster rates is a consequence of the larger fraction of high-angle grain boundaries introduced at 373 K since it is well established that the fractions of high-angle boundaries present in materials after ECAP processing depends critically on the precise pressing conditions. [30]

There have been numerous attempts to achieve superplastic elongations in Al-2024 and modified Al-2024 alloys. The reported data are summarized in Figure 7 where the solid points represent the results achieved after ECAP and the open points denote various thermo-mechanical processing (TMP) including through the use of powder metallurgy (PM) processing. [31–34]

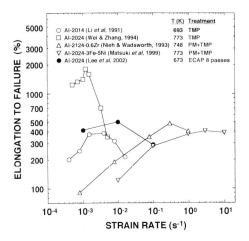

Figure 7: Elongation to failure versus strain rate in tensile testing for various Al-2024 alloys after ECAP (solid points) [21] and after thermomechanical processing (TMP) (open points) [31–34]: PM denotes powder metallurgy processing

It is apparent from inspection of Figure 7 that, although good tensile ductilities are achieved in the two alloys prepared using powder metallurgy processing, the two ingot metallurgy materials shown in Figure 7 after TMP give high elongations but these elongations occur at low strain rates and they do not provide a capability for rapid superplastic forming. Furthermore, the solidus temperature of this alloy is ~775 K and some of the data documented in Figure 7 were obtained at temperatures that are extremely close to this temperature and therefore not practical for use in industrial forming operations. By contrast, processing by ECAP is a simple procedure, it can be used for the processing of a wide range of alloys without requiring the development of specific TMP procedures, it requires no modification of the alloy through the addition of alloying elements, and superplastic elongations are achieved at rapid rates at testing temperatures which are realistic for use in forming operations.

5.3 Spray-cast Al-7034 Alloy

The Al-7034 alloy used in this investigation was produced commercially by spray-casting, where this alloy was selected because there is evidence that spray-casting produces a higher density of dispersoid particles than may be attained using conventional casting procedures. [35] The alloy contained, in wt.-%, 11.5 % Zn, 2.5 % Mg, 0.9 % Cu, and 0.2 % Zr and it was received in an extruded condition with an initial grain size of ~2.1 μm and with a distribution of rod-shaped precipitates having average lengths of ~0.48 μm and widths of ~0.07 μm. These precipitates were identified as the $MgZn_2$ η-phase. The alloy was processed by ECAP at 473 K for either six or eight passes and then tested in tension at temperatures in the range from 573 to 698 K. The average grain size after ECAP was ~0.3 μm and the grains were then in a reasonably equiaxed configuration with a high fraction of high-angle grain boundaries. Inspection showed the rod-shaped precipitates were broken by ECAP and in the as-pressed condition there was an array of spherical precipitates with an average size of ~0.025 μm.

The results from the tensile testing are shown in Figure 8. Thus, high superplastic ductilities were recorded at testing temperatures at and above 673 K, with a maximum elongation of ~1100 % after pressing through eight passes and "testing at 673 K with an initial strain rate of 3.3 × 10^{-2} s^{-1}. The displacement of the peak elongation to a faster strain rate with an increasing number of passes of ECAP is attributed again to an increase in the fraction of high-angle boundaries with increasing strain in ECAP. [30] Thus, as in the other "Al-"based alloys, the spray-cast Al-7034 alloy also exhibits superplastic elongations at rapid strain rates.

Figure 8: Elongation to failure versus strain rate for a spray-cast Al-7034 alloy after ECAP at a temperature of 473 K for six or eight passes

6   Summary and Conclusions

Experiments were conducted to examine the feasibility of achieving superplasticity at rapid strain rates in three different aluminum-based alloys: an Al-3 % Mg-0.2 % Sc alloy, a commercial wrought Al-2024 alloy and a commercial spray-cast Al-7034 alloy. All three alloys exhibit good to excellent superplastic properties after processing by ECAP.

7   Acknowledgement

This work was supported in part by the Light Metals Educational Foundation of Japan and in part by the U.S. Army Research Office under Grant No. DAAD19-00-1-0488.

8   References

[1]   T. G. Langdon, Metall. Trans. A 1982, 13A, 689
[2]   R. Z. Valiev, R. K. Islamgaliev, I. V. Alexandrov, Prog. Mater. Sci. 2000, 45, 103
[3]   V. M. Segal, V. I. Reznikov, A. E. Drobyshevskiy, V. I. Kopylov, Russian Metall. 1981, 1, 99
[4]   M. Furukawa, Z. Horita, M. Nemoto, T. G. Langdon, J.'Mater. Sci. 2001, 36, 2835
[5]   T. G. Langdon, Mater. Sci. Eng. 1994, A174, 225
[6]   A. Ball, M. M. Hutchison, Metal Sci. J. 1969, 3, 1
[7]   T. G. Langdon, Acta Metall. Mater. 1994, 42, 2437

[8] L. K. L. Falk, P. R. Howell, G. L. Dunlop, T. G. Langdon, Acta Metall. 1986, 34, 1203
[9] R. Z. Valiev, T. G. Langdon, Acta Metall. Mater. 1993, 41, 949
[10] P. K. Chaudhury, F. A. Mohamed, Acta Metall. 1988, 36, 1099
[11] S. Yan, J. C. Earthman, F. A. Mohamed, Phil. Mag. A 1994, 69, 1017
[12] F. A. Mohamed, M. M. I. Ahmed, T. G. Langdon, Metall. Trans. A 1977, 8A, 933
[13] <?twb0.20w>Y. Ma, M. Furukawa, Z. Horita, M. Nemoto, R. Z. Valiev, T. G. Langdon, Mater. Trans. JIM 1996, 37, 336
[14] Y. Iwahashi, J. Wang, Z. Horita, M. Nemoto, T. G. Langdon, Scripta Mater. 1996, 35, 143
[15] M. Furukawa, Y. Iwahashi, Z. Horita, M. Nemoto, T. G. Langdon, Mater. Sci. Eng. 1998, A257, 328
[16] M. Furukawa, Z. Horita, T. G. Langdon, Mater. Sci. Eng. 2002, A332, 97
[17] K. Oh-ishi, Z. Horita, M. Furukawa, M. Nemoto, T. G. Langdon, Metall. Mater. Trans. A 1998, 29A, 2011.
[18] S. Komura, M. Furukawa, Z. Horita, M. Nemoto, T. G. Langdon, Mater. Sci. Eng. 2001, A297, 111
[19] H. Hasegawa, S. Komura, A. Utsunomiya, Z. Horita, M.'Furukawa, M. Nemoto, T. G. Langdon, Mater. Sci. Eng. 1999, A265, 181
[20] P. B. Berbon, S. Komura, A. Utsunomiya, Z. Horita, M.'Furukawa, M. Nemoto, T. G. Langdon, Mater. Trans. JIM 1999, 40, 772
[21] S. Lee, M. Furukawa, Z. Horita, T. G. Langdon, Mater. Sci. Eng. 2002, A342, 295
[22] Y. Iwahashi, Z. Horita, M. Nemoto, T. G. Langdon, Acta Mater. 1998, 46, 3317
[23] Y. Iwahashi, Z. Horita, M. Nemoto, T. G. Langdon, "Metall. Mater. Trans. A, 1998, 29A, 2503
[24] S. Komura, P. B. Berbon, M. Furukawa, Z. Horita, M.'Nemoto, T. G. Langdon, Scripta Mater. 1998, 38, 1851
[25] S. Komura, Z. Horita, M. Furukawa, M. Nemoto, T. G. Langdon, J. Mater. Res. 2000, 15, 2571
[26] S. Komura, Z. Horita, M. Furukawa, M. Nemoto, T. G. Langdon, Metall. Mater. Trans. A 2001, 32A, 707
[27] H. Akamatsu, T. Fujinami, Z. Horita, T. G. Langdon, Scripta Mater. 2001, 44, 759
[28] Z. Horita, M. Furukawa, M. Nemoto, A. J. Barnes, T. G. Langdon, Acta Mater. 2000, 48, 3633
[29] E. M Taleff, G. A. Henshall, T. G. Nieh, D. R. Lesuer, J.'Wadsworth, Metall. Mater. Trans. A 1998, 29A, 1081
[30] S. D. Terhune, D. L. Swisher, K. Oh-ishi, Z. Horita, T. G. Langdon, T. R. McNelley, Metall. Mater. Trans. A 2002, 33A, 2173
[31] X. P. Li, L. Covelli, V. Tagliaferri, Y. W. Liu, J. Mater. Sci. Lett. 1991, 10, 585
[32] Z. Wei, B. Zhang, J. Mater. Sci. Lett. 1994, 13, 1806
[33] T. G. Nieh, J. Wadsworth, Scripta Metall. Mater. 1993, 28, 1119
[34] K. Matsuki, T. Aida, J. Kusui, Mater. Sci. Forum 1999, 304-306, 255
[35] Z. C. Wang, P. B. Prangnell, Mater. Sci. Eng. 2002, A328, 87

Production of Superplastic Mg Alloys Using Severe Plastic Deformation

Zenji Horita[1], Kiyoshi Matsubara[1], Yuichi Miyahara[1] and Terence G. Langdon[2]
[1]Kyushu University, Fukuoka, Japan
[2]University of Southern California, Los Angeles, U.S.A.

1 Abstract

Equal-Channel Angular Pressing (ECAP) was conducted to refine grain size to the submicrometer level in three different Mg alloys having nominal compositions of Mg-0.6wt%Zr, Mg-9wt%Al and Mg-7.5wt%Al-0.2wt%Zr. It is shown that the grain refinement was successfully achieved when the alloys were subjected to extrusion prior to ECAP. Thermal stability of the fine-grained structures was examined by conducting static annealing experiment at various temperatures. Transmission electron microscopy was used for the observation of microstructures. Tensile tests were performed at elevated temperatures and it is shown that high-strain-rate and/or low-temperature superplasticity was attained in the alloys subjected to ECAP.

2 Introduction

Grain refinement may be achieved in metallic materials through the process of severe plastic deformation (SPD) [1,2]. With this process, the grain size is reduced to the submicrometer range or even to the nanometer range. Several different procedures are available for the SPD process but the most popular one may be Equal-Channel Angular Pressing (ECAP) where a sample is pressed through an L-shaped channel made within a die [3]. An important feature of the ECAP process is that the cross-sectional dimensions of the sample remain unchanged after pressing and, unlike materials produced using powder metallurgy, it is possible to apply to bulk cast materials without introducing any residual porosity. It has been shown that the ECAP process can scale up to a large size for practical applications [4] and therefore it provides a great potential for producing superplastic materials on a large scale. Furthermore, because ultrafine grains are produced, it is anticipated that superplasticity occurs at higher strain rates [5] and/or lower temperatures [6].

There are many applications of the ECAP process but most have been carried out using Al alloys [7]. There are also some applications of the ECAP process to Cu, Fe, Ni and Mg alloys [8-11] but the related reports are rather limited by comparison with Al alloys. In this study, the ECAP process is applied to Mg alloys to produce microstructures having submicrometer grain sizes. Because Mg alloys are inherently brittle due to the limited number of dislocation slip systems, the ductility improvement is an important requirement for the alloys. Thus, this study attempts to improve the ductility through the grain refinement and subsequent introduction of superplasticity into the alloys. The thermal stability of the fine-grained structures produced by ECAP is first examined by conducting static annealing experiments. The tensile ductility is then examined for the advent of superplasticity at low temperatures and high strain rates.

3 Experimental

Three different Mg alloys having nominal compositions of Mg-0.6wt%Zr, Mg-9wt%Al and Mg-7.5wt%Al-0.2wt%Zr were prepared by melting and casting procedures. Ingots of the alloys were subjected to extrusion at 623 K at a rate of 5 mm s^{-1} to give rods having diameters of 10 mm with a reduction ratio of 1/36. The grain sizes after this extrusion were measured as ~11, ~12 and 21 µm, respectively. ECAP processing was conducted using a die having a channel angle of 90° at 573 K up to a maximum of 4 passes for the Mg-0.6wt%Zr alloy and at 473 K up to a maximum of 2 passes for the Mg-9wt%Al and Mg-7.5wt%Al-0.2wt%Zr alloys. Each rod was held in the die at the testing temperature for ~10 minutes prior to the first pass and for ~1 minute prior to the next pass. The rod was rotated by 90° about the longitudinal axis for the following pass. After ECAP, some samples were sliced perpendicular to the longitudinal axes to thicknesses of ~0.4 mm and annealed for 1 hour at selected temperatures in the range from 373 to 573 K in order to examine the thermal stability of the ECAP structure.

Microstructural examination was conducted using optical microscopy (OM), transmission electron microscopy (TEM) and analytical electron microscopy (AEM). To evaluate the tensile properties of the alloy, tensile specimens were prepared with gauge lengths of 5 mm and cross-sectional areas of 2×3 mm^3 with the tensile axes parallel to the longitudinal axes of the rods. Some tensile specimens were also prepared from the as-cast alloys and the ECAP alloys without extrusion, for comparison. Tensile tests were conducted over a range of temperatures from 373 to 573 K with initial strain rates from $1.0 \cdot 10^{-4}$ to $3.3 \cdot 10^{-1}$ s^{-1} using a testing machine operating at a constant rate of cross-head displacement.

4 Results and Discussion

Figure 1 shows a TEM micrograph of the Mg-9wt%Al alloy after extrusion followed by ECAP through 2 passes at 473 K: a selected area electron diffraction (SAED) pattern taken from a region of 6.2 µm is included. It is apparent that many of the grain boundaries are ill-defined and diffuse in appearance. Similar boundaries were reported in Al alloys [12] and they are interpreted as representative of high-energy non-equilibrium boundaries. The SAED pattern shows that the diffracted beams form rings which are indicative of the presence of boundaries having high angles of misorientation. The average grain size was measured as ~0.7 µm. Similar microstructures were also observed in the M-0.6wt%Zr and Mg-7.5wt%Al-0.2wt%Zr alloys after extrusion followed by ECAP through 2 passes, having the average grain sizes of ~1.3 µm and ~0.8 µm, respectively. It should be noted that the microstructures consisted not only of fine grains with high angle boundaries but also of subgrains with low angle boundaries. The area fraction of the former grains increased as the number of ECAP pass increased.

The variations of grain size with the static annealing temperatures are plotted in Fig. 2. It is apparent that the fine-grained structure produced by ECAP is stable up to ~573 K for the Mg-0.6wt%Zr alloy and up to ~473 K for the Mg-9wt%Al and Mg-7.5wt%Al-0.2wt%Zr alloys. The inclusion of Zr is effective in the Mg-0.6wt%Zr alloy for retaining the fine grains but it is not in the Mg-7.5%Al-0.2wt%Zr alloy. This may be due to the lower amount in the Mg-7.5%Al-0.2wt%Zr alloy but there are different types of contribution of Zr to the formation of particles. Whereas Zr-rich particles formed in the Mg-0.6wt%Zr alloys, AEM analysis revealed that Zr surrounded β phase (Mg$_{17}$Al$_{12}$) particles. Figure 3 shows a TEM micrograph of the Mg-9wt%Al alloy after

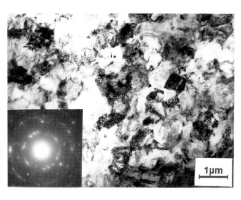 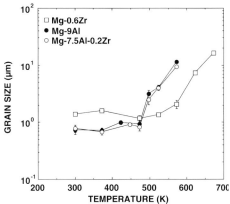

Figure 1: TEM micrograph and SAED pattern for Mg-9wt%Al alloy after extrusion followed by ECAP through 2 passes at 473 K

Figure 2: Grain size versus static annealing temperature for three alloys after extrusion followed by ECAP

2 passes at 473 K followed by static annealing at 498 K for 1 hour. A comparison with Fig.1 shows that grain growth occurs with the average grain size from ~0.7 μm to ~3.1 μm. The grain boundaries are now well-defined and there are less dislocations but many particles visible within the grains.

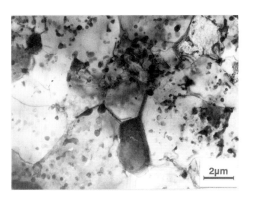 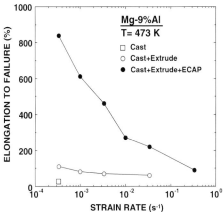

Figure 3: TEM micrograph for Mg-9wt%Al alloy subjected to ECAP followed by annealing at 498 K for 1 hour

Figure 4: Elongation to failure versus initial strain rate for Mg-9wt%Al alloy after casting, after extrusion and after extrusion and ECAP

Figure 4 plots the elongation to failure against the initial strain rate after tensile testing at 473 K. The tensile specimens were prepared from the Mg-9wt%Al after extrusion followed by ECAP through 2 passes at 473 K and, for comparison, after as-casting and after extrusion without ECAP. It should be noted that no data are available from cast samples subjected to ECAP without the extrusion step because these samples broke on the first passage through the die. It is apparent that there is a significant improvement in the ductility when the samples are subjected to ECAP after extrusion. Thus, this comparison demonstrates the importance of a two-step process of extrusion and ECAP in order to obtain superplastic elongations. The results of tensile tests with an initial strain rate of $3.3 \cdot 10^{-4}$ s^{-1} are shown in Fig.5 for the Mg-7.5%Al-0.2%Zr alloy subjected to 2 passes at 473 K after extrusion. For comparison, Fig.5 includes the results for the as-extruded condition. Again, there is a significant improvement of the ductility when the samples are subjected to ECAP. Great improvement in the elongation to failure was also confirmed in the Mg-0.6%Zr alloy.

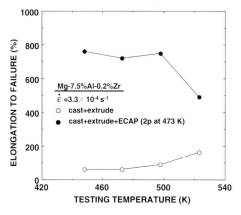
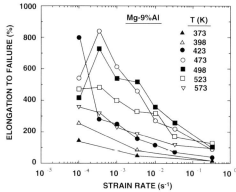

Figure 5: Elongation to failure plotted against testing temperature for Mg-7.5%Al-0.2%Zr alloy subjected to 2 passes at 473 K after extrusion

Figure 6: Elongation to failure versus initial strain rate for Mg-9%Al alloy after extrusion followed by ECAP

The elongation to failure is further plotted against the initial strain rate in Fig.6 for the Mg-9%Al alloy where all samples were prepared by the two-step process of extrusion and ECAP and tensile testing was conducted at temperatures from 373 to 573 K. Five significant conclusions may be reached from inspection of Fig.6. First, the elongations generally increase with decreasing initial strain rate except at temperatures from 473 to 523 K where peak elongations are achieved at $3.3 \cdot 10^{-4}$ s^{-1} and the elongations decrease when the strain rate is reduced to $1 \cdot 10^{-4}$ s^{-1}. Second, the maximum elongation achieved at each temperature tends to increase with increasing temperature up to 473 K, corresponding to the temperature used for ECAP, but thereafter, at even higher temperatures and especially at 523 and 573 K, the elongations are reduced. The reduction in the elongations to failure at 523 and 573 K is a consequence of the rapid grain growth occurring above ~500 K as documented in Fig. 2. Third, close inspection of the two curves for 473 and 498 K shows that high elongations are achieved at the faster strain rates when using a testing temperature of 498 K but the situation is reversed at strain rates at and below ~$1.0 \cdot 10^{-3}$ s^{-1}. The high elongations achieved at 498 K at the rapid strain rates are attributed to the short-term nature of the tensile testing and the consequent limited grain growth occurring within the specimens. Fourth, an elongation of ~360% was achieved at 498 K when testing with an initial strain rate of $1.0 \cdot 10^{-2}$ s^{-1} and this result is significant because the strain rate is within the range generally associated with high strain rate superplasticity. Fifth, an elongation of ~800% was attained at 423 K with an initial strain rate of $1.0 \cdot 10^{-4}$ s^{-1}. This testing temperature corresponds to 0.55 T_m, where T_m is the absolute melting point of the alloy and this is well within the range associated with low temperature superplasticity. Thus, there is evidence in Fig. 6 for superplasticity at both high strain rates and low temperatures.

Figure 7 plots the elongation to failure as a function of the initial strain rate for the Mg-0.6%Zr alloy subjected to different numbers of ECAP pass including the samples after extrusion but without ECAP. Tensile testing was conducted at 573 K. Whereas the elongation to failure appears to be almost the same at a strain rate of $3.3 \cdot 10^{-4}$ s^{-1}, it is well enhanced at higher strain rates with an increase in the number of ECAP pass. This should be attributed to an increased fraction of fine grains having high angle boundaries. Figure 8 shows double logarithmic plot of the flow stress against the initial strain rate. The flow stress was taken at a strain of 0.1. The level of the flow stress is significantly decreased when ECAP is applied and this increase is more prominent as the number of ECAP pass is increased. The strain rate sensitivity of ~0.4 is achieved at lower strain rates and this is consistent with Fig.7 where higher elongation to failure is observed at lower strain rates.

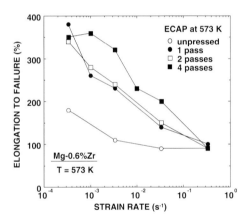 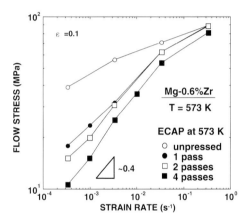

Figure 7: Elongation to failure plotted against testing temperature for Mg-0.6%Zr alloy subjected to different numbers of ECAP pass including samples after extrusion but without ECAP

Figure 8: Double logarithmic plot of the flow stress against the initial strain rate for Mg-0.6%Zr alloy subjected to different numbers of ECAP pass including samples after extrusion but without ECAP

5 Summary and Conclusions

1. The grain sizes of Mg-0.6wt%Zr, Mg-9wt%Al and Mg-7.5wt%Al-0.2wt%Zr alloys were reduced to ~1.3 µm, ~0.7 µm and 0.8 µm, respectively, using the process of Equal-Channel Angular Pressing (ECAP). This grain refinement was feasible for the alloys subjected to extrusion prior to ECAP.
2. Static annealing experiments showed that the fine-grained structures produced by ECAP were retained up to 473 K for the Mg-9wt%Al and Mg-7.5wt%Al-0.2wt%Zr alloys and up to 573 K for the Mg-0.6wt%Zr alloy.

3. Elongation to failure increased with increasing numbers of ECAP pass. This was attributed to an increasing fraction of high-angle grain boundaries.
4. Low-temperature superplasticity and high-strain-rate superplasticity were both attained in the fine grained alloys

6 Acknowledgements

This work was supported in part by the Mitsubishi Foundation, in part by the Light Metals Educational Foundation of Japan, and in part by the U.S. Army Research Office under Grant No. DAAD19-00-1-0488.

7 References

[1] R.Z. Valiev, R.K. Islamgaliev, I.V. Alexandrov, Prog. Mater. Sci., 2000, 45, 103–189
[2] Y.T. Zhu, T.G. Langdon, R.S. Mishra, S.L. Semiatin, M.J. Saran and T.C. Lowe (eds.): Ultrafine Grained Materials II, The Minerals, Metals & Materials Society, Warrendale, PA, (2002)
[3] V.M. Segal, V.I. Reznikov, A.E. Drobyshevskiy, V.I. Kopylov, Russian Metall., 1981, 1, 99–105
[4] Z. Horita, T. Fujinami and T.G. Langdon: Mater. Sci. Eng., 2001, A318, 34–41
[5] Y. Ma, M. Furukawa, Z. Horita, M. Nemoto, R.Z. Valiev, T.G. Langdon, Mater. Trans. JIM, 1996, 37, 336–339
[6] S. Ota, H. Akamatsu, K. Neishi, M. Furukawa, Z. Horita and T.G. Langdon: Mater. Trans. JIM , 2002, 43, 2364–2369
[7] S. Komura, Z. Horita, M. Furukawa, M. Nemoto and T.G. Langdon: Metall. Mater. Trans., 2001, 32A, 707–716.
[8] K. Neishi, T. Uchida, A. Yamauchi, K. Nakamura, Z. Horita and T.G. Langdon: Mater. Sci. Eng., 2001, A307, 23–28
[9] Y. Fukuda, K. Oh-ishi, Z. Horita and T.G. Langdon, Acta Mater., 2002, 50, 1359–1368
[10] K. Neishi, Z. Horita and T.G. Langdon: Mater. Sci. Eng., 2001, A325, 54–58
[11] M. Mabuchi, H. Iwasaki, K. Yanase and K. Higashi: Scripta Mater., 1997, 36, 681–686
[12] K. Oh-ishi, Z. Horita, D.J. Smith and T.G. Langdon, 2001, 16, 583–589

High Strain Rate Superplasticity in an Micrometer-Grained Al-Li Alloy Produced by Equal-Channel Angular Extrusion

M. M. Myshlyaev[1,2], M. M. Kamalov[1], M. M. Myshlyaeva[1]
[1] Institute of Solid State Physics, Russian Academy of Sciences, Chernogolovka
[2] Baikov Institute of Metallurgy and Material Science, Russian Academy of Sciences, Moscow

1 Introduction

Materials scientists, designers, and metal physicists have been showing recently acute interest in aluminum-lithium alloys due to the unique combination of their properties, namely, an increased elastic modulus, sufficiently high strength and low density. These features enable one to reduce appreciably the weight of space equipment with all concomitant advantages. Nowadays work is in progress aimed at improvement of properties of these alloys, also by forming in them nano- and microcrystalline structure via intensive plastic deformation. It is general knowledge that equal-channel angular extrusion (ECA-extrusion) is one of the most promising methods of achieving this goal. The following paper deals with precisely this method of forming a fine-grained structure.

The object of the study was the most advanced lightest (density 2.47 g cm^{-3}) corrosion resistant weldable alloy Al–5.5%Mg–2.2%Li–0.12%Zr. When employed in welded constructions, it reduces the weight by 20–25 % and increases the rigidity by 6 %. It is superplastic and is widely used to fabricate workpieces of complex profile. Typical characteristics of its superplasticity (SP) are as follows: strain to failure is 350 % and the coefficient of strain rate sensitivity of stress is 0.45 at a strain rate of $5 \cdot 10^{-3}$ s^{-1} at $T = 480$ °C [1].

2 Procedures

The rods (20 mm in diameter, 70 mm in length) were produced by sequentially ECA-extrusion the material for 10 passes at 370 °C. Samples for ECA-extrusion were cut from a hot-rolled plate with a recrystallized structure. The extrusion was conducted in air.

Structure and phase state were studied by an electron microscopy (JEM–100CX). Three sample sections, i.e. normal to the rod axis and two mutually perpendicular and parallel to the rod axis were examined.

Flat samples for mechanical tests were prepared from the ECA-extruded rods. They were 0.85 mm thick with gauge length 5 mm. The frontal and side sample surfaces were carefully polished. The layers of disturbed structure brought about by the sample preparation were removed. The differences in thickness and width along the gauge axis did not exceed 0.01 mm. The symmetry axis along the samples was set in parallel with the axis of the rods.

The samples were strained on the Instron testing machine under conditions of uniaxial tension along their axis at preset temperatures and tension rates. The measurements of the applied stress and sample elongation were within the tolerance of 0.25 % and 1 % respectively. In the course of the test the temperature was kept constant to within ± 2 K.

3 Results and Discussion

The investigations showed. Rods demonstrated grained structure. About 50 % of grains measured from 0.5 to 3 µm, grains measuring from 3 to 5 µm made up 30–40 %, from 5 to 8 µm – 10–20 %. Normally, the grains exhibited subgrains containing both individual dislocations and dislocation cells and tangles. The subgrain misorientations were 2–8°. The subgrain boundaries consisted of dislocations. Often they were quite regular dislocation networks and single-row walls. They measured from 0.3 to 2 µm, depending on the grain size. Fractured and broken subgrain boundaries were frequent. Dislocation motion and migration of subgrain boundaries were observable when examining the structure. In individual cases bent extinction contours were observed that is suggestive of the occurrence of internal stresses. The rods demonstrated numerous particles of the Al_2LiMg phase of various sizes and configurations and small particles of the δ' (Al_3Li) phase. The formers were found in the grain and subgrain interiors and at their boundaries as well as in dislocations. In the latter case particles were small. The characteristic pattern is illustrated in Fig. 1.

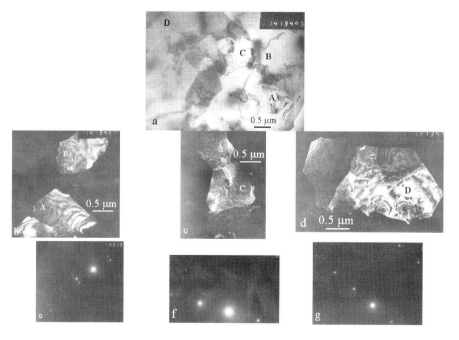

Figure 1: Structure of ECA extruded rods. A–BF. Al2LiMg particles and structural fragments are seen. Positions of A–D grains are marked of; b, c, d – DF images of A–D grains respectively obtained by using short range of the reflections. The images of A and B grains have been obtained at a simultaneous passage of two diffracted beams through a diaphragm; e, f, g – MDP from A, C and D grains respectively. It is seen they are pertinent to different axes of zones, i.e. they correspond to strongly misoriented grains

Studies of the dependence of deformation up to failure on the initial strain rate and testing temperature (T) (Figs. 2, 3) showed that the range of 365–400 °C and $\dot{e}_{in} = 1.7 \cdot 10^{-2}\,s^{-1}$ were the most optimum to attain the largest strain. The greatest value of the attained strain was 1900 %.

Note (see Introduction) that this alloy when not subjected to ECA-extrusion showed the lowest SP strain (350 %) and its SP manifested itself at a much higher temperature (480°) and a noticeably smaller strain rate ($5 \cdot 10^{-3} s^{-1}$).

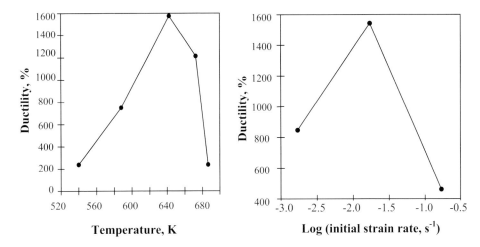

Figure 2: Strain to failure vs temperature. Initial strain rate $\dot{e}_{in} = 1.7 \cdot 10^{-2} s^{-1}$.

Figure 3: Strain to failure vs initial strain rate at $T = 370$ °C.

The diagrams describing the connection of true stress σ with true strain e were experimentally obtained (Fig. 4). They showed three stages of plastic deformation. The first one was rather continuous stage of deformation hardening. The second stage was characterized by constancy σ. The third one was a stage of monotonous fall of σ with increase in e value. This stage was the most continuous in true strain and, consequently, in elongation. To determine the true strain rates in these stages we obtained a dependence of on e using the same testing conditions (Fig 5). This dependence showed that strain rates $10^{-2} s^{-1}$ and $10^{-3} s^{-1}$ corresponded to the first and the third stages, correspondingly. The former indicated strain rate is characteristic for SP deformation (SPD) at the expense of sliding inside grains [2]. The last indicated strain rate is typical for SPD of fine-grained materials, when SP is conditioned by grain boundary sliding [3, 4].

The analysis of the collected experimental data with take into account the ones presented in literature showed that the connection among , σ and T can be well described by the known relationship:

$$\dot{e} = \dot{e}_0 \exp(-U/kT) = A \sigma^n T^{-1} \exp(-U/kT), \tag{1}$$

where $n = 2$, and U – activation energy of SPD, k – Boltzmann constant, A – constant. Estimation of value n and U were evaluated using standard techniques. According to our experiments $n = 2.23$ and $U = 1.4$ eV in the first stage and $n = 2.3$ and $U = 0.98$ eV in the third stage. The experimental values n coincide with rather a high accuracy in the value $n = 2$ in Eqn. (1). Using these values of n and U values of parameters $\dot{e}_0 = 5 \cdot 10^{10} s^{-1}$ and $A = 1.6 \cdot 10^6$ K·mm²·Mpa⁻¹·s⁻¹ were calculated. The aforementioned of SPD activation energy value of $U = 1.4$ eV corresponds to self-diffusion energy (1.4–1.5 eV) inside of grain [5]. The value of $U = 0.98$ eV corresponds to grain boundary self-diffusion energy $Q = W + R_{gb} = 0.99$ eV, where $W = 0.8$ eV [5] – vacancy formation energy and $R_{gb} = 0.19$ eV [6] – vacancy migration energy along grain boundaries or dislocations (pipe diffusion). The obtained different values of activation energy which correspond to the first and third stages point to the presence of plastic deformation under different mechanisms during these stages. Thus, the second stage is a transitory one and transforms from one mechanism to another.

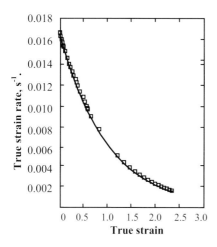 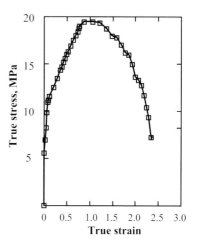

Figure 4: True stress vs true strain. Initial true strain rate $\dot{e}_{in} = 1.7 \cdot 10^{-2} s^{-1}$, deformation temperature T = 395 °C

Figure 5: True strain rate vs true strain. Initial true strain rate $\dot{e}_{in} = 1.7 \cdot 10^{-2} s^{-1}$, deformation temperature T = 395 °C.

TEM studies were carried out to investigate structure of the samples subjected to tensile straining. The first stage shows overall continuous rearrangement of structure with domination of hardening processes over dynamic recovery ones, and active sliding inside grains. As a result, the grains became elongated in the strain direction. Elongation of grains decreased during the deformation in the third stage. By the end of the stage grains became nearly equiaxed. Important circumstances attract attention in the whole stage: absence of dislocations in many grains and the presence of fine Al$_2$LiMg particles in them and in grain boundaries. The important fact is also that Al$_2$LiMg particles make chains in grains and situated as grain boundary profiles (Fig. 6). All these points to continuous and overall dynamic recrystallization with grain boundary sliding and migration. One should note that the described herewith in agreement with aforesaid activation energies.

4 Conclusion

The structure and mechanical behavior of the ECA pressed 1420 alloy have been studied in SP conditions. Three stages of SPD have been shown. The data showing intra-grain sliding during the hardening stage and dynamic recrystallization with participation of grain boundary sliding and migration during the stage of the decrease of true stress have been obtained. It has been shown the elongation up to 1900 % corresponds to alloy, and $n \approx 2$ and $m \approx 0.45$ for both stages.

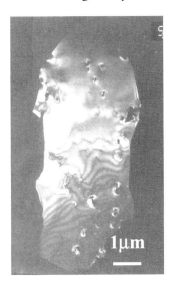

Figure 6: Characteristic partitioning of Al_2LiMg particles. DF. The particles are seen at the grain boundaries and in the grain interiors

5 Acknowledgements

The support from the Russian Foundation for Basic Research (Projects 01-02-16505, 02-02-81021, 02-02-96413) is greatly appreciated. We are also grateful to Dr. R. K. Islamgaliev, Ufa State Technical University of Aviation, who kindly supplied us with the ECA-extruded rods used in the investigation.

6 References

[1] I. Ya. Novikov, V. K. Portnoi, I. L. Konstantinov, N. I. Kolobnev, Physical Metallurgy of Aluminium Alloys, Nauka, Moscow, 1985, p. 245
[2] V. A. Likhachev, M. M. Myshlyaev, O. N. Sen'kov, Laws of the Superplastic Behavior of Aluminum in Torsion, Lawrence Livermore National Laboratory, Livermore, 1987, p. 45
[3] M. V. Grabskii, Structural Superplasticity of Metals, Metallurgia, Moscow, 1975, p. 270
[4] O. A. Kaibyshev, Plasticity and Superplasticity of Metals, Metallurgia, Moscow, 1975, p. 279
[5] J. Friedel, Dislocations, Pergamon Press, Oxford, 1964, p. 467
[6] J. P. Stark, Diffusion in Solids, Energia, Moscow, 1980, p. 217

Diffusion-Controlled Processes and Plasticity of Submicrocrystalline Materials

Yu.R. Kolobov, K.V. Ivanov, G.P. Grabovetskaya, E.V. Naidenkin,
Institute of Strength Physics and Materials Science, Tomsk, Russia

1 Introduction

Purposeful formation of such structural-phase states in polycrystalline materials which provide for high ductility is one of the urgent issues of material science. It is well-known that achievement of high ductility and superplasticity in polycrystalline materials is related to the development of grain boundary sliding which under superplasticity conditions contributes significantly (up to 80 %) to the overall deformation [1]. It is established that the activation of grain boundary sliding in metal materials may occurs in the presence of grain boundary diffusion fluxes of impurity from an external source, e.g. coating. Under the above conditions a significant decrease of resistance to deformation and an essential increase of ductility as well as superplastic flow under certain conditions are observed for metal polycrystals [2, 3].

In recent years submicrocrystalline (SMC) materials (grain size 0.1–0.3 µm) produced by severe plastic deformation have been extensively investigated [4]. This is motivated by the fact that SMC materials differ significantly from their coarse-grained counterparts in physicochemical, mechanical, etc., properties [4]. Thus, in these materials many mechanical properties of practical significance, e.g. ultimate strength, yield strength, fatigue resistance, etc., are found to improve [4, 5]. Moreover, the deforming SMC metals and alloys are found to exhibit in the range of elevated temperatures high strain-rate and/or low-temperature superplasticity [6]. In the case of SMC materials, transition to a superplastic state may occur at temperatures which are lower by several hundreds of degrees relative to the respective superplastic fine-grained counterparts (grain size of a few micrometers), with the rate of superplastic flow in the former case being higher by several orders of magnitude. However, up to now the superplastic behaviour of SMC metallic alloys has been investigated for a limited number of alloys which exhibit superplasticity in fine-grained state without treatment by severe plastic deformation. In this connection, it is of great interest to investigate the possibility of SMC materials, which are not superplastic in fine-grained state, passing to a superplastic state.

It is well known that grain boundary state plays a key role in the effects which are due to the activation of grain boundaries by diffusion fluxes of impurity from an external source. Due to the increased fraction of grain boundary phase in SMC materials [4], the activation of grain boundaries by diffusion fluxes of impurity and the related change of mechanical properties in these materials may have special features. In particular, due to the increased diffusivities in SMC materials relative to the coarse-grained counterparts [7], the effect of creep acceleration by grain boundary diffusion fluxes of impurity from the environment or an internal source may be manifested at less elevated temperatures. In superplastic alloys, mostly polyphase ones, the role of internal sources may be performed by grain boundary secondary phases dissolving during deformation.

In view of the above, of great interest is an investigation of the characteristic features of plastic deformation development at elevated temperatures as well as of the mechanical properties of SMC metals and alloys under the action of grain boundary fluxes of impurity from the environment or an internal source.

2 Experimental

The materials and experimental procedures are described in detail elsewhere [4, 8–10]. Commercial-grade nickel and copper samples as well as samples of the Al-5.5%Mg-2.2%Li-0.12%Zr and Al-5%Mg-2.2%Li-0.12%Zr-0.2%Sc alloys were studied. The coarse-grained pure metals and Al-Mg-Li alloys investigated had grain size of about 20 and 10 μm, respectively. Submicrocrystalline state in the above metals (d~0.3 μm) and ultrafine-grained state in the alloys (d~1 μm) were formed by equal-channel angular pressing.

A diffusion impurity layer (copper for nickel and aluminum for copper) 10 μm in thickness was electrodeposited on the sample surface. The materials were tested in tension and creep in vacuum in a wide range of temperatures. The structure evolution during testing was controlled by optical metallography and electron microscopy.

3 Results and Discussion

It is established that the creep curves of SMC copper at 373–473 K have the usual three-stage shape which is characteristic for plastic deformation of polycrystalline materials. The curves obtained at 473K under the load of less than 140 MPa were an exception since they showed five stages. In the latter case the deformation to failure (ductility) of copper increases sharply (from 10 up to 50 %; Table 1). Such creep behaviour is assumed to be due to the dynamic recrystallization during the creep.

Let us consider the effect of diffusion fluxes of aluminium on the creep of SMC Cu. The effect of creep acceleration is defined as the ratio $\dot{\varepsilon}_2/\dot{\varepsilon}_1$ ($\dot{\varepsilon}_1$ and $\dot{\varepsilon}_2$ are the creep rates of Cu in vacuum and under diffusion contact with Al, respectively). The value obtained for SMC copper under the influence of grain boundary diffusion fluxes of aluminium at 423–473 K is significantly lower relative to the coarse-grained counterpart (573–673 K; Fig. 1a). That the above effect is manifested at less elevated temperatures may be due to the significant increase of the diffusion coefficients of aluminium in SMC copper relative to the coarse-grained counterpart [7].

It is shown that the dependence of steady-state creep rate on applied stresses is linear in all the investigated temperature and creep rate intervals for the creep of copper in vacuum and under the diffusion contact with aluminium (Fig. 2). At 423 K the grain boundary diffusion fluxes results in a five-fold increase of the creep rate. The strain rate sensitivity m ($m = d\lg\sigma/d\lg\dot{\varepsilon}$) increases from 0.08 for the copper creep in vacuum up to 0.1 for the copper creep under the effect of the grain boundary diffusion fluxes. The ductility of SMC copper under the diffusion fluxes of aluminium remains the same (~10 %; Table 1). Of particular interest is the dependence of steady-state creep rate on applied stresses obtained for the same material at 473 K. Although the effect of creep acceleration of SMC copper by grain boun-dary diffusion fluxes at 473 K is less strongly pronounced than that obtained at 423 K, the strain-rate sensitivity

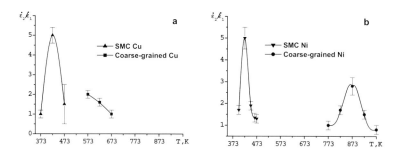

Figure 1: Dependence of the value of creep acceleration effect on the deformation temperature

($m = 0.5$) obtained for the creep of copper under diffusion fluxes greatly exceeds that for creep in vacuum ($m = 0.19$; Table 1). Generally, one associates such an increase of m with the change of the main deformation mechanism from the intragrain dislocation slip to the grain boundary sliding and transit of material to the superplastic state. The ductility of copper under the diffusion fluxes of aluminium is the same as that of copper during the creep in vacuum and amounts to 40–50 %. However, the creep curves of SMC copper with aluminium coating show the usual three-stages as distinct from the curves obtained for the creep in vacuum which, as mentioned above, show five-stages. It suggests that the action of diffusion fluxes on the grain boundaries affects the development of plastic deformation by increasing the contribution to the overall deformation of grain boundary sliding. In view of the above, we can assume that in principle superplastic state can be attained in SMC copper by the action of grain boundary diffusion fluxes of aluminium from an external source.

Table 1: Parameters of the creep of submicrocrystalline (SMC) and coarse-grained (CG) copper and nickel. Diffusion impurity is indicated in brackets.

Material/State	Test temperature, K	Steady-state creep rate, s^{-1}	Deformation to failure, %	m	$Q_c \pm 20$, kJ/mol
Cu / SMC	423	$5.7 \cdot 10^{-6}$	10	0.08	–
Cu(Al) / SMC	423	$2.7 \cdot 10^{-5}$	10	0.10	–
Cu / SMC	473	$3.3 \cdot 10^{-6}$	~50	0.19	–
Cu(Al) / SMC	473	$5.3 \cdot 10^{-6}$	~50	0.50	–
Ni / SMC	423	$3.6 \cdot 10^{-6}$	4	0.11	115
Ni(Cu) / SMC	423	$1.8 \cdot 10^{-5}$	8	0.23	70
Ni / CG	873	$1.1 \cdot 10^{-6}$	32	0.13	274
Ni(Cu) / CG	873	$3.1 \cdot 10^{-6}$	45	0.21	168

The essential effect of grain boundary diffusion fluxes on the creep of coarse-grained and SMC materials can be further illustrated by the results of creep tests obtained for nickel under grain boundary diffusion of copper. The effect of creep activation in SMC Ni under the action of diffusion fluxes of copper is observed in the temperature interval of 398–473 K, which is significantly lower relative to the coarse-grained counterpart (773–973 K; Fig. 1b). The shift of

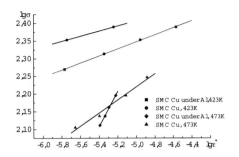 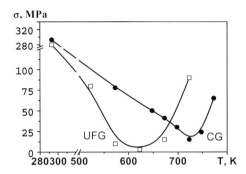

Figure 2: Dependence of flow stress on the creep rate

Figure 3: Temperature dependence of superplastic flow stress obtained for the Al-Mg-Li-Zr alloy

temperature of the effect manifestation is $\Delta T = 450$ K. In the case of SMC state the deformation to failure grows by two times and the strain rate sensitivity increases from 0.11 to 0.23 (Table 1).

By analogy with coarse-grained materials, one may assume that the effect of decrease of creep resistance observed for SMC metals under the action of grain boundary diffusion fluxes is associated by the activation of grain boundary sliding and by the growth of its contribution to the overall deformation. It is well known that deformation by intragranular dislocation slip occurring at elevated temperatures is controlled by volume diffusion, while grain boundary sliding process is controlled by grain boundary diffusion. For this reason, the growth of the contribution to the overall deformation of grain boundary sliding results, as a rule, in a decrease of the apparent activation energy of creep (Q_c) from values close to the activation energy of volume diffusion (Q_v) down to the respective values of grain boundary diffusion (Q_b). In this connection, we examined the temperature dependency of the apparent activation energy of creep obtained for SMC nickel in vacuum and under the diffusion of copper atoms from the coating in the temperature interval $T \leq 0.3\ T_{melt}$.

It has been found that during the creep of SMC nickel under the grain boundary diffusion of copper at 423 and 443 K (the effect of creep activation is observed) the Q_c value is close to the activation energy of grain boundary diffusion of copper in SMC nickel (~43 kJ/mol [7]), while at 473 K (an inverse effect is observed, i.e. an increase of creep resistance at low creep rates) Q_c is practically equal to the respective Q_c value for the creep in vacuum (Table 1). The data obtained speak in favour of the above contention that the effect of creep activation of SMC nickel by grain boundary diffusion fluxes of copper is caused by the activation of grain boundary sliding and by its growing contribution to the overall deformation.

It is well known that the process of grain boundary sliding is controlled by diffusion along grain boundaries [1,2]. It suggests that the shift of the maximum of creep activation effect in the case of SMC Ni to lower temperatures is due to the increase of grain boundary diffusion coefficient of Cu in SMC material relative to coarse-grained one. The direct measurements of the grain boundary diffusion coefficients [7] revealed the increase of D^{Cu} in SMC Ni by several orders of magnitude in comparison with coarse-grained one.

Let us consider one more diffusion-controlled process, i.e. superplastic flow. It is common knowledge that the main deformation mechanism of superplastic flow is a grain boundary sliding which is controlled by grain boundary diffusion [1]. Using the experimental data on the temperature dependence of flow stress obtained for ultrafine-grained alloys produced by severe plastic deformation and the coarse-grained counterparts, one can compare and evaluate the diffusion parameters of the material in the indicated structural states.

Figure 3 demonstrates the temperature dependence of flow stress obtained for the ultrafine-grained and coarse-grained superplastic Al-Mg-Li-Zr alloys. These curves show minimums corresponding to the optimal temperature, at which the material in the above two states manifests superplastic properties, and to the maximal elongation to failure. In the case of ultrafine-grained alloy, the temperature of superplastic properties manifestation is evidently shifted to region of less elevated temperatures. It is widely known that superplastic deformation of alloys is described satisfactorily by the following equation [1]:

$$\dot{\varepsilon} = A \frac{D^{\text{eff}}}{kT} \sigma^2 \frac{1}{d^2}, \tag{1}$$

where A is a constant; D^{eff} is the effective diffusion coefficient; σ is the applied stress; d is the grain size. Given that the grain size ($d_{cg} \sim 10$ μm and $d_{ufg} \sim 1$ μm), the stresses (values from Fig. 3) and the rates of superplastic deformation of coarse-grained and ultrafine-grained alloys ($\dot{\varepsilon}_{cg} = 5 \cdot 10^{-4}$ s^{-1} and $\dot{\varepsilon}_{ufg} = 10^{-2}$ s^{-1}) and assuming the A coefficient to be independent of structure, one can evaluate the ratio of the effective values:

$$\frac{D^{\text{eff}}_{cg}(T = 723 \text{ K})}{D^{\text{eff}}_{ufg}(T = 623 \text{ K})} \sim 1. \tag{2}$$

Thus, the effective diffusion coefficients are approximately equal despite of the difference in the temperature of 100 K.

Assume that the activation energy in the Arrhenius law governing the effective diffusion coefficient is characteristic for the superplastic deformation of the Al-Mg-Li-Zr alloy ($Q = 145$ kJ/mol). Hence, we obtain:

$$D^{\text{eff}}_{cg}(623 \text{ K}) = 2 \cdot 10^{-2} D^{\text{eff}}_{cg}(723 \text{ K}). \tag{3}$$

From equations (3) and (4) follows that the effective diffusion coefficient in the ultrafine-grained alloy exceeds by approximately two orders of magnitude the respective value in the coarse-grained counterpart. In the case of Al-Mg-Li-Zr-Sc alloy in ultrafine-grained state, similar estimation reveals that D^{eff} grows by one order of magnitude.

4 Conclusions

1. The effect of creep enhancement of submicrocrystalline nickel and copper, which is due to the action of grain boundary diffusion fluxes from the surface, is observed at less elevated temperatures relative to coarse-grained state. Moreover, the apparent activation energy of creep in submicrocrystalline nickel decreases down to the energies of grain boundary diffusion of copper in submicrocrystalline nickel.

2. Observations of the structurally induced temperature decrease of creep activation by means of diffusion flux measurements, and of the evidence of superplasticity have revealed that the effective diffusion coefficients obtained for the test materials in submicrocrystalline and ultrafine-grained states (produced by severe plastic deformation) exceed the respective value for coarse-grained state by several orders of magnitude.

5 Acknowledgments

The authors express their gratitude to Prof. R.Z. Valiev (UGATU, Ufa, Russia) for supplying submicrocrystalline materials for the experiments and for fruitful discussion. We also thank Prof. M. Zehetbauer for his valuable comments on the results discussed in the paper. The work is financially supported by the Russian Foundation of Basic Research (grant No. 03–02–16955, 02-02-06895), INTAS (grant No. 01–0320), ISTC (grant No. 2398p) and CRDF (TO 016-02).

6 References

[1]   O.A. Kaibyshev, R.Z. Valiev, Grain boundaries and properties of metals, Metallurgy, Moscow, 1987 (in Russian)
[2]   Yu.R. Kolobov, Diffusion-controlled processes on grain boundaries and plasticity of metal polycrystals, Nauka publ., Novosibirsk, 1998 (in Russian)
[3]   Yu.R. Kolobov, I.V. Ratochka, J. Mater. Sci. Technol., 1995, 11, 38–40
[4]   Yu.R. Kolobov, R.Z. Valiev, G.P. Grabovetskaya, et al., Grain boundary diffusion and properties of nanostructured materials, Nauka publ., Novosibirsk, 2001 (in Russian)
[5]   H.W. Höppel, R.Z. Valiev, Zeitschrift für Metallkunde, 2002, 93, 641–648
[6]   S.X. McFadden, R.S. Mishra, R.Z. Valiev, et al. Nature, 398, 684–686
[7]   Yu.R. Kolobov, G.P. Grabovetskaya, M.B. Ivanov, A.P. Zhilyaev, R.Z. Valiev, Scripta Materialia, 2001, 44, 873–878
[8]   K.V. Ivanov, I.V. Ratochka, Yu.R. Kolobov, Nanostr. Materials, 1999, 12, 947–950
[9]   Yu.R. Kolobov, G.P. Grabovetskaya, K.V. Ivanov, M.B. Ivanov, Interface Science, 2002, 10, 31–36
[10] G.P. Grabovetskaya, K.V. Ivanov, Yu.R. Kolobov, Ann. Chim. Sci. Mat., 2002, 27, 89–98

Superplastic Behavior of Deformation Processed Cu-Ag Nanocomposites

S. I. Hong, Y. S. Kim and H. S. Kim
Department of Metallurgical Engineering, Chungnam National University, Taedok Science Town, Taejon, Korea

1 Introduction

Deformation processed Cu base nanocomposites possess high strength in excess of 1 GPa with a conductivity of 60–70 % IACS [1]. Recently, Hong and Hill [1] examined the microstructural stability of Cu-Ag nanocomposites at high temperatures and reported that extensive recrystallization occurred following heat treatment at 400 °C. They also observed that strength level decreased substantially for Cu-Ag nanocomposite heat treated at 400 °C. The room temperature ductility, however, did not improve even after the extensive recrystallization occurred upon annealing at 400 °C. One interesting observation by Hong and Hill [1] was that heat treating at 100 or 200 °C significantly reduced the ductility. High strength materials manifest limited tensile elongation due to flow localization even though some of them can be quite tough. The recovery/recrystallization occurs at and above 400 °C and the deformation behavior after recrystallization is expected to differ from that of the as-drawn Cu-Ag nanocomposites. However, much of the work on Cu-Ag nanocomposites has focussed on room temperature mechanical properties and its relation to the microstructures [1–4].

The observation of fine two-phase microstructure in Cu-16 vol.% Ag nanocomposites [1-6] at high temperatures suggests that it may exhibit superplastic behavior. Superplastic behavior has been reported for a range of α/β Cu alloys [6,7] including brasses, aluminum bronzes and nickel-silvers. For the binary brasses and aluminum bronzes the duplex superplastic microstructure was obtained by hot working the material in the $\alpha+\beta$ phase field or during cooling from the β to the $\alpha+\beta$ fields [6,7]. The optimum deformation temperature for the brasses was 600 °C with maximum tensile elongations of 500 % being obtained for an alloy with approximately equal volume fraction of the two phases [6,7]. For the aluminum bronzes the optimum deformation temperature was 700 °C with the elongations over 1000 % [8]. In these binary α/β Cu alloys, the grain growth was not strongly inhibited because the composition difference between two phases is relatively small [8]. In nickel-silvers, the optimum deformation temperatures were 570–600 °C with tensile elongations of 200–680 [6].

The superplastic duplex microstructure can be obtained in filamentary Cu-Ag nanocomposites by heat treatments and/or thermo-mechanical processing. Actually Cline and Lee [9] observed the superplastic behavior in eutectic Cu-72 wt.% Ag at temperatures between 650–730 °C with the maximum elongation of 500 %. They also observed that the equiaxed structure is stronger and more ductile than the lamellar structure at room temperature in Cu-72 wt.% Ag [9]. The major objective of this research is to correlate the high temperature deformation behaviors of Cu-Ag nanocomposite wires and the microstructural evolution.

2 Experimental Methods

Cu-24 wt.% Ag (Cu-16 at.% Ag) alloy was cast using a horizontal continuous casting machine. The alloy was subsequently hot forged at 450 °C, annealed at 450 °C for 10–20 hours, and then cold drawn to a reduction in area (RA) of 98.5 % (draw ratio η = 4.2) with intermediate annealing [1,2]. The tensile specimens for this study, consisting of 2 mm gauge width and thickness and 6 mm gauge length were machined from Cu-Ag nanocomposite wires. Tensile testing was carried out using a United testing machine with a three-zone split-furnace at temperaures between 400 and 600 °C. Some tensile tests were stopped before fracture to examine the change of surface morphologies at high temperatures using SEM. TEM specimens were prepared by mechanically thinning, dimpling and ion milling on a liquid nitrogen stage. To ensure the samples were adequately cooled, the specimen rotation drive rod was submersed in liquid nitrogen for one hour prior to ion milling. TEM observations were carried out using Phillips CM30 and Jeol JEM2010 electron microscopes.

3 Results and Discussion

Fig. 1 shows the three dimensional view of as-drawn Cu-Ag nanocomposites. In this figure, the linear white phase represents Ag-rich phase and the black phase represents Cu-rich phase. Hong and Hill [1] reported that the linear Ag-rich phase observed by SEM consists of two different morphologies: relatively thick silver lamella (20–80 nm in thickness) and Cu matrix with numorous fine silver filaments (0.8–6 nm in diameter). The alignment of Ag-rich phase with wire axis is readily apparent whereas the Ag-rich phase take the cellular morphology in the transverse section [1]. Fig. 2a and 2b show the general TEM microstructural features of the longitudinal (2a) and transverse (2b) sections of as-drawn Cu-Ag wires, respectively. As seen in Fig. 2a and 2b, the microstructure of these wires consist of very fine silver filaments (indicated by "A"), copper-rich α phase (indicated by "B"), and relatively thick silver lamellae (indicated by "C"). In Fig. 2b, the fine silver filaments appear as small dots (see the region indicated by "A") since the filaments are perpendicular to the image plane while the silver lamellae appear as thin dark strips ("C"). The bright areas are copper-rich α phase ("B"). As shown in Fig. 2b, the regions containing silver filaments were separated by dark channels of silver-rich lamellae. The thick-

Figure 1: Three dimensional view of as-drawn Cu-Ag nanocomposites. Drawing strain η = 4.2

ness of the lamellae was 20–80 nm. Since these samples are highly textured, the contrast between Cu grains was not pronounced [1].

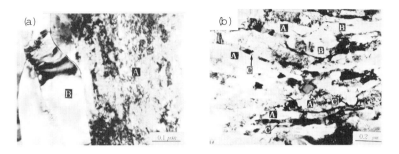

Figure 2: Longitudinal (a) and transverse (b) sections of Cu-Ag nanocomposites observed using TEM

Fig. 3a and 3b are TEM micrographs of longitudinal sections of Cu-Ag heat-treated at 400 °C and 500 °C. These figures show that extensive recrystallization occurred following heat treatment at 400 °C. Although most regions containing silver filaments were replaced by newly recrystallized grains, some regions did not recrystallize (labeled "A" in Fig. 5a). At 500 °C, recrystallization occurred more extensively and no uncrystallized region with filaments were observed. The grain size (500 nm) in Cu-Ag heat treated at 500 °C were found to be larger than that (350 nm) at 400 °C.

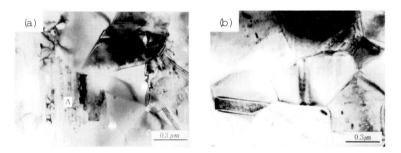

Figure 3: Microstructure of the longitudinal sections of Cu-Ag wires heat treated at 400 °C (a) and 500 °C(b)

Fig. 4 shows the variation of ductility as a function of strain rate at various temperatures. The strain rate at which the largest elongation was observed increased with increase of temperature. The superplastic elongations of Cu-16 vol.% Ag nanocomposites observed in this study is larger than those of eutectic Cu-60 vol.% Ag observed by Cline and Lee (500 % elongation at 675 °C [9]. Examples of specimen deformed to failure at 600 °C at various strainrates are shown in Fig. 5. Fig. 6 shows a plot of maximum flow stress versus the strain rate at various temperatures. The slope of the curve in Fig. 6 gives the value of m, the strain rate sensitivity. The maximum value, m is close to 0.5 above 500 °C.

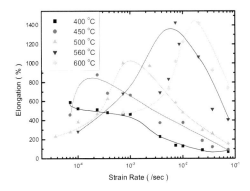

Figure 4: Variation of ductility as a function of strain rate at various temperatures

Figure 5: Tensile specimens before and after superplastic deformation at various strain rates at 600 °C

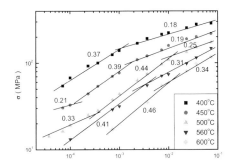

Figure 6: Log stress versus log strain rate curves

Fig. 7(a)-(c) shows the surface morphologies near fracture surface for Cu-Ag strained until fracture at various train rates. As shown in this figure, the initial filamentary microstructure (Fig. 1) was totally altered to the typical two-phase microstructure. It should be noted that Ag phase (white phase) is not elongated parallel to the loading axis, but mostly lies perpendicular to the loading axis in the longitudinal section. In Fig. 7(d), the surface morphologies of Cu-Ag in the grip region which was kept at 600 °C until final fracture. It should be noted that the general fibrous structure parallel to the drawing axis was retained in the grip region at 600 °C although the break-up and spheroidization of Ag filaments and/or lamellae were quite apparent. As the deformation proceeds, the elongated silver lamellae and/or filaments gradually broke up and rearranged into chevron patterns. One interesting observation is that the chevron patterns becomes coarser as the strain rate decreases. The filamentary microstructure is not stable because of its high interface energy and high cold-work energy and, therefore, at high temperatures, filaments and lamellae tend to break up by spheroidization, grooving and/or recrystallization [10–12]. At low strain rate, more stable structure with a lower interfacial energy can be attained and the chevron pattern becomes coarser.

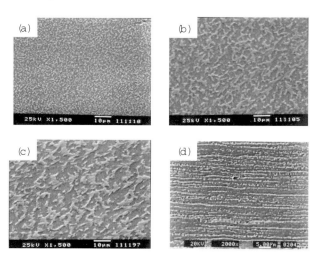

Figure 7: Longitudinal sections of Cu-Ag strained until fracture at 600 °C. (a) strain rate = $3 \cdot 10^{-1}$/sec, (b) strain rate = $2 \cdot 10^{-2}$/sec, (c) strain rate = $7.3 \cdot 10^{-4}$/sec, (d) not strained (from grip)

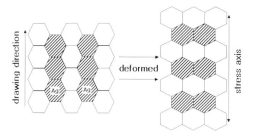

Figure 8: The rearrangement of Ag phase by a grain-switching mechanism

The rearrangement of Ag phase can be explained by a grain-switching mechanism [13] as schematically shown in Fig. 8. In Fig. 8(a), Ag lamella at high temperatures can be assumed to consist of a string of grains (see Fig. 4). Following a grain-switching event, Ag grains are rearranged to appear approximately perpendicular to the loading axis. To accommodate the switched grains in their new configuration, a large localized concentration of strain develops in the surrounding matrix. The local strain would be relatively easily relaxed by plastic flow and diffusional flow than in other Cu-base alloys since the diffusivity of Ag in Cu is high [14,15] and the Ag phase is softer [16].

4 Conclusions

The elongated silver lamellae and/or the region with numerous filaments break up and rearrange into chevron patterns which mostly lie perpendicular to the loading axis during superplastic deformation. The morphological change of Ag phase can be explained in terms of grain switching model. The chevron patterns becomes coarser as the strain rate decreases. The filamentary microstructure is not stable because of its high interface energy and high cold-work energy and, therefore, at high temperatures, filaments and lamellae tend to break up by spheroidization, grooving and/or recrystallization. At low strain rate, more stable structure with a lower interfacial energy can be attained and the chevron pattern becomes coarser. The elongation of 1425 % was obtained at 600 °C at the strain rates of $2 \cdot 10^{-2}$/sec and the elongation up to 600 % was obtained at 400 °C at the strain rate of $1 \cdot 10^{-4}$/sec.

5 Acknowledgments

The authors acknowledge the support from the Korea Research Foundation (2001-041-E00431).

6 References

[1] S. I. Hong and M. A. Hill, Acta Metall. Mater. 1998, 46, 4111
[2] Y. Sakai and H. J. Schneider-Muntau, Acta Metall. Mater. 1997, 45, 1017
[3] G. Frommeyer and Wassermann, Acta Metall. 1975, 23, 1353
[4] A. Bengalem and D. G. Morris, Acta Metall. Mater. 1997, 45, 397
[5] S. I. Hong and M. A. Hill, Mater. Sci. Eng. 1999, A264, 151
[6] D. W. Livesey and N. Ridley, Metall. Trans. 1982, 13A, 1619
[7] S. A. Shei and T. G. Langdon, Acta Metall. 1978, 26, 1153
[8] K. Higashi, T. Ohnishi and Y. Nakatami, Scripta Metall. 1985, 19, 821
[9] H. E. Cline and D. Lee, Acta Metall. 1970, 18, 315
[10] J. C. Malzhan Kampe, T. H. Courtney and Y. Leng, Acta Metall. 1989, 37, 1735
[11] S. I. Hong and M. A. Hill, Mater. Sci. Eng. 2000, A281, 189
[12] S. I. Hong and M. A. Hill, Scripta Mater. 2000, 42, 737
[13] M. F. Ashby and R. A. Verrall, Acta Metall. 1973, 21, 149
[14] O. Kubaschewski, Trans. Faraday Soc. 1950, 46, 713
[15] J. R. Cahoon and W. V. Youdelis, Trans Met. Soc. AIME, 1967, 239, 127
[16] G. Frommeyer and G. Wassermann, Acta Metall. 1975, 23, 1353

Features of Microstructure and Phase State in an Al-Li Alloy after ECA Pressing and High Strain Rate Superplastic Flow

M.M. Myshlyaev [1,2], A.A. Mazilkin[1] and M.M. Kamalov[1]
[1] Institute of Solid State Physics, Russian Academy of Sciences, Chernogolovka, Russia
[2] Baikov Institute of Metallurgy and Material Science, Russian Academy of Sciences, Moscow, Russia

1   Introduction

Aluminum-lithium alloys recently give rise to a big interest since they possess a remarkable combination of physical and mechanical properties. These properties can be substantially improved by methods of severe plastic deformation (SPD), e.g. by equal channel angular (ECA) pressing technique [1, 2]. The material after such a treatment is considered to characterize not only by small grain size but also by high level of internal stress, non-equilibrium grain boundaries, changed phase composition. It is of interest to study the alloy structure formed by ECAP, its evolution at deformation and also the thermostability of the structure obtained by SPD techniques.

2   Experimental

In this work aluminum-lithium alloy (Al-5,5% Mg-2,2%Li-0,12%Zr) was chosen for the investigations. In order to obtain submicrocrystalline structure the alloy billets were subjected to intensive plastic deformation by ECA pressing technique. The produced material was tested by axial tension with a constant rate of 0.5 mm/min at the temperature $T = 370$ °C. During the tests samples demonstrated superplastic flow with the largest lengthening ~ 1900 %. Structural features and phase composition of the alloy were investigated by transmission electron microscopy using JEM-100CX microscope. Characteristics of subboundaries (structure, types of the dislocations) were determined by $g \cdot b$ and of trace analysis. *In-situ* experiments on heating of the samples in the microscope column were conducted. X-ray diffraction analysis of the studied alloy was carried out on SIMENS diffractometer with CuK_α radiation.

3   Results and Discussion

The mean grain size of 1,5–2 µm corresponded to the initial state of the alloy after the ECA pressing though larger grains with size up to 8 µm were observed in the structure. Grains had equiaxial shape; inside the grains there existed a developed substructure, characterized by a presence of individual dislocations, piles-up of dislocations and dislocation subboudaries. So, taking into account the presence of subboundaries, the material can be described as one with submicrocrystalline structure. Average dislocation density was about 10^9 cm^{-2}. Precipitates of S_1 (Al_2LiMg) and δ' (Al_3Li) phases were detected in the structure. The S_1-phase precipitated as colonies on grain boundaries, in triple junctions and also inside the grains (Fig. 1).

Figure 1: DF image of ECA pressed alloy

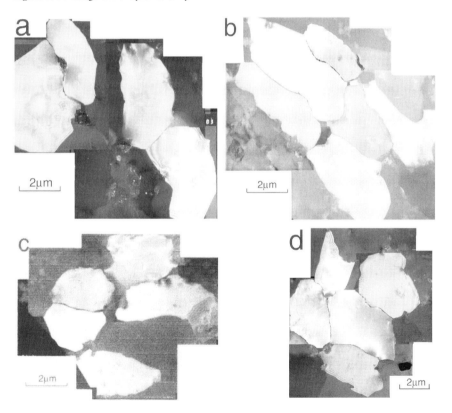

Figure 2: Structure of the alloy in the course of tensile tests: $\varepsilon = 110\%$ (a), 300 % (b), 900 % (c), 1200 % (d)

Set of photographs (Fig. 2) illustrates changes in the alloy structure after different degrees of tensile deformation. It was aforementioned that equiaxial grain structure was a distinctive feature of the initial state of the alloy. In samples deformed to $\varepsilon = 110\,\%$ and $300\,\%$ (Fig. 2, a and b) predominance of extended grains is noticeable. Samples subjected to higher degrees of deformation ($900\,\%$ and $1200\,\%$) demonstrated the structure, in which the equiaxial grains are observed again (Fig. 2, c and d). Herewith the average grain size increases slightly. The value of dislocation density decreases to 10^6–10^7cm^{-2}. There were no visible changes in size, shape and relative volume fraction of intermetallic compounds.

The structure of dislocation subboundaries observed inside the grains was investigated in detail. It should be noted that subboundary structure was rather regular though it was formed during significant plastic deformation. It obviously indicates that along with the dislocation glide the non-conservative processes also took an active part in forming of the subboundaries. Examples of two such subboundaries are represented in Figure 3.

The subboundary in Fig. 3a is a tilt dislocation wall. Foil area corresponded to the grain in which the wall is situated coincides with the (110) plane. By the trace analysis the plane in which the subboundary lies was determined to be parallel to (013). The technique of $g \cdot b$-analysis allowed to determine that dislocation Burgers vector was $a/2\,[1\,1\,0]$; unit vector along the dislocation line was parallel to the $[\bar{3}\,\bar{3}\,\bar{1}]$ direction. Therefore the wall consists of mixed dislocations, and the angle between the Burgers vector and the dislocation line is ~47°.

Figure 3b demonstrates another type of subboundary; it is a hexagonal dislocation net formed by dislocations of three families. Usage of the same procedure as in previous case permitted to determine that the plane of given subboundary is parallel to $(1\,2\,1)$ crystallographic plane. In order to determine dislocation Burgers vectors the value of $g \cdot [b \times u]$ product was calculated. The dislocation Burgers vectors and corresponding unit vectors along the dislocation lines were $b_1 = a/2\,[1\,1\,0]$, $u_1 = [1\,2\,3]$, $b_2 = a/2\,[0\,1\,1]$, $u_2 = [1\,0\,1]$ and $b_3 = a/2\,[1\,0\,1]$, $u_1 = [3\,2\,1]$. The found Burgers vectors also meet the requirements for their amount to be equal to zero in the point of the dislocations intersection. All the dislocations being included in the network are mixed with a significant edge component.

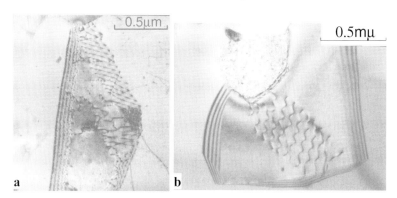

Figure 3: Microphotographs of the dislocation subboundaries: a. dislocation wall; b. dislocation network

It was already mentioned that the presence of two intermetallic phases known for Al-Li-Mg system [3, 4] was discovered in the structure of the studied alloy. These are equilibrium cubic S_1-phase and non-equilibrium δ'-phase with cubic structure of $L1_2$ type. The precipitates of δ'-

phases are extremely dispersed and often their presence is revealed only by reflections on electron diffraction patterns. S_1-phase precipitates as colonies which size is about 0.2–0.3 μm. If the size of the colonies is regarded, the volume fraction of this phase makes up about 15 %. However, the X-ray analysis of the alloy gives the phase fraction less than a percent (Fig. 4). Positions of the reflections from the Al_2LiMg phase and aluminum are indicated in the bottom part of the figure.

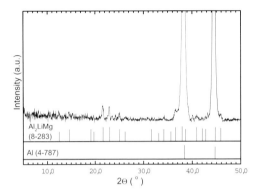

Figure 4: X-ray diffraction data for aluminum-lithium alloy, CuK_α radiation

Dark field electron-microscopic image of S_1-phases colony and corresponding diffraction pattern are shown in Fig. 5. The dark field image is received in one of the phase reflections. It can be seen in the image that the phase colony consists of randomly orientated tiny crystals which size is ~ 25 nm. Region of the material between the crystals is supposedly depleted solid solution of Li and Mg in aluminum matrix.

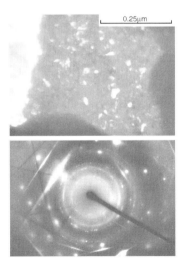

Figure 5: DF image of S_1-phase colony and corresponding diffraction pattern

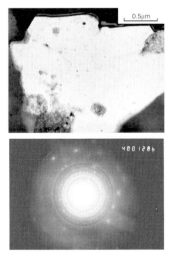

Figure 6: DF image and electron diffraction pattern of S_1-phase after *in-situ* heating

In-situ experiments on samples heating in the microscope column are carried out to study thermal stability of the alloy structure at different annealing temperature up to the temperature of the tensile tests. Thermoactivated disintegration of dislocation subboundaries and decreasing of dislocation density due to dislocation annihilation, their migration to the grain boundaries or sample surface are observed during the heating. Changes in size and shape of part of S_1-phase colonies are also recorded. The colonies become larger with size about 1 μm and regularly shaped as can be seen in Fig. 6. The size of the crystals forming the colonies practically is not changed.

Thermostability of the alloy has been also studied in the course of isothermal annealing. It is shown that at the temperature $T = 370°C$, which is equal to the temperature of mechanical tests, nonuniform grain growth takes place: against the stable submicrocrystalline structure a significant enlargement of separate grains occurs. Fraction of such grains increases with increasing of annealing time. So, the anomalous grain growth at annealing of the Al-Li alloy after ECA pressing is observed.

It is interesting to note that the obtained results, indicating existence of at least two stages of deformation of ECA pressed microcrystalline aluminum-lithium alloy, are in agreement with the study of its mechanical behavior [5]. The authors of [5] came to the conclusion of multistage character of plastic deformation by examining the true stress *vs.* true strain dependence (Fig. 7). It is obvious from the plot in Fig. 7 that the deformation is characterized by two main stages. The first one is the stage of hardening that is followed by the softening stage. For these two stages the value of activation energy was determined. For hardening stage activation energy turn out to be equal to 1.4 eV, for softening one it was 1 eV. Using these activation energy values, the suggestions were made about mechanisms controlled the superplastic deformation. On the first stage the deformation is supposed to be controlled by self-diffusion in the grain bulk that corresponds to the intragranular dislocation slide; on the second, it is the self-diffusion along the grain boundaries that corresponds to the grain boundaries sliding.

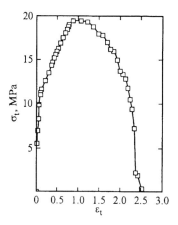

Figure 7: Tensile true stress–true strain diagram of the aluminum–lithium alloy

The study of grain structure evolution of the alloy confirms validity of the conclusions concerning the deformation controlling mechanisms. It was observed that on the stage of hardening

(ε = 110 % and 300 %) the structure consisted mainly of extended grains. The density of dislocations inside the grains on this stage of deformation is rather high. These structural features indicate that the most probable deformation mechanism is the intragranular dislocation. The observations of alloy structure on the softening stage (ε = 900 % and 1200 %) showed that grain gained equiaxial shape but the dislocation density decreased significantly, i.e. it would be quite correct to expect that the deformation on this stage is fulfilled by grain boundary slide.

The results of electron-microscopic studies of dislocation subboundaries showed that mainly non-equilibrium subboundaries of different kind constituted the substructure of the grains. The non-equilibrium character of subboundaries may be caused by various factors. First, it is breaking in their structure regularity. Second, it is because the dislocations forming dislocation walls have rather great screw component, and the dislocations constituted dislocation networks have a significant edge component. Non-equilibrium subboundaries, possessing increased energy, demonstrate thermoactivated destruction at samples heating by migration of their dislocations to high angle grains boundaries [6]. On the other hand such non-equilibrium subboundaries generate long-range stress fields in grains which in turn can cause emergence of the dislocations from other subboundaries i.e. practically their destruction [7].

4 Acknowledgements

The support from Russian Foundation for Basic Research (Projects 01-02-16505, 02-02-81021 and 02-02-96413) is greatly appreciated.

5 References

[1] V.M. Segal, V.I. Reznikov, A.E. Drobyshevskiy, V.I. Kopylov, Russian Metallurgy 1981, 1, 99
[2] V.M. Segal, Mater. Sci. Eng. 1995, A197, 157
[3] Mondolfo L.F., Aluminum Alloys: Structure and Properties, Butterworth & Co Ltd, London, 1976, p. 566
[4] J. N. Fridlyander, K. V. Chuistov, A. L. Beresina, N. I. Kolobnev, Aluminum-Lithium Alloys (Structure and Properties) Naukova Dumka, Kiev, 1992, p. 192
[5] M. M. Myshlyaev, V. V. Shpeizman, M. M. Kamalov, Physics of the Solid State, 2001, 43, 2099–2104
[6] J.C.M. Li, in Electron Microscopy and Strength of Crystals N.Y., 1963, p. 713–779
[7] J.P. Hirth and J.Lothe, Theory of Dislocations, Wiley, N.Y. 1968

Microstructure Refinement and Improvement of Mechanical Properties of a Magnesium Alloy by Severe Plastic Deformation

A. Mussi, J.J. Blandin, E. F. Rauch
Génie Physique et Mécanique des Matériaux (GPM2) Saint-Martin d'Hères, France
Institut National Polytechnique de Grenoble (INPG) Saint-Martin d'Hères, France

1   Abstract

Equal channel angular extrusions (ECAE) have been performed on a magnesium alloy (AZ91) to improve its mechanical properties at room and high temperature. Two conditions of extrusions have been selected: at constant and varying temperature. In both cases, the structure is drastically refined and cells of about 0.3 µm could be achieved. These microstructures exhibit high yield stresses at room temperature and superplastic properties at lower temperatures than those usually reported for magnesium alloys. A specificity of this alloy is the ability to contain a large amount of β phase ($Al_{12}Mg_{17}$). A particular attention is given to the role of this intermetallic phase during both the ECAE process and the superplastic deformation. Moreover, the effect of deformation (ECAE, superplasticity) on the β precipitation is also discussed.

2   Introduction

Despite of their low density ($\approx$ 1.8 g/cm^{-3}), the use of magnesium alloys in industry is still limited. To enhance the attraction for magnesium alloys, their mechanical properties at room temperatures as well as their ability to be formed at high temperatures, can be improved by grain refinement. Such refinement may be achieved by conventional extrusion or rolling. In the case of magnesium alloys, grain sizes of about 5 µm can be produced with such procedures [1]. Severe plastic deformation processed with Equal Channel Angular Extrusion (ECAE), has been extensively applied in aluminum alloys, leading to submicronic grain sizes [2]. It is now well known that in this process, the material is highly sheared in a die containing two intersecting channels of identical cross sections [3]. Moreover, as the sample geometry remains the same after each extrusion, this process can be repeated several times. The technique has been recently applied successfully to magnesium alloys [4]. The aim of this paper is to study the mechanical properties at room and high temperature of an AZ91 alloy subjected to ECAE.

3   Experimental procedure

The composition of the present AZ91 magnesium alloy is Mg-9.1Al-0.9Zn-0.2Mn (wt. %). Before extrusion, the alloy is homogenized at 413 °C during 19 h, leading to a mean grain size of about 50 µm. The channels of the ECAE device are at 90° the one from the other and have a square section. In the present investigation, a B route was selected since it has been shown that it was the most effective route to produce an equiaxed submicronic structure in an aluminum alloy [2]. The crosshead velocity was 5 mm/min and a MoS_2 lubricant was used to limit friction. The samples were parallelepiped bars of 45 mm long with a cross-section of 10 mm × 10 mm.

After extrusion, thin foils were extracted thanks to a Precision Ion Polishing System (PIPS) with a 7° ion beam angle and 5 keV. Microstructure characterizations were mainly performed by TEM with a JEOL 3010 operating at a 300 kV and SEM. Tensile tests were carried out at room temperature and 250°C on specimens of 8 mm gauge length, 2 mm gauge width and 1.5 mm gauge thick. Tests were carried out at 10^{-3} s^{-1}.

4 Results

4.1 Effects of Thermo-Mechanical Treatments on the Microstructures

The first extrusion for AZ91 could not be performed below 265 °C since fracture was systematically observed at lower temperatures. Therefore, extrusions were processed at that temperature for all the successive passes. This procedure is referred as CT (constant temperature) in the following. After 8 extrusions, the resulting structure size is about 1 µm (figure 1). During this process, the measured load decreased continuously. A smaller structural size being expected if the load could be kept constant, it was attempted to perform the extrusions at a decreasing temperature for each pass.

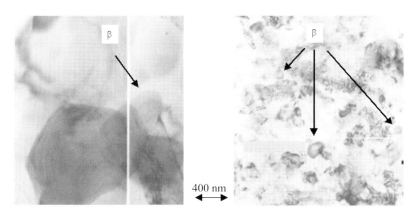

Figure 1: TEM analysis of AZ91 ECAE processed at 265 °C 8 times (CT)

Figure 2: TEM analysis of AZ91 ECAE processed with a decrease of temperature (DT) 8 passes

With ECAE performed at a temperature decreasing from 265 °C for the first extrusion down to 150 °C for the last one, no fracture was observed. Hereafter, this process is called DT (decreasing temperature). Figure 2 displays a TEM observation of the microstructure after 8 passes according to this second procedure. TEM micrograph reveals a structure size of 0.3 µm (Figure 2). Note that the observed structural sizes cannot unambiguously be considered as the grain size.

4.2 Mechanical Properties at Room Temperature

Figure 3 shows the stress/strain curves obtained at room temperature for the two ECAE procedures (8 passes) and the corresponding mechanical properties are detailed in table 1.

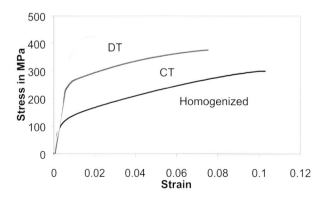

Figure 3: Tensile test comparisons at room temperature

Table 1 : Mechanical properties at room temperatures

	Yield stress (MPa)	Ultimate tensile stress (MPa)	Elongation to failure
Homogenized	100	300	9.6 %
CT	240	375	6.9 %
DT	375	425	2.2 %

According to the Hall-Petch law, the yield stress is related to the grain size through:

$$\sigma = \sigma_0 + \frac{k}{\sqrt{d}} \qquad (1)$$

where d is the grain size, σ_0 is a constant, and k is the Hall-Petch coefficient. For magnesium alloys, a value of k about 210 MPa µm$^{-1/2}$ was reported [5]. For ECAE-processed samples, the use of the Hall-Petch relation is delicate since it requires on the one hand, a precise measurement of the grain size, on the other hand, to assume that all the other microstructural features are constant, including the dislocation density. Nevertheless, if it is assumed that the Hall-Petch relationship can be used in the present investigation, it leads to grain size of 1.5 µm (respectively 0.5 µm) for the CT (respectively DT) conditions. These values are larger than the ones derived from the TEM observations and suggest that the structure is composed of grains mixed with cells.

4.3 Mechanical Properties at High Temperatures: Superplastic Behavior

A finer grain size is known to promote superplastic deformation. Indeed, an AZ91 structure refined up to 5 µm from conventional thermo-mechanical treatments, allows a high elongation to failure of 340 % at 300 °C and 2.10^{-4} s^{-1} [1]. After ECAE, the structure refinement is intense (figures 1 and 2) which may promote superplastic deformation at lower temperature and/or higher strain rate. Figure 4 displays the stress – strain curves corresponding to tensile tests performed at 10^{-3} s^{-1} and 250 °C. In the homogenized state, the ductility of the alloy is very limited and a large flow stress is measured. After extrusion, a significant stress decrease is detected confirming grain size sensitivity. The alloy extruded in CT conditions shows an elongation of 190 % whereas this value increases up to of 490 % for AZ91 for the DT conditions (figure 4). These curves confirm also that the AZ91 structure is finer in DT than CT conditions. Furthermore, strain hardening is detected during superplastic deformation, which can be probably related to dynamic grain growth.

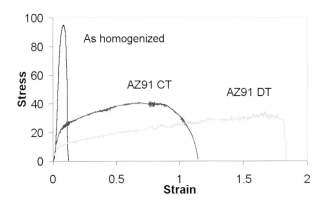

Figure 4: Stress/strain curves obtained at 250 °C and 10^{-3} s^{-1}

5 Role of the $Al_{12}Mg_{17}$ Phase

One specificity of the AZ91 alloy is the ability to contain a large fraction of intermetallic $Al_{12}Mg_{17}$ β phase as a result of the important amount of aluminum and this phase can interact with ECAE and superplastic deformation.

5.1 Interaction with ECAE

Since ECAE passes were carried out between 265 °C and 150 °C on a homogenized microstructure, β precipitation is expected. This is confirmed by the TEM observations shown in figures 1 and 2. The β precipitates are larger for the CT conditions, due to higher processing temperatures. A streaking feature about these precipitates is their nodular shape. Usually, $Al_{12}Mg_{17}$ precipitates are flat (lath shapes [6]). However, in the present investigation, the precipitates were probably intensively sheared during each ECAE pass. Moreover, due to their particular small size (≈ 50 nm), some of them could be also dissolved due to their high surface energy, as already reported in other studies dealing with severe plastic deformation techniques [7,8].

Another consequence of this precipitation is the depletion in aluminum of residual solid solution, which may affect the flow stress of the alloy due to the strong interaction between dislocations and the aluminum solutes.

5.2 Interaction with Superplastic Properties

Figure 5.a displays a SEM observation of the gauge length of a sample extruded in DT conditions and superplastically deformed at 250 °C and 10^{-3} s^{-1} (e = 1.7). An important volume fraction of large β precipitates is detected and this seems to be preferentially located nearby grain boundaries. The mechanical properties at high temperature of this intermetallic phase remain poorly documented, but some authors suggested that $Al_{12}Mg_{17}$ could be soft for temperatures larger than 200 °C [5]. If it is the case, this phase could have a beneficial effect on cavitation resistance during superplastic deformation of the AZ91 alloy.

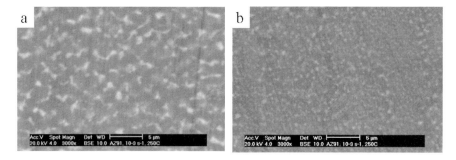

Figure 5: a) β precipitates in AZ91 for the DT conditions deformed at 250 °C and 10^{-3} s^{-1}; b) β precipitates in an equivalent static state

Another noticeable feature about the interaction of the β phase with superplastic properties is that superplastic deformation affects drastically the β distribution through the microstructure. Figure 5.b displays a SEM micrograph of the head of the sample of which the gauge length was shown in figure 5.a. The β phase is much finer than in the gauge length which means that the $Al_{12}Mg_{17}$ had undergone dynamic phase growth.

6 Conclusion

An AZ91 alloy was ECAE processed at constant and decreasing temperatures. In both cases, the microstructure was sharply refined and structure size of about 0.3 µm was obtained when a continuous reduction of the ECAE temperature was selected. As a consequence, particularly high yield stresses were measured as well as good superplastic properties. In this alloy, the intermetallic $Al_{12}Mg_{17}$ phase play an important role, not only during the ECAE process but also during superplastic deformation.

7 References

[1] Kubota, K.; Mabuchi, M.; Higashi, K., J. Mater. Sci., 1999, 34, 2255
[2] Dupuy, L.; Blandin, J.J.; Rauch, E. F., Mater. Sci. Technol., 2000, 16, 1256
[3] Segal, V. M., Mat. Sci. Eng., 1995, A197, 157
[4] Mabuchi, M.; Ameyama, K.; Iwasaki, H.; Higashi K., Acta Metall., 1999, 47, 2047
[5] Lukac, P.; Trojanova, Z.; Mathis, K., in Euromat 2000, 'Advances in mechanical behaviour, plasticity and damage', (ed. Miannay, D.; Costa, P.; François, D.; Pineau, A.) 2000, p. 1291
[6] Murayama, M.; Horita, Z.; Hono, K., Acta Mater., 2001, 49, 21
[7] Celotto, S., Acta. Mater., 2000, 48, 1775
[8] Languillaume, J.; Kapelski, G.; Baudelet, B., Acta. Mater., 1997, 45, 120

Grain Refinement and Superplastic Properties of Cu-Zn Alloys Processed by Equal-Channel Angular Pressing

Koji Neishi[1], Zenji Horita[1] and Terence G. Langdon[2]
[1]Kyushu University, Fukuoka, Japan
[2]University of Southern California, Los Angeles, U.S.A.

1 Abstract

Two Cu-based alloys, containing either 40 wt% or 42 wt% Zn, were subjected to equal-channel angular pressing (ECAP) in order to achieve significant grain refinement. The alloys were annealed initially at 1123 K to give a single phase and subsequently at 633 K to give a two-phase structure. Grain refinement was achieved through a combination of severe plastic straining using ECAP and the phase transformation which occurs on passing from a single to a dual-phase structure. An ultrafine grain size was achieved after only one pass of ECAP and there was little or no grain coarsening even after annealing for 1 hour at 623 K. Tensile testing showed that the Cu-40 % Zn alloy was superplastic at a temperature of 623 K, corresponding to a homologous temperature of ~0.53 T_m where T_m is the melting temperature of the alloy, and there was evidence for superplastic flow both at low temperatures and at reasonably high strain rates.

2 Introduction

Superplasticity requires a small grain size that is stable at reasonably high homologous temperatures [1]. To date, most of the superplastic data relate to materials having grain sizes of the order of ~5 µm tested at temperatures significantly above ~0.5 T_m, where T_m is the absolute melting temperature. However, there is evidence that the grain sizes of bulk materials may be reduced to the submicrometer or even the nanometer size using processes involving the introduction of severe plastic deformation [2], and this raises the possibility that it may be feasible to achieve superplastic elongations both at high strain rates and at relatively low homologous temperatures [3].

The present investigation was initiated to investigate this possibility using two copper-based alloys with zinc contents of either 40 wt% or 42 wt%. Earlier reports described the testing of a Cu-40% Zn alloy [4] and the occurrence of low temperature superplasticity in a Cu-Zn-Sn alloy [5].

3 Experimental

Ingots of two Cu alloys containing 40 wt%Zn and 42 wt%Zn were produced using a melting and casting procedure and, following extrusion at 973 K to rods with diameters of 10 mm, they were supplied by TOTO, Ltd. (Kitakyushu, Japan). These rods were cut into lengths of 60 mm and they were heated to 1123 K, held for 1 hour and quenched into iced water to make the alloys a β single phase. The ECAP was conducted for 1 pass at 633 K or 673 K using a die having a channel angle of 90° so that a single pass gives a strain of ~1 [6]: the principles of processing by ECAP were first described by Segal *et al.* [7] and a recent report described these principles in detail [8]. Before pressing, the samples were held in the die either for 2 or 5 minutes where these times were long enough to reach the designated temperatures. The two different holding times were adopted in order to check any change in the fraction of the β phase transforming to the α phase and a subsequent change in the grain size. After pressing, the samples were cut perpendicular to the longitudinal axis to thicknesses of ~5 mm and subjected to static annealing at 623 K for 1 hour. Microstructural observations were conducted on the samples, both after ECAP and after a combination of ECAP and annealing, using optical microscopy and orientation image microscopy (OIM). The area fractions of α and β phases were analyzed from the OIM images.

To evaluate the occurrence of superplasticity, tensile specimens were machined from the as-pressed rods with the tensile axes parallel to the longitudinal axes and with gauge lengths of 5 mm and cross-sections of 2 × 3 mm². Tensile tests were conducted at 623 K with initial strain rates of $1.0 \cdot 10^{-4}$ to $1 \, s^{-1}$ using a machine operating at a constant rate of cross-head displacement.

4 Results and Discussion

Figure 1 shows the microstructures observed by optical microscopy after ECAP for the Cu-40% Zn and Cu-42% Zn alloys. All samples were held for 5 minutes before ECAP. Grain refinement appears to be more pronounced as the ECAP temperature is decreased for both alloys. Comparison reveals that the grain size is more refined in the Cu-40%Zn alloy than in the Cu-42%Zn alloy. These trends are more clearly demonstrated in Table 1 where the grain sizes are documented from quantitative measurements using OIM. It is noted that, for the Cu-40 % Zn alloy subjected to ECAP at 673 K, the grain size was measured from optical micrographs. There

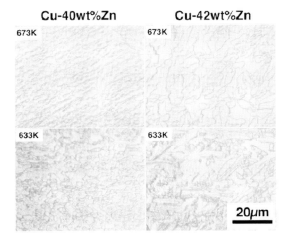

Figure 1: Optical micrographs of the Cu-40%Zn and Cu-42%Zn alloys processed by ECAP

appears to be no significant change in the grain size when the holding time is shortened to 2 minutes.

Table 1: Grain sizes after ECAP and after ECAP plus annealing at various conditions

	ECAP		Grain Size	
	Temperature (K)	Holding time (min)	As-ECAP (µm)	ECAP+Anneal (623 K, 1 h) (µm)
Cu-40%Zn	633	2	1.0	1.3
	633	5	1.0	1.7
	673	5	1.4	1.5
CU-42%Zn	633	2	1.7	1.5
	633	5	1.5	1.5
	673	5	–1.0	–

In order to check the thermal stability of the fine-grained structures produced by ECAP, the samples were annealed at 623 K for 1 hour. Figure 2 compares OIM images of the Cu-40% Zn alloy after ECAP and after ECAP followed by annealing, where the samples were pressed at 633 K with a holding time of 5 minutes. The bright areas represent the α phase and the gray areas correspond to the β phase. Three features arise from the annealing after ECAP: (1) the area fraction of the β phase is decreased, (2) there is some grain growth and (3) the grain orientation of each phase becomes random. The grain sizes measured using OIM analysis are included in Table 1 for all samples except for the Cu-42% Zn alloy subjected to ECAP at 673 K with a holding time of 5 minutes.

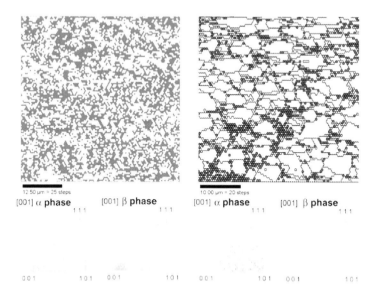

Figure 2: OIM images of the Cu-40%Zn alloy after ECAP and after ECAP followed by annealing

Although grain growth occurs by annealing at 623 K for 1 hour, the extent of the growth remains relatively small. It is therefore anticipated that superplastic elongations may be achieved for the present alloys. Figure 3 plots the elongation to failure against the initial strain rate obtained after tensile testing at 623 K for the Cu-40% Zn alloy subjected to ECAP at 633 K with holding times of 2 and 5 minutes. The elongation to failure increases with decreasing strain rate and it reaches a maximum of 740 % at a strain rate of $3.3 \cdot 10^{-4}$ s^{-1} for a holding time of 2 minutes and continuously increasing to 1000 % within the strain rates covered in this study for a holding time of 5 minutes. Although the maximum elongation is lower for the sample with a holding time of 2 minutes, the elongation to failure at higher strain rates is greater than for the sample with a holding time of 5 minutes. This difference may be due to the fraction of the α phase. Observations by OIM revealed that the α phase occupied 84 % for a holding time of 5 minutes and 55 % for a holding time of 2 minutes. Therefore, more α/β boundaries are available for the samples with a holding time of 2 minutes.

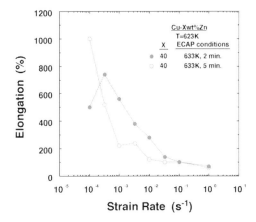

Figure 3: Elongation to failure against initial strain rate after tensile testing at 623 K for the Cu-40%Zn alloy

Figure 4 shows the appearance of specimens after deformation for the samples pressed at 633 K with a holding time of 2 minutes: the upper specimen is untested. These specimens reveal clear evidence for the occurrence of superplastic deformation, especially in the vicinity of strain rates of 10^{-3} and 10^{-4} s^{-1}. The testing temperature of 623 K is equivalent to a homologous temperature of ~0.53 T_m, so that, when considering also the results in Fig. 3, superplasticity is attained at a relatively low temperature and, in addition, the occurrence of an elongation of 280 % at a strain rate of $1.0 \cdot 10^{-2}$ s^{-1} shows the potential for achieving high strain rate superplasticity in this alloy.

5   Summary and Conclusions

1. Application of ECAP to Cu-40wt%Zn and Cu-42wt%Zn alloys at a phase transforamtion temperature led to grain sizes of 1–2 μm.
2. The fine-grained structures were stable at 623 K.

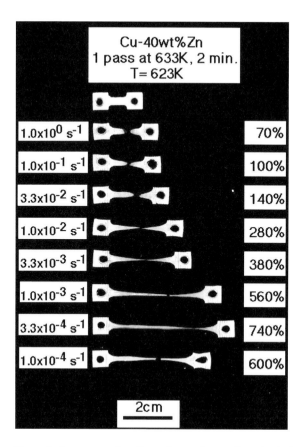

Figure 4: Appearance of tensile specimens after tensile testing including untested specimen

3. The fine-grained Cu-40%Zn alloy exhibited superplastic elongations of ~1000 % at 623 K with an initial strain rate of $1.0 \cdot 10^{-4}$ s^{-1} and ~280 % at 623 K with an initial strain rate of $1.0 \cdot 10^{-4}$ s^{-1}.
4. Low-temperature and high-strain-rate superplasticity were atained in the ECAP-processed Cu-40%Zn alloy.

6 Acknowledgements

We are grateful to Mr. T.Fujinami for his helpful assistance. This work was supported by the U.S. Army Research Office under Grant No. DAAD19-00-1-0488.

7 References

[1] T.G. Langdon, Metall. Trans. A, 1982, 13A, 689–701
[2] Y. Ma, M. Furukawa, Z. Horita, M. Nemoto, R.Z. Valiev, T.G. Langdon, Mater. Trans. JIM, 1996, 37, 336–339
[3] R.Z. Valiev, R.K. Islamgaliev, I.V. Alexandrov, Prog. Mater. Sci., 2000, 45, 103–189
[4] K. Neishi, Z. Horita, T.G. Langdon, Scripta Mater., 2001, 45, 965–970
[5] K. Neishi, T. Uchida, A. Yamauchi, K. Nakamura, Z. Horita, T.G. Langdon, Mater. Sci. Eng., 2001, A307, 23–28
[6] Y. Iwahasi, J. Wang, Z. Horita, M. Nemoto, T.G. Langdon, Scripta Master., 1996, 35, 143–146
[7] V.M. Segal, V.I. Reznikov, A.E. Drobyshevskiy, V.I. Kopylov, Russian Metall., 1981, 1, 99–105
[8] M. Furukawa, Z. Horita, M. Nemoto, T.G. Langdon, J. Mater. Sci., 2001, 36, 2835–2843

XIV Mechanisms of Diffusion Related Processes in Nanocrystalline Materials

Diffusion in Nanocrystalline Metals and Alloys – A Status Report

Roland Würschum[1], Simone Herth[2], and Ulrich Brossmann[1]
1 Technische Universität Graz, Institut für Technische Physik, Graz
2 Forschungszentrum Karlsruhe, Institut für Nanotechnologie, Karlsruhe, Germany

1 Abstract

Diffusion is a key property determining the suitability of nanocrystalline materials for use in numerous applications, and it is crucial to the assessment of the extent to which the interfaces in nanocrystalline samples differ from conventional grain boundaries. The present article offers an overview of "diffusion in nanocrystalline metals and alloys. Emphasis is placed on the interfacial characteristics that affect diffusion in nanocrystalline materials, such as structural relaxation, grain growth, porosity, and the specific type of interface. In addition, the influence of intergranular amorphous phases and "intergranular melting on diffusion is addressed, and the atomistic simulation of GB structures and "diffusion is briefly summarized. On the basis of the available diffusion data, the diffusion-mediated processes of deformation and induced magnetic anisotropy are discussed.

2 Introduction

Since the pioneering work performed on nanocrystalline materials nearly twenty years ago, [1] diffusion in these novel types of materials has attracted permanent interest, largely because material transport belongs to the group of physical properties differing most in nanocrystalline materials in comparison to their coarse-grained or single-crystalline counterparts. In nanocrystalline specimens, rapid mass transport can occur because of the high number density of crystallite interfaces and the fact that diffusion along interfaces is usually much faster than in crystals.

Diffusion is a determining feature of a number of application-oriented properties of nanocrystalline materials, such as enhanced ductility, diffusion-induced magnetic anisotropy, enhanced ionic mass transport, and improved catalytic activity. [2,3] Moreover, diffusion in nanocrystalline materials is also relevant to the basic physics of interfaces. Since interface diffusion is highly structure sensitive, diffusion studies can provide valuable insight into the question of the extent to which interfaces in nanocrystalline materials differ from conventional grain boundaries (GBs).

The present article is an up-to-date review of diffusion as it occurs in nanocrystalline metals and alloys. Following a brief description of GB diffusion models (Section 2) and an introductory overview of experimental results (Section 3.1), focus will be laid on the effects of structural relaxation and grain growth (Section 3.2), porosity and different types of interfaces (Section 3.3), intergranular amorphous phases (Section 3.4) and intergranular melting (Section 3.5). Recent literature on atomistic simulations of GB structures and diffusion will be briefly summarized in Section 4. An assessment of the diffusion-mediated processes of deformation and induced magnetic anisotropy will be given in Section 5 on the basis of the available diffusion data.

For a summary of diffusion data that includes hydrogen diffusion in nanocrystalline metals and alloys as well as diffusion in nanocrystalline ceramics, the reader is referred to our previous review article. [4] Furthermore, reviews of GB diffusion in coarse-grained materials and bicrystals have been published recently by Mishin and Herzig. [5,6]

3   Modeling

Diffusion coefficients are usually determined from the penetration profile of tracer atoms measured in a diffusion experiment. In the general case of interface diffusion in polycrystalline materials, two simultaneous diffusion processes must be taken into account: rapid diffusion in the crystallite interfaces (GB diffusion coefficient D_{GB}), and diffusion from the interfaces and specimen surface into the volume of the crystallites (diffusion coefficient D_V). According to Harrison's classification scheme, three different diffusion regimes – denoted by A, B, and C – are observed in polycrystalline materials; in each of these regimes, evaluation of the diffusion "profiles is relatively straightforward (see Kaur et al.). [7] These diffusion regimes are characterized by appropriate ratios between the diffusion length L_V in the crystallites ($\sim D_V$ t) 0.5, with diffusion time t) and the crystallite diameter d or the interface thickness δ.

Roughly speaking, in the regimes of type A ($L_V > d$) and C ($L_V < \delta$), simple Gaussian diffusion profiles develop, from which DGB (type C) or an average value of DGB and D_V (type A) can be determined directly. In the type-B regime ($d > L_V > \delta$), in addition to the Gaussian part characterizing the volume diffusion, a tail is observed, from which the product $s\delta D_{GB}D_V^{0.5}$ (segregation factor s) may be determined. Transitions between the regimes A and B, or B and C can be treated numerically. [8–10] Moreover, an immobilization of diffusing atoms from fast interfacial diffusion paths in the crystallites that arise from the migration of GBs in the wake of grain growth can be taken into account quantitatively. [11]

For fine-grained materials, Mishin and Herzig proposed an extension of the aforementioned scheme, distinguishing between different ratios between the diffusion length along the interfaces and the crystallite diameter (see Kaur et al.). [7] For a number of recent measurements of diffusion in nanocrystalline materials, it has turned out to be useful to extend the GB diffusion model to include two types of interfaces manifesting low and high diffusivity (see Section 3.3). Between fast and slow interface diffusion, relationships can be derived that are analogous to those valid for the case of type-B kinetics of GB diffusion. [10,12,13] In another extension, Peteline and co-workers [14] considered the possibility of diffusion along the triple-line intersections of interfaces in addition to diffusion within the interfaces and crystallites.

4   Diffusion Measurements

4.1   Overview

An overview of diffusion measurements that have been performed on nanocrystalline metals and alloys is given in Table 1. The first diffusion study in this field was published by Horváth et al. in 1987, [15] in which self-diffusivities in nanocrystalline Cu (n-Cu) were measured to be about three orders of magnitude larger than in grain boundaries. Subsequently, further findings of extraordinarily high diffusivities were reported in nanocrystalline metals (Table 1), such as the diffusivity of Ag in n-Cu. [16] In the period following this initial era, however, it was realized that factors like structural relaxation, grain growth, and residual porosity must be taken into account in order to obtain an unambiguous assessment of the diffusion in nanocrystalline metals. As a consequence of experimental advances in preparing porosity-free nanocrystalline metals of high purity, grain growth became even more prevalent during the diffusion annealing performed in subsequent studies, because of the absence of pinning centers for interfaces. [17–20] However, thermally stable nanostructures could be achieved for alloys, which permitted the study of diffusion without grain boundary migration in a number of cases. [10,13,21–24] As outlined in the following, most of the recent studies taking these factors into consideration (Sections 3.2, 3.3) have concluded that the diffusivities in relaxed interfaces of nanocrystalline materials are similar to or only slightly higher than GB diffusivities or the values extrapolated from high-temperature GB diffusion data. However, a particular situation appears to apply to nanocrystalline alloys prepared by crystallization, in which lower interfacial diffusivities are observed, presumably as a result of the presence of residual intergranular phases (Section 3.4).

A wide spectrum of well-established experimental techniques has been applied to the measurement of diffusion in nanocrystalline materials, including the radiotracer method with sputter or mechanical sectioning, electron-beam microanalysis, Auger electron or secondary ion-mass spectroscopy with depth profiling, Rutherford backscattering and nuclear magnetic resonance. These techniques are described in textbooks on diffusion (see, e.g., Kaur et al.) [7] or in the respective references quoted in Table 1.

4.2 Structural Relaxation and Grain Growth

Since the conditions prevalent during the synthesis of "nanocrystalline materials are far from thermodynamic equilibrium, the initial structure of interfaces in such samples may depend sensitively on the time-temperature history. For nanocrystalline metals prepared by crystallite condensation in an inert-gas atmosphere with subsequent crystallite compaction, extensive experimental evidence exists that structural relaxation occurs at slightly elevated temperatures (see Würschum et al.). [4,25] That this structural relaxation exerts an influence on the diffusion behavior can be concluded from the decrease in self-diffusion coefficient with increasing time of diffusion annealing, as measured, for instance, in nanocrystalline Fe prepared by the cluster condensation and compaction route and by severe plastic deformation. [19] Similar observations were recently made by Kolobov et al. [26] on nanostructured Ni prepared by severe plastic deformation. In that case, the diffusivity of Cu at 423 K was found to decrease by more than three orders of magnitude upon pre-annealing the sample at 523 K prior to the onset of grain growth. [26] Both in n-Fe (Fig. 1) [19] and in n-Ni, [26] the interfacial diffusion coefficients in the relaxed state appear to be similar to or only slightly higher than the values expected for conventional GBs in coarse-grained materials from an extrapolation of high-temperature data.

The enhanced diffusivity observed in nanocrystalline metals prior to relaxation may indicate a non-equilibrium structure of the interfaces. Non-equilibrium GBs are known in the case of cold-worked metals, where they exhibit a gradual relaxation behavior or a metastable character. [27] The enhanced diffusivities found in such samples are ascribed to the absorption of lattice

Table 1: Overview of diffusion studies of nanocrystalline metals and alloys. Part A (metals): Diffusion coefficients D at 25 % of the melting temperature TM, according to a linear interpolation of the measurement data (diffusing elements given in parentheses). D-values obtained by linear extrapolation are placed in parentheses. Part B (alloys): Activation energy Q and pre-exponential factor dD_0 or D_0 of diffusion. The model used in analysis is specified as follows: GB diffusion kinetics of type A, B, or C; models considering two different types of interfaces (2 IF); brick-layer model (BL); or Boltzmann-Matano technique (BM). Ta: pre-annealing temperature (AT: ambient temperature); Td: temperature interval of diffusion; d: crystallite size prior to diffusion. NS: no statement given in quoted reference. Synthesis routes: crystallite condensation and compaction (CC), severe plastic deformation (SPD), sputtering of films (MS), electrodeposition (ED), melt-spinning and crystallization (CRY), melt-spinning and hot compaction (MS/HC), mechanical alloying and hot compaction (MA/HC). Measurement methods: radiotracer (RT), electron-beam microanalysis (EBMA), Auger electron (AES) or secondary ion-mass spectroscopy (SIMS), Rutherford backscattering (RBS), nuclear magnetic resonance (NMR).

Part A: Metals	Synthesis	T_a [K]	T_d [K]	D (0.25 T_M) [m^{2}s^{-1}]	Model	d [nm]	Method
Pd (Fe)[17] [a]	CC	373 [f]	423–523	2.8×10^{-20}	C	50 [c]	RT
Pd (Fe) [18] [a]	SPD	293–673	371–623	3×10^{-20}	C	80 [c]	RT
Fe (Fe) [19][a]	CC	403 [f]	452–499	1×10^{-21} [b]	C	31 [c]	RT
Fe (Fe) [19][a]	SPD	527	527	3×10^{-19} (0.29T_M) [b]	C	NS	RT
Literature data:							
Cu (Cu) [15]	CC	AT	293–393	9.2×10^{-19}	C	8 [d]	RT
Cu (Ag) [32]	MS	NS	393–428	NS	C	20 [d]	AES
Cu [33]	CC	AT	293–420	NS		11	NMR
Cu (Ag) [16]	CC	AT	303–373	1.6×10^{-18}	C	8	EBMA
Cu (Au) [34]	CC	AT	373	4.8×10^{-22} (0.27T_M)	C	10	AES
Cu (Bi) [34]	CC	AT	293–413	3×10^{-20} [b]	C	10	RBS
Cu (Sb) [35]	CC	AT	323–373	2.4×10^{-21} [e] 1.3×10^{-22} [e]	2 IF	50	RBS
Ni (Ni) [12]	CC	773 [f]	293–473	5.9×10^{-18} [e] 1.6×10^{-19} (0.27T_M) [e]	2 IF	70	RT
Ni (Cu) [26]	SPD	398	398–448	9.6×10^{-15} (0.245T_M) [b]	C	300 [d]	SIMS
Ni (Cu) [26]	ED	NS	423	3.8×10^{-17} (0.245T_M)	C	30	SIMS
Fe (B) [36]	CC	AT	293–383	(4.5×10^{-18})	C	7	SIMS
Grain boundaries (GBs) in coarse-grained metals and bicrystals:							
fcc, general [37]				(1.7×10^{-21}) [g]			
bcc, general [37]				(7.1×10^{-23}) [g]			
Ag [38]				1.2×10^{-16} (0.29 T_M)			
Σ-tilt-GB Ag/Au [39]				(2.7×10^{-19}) [g]			

Part B: alloys	Synthesis	T_a [K]	T_d [K]	δD_0 or D_0	Q [eV]	Model	D [nm]	Method
Fe$_{73.5}$Si$_{13.5}$B$_9$–Nb$_3$Cu$_1$ (Fe) [21][a]	CRY	810–818	628–773	NS	1.9	BL	13 [d]	RT
Fe$_{73.5}$Si$_{13.5}$B$_9$–Nb$_3$Cu$_1$ (Ge) [22][a]	CRY	817	735–783	1.14×10^{-8} [m^{2}s^{-1}]	2.91	B	13 [d]	RT
Fe$_{90}$Zr$_{10}$ [13][a]	CRY	878–895	573–767	NS	NS	2 IF	29 [d]	RT
Fe$_{90}$Zr$_7$B$_3$(Fe) [22][a]	CRY	873	593–773	2.78×10^{-7} [e] [m^{2}s^{-1}]	1.70 [e]	2 IF	18 [d]	RT
			623–741	6.84×10^{-14} [e] [m^{3}s^{-1}]	1.64 [e]			
Nd$_{14.2}$Fe$_{80.8}$B$_5$ (Fe) [23][a]	MS/HC	1048	689–951	1.53×10^{-11} [m^{3}s^{-1}]	1.74	B	100 [d]	RT
Literature data :								
γ-Fe$_{61.2}$Ni$_{38.8}$ (Fe) [10,24]	MA/HC	1123	636–1013	4.2×10^{-3} [e] [m^{2}s^{-1}] 3.4×10^{-3} [e] [m^{2}s^{-1}]	1.93 [e] 1.53 [e]	A,AB, B, 2 IF	80–100 [d]	RT
Al$_{91.9}$Ti$_{7.8}$Fe$_{0.3}$(Cu) [40]	MA/HC	673	371–571	2.4×10^{-12} [m^{2}s^{-1}]	0.364	C	22	SIMS
Al$_{96.8}$Mg$_3$Sc$_{0.2}$/Al$_{99.8}$Sc$_{0.2}$ [41]	SPD	AT	523–723	NS	1.02 [h]	BM	200 [c]	EPMA

[a] Data from authors\9 group. [b] Variation with temperature of preanneal or with diffusion time observed. [c] Crystallite growth upon diffusion annealing observed. [d] Crystallite size constant upon diffusion annealing. [e] 2-IF diffusion model (values quoted in 1st and 2nd lines refer to diffusion along two types of interfaces). [f] Temperature of crystallite compaction. [g] Related to grain-boundary thickness δ = 1 nm. [h] Interdiffusion

dislocations at GBs. [27,28] In the case of nanocrystalline metals, in which dislocation activity ceases due to the small crystallite size, [29] enhanced diffusivities may arise from local excess free volume, which, in the case of inert-gas-condensed samples, may remain following the high-pressure compaction step. Evidence for the existence of vacancy-type interfacial free volumes and their variation upon annealing at slightly elevated temperatures could be derived from positron annihilation studies. [30] This situation resembles the well-known structural relaxation in amorphous alloys, where the annealing out of excess volume gives rise to a decrease in diffusivity (see, Frank). [31] Unlike the case of amorphous alloys, the microstructure of which remain stable over a relatively wide temperature range below the onset of crystallization, the relaxed structure of nanocrystalline metals is highly prone to interface migration and grain growth. This is particularly true for porosity-free nanocrystalline metals of high purity, in which pinning centers for interfaces have largely been eliminated. [17] In that case, the assessment of the diffusion behavior is affected by the concomitant GB migration.

The occurrence of crystallite growth during diffusion gives rise to a decrease in the fraction of interfaces and, as a consequence of growth-induced interface migration, to a slowing-down of tracer diffusion, since the tracer atoms are immobilized by incorporation at lattice sites of the crystallites. Although this leads to deviations from Gaussian profiles, the diffusivities derived from fits of Gaussians to the diffusion tails according to type-C kinetics may still lie fairly close

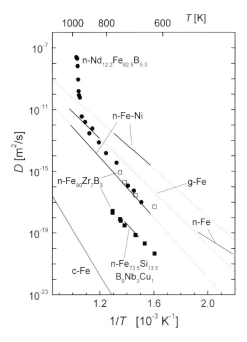

Figure 1: Arrhenius plots of 59Fe-tracer diffusivities in nanocrystalline Fe [19] and the Fe-rich nanocrystalline alloys Fe73.5Si13.5B9Nb3Cu1, [21] Fe90Zr7B3 [22] and Nd12.2Fe81.8B6 [22] (interface thickness d = 1 nm). For Fe90Zr7B3 the diffusivities in two types of interfaces (□ and ■) are shown. The data of n-Fe refer to relaxed GBs. Literature data of Fe diffusion in the ferromagnetic phase of crystalline a-Fe (c-Fe), [45] in grain boundaries of Fe (g-Fe) [46] and in nanocrystalline g-Fe–Ni [10,24] are shown for comparison (extrapolations are dotted).

to the actual values of D_{GB}. This follows from the fact that the tails of the diffusion profiles are least affected by grain growth, because the tails are governed by those GBs that migrate at the lowest rate. [20] Here, it is useful to note that regions with low or negligible GB migration are commonly observed during inhomogeneous grain growth, which frequently prevails in nanocrystalline samples. [4,20]

The complications introduced by grain growth in nanocrystalline metals are even more serious when attempting to ascribe a physical interpretation to experimental values for the activation energies of diffusion, because grain growth seriously limits the temperature range of such studies. For this reason, only characteristic diffusion coefficients are quoted for n-metals in Table 1, rather than activation energies, as well.

An enhanced stability of nanocrystalline metals with respect to grain growth may obtain in the case of thin films, as studied recently by Erdélyi et al. [32] In n-Cu-films, the diffusion of Ag in the low-temperature regime of type-C kinetics was found to be consistent with the GB diffusion and segregation data obtained for conventional GBs in the type-B regime. [32]

4.3 Porosity and Different Types of Interfaces

In a number of nanocrystalline alloys, it has been possible to carry out diffusion studies over a wide temperature range without complications arising from grain growth (Section 3.2). Despite the stable microstructure, the diffusion characteristics of such nanocrystalline alloys may be quite complex, owing to the presence of various types of interfaces (this Section), intergranular amorphous phases (Section 3.4), or the occurrence of intergranular melting (Section 3.5).

A range of interface types may represent a characteristic feature of nanocrystalline materials, particularly in those cases in which bulk samples are prepared from powders consisting of agglomerates of nanocrystallites. The interfaces between the agglomerates (i.e., inter-agglomerate boundaries) as well as residual porosity between the agglomerates provide pathways of high diffusivity, whereas a lower diffusivity occurs along the interfaces between the nanocrystallites located within the agglomerates (so-called intra-agglomerate boundaries). The first diffusion studies taking these two types of interfaces into account were performed by Bokstein et al. [12] on cluster-assembled nanocrystalline Ni with a density of 92–93 % of the bulk density and more recently by Divinski et al. [24] on 98 %-dense γ-Fe61.2Ni38.8 prepared by mechanical alloying and hot compaction (Fig. 1). In both cases, the diffusion profiles were analyzed using relationships analogous to those of type-B GB diffusion kinetics, taking into consideration diffusion fluxes from the inter-agglomerate to the intra-agglomerate boundaries. Both in nanocrystalline Ni [12] and γ-Fe61.2Ni38.8, [24] the diffusion coefficients of the intra-agglomerate boundaries were found to be characteristic of conventional GBs, whereas diffusivities several orders of magnitude higher were deduced for the inter-agglomerate boundaries. That the diffusion coefficients of the intra-agglomerate boundaries in γ-Fe61.2Ni38.8 are very similar to those of GBs in coarse-grained γ-Fe could also be deduced from an extension of the diffusion studies to the type-B and A regimes, where volume diffusion from the intra-agglomerate boundaries into the crystallites becomes increasingly dominant. [10]

In addition, evidence for the presence of various interface types has been obtained in nanocrystalline Fe-based alloys prepared by crystallization. [13,22] In these highly dense alloys, residual intergranular amorphous phases rather than inter-agglomerate boundaries give rise to additional interfacial diffusion paths, as outlined in the following section.

4.4 Intergranular Amorphous Phases

Nanocrystalline Fe-based alloys prepared by crystallization are of high technical relevance due to their superior soft-magnetic properties. [42,43] The two basic types of alloys are Fe73.5Si13.5B$_9$Nb$_3$Cu$_1$ (VITROPERM^{TM}) and Fe90Zr$_7$B$_3$ (NANOPERM^{TM}), consisting of D0$_3$-Fe$_3$Si or α-Fe nanocrystallites, respectively, with sizes of 10 to 15 nm embedded in a residual amorphous matrix. Larger crystallite sizes of ca. 30 nm with a two-phase structure of α-Fe and Fe-Zr nanocrystallites are formed upon crystallization of Fe90Zr10. [13] Each of these three Fe-based alloys manifests Fe-tracer diffusivities that are substantially lower than those of GBs in α-Fe (Fig. 1). [13,21,22,44] This particular feature strongly differs from the diffusion behavior observed in other nanocrystalline metals and alloys (Sections 3.2, 3.3). The reduced interfacial diffusivities, which are similar to (Fe73.5Si13.5B$_9$Nb$_3$Cu$_1$) or even lower (Fe90Zr$_7$B$_3$, Fe90Zr10) than those of the initial amorphous phases, are considered to arise from the residual intergranular amorphous phases.

In addition to the slow interface diffusivity, indications for a second fast diffusion process have been deduced from the tails of the diffusion profiles measured in nanocrystalline Fe90Zr$_7$B$_3$ and Fe90Zr10. Adopting the simple picture of the existence of two types of interfaces with high and low interface diffusivities – which can be described with diffusion kinetics analogous to those of type B (Section 3.3) – we find the fast process to be characteristic of diffusion along conventional GBs (Fig. 1). [13,22,44–46] No fast interfacial paths are found in nanocrystalline Fe73.5Si13.5B$_9$Nb$_3$Cu1, [21] presumably due to the higher volume fraction of the intergranular amorphous phase in comparison to the Fe-Zr alloys.

4.5 Intergranular Melting

Diffusion studies of nanocrystalline alloys have been recently extended to Nd$_2$Fe14B-based systems, which manifest an intergranular melting transition of Nd-enriched GBs. [47] These studies are motivated by the important role played by intergranular melting in the technological application of this permanent magnetic material. The powder-metallurgical processing of Nd$_2$Fe14B takes advantage of intergranular melting in order to induce crystallographic and thus magnetic texture (Grünberger et al., [47] see Section 5).

The diffusion studies performed on n-Nd$_2$Fe14B shed light on a novel type of diffusion behavior in nanocrystalline materials. Radiotracer measurements carried out with the isotope ^{59}Fe show a substantial increase of the interface diffusion coefficient D_{GB} above the intergranular melting [22,23] (Fig. 1). The increase of D_{GB} towards the diffusivity in bulk melts occurs over an extended temperature range, rather than abruptly. Such a gradual change could result either from confinement or from an initially local melting. Both effects give rise to reduced long-range diffusivities compared to bulk melts, as long as the dimensions of the molten region are small or interconnected liquid regions have not yet formed during the initial stage of melting. Below the intergranular melting transition, GB diffusivities in n-Nd$_2$Fe14B are found to be similar to those of α-Fe, within a framework of GB diffusion kinetics of type B with an assumed volume self-diffusivity as in α-Fe (Fig. 1).

5 Atomistic Simulations

With the increase of computational power during the last years, atomistic simulations have become a valuable tool for gaining additional insights into interfacial diffusion processes in bicrystals (for reviews see Mishin et al.) [5,48] and nanocrystalline materials. [29]

In coincidence GBs of metals, which represent special GBs, diffusion is found to be mediated by point defects with reduced activation energies of formation and migration compared to the lattice. In contrast to lattice diffusion in metals, where vacancy-mediated diffusion prevails, computer simulations show that GB diffusion can be dominated either by vacancy or by interstitial-related mechanisms, depending on the GB structure. [49,50] The picture of single point defects in ordered GBs appears to be restricted to low and moderate temperatures, where the thermal motion of atoms induces only limited distortions of the GB structure. At high temperatures – according to recent molecular dynamics simulations performed by Keblinski et al. [51] – high-energy GBs may "undergo a transition from a solid to a confined liquid state, accompanied by a decrease in the activation energy for diffusion.

Nanocrystalline structures were also studied by Wolf and co-workers in a number of computer simulations. [29,52,53] They find that low-angle and special boundaries are absent in "nanocrystalline materials; according to these studies, nanocrystalline materials contain only high-energy GBs, with atomic structures like those of high-energy GBs in bicrystals. Therefore, nanocrystalline materials are characterized by a narrow distribution of interfacial widths and energy densities. As far as diffusion is concerned, simulations of GB diffusion creep in a nanocrystalline Pd model structure at high temperatures of 900 K and above reveal a Coble-creep-type behavior, with a fast liquid-like self-diffusivity as found in high-energy GBs at high temperatures. [29] So far, an unambiguous comparison with experimental diffusion data is hampered by the drastically different temperature regimes accessible in computer simulations ($T > 900$ K) and experimental studies ($T < 600$ K; see Table 1, part A).

6 Comparison with Diffusion-Mediated Processes of Deformation and Induced Magnetic Anisotropy

Diffusion is particularly relevant for the deformation behavior of nanocrystalline metals and alloys, because, at small crystallite sizes, conventional dislocation mechanisms may be suppressed and grain-boundary-mediated deformation processes may become dominant. [29] Conclusions regarding the deformation mechanisms of nanocrystalline metals and alloys can be drawn by comparing the diffusion and deformation characteristics of such materials.

In the above-mentioned molecular-dynamics simulations of the high-temperature deformation of nanocrystalline Pd, Yamakov et al. [29] observed the "steady-state strain rate to obey Coble-creep behavior, as "characteristic of a GB-diffusion-mediated deformation. Grain-boundary sliding was found to be an accomodation mechanism for the Coble creep, with the entire deformation process being controlled by GB diffusion. [29]

The occurrence of a GB-diffusion-mediated deformation process is also the conclusion reached by room-temperature creep experiments performed by Wang et al. [54] on electrodeposited nanocrystalline Ni. At small grain sizes of 6 nm, the strain rate $d\varepsilon/dt \sim \sigma^n$ is found to be related to the stress σ by a stress exponent n of 1.2 – a value lying between that of Coble creep

(n = 1) and GB-diffusion-controlled GB sliding (n = 2). [54] At grain sizes between 20 nm and 40 nm, higher values of n up to 5.3 indicate a transition to dislocation creep with increasing crystallite size. [54] Creep studies carried out on electrodeposited n-Cu by Cai et al. [55] also find a linear relation between strain rate and stress (i.e., n = 1), as characteristic of GB-diffusion-controlled Coble creep. This is further supported by the observed increase of the strain rate of n-Cu with temperature, from which an activation energy for diffusion creep of 0.72 eV [55] – rather similar to the activation energy for GB self-diffusion in coarse-grained Cu [56] – is derived. For nanostructured Ni prepared by severe plastic deformation, Kolobov et al. [57] deduced an apparent creep activation energy of 1.2 eV. Although this value again nearly equals the activation energy for GB self-diffusion in coarse-grained Ni, [58] in this case creep mechanisms in addition to GB diffusion may be operative, owing to the larger grain size of 300 nm.

An important application-relevant issue concerning the high-temperature deformation of nanocrystalline materials is the processing and magnetic anisotropy of soft-magnetic and hard-magnetic nanocrystalline alloys. A high-temperature treatment of magnetic alloys placed under mechanical load or in an external magnetic field may give rise to diffusion-mediated magnetic anisotropy or texture, which may be desirable for tailoring the hysteresis loop. [42,47,59] Recently obtained "diffusion data and their comparison with deformation characteristics may offer insight into the microscopic processes underlying the magnetic anisotropy induced during the processing of magnetic nanocrystalline alloys.

Based on a model developed by Néel, the induced anisotropy in the soft-magnetic nanocrystalline alloy $Fe_{73.5}Si_{13.5}B_9Nb_3Cu_1$ can be considered to arise from atomic pair ordering. [44,59] According to Emura et al., [60] the activation energy of 2.9 eV measured for the formation of magnetic anisotropy in the nanocrystalline state of $Fe_{73.5}Si_{13.5}B_9Nb_3Cu_1$ is higher than the corresponding value, 1.92 eV, of the amorphous state. [60] In addition, the magnetic anisotropy was found to be induced more efficiently during crystallization (so-called one-step annealing) than by a two-step annealing procedure of crystallization followed by anisotropy formation. [61] A comparison of both results with the diffusion data of n-$Fe_{73.5}Si_{13.5}B_9Nb_3Cu_1$ (Table 1, part B) shows that, for the following reasons, the diffusivity of Fe can be excluded as the controlling factor for anisotropy formation: [21] i) The Fe diffusivity increases upon crystallization, in contrast to the efficiency of anisotropy formation. ii) The activation energy of 1.9 eV for Fe diffusion in the nanocrystalline state is substantially lower than the activation energy of 2.9 eV for anisotropy formation. iii) On the time scale of anisotropy formation, the Fe diffusion length substantially exceeds the crystallite diameter, which is considered to be an upper bound for the characteristic diffusion length required for anisotropy formation by atomic pair ordering. Instead of by Fe diffusion, the anisotropy formation appears to be controlled by diffusion of Si. Indeed, the diffusion of Ge in $Fe_{73.5}Si_{13.5}B_9Nb_3Cu_1$, which is considered to resemble Si diffusion, is found to be much slower than Fe diffusion, with an activation energy similar to that of anisotropy formation. [22] From a detailed analysis of Ge diffusion profiles, it can be concluded that the limiting factor for anisotropy formation in n-$Fe_{73.5}Si_{13.5}B_9Nb_3Cu_1$ is Si diffusion in the nanocrystallites, rather than through the intergranular amorphous matrix, indicating that the development of magnetic anisotropy requires diffusion-mediated Fe-Si pair ordering to occur within the nanocrystallites. This is unlike the case of diffusion creep, which appears to be governed by GB diffusion. In fact, in nanocrystalline Fe-Si-B, prepared by crystallization with a composition similar to that of VITROPERM^{TM}, a substantially lower activation energy of 1.5 eV is found for diffusion creep, [62] which is characteristic of Fe diffusion in Fe-rich amorphous alloys.

During the processing of nanocrystalline Nd_2Fe14B permanent magnets, hot deformation is applied in order to induce a magnetic texture. Although the deformation of these materials is promoted by the presence of a liquid grain-boundary phase, which forms as a consequence of an excess content of Nd, the stress exponent n of the strain rate and the activation energy of deformation does not depend on whether the grain-boundary phase is liquid or solid. [47] This contrasts with the aforementioned Fe self-diffusion studies of n-Nd_2Fe14B (Section 3.5, Table 1), which show a pronounced increase of DGB upon intergranular melting. Moreover, the activation "energy of 2.9 eV for deformation [47] is higher than the value of 1.74 eV for grain-boundary diffusivity (Table 1). These discrepancies between the characteristics of deformation and Fe diffusion confirm prior conclusions that, in this case, the deformation process is controlled by a solution-precipitation process occurring at GBs, rather than by GB diffusion, [47] a notion further supported by preliminary Nd self-diffusion studies. [63]

7   Acknowledgement

The authors are indebted to C. Krill for carefully reading the manuscript.

8   References

[1]   R. Birringer, H. Gleiter, H.-P. Klein, P. Marquardt, Phys. Lett. A 1984, 102, 365
[2]   Nanomaterials: Synthesis, Properties, and Applications (Eds: A. S. Edelstein, R. C. Cammarata), Institute of Physics, Bristol 1996
[3]   Nanostructured Materials: Processing, Properties and Applications (Ed: C. C. Koch), Noyes Publications, William Andrew Publishing, Norwich, NY 2002
[4]   R. Würschum, U. Brossmann, H.-E. Schaefer, in Nanostructured Materials: Processing, Properties and Applications (Ed: C. C. Koch), Noyes Publications, William "Andrew Publishing, Norwich, NY 2002, pp. 267–299
[5]   Y. Mishin, C. Herzig, Mater. Sci. Engin. A 1999, 260, 55
[6]   Y. Mishin, Diffus. Deffect Data, Pt. A 2001, 194–199, 1113
[7]   I. Kaur, Y. Mishin, W. Gust, Fundamentals of Grain and Interphase Boundary Diffusion, John Wiley, Chichester 1995
[8]   I. A. Szabó, D. L. Beke, F. J. Kedves, Phil. Mag. A 1990, 62, 227
[9]   I. V. Belova, G. E. Murch, Phil. Mag. A 2001, 81, 2447
[10] S. V. Divinski, F. Hisker, Y.-S. Kang, J.-S. Lee, C. Herzig, Z. Metallkd. 2002, 93, 256
[11] a) A. M. Glaeser, J. W. Evans, Acta Metall. 1986, 34, 1545. b) F. Güthoff, Y. Mishin, C. Herzig, Z. Metallkd. 1993, 84, 584
[12] B. S. Bokstein, H. D. Bröse, L. I. Trusov, T. P. Khvostantseva, Nanostruct. Mater. 1995, 6, 873
[13] R. Würschum, T. Michel, P. Scharwaechter, W. Frank, H.-E. Schaefer, Nanostruct. Mater. 1999, 12, 555._14]a) L. M. Klinger, L. A. Levin, A. L. Peteline, Diffus. Deffect Data, Pt. A 1997, 143–147, 1523. b) A. L. Peteline, S. Peteline, O. Oreshina, Diffus. Deffect Data, Pt. A 1996, 194–199, 1265
[14] J. Horváth, R. Birringer, H.Gleiter, Sol. Stat. Comm. 1987, 62, 319
[15] S. Schumacher, R. Birringer, R. Strauß, H. Gleiter, Acta Metall. 1989, 37, 2485

[16] R. Würschum, K. Reimann, S. Gruß, A. Kübler, P. Scharwaechter, W. Frank, O. Kruse, H. D. Carstanjen, H.-E. Schaefer, Phil. Mag. B 1997, 76, 407
[17] R. Würschum, A. Kübler, S. Gruß, P. Scharwaechter, W. Frank, R. Z. Valiev, R. R. Mulyukov, H.-E. Schaefer, Ann. Chim (Paris, Fr) 1996, 21, 471
[18] H. Tanimoto, P. Farber, R. Würschum, R. Z. Valiev, H.-"E. Schaefer, Nanostruct. Mater. 1999, 12, 681
[19] R. Würschum, K. Reimann, P. Farber, Diffus. Deffect Data, Pt. A 1997, 143–147, 1463
[20] R. Würschum, P. Farber, R. Dittmar, P. Scharwaechter, W. Frank, H.-E. Schaefer, Phys. Rev. Lett. 1997, 79, 4918
[21] S. Herth, R. Würschum, Ph. D. Thesis, University Stuttgart 2003
[22] M. Eggersmann, F. Ye, S. Herth, O. Gutfleisch, R. Würschum, Interface Sci. 2001, 9, 337
[23] S. V. Divinski, F. Hisker, Y.-S. Kang, J.-S. Lee, C. Herzig, Z. Metallkd. 2002, 93, 265
[24] R. Würschum, Habilitation Thesis, Univ. Stuttgart 1997
[25] Y. R. Kolobov, G. P. Grabovetskaya, M. B. Ivanov, A. P. Zhilyaev, R. Z. Valiev, Scripta Mater. 2001, 44, 873
[26] L. G. Kornelyuk, A. Yu Lozovoi, I. M. Razumovskii, Diffus. Deffect Data, Pt. A 1997, 143–147, 1481
[27] R. Z. Valiev, I. M. Razumovskii, V. I. Sergeev, Phys. Stat. Sol. A 1993, 139, 321
[28] V. Yamakov, D. Wolf, S. R. Phillpot, H. Gleiter, Acta "Mater. 2002, 50, 61
[29] R. Würschum, H.-E. Schaefer, in Nanomaterials: Synthesis, Properties, and Applications (Eds: A. S. Edelstein, R. C. Cammarata), Institute of Physics, Bristol 1996, pp. 277–301
[30] W. Frank, Diffus. Deffect Data, Pt. A 1997, 143–147, 695
[31] Z. Erdélyi, Ch. Girardeaux, G. A. Langer, D. L. Beke, A. Rolland, J. Bernardini, J. Appl. Phys. 2001, 89, 3971
[32] W. Dickenscheid, R. Birringer, H. Gleiter, O. Kanert, B. Michel, B. Günther, Solid State Commun. 1991, 79, 683
[33] H. J. Höfler, R. S. Averback, H. Hahn, H. Gleiter, J. Appl. Phys. 1993, 74, 3832
[34] I. L. Balandin, B. S. Bokstein, V. K. Egorov, P. V. Kurkin, Diffus. Deffect Data, Pt. A 1997, 143–147, 1475
[35] H. J. Höfler, R. S. Averback, H. Gleiter, Phil. Mag. Lett. 1993, 68, 99
[36] W. Gust, S. Mayer, A. Bögel, B. Predel, J. de Physique 1985, 46, 537
[37] J. Sommer, C. Herzig, J. Appl. Phys. 1992, 72, 2758
[38] Qing Ma, R. W. Balluffi, Acta Metall. Mater. 1993, 41, 133._40 Y. Minamino, S. Saji, K. Hirao, K. Ogawa, H. Araki, Y. Miyamoto, T. Yamane, Mater. Trans., Jap. Inst. Metals (JIM) 1996, 37, 130
[39] T. Fujita, H. Hasegawa, Z. Horita, T. G. Langdon Diffus. Deffect Data, Pt. A 2001, 194–199, 1205
[40] G. Herzer, Scripta Metall. Mater. 1995, 33, 1741
[41] A. Makino, A. Inoue, T. Masumoto, Mater. Trans. JIM 1995, 36, 924
[42] S. Herth, T. Michel, H. Tanimoto, M. Eggersmann, R. Dittmar, H.-E. Schaefer, W. Frank, R. Würschum, Diffus. Deffect Data, Pt. A 2001, 194–199, 1199
[43] M. Lübbehusen, H. Mehrer, Acta Metall. Mater. 1990, 38, 283
[44] J. Bernardini, P. Gas, E. D. Hondros, M. P. Seah, Proc. Roy. Soc. Lond. A 1982, 379, 159
[45] W. Grünberger, D. Hinz, A. Kirchner, K.-H. Müller, L. Schultz, J. Alloys Comp. 1997, 257, 293
[46] Y. Mishin, C. Herzig, J. Bernadini, W. Gust, Internat. Mater. Rev. 1997, 42, 155

[47] Qing Ma, C. L. Liu, J. B. Adams, R. W. Balluffi, Acta Metall. Mater. 1993, 41, 143
[48] M. R. Sørensen, Y. Mishin, A. F. Voter, Phys. Rev. B 2000, 62, 3658
[49] P. Keblinski, D. Wolf, S. R. Phillpot, H. Gleiter, Phil. Mag. A 1999, 79, 2735
[50] P. Keblinski, D. Wolf, S. R. Phillpot, H. Gleiter, Scripta Mater. 1999, 41, 631
[51] P. Keblinski, D. Wolf, H. Gleiter, Interface Sci. 1998, *6*, 205
[52] N. Wang, Z. Wang, K. T. Aust, U. Erb, Mater. Sci. Engin. A 1997, 237, 150
[53] B. Cai, Q. P. Kong, L. Lu, K. Lu, Scripta Mater. 1999, 41, 755
[54] T. Surholt, C. Herzig, Diffus. Deffect Data, Pt. A 1997, 143–147, 1391
[55] Y. R. Kolobov, G. P. Grabovetskaya, K. V. Ivanov, M. B. Ivanov, Interface Sci. 2002, 10, 31
[56] I. Kaur, W. Gust, Handbook of Grain and Interphase Boundary Diffusion Data, Ziegler, Stuttgart 1989.&feks;
[57] B. Hofmann, H. Kronmüller, J. Magnetism Magn. Mater. 1996, 152, 91
[58] M. Emura, A. M. Severino, A. D. Santos, F. P. Missell, IEEE Trans. Magn. 1994, 30, 4785
[59] A. Lovas, L. F. Kiss, B. Varga, P. Kamasa, I. Balogh, I. Bakonyi, J. de Physique. IV 1998, *8*, 291
[60] a) Xiao Menglai, Q. P. Kong, Scripta Mater. 1997, 36, 299. b) Q. P. Kong, B. Cai, M. L. Xiao, Mater. Sci. Engin. A 1997, 234, 91
[61] W. Sprengel, V. Barbe, S. Herth, O. Gutfleisch, H.-E. Schaefer, R. Würschum, Proc. 2nd Int. Conf. on Nanomaterials by Severe Plastic Deformation (Eds: M. J. Zehetbauer, R. Z. Valier), Wiley-VCH, Weinheim, in press

Self-Diffusion of ^{147}Nd in Nanocrystalline Nd$_2$Fe$_{14}$B

W. Sprengel[1], V. Barbe[1], S. Herth[1,2], T. Wejrzanowski[1,3], O. Gutfleisch[4], P.D. Eversheim[5], R. Würschum[6], and H.-E. Schaefer[1]

[1] Institut für Theoretische und Angewandte Physik, Universität Stuttgart, Stuttgart, Germany
[2] Institut für Nanotechnologie, Forschungszentrum Karlsruhe GmbH, Karlsruhe, Germany
[3] Faculty of Materials Science, Warsaw University of Technology, Warsaw, Poland
[4] Leibniz-Institut für Festkörper- und Werkstoffforschung Dresden, Dresden, Germany
[5] Helmholtz Institut für Strahlen- und Kernphysik, Universität Bonn, Bonn, Germany
[6] Institut für Technische Physik, Technische Universität Graz, Graz, Austria

1 Abstract

For nanocrystalline Nd$_2$Fe$_{14}$B the self-diffusivity of ^{147}Nd was studied by the radio tracer technique. In the case of ^{59}Fe diffusion the temperature dependence of the product of the interface thickness and the interface diffusion coefficient, δD_{GB}, shows a sharp increase for temperatures above 951 K for which the existence of a liquid grain-boundary phase has been reported. Below 951 K the temperature dependence of the ^{59}Fe diffusivity can be described by an Arrhenius type behavior. In the same temperature range the ^{147}Nd diffusivity also follows an Arrhenius law, however, is slightly higher than that of ^{59}Fe.

2 Introduction

The intermetallic compound Nd$_2$Fe$_{14}$B forms the base material for high-performance permanent magnets. By reducing the grain size so that each grain is a particle with a single magnetic domain and by decoupling of neighboring grains with a paramagnetic, Nd-rich intergranular phase, the magnetic energy product can be maximized [1]. In the resulting nanocrystalline structure, interfaces play a significant role for tailoring the magnetic properties of these materials. However, little is known about the interface structure and atomic processes such as diffusion therein. Furthermore, nanocrystalline (n-) Nd-rich Nd$_2$Fe$_{14}$B exhibits the phenomenon of intergranular phase melting at temperatures far below the melting temperature of the bulk material [2]. These aspects are pivotal for the processing and the understanding of the physics of these high-performance permanent magnets. Therefore, studies of grain boundary diffusion in Nd-rich n-Nd$_2$Fe$_{14}$B are highly desirable.

In general, diffusion of atoms in nanocrystalline structures can occur either in the grains (volume diffusion with the diffusivity D_V) or in the interfaces (grain boundary diffusion D_{GB}). Depending on the ratio between the grain size d and the penetration $(D_V t)^{1/2}$ into the volume of the grains after the diffusion time t, three different cases can be distinguished [3]. In the so called type A regime, where the condition $(D_V t)^{1/2} \gg d$ is valid, an effective diffusivity D_{eff} resulting from volume and grain boundary diffusion is measured. In the opposite case, where the diffusion length in the grains can be neglected in comparison to the grain boundary width δ $((D_V t)^{1/2} \ll \delta)$, the grain boundary diffusivity D_{GB} can directly be determined. This is called the type C regime. In both cases, types A and C, the concentration-penetration curve $c(x)$ is described by a Gaussian distribution

$$c(x,t) = c_0 \exp\left(-\frac{x^2}{4Dt}\right) \qquad (1)$$

if diffusion occurs from a thin layer of source material (thin film solution). Here, the logarithm of the specific activity c is proportional to the square of the penetration depth x from which the respective diffusion coefficient can be obtained. The third case, the type B regime, is an intermediate case between type A and C, where $\delta \ll (D_V t)^{1/2} \ll d$ is valid. For this regime only the product in the form of

$$s\delta D_{GB} = 1.322 \sqrt{\frac{D_V}{t}} \left(\frac{\partial \ln c(x,t)}{\partial x^{6/5}}\right)^{-5/3} \qquad (2)$$

can be determined where s denotes the segregation factor and where data for the volume diffusivity D_V are required for a determination the grain boundary diffusivity D_{GB}. In the type B regime the $c(x)$ curves consist of two parts where the part of deep penetration caused by the grain boundary diffusion will give a straight line in a log c vs. $x^{6/5}$ plot according to Eq.(2). The reader is referred to the literature for details of the theory of grain boundary diffusion [4] and diffusion in nanocrystalline materials [5].

The present paper will report on a study of Nd self-diffusion in Nd-rich nanocrystalline $Nd_2Fe_{14}B$ intermetallic compounds by making use of the radiotracer technique. The results are ascribed to grain boundary diffusion processes and will be compared to ^{59}Fe self-diffusion data obtained for Nd-rich n-$Nd_2Fe_{14}B$ [6].

3 Experimental Procedure

Cylindrical Nd-rich n-$Nd_2Fe_{14}B$ rods with Nd excess of 2.4 at% (height 8 mm, diameter 8 mm) were prepared by hot-pressing rapidly quenched commercial powders (Magnequench Inc.) at about 1000 K for 2 min. The microstructure characterized subsequently by x-ray diffraction and transmission electron microscopy revealed $Nd_2Fe_{14}B$ grains with an average size of about 80 to 100 nm embedded in a Nd-rich intergranular phase. For the diffusion experiments discs of 2 to 3 mm height were cut from these rods. The disc surface was mirror polished by standard metallographical procedures. For the diffusion studies the radiotracer method was used employing the radioactive isotope ^{147}Nd. The isotope ^{147}Nd (half-life 11 d) was prepared by neutron activation of ^{146}Nd in the nuclear reactor of GKSS Geesthacht, Germany and then implanted into the specimen below the surface at the implantor facility of Bonn University, Germany. The implantation energy of 80 keV resulted in a mean implantation depth for the $^{147}Nd^+$ ions of 15 nm with a width of 15 nm as determined from an activity profile prior to diffusion annealing measured by ion beam sputtering. The short-time diffusion annealings were performed in a mirror furnace with extremely high heating rates. After diffusion annealing mechanical precision grinding was used for sectioning. The specific activity of each section was determined

4 Results and Discussion

Tracer concentration-profiles $c(x)$ for the ^{147}Nd diffusion have been measured in the temperature range from 773 K to 968 K. The diffusion profiles yield straight lines in a log c vs. $x^{6/5}$ plot for deep penetration (Fig. 1). This behavior clearly indicates a grain boundary diffusion process of type B kinetics with small penetration into the grains in the order of several nm. It should be mentioned that from the initial part of the profile in general direct information about the volume diffusivity D_V can be obtained. However, in the present case the part in the range of several µm below the surface which is characterized by a steep slope cannot be used for the analysis of D_V because of an instrumental broadening. This broadening is due to the coarseness of the abrasive film used in the grinding procedure for diffusion profiles which exceed several hundred µm in length. The straight lines indicated in Fig. 1 have been fitted to the data according to Eq.(2). From the diffusion profile the ratio $s\delta D_{GB} D_V^{-1/2}$ is obtained. The temperature dependence of this ratio determined for the 147Nd diffusion as function of the reciprocal temperature T^{-1} is shown in Fig. 2. In this representation no assumptions about the volume diffusivity and its temperature dependence are required. As for Nd-rich Nd$_2$Fe$_{14}$B an enrichment of Nd in the grain boundary has been observed [7] a temperature independent value of $s = 3$ for the segregation factor s has been assumed.

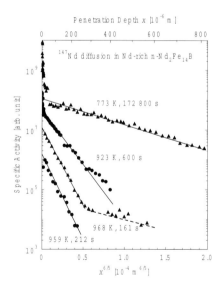

Figure 1: The specific activity of ^{147}Nd as function of the penetration depth x in Nd-rich n-Nd$_2$Fe$_{14}$B after diffusion annealing for different temperatures and annealing times. The diffusion profiles measured under conditions of the type B regime give straight lines in the representation of the logarithm of the specific activity c vs. $x^{6/5}$ (for details see text). The diffusion profile measured for the highest temperature at 968 K shows for deep penetration (dashed line) a deviation. This behavior is ascribed to diffusion in the liquid grain boundary phase.

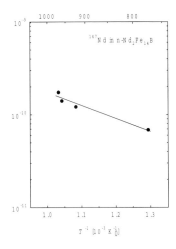 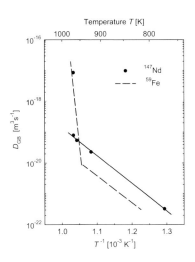

Figure 2: Temperature dependence of the ^{147}Nd diffusion behavior in n-Nd$_2$Fe$_{14}$B represented by the ratio of the product of grain boundary width and grain boundary diffusivity, δD_{GB}, to the square root of the volume diffusivity $D_V^{1/2}$ as a function of the reciprocal temperature T^{-1}. The values for the ratio $\delta D_{GB} D_V^{-1/2}$ were directly obtained from the type B diffusion profiles assuming a temperature independent segregation factor $s = 3$.

Figure 3: Temperature dependence of the product of grain boundary width δ and grain boundary diffusivity D_{GB} for ^{147}Nd (●, —, present work) and ^{59}Fe (---, Ref. 6) diffusion in n-Nd$_2$Fe$_{14}$B. In both cases the volume diffusivities of Fe and Nd in Nd$_2$Fe$_{14}$B have been described by the volume diffusivity D_V of Fe in bcc-Fe [8].

For further analysis of the temperature dependence of the Nd grain boundary diffusion knowledge about the volume diffusivity D_V of Nd in Nd$_2$Fe$_{14}$B is required. Unfortunately, data for the volume diffusivity of Nd in Nd$_2$Fe$_{14}$B or related intermetallic compounds are lacking [9]. For a preliminary analysis the temperature dependence of the Fe self-diffusivity in bcc-Fe as reported in Ref. 6 has been used as a first approximation for the Nd volume diffusivity in Nd$_2$Fe$_{14}$B. The resulting temperature dependence of the product of the grain boundary width and the grain boundary diffusivity, δD_{GB}, is shown in Fig. 3. The values below the transition temperature can be described with an Arrhenius law in the form of

$$\delta D_{GB}^{Nd}(T) = 1.35 \times 10^{-10} \exp\left(\frac{-1.78 \text{ eV}}{k_B T}\right) \text{m}^3 \text{s}^{-1} \quad . \tag{3}$$

The diffusion profile of the specimen measured at 968 K which is above the transition temperature of 951 K of the intergranular phase showed type B regime for the grain boundary diffusivity. Additionally, due to the initially very high specific activity of the ^{147}Nd tracer, fast diffusion in the liquid intergranular phase is also observed. The corresponding specific activities, marked by a dashed line in Fig. 1, were still significantly above the background. For the liquid Nd-rich intergranular phase a value for $\delta D^{Nd} = 8.6 \cdot 10^{-18}$ m^{3}s^{-1} is determined for 968 K if out-diffusion into the solid grain boundary is neglected. Assuming a width of $\delta = 10$ nm for the interfacial liquid film [7] a diffusivity of $D^{Nd} = 8.6 \cdot 10^{-10}$ m^2 s^{-1} is determined. This value is close to values determined for diffusion in metallic melts which are in the range of 1 to $6 \cdot 10^{-9}$ m^{2}s^{-1} [10] supporting the observation of intergranular phase melting.

^{59}Fe diffusion studies in Nd-rich n-Nd$_2$Fe$_{14}$B in the temperature range from 800 K to 960 K, i.e., below and above the melting transition of the Nd-enriched intergranular phase at 951 K [2] have been reported earlier [6] and are shown in Fig. 3 as dashed lines. The experiments below the transition temperature were performed in the regime of type B kinetics. For the analysis of the temperature dependence of the grain boundary diffusion below the transition temperature the volume self-diffusion data of bcc-Fe [8] were used for D_V because volume diffusion data of Fe in Nd$_2$Fe$_{14}$B are lacking [9]. The temperature dependence of δD_{GB} can be described by an Arrhenius relation [6]. From this relation the activation enthalpy for ^{59}Fe grain boundary diffusion and the pre-exponential factor can be deduced giving the Arrhenius relationship

$$\delta D_{GB}^{Fe}(T) = 1.53 \times 10^{-11} \exp\left(\frac{-1.74 \text{ eV}}{k_B T}\right) \text{m}^3\text{s}^{-1} . \tag{4}$$

Above the intergranular melting temperature at 951 K the temperature dependence of δD_{GB} shows a substantial increase which is attributed to the occurrence of a liquid intergranular phase giving rise to enhanced atomic transport [5]. This increase is not sharp but occurs over a temperature range of 20 to 30 K. This may indicate an inhomogeneous intergranular melting process.

In the investigated temperature range below the intergranular melting transition temperature the activation energy for the ^{147}Nd grain boundary diffusion of $Q = 1.78$ eV is similar to the activation energy of $Q = 1.74$ eV determined for ^{59}Fe grain boundary diffusion in n-Nd$_2$Fe$_{14}$B. However, the absolute values of the ^{147}Nd grain boundary diffusivities are about one order of magnitude higher than the Fe diffusivities [6, 7]. This is puzzling on first glance as for an atom like Nd which is much larger than an Fe atom, one should expect the opposite behavior. However, this result is a preliminary one. Due to the lack of volume diffusivity data in both cases Fe diffusion in bcc-Fe has been used to approximate as well the Fe as the Nd volume self-diffusion in Nd$_2$Fe$_{14}$B. In an analogy to self-diffusion in other intermetallic compounds [11] the volume diffusivity of the minority Nd in Nd$_2$Fe$_{14}$B is expected to be slower than that of Fe as Nd lacks the possibility to diffuse via its own sublattice. In the case that the actual Nd volume diffusion in Nd$_2$Fe$_{14}$B is much slower than the Fe diffusion, the resulting grain boundary diffusivity of Nd would be lower and could even be lower than the Fe diffusivity.

The occurrence of the intergranular phase melting seems to be essential for the processing of Nd-rich Nd$_2$Fe$_{14}$B permanent magnets by hot deformation to induce magnetic anisotropy. The liquid phase is necessary to prevent crack propagation. The activation energy for the deformation process was determined to be 2.9 eV [2]. From this value it is obvious that for deformation diffusion is not the rate controlling process as the activation energies for Fe as well as for Nd diffusion of about 1.7 eV are significantly lower.

Diffusion in nanocrystalline Nd-rich Nd$_2$Fe$_{14}$B permanent magnets is a complex process involving volume diffusion, grain boundary diffusion and diffusion in the intergranular phase. In the present work the Nd interfacial self-diffusivity in nanocrystalline Nd$_2$Fe$_{14}$B was studied by making use of an approximation of the Nd volume self-diffusivity. It turns out that the value for interfacial self-diffusivity of Nd, in nanocrystalline Nd$_2$Fe$_{14}$B is higher than the value for Fe, although the Nd atom is larger than the Fe atom. For a complete understanding of the interface structure and the interface diffusion in nanocrystalline Nd$_2$Fe$_{14}$B the determination of the Nd volume diffusivity in coarse grained $Nd_2Fe_{14}B$ is highly desirable. Specific studies of the atomic structure of interfaces in n-$Nd_2Fe_{14}B$, e.g., by investigating the sizes of interfacial free volumes by positron lifetime spectroscopy [12,13], or their chemical surroundings by coincident measurements of the positron electron annihilation photons [14] may yield further helpful information.

5 Acknowledgements

Technical support by GKSS Geesthacht for neutron activation of ^{146}Nd is appreciated. The work was financially supported by Deutsche Forschungsgemeinschaft (WU 191/4-1) and the European Community (Contract No. HPMT-CT-2001-00224).

6 References

[1] D. Goll, H. Kronmüller, Naturwissenschaften 2000, 87, 423–438
[2] W. Grünberger, D. Hinz, A. Kircher, K.-H. Müller, L. Schultz, J. Alloys Compounds 1997, 257, 293–301
[3] L.G. Harrison, Trans. Faraday Soc. 1961, 57, 1191–1199
[4] I. Kaur, Y. Mishin, W. Gust, Fundamentals of Grain and Interface Boundary Diffusion, Wiley & Sons, Chichester, New York, 1995
[5] R. Würschum, S. Herth, U. Brossmann, this volume
[6] M. Eggersmann, F. Ye, S. Herth, O. Gutfleisch, R. Würschum, Interface Science 2001, 9, 337–341
[7] S. Herth, Ph.D. thesis, Stuttgart University, Germany 2003
[8] M. Lübbehusen, H. Mehrer, Acta Metall. Mater. 1990, 38, 283–292
[9] H. Mehrer (Ed.), Diffusion in Solid Metals and Alloys, Landoldt Börnstein III/ 26, Springer, Berlin, Germany 1990
[10] W. Gust, S. Mayer, A. Bögel, B. Predel, J. Phys. Colloq. 1995, 46, 537–544
[11] W. Sprengel, M.A. Müller, H.-E. Schaefer, Intermetallic Compounds, Vol. 3, Principles and Practice, (Eds.: J.H. Westbrook, R.L. Fleischer), J. Wiley and Sons, Chichester, UK, 2002, pp. 275–293
[12] R. Würschum, E. Shapiro, R. Dittmar, H.-E. Schaefer, Phys. Rev. B, 2000, 62, 12021–12027
[13] L. Pasquini, A.A. Rempel, R. Würschum, K. Reimann, M.A. Müller, B. Fultz, H.-E. Schaefer, Phys. Rev. B 2001, 63, 134114-1 – 134114-7
[14] A.A. Rempel, W. Sprengel, K. Blaurock, K. Reichle, J. Major, H.-E. Schaefer, Phys. Rev. Lett. 2002, 89, 185501-1 – 185501–4

Theoretical Investigation of Nonequilibrium Grain Boundary Diffusion Properties

V. N. Perevezentsev

Blagonravov Nizhni Novgorod Branch of Mechanical Engineering Research Institute, Russian Academy of Sciences, Nizhny Novgorod, Russia

1 Introduction

Despite of a number of works devoted to the theory of grain boundary (GB) diffusion, the mechanisms of diffusion in general GBs (i.e., in boundaries with a disordered atomic structure [1]) are open to discussion [2,3]. Theoretical models based on the concept of the mechanism of diffusion as a motion of localized GB vacancies contradict the results of computer simulation of GB atomic structure. According to these calculations, a vacancy in a general GB is unstable and is delocalised as a result of atoms relaxational displacements [4–5]. In recent years, interest in this problem increased because of anomalously high coefficients of diffusion along GBs were revealed in submicrocrystalline materials obtained by severe plastic deformation [6]. It has been shown [7–9] that coefficients of self- and heterodiffusion in such materials may be one order of magnitude (or even more) higher than in the coarse-grain state. These anomalies of NMC diffusion are assumed to be associated with the nonequilibrium structure of GBs [4].

In this paper, we consider a new mechanism of selfdiffusion in GBs with a disordered atomic structure. This mechanism is based on the concept of metastable vacancies arising as a result of thermal fluctuations of density and disappearing upon subsequent relaxational rearrangement of the atomic structure of the GB.

2 Description of the Model

Let us consider general GB as a thin amorphous layer with thickness $d_0 \sim 2a$ (a is the interatomic spacing) located between two mutually misoriented crystallites. Let F_b^o and v_b^o be the free energy and the volume of GB atoms corresponding to a minimum of the boundary energy (we assume that, in the first approximation, the energy and the volume of the boundary are uniformly distributed over constituting atoms). Then, at a constant number of atoms, the free energy of a GB atom F_b as a function of its volume v_b can be represented in the form of an expansion in powers of $(v_b - v_b^o)$. To an accuracy of second order terms and taking into account that the linear term of the expansion is absent, we can write

$$F_b = F_b^o + K'(v_b - v_b^o)^2 / 2, \qquad (1)$$

where the coefficient $K' = \partial^2 F_b / \partial v_b^2$ is related to the compressive coefficient of the material $1/K_b = -(1/v_b) \cdot (\partial v b / \partial P)$ (where $P = \partial Fb / \partial vb$ is the pressure): $K' = K_b / v_b$. The second term in equation (1) describes the increase in the elastic energy of GB atoms upon the deviation of v_b from v_b^o. Let us analyse changes in free energy ΔF of the system for the case where lattice vacancies enter GB. Let in the initial state there be n lattice vacancies located in layers of single atomic thickness on both sides of the boundary. The free energy of initial state can be written (per unit area of the boundary) as

$$F_1 = N_b F_b^o + n F_{vf}^c + kT[n\ln(n) + (N_c - n)\ln(N_c - n) - N_c \ln N_c], \qquad (2)$$

where k is the Boltzmann constant; N_b and N_c are the number of atoms in the GB and the number of sites in atomic layers of the grains, respectively; and F_{vf}^c is the free energy of vacancy formation in the crystal lattice. The last term in equation (2) describes the configurational entropy of lattice vacancies.

Upon a jump of a single vacancy from the crystal lattice to the GB, one atom of the boundary vanishes and one GB vacancy is created. In the absence of a relaxational rearrangement of the atomic structure, the energy consumption on the formation of a vacancy in the GB and in the monocrystal differ only by the free energy of the GB atom F_b^o and, consequently, the energy of an unrelaxed GB vacancy is equal to $(F_{vf}^o - F_b^o)$. As was noted above, the vacancy in a disordered atomic structure of general GB is unstable. Relaxational displacements of atoms lead to its delocalization. The transition of n vacancies from the bulk of the grain into the boundary and their delocalization lead to an increase of excess GB atomic volume $\Delta v_b = v_b - v_b^o = n\, v_b^o / N_b$ and, in view of equation (1), to an increase in energy of GB atoms. Change in elastic energy ΔE_e^b per unit area of the boundary is

$$\Delta E_e^b = (N_b - n) \frac{K_b}{2 v_b^o} \left(\frac{n v_b}{N_b} \right)^2 .$$

Thus, the free energy F_2 of the system after the passage of n lattice vacancies into the GB and their delocalization can be written as

$$F_2 = (N_b - n)\left[F_b^o + \frac{1}{2} K_b \left(\frac{n}{N_b} \right)^2 v_b \right]. \qquad (3)$$

When writing equation (3), we neglected the change in the vibration entropy of GB atoms. Corresponding change in free energy $\Delta F = F_2 - F_1$ of the system is (at $N_b \gg n$):

$$\Delta F \approx \frac{1}{2} K_b v_b \frac{n^2}{N_b} - n\left(F_{vf}^c + F_b^o \right) - kT\left[n\ln(n) + (N_c - n)\ln(N_c - n) - N_c \ln N_c \right]. \qquad (4)$$

Using equation $(\partial \Delta F / \partial n) = 0$, we can find the equilibrium number of vacancies delocalized in the GB or, with allowance for the relation $\Delta v_b = n v_b / N_b$, the equilibrium excess atomic volume Δv_b^* corresponding to the minimum of the free energy of the system under consideration. On the assumption $\delta_o \cong 2a$ and, consequently, $N_c / N_b \approx 1$, we obtain

$$\left(\frac{\Delta v_b^*}{v_b^o} \right) \approx \frac{F_{vf}^c + F_b^o + kT \ln\left(\Delta v_b^* / v_b^o \right)}{K_b v_b^o} . \qquad (5)$$

At $\Delta v_b < \Delta v_b^*$, an energetically favourable process is the absorption of a lattice vacancies by the GB, while at $\Delta v_b > \Delta v_b^*$, emission of vacancies from the boundary into the grain bulk (the mechanism of the formation of vacancies is considered below). The GB in equilibrium with lattice vacancies, i.e., the boundary at $\Delta v_b = \Delta v_b^*$, will below be called equilibrium boundary.

Now, we consider the mechanism of GB self-diffusion. Let each GB atom have an excess volume Δv_b. As a result of thermal fluctuations, a local increase in the density of the material of the GB can occur. If such fluctuation involves a group of m atoms, where $m = \Omega v / \Delta v_b$, and leads to a decrease in the atomic volume v_b to v_b^o, then, an unrelaxed vacancy of volume Ωv (we assume below that $\Omega_v = v_b^o$) will appear. In this case, a release of the elastic energy of GB atoms involved in this fluctuation occurs and the free energy increases by the energy of the unrelaxed vacancy $\left(F_{vf}^c - F_b^o\right)$. The entropy of this group of atoms also changes by the magnitude of the configurational entropy of the vacancy in the group of m atoms, i.e., by $\Delta S = k \ln(m)$. Therefore, the free energy of formation of a GB vacancy F_{vf}^b, according to the mechanism considered, can be represented in the form

$$F_{vf}^b = \left(F_{vf}^c - F_b^o\right) - m\frac{1}{2}K_b\frac{(\Delta v_b)^2}{v_b} - kT\ln(m). \tag{6}$$

The corresponding expression for the free activation energy of GB diffusion $F_d^b = F_{vf}^b + F_{vm}^b$, where F_{vm}^b is the activation energy for a jump of one of the nearest GB atoms into the newly formed vacancy, with allowance for the relation $m = v_b^o / \Delta v_b$ has the form

$$F_d^b = \left(F_{vf}^c - F_b^o + F_{vm}^b\right) - \frac{1}{2}K_b\Delta v_b fkT\ln\left(\Delta v_b / v_b^o\right). \tag{7}$$

For an equilibrium GB $\Delta v_b = \Delta v_b^*$ and, consequently, with allowance for Eqs. (5) and (7)

$$F_d^b = \frac{1}{2}\left[F_{vf}^c - 3F_b^o + kT\ln\left(\Delta v_b^* / v_b^o\right)\right] + F_{vm}^b. \tag{8}$$

3 Coefficients of Self-Diffusion in Equilibrium and Nonequilibrium GBs

The expression for the coefficient of GB diffusion has the form

$$D_b = D_{bo}\exp\left(-\frac{H_{vf}^c - 3H_b^o + 2H_{mv}^b}{2kT}\right), \tag{9}$$

where pre-factor

$$D_{bo} = \frac{1}{Z_b}a_b^2\omega_b\sqrt{\frac{v_b^o}{\Delta v_b^*}}\cdot\exp\left(\frac{S_{vf}^c - 3S_b^o + 2S_{mv}^b}{2k}\right). \tag{10}$$

Here, Z_b is the coordination number; ω_b is the Debye frequency of atomic vibrations in the boundary; H_{vf}^c, H_b^o and H_{mv}^b are the enthalpies and S_{vf}^c, S_b^o and S_{mv}^b are the entropies in the corresponding expressions for the free energy $F = H - TS$.

Let us estimate the activation energy of GB diffusion $Q_b = \left(H_{vf}^c - 3H_b^o + 2H_{mv}^b\right)/2$ using the following characteristic values of the parameters: $H_{vf}^c \approx 10kT_m$ (T_m is the melting temperature)

[10]; $H_b^o = \gamma_b^o \upsilon_b^o / \delta_o \cong kT_m$ (γ_b^o is the specific surface enthalpy of GB, and $\gamma_b^o a_b^2 \cong 2kT_m$); and $H_{vm}^c \approx 8kT_m$ [12]. Since, there is no available data for H_{mv}^b in the literature, we will use for estimation the assumption that $H_{vm}^b \approx H_{vm}^c$. Substituting these values into equation (10), we obtain $Q_b \approx 11,5\ kT$.

Now, we estimate the magnitude of the relative excess volume of an equilibrium GB ($\Delta \upsilon_b^* / \upsilon_b^o$ equation (5)) and the pre-factor D_{b0} (10). We assume that, on the order of magnitude, $S_{vf}^c \cong 2k$ [12], $S_b^o \cong 1k$ [11] and $S_{mv}^b \cong 1k$. We also assume that $Z_b = 6$, $\omega_b \sim 10^{13}$ s^{-1} and $T = T_m / 2$. With these values of the parameters, we obtain $\Delta \upsilon_b^* / \upsilon_b^o \cong 4 \cdot 10^{-2}$ and $D_{bo} \sim 6 \cdot 10^{-2}$ sm^2/s. Values of Q_b and D_{bo} on the order of magnitude correspond to experimental values [2,10].

The activation energy of diffusion in nonequilibrium GB ($\Delta \upsilon_b > \Delta \upsilon_b^*$) can be obtained by substituting $\Delta \upsilon_b = \Delta \upsilon_b^* + \Delta \tilde{\upsilon}_b$, where $\Delta \tilde{\upsilon}_b$ is the nonequilibrium excess atomic volume into equation (6). The corresponding expression for the diffusion coefficient $\tilde{D}_b$ in a nonequilibrium GB can be written in the form

$$\tilde{D}_b = D_b \left(\frac{\Delta \upsilon_b^*}{\Delta \upsilon_b^* + \Delta \tilde{\upsilon}_b} \right)^{1/2} \exp \left(\frac{K_b \Delta \tilde{\upsilon}_b}{2kT} \right), \qquad (11)$$

where D_b is the coefficient of diffusion in the equilibrium GB.

Thus, coefficient of diffusion in a nonequilibrium GB ($\Delta \tilde{\upsilon}_b > 0$) can increase as compared to its equilibrium state. In particular, at $T = T_m / 2$, the increase in the diffusion coefficient $\tilde{D}_b$ by an order of magnitude is achieved at $\Delta \tilde{\upsilon}_b \cong 2 \cdot 10^{-2} \upsilon_b^o$.

Nonequilibrium excess GB volume may appear as a result of generation of noneqilibrium vacancies and their delocalization due to GB defect structure recovery processes (climb of dislocations appeared in GB during plastic deformation, annihilation of dislocations dipoles or sessile dislocation loo`ps, partial annihilation of sessile components of dislocations, which get into a reaction with formation of orientational misfit dislocations and so on).

4 Recovery of Diffusion Properties of Nonequilibrium GB during Annealing

Let us suggest that initially GB has nonequilibrium relative excess volume $\Delta \tilde{\upsilon}_b^0 / \upsilon_b^0$ and analyse kinetics of relaxation of nonequilibrium volume and coefficient of GB diffusion. Decreasing of $\Delta \tilde{\upsilon}_b$ at annealing in case of infinite boundary may be connected with termofluctuational creating of localized GB vacancies and their jumps to volume of grains. Frequency of such events (per unit of GB square) is equal to

$$\dot{N}^- = \frac{\omega_b}{a^2} \exp \left(-\frac{F_{vf}^b + F_b^o + \Delta F_m}{kT} \right) \qquad (12)$$

where ΔF_m is free activation energy of atom jump from volume of grain into localized GB vacancy; ω_b – frequency of oscillations of GB atoms. Using (5) and (7) one can write (12) as

$$\dot{N}^- = \frac{\omega_b}{a^2} A_o \exp \left(\frac{K_b \Delta \tilde{\upsilon}_b}{2kT} \right),$$

where

$$A_o = \frac{\omega_b}{a^2} \exp\left(-\frac{F_{vf}^b + \Delta F_m - 0.5K_b \Delta \tilde{v}_b^*}{kT}\right) \tag{13}$$

Frequency of reverse jumps of vacancies into GB is

$$\dot{N}^+ = \omega a C_1 \exp\left(-\frac{\Delta F_m}{kT}\right) \tag{14}$$

where ω is a frequency of oscillations of atoms in crystalline lattice; $C_1 a$ – number of vacancies in near boundary layer with thickness a. Hence, the kinetic equation, describing evolution of nonequilibrium volume $\Delta \tilde{v}_b^0 / v_b^0$ during isothermal annealing, can be written as

$$\frac{\delta}{\Omega} \frac{\Delta \dot{\tilde{v}}_b}{v_b} = \dot{N}^+ - \dot{N}^- = -A_o \exp\left(-\frac{K_b \Delta \tilde{v}_b}{2kT}\right) + BC_1 \tag{15}$$

From the condition $\dot{N}^+ = \dot{N}^-$, we find the equilibrium concentration of vacancies near GB:

$$C_1^* = \frac{A_0}{B} = C_0 \exp\left(\frac{K_b \Delta v_b^*}{2kT}\right), \tag{16}$$

where $C_1^* = \dfrac{A_0}{B} = C_0 \exp\left(\dfrac{K_b \Delta v_b^*}{2kT}\right)$

is the equilibrium concentration of vacancies in the monocrystal.

To find function $C_1(t) = C_{(x=0,t)}$ in (15) it is necessary to calculate concentration profile of lattice vacancies $C(x,t)$. Hence it is necessary to solve diffusion equation at following boundary conditions for vacancies flow:

$$-D_v \left.\frac{\partial C_{(x,t)}}{\partial x}\right|_{(x=0)} = A_o \exp\left(\frac{K_b \Delta \tilde{v}_b}{2kT}\right) - BC_{(x=0;t)} \tag{17}$$

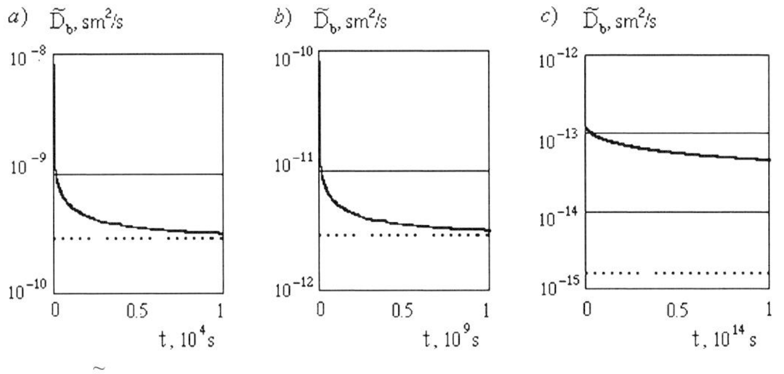

Figure 1: $\tilde{D}_b(t)$ at different temperatures of annealing: $T = 0.5T_m$ (a); $T = 0.4T_m$ (b); $T = 0.3T_m$ (c); diffusion coefficient of equilibrium GB – dotted line

(D_v is a coefficient of diffusion in a volume of grain).

Solution of such system of equations allow us to obtain dependencies $C_{(x,t)}$, $\Delta \tilde{v}_b(t)$ and $\tilde{D}_b(t)$ at different temperatures of annealing. Results of $\tilde{D}_b(t)$ calculations at initial value of $\Delta \tilde{v}_b^0 / v_b^0 = 2 \cdot 10^{-2}$ and annealing temperatures $T/T_m = 0.4; 4; 0.5$ for above-mentioned values of parameters are shown on figure 1. Characteristic relaxation time τ (time, when D_b decrease in e times) on temperature is shown in figure 2. It is evident that time of life for nonequilibrium state is essentially defined by annealing temperature. Thus, $\tau \sim 15$ s at $T/T_m = 0.5$; $\tau = 10^5$ s at $T/T_m = 0.4$. Nonequilibrium state of GBs can be retained for a long time at relatively low temperatures ($T \leq 0,4 Tm$).

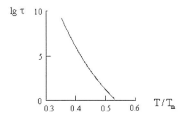

Figure 2: The dependency of characteristic relaxation time τ on homological temperature of annealing

5 Possible Applications to Nanocrystalline Materials.

As it is known, GBs in nanocrystalline materials, produced by severe plastic deformation, contain extremely high density of dislocations ($\rho \sim 10^6$ sm^{-1} [6]). Their climb and annihilation during heating leads to a formation of nonequilibrium excess boundary volume, relative value of which $\Delta \tilde{v}_b / v_b^0 = \alpha \rho b \approx 2 \cdot 10^{-2}$ ($\alpha \approx 0,6$ – geometrical coefficient) is close to experimental estimation of $\Delta \tilde{v} / v$ from dilatometric investigations [6]. The state of GBs at relatively low temperatures may be nonequilibrium for a sufficiently long time and influence on different processes in nanocrystalline materials, which are controlled by GB diffusion (anomalous grains growth, precipitation on GBs and so on). During plastic deformation nonequilibrium state of grain boundary may be maintained due to above-mentioned mechanism (climbing and partial annihilation of lattice dislocations coming to GB) and in case of intragranular deformation absences (very small size of grains) – due to generation of nonequilibrium vacancies during GB sliding along curved grain boundaries.

The diffusion along nonequilibrium GBs may play considerable role in deformation behavior of nano and microcrystalline materials. In particular, decreasing of activation energy of GB diffusion (on $0.5 k_b \Delta \tilde{v}_b$ value accordingly to the theory) may lead to activization of GB sliding at essentially low temperatures ($T \geq 0,25 \, T_m$). It is possible to suppose that low-temperature superplasticity in submicrocrystalline alloys is connected with such effect. Besides, transition of GBs into nonequilibrium state during deformation may be considered as factor promoting achievement of high strain rate superplasticity.

GB nonequilibrium state influence on deformation behaviour of nanocrystalline materials (on the dependency of plastic flow stress on strain rate, temperature and grain size) will be considered in subsequent works.

6 Summary

- The change of the free energy is analyzed for the case of transition of lattice vacancies into GB and their subsequent delocalization in the disordered atomic structure of the boundary. An expression is obtained for the relative excess volume of a relaxed GB, which is in equilibrium with lattice vacancies.
- New mechanism of GB diffusion is suggested. As an elementary act of diffusion, the formation of a metastable GB vacancy as a result of a local fluctuation of the density and a jump of one of the nearest GB atoms into this vacancy are considered.
- Expressions are obtained for the activation energy and the coefficient of self-diffusion in equilibrium and nonequilibrium GBs. A fundamental relationship between the activation energy for GB diffusion and the energy of formation of lattice vacancies is established.
- Kinetics of recovery of nonequilibrium GB diffusional properties is analysed.

7 Acknowledgments

This work was supported in part by the Russian Foundation for Basic Research (project No. 02-03-33043) and U.S Civilian Research and Development Foundation (grant no. RE2-2230).

8 References

[1] A. N.Orlov, V.N.Perevezentsev, V.V.Rybin, Grain Boundaries in Metals, Moscow: Metallurgy, 1980 (in Russian), p. 156
[2] I.Kaur, W.Gust, Fundamentals of Grain and Interface Boundary Diffusion, Ziegler, Stuttgart, 1989
[3] V. Naundorf, M.P.Macht, A.S.Bakai, et al., J. Noncryst. Solids, 1999, 250–252, 679–683
[4] H.Gleiter, Mater. Sci. Eng., 1982, 52, 91–131
[5] R.W.Ballufi, T.Kwak, P.D.Bristowe, et al., Scr. Metall., 1981, 15, 951–956
[6] R.Z.Valiev and I.V.Aleksandrov, Nanostructured Materials Produced by Severe Plastic Deformation, Moscow: Logos, 2000 (in Russian), p. 272
[7] R.Z.Valiev, T.M.Razumovskii, V.I.Sergeev, Phys. Stat. Sol. (a), 1993, 139, 321–328
[8] R.Wurshum, A.Kubler, S.Gruss at al., Ann. Chim. Fr., 1996, 21, 471–476
[9] Yu.R.Kolobov, R.Z.Valiev, G.P.Grabovetskaya et al., Grain boundary diffusion and properties of nanostructured materials, Novosibirsk, Nauka, 2001, (in Russian), p. 232
[10] B.S.Bokshtein, S.Z.Bokshtein, A.A.Zhukhovitskii, Thermodynamics and Kinetics of Diffusion in Solids, Moscow: Metallurgiya, 1974 (in Russian)
[11] J.W.Prowan, O.A.Bamiro, Acta Metall., 1977, 25, N 3, 309–319
[12] Physical Metallurgy, R.W.Cahn and P.Haasen, Eds., New York, North Holland Physics, 1983

On Annealing Mechanisms Operating in Ultra Fine-Grained Alloys

T. Sakai[1], H. Miura[1], A. Belyakov[2] and K. Tsuzaki[2]

[1] Department of Mechanical Engineering and Intelligent Systems, University of Electro-Communications, Chofu, Tokyo, Japan
[2] Steel Research Center, National Institute for Materials Science, Tsukuba, Ibaraki 305-0047, Japan

1   Abstract

Annealing processes of an ultra fine-grained (UFG) 304 stainless steel with an average grain size of about 0.3 µm were studied at 973 K. The strain-induced UFG structure is essentially stable against recrystallization. The sequential annealing processes are recovery and transient recrystallization, followed by normal grain growth. The annealing behavior is discussed with reference to classical recrystallization.

2   Introduction

Microstructures have great effects on the mechanical properties of metallic materials. Ultra fine-grained metals and alloys are believed to have some advantages for the improvement of mechanical behavior. Recently, submicrocrystalline structures have been developed in various metallic materials by large strain deformation at relatively low temperatures [1–3]. The authors utilized a warm multiple multi-axial deformation to obtain the ultra fine-grained microstructures in some metallic materials [4–6]. Multi-axial deformation promoted rapid formation of many intersecting subboundaries. A gradual rise in misorientations across the strain-induced subboundaries with increasing strain finally led to the evolution of ultra fine-grained microstructures. Such materials processed by severe large deformation are characterized by high internal stresses, which are associated with a high dislocation density evolved during cold-to-warm deformation as well as with a non-equilibrium state of the strain-induced grain boundaries [3, 5, 7]. Upon heating, these high internal distortions may lead to rapid development of some discontinuous grain coarsening like a primary recrystallization. On the other hand, materials with a high density of high-angle (sub)boundaries have been suggested to be essentially resistant to discontinuous recrystallization [2, 8]. A detailed understanding of the restoration mechanism operating in the strain-induced submicrocrystalline structures during annealing, however, is complicated by a scarcity of experimental data [2, 3].

The aim of the present work is to study the annealing behavior of a 304 stainless steel with an ultra fine-grained structure, which was developed by large strain multiple deformation. The restoration mechanisms operating during annealing of such fine-grained matrices are discussed with reference to the nucleation mechanism that takes place during classical primary recrystallization.

3 Experimental Procedure

A 304 type austenitic stainless steel (0.058%C, 0.7%Si, 0.95%Mn, 0.029%P, 0.008%S, 8.35%Ni, 18.09%Cr, 0.15%Cu, 0.13%Mo and the balance Fe, mass%) was used as the starting material. The ultra fine-grained structure with an average grain size of about 0.3 µm was developed by multi-pass compression at 873 K with sequential (from pass to pass) changing the compression axis. The details of the processing are described elsewhere [5]. These deformed specimens with final rectangular dimension of about 5.0 : 4.2 : 3.5 mm were annealed in ambient atmosphere at temperature of 973 K.

The TEM observations were carried out using a JEOL JEM-2000FX operating at 200 kV. The average grain size was measured by the linear intercept method, and the twins were omitted from measurements. The restoration kinetics was studied by means of the fractional softening (X) calculated by the equation: $X = (Hv_\varepsilon - Hv_t)/(Hv_\varepsilon - Hv_0)$ [9]. Here $Hv_\varepsilon = 3800$ MPa is the hardness for as-processed ultra fine-grained sample, Hv_t and Hv_0 are those for partially and fully annealed samples, respectively. Hv_0 was 1850 MPa for a fully annealed sample with an average grain size of 25 µm.

4 Results and Discussion

Fig. 1 shows typical ultra fine-grained microstructures evolved in the 304 stainless steel after a large strain deformation (Fig. 1a) followed by annealing at 973 K (Figs. 1b and 1c). Referring to the as-processed state, it is clearly seen that the annealing for a relatively short time of 0.45 ks does not lead to any significant microstructural changes. However, it should be noted that the microstructure subjected by early annealing demonstrates some increasing level of heterogeneity that developed on a substructural scale as compared to the as-deformed state. Namely, a certain amount of fine grains accommodates a high density of interior dislocations, whereas other grains are almost free of dislocations. Such dislocation free grains may be considered as potential recrystallization nuclei, which are able to grow during annealing for rather long time.

Upon further annealing, the average grain size increases and grain boundaries become sharp and clear observable. After long-time annealing for 28.8 ks (Fig. 1c), dislocation substructures, which were inherent in the as-processed state, disappear completely, and relatively coarse recrystallized grains homogeneously develop including several twins. It should be noted that the final microstructure evolved after long-time annealing can still be considered as a fine-grained one. The average grain size in Fig. 1c is about 3 µm. This suggests that any recrystallization mechanisms based on a drastic grain growth did not operate during annealing of the present steel with a strain-induced fine-grained structure.

Let us consider the structural mechanisms operating in the fine-grained steel upon annealing in more details. Fig. 2 presents the Avrami plot of structural softening (X) and a relationship between the average grain size (D) and the annealing time (t) at 973 K. Contrary to the classical primary recrystallization, the softening kinetics does not vary remarkably with the annealing time. At early annealing the slope of the Avrami plot is about 0.3. Then, at intermediate annealing time, there is a stepwise increase in softening with the Avrami exponent of about 1 for a quite short period. Upon further annealing the softening kinetics slows down and then results in minor softening with an exponent of about 0.2.

Figure 1: Typical microstructures developed in submicrocrystalline 304 stainless steel processed by (a) large strain deformation to $\varepsilon = 6.4$ at 873K, followed by annealing at 973 K (b) for 0.45 ks and (c) for 28.8 ks

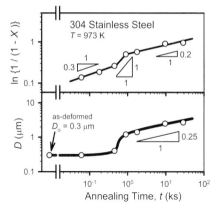

Figure 2: Avrami plot of fractional softening (X) and the changes in grain size (D) during annealing of ultra fine-grained 304 stainless steel at 973 K

The changes in the average grain size with annealing correspond roughly to the softening curve. There is almost no grain coarsening at the beginning of the treatment. Such behavior suggests that the recovery can be a structural restoration mechanism operating at early annealing. Then, the average grain size steeply increases. This grain size change with sharp softening is indicative of some recrystallization process. The latest annealing stage is characterized by a relatively slow grain coarsening, which is similar to the normal grain growth process. The average grain size follows a power law function of annealing time with a grain growth exponent of about 0.25. The latter is in good agreement with that of 0.1 ~ 0.5, which was reported in many studies on the normal grain growth of single-phase alloys [10].

The results described above suggest that the annealing behavior of the strain-induced ultra fine-grained microstructure can be characterized by the presence of three sequential time intervals. The fastest operating process is a conventional recovery, which does not require any incubation period and develops just after heating. Following the recovery, a kind of recrystallization process results in a stepwise grain coarsening accompanied by a limited softening. Finally, a conventional normal grain growth takes place during long-time annealing. The interesting point is the transient recrystallization, which is characterized by a rather slow kinetics and by a limited softening taking place in a short period of time. The Avrami exponent of 1 in Fig. 2 seems to be small for the primary recrystallization, e.g. the exponent for classical primary recrystallization is 4, or it is 3 for the site saturation considerations [10]. Moreover, the softening provided by the transient recrystallization is not clear detectable. Strictly speaking, the fractional softening data in Fig. 2 might be roughly approximated by a single straight line. This suggests that the studied transient recrystallization is based on a different structural mechanism from the primary recrystallization.

The classical recrystallization mechanism responsible to nucleation and long-distance growth of the nuclei has been discussed as a strain induced grain boundary migration [10, 11]. Here, any internal interfaces providing high misorientations can be considered as the grain boundaries. The strain induced grain boundary migration involves the local bulging of a high-angle boundary, leaving a dislocation free region behind the moving boundary. The driving force (P_1) for this process is the stored deformation energy, which may be roughly related to the difference in dislocation densities ($\Delta\rho$) between the recrystallization nucleus and the work hardened grain as

$$P_1 \sim G b^2 \Delta\rho \tag{1}$$

where G is the shear modulus and b the Burgers vector [10-12]. On the other hand the local bulging of grain boundaries can be suppressed by several dragging forces. One of them results from the increase in total surface area of grain boundaries and can be written as

$$P_2 \sim 2\gamma/R \tag{2}$$

where γ is the specific grain boundary energy and R is the mean radius of bulge curvature [10-12]. The recrystallization nucleus will grow if $P_1 > P_2$. This leads to the critical conditions for primary recrystallization:

$$\Delta\rho_{crit} > 2\gamma/R G b^2 \quad \text{or} \quad R_{crit} > 2\gamma/\Delta\rho G b^2 \tag{3}$$

It is clearly seen from Eq. (3) that the critical value of the dislocation density difference and the bulge curvature are in inverse proportion to each other, i.e. the smaller nucleus curvature must correspond to the higher dislocation heterogeneity and/or vice versa.

The primary recrystallization nuclei have been reported to have radii of about 1 µm [11]. This is a reasonable dimension because conventional cold worked microstructures are characterized by highly elongated (or pan-caked) grains, and some shear bands and transition bands, which are also preferential nucleation sites, consist of long and relatively straight high-angle (sub)boundaries [10, 11]. Therefore, the recrystallization progress depends on the stored energy due to increased dislocation density. However, the studied strain-induced steel has very fine grains of about 0.3 µm in size, which is three times smaller than conventional recrystallization nuclei. The primary recrystallization in the fine-grained materials require a higher stored energy than that in conventional cold worked ones. On the other hand, the interior dislocation density in submicrocrystalline materials processed by severe deformation tends to decrease with increasing strain [4–6]. For the present material, the average interior dislocation density decreases in about two thirds of its maximal value at large cumulative strains [5]. It can be concluded that the primary recrystallization has a high dragging force to develop in strain-induced fine-grained materials.

Finally, let us consider relationship between the hardness (Hv) and the grain size (D) (Fig. 3). Fig. 3 also represents the data obtained by annealing at 1073 K and 1173 K. The recovery at the fast annealing stage results in the softening without any change of the grain size. The grain coarsening takes place as the transient recrystallization develops during further annealing. The hardness of the annealed samples (open marks) at the beginning of the transient recrystallization is lower than that of the dynamically recrystallized (DRX) samples (solid marks). At completing the transient recrystallization, the hardness becomes almost identical with the DRX samples. At the late annealing stages, the samples with the rather large annealed grains, that develop with normal grain growth, demonstrate the same hardness levels as those reported in the literature. Therefore, the softening during the transient recrystallization and the normal grain growth can be explained by a continuous grain coarsening. A conventional recovery creates the appropriate conditions for the development of the transient recrystallization [15]. Then, a certain amount of fine grains starts to grow and consumes neighboring grains. The ability of certain grain to grow rapidly can be motivated by some structural inhomogeneity that is inherent in the strain-induced state and the variety of the recovery kinetics in different grains.

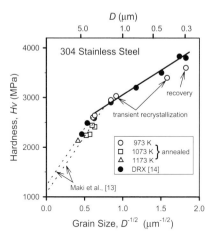

Figure 3: Relationship between the hardness (Hv) and the grain size (D) evolved under annealing (open marks), static recrystallization [13], and dynamic recrystallization (DRX) (solid marks) [13, 14].

5 Conclusions

The structural softening mechanisms operating upon annealing of a submicrocrystalline 304 stainless steel processed by severe plastic working were studied. The main results were:

1. In spite of the apparently high stored energy resulting from severe large deformation, the strain-induced submicrocrystalline steel is essentially stable against primary recrystallization.
2. The annealing behavior can be characterized by the operation of the following major sequential processes: recovery, transient recrystallization, and normal grain growth.
3. At early annealing, the transient recrystallization following recovery results from the homogeneous nucleation that operates in uniformly distributed strain-induced ultra fine grains. Such annealing process is considered as a continuous phenomenon.

6 References

[1] Y. Iwahashi, Z. Horita, M. Nemoto, T. Langdon, Acta. Mater. 1997, 45, 4733–4741
[2] F.J. Humphreys, P.B. Prangnell, J.R. Bowen, A. Gholinia, C. Harris, Phil. Trans. R. Soc. Lond. 1999, 357, 1663–1681
[3] R.Z. Valiev, R.K. Islamgaliev, I.V. Alexandrov, Progr. Mat. Sci. 2000, 45, 103–189
[4] A. Belyakov, W. Gao, H. Miura, T. Sakai, Metall. Trans. A 1998, 29A, 2957–2965
[5] A. Belyakov, T. Sakai, H. Miura, Mater. Trans., JIM 2000, 41, 476–484
[6] A. Belyakov, T. Sakai, H. Miura, K. Tsuzaki, Phil. Mag. A 2001, 81, 2629-2643
[7] A. Belyakov, T. Sakai, H. Miura, R. Kaibyshev, Phil. Mag. Let. 2000, 80, 711–718
[8] F.J. Humphreys, Acta Mater. 1997, 45, 4231–4240
[9] T. Sakai, in Recrystallization and Related Phenomena (Ed.: T.R. McNelley), TMS, 1996, p. 137
[10] F.J. Humphreys, M. Hatherly, Recrystallization and Related Annealing Phenomena, Pergamon Press, Oxford, 1996, p. 127
[11] H.P. Stuwe, in Recrystallization of Metallic Materials (Ed.: F. Haessner), Verlag, Stuttgart, 1978, p. 11
[12] T. Sakai, M. Ohashi, Materials Science Forum 1993, Vol. 113–115, p. 521
[13] T. Maki, T. Akasaka, K. Okuno, I. Tamura, Trans. ISIJ 1982, 22, 253–261
[14] A. Belyakov, T. Sakai, H. Miura, R. Kaibyshev, ISIJ Inter. 1999, 39, 592–599
[15] A. Belyakov, T. Sakai, H. Miura, R. Kaibyshev, K. Tsuzaki, Acta Mater. 2002, 50, 1547–1557

XV Application of SPD Materials

Commercialization of Nanostructured Metals Produced by Severe Plastic Deformation Processing

Terry C. Lowe and Yuntian T. Zhu
Los Alamos National Laboratory, Los Alamos, NM, USA

1 Abstract

The promise of nanotechnology is increasingly being realized as governments, universities, public and private research laboratories, and the various industrial sectors devote resources to this emerging area. Estimates for the economic impact of nanotechnology on existing global markets exceed $700 billion by the year 2008. Nanomaterials are projected to be one of the earliest components of nanotechnology to appear in commercial applications. Amongst the emerging new nanomaterials, bulk nanostructured metals produced by severe plastic deformation (SPD) have shown promise in a wide range of application areas. In this paper, we overview developments in severe plastic deformation technology, emphasizing progress since the international workshop „Investigations and Applications of Severe Plastic Deformation" held 2–8 August 1999 in Moscow, Russia. Then, we overview some of principal areas of application for SPD metals and alloys.

2 Introduction

The science and technology of materials subject to severe plastic deformation (SPD) has evolved largely during the past 30 years. During this period, substantial advances have been made. Most of the body of knowledge on SPD was represented in the first international workshop to focus on the topic, "Investigations and Applications of Severe Plastic Deformation" held 2–8 August 1999 in Moscow, Russia. The purpose of this paper is to summarize progress in the field since this workshop, focusing largely on the commercial potential and markets for SPD technology.

It is appropriate to first consider SPD produced metals in the broader context of nanotechnology. Because of the high level of interest and excitement surrounding nanotechnology, non-government investment in nanotechnology is matching global government investment, even in early stage research and development, thus creating a notable drive for rapid commercialization. [1] Nanomaterials in particular are expected to be among the earliest areas of nanotechnology to develop and therefore, also among the earliest to be commercialized. [1] Thus, the commercial interest in nanomaterials is accelerating research and development in this area and facilitating the early introduction of SPD-processed metals into commercial markets. One indicator of maturity of the nanomaterials technology is the recent publication of the first handbook on nanostructured material processing. [2]

SPD has emerged as a promising method for producing bulk nanomaterials by refining the grains of conventional metals and alloys to be submicrocrystalline (grain diameter of 100 nm–1000 nm) or nanocrystalline (grain diameter of 1–100 nm). However, it is not solely the ability to create nanoscale grain sizes that characterizes SPD processed metals as nanomaterials. It is also that severe plastic deformation processing provides a means to control additional nanoscale features such as grain boundary and sub-boundary structures.

Since 1999 there has been a proliferation of conferences and topical workshops on nanotechnology and in parallel, a steadily growing number of symposia on ultrafine-grained (UFG) metals. The topical symposia on UFG metals have highlighted SPD as an approach to grain refinement into the nanoscale regime. [3,4] In addition, knowledge pertaining to large strain deformation has evolved within additional technological contexts including metal forming, machining, ball milling of powders, wear of high friction surfaces, localized phenomena near the tips of propagating cracks, fatigue, and superplasticity. Research in all of these areas has contributed to the overall knowledge of how severe plastic deformation creates the characteristic nanostructures found in SPD processed crystalline materials. What distinguishes SPD research from some of these other nanomaterials research areas is the explicit focus on achieving nanoscale control of internal structure in bulk materials.

Though interest in SPD has been fueled by the swell of interest in nanotechnology, research on SPD pre-dates the emergence of nanotechnology. SPD processing exists as a technology area independent of nanomaterials. The maturity of severe plastic deformation technology is of particular interest to those seeking its commercialization. Thus this work addresses the question of maturity through compilation and analysis of the body of knowledge that is available in the academic and patent literature.

3   Literature and Patent Analysis

The approach used in this paper is identical to that used previously by Lowe et al. [5] Library databases covering materials and physics were searched for keywords, including variations of "severe plastic deformation" and "equal channel angular." The search focused on archival journal publications, but the databases that were searched had some information on conference proceedings and reports. The databases searched included METADEX, INSPEC, Engineering Index, and U.S.D.O.E Energy References. The search period was restricted to January 1990 to December 2002. Prior literature analysis by Lowe, et al. covering the period from 1980 to 1999 showed very little research related to SPD processing during the 10 years prior to 1990. Because there is inevitable lag between the time of publication and the entry of publication data into databases, the representation of the literature for the last six months of the year 2002 is only partially complete. It is also noteworthy, that because of the length of the peer review process used for most journals, the results of this analysis reflect a state of knowledge that is potentially 6–18 months behind the current frontiers of research on severe plastic deformation. All references found were compiled into a single master database to allow for elimination of duplicates and enable detailed analysis. The compiled list of literature was far too long for each paper to be individually referenced in this paper, but selected examples of references from several sub-topic areas are included.

Patent publications on severe plastic deformation technology were searched using electronic databases from the U.S. Patent and Trademark Office, DEPATISnet from the German Patent and Trademark Office, and the Delphion Research network tools accessing its INPADOC database, which includes filings from 65 patent offices worldwide. The same keywords as used for the literature analysis were used in the queries of the patent databases. Because the titles and descriptors used to write patents are often designed to be distinct from prior patents in the same technology area, it can be difficult to identify patents using the keyword search approach used in this work. Thus, the results obtained in this work should be regarded as indicative of the level of patent activity in SPD only, and not a complete or definitive representation of all known SPD intellectual property. The specific patent databases searched included Granted U.S. Patents, U.S. Patent Applications, Granted European Patents, European Patent Applications, INPADOC (from Delphion), World Intellectual Property Organization (WIPO) Patent Cooperation Treaty (PCT) Publications, and Patent Abstracts of Japan. Because patents typically issue more than one year after their initial application, the state of technology represented by searching only issued patents is behind the current frontier by a year or more. For this reason, the search for SPD-related technology also included patent applications.

4 Results

4.1 Literature Analysis

A total of 828 distinct publications on severe plastic deformation were identified between 1990 and the present. Of these, 805 appeared in archival journals and 23 in conference proceedings. Information obtained on conferences was far from complete since there have been at least 15 conferences or symposia which have included severe plastic deformation as a topic. [6] Of these, most have published proceedings or conference reports, for example. [3–5] The proceeding of [3,4] alone contain 116 papers. Because the literature search method did not seek to include conference proceedings, and because access to the proceedings content is limited, we have not analyzed this source of information on SPD.

Figure 1 shows the steady increase in publication from 1990 through 2000, with the rate of publication nearly doubling between 2000 and 2001. The cumulative total number of SPD publications previously identified by Lowe et al. [5] from 1980 through 1999 was 197. In 2001 alone 205 papers on SPD were published.

The distribution of topical areas of these papers illustrates the current level of development of severe plastic deformation technology. These topical areas correspond with the natural evolutionary stages of research in materials science and materials development. They are: 1) synthesis and processing, 2) characterization of microstructures, 3) characterization and measurement

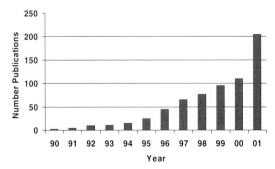

Figure 1: Growth in the number of technical journal publications on severe plastic deformation

of properties, 4) modeling, and 5) application of SPD materials. These five topical areas typically evolve sequentially within the synthesis–structure–properties triad that characterizes materials science investigations. First the method to synthesize a new material is pioneered. Then its microstructure is characterized. As the synthesis technique is validated, samples are produced for more extensive characterization of properties. Modeling and development toward applications typically follow only after sufficient data is available for analysis and experience with the new material has evolved.

The number of publications addressing each of the five sub-topical areas pertaining to severe plastic deformation materials is summarized in Table 1. Contrast is shown between the body of knowledge that could be identified in August of 1999 and in December of 2002. Note that most papers emphasize more than one of the topical areas. For example, over half analyze microstructures.

Table 1: Distribution of topical emphases in publications on severe plastic deformation from 1990–2002

Topic	Number of publications as of August 1999	Number of publications as of December 2002
Microstructures	144	490
Properties	78	445
Synthesis and processing methods	44	255
Modeling	18	102
Applications and products	18	73

Aside from the general increase in research in all areas since 1999, the relative proportion of research in the respective areas is most notable. In 1999, three-quarters of the 197 total publications on SPD reported microstructures of SPD materials, but only 40 % of the publications reported measurements of any properties. By 2002, 59 % of the published research on SPD addressed microstructures while 54 % reported data on one or more additional properties. Thus, an increasing proportion of research is going beyond the early stage of just reporting microstructures. However, there is still relatively little work on key engineering properties such as fatigue resistance and fracture toughness. Only 13 papers focused specifically on fatigue, for example Vinogradov et al., [7] and only six specifically addressed fracture properties (for example Xia et al.). [8] However, 132 papers reference fatigue and 151 reference fracture. Thus, there is clearly interest in the topic, but as of yet, only a small number of significant contributions to these areas.

In 1999 less than a quarter of the publications described innovations in synthesis and processing of SPD materials. By 2002 this proportion grew to over 30 %. The proportion of research including modeling and simulation has also grown from 9 % in 1999 to over 12 % in 2002. This is a positive indication of increasing depth of research and a growing ability to quantify our understanding of SPD processing and resulting materials properties.

The number of publications referencing applications or commercialization of SPD has increased, but has not grown proportionately with the other areas. This is not surprising since the databases searched include largely academic archival journals that focus on reporting of scientific frontiers. Conclusions about the status of applications are better drawn from the analysis of patents and the presence of nanostructured metals in commercial markets, as discussed in the next section.

In Table 2 the papers from the literature are categorized according to the material studied. Note that some theoretical and modeling papers reference no material at all. In other cases, results for several materials are referenced in a single paper. Contrast is shown between the body of knowledge that existed in August of 1999 and in December of 2002.

Table 2: Distribution of materials studied in publications on severe plastic deformation from 1990–2002

Material	Number of publications August 1999	Number of publications December 2002
Aluminum or aluminum alloys	58	369
Copper or copper alloys	48	182
Nickel or nickel alloys	28	84
Iron or iron alloys	24	160
Titanium or titanium alloys	10	72
Intermetallics	11	53
Semiconductors	7	9
Composites	7	59
Palladium or palladium alloys	7	14
Magnetic materials & rare earths	5	39
Polymers	3	15
Other metals: Ag, Nb, Co, Zn, Mg, Li,	11	96

As was true up to 1999, the largest body of research continues to focus on aluminum and its alloys. Aluminum and copper are readily processed by SPD at relatively low temperatures and are available in grades suitable for fundamental research. The proportionately larger increase in the amount of work on aluminum than on copper is due in part to the increasing number of aluminum alloys that are being explored. There is also substantial interest in developing ultrafine grain forms of aluminum alloys to exploit their superplastic properties. Sixty-nine papers have been published on superplasticity in SPD-processed aluminum alloys (for example Kim et al.). [9]

The breadth of materials studied has grown in recent years, with particularly large increases in the amount of work on steels, titanium, intermetallics, and composites. The increase in work on steels is largely from a growing number of researchers working on severe plastic deformation technology in South Korea, China, and Japan. Interest in titanium appears to be driven by a combination of factors. Titanium can be formed superplastically, as discussed in 20 of the 72 papers on titanium and its alloys (for example Sergueeva et al.). [10] Titanium is also a relatively expensive metal valued for use in long-life applications because of its corrosion resistance and its bio-compatibility. Thus, there are technological drivers for interest in the ability to further enhance the properties of titanium by SPD. Interest in intermetallics has focused on Ni_3Al

(18 papers) and TiAl (19 papers). This work is of particular scientific interest since it shows the effects of SPD on locally-ordered structures, and bridges with work on amorphous metals and novel processing methods. There is clearly interest in superplastic properties of intermetallics as well. Twenty-two of the 53 papers on intermetallics address their superplastic behavior (for example McFadden et al.). [11]

SPD processing has been successfully applied to non-metals, including polymers and semiconductors. Equal channel angular pressing has been shown to be effective for controlling molecular and macro scale structures in polyethylene (for example see Zhiyong et al.), [12] polycarbonate (for example see Xia et al.), [8] and several metal and polymer nanocomposite systems (for example see Islamgaliev et al.). [13] The work on polymers is promising and is likely to continue to grow. Exploratory work on SPD processing of semiconductor materials has not grown significantly since 1999.

4.2 Patent Analysis

Thirty-three patent actions, including 14 issued patents and 19 applications or other actions were found in the patent literature. The progression of issued patents and other patent-related activities to protect intellectual property is shown in Figure 2 below. The first SPD patents, issued in the U.S., appeared in 1996. Since then the level of patent activity has grown, with a total of 11 patent actions in 2001 and 12 in 2002. Patent actions include such events as new patent applications, foreign filings, and continuations of existing patents. The number of pending patent actions exceeds the total number of patents in existence. Thus we can expect substantial growth in the amount of intellectual property pertaining to SPD over the next several years.

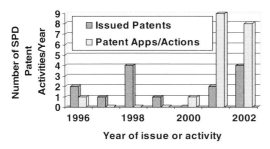

Figure 2: Severe plastic deformation patent activity

The patents and applications that have been issued can be categorized as pertaining to an apparatus, a method, or a specific product. This categorization is made by reviewing the detailed claims for issued patents. For other patent actions such as applications, it is only possible to judge the content of the action from the title of the action. Table 3 shows the distribution of intellectual property within these categories.

The majority of intellectual property pertaining to severe plastic deformation pertains to specific methods of processing. Specific products and apparatus to produce SPD materials and products are less common. The products patented include particular metals and alloys and particular product forms, such as sputtering targets. Particular machines and devices are patented as well. Variations of the original equal channel angular pressing method of Segal [14] have

Table 3: Distribution of intellectual property type pertaining to severe plastic deformation

	Apparatus	Method	Product
Patents	5	9	5
Patent applications, non-US filings	0	17	7

been patented, but the most recent emphasis has been on methods that are suitable for continuous production, for example see Zhu and Lee et al. [15,16]

5 Markets for SPD Metals

Markets for materials processed by SPD exist in virtually every product sector where superior mechanical properties, in particular strength, strength/weight ratio, and fatigue life are critical design parameters. Products which depend upon achieving uniform fine grain sizes are also impacted by the prospective availability of SPD metals. For example, ultrafine grain sizes are attractive independent of the their mechanical properties for such applications as pre-forms for forgings, targets for sputtering of precious metals, and sheet for superplastic forming.

Formal market analyses conducted by nanostructured metals companies such as Metallicum have identified over 100 specific markets for SPD metals in aerospace, transportation, medical devices, sports products, food, and chemical processing, electronics, and conventional defense. However, the accessibility of these markets depends upon several factors. First, the ability to introduce new materials into an existing product sector relies on the inclination for the sector to adopt new materials and technologies. For example, the sporting goods products industry thrives on the introduction of new materials and technologies to consumers. In contrast, transportation and aerospace industries are more deliberate in how they evolve to new materials. Highly regulated industries, such as nuclear power generation are among the slowest to evolve.

In all cases, there must be a significant economic, performance, or regulatory drivers for adopting SPD metals. Since new materials typically demand a premium price, markets that are cost insensitive are most attractive for metals nanostructured by SPD. For example, the medical device markets use small volumes of materials in high value devices such as prosthetics. For these products the cost of the material may be less than 2 % of the overall product cost. Furthermore, competition amongst medical device manufacturers tends to drive continuous innovation, including the adoption of new materials that give medical devices distinct competitive advantages.

The automotive industry is a contrasting case in point. It seeks superior materials for many applications, including those driven by a need to reduce vehicle weight, enhance crash resistance, and more readily comply with emissions and safety requirements. However, price sensitivity of material costs in automobile manufacturing also demands that SPD metal production technology be sufficiently mature and highly optimized. Furthermore, the multi-year design and large-volume materials procurement cycles make the adoption of high strength SPD metals for automobiles a more difficult market. Whereas there is a tendency to believe that superior properties of SPD materials allows them to compete with existing conventional metals, the reliability and maturity of the metal production technology remains a substantial barrier to penetrating any market in which alternative competing materials are already produced in large volume at competitive prices.

The first appearance of severe plastic deformation processed metal in a commercial product has been for sputtering targets. Honeywell International Inc. uses equal channel angular extrusion to produce aluminum, copper, and copper alloy sputtering targets. They advertise average grain sizes in the aluminum targets on the order of 500 nm leading to extended target lifetime, low arcing levels, and more uniform sputtered coatings. Increase in target lifetime is on the order of 20 % over conventional aluminum targets which have a 50 micrometer grain size. As an aside, it is notable that ECAE is a registered trademark of Honeywell.

Metallicum, a company that operates facilities in New Mexico in the U.S., is using several severe plastic deformation processes to develop titanium-based products for medical implant markets. The initial focus of the company is on processing commercial purity titanium to impart properties that are comparable to, or in some instances exceed, those of titanium alloys. This company\9s special emphasis is to provide highly biocompatible materials with superior fatigue and fracture resistance for medical devices.

The U.S. Department of Energy Office of Industrial Technologies is funding a joint university–industry effort to develop severe plastic deformation processing of aluminum. A particular focus of this effort has been to produce continuous length, large cross-section products. This collaboration has been successful implementing a continuous process to fabricate 12.7 mm square cross-section 6061 aluminum bar in lengths up 1.5 meters. [17] The industrial partner in this project, Intercontinental Manufacturing Company, is evaluating the use of severe plastic deformation to produce ultrafine grain aluminum and magnesium forging pre-forms with enhanced forgeability. They are using ECAP to produce 100 mm square cross-section forging pre-forms up to 350 mm in length. [18] The joint effort is pioneering the development of aluminum stock that can be forged at lower temperatures and at lower cost than is possible with conventional aluminum alloys.

6 Conclusion

Severe plastic deformation technology remains at an early stage of development. However, the pace of work in this area has accelerated dramatically since 1999. The maturity of the field is evolving as a growing number of researchers investigate facets of SPD technology.

The first SPD-processed metals are available as commercial product in just a few market sectors. Based on the patent activity, additional products and markets, encompassing a larger array of metals appear eminent. Large volume production of SPD metals is not likely to emerge until the economics of continuous processing methods that are just now emerging are well established and greater penetration of key markets is established.

7 References

[1] P. Hoister, T. E. Harper, The Nanotechnology Opportunity Report, vol. 1, CMP-Cientifica. 239, Las Rozaz, Spain 2002
[2] C. C. Koch, Nanostructured Materials—Processing, Properties, and Applications, Noyes Publications and Williams Andrews Press, Westwood, NJ 2002

[3] Ultrafine Grain Materials (Eds: R. S. Mishra et al.), The Minerals, Metals and Materials Society, Warrendale, PA 2000, p. 434.
[4] Ultrafine Grained Materials II. (Eds: Y. T. Zhu et al.), The Minerals, Metals, and Materials Society, Warrendale, PA 2002, p. 685
[5] T. C. Lowe, Y. T. Zhu, S. J. Semiatin, D. R. Berg, in Investigations and Applications of Severe Plastic Deformation, (Eds: T. C. Lowe, R. Z. Valiev), Kluwer Academic Publishers, Norwell, MA 1999
[6] R. Z. Vailev, personal communication
[7] A. Y. Vinogradov, V. V. Stolyarov, S. Hashimoto, R. Z. Valiev, Mater. Sci. Eng. A 2001, A318, 163
[8] Z. Xia, H.-J. Sue, A. J. Hsieh, J. Appl. Polym. Sci. 2001, 79, 2060
[9] W. J. Kim, J. K. Kim, T. Y. Park, S. J. Hong, D. J. Kim, Y. S. Kim, J. D. Lee, Metall. Mater. Trans. A 2002, 33, 3155
[10] A. V. Sergueeva, V. V. Stolyarov, R. Z. Valiev, A. K. Mukherjee, Scr. Mater. 2000, 43, 819
[11] S. X. McFadden, R. Z. Valiev, A. K. Mukherjee, Mater. Sci. Eng. A 2001, A319–321, 849
[12] X. Zhiyong, S. Hung-Jue, A. J. Hsieh, J. W. L. Huang, J.'Polym. Sci. B 2001, 39, 1394
[13] R. K. Islamgaliev, W. Buchgraber, Y. R. Kolobov, N. M. Amirkhanov, A. V. Sergueeva, K. V. Ivanov, G. P. Grabovetskaya, Mater. Sci. Eng. A 2001, A319-321, 872
[14] V. M. Segal, Plastic Deformation of Crystalline Materials, USPTO, #5513512, May 7, 1996
[15] Y. T. Zhu, T. C. Lowe, Method for Producing Fine-Grained Materials Using Repetitive Corrugation and Straightening, USPTO, #6197129, March 6, 2001
[16] J.-C. Lee, H. K. Seok, J. W. Park, Y. H. Chung, H. J. Lee, Continuous Shear Deformation Device, #6370930, April 16, 2002
[17] R. Srinivasan, personal communication
[18] P. Chaudhury, personal communication

The Main Directions in Applied Research and Developments of SPD Nanomaterials in Russia

V. A. Fokine

Editor of the Journal "Russia and World: Science and Technology", Director General of the F&F Consulting Co., Moscow, Russia

1 Russia and World: Science and Technology

"Russia and World: Science and Technology" is an international periodical on Russian technologies, new developments and high-tech products from different regions of the country. It has been issued in two languages for nearly 10 years and distributed in more than 20 countries.

The distinguishing feature of the journal supported by the Ministry of Industry, Science and Technologies of the Russian Federation is that it publishes only firsthand information. The new developments to be covered have been given careful consideration by its experts, both those on the payroll and doing specific work under contract, who are acknowledged as authorities in their fields of knowledge. As a rule, publishing information is preceded by a visit to the developer-company and discussion of the key points of the offered developments on the spot.

By now the scientific potential of Russia has been given very different, and sometimes polar, appraisals.

Some view Russia as a gigantic, unlimited source of high-technologies capable of bringing multibillion profits. Some experts estimate the cost of intellectual property in Russia at about $ 400 billion. Others, on the contrary, believe that during the last several years the high-technology potential of the country has been practically wasted away, while available technologies and developments are related, as a rule, to defense industry and can hardly be built-in into modern technological processes.

The very existence of such polarity in assessments testifies to the fact that, despite openness of Russia during the last decade, neither foreign, nor even most Russian experts have performed profound research into its scientific potential.

In opinion of the editorial board of the journal, the Russian market of new technologies can be compared to the ocean with shoals of fish living at specific depths and in specific areas. And it requires skills, knowledge and luck to catch fish in this mass of water.

Such a good catch was made when the editorial board got profoundly enough familiarized with the recent achievements in the area of bulk nanostructured materials.

From that point on attention of experts in material science was focused on this line of research, and specifically on the techniques of severe plastic deformation (SPD) used for their production, which can be explained by the following:

1. Many of the pioneer works on SPD nanostructured materials were performed in Russia, moreover, at present investigations into this subject are being actively pursued in the country. That verifies the big amount of original papers, special issues and patents in Russia and in the world (see, e.g., [1–7]).
2. Being oriented to applied results, the editorial board came to the conclusion that in case of positive development nanostructured materials become very promising for use in a number of branches of science and technology. Featuring a unique atomic-crystalline structure, determined by fine grain sizes, very long intercrystalline boundaries, very high density of other atomic-crystalline defects, the materials reveal extraordinary properties. It is these properties that make nanostructured materials promising for different structural applications, for example, fabrication of superstrong articles from metals, plastic ceramics, wear-free products, etc. in such new areas of technology as micromechanical systems, biomedical implants and others.
3. With regard to complexity of the problem and the economic situation in Russia, real applied results can be obtained in the course of efficient interdisciplinary and international technical and technological cooperation to be organized in this area.

A special stress should be made on technical and technological collaboration, because cooperation in basic research has been carried out on a wide scale, in many respects due to active efforts of Professor R. Valiev, a pioneer of SPD nanomaterials research and co-chairman of this conference.

The first step the F&F Consulting firm made was performing analysis of the situation in investigations into obtaining bulk nanostructured materials in Russia.

In particular, the prepared review "Works in Russia in the field of bulk nanostructured materials" included analysis of the three main ways of obtaining them:

- using severe plastic deformation (SPD);
- nanocrystallization of amorphous alloys (NAA);
- nanopowders compacting (NPC).

Nanocrystallization of amorphous alloys and nanopowders compacting are not among the subjects of close consideration by the conference Thus, emphasis should be made on the leading Russian scientific centers engaged in SPD research that have attained results that are assessed as very interesting in the view of their translation into applied works.

As for SPD R&D centers (Fig. 1), these are, primarily, the Institute of Physics of Advanced Materials of the Ufa State Aviation Technical University, Tomsk Institute of Material Science, Nizhniy Novgorod Branch of Institute of Mechanical-Engineering Science, Institute of Problems of Metals Superplasticity of the RAS (Ufa), Institute for Physics of Metals of the RAS and Urals State Technical University (Yekaterinburg), Ulyanovsk State University, Moscow Institute of Steel and Alloys, Moscow Institute of Metallurgy and Material Science. Interesting investigations are underway in St. Petersburg and Sarov as well.

Among the most important directions of fundamental and applied research conducted by the above R&D centers are the following topics:

1. development of nanostructured TiNi alloys with shape memory effect;
2. development of nanostructured Ti for medicine;
3. development of nanostructured Al alloys for superplastic forming;
4. development of methods to process an ultrafine-grained (UFG) structure in stainless steels;
5. application of multiple forging for producing a TiAl intermetallic compound with an UFG structure;
6. development of SPD processing to produce an UFG structure in long-sized rods and sheets;
7. development of SPD processing for producing an UFG structure in low-carbon MnNi steels;
8. fabrication of Al-based nanocomposites by SPD techniques;
9. fabrication of Cu-based nanocomposites by SPD techniques;
10. production of electric contacts and parts for switching equipment with increased service life using SPD methods;

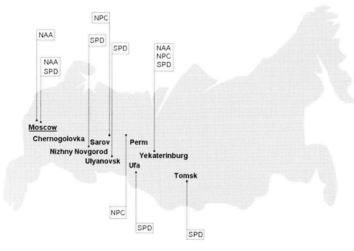

NAA - Nanocrystallization of Amorphous Alloys
NPC - NanoPowders Compacting
SPD - Severe Plastic Deformation

Figure 1: The major Russian centers of SPD nanomaterials research

11. processing of materials for MEMS by SPD methods;
12. fabrication by SPD techniques of nanostructured silicon radiating in the visible spectrum.

Undoubtedly, in conditions of limited resources in Russia progress in promoting these works to a great extent depends on the Ministry of Industry and Science of Russia that lately has been paying special attention to coordination and support of works in the field of nanotechnology.

As to the F&F Consulting Co., the firm sees its main task in initiating interest of the society to practical application of fundamental results attained in different branches of science and technology. To do so the F&F uses all available channels: information is deployed on Web-sites, it is published in scientific and technical publications, presented in popular sci-tech magazines, at National workshops in the field of innovations.

A range of developments in SPD being done in Russia are close to reaching the stage of applied research. It is the synergies of profound knowledge of specific fundamental properties of nanostructured materials and a significant amount of know-how accumulated by manufacturers that can ensure the desired effect. In this connection it is appropriate to single out two developments: works on nanostructured titanium for medical applications and nanostructured aluminium alloys for superplastic forming.

Both works were carried out jointly by the IPAM USATU in Ufa under the direction of Professor R. Valiev and Design and Technology Bureau Iskra (the project manager Dr. V.V. Latysh). The first work was performed with participation of Prof. Kolobov (IPAM RAS, Tomsk) in cooperation with Drs. Y. Zhu and T. Lowe from Los-Alamos National Laboratory, USA.

In the first case the researchers mastered stocks of very strong nanostructured Ti for implants fabrication for orthopedic medicine (Fig. 2). They accumulated a significant amount of know-how.

They created a tangible resource and the samples are really interesting. Next year they plan to start a large-scale production of nanostructured metals with output of 5 tons.

Specialists, and not only they, realize how hard it would be to get these products through all the medical trials and certifications. An integrated program of works on SPD nanomaterials supported at the federal or international level could become one of possible solutions that would facilitate progress in this area.

Figure 2: A device for correction and fixation of the spine column made of high-strength nanostructured Ti

Figure 3: View of a piston fabricated from a nanostructed Al alloy which is supposed to be used in small overall internal combustion engines (it is shown in the figure behind the piston)

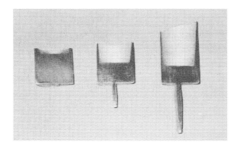

Figure 4: View of a fitting produced from an A1 alloy by high strain rate superplastic forming

In the second case, on the laboratory level, the developers mastered methods of fabrication of complex-shape articles, namely pistons of small-size engines of internal combustion (Fig. 3) and articles of fitting type (Fig. 4), by superplastic forming using nanostructured aluminium alloys.

The developers carried out detailed investigations of the produced material and discovered very interesting possibilities for using nanostructured alloys. In particular, a pronounced reduction in the stamping specific force, in forming temperature and others. At this stage it is especially important to identify the range of products that can find demand in the market and that can be maximally efficiently fabricated from SPD materials. Researchers in the Ulyanovsk University and in Ufa University have also carried out very interesting works on SPD-processed nanometals to be used in microsystems.

Note should be made that such materials featuring high plasticity and strength characteristics can be promising for use as membranes and springs, silicon substitutes, as well as to produce shafts and parts of microengines and in other devices of several dozens of micrometers in size.

The developers mastered laboratory-level modes for production of blanks of nickel plates with ultra-fine grain structure with sizes $0.2 \times 20 \times 200$ mm. It can be said that use of such parts in MEMS has no analogs in world practice and can bring about a notable economic effect and improvement of the consumer properties of the products.

At the same time, it should be stressed that solution of a range of fundamental problems related to use of such products is impossible without close cooperation with producers and developers of microsystems.

On the whole, despite the existing difficulties, the general line of research into SPD in Russia testifies to the fact that in the nearest future we can expect sufficiently active development of works on practical application of bulk nanomaterials as new structural and functional materials of a new generation, as well as on commercialization of obtained results.

A special workshop held on February 5, 2003 within the framework of the Moscow International Salon of Innovations and Investments organized by the Ministry of Industry and Science of Russia with participation of the IPAM (Ufa) and F&F Consulting Co. was devoted to this particular subject. This successful event has evidenced that international technical and technological cooperation, including conducting R&D by order from foreign partners, as well as using international experience in innovations commercialization can play an important role in SPD technologies development and commercialization.

2   Conclusions

In recent years the journal "Russia and World: Science and Technology" has kept close track of new developments in basic and applied research of SPD-produced nanomaterials. These works are actively pursued in Russia and appear to be very promising for efficient commercialisation. The journal provides all-round information support to the investigations on SPD, and points out that in order to enhance these investigations it is highly important to co-ordinate all the works in this field ongoing in the Russian Federation and develop international co-operation between Russian and foreign researchers.

3   References

[1] R. Z. Valiev, N. A. Krasilnikov, N. K. Tsenev, Mater. Sci. Eng. 1991, A137, 35–40
[2] R. Z. Valiev, A. V. Korznikov, R. R. Mulyukov, Mater. Sci. Eng. 1993, A168, 141–148
[3] Proceedings of the NATO ARW "Nanostructured Materials: Science and Technology" (St. Petersburg, Russia), NATO Sci. Series, eds. G.-M. Chow and N. I. Noskova, Kluwer Publ., 50, (2000), p. 456
[4] Proceedings of the NATO ARW "Investigations and Applications of Severe Plastic Deformation" (Moscow, Russia), NATO Sci. Series (Eds. T. C. Lowe and R. Z. Valiev), Kluwer Publ., 80, (2000), p. 394
[5] Ultrafine-Grained Materials Processed by Severe Plastic Deformation (Ed.: R. Z. Valiev), Ann. Chim. (special issue) 1996, 21, 369–554
[6] R. Z. Valiev, R. K. Islamgaliev and I. V. Alexandrov, Prog. Mater. Sci. 2000, 45, 103–189
[7] Phys. Met. Metallogr. (special issue) 2002, 94, Suppl. 1 (in Russian)

Developing of Structure and Properties in Low-Carbon Steels During Warm and Hot Equal Channel Angular Pressing

S.V. Dobatkin[1,2], P.D. Odessky[3], R.Pippan[4], G.I. Raab[5], N.A. Krasilnikov[5], A.M.Arsenkin[2]
[1]Baikov Institute of Metallurgy and Material Science, Russian Academy of Sciences, Moscow, Russia
[2]Moscow State Steel and Alloys Institute (Technological University), Moscow, Russia
[3]Institute of Building Constructions, Moscow, Russia
[4]Erich Schmid Institute of Material Science, Austrian Academy of Sciences, Leoben, Austria
[5]Ufa State Aviation Technical University, Ufa, Russia

1 Introduction

Ultrafine-grained (UFG) materials obtained by SPD at lower temperature call for special interest because of their unusual properties [1,2]. Equal channel angular pressing (ECAP) is currently one of the most advanced way that can produce UFG materials (grain size – less than 1 µm)[1–4]. The most investigations were carried out, basically, on pure metals and plastic alloys [3,4]. Industrial steels are investigated a little. Certainly, it is impossible to use widely methods of severe plastic deformation in an industry now. But on our sight, it is necessary to study limiting structural states of industrial materials and properties, appropriate to them. There are some investigations dealt with low-carbon steels [5–8], but connection between structure after ECAP and mechanical properties, especially impact toughness is studied now not enough.

The purpose of this paper is to study of structure and mechanical properties of low-carbon 0,17C; 0,2CMnSiV and 0,25CMnSi steels after equal channel angular (ECA) pressing in warm and hot deformation conditions.

2 Experimental Procedure

As materials of research were chosen low carbon steels having rather low YS and used, basically, as elements of designs (Tab. 1).

Table 1: Chemical composition of low-carbon steels studied

Steel	Alloying element, wt.%					
	Si	Mn	V		P	S
0,17C	0,17	0,18	0,32	-	<0,04	<0,05
0,2CMnSiV	0,21	0,78	0,89	0,16	<0,030	<0,025
0,25CMnSi	0,23	0,75	1,24	-	0,015	0,031

ECA pressing has been carried out using the samples of 20 mm in diameter and 80 mm in length at the temperatures 500–550 °C with an angle of intersection of channels 90° and N (number of passes) = 4 and at the temperature 750 °C with channel intersection angles 90° and 110° with $N = 8$ and 12, correspondingly. We could not use cold ECA pressing due to specific equipment possibilities. The limit temperature of 500 °C for warm deformation was selected with account for the capability of the deformation tool

The structure analyses has been carried out using the transmission electron microscope JEM-100X. Microhardness was defined using the PMT-3 device with 50gf load.

3   Results and Discussion

To obtain nanocrystalline structure in metals and alloys using equal channel angular pressing (ECAP) the deformation should be cold [2,3]. However, with decreasing ECAP temperature the strain (the number of passes), that is required to form high angle boundaries, increases [9]. To attain high strains at lower temperatures it is necessary to reduce friction in the entry channel of ECAP die as much as possible, and to apply back pressure in the exit channel [1].

The ECAP die used in the present work allows deformation of low carbon steels 0.17C, 0.2C-Mn-Si-V and 0.25C-Mn-Si at 500–550 °C with N = 4 and channel angle of 90°.

In warm ECAP, the initial ferrite grains elongate with increasing number of passes and align at a certain angle with respect to specimen axis (Table 2). The amount of pearlite does not change.

Table 2: Cross sizes of ferritic-pearlitic structure components of 0,17 %C steel after ECA-pressing (t = 500 °C, φ = 90°, N = 4) [1]

Place of measurement	D_F, μm	D_P, μm
Surface	13,4 ± 2,0	6,4 ± 1,2
Center	17,9 ± 2,6	8,5 ± 1,9
Surface	12,7 ± 2,1	5,3 ± 1,0

[1]The initial mean sizes of ferrite and pearlite regions correspond to 30 μm and 17 μm.

TEM observations of 0.17C steel ECA pressed at T = 500 °C and N = 4 reveal a mixed microstructure inside the initial elongated grains, which consists of dynamically recovered structure with low angle boundaries of subgrains and dynamically recrystallized (submicrocrystalline) grains with high angle boundaries (Fig.1).

The average size of microstructure elements is 0.35 μm. Also the nucleation of new grains between cementite laths was observed. In this case the grains were of 0,1–0,2 μm in size. The fragmentation of cementite laths, their coalescence and spheroidization were also detected.

EBSD analysis confirmed that two types of microstructure, one with low- and one with high-angle misorientation of grain boundaries, exist inside the elongated ferrite grains. As shown in Fig. 2, the increase in the number of ECAP passes from N = 1 to N = 4 leads to refinement of microstructure elements and to increasing fraction of grains, i.e., elements with high-angle misorientation. However, even at N = 4 only partially submicrocrystalline microstructure was revealed by both TEM and EBSD. It can be assumed that more passes are needed to obtain a fully submicrocrystalline microstructure in warm ECAP.

TEM and EBSD of 0.2C-Mn-Si-V and 0.25C-2Mn-Si steels reveal similar microstructure – a mixture of subgrains and grains with the size of microstructure elements of 0.3–0.5 μm.

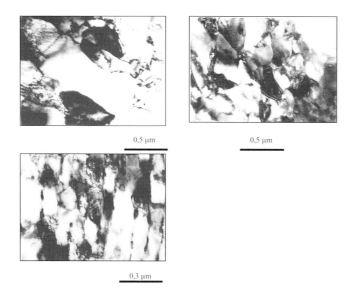

Figure 1: Structure of 0,17 % C steel after ECA pressing with $t = 500$ °C, $\varphi = 90°$; $N = 4$

This microstructure results in significant strengthening of steels: YS reaches 840 MPa for 0.17C steel and exceeds 1000 MPa for two other low alloyed steels (Table 3). EL in all cases is 10–15 %. Regretfully, the microstructure of the described type leads to a very low values of impact toughness KCV both at room temperature and at –40 °C (Table 3). What could be the reasons for such low impact toughness? Probably, the mixed microstructure with high dislocation density inside subgrains, non-equilibrium structure of grain boundaries and small size of microstructure elements produce a combined effect that manifests itself through close values of YS and TS.

What are the feasible ways to increase the impact toughness in this case?

We believe that these are as follows:

1. Increase in warm ECAP strain (number of passes).
2. Reheating after warm ECAP.
3. Hot ECAP.

With ECAP die used in this work it is impossible to increase the number of passes during pressing.

The reheating after warm ECAP was aimed to form ultrafine grained structure with average grain size of 1 µm. In 0.17C steel, the reheating to 600–700 °C after ECAP at 500 °C with $N = 4$ resulted in grain size of 7–8 µm (Fig.3). The decrease of reheating temperature down to 550 °C also did not allow for UFG structure. Instead, a mixture of recrystallized grains (0.4–0.5 µm) and subgrains (0.25–0.65 µm) was observed. With this microstructure the impact toughness KCV^{+20} of 0.17C steel increased fivefold as compared to that after warm ECAP (Table 3).

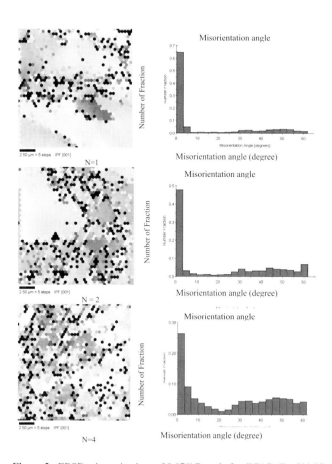

Figure 2: EBSD – investigations of 0,17%C steel after ECAP (T = 500 °C, Φ = 90°)

Table 3: Mechanical properties of low carbon steels after ECA pressing

No	Steel	T, ° of ECA pressing	Φ, angle of channel intersection	N, number of passes	T, ° of heating (30 min)	UTS, MPa	YS, MPA	EL,%	RA,%	KCV,MJ/m² +20 °C	–40 °C
1	0,17%C	500	90°	4	-	850	840	10	52	0,39	–
					550	–	–	–	–	1,98	–
2	0,2%C-Mn-Si-V	550	90°	4	-	1120	1110	8	40	0,55	0,15
		750	110°	8	-	850	820	15	45	2,52	–
			90°	4	-	975	905	13	36	2,0	1,2
3	0,25%C-Mn-Si	550	90°	4	-	1005	1000	11	44	0,21	0,14
		750	110°	8	-	875	870	12	34	2,19	1,65

Hot ECAP was performed at $T = 750$ °C with the channel angle of 110° and $N = 8$, as well as with 90° at $N = 4$. In the first case, a mixture of recrystallized grains of 0,3–6 μm and subgrains of 0,5 μm was obtained.

Steels 0.2C-Mn-Si-V and 0.25C-2Mn-Si exhibit strengthening up to $YS > 800$ MPa and $EL = 10$–15 % at fairly high impact toughness both at +20 °C and –40 °C (Table 3) after hot ECAP with the channel angle $\varphi = 110°$. Hot ECAP with $\varphi = 90°$ and $N = 4$ results in mostly recovered microstructure that leads to even higher strength ($YS = 905$ MPa) at $EL = 13$ % and high impact toughness (Table 3).

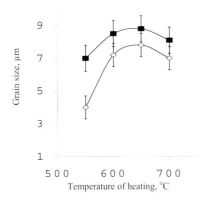

Figure 3: Grain size during heating of low carbon steels after warm ECA pressing, ■ - 0.25C-2Mn-Si, ◇ - 0.17C

Thus, neither reheating after warm ECAP nor hot ECAP permit for UFG structure with grain size of about 1 μm in the studied low carbon steels. However, mostly subgrain microstructure obtained in both cases allowed for significant increase in impact toughness, including that at –40 °C, as compared to as warm ECA pressed microstructure, with preserving its high strength (Table 3).

This paper doesn't give us the answer: do we need the low – carbon steels with the grain size less than 1 μm?. In another words, could we get high impact toughness because the close values of YS and TS. For answering we should test the low carbon steels with the grain size less then 1 μm and equilibrium high angle boundaries.

4　　Conclusions

1. Warm ECAP of low carbon steels (0.17C, 0.2C-Mn-Si-V, 0.25C-2Mn-Si) leads to mixed microstructure of grains and subgrains with average size of 0.3–0.5 μm and high dislocation density. This partially submicrocrystalline microstructure results in significant strengthening ($YS > 800$ MPa for 0.17C steel, $YS > 1000$ MPa for low alloyed steels), but also in low impact toughness.

2. Reheating of 0.17C steel at 550 °C after ECAP allows for a fivefold increase in impact toughness due to microstructure comprising recrystallized grains of 0.4–5 µm and subgrains of 0,25–0.65 µm.
3. Hot ECAP of low carbon steels (0.2C-Mn-Si-V and 0.25C-2Mn-Si) at 750 °C results in mostly subgrain microstructure that leads to high strength ($YS > 900$ MPa) and impact toughness ($KCV^{+20} > 2$ MJ/m^2).

5   References

[1] V.M.Segal,V.I.Reznikov,A.E.Drobyshevsky,V.I.Kopylov, Metally,1981, 1, 115–123
[2] "Investigations and Applications of Severe Plastic Deformation", (Eds.T.C.Lowe and R.Z.Valiev), Kluwer Academic Publishers,Dordrecht,2000
[3] R.Z.Valiev,A.V.Korznikov,R.R.Mulyukov, Mater.Sci.Eng. 1993,A168, 141–148
[4] Y.Iwahashi, Z.Horita, M.Nemoto and T.G.Langdon, Acta Mater. 1998, 46, 3317–3331
[5] D.H.Shin, Y.-S.Kim and E.J.Lavernia, Acta Mater. 2001, Vol.49, 13, 2387–2393
[6] J.Kim, I.Kim and D.H.Shin, Scripta Mater. 2001, 45, 421–426
[7] Y.Fukuda, K.Oh-ishi, Z.Horita and T.Langdon, In Proc. Of Inter. Symposium on Ultra-fine Grained Steels (ISUGS 2001), ISIJ, 2001, 156–159
[8] S.V.Dobatkin, P.D.Odessky, N.A.Krasilnikov et al. In Proc. Of the First Joint Inter. Confer. on Recrystallization and Grain Growth, (Eds. G.Gottstein and D.A.Molodov), Springer-Verlag, 2001, 543–548
[9] S.V. Dobatkin. In "Investigations and Applications of Severe Plastic Deformation". Eds T.C. Lowe and R.Z. Valiev. Kluwer Academic Publishers, 2000, 13–22

Mechanical Properties of Severely Plastically Deformed Titanium

L. Zeipper[1], M. Zehetbauer[2], B. Mingler[2], E. Schafler[2], G. Korb[1], H. P. Karnthaler[2]
[1] ARC Seibersdorf research GmbH, Seibersdorf, Austria
[2] Institute of Materials Physics, University of Vienna, Vienna, Austria

1 Introduction

Ultrahigh-strength commercially pure titanium (CP-Ti) as well as titanium alloys (e.g. Ti6Al4V ELI) have a great impact on medical implants and other high-tech devices. By Severe Plastic Deformation (SPD) ultrafine-grained and nanostructured materials are achieved during a "top down" approach, starting from conventional coarse-grained metals and alloys. As far as batch processing is concerned, Equal Channel Angular Pressing (ECAP) is one of the most promising methods so far. By creating ultrafine-grained structures in commercially pure titanium, titanium alloys for medical use can be replaced without loosing the desired mechanical properties [1]. This will result in an increasing biocompatibility [2] and an adjustable level of ductility for post-processing.

2 Experimental

2.1 Samples Preparation

2.1.1 ECAP Samples

An equivalent VT1-O had to be chosen because of the lack of a 1pass UNS grade 2 titanium. CP-Titan grade 2 samples in a mill-annealed condition of diameter 20 mm (1pass) to 40 mm (0, 4, 8 & 10pass) have been ECAP processed in Ufa/Russia at 450 °C–400 °C at a pressing speed of 6mm/s for 1,4,8 and 10 passes by route Bc (90° turn after each pass) [3] and analysed in Austria. The chemical composition is indicated in Table 1.

Table 1: Chemical composition of the initial CP-Titanium grade 2:

	Fe	C	O	N	H	Si	others
	[wt.%]	[wt%]	[wt.%]	[wt%]	[wt%]	[wt%]	[wt%]
UNS: R50400 (0 & 8pass)	0,3	0,1	0,25	0,03	0,0125		
Russian: VT1-O (1pass)	0,3	0,07	0,2	0,04	0,01	0,1	0,3

2.1.2 Samples by Cold-Rolling

CP-Titan grade 2 samples in a mill-annealed condition were cold-rolled up to a reduction of 54–83 %, that means a true strain ε_{true} of 0,77–1,77 (Fig.2). The rolling procedure was carried out on a laboratory scale rolling mill with small steps of about e_{true} = 0,014. The 54 % sample was prepared to carry out standard tensile tests (Fig.3). The 83 % samples are showing the microstructure being typical for heavy cold deformation, where first major cracks appeared at the edges of the sheets and the procedure needed to be stopped. This amount of deformation is approximately comparable to a 1pass ECAP step (true strain ≈ 1,13 [4]). One can see right here, that even a slightly heavier deformed sample show larger and strongly macro-textured structural elements than observed in ECAP samples (Fig.2). However, one must state that the temperatures of deformation were different and further investigations at elevated temperature are ongoing.

2.2   Results and Discussion

2.2.1   3D Ultrasonic Measurements

First quality controls after ECAP processing used 3D Ultrasonic Impulse-Echo Testing (Fig.1) for Ø 20mm CP-Ti billets (post-processed by cutting to Ø 16 mm; scanning interval: 0,25 mm; resolution: ≥ 70 µm at 50 MHz in dist. H_2O). The Ø 40 mm ECAP CP-Ti sample (post-processed by cutting to Ø 32 mm; scanning interval: 0,25 mm; resolution: ≥ 140 µm at 25 MHz in dest. H_2O) was analysed in the same way. No cracks due to the SPD processing were observed (see left hand side, left picture). Back Wall Echos showed the marks of cutting at the billets' surface.

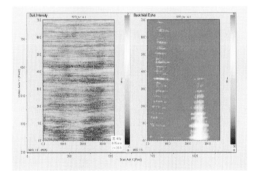 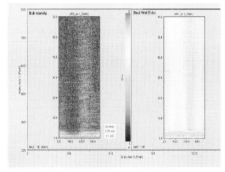

Figure 1: 3D ultrasonic analysis of Ø 40 mm 10pass (left hand side) and Ø 20 mm 8pass (right hand side) nano-SPD CP-Ti cut to diameters 32mm and 16mm

2.2.2   Light Microscopy (LIMI) & Transmission Electron Microscopy (TEM)-Microstructure

Optical LIMI and TEM investigations of different microstructures of CP-Ti, 83% cold-rolled, 1pass, 8pass SPD CP-Ti were carried out. The micrographs were prepared with conventional equipment, etched with a agent of 15 ml HNO_3, 5 ml HF, 80 ml H_2O and analysed by an Axioplan light-microscope. For TEM picture details please see [5]. In the bottom left hand picture of Fig.2, a 1pass ECAP sample is shown with a mean transversal structural size of about 300 nm and a length of up to 2 µm [6–8]. The two LIMI pictures with the TEM inserts (Fig.2) show the homogenisation of the structure due to further passes from 1 to 8 pass CP-Ti. This is also indicated by SAED patterns [5]. Additionally, compared to 1pass samples a sharper relief is visible

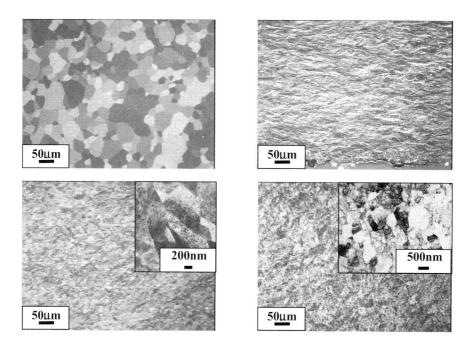

Figure 2: LIMI pictures with TEM inserts - Upper left hand side: coarse-grained, mill-annealed CP-Ti; Upper right hand side: 83 % cold rolled CP-Ti; Bottom left hand side: 1 pass ECAP CP-Ti; Bottom right hand side: 8 pass ECAP CP-Ti

in etched 8pass LIMI micrographs. Comparing TEM with LIMI, this can be attributed to a increased number of high angle grain boundaries and regions with low interior dislocation densities. Consequently, a considerably more distinct etching takes place (Fig.2).

2.2.3 Mechanical Properties

From the tensile properties of 8 and 10pass samples it can be seen that the elongation after fracture reaches levels similar to that ones of the annealed state. Additionally their yield strength and ultimate tensile values are increased by 60 % and 40 %, when compared to the initial coarse-grained state. After the first ECAP pass the elongation after fracture (A5, gage length equals 5 times gage diameter) decreases dramatically (–32 %). It can be clearly seen in Fig.3, that not only A5 but also Ag (the uniform elongation before necking) decreased even more pronounced (–60 %). By applying 7 additional passes both values increase again and reached a saturation level at about 70 % (Ag) and 85 % (A5) of the initial coarse-grained value. This fact, that both strength and ductility increase is a paradoxon, which is famous for SPD materials [9]. Therefore there is a great need to analyse this abnormal behaviour in more detail by experiments [10] and modeling [11–14].

To demonstrate the technological relevance of SPD Fig.3 shows data of a 53 % cold rolled Ti sample, which exhibits a substantially lower ductility. Due to a combination of ECAP plus cold rolling mechanical properties of the Ti6Al4V ELI alloy can be reached in unalloyed CP-Titanium [15].

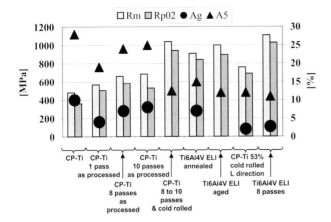

Figure 3: Strength and Ductility measured in tensile-tests according to EN 10002-Teil 1 (excl. Ti6Al4 ELI aged: [16], 8pass CP-Ti & cold rolling: [15]) Tensile tests were carried out with round small proportional specimen at room temperature, 6 MPa/s up to $Rp_{0,2}$ and 5 %/min afterwards. A 20mm clip-gage was used to determine the exact strain values.

If technological post-deformation techniques like rolling, forging or deep drawing are applied to SPD CP-Titanium, a great difference will be observed between cold rolled initial CP-Ti and ECAP processed SPD CP-Ti. Using Z (the reduction of area), it can be assumed, that the latter has got the same or even slightly increased formability (Z = 50–60 %) than coarse-grained, annealed CP-Ti (Fig.4), although the strength has been dramatically increased due to SPD (Fig.3). That means that sheets manufacturing can be carried out roughly in the same way as it is used for coarse-grained, low-strength CP-Ti. However, texture will play an important role (see weak dotted lines in the right hand side picture in Fig.4) and needs to be clarified in more detail. 53 % cold rolled CP-Ti exhibits a low formability of Z = 36 %.

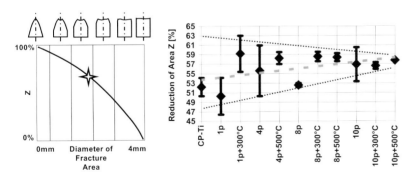

Figure 4: Z as a measure for formability; Left hand side: general dependence of the reduction of area versus diameter of fracture area (side-view of the fracture area of a tensile test specimen on top of the diagram) [17]; Right hand side: formability of different pre-deformed CP-Ti samples measured Z in tensile tests (see Fig.3) and the deviation of perpendicular cross section diameters by error bars indicating non uniform deformation

3 SEM Fracture Analysis

The microstructure has a great influence on the fracture behaviour. Therefore it is assumed that fracture surfaces can reveal new insights into the microstructure-properties relationship.

From Fig.5-7 we can see, that the dimple size distribution per equal area, gets more homogeneous by additional ECAP passes. That is clearly confirmed by the standard deviation σ.

If we compare the yield strength values (Fig.3) with the mean dimple sizes (Fig.5–7) we find a ratio of $Rp_{0,2}$ increase to dimple size decrease of approximately 0,66 (tab.2). The dimple size strongly correlates with the yield strength in both, coarse-grained and SPD CP-Ti states.

Table 2: Yield strength and dimple size evolution after different numbers of ECAP passes:

	$Rp_{0,2}$ [%]	Dimple Size [%]	$Rp_{0,2}$/Dimple Size
0pass to 1pass	+40	-58	0,68
1pass to 8pass	+16	-25	0,64

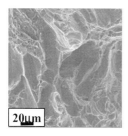 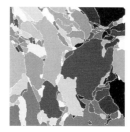 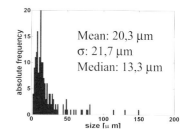

Figure 5: Fracture surface of coarse-grained CP-Ti and area equivalent circle diameter analysis (Matlab routine)

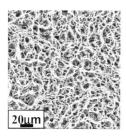 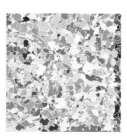 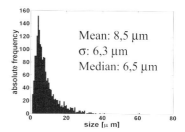

Figure 6: Fracture surface of 1pass SPD CP-Ti and area equivalent circle diameter analysis (Matlab routine)

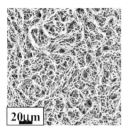 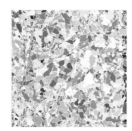 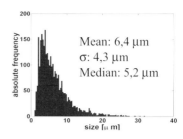

Figure 7: Fracture surface of 8pass SPD CP-Ti and area equivalent circle diameter analysis (Matlab routine)

4 Conclusions

- It has been found, that no major cracks (≥ 75 µm) are visible in the as processed ECAP CP-Ti billets up to 8passes.
- From 1pass to 8pass a homogeneous microstructure formation with an increasing amount of high angle grain boundaries and recovered grains was visible. This homogenisation is strongly linked to the 200–300 nm (width) plate like boundary structures after 1pass (Fig.2 left hand bottom), which seems to be the limit in the achievable mean globular grain-size after an increasing number of ECAP passes at a given temperature and die angle (90°).
- The tensile properties increase (strength & ductility) with increasing ECAP passes. The ductility profits from the stress homogenization due to the developing, uniform "multi-modal" structure. The influence of these features, however, is not clarified so far and numerical models need to explain this behaviour in more detail. Furthermore, the role of impurities needs to be taken into consideration [18].
- The formability at room temperature, measured by tensile parameters, does not change significantly after ECAP, when compared to the annealed initial state. Cold rolled CP-Ti, starting from annealed material is about 35 % less ductile. Reasons might be the different temperature, the texture, another deformation path, and others. These dependencies are under current investigation.
- Fracture analysis exhibits a dependence of the yield strength on the dimple size. Its ratio stays constant for the whole fracture surfaces, from 0pass to 8pass ECAP CP-Ti. That means that important micro- or nanostructural features concerning the SPD materials' mechanical behaviour may be closely linked to the fracture dimples. Additional experimental techniques will be used to check out this linkage.
- The fracture mechanisms in tension seem to act similary in coarse-grained and SPD materials, although the stress levels are shifted dramatically. The enhanced ductility may be pretty well linked to the higher number of grain boundaries per area and the resulting well distributed stresses at an elevated level. If the type of boundaries plays a big role, or if it is just the general geometric distribution of structural elements remains unclear. Further investigations by means of experiments and numerical models will help us to get to know more about SPD and the post-deformation behaviour of SPD materials.

5 Acknowledgement

The author is indebted to Prof. R.Z. Valiev and his team at the Ufa State Technological University, Dr. P. Hahn and W. Costin from ARC Seibersdorf for their contribution with special knowledge and assistance. Furthermore the author acknowledges with gratitude the very necessary contributions of the technicians M. Rohrer, R. Blach and Ing. H. Lichtl as well as many others involved.

6 References

[1] V.V. Stolyarov, V.V. Latysh, R.Z. Valiev, Y.T. Zhu, T.C. Lowe in Investigations and Applications of Severe Plastic Deformation, NATO Conference, Kluwer Academic Publisher, Netherlands, 2000, 367–372
[2] D. Kuroda et al., Mater. Sci. Eng. A 1998, 243, 244–249
[3] V.V. Stolyarov, Y.T. Zhu, I.V. Alexandrov, T.C. Lowe, R.Z. Valiev, Mater. Sci. Eng. A 2001, 299, 59–67
[4] Y. Iwahashi, J. Wang, Z. Horita, M. Nemoto, T.G. Landong, Scripta Mater. 1996, 35, 143–146
[5] B. Mingler, L. Zeipper, H.P. Karnthaler, M. Zehetbauer, this issue
[6] I. Kim, J. Kim, D.H. Shin, X.Z. Liao, Y.T. Zhu, Scripta Mater. 2003, 48, 813–817
[7] I. Kim, J. Kim, D.H. Shin, C.S. Lee, S.K. Hwang, Mater. Sci. Eng. A, 2003, 342, 302–310
[8] D.H. Shin, I. Kim, J. Kim, Y.T. Zhu, Mater. Sci. Eng. A, 2002, 334, 239–245
[9] R.Z. Valiev, T.C. Lowe, A.K. Mukherjee, JOM 2000, 27–28
[10] T. Ungar, I. Alexandrov, M. Zehetbauer, JOM 2000, 34–36
[11] M. Zehetbauer, Acta Metall. Mater. 1993, Vol. 41, No. 2, 589–599
[12] M. Zehetbauer, P. Les, Metall. Mater., 36, 1998, 153–161
[13] S. Balasubramanian, L. Anand, Acta Mater. 2002, 50, 133–148
[14] H.S. Kim, Y. Estrin, M.B. Bush, Acta Mater. 2000, 48, 493–504
[15] V.V. Stolyarov, Y.Th. Zhu, I.V. Alexandrov, T.C. Lowe, R.Z. Valiev, Mater. Sci. Eng. A 2003, 343, 43–50
[16] R. Boyer, G. Welsch, W.W. Collings, Materials Properties Handbook: Titanium Alloys, 2nd ed., ASM International, Materials Park OH, 1998
[17] K. Lange, Umformtechnik, Band1: Grundlagen, 2nd ed., Springer, Berlin Heidelberg, 1984, Studienausgabe 2002
[18] Yu.R. Kolobov, O.A. Kashin, E.E. Sagymbaev, E.F. Dudarev, L.S. Bushnev, G.P. Grabovetskaya, G.P. Pochivalova, N.V. Girsova, V.V. Stolyarov, Russian Physics Journal 2000, Vol. 43, No. 1, 71–77

Ways to Improve Strength of Titanium Alloys by Means of Severe Plastic Deformation

A.A. Popov
Ural State Technical University, UGTU-UPI, Ekaterinburg, Russia

1 Introduction

Titanium alloys are regarded as materials with a high strength-to-weight ratio. As a rule, to obtain high-strength characteristics in such alloys, a two-phase structure is formed and precipitation of the second phase, which is comparatively fine-structured, is regulated by means of creating the corresponding dislocation structure in the matrix. At modern industrial facilities the achievable level of hardening for titanium alloys can be as high as 1400–1450 MPa, while in laboratory conditions such alloys can be hardened up to 1500–1600 MPa, with satisfactory plastic characteristics retained [1–2]. Usually, such properties are observed in materials with relatively coarse (30–50 µm) β-grains. When low-alloyed metals are used (with the molybdenum equivalent less than 4–5 %) then the size of relatively globular phases obtained by thermostrengthening is reported to be of 4–5 µmBy forming an ultrafine-grained structure, severe plastic deformation methods, e.g. high pressure torsion (HPT) or equal channel angular pressing (ECAP) [3], should enhance high strength which otherwise would be impossible to obtain by traditional thermomechanical treatment. As this takes place, fatigue characteristics are expected to get improved as well.

2 Results and Discussion

For commercially pure titanium (VT1-0 and Grade 2 Ti) it has been shown [4] that the strength can reach a value that is two or three times larger than the initial values of σ ($\sigma_{0,2}$) in the coarse-grained state. Thus, for instance, if commercially pure titanium would normally have the tensile strength of 450–500 MPa, then after severe deformation by high pressure torsion it can reach up to 1000–1100 MPa [4]. The structure thus formed is fragmented with grain boundaries of 70–80 nanometers in size (Figure1). Predominantly, these fragments have high-angle boundaries and are of an equiaxial shape.

Such structure is formed as a result of the dislocation and disclination mechanisms of severe deformation. Many of new boundaries have a deformation origin and possess long-range stress fields, which is attested by diffraction contrasts revealed at such boundaries [4]. This structure is characterized with low thermal stability and, when heated above 300 °C, intense grain growth is observed. We believe that the original fragments mainly have random orientations at high angles (see the electron diffraction pattern in Figure 1a). At the same time, heating above 450 °C led to the enlarging of grains up to 550–600 nm in size at an average and as this occurred all the boundaries would became equiaxial.

When alloys are subjected to severe plastic deformation, the strengthening effect is considerably less than for pure metals. This is associated both with a large initial strength and with a par-

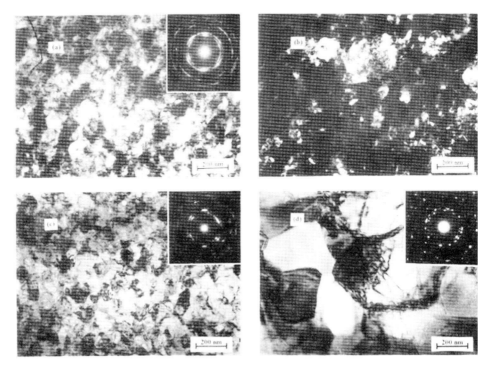

Figure 1: Micrographs of structure of commercially pure Ti after SPD (a) SPD-processed state, bright field image, (b) SPD-processed state, dark field image, (c) annealing 300 °C for 2 hours, (d) – annealing 350 °C for 2 hours

tial metastable phase decomposition during the following heat treatment. For the Ti6Al4V alloy we found that severe plastic deformation increases the level of strength by not more than 50 % (i.e. from 900 up to 1300 MPa). At the same time, it should be taken into account that by applying heat treatment the strength of these alloys can be increased up to 1200 MPa, and their plastic properties would remain comparatively high in this case. Besides, the thermal stability of the alloy structure subjected to severe plastic deformation is considerably lower than that for the heat-treated alloy.

Table 1 shows strength values for several titanium-based alloys after various standard heat treatments. Assuming that the different hardening mechanisms contribute to the strength of the alloy in an additive way, one can write an expression for the yield strength [5]:

$$\sigma_{0.2} = \sigma_{s.s.} + \Delta\sigma_b + \Delta\sigma_{def} + \Delta\sigma_{s.ph.}$$

where $\sigma_{s.s.}$ – solid solution strength; $\Delta\sigma_b$ – grain boundary strength; $\Delta\sigma_{def}$ – dislocation deformation strength; $\Delta\sigma_{s.ph}$ – strength of the second phase particles.

Unfortunately, in the literature too little data exist on actual values of the yield strength for titanium-based alloys and therefore such equation could be only taken with some uncertainty for the tensile strength.

The strength components were evaluated by standard expressions applied in the strength theory for polycrystalline alloys. As follows from the results presented, severe plastic deformation achieves an appreciable benefit in strength only for commercially pure titanium, when strain-hardening mechanisms cannot involve the second phase precipitation. When the alloy with an significant amount of second phase is subjected to severe deformation, then the hardening effect of grain refining considerably exceeds the hardening effect of aging. This could be explained on the basis that the stress fields which result from deformation, become considerable obstacles to the nucleation of the second phase thus hindering the process of decomposition of the metastable solid solution.

Table 1: Strength properties of different Ti-alloys after different mechanical and thermal treatment. $\sigma_{s.s.}$ is the solid solution strength; $\Delta\sigma_b$ the grain boundary strength; $\Delta\sigma_{def}$ stands for deformation strength, and $\Delta\sigma_{s.ph}$ for the strength of the second phase particles

Alloy	Treatment (resulting structural state)	$\sigma_{s.s}$	$\Delta\sigma_b$	$\Delta\sigma_{def}$	$\Delta\sigma_{s.ph}$	$\Sigma\sigma$
V1-0	Annealing (coarse grains)	450	–	–	–	450
	+cold deformation 50 %	450	–	150	–	600
	Annealing, fine grains ($D = 10$ μm) + cold deformation 50 %.	450	300	100	–	850
	SPD + Cold Deformation	450	450	150	–	1050
6-4	Annealing (80%α+20%β)	650	150	-	150	950
	+cold def. 30 %	650	150	100	150	1050
	+aging at 500 °C	630	150	50	300	1150
	Annealing+SPD	650	400	100	150	1300
15-3-3	Quenching from 900 °C (100% of β-phase)	900	–	–	–	900
	+ aging at 500 °C	900	–	–	450	1350
	Cold def. 50% + aging at 500 °C	900	–	100	450	1450
	SPD +aging at 500 °C	900	250	100	200	1450
V22	Quenching from 950⁰ (100 % of β-phase)	850	–	–	–	850
	+ cold def. 50 %	850	–	100	–	950
	+ aging at 500 °C	800	–	100	500	1400
	Annealing (50%α+50%β)	800	50	–	50	950
	Annealing +SPD	800	400	100	50	1350
	Quenching+SPD	850	350	100	–	1300
	Quenching+SPD+aging	850	300	100	100	1350

An exception to this is the alloys with the matrix being unstable under deformation. This effect is the most clearly pronounced in the alloys with a molybdenum equivalent of 10–12 (Figure 2), when strain-induced fine martensite is formed and when its decomposition, which

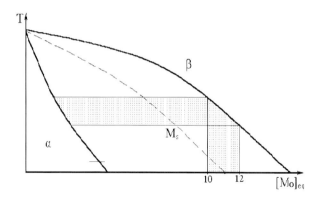

Figure 2: Metastable phase diagram of Ti - β-stabilized alloys

Figure 3: TEM micrograph of SPD-processed titanium alloy VT22, bright field image ; ×100000

follows the heat treatment, leads to a nearly nano-grained state with a phase ratio approaching 1:1.

As the example for V22 alloy (Figure 3) has shown, a strength level of 1850–1900 MPa can be achieved, with plastic characteristics remaining sufficiently good. Such properties are obtained via warm deformation of the structure containing 10–15 % of the rhombic phase, followed by aging at temperatures 100–150 °C below the deformation temperature. Such deformation provides the forming of the β-phase with tetragonal distortions, which is characterized by improved strength and high susceptibility to subsequent thermostrengthening.

3 Conclusions

The efficiency of severe plastic deformation is considerable for commercially pure titanium and for Ti alloys which are sensitive to phase transformation under deformation. In cases where thermal treatment can be applied to improve the strength of alloys by means of precipitation of second phase particles, the application of severe plastic deformation is less effective.

4 References

[1] Proceedings of the 9th Intern. Conf. of Titanium: Science and Technology. S. Petersburg. 1999
[2] V.K. Aleksandrov, N.F. Anoshkin, A.P. Belozerov, et.al., Half-finished Products of Titanium Alloys. M. 1996, 581
[3] R.Z. Valiev, R.K.Islamgaliev, I.V. Alexandrov, Progress Mater. Sci., V. 45, 2000, 103–189
[4] A.A. Popov, I.Yu. Pyshmintsev, S.L. Demakov, et al., Scripta Mater., V. 37, No.7, 1997, 1089–1094
[5] M.I. Goldshein, V.S. Litvinov, M.F. Bronfin, Metallophysics of high-strength alloys. .: Metallurgia, 1986, 312

Structures, Properties, and Application of Nanostructured Shape Memory TiNi-based Alloys

V.G. Pushin
Institute of Metal Physics, Ural Division of Russian Academy of Sciences, Ekaterinburg, Russia

1 Introduction

The TiNi-based alloys in the group of metallic alloys with thermoelastic martensitic transformations (MT) are distinguished by a combination of attractive characteristics such as strength and plasticity, unique shape memory effect (SME), superelasticity, high reliability, endurance, weldability, corrosion resistance, biological compatibility [1–3]. Moreover, these alloys have relatively simple chemical compositions and technological stages of production. These circumstances ensure their wide and exclusive application as a functional material of new generation in fields like medicine [1–3] and others.

On the other hand, recent studies have shown that nanocrystalline materials (NCM) with grain sizes less than 100 nm, and submicrocrystalline (SMC) materials with a mean grain size of a part of micron can demonstrate properties significantly different from those of corresponding coarse-grained ones [4–6]. Their superior properties include a combination of high strength and ductility, high strain rate and low temperature superplasticity [4–8]. Various methods of production of NCM and SMC materials have been developed [4–12], but only for commercial alloys. Methods of severe plastic deformation (SPD) [6, 8], e.g. high pressure torsion (HPT) and equal channel angular pressing (ECAP), are of special interest.

The aim of this paper is to report our own investigations on shape memory B2-titanium alloys, that are produced by methods of severe plastic deformation (SPD), superrapid quenching (SRQ), and combinations of cold plastic deformation and thermal treatments. The present work deals with a study of initial substructures, structural evolution and their recovery during subsequent annealing, as well as with peculiarities of MT in such alloys and related phenomena.

2 Experimental

In the present work, binary and multicomponent (including Cu, Fe, Co) TiNi-based alloys were investigated. SRQ-alloys were produced by melt spinning at cooling rates of 10^4–10^7 K/s in an inert gas atmosphere as ribbons 30–100 µm thick.. Samples for severe plastic torsion straining under high pressure (HPT) were discs, 10–12 mm in diameter and 0.3–0.5 mm in thickness. The other group of SPD TiNi specimens, 18 mm in diameter and 100 mm in length, was obtained by means of equal channel angular pressing (ECAP) [12]. True logarithmic strains of about 5–7 were imposed. Also, combinations of iterative large plastic deformation (RPD by rolling and drawing) and annealing were used.

The structural states and phase composition of specimens were examined by X-ray diffraction (XRD, using DRON-3M and DRON-4M diffractometers), transmission electron microscopy (TEM) and selected area electron diffraction (SAED), using a JEOL 200CX TEM, operated at 200 kV, including in situ heating and cooling experiments. The nanocrystallization temperatures and the critical points of MT of the alloys were determined from the curves of the temperature dependence of electrical resistance $\rho(T)$ and magnetic susceptibility $\chi(T)$. To determine the deformation characteristics of SME the effect of strain by bending on shape restoration of the alloys was studied. Tensile tests and measurements of microhardness were performed at room temperature.

3 Results and Discussion

3.1 Structure and Phase Transformations in TiNi-Polycrystalline Alloys

It is known that cooling below the M_S' temperature of a number of the B2 TiNi-based alloys in single crystalline and usual coarse-grained (with a mean grain size of 50–80 µm) states leads to the formation of the R-martensite with a rhombohedrical (or hexagonal) symmetry [1–3]. The rhombohedrical lattice parameters of the R-martensite are $a_R \cong 0.9$ nm, $\alpha_R = 89.5°–89.0°$, and $a_H \cong 0.734$ nm, $c_H \cong 0.528$ nm – for the hexagonal structure [3]. The volume change $\Delta V/V$ upon B2 → R MT is –(0.01–0.02) %. However, after transition the linear distortions along and perpendicular to the axis of the rhombohedrical extension $[111]_{B2}$ are anisotropic and their magnitudes became larger by two order of magnitude after cooling in cryogenic field ($\Delta c/c = +1.0$ %, $\Delta a/a = -1.0$ %). Therefore, the grains of an intermediate size may contain tens of packets that consist of crystals twinned pairwise on the habit and twinning planes $\{110\}_{B2}$ and $\{100\}_{B2}$. The latter variant is much more rare.

In binary and multicomponent TiNi-based alloys at certain concentration of a third elements (e.g. ≤ 3 at.% Fe, ≤ 7 % Co, ≤15 % Cu), cooling below the M_S temperature leads to the formation of the monoclinic B19'-martensite. A unit cell of B19'-martensite in TiNi has lattice parameters at room temperature close to $a = 0.289$ nm, $b = 0.412$ nm, $c = 0.462$ nm, and $\beta = 97°$ [3]. In TiNi-based alloys at certain concentration of such elements as Cu, Pd, Pt, Au, cooling to below the M_S point leads to the formation of orthorhombic B19-martensite [3]. $\Delta V/V$ upon B2 → B19' and B2 → B19 MT is –(0.03–0.04) % [3]. The linear distortions along axises of $<100>_{B2}$ and $<110>_{B2}$ are also anisotropic and greater ($\Delta a/a = -1.0$ %, $\Delta b/b = -2.0$ %, $\Delta c/c = 1.5$ %). Note the variety of morphological variants of the B19'- and B19-martensites. A number of lentiqular and platelike crystals is observed, but more frequently, especially near and below the M_f temperature, they are grouped into packets of pairwise twinned crystals. They can contain various internal defects, such as twins and stacking faults. Thus, the spatial self-accommodation of anisotropic elastic stresses caused by the thermoelastic MT is achieved on a microscopic level due to the formation of microtwins and on a mesoscopic level (within a grain) due to the formation of coherent crystals of packed-pyramidal morphology.

3.2 Submicrocrystalline TiNi-Based Alloys

The methods of production of a bulk SMC materials can be classified into three groups based on their approaches: SRQ, SPD, and combinations of RPD and subsequent annealing. Firstly, all investigated TiNi-based alloys can be produced using the SRQ-technique, by melt-spinning [9-11]. Secondly, the TiNi alloys were processed using also severe ECAP [12]. It is shown that both SRQ and ECAP results in formation of equi-axed granular SMC structure. The SAED patterns exhibit rings of spots, which indicate that within the field of view are contained many ultrafine grains with random high-angle misorientations. The average grain size measured by TEM was 0.2–0.6 μm depending the treatment regimes. In the TEM images of ECAP samples some grains are visible but many grain boundaries are poorly defined. Their diffraction contrast is not uniform and it changes in a complex way due to high distortions localized both inside grains and on their wavy boundaries. Moreover, these grains has subgrain structure with high density of dislocations observed on the background of bent and broadened extinction contours. According to TEM data, sufficiently intense fundamental B2 reflections broadened at their bases are discovered in XRD patterns, testifying to high internal stresses and distortions in the alloys after severe ECAP in comparison with SRQ-alloys. Thirdly, TiNi-based alloys were processed using moderate plastic deformation coupled with intermediate annealing (with a reductions of about 10 % per pass at deformation and annealing temperatures decreased from 700 °C down to room temperature). Such combined method of RPD permits one to obtain sheets, bands or wires with structure of non-equiaxed SMC grains. The TEM study of samples after SRQ, ECAP or RPD and annealing at 200–500 °C shows that grains change only little in size, but the density of dislocations decrease within every grain. As the annealing temperature for TiNi is increased to 500 °C, most of ultrafined grains are nearly free from lattice dislocations, the mean grain size becomes at least twofold greater, the grain boundaries become well defined, suggesting some recovery of structure.

It was found that the critical temperatures of B2 → R MT weakly change, whereas those of the MT B2(R) → B19' and B2 → B19 decrease more substantially (by 20°–40°). Using in situ low-temperature TEM, we established that MT in the as-processed TiNi began with the formation of thin single-crystal R-, B19-, or B19'-martensite plates nucleated both inside the ultramicrograins and heterogeneously at their boundaries. The MT is completed by the formation of a single-packet twinned structure (Fig.1,2). The coherent inter-packet boundaries of the martensite are stepwise or even wavy rather than strictly planar. In the coarsest grains, the martensite can consist of groups of two or three packets, whereas in finer grains it is present as single-packet group. The packets of twinned crystals in neighbouring grains make large angles close to 60° or 90° with one another. Comparing the average sizes of the R-, B19- and B19'-martensite

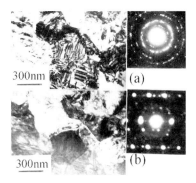

Figure 1: TEM bright- (a) and dark-field (b) micrographs and SAED patterns of R-martensite in Ti49.5Ni50.5 at 0 °C as-processed by ECAP

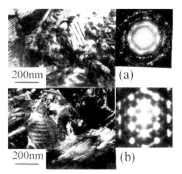

Figure 2: TEM bright- (a) and dark-field (b) micrographs and SAED patterns of B19'-martensite in $Ti_{49.5}Ni_{50.5}$ at −100 °C as- processed by ECAP

crystals formed in as-processed and usual undeformed polycrystalline alloys of given composition, we should note that the plate martensite crystals in as-processed state are substantially smaller both in length and in thickness. Finally, the SMC alloys as-processed by SRQ, severe ECAP, or RPD are characterized by a crystallographic and microstructural texture in martensite and austenite states.

3.3 Amorphous and Nanostructured TiNi-Based Alloys

Binary Ti-Ni (with Ti-content more 55 at. % or less 45 at. %) and quasi-binary TiNi-TiCu alloys (containing more 20–22 at.% Cu) produced by melt spinning at quenching rates $V_q = 10^6$–10^7 K/s are amorphous [9]. It is shown that all the alloys TiNi and TiNiFe after HPT are amorphous or amorphous-nanocrystalline (Fig.3 a, b) [12]. For these alloys, typically wide diffuse maxima are observed in XRD patterns, and halos near the angular positions of the fundamental reflections to form are observed in the SAED patterns. The amorphous alloys are stable at low temperatures. Nanostructured materials can be made from amorphous ones using nanocrystallization annealing of SPD TiNi-based alloys by heating up to temperatures between starting temperature of nanocrystallization $T_s \approx 230$–240 °C and finishing one $T_f \approx 300$ °C. Their grain size is very small and on average equals to 10–20 nm, 20–30 nm, 60–70 nm after annealing for 20 min at 250 °C and 300 °C, and for 5 min at 500 °C, respectively (Fig.3,4). SRQ alloys are more stable and undergo nanocrystallization at heating to temperatures 450–500 °C. Using in situ low-temperature TEM and XRD, we established that in nanostructured HPT and SRQ TiNi-based alloys cooled below the M_s' or M_s temperatures nucleation and growth of R-, B19'- or B19-

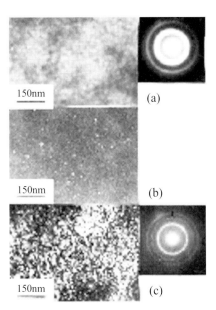

Figure 3: Bright- (a, c) and dark-field (b) TEM micrographs and SAED patterns of TiNi-based alloys as-processed by HPT ($n = 10$) at room temperature in amorphous state (a, b) and nanostructured state after annealing at 250 °C, 20 min (c).

martensites occur by "single nanocrystal – single nanocrystal" mechanism without internal microtwins and stacking faults [9–12]. Therefore, it is apparent that in this case the quasi-isotropic spatial compensation of elastic stresses is attained outside a single nanograins, in a group of many neighbouring nanograins which form the self-accommodating elastic system.

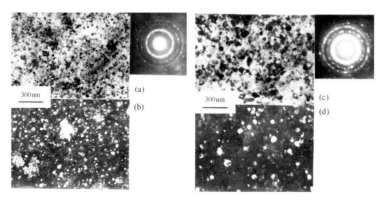

Figure 4: Bright- (a, c) and dark-field (b, d) TEM micrographs and SAED patterns of nanostructured R- (a, b) and B19'-martensites (c, d) in TiNi-based alloys as-processed by HPT ($n = 5$) at room temperature and annealed at 300 °C for 20 min (a, b) or at 500 °C for 5 min (c, d) taken at cooling temperature of 0 °C (a, b) and –100 °C (c, d)

3.5 Mechanical Properties of Nanostructured TiNi-Based Alloys

The room temperature tensile tests of TiNi-based alloys in quenched coarse-grained state, after ECAP, HPT, and RPD showed that the alloys in the as-processed condition exhibit maximum strengthening in the nanocrystalline state after annealing at 200–300 °C (ultimate strength σ_u = 2.5–2.6 GPa) and in the submicrocrystalline ones (σ_u =1.5–1.7 GPa). The measurements of the shape memory effects (SME) showed that a complete shape recovery (S = 95–100 %) occurred after a bending deformation γ below 5–7 %. The thermal cycling of the nanocrystalline TiNi-based alloys preliminarily bent in the B19'-martensitic state revealed a spontaneous reversible shape memory effect (RSME), which may reach 10–15 % of the conventional "one-way" SME. Such RSME in ECAP TiNi alloys is practically absent. These materials have high recovery stresses (up to 1 GPa) generated by memory elements. The microhardness of TiNi-based alloys in the usual coarse-grained and nanostructured states is near 3.0 and 5.0 GPa, accordingly.

3.6 Application

Numerous engineering and medical applications of TiNi materials with SME have been developed to date. Here we illustrate a few of our own inventions. In Russia, fire detectors, which we designed from high strength TiNi-based alloys, are successfully mounted and exploited for several years. In the late 1980's, we designed unique facilities for endoscopic extraction of concrements from hollow organs (ureter, gall ducts, etc.), their dilatation and endoscopical elecrosurgery in urology, proctology and gastroenterology using self-spreading instruments made of high strength TiNi wires with SME produced by cold drawing [3] (Fig.5). These facilities are widely used in Russian urology, surgery and gastroenterology and demonstrated in China, France, Germany, Israel and USA. For weakly invasive osteoplasty we designed a number of new devices with SME for treating fractures of femur neck [3]. The recent fabrication of nanostructured TiNi alloys with enhanced functional properties is expected to be very efficient for further developments and new practical use.

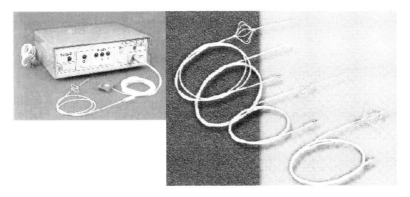

Figure 5: Facilities for endoscopic extraction of concrements from hollow organs (ureter, gall ducts, ets), their dilation and endoscopical elecrosurgery in urology, proctology and gastroenterology with self-spreading instruments from high strength TiNi wires with SME

4 Conclusions

Various methods of production of submicrocrystalline TiNi-based alloys, including superrapid solidification from melt, severe plastic deformation, and combined method of repeated plastic deformation and annealing, have been developed during recent years. Moreover, nanocrystalline TiNi-based alloys can be processed using the techniques of the super-rapid solidification or severe plastic torsion straining under high pressure providing a fully amorphous state with their further annealing. These TiNi-based alloys exhibit high strength behaviour, thermoelastic martensitic transformations and enhanced related phenomena, including shape memory effects. In nanostructured alloys cooled below the martensitic point the nucleation and growth of martensite occur by "B2-austenite single nanocrystal – martensite single crystal" mechanism without twinning. In SMC alloys (with B2-grain size more than 100–200 nm) martensite has twinned single-packet morphology. It is concluded that nanostructured TiNi-based alloys have remarkable microstructures and mechanical properties and that there exist possibilities for using them as high strength shape memory materials with high recovery stress in the fields of medicine and engineering.

5 Acknowledgement

The present work was supported in part by grants of INTAS, no. 99-01741, 01-0320, RFBR no. 02-02-16420, and ISTC no. 2398r-2002.

6 References

[1] Miyazaki S., Otsuka K., ISTJ International 1989, 29, 353–377
[2] Pushin V.G., Kondratjev V.V., Phys Met. Metallogr. 1994, 7,. 497–511
[3] Pushin V.G., Phys. Met. Metallogr. 2000, 90, S68–S95
[4] Gleiter H., Progress Mater. Sci. 1989, 33, 223–315
[5] Koch C.C., Cho Y.S., Nanostructured Materials 1992, 1, 207–212
[6] Valiev R.Z., Korznikov A.V., Mulyukov R.R., Mat. Sci. Enginer. 1993, A168, 141–148
[7] Siegel R.W., Fougere G.E., Nanostructured Materials 1995, 6, 205–216
[8] Valiev R.Z., Islamgaliev R.K., Alexandrov I.V. Progress Mater. Sci. 2000, 45, 103–189
[9] Pushin V.G., Volkova S.B., Matveeva N.M., Phys. Met. Metallogr. 1997, 83, 275–288, 435–443
[10] Pushin V.G., Popov V.V., Kuntsevich T.E., Phys. Met. Metallogr. 2001, 91, 374–382, 486–493
[11] Pushin V.G., Kourov N.I., Kuntsevich T.E., Phys. Met. Metallogr. 2001, 92, 58–69
[12] Pushin V.G., Stolyarov V.V., Valiev R.Z., Kourov N.I., Kuranova N.N., Prokofiev E.A., Yurchenko L.I., Ann. Chim. Sci.Mat. 2002, 27, 77–88

Microstructure and Properties of a Low Carbon Steel after Equal Channel Angular Pressing

J. Wang[1], C. Xu[2], Y. Wang[1], Z. Du[1], Z. Zhang[1], L. Wang[1], X. Zhao[1] and T. G. Langdon[2]
[1]School of Metallurgical Engineering, Xi'an Univ. of Arch. & Tech., Xi'an, P. R. China
[2]Depts. of Mech. Eng. and Mater. Sci., Univ. of Southern California, Los Angeles, USA

1 Introduction

Severe Plastic Deformation (SPD) processing, especially Equal-Channel Angular Pressing (ECAP) of steel and other iron-based alloys, has achieved increasing interest recently because of the potential for obtaining ultrafine grains and improving the properties [1–8]. Shin and coworkers [1–5] have conducted extensive investigations on ECAP processing of a 0.15 wt% C commercial carbon steel, usually at 623 K using route C. A near-equiaxed ultrafine grain structure of ~0.2 µm resulted from 4 passes of ECAP [1] and this led to a substantial increase of the yield stress (YS) to ~937 MPa and the ultimate tensile strength (UTS) to ~943 MPa with reasonable elongations of ~10 % [2,3]. Investigations by Fukuda et al. [8], conducted at room temperature using route B_c, achieved an essentially equiaxed microstructure with high grain boundary misorientations and an average grain size of ~0.2 µm with a UTS of ~800 MPa after ECAP through only three passes. These latter experiments used a carbon steel with a fairly low carbon content of 0.08 % C and there was an extensive region of strain hardening and a reasonably high elongation to failure in tensile testing at room temperature.

Although a number of investigations are therefore available on the application of ECAP to low-carbon steels, the results reveal some contradictions. In the work of Shin et al. [1] an equiaxed microstructure was developed after 2 passes of ECAP using route C while in the investigation of Fukuda et al. [8] the extensive equiaxed microstructure only formed after 3 passes of ECAP using route B_c in a steel with only 0.08 wt% C. On the other hand, an analysis of shearing characteristics [9] plus experimental observations on aluminum and aluminum-based alloys [10-12] suggest that route B_c is more effective than route C for forming equiaxed homogeneous microstructures and new high-angle boundaries. The present investigation was conducted in part to examine the essential features behind this contradiction.

Although Shin and coworkers [4] showed the critical effect of processing temperature on microstructure refinement, nearly all of the ECAP processing on carbon steel has been conducted at elevated temperatures (significantly higher than room temperatures) [1–4,7] with only a small number of tests at room temperature [6,8]. Another noteworthy point is that, regardless of the deformation routes and temperatures used, ECAP processing of carbon steel is seldom taken beyond 4 passes. This implies there is an upper limit for the ECAP of carbon steel since 4 passes of ECAP, giving an accumulated strain of ~4 [13], is only at the very initial stages of severe plastic deformation and, for microstructure refinement, an accumulated strain over ~4–7 is generally needed [14]. Thus, the present investigation was also designed to process a commercial plain carbon steel at room temperature to a higher number of passes and to investigate the structure and property evolution of the steel during ECAP.

2 Experimental Materials and Procedures

A hot-rolled commercial low carbon steel, containing 0.15 wt% C, was used for the present experiments. The composition of this steel is shown in Table 1. Figure 1 shows a typical optical microstructure. Square bar-shaped samples were used for the present investigation to permit easy rotation by 180° between passes and to prevent any undesirable rotations of the samples during ECAP. Samples were cut from the hot-rolled plate along the longitudinal direction with sizes of 15×15×60 mm for ECAP processing.

Table 1: Composition of the experimental alloy (wt%)

Element	C	Si	Mn	P	S	Cr	Ni	Cu	Fe
Content	0.15	0.17	0.52	0.019	0.021	0.10	0.10	0.10	Bal.

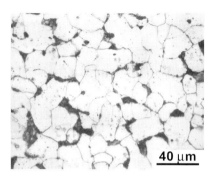

Figure 1: Optical microstructure of the as-received low carbon steel

ECAP was conducted at room temperature using route C with a rotation of 180° between passes. In the present study, the intersecting angle between the two equal channels was $\pi/2$.

Transmission electron microscopy was used for the structure characterization on the transverse cross-sections of the samples before and after ECAP from 1 to 11 passes. The foils were thinned with a twin-jet polisher using a solution of 5 % $HClO_4$, and 95 % C_2H_5OH. A JCM-200CX electron microscope operating at 160 kV was used to examine the foils. The aperture size for selected area diffraction (SAD) was 2.5 µm in diameter.

Tensile specimens were cut from the ECAP bars along the longitudinal direction. Tensile tests were conducted at room temperature using an Instron machine with the cross-head speed adjusted to give an initial strain rate of $1.0 \cdot 10^{-3}\,s^{-1}$.

3 Experimental Results

3.1 Microstructures

Figure 2. shows typical TEM microstructures, plus the corresponding SAD patterns from an area of 2.5 µm in diameter, taken on the transverse sections after ECAP at room temperature.

After one pass, the microstructure consists of nearly parallel bands of irregular elongated structures, with dislocation tangles slanting inside the bands and the bands divided into short sections. The band width is ~0.3–0.4 μm. The SAD pattern is characterized by diffusive spots in an array of a single crystal net pattern, indicating lattice distortion or small-angle misorientations between the bands. Although there appears to be some equiaxed tangled structures after 2 passes, the microstructure is essentially banded and elongated with a width of ~0.3–0.4 μm and with similar characteristics as after one pass.

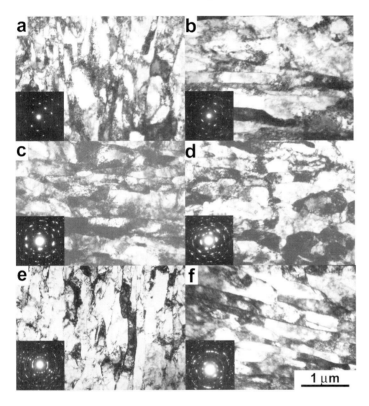

Figure 2: Typical TEM microstructures on transverse cross-sections of samples of the low carbon steel after ECAP for (a) 1, (b) 2, (c) 6, (d) 8, (e) 10 and (f) 11 passes

The diffraction spots in the SAD single crystal net pattern after 2 passes are spread into arcs indicating higher angle, but still relatively low angle, misorientations between the bands. This trend of microstructural evolution extends up to 6 passes when the structure is characterized with a high density of dislocation tangles. After 8 passes, the dislocation density starts to decrease and the diffraction spots in the SAD patterns are so spread that they form nearly discontinuous circles in the high index region; thus, it is then difficult to find the trace of low index spots in the single crystal net pattern. This evolution indicates the low-to-high transition of the misorientation angles between different grains. Some form of recovery of defects, evidenced by the decreasing dislocation density, has thus occurred.

Compared with microstructures reported previously, the distinctive difference after 10 and 11 passes is that it is impossible to find the trace of spots in the array of a single crystal net pattern, even in the low index region. All the diffraction spots, with typical azimuth spreading, arrange themselves randomly to form rings indicating high-angle misorientations. The microstructure still consists of nearly parallel bands of elongated structures, with a band width of ~0.2–0.3 µm but these structure bands are now much lower in dislocation density with much straighter and better defined boundaries, especially after a total of 11 passes. This indicates the occurrence of more recovery although it is clear that the structure remains in a very non-equilibrium state.

3.2 Mechanical Properties after ECAP

Figure 3 gives stress-strain curves of the as-received material and after ECAP. The as-received sample exhibits yielding and strain hardening behavior with a relatively large elongation typical of a conventional low carbon steel. All the samples processed by ECAP exhibit similar behavior with rapid softening after yielding and no evidence of strain hardening after yielding. For samples subjected to 2, 3 and 4 passes, the yielding strength reaches ~800 MPa. For samples pressed through 6 and 8 passes, the yielding strength reaches ~950 MPa and for the sample pressed through 10 passes the yielding strength reaches to higher than 1200 MPa. Compared with the results of Shin et al. [2,3,5] and Fukuda et al. [8], these data indicate that an increase in the number of ECAP passes has a very substantial effect in increasing the strength of carbon steel and this increase in strength occurs to rather a high level.

Figure 4 shows the increase of yielding strength with ECAP passes, together with the change of tensile elongation. After an initial decrease of tensile elongation at low numbers of ECAP passes, there is a subsequent increase when ECAP is continued to more than 4 passes. This trend has similarities to that observed by Fukuda et al. [8] except that they pressed only to a total of 3 passes.

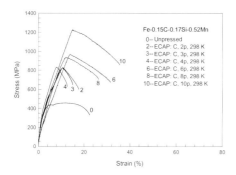

Figure 3: Stress-strain curves of as-received and ECAP samples

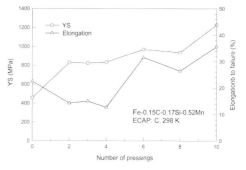

Figure 4: Dependence of yielding strength and elongation on number of ECAP passes

4 Discussion

From the above experimental results, the features of ferrite microstructure evolution in processing by ECAP can be summerized as follows. (i) Nearly parallel bands of elongated substructures are dominant in the ferrite microstructure after ECAP from 1 to 11 passes, with a slight decrease in the band width from ~0.3–0.4 to ~0.2–0.3 µm. (ii) The microstructure is highly non-equilibrium as demonstrated by the high density of dislocation tangles and the azimuth spreading of the diffraction sports; the dislocation density increases with ECAP from 1 to 6 passes, and decreases when ECAP continues beyond 6 passes indicating the structure is then changing towards a metastable state. (iii) The SAD patterns change first from single crystal net pattern with increasing azimuth spreading of the spots corresponding to 1 to 8 ECAP passes and then to a discontinuous ring pattern with extensive azimuth spreading of the spots in samples corresponding to 10 and 11 passes. These features are similar to those of high purity aluminum processed with the same route C at room temperature for the initial 3 passes [9,10]. A difference appears as the ECAP is taken to 4 passes or higher when very few elongated grains were observed after 4 passes in 99.99 % aluminum but the elongated band structure persists up to 11 passes in the 0.15 wt% C plain carbon steel. The formation of equiaxed grains during ECAP of 99.99 % aluminum may be attributed to the possible development of *in situ* continuous recrystallization during ECAP.

In the case of commercial carbon steel, the present observations show that only limited recovery occurs at room temperature. This is further confirmed in Figure 5. This implies that room temperature is not high enough for the activation of recrystallization in the experimental steel even during the severe deformation of ECAP and with the help of any deformation heating. Thus, Shin *et al.* [3] reported that recrystallization of 0.15 wt% C commercial plain carbon steel after ECAP to 4 passes using route C takes place at ~812 K.

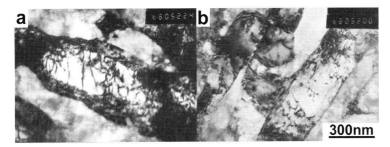

Figure 5: Typical band structures for samples subjected to ECAP for (a) 6 passes and (b) 10 passes

In the investigation of Shin *et al.* [1], an equiaxed structure was achieved after 2 and 4 passes of ECAP at 623 K using route C. Compared with the strongly banded microsturcture obtained in this investigation processed at room temperature, it is concluded that the processing temperature plays a very dominant role in the formation of the structure. In the investigation of Fukuda *et al.* [8], an equiaxed structure was achieved after 3 passes of ECAP of a 0.08 wt% C steel at room temperature using route B_c. Compared with the present strongly banded microstructure using ECAP through route C, it is concluded that the processing route also plays a dominant role in the formation of this processed structure and route Bc is more satisfactory than route C for the formation of an equiaxed structure in carbon steel.

5 Summary and Conclusions

1. Equal-Channel Angular Pressing was successfully carried out on a commercial low-carbon steel at room temperature up to an equivalent true strain of ~11 via route C.
2. Nearly parallel bands of elongated substructures dominated the ferrite microstructure after ECAP from 1 to 11 passes, with a slight decrease of the band width from 0.3–0.4 to 0.2–0.3 µm.
3. The microstructure was highly non-equilibrium after ECAP; the dislocation density increased with ECAP from 1 to 6 passes, and decreased when ECAP continued beyond 6 passes, indicating the structure changed towards a metastable state.
4. The SAD pattern changed first from a single crystal pattern with increasing azimuth spreading of the spots from 1 to 8 ECAP passes, and then to a discontinuous ring pattern with extensive azimuth spreading in samples after 10 and 11 ECAP passes.
5. After 10 passes of ECAP at room temperature using route C, the YS of 0.15 wt% C low carbon steel increases to a high level of over 1200 MPa but with good ductility.

6 Acknowledgements

The work in Xi'an was supported by the National Nature Science Foundation of China under Grant No. 59974018. The work in Los Angeles was supported by the U.S. Army Research Office under Grant No. DAAD19-00-1-0488.

7 References

[1] D.H. Shin, B.C. Kim, Y-S. Kim, K-T. Park, Acta Mater., 2000, 48, 2247–2255
[2] D.H. Shin, C.W.Seo, J.Kim, K-T. Park, W.Y.Choo, Scripta Mater., 2000, 42, 695–699
[3] K-T. Park, Y-S. Kim, J.G. Lee, D.H. Shin, Mater. Sci. Eng., 2000, A293, 165–172
[4] D.H. Shin, J-J. Pak, Y.K. Kim, K-T. Park, Y-S. Kim, Mater. Sci. Eng., 2002, A325, 31–37
[5] H.K Kim, M-I. Choi, C-S. Chung, D.H. Shin, Mater. Sci. Eng., 2003, A340, 243–250
[6] S.L. Semiatin, D.P. DeLo, E.B. Shell, Acta Mater., 2000, 48, 1841–1851
[7] W.J. Kim, J.K. Kim, W.Y. Choo, S.I. Hong, J.D. Lee, Mater. Lett., 2001, 51, 177–182
[8] Y. Fukuda, K. Oh-ishi, Z. Horita, T.G. Langdon, Acta Mater., 2002, 50, 1359–1368
[9] M. Furukawa, Y. Iwahashi, Z. Horita, M. Nemoto, T.G. Langdon, Mater. Sci. Eng., 1998, A257, 328–332
[10] Y. Iwahashi, Z. Horita, M. Nemoto, T.G. Langdon, Acta Mater., 1998, 46, 3317–3331
[11] K. Oh-ishi, Z. Horita, M. Furukawa, M. Nemoto, T.G. Langdon, Metall. Mater. Trans., 1998, 29A, 2011–2013
[12] A. Gholinia, P.B. Prangnell, M.V. Markushev, Acta Mater., 2000, 48, 1115–1130
[13] Y. Iwahashi, J. Wang, Z. Horita, M. Nemoto, T.G. Langdon, Scripta Mater., 1996, 35, 143–146
[14] R.Z. Valiev, R.K. Islamgaliev, I.V. Alexandrov, Prog. Mater. Sci., 2000, 45, 103–189

Formation of Submicrocrystalline Structure in Large-Scale Ti-6Al-4V Billets during Warm Severe Plastic Deformation

[1]S.V. Zherebtsov, [1]G.A. Salishchev, [1]R.M. Galeyev, [1]O.R. Valiakhmetov, [2]S.L. Semiatin
[1]Institute for Metals Superplasticity Problems, Ufa, Russia
[2]Air Force Research Laboratory, Wright-Patterson Air Force Base, USA

1 Introduction

Materials with a submicrocrystalline (SMC) structure have an average grain size less than 1 µm and show enhanced mechanical properties such as increased strength and fatigue resistance [1]. They also exhibit superplastic behavior at temperatures much below the temperature range typical for materials with micron-sized grains [1, 2], thereby leading to a decrease in processing tool costs and material savings due to reduced contamination [3]. Specifically, low-temperature superplasticity in titanium alloys can be used to produce structural components by superplastic forming (SPF) or isothermal forging at much lower temperatures (600–700 °C) than those used presently (850–950 °C) [4]. However, this requires methods by which an SMC structure can be developed in large-scale billets. A submicrocrystalline structure can be produced in bulk material by severe plastic deformation (SPD) using methods such as equal channel angular extrusion [1] or multi-step isothermal forging [5]. In either case, the development of the SMC structure requires lower working temperatures than those commonly used in the conventional manufacture of semi-finished products. In addition to determining the deformation-temperature regime, the initial (perform) microstructure should also be established, for it can have a substantial effect on the kinetics of grain refinement. For two-phase (alpha/beta) titanium alloys, the perform microstructures are typically globular, bimodal, or lamellar. After hot- or warm- working, the characteristics of these alloys and their final microstructures will be different. However, there is no information in the literature concerning the influence of the initial microstructure on the development of a homogeneous SMC structure during SPD.

The objective of the present work was to establish a method for creating an SMC structure in large-scale titanium billet products by means of multi-step isothermal forging. In particular, the effect of SPD processing conditions and initial microstructure on the formation of a homogeneous SMC structure in large-scale billet of the titanium alloy Ti-6Al-4V was to be determined.

2 Materials and Methods

The alpha/beta titanium alloy Ti-6Al-4V was used to establish methods for producing a SMC structure. It had a measured composition (in weight percent) of 6.3 Al, 4.1 V, 0.18 Fe, 0.182 O, and 0.03 Si. The material had a beta transus temperature (at which alpha + beta → beta) of 995 °C.

Isothermal compression tests were conducted to establish the effect of processing parameters and initial microstructure on the formation of an SMC structure. To establish plastic-flow behavior and optimal processing parameters, material water quenched from the β-region (1010°C) was used; this material had a beta grain size of 250 μm. Initially, cylindrical samples (⌀ 10 mm × 15mm) were subjected to uniaxial compression to a height reduction of 70–75 % at an initial strain rate 10^{-3} s^{-1} in the temperature range 450–800 °C. Subsequent work consisted of sequential deformation of prismatic samples (16 × 18 × 20 mm^3) along three orthogonal directions, or so-called 'abc' deformation. Prior to each rotation, a prismatic shape was restored to the sample by machining the curved faces. The initial strain rate and the true strain per deformation step were 10^{-3}s^{-1} and ~0.4, respectively. The true strain was calculated as ln(h$_o$/h), where h$_o$ and h are initial and final height, respectively.

Compression tests were also performed to determine the influence of various types of initial microstructure on the processing parameters needed for the formation of an SMC structure. The initial microstructures were the following: a martensitic structure (formed by water quenching following beta annealing at 1010 °C), a coarse lamellar structure (formed during air cooling following beta annealing at 1010 °C), a globular-alpha-beta structure (developed via multi-step isothermal forging at 700 °C), and a bimodal structure (formed during multi-step isothermal forging at 950 °C followed by air cooling).

The ability to scale up the multi-step isothermal forging approach to form an SMC structure in a large Ti-6Al-4V billet with SMC was demonstrated via prototype production trials in a 16 MN hydraulic press equipped with a hot die set.

3   Results and Discussion

3.1   Microstructure Evolution in Beta-Annealed-and-Water-Quenched Ti-6Al-4V

The deformation of beta-annealed-and-water-quenched Ti-6Al-4V at temperatures between 450 and 800 °C led to a distinctly globularized structure of equiaxed α- and β-phase grains. For deformation temperatures of 650 °C and below, the size of the grains was in the submicron range (Fig. 1). Transmission electron microscopy (TEM) of samples with SMC structure revealed a significant amount of the fringe-diffraction contrast, indicative of elastic lattice distortion, as well as an increased dislocation density in a large number of grains (Fig. 1b). Based on these initial results, sub-

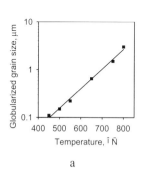
a

b

Figure 1: Hot compression results for beta-annealed-and-water quenched Ti-6Al-4V: (a) Dependence of the globularized grain size on deformation temperature for samples compressed to a height reduction of 70 % at a strain rate of 10^3 s^{-1} and (b) TEM micrograph for a sample compressed 70 % at 550 °C and 10^{-3} s^{-1}

sequent warm working trials were performed at 550 °C and compared to the microstructure and mechanical behavior of samples processed at 800 °C.

Using the method of multi-step isothermal forging under *laboratory* conditions (Fig. 2), the plastic-flow behavior in terms of stress versus cumulative strain (S-Σe) was readily determined. Such results for deformation at 550 and 800 °C and 10^{-3} s^{-1} are shown in Fig. 2 b,c. The cumulative S-Σe curves had a similar shape for both temperatures. The curves exhibited a peak stress, flow softening, and steady-state flow at large strains. At 550 °C, however, the alloy underwent a substantially greater amount of flow softening than at the higher temperature. Furthermore, the stress-strain data revealed that during each increment of deformation the flow stress decreased and became almost constant. The strain-rate-sensitivity m at 550 °C and 10^{-3} s^{-1} was equal to 0.17 for e = 0.4 and 0.35 for Σe = 0.9. At 800 °C and 10^{-3} s^{-1}, the values of m were 0.24 for e = 0.4 and 0.42 for Σe = 0.9. The presence of the steady flow stage on the true stress-strain curves and the values of the strain-rate sensitivity m demonstrated that plastic flow had evolved to a superplastic mode during in the process of "abc" forging.

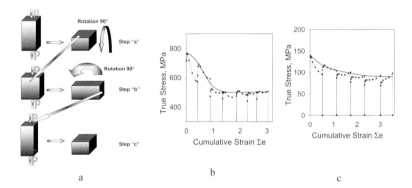

Figure 2: Multi-step deformation of beta-annealed-and-water-quenched Ti-6Al-4V: (a) deformation sequence and cumulative *S-Σe* curves for 'abc' deformation at 10^{-3} s^{-1} and temperatures of (b) 550 °C or (c) 800 °C.

The present results indicated that microstructure evolution during deformation of the titanium alloy at temperatures of 550 and 800 °C occurs as follows. During the initial stage of plastic deformation, strain hardening occurs due to the increased density of dislocations in the α-plates and β-matrix. As the strain increases further, the majority of plates rotate in the direction of deformation. In unfavorably-oriented colonies, deformation comprises intense shear strain and bending of the plates. At a finer scale, transverse sub-boundaries are formed within in α- and β-phases. As deformation increases further, the misorientation of these sub-boundaries increases as dislocations are absorbed into the sub-boundary walls, thereby leading to high-angle boundaries. Concurrently, the semi-coherent interphase α/β boundaries are transformed to non-coherent ones. Due to the transformation of the interphase and development of intra-lamellar boundaries, mass transfer becomes more rapid, and grooves are formed on the surface of the α-plates, leading to the segmentation of the alpha plates. The fragmented β-interlayers and α-plates are thus spheroidized. As the globularization process proceeds, plastic flow via grain-boundary sliding (accommodated by glide and climb of dislocations within the phases) becomes activated, the flow stress decreases, the m value increases, and the overall homogeneity of the structure is improved. As a whole, the present results agree with those published earlier [6, 7].

It is important to note that the development of superplastic flow during the final stages of the formation of an SMC structure contributes to an increase in the homogeneity of the final microstructure during warm deformation at 550 °C. After deformation to $\Sigma e = 3$, a homogeneous microstructure with globular grains of α- and β-phases, the mean size of which is ~0.4 µm, was formed.

3.2 Effect of Preform Microstructure on the Formation of an SMC Structure

Post-deformation metallography and TEM revealed a noticeable effect of perform microstucture on the formation of an SMC structure in Ti-6Al-4V. After compression of the alloy with the initial martensitic and fine-grained globular-alpha-beta microstructures to a height reduction of 70 % at 550 °C, a homogeneous globular SMC structure with a grain size of ~0.3 µm was formed (Fig. 3a, b). As a result of deformation at 550 °C, SMC grains of ~0.3 µm were formed in the samples with the coarse-lamellar starting structure as well. However, the microstructure was noticeably heterogeneous in this case because the lamellar component was preserved to a height reduction of ~30 % (Fig. 3c). For the bimodal (globular-lamellar) perform microstructure, only the lamellar component underwent transformation to produce grains refined to an SMC size; the coarse α-globules only elongated slightly in the direction of metal flow (Fig. 3d).

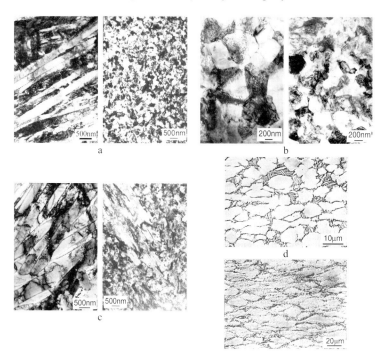

Figure 3: Microstructures developed in Ti-6Al-4V before (left - a, b, c, and top - d) and after (right - a, b, c, and bottom - d) a height reduction of 70 % at 550 ° and a strain rate of $10^{-3}\,s^{-1}$. The initial microstructures were: (a) martensitic, (b) globular, (c) coarse-lamellar, and (d) bimodal

3.3 Process Validation/Scale-Up

The laboratory-scale uniaxial upset and 'abc' forging trials demonstrated that the development of a homogeneous SMC microstructure in Ti-6Al-4V should be performed via warm working at a temperature equal to or below 700 °C (for a strain rate of 10^{-3} s^{-1}). In addition, the preform microstructure should consist preferably of martensitic or globular-alpha-beta. Because the critical strain for the initation of globularization increases sharply at lower temperatures, a total strain of not less than $\Sigma e = 3$ should be imposed within the whole volume of the workpiece. Using these guidelines, a homogeneous SMC structure was produced in a large Ti-6Al-4V billet measuring 150 mm diameter and 200 mm length via multi-step isothermal forging (Fig. 4a). Metallographic analysis of the billet indicated that its macrostructure was homogeneous across the entire section (Fig. 4b), and its microstructure comprised a uniform distribution of globular of α- and β-phase particles whose size did not exceed 0.5 µm (Fig. 4c, d).

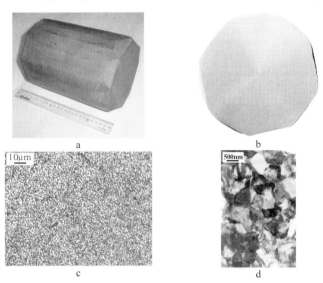

Figure 4: Ti-6Al-4V SMC billet, measuring 150 mm diameter and 200 mm length: () as-processed billet, (b) macrostructure, (c) microstructure, and (d) TEM micrograph

Table 1: Mechanical properties of samples cut out from the Ti-64 SMC billet

Direction of sampling	YS, [MPa]	UTS, [MPa]	EL [%]	RA [%]
Radial	1350	1360	7	62
Tangential	1335	1355	7	61

The mechanical properties of the SMC billet were also excellent. After multi-step isothermal forging, the ultimate tensile strength was very high (1360 MPa), and the total elongation was 7 % (Table 1). The property uniformity along the radial and tangential directions was also very uniform. These values can be compared to those in conventionally-processed (microcrystalline)

Ti-6Al-4V. After solution treatment and aging, conventional Ti-6Al-4V alloy typically has an ultimate tensile strength no higher than 1100 MPa and an elongation of approximately 8 %.

4   Summary

1. The process of successive compression of alpha/beta titanium alloys (such as Ti-6Al-4V) along three orthogonal directions at 550 °C and $10^{-3}\,s^{-1}$ to a cumulative strain of $\Sigma e = 3$ leads to the formation of a homogenous submicrocrystalline structure (with a grain size of ~0.3 µm). Superplastic flow during the final stages of deformation contributes to the formation of a homogeneous SMC structure.
2. The formation of a homogeneous SMC structure is readily promoted by using a perform microstructure which is either fully martensitic or globular. The present work suggests that the SMC structure can not be obtained with a starting bimodal microstructure.
3. Using the method of multi-step forging, a large-scale billet (∅ 150 mm × 200 mm) with a homogeneous microstructure and a grain size less than 0.5 µm was produced. The refined grain size provided a substantial increase in strength without a loss in ductility. Its strength and ductility were almost identical in the tangential and radial test directions.

5   Acknowledgements

The present work was supported by the Air Force Office of Scientific Research (AFOSR) and the AFOSR European Office of Aerospace Research and Development (AFOSR/EOARD) within the framework of ISTC partner project 2124p. The encouragement of the AFOSR program managers (Drs. C.H. Ward and C.S. Hartley) is greatly appreciated.

6   References

[1] Valiev R.Z., Islamgaliev R.K., Alexandrov I.V., Progress in Materials Science 2000, 45, p. 102–189
[2] O.A. Kaibyshev, Superplasticity of Alloys, Intermetallides and Ceramics, Springer Verlag, Berlin, 1992
[3] Froes F., Yolton C., Chesnutt J., Hamilton C., in: The Forging and Properties of Aerospace Materials. The Chameleon Press, London, 1978, p. 371–398
[4] Salishchev G.A., Galeyev R.M., Valiakhmetov O.R., Safiullin R.V., Lutfullin R.Ya., Senkov O.N., Froes F.H., Kaibyshev O.A., Mat. Tech. & Adv. Perf. Mater., 15.2 (2000) 133–135
[5] Patent PCT/US97/18642, WO 9817836 A1, 1998
[6] Salishchev G.A., Valiakhmetov O.R., Galeev R.M., Journal of Materials Science 1993, 28, p. 2898–2902
[7] Semiatin S.L., Seetharaman V., Weiss I., in: The Advances in the Science and Technology of Titanium Alloy Processing (Eds.: I Weiss, R. Srinivasan, P.J. Bania, D. Eylon, S.L. Semiatin), The Minerals, Metals and Materials Society, 1997, p. 3–73

Authors

Akamatsu, H. 459
Alberdi, J. M. 65
Aldazabal, J. 65
Alexandrov, I. V. 183, 207, 245, 271, 309, 315, 465, 609
Alkorta, J. 491
Ameyama, K. 571
Anwand, W. 407
Apps, P. J. 138
Arsenkin, A. M. 798, 804

Bacroix, B. 564
Baik, S. C. 233, 257
Barbe, V. 767
Barbé, L. 530
Baró, M. D. 387
Barta-Schreiber, N. 677
Bassani, P. 145
Bastarash, E. N. 158, 420
Belyakov, A. 558, 780
Bengus, V. Z 151, 207
Bernard, F. 539
Berveiller, M. 101
Betzwar-Kotas, A. 636
Beygelzimer, Y. 511
Billard, S. 564
Blandin, J. J. 740
Blavette, D. 118
Bonarski, J. 309, 315
Bowen, J. R. 138, 257
Brauer, G. 407
Brossmann, U. 755
Brunner, D. 131
Budilov, I. N. 271

Champion, Y. 545
Chinh, N. Q. 87
Chun, B. S. 239
Cieslar, M. 407, 630
Cizek, J. 407, 630

Dahlgren, M. 18
de Messemaeker, J. 332
Derevyagina, L. S. 37
Derlet, P. M. 599
Dirras, G. 564
Ditenberg, I. A. 381
Dobatkin, S. 158, 170, 420, 798, 804
Dotsenko, T. V. 183
Du, Z. 829
Dubravina, A. A, 465
Dupuy, L. 297
Dvorak, J. 200

Enikeev, N. A. 245
Estrin, Y. 233, 257
Eversheim, P. D. 767

Fátay, D. 420
Fecht, H.-J. 30, 453
Firstov, S. A. 72
Fokine, V. A. 798
Fondere, J. P. 564
Froyen, L. 332, 585
Fujiwara, H. 571
Furukawa, M. 459, 701

Gaffet, E. 539
Galeyev, R. M. 835
Garcia Oca, C. 623
González, P. A. 251
Gonzalez Doncel, G. 623
Grabovetskaya, G. P. 722
Gröger, V. 636
Grössinger, R. 18
Gubicza, J. 420
Guillet, A. 118
Gunderov, D. V. 165
Gust, W. 642
Gutfleisch, O. 767
Hahn, H. 3

Hamaji, H.   375
Han, K.   95
Hara, T.   558
Hashimoto, S.   363, 663
Hasnaoui, A.   599
Heason, C. P.   498
Hebesberger, T.   447
Hellmig, R. J.   233, 257
Herth, S.   755, 767
Hidaka, H.   345
Higashida, K.   517
Hollang, L.   131
Holste, C.   131
Holzer, D.   18
Hong, S. I.   239, 245, 728
Höppel, H. W.   677
Horita, Z.   87, 459, 701, 711, 746
Houbaert, Y.   530
Houtte, P. Van   226
Huang, X.   323
Hÿtch, M. J.   545

Ignatenko, L. N.   357
Inomoto, H.   571
Ishmaku, A.   95
Islamgaliev, R. K.   407, 630
Ivanisenko, Yu.   453
Ivanov, K. V.   722

Juul Jensen, D.   257

Kaloshkin, S. D.   579
Kamalov, M. M.   717, 734
Karnthaler, H. P.   80, 339, 351, 369, 810
Kautz, M.   677
Kawasaki, K.   345
Kazykhanov, V.   351
Kestens, L.   530, 585
Khatibi, G.   636
Khmelevskaya, I. Yu.   170
Kim, B.-K.   387
Kim, H. S.   233, 239, 245, 257, 728
Kim, Y. S.   728
Kimura, Y.   558

Klöden, B.   303
Knabl, W.   648
Kohout, J.   435
Kolobov, Yu. R.   722
Koneva, N. A.   263, 357
Kopylov, V. I.   37
Korb, G.   810
Korneva, A. V.   177
Korotaev, A. D.   381
Korznikov, A. V.   177
Korznikova, G. F.   177
Koyama, T.   363
Kozlov, E. V.   263, 357
Krajczyk, L.   44
Krallics, G.   183, 271
Krasilnikov, N. A.   798, 804
Krasilnikov, N. A.   158
Krummeich-Brangier, R.   101
Kuzel, R.   407, 630

Langdon, T. G.   87, 387, 459, 677, 701, 711, 746, 829
Langlois, C.   545
Langlois, P.   545
Lapovok, R. Ye.   551
Laptev, A. I.   579
Lartigue, S.   545
Lee, H. R.   239
Lee, S-H   479
Lendvai, J.   190
Litovchenko, I. Yu.   381
López, G. A.   642
Lowe, T. C.   789
Luis, C.   251
Lukác, P.   190, 413

Macalik, B.   44
Máthis, K.   190
Matsubara, K.   711
Mazilkin, A. A.   734
Mimaki, T.   363
Minamino, Y.   479
Mingler, B.   369, 810
Minkow, A.   453
Mironov, S. Yu.   523, 691

Mittemeijer, E. J.   642
Miura, H.   375, 780
Miyahara, Y.   711
Miyamoto, H.   363
Morikawa, T.   517
Morris, D. G.   623
Morris-Muñoz, M. A.   623
Mughrabi, H.   677
Mussi, A.   190, 740
Myshlayev , M. M.   523, 691, 717, 734
Myshlyaeva, M. M.   717

Naidenkin, E. V.   722
Natsik, V. D.   207
Neishi, K.   746
Nurislamova, G. V.   387
Nyilas, K.   420

Odessky, P. D.   798, 804
Oertel, C.-G.   303
Orlov, D.   511
Ovchinnikov, S. V.   381

Pinzhin, Yu. P.   381
Pakiela, Z.   194
Panin, A. V.   37
Panin, V. E.   37
Paris, S.   539
Park, K.-T.   616
Pashinska, Y.   511
Pekarskaya, E. E.   357
Perevezentsev, V. N.   773
Pippan, R.   447, 648, 654, 684, 691, 798, 804
Podolskii, A. V.   207
Popov, A. A.   817
Popov , A. G.   165
Popova, N. A.   263, 357
Prangnell, P. B.   138, 498
Principi, G.   579
Prochazka, I.   407, 630
Prokofjev, E. A.   170
Prokoshkin, S. D.   170
Pushin, V. G.   822

Raab, G. I.   271, 471, 798, 804
Rauch, E.   190
Rauch, E. F.   297, 740
Reis, A. C. C   530, 585
Rentenberger, C.   80
Riehemann, W.   413
Romanov, A. E.   215
Rombouts, M.   585
Rostova , T. D.   158
Roven, H. J.   591
Rybacki, E.   303

Sabar, H.   101
Saito, Y.   479
Sakai, T.   375, 780
Sakai, Y.   558
Salishchev, G. A.   523, 691, 835
Sato, R.   18
Sauvage, X.   118
Schaefer, H.-E.   767
Schafler, E.   426, 435, 810
Schedler, B.   654
Schegoleva , N. N.   165
Seefeldt, M.   226
Semenova, I. P.   183
Semiatin, S. L.   835
Seo, M. H.   257
Sergueeva, A. V.   453, 465
Sevillano, J. G.   65, 491
Shelekhov, E. V.   579
Shevchenko, N. V.   381
Shin, D. H.   616
Sklenicka, V.   200
Skrotzki, W.   303
Smirnov, S. N.   151, 207
Soshnikova, E. P.   471
Spataru, T.   579
Sprengel, W.   767
Spuskanyuk, A.   511
Stanek, M.   413
Stolyarov, V. V.   125, 165, 170, 207
Straumal, B. B.   642
Stulíková, I.   630
Stüwe, H. P.   55, 435, 447
Surikova, N. S.   381
Sus-Ryszkowska, M.   194

Suszynska, M.   44
Svoboda, M.   200
Synkov, S.   511
Szeles, Z.   183
Szpunar, J. A.   387

Tabachnikova, E. D.   207
Takaki, S.   345
Tamm, R.   303
Tarkowski, L.   315
Tasca, L.   145
Tcherdyntsev, V. V.   579
Thiele, E.   131
Thomson, P. F.   551
Todaka, Y.   505
Tolleneer, I.   530
Tóth, L. S.   281
Trojanová, Z.   190, 413
Trubitsyna, I. B.   158, 170
Tsuchiya, K.   505
Tsuchiyama, T.   345
Tsuji, N.   479
Tsuzaki, K.   558, 780
Tyumentsev, A. N.   381

Umemoto, M.   505
Ungár, T.   395, 420

Valiakhmetov, O. R.   835
Valiev, R. Z.   37, 80, 109, 125, 158, 207, 271, 339, 381, 453, 465
van Humbeeck, J.   332
van Swygenhofen, H.   599
Varyukhin, V.   511
Vedani, M.   145
Verlinden, B.   332
Vinogradov, A.   158, 363, 663
Vorhauer, A.   435, 447, 648, 684

Wadsack, R.   654
Waitz, T.   339, 351
Wang, J.   829
Wang, L.   829

Wang, Y.   829
Wcislak, L.   303
Weidenfeller, B.   413
Weiss, B.   636
Wejrzanowski, T.   767
Werenskiold, J. C.   591
Winter, G.   323
Wottle, I.   636
Würschum, R.   755, 767

Xin, Y.   95
Xu, C.   677, 701, 829

Yavary, A. R.   165

Zakharov, V. V.   158
Zehetbauer, M.   369, 426, 435, 810
Zeipper, L.   369, 426, 810
Zhang, Z.   829
Zhao, X.   829
Zhdanov, A. N.   263, 357
Zherebtsov, S. V.   691, 835
Zhernakov, V. S.   271
Zhilyaev, A. P.   387
Zhu, Y. T.   789
Zimprich, P.   636

Subject Index[*]

A

AA6082   145
AA8079   498
Accumulated roll bonding   479, 498, 530
acommercial Al-Mg-Mn-Alloy   251
Aging behaviour   145
AISI304L stainless steel powder   571
Al-3Mg   623
Al-Cu-Fe quasicrystal   579
Al-Li alloy   717, 734
Alloy   170
   AA6082   145
   AA8079   498
   Al-Li   717, 734
   Al-Mg-Mn   251
   Al-Mg-Si   183
   Al-Mg-Sc-Zr   420
   Al-Sc   158
   Al-Zn   642
   Al-3Mg   623
   aluminum   459, 701
   AZ91   190
   Cu-Ag   728
   Cu-Zn   746
   CP-Al   498
   Cu-Ag   728
   Cu-Zn   746
   Fe-based   530
   magnesium   740
   magnetic   177
   nanocrystalline   755
   nanostructured   125
   Nd-Fe-B   165, 767
   NiAl   303
   NiTi   339, 345
   powders   585
   superplastic   711
   Ti-6Al-4V   835
   TiNi   822
   titanium   817
   ultra fine grained   780
Alloying, mechanical   579
Al-Mg-Mn, acommercial alloy   251
Al-Mg-Sc alloys   158
Al-Mg-Sc-Zr   420
Al-Mg-Si   183
Al-Sc alloys   158
Aluminum   87, 233
   alloy   459, 701
   nanocrystalline   564
   polycrystals   226
   pure   200, 684
   severe deformation   138
Al-Zn system   642
Amorphous alloy   165, 339
Analysis, X-ray   420
Anelastic Properties, Magnesium   413
Angular extrusion   281, 297, 332, 717
Anisotropy, thermal   151
Annealed microstructures   558
Annealing mechanisms, alloy   780
Annealing treatment   623
Application of nanocrystalline structures   18
Applied research, russian   798
Armco iron   453
Atomistic modelling   599
Austenitic stainless steel   375
AZ91   190

B

Back-pressure   551
BCC-polycrystals   72
Boundary type   357
Boundary characteristics, metal   323
Boundary diffusion   773

---

[*] The page numbers refer to the first page of the respecting article

Broadened X-ray diffraction peaks   395
Bulk Al-Cu-Fe quasicrystals   579
Bulk nanomaterials   789
Bulk ultra fine grained material   479

C

Carbon steel   505
   nanostructured   616
Chromium, nanostructured   654
Cold deformation   523
Cold-rolled metal   517
Cold torsion   226
Complex loading   177
Composite   30
   nanolamellar   101
   grain model   263
Compression, uniaxial   691
Contrained groove pressing   491
Copper   87, 245, 315, 363, 465
   deformed   257
   nanoparticle   44
   nanostructured   545
   pure   511, 684
   ultra fine grained   407
CP-Al   498
Creep behaviour   200
Crystalline alloys   165
Crystallographic microstructure   315
Crystallographic texture   315
Cu-Ag nanocomposite   728
Cu-Foils, fine grained   636
Cu-Zn alloy   746
Cyclic deformation   663, 677

D

Defects, distribution   407
Deformation   728
   Al-Mg-Sc-Zr   420
   behaviour   233, 245
   cold   523
   copper   257, 465
   iron   345
   large strain   80
   path   684
   plastic   663
   single crystal   363
   substructure   72
   type   684
Deformed material   207
Deformed Ti   369, 810
Degradation, plastic   663
Dense nanostructured material   539
Densification   551
Dependence, thermal stability   630
Development, copper   363
Development, microstructure   345
Die   459
Diffusion   755, 767
Diffusion controlled processes   722
Diffusion properties, grain boundary   773
Disclination   207, 226
Dislocation   357
Dislocation-based model   245
Distribution of defects, spatial   407
Draw hardening   101

E

ECAP, Nickel   131
ECAP processing   145
ECA pressing   734
Effect, microscopic   65
Electrodeposited Cu-Foil   636
Enhancing, fatigue properties   677
Equal channel angular extrusion   131, 138, 183, 190, 200, 226, 233, 239, 245, 251, 257, 271, 315, 363, 369, 459, 471, 551, 591, 616, 746
Equivalent strain   55
Evolution, Al-3Mg   623
Extrusion, angular   138, 281, 297, 332, 717
Extrusion, twist   511

F

Fatigue, metal   663, 677
Fe-based
   alloys   530
   powders   585
FeCoCr   177

Fe powders, milled   558
Fine grained alloys   780
Fine grained Cu-Foils   636
Fine grained metals   630
Finite element method, simulation   271
Flow, plastic   37
Flow stress   131
Formation, nanostructure   30, 381
Forming capability, superplastic   701
Fragmentation, grain   226

G

Glasses, soda lime silicate   44
Grain, copper   407
Grain boundary diffusion   773
Grain boundary phase transition   642
Grain boundary state   722
Grain type   357
Grained Cu-Foil   636
Grained material   479
Grain fragmentation   226
Grain model   263
Grain refinement   303, 387, 630, 663
   alloy   530
     Cu-Zn alloy   746
Grain size   345
Grain structure, steel powder   571
Groove pressing   491

H

Hardening, draw   101
Hard magnetic alloy FeCoCr   177
Hard-to-deform material   471
Heavily deformed metal   323
High-energy nanostructures   381
High pressure torsion   158, 303, 387, 407,
   420, 447, 453, 465
High strain rate   717, 734
Hot isostatic pressing   564
Hot pressing   804
HPT   339
HRTEM investigations, alloy   339
Hydrostatic pressure   435

I

IF Steel   332
Internal Stress   151
Investigations, TEM   369
Iron, nanostructured   453
Iron, polycrystalline   345
Isostatic pressing   564

J

Joint disclinations   357

L

L12 intermetallics   80
Large scale billets, Ti-6Al-4V   835
Large strain   80
Low carbon steel   804, 829
   nanostructured   616

M

Magnesium alloy   740
Magnesium, nanocrystalline   413
Magnesium particles   551
Magnetic alloy   177
Magnetic material   18
Material
   deformed   207
   hard-to-deform   471
   magnetic   18
   nanocrystalline   194
   nanostructured   3, 37, 118, 539
   tungsten based   648
   UFG   263, 357
Mechanical alloying   579
Mechanical behaviour, titanium   523
Mechanically activated powder metallurgy   539
Mechanically milled Fe powders   558
Mechanical properties
   copper   257, 545
   iron   453
   magnesium   740
   titanium   810

Mesoscopic Effect 65
Metal
  cold rolled 517
  deformed 323, 663
  fine grained 630
  nanocrystalline 599, 755
  nanostructured 207, 789
  ultra fine grained 677
Metallurgy, powder 539, 545
Metastable alloys 125
Mg alloys 711
Micrometer grained alloy 717
Microscopic effect 65
Microstructural analysis, steel 332
Microstructural evolution 375, 387
Microstructural heterogeneity 194
Microstructural properties, copper 257
Microstructural stability, carbon steel 616
Microstructure
  Al-Li 734
  alloy 623
  Al-Mg-Sc-Zr 420
  annealed 558
  chromium 654
  crystallographic 315
  development 345, 363
  evolution 426
  intermetallics 80
  low carbon steel 829
  NiTi 345
  refinement, magnesium alloy 740
  tungsten 648
Model, structural 381
Modelling
  atomistic 599
  deformation 233
  grain fragmentation 226
  process 239
Multiscale studies, SPD materials 599
Multiscale transition method 101

N

Nanocomposites, Cu-Ag 728
Nanocrystalline
  alloys 755
  aluminum 564
  Fe based alloy powders 585
  magnesium 413
  material 194, 599, 755
  NiTi alloys 339
  structure 18, 158
Nanocrystallisation 505
Nano grain structure 571
Nanolamellar composite 101
Nanoplatelets, formation 95
Nanostructure, intermetallics 80
Nanostructured
  chromium 654
  copper 545
  iron 453
  low carbon steels 616
  material 3, 37, 118, 539
  metal 207, 789
  metastable alloy 125
  shape 822
  titanium 151
Nanostructure, formation 30
Nanostructure, high energy 381
NdFeB 165, 767
NiAl 303
Nickel 131, 387
NiTi 339, 351
Nonequilibrium grain boundary
  diffusion 773

P

Paradoxes 109
Particles, magnesium 551
Particles, second-phase 138
Particles 375
Peak analysis, X-ray 420
Peaks, X-ray diffraction 395
Periodical 798
Phases, sintering 579
Phase state, alloy 734
Phase transformation 118, 125, 165, 345
Phase transition 642
Planar simple shear 297
Plastic flow 37
Plasticity 722

Plasticity, titanium 151
Plates, processing 491
Polycrystalline iron 345
Polycrystals 72
   aluminum 226
Porousless 609
Positron annihilation spectroscopy 407
Powder
   Fe based 585
   formation 579
   steel 571
   metallurgy 539, 545
Pressing 734, 804
   angular 183, 190, 200, 226, 233, 239, 245, 251, 257, 271, 315, 363, 369, 459, 471, 551, 591, 616, 746
   constrained 491
   hot 564
   simulation 271
Pressure, hydrostatic 435
Printed wiring boards 636
Proccessed materials, nanocrystalline 194
Processing parameters, torsion 447
Process modeling 239
Properties
   alloy 822
   copper 257
   low carbon steel 804, 829
   BCC-polycrystals 72
Pure aluminum 684
Pure copper 511, 684

Q

Quasicrystals, Al-Cu-Fe 579

R

Rapidly quenched alloys 165
Roll bonding, accumulated 479, 498, 530

S

Scale levels of plastic flow 37
Second phase particles 138, 375
Self diffusion 767

Sensitivity, temperature 207
Severe deformation
   aluminium alloy 138
   steel 375
   Al-3Mg 623
   material 207
   metal 663
Severe plastic deformation 194, 251, 263, 281, 309, 357, 381, 426, 435, 505, 511, 517, 648, 654, 701, 711, 740
   equivalent strains 55
   material 609
   paradox 109
   processing 491
   phase transformation 118
   shape memory alloy 170
Shape memory alloy 170, 822
Shear, planar 297
Shear band 517
Silver nanoparticle 44
Single crystal, copper 363
Sintering 579
Soda lime silicate glasses 44
Spatial distribution, defects 407
Spectroscopy, Positron annihilation 407
Stability
   carbon steel 616
   thermal 654
Stage IV 65
Steel
   austenitic 375
   carbon 505
   IF 332
   low carbon 804, 829
   powder 571
Strain range 87
Strain hardening 95
Strain measurement 591
Strain rate 131
Straining, alloy 420
Strength
   metal 599
   titanium 817
Strengthening, grain 263
Stress, internal 151
Stress strain relationship 87

Structural model 381
Structure
    alloy 170, 822
    deformed 517
    dislocation 357
    evolution, titanium 523
    formation 158
    metal 72
    nanocrystalline 18
    phase transformations 125
    steel 804
    submicrocrystalline 177, 691, 835
Submicrocrystalline
    deformation 789
    material 722
    structure, titanium 523, 691
    structure 158, 177, 835
Successive uniaxial compression 691
Superplastic
    Cu-Zn alloy 746
    forming capability 701
    Mg alloys 711
Superplasticity 642, 717, 728, 734

T

TEM investigations 369
Temperature rate 131
Temperature strain rate 207
Tensile properties, carbon steel 616
Texture
    changes 309
    development 233, 315
    evolution 281, 297, 303
    modeling 498
Thermal anisotropy 151
Thermal stability
    chromium 654
    iron 453
    metal 630
    tungsten 648

Thermal treatment 426
Thermo-mechanical properties, Cu-foil 636
Ti-6Al-4V alloyy 835
Ti-Ni-based shape memory alloys 170, 822
Titanium
    deformed 309
    nanostructured 151
    submicrocrystalline 523
Titanium 315, 369, 426, 691, 810, 817
Torsion 158
    cold 226
    high pressure 387, 447, 453, 465
    high strain 303
Transition method, multiscale 101
Tungsten 271, 648
Twist extrusion 511

U

UFG materials 263, 357
Ultra fine grained
    alloy 780
    copper 407
    Cu-Foil 636
    material 479
    metal 630, 677
Ultra grain refinement, alloy 530
Uniaxial compression 691

W

Warm pressing 804
Wiring boards, printed 636

X

X-ray diffraction 395
X-ray peak profile analysis 420